THERMAL RADIATION
HEAT TRANSFER

THERMAL RADIATION HEAT TRANSFER

Second Edition

Robert Siegel

Fluid Mechanics and Acoustics Division
NASA Lewis Research Center

John R. Howell

Professor of Mechanical Engineering
University of Texas at Austin

Hemisphere Publishing Corporation, New York
A subsidiary of Harper & Row, Publishers, Inc.

Cambridge   Philadelphia   San Francisco   Washington
London   Mexico City   São Paulo   Singapore   Sydney

THERMAL RADIATION HEAT TRANSFER, Second Edition

7 8 9 0   B C B C   8 9 8 7 6

This book was set in Press Roman by Communication Crafts Ltd. The editors were Diane Heiberg, Christine Flint, and Edward Millman. BookCrafters Inc. was printer and binder.

Library of Congress Cataloging in Publication Data

Siegel, Robert, date.
    Thermal radiation heat transfer.

    (Series in thermal and fluids engineering)
    Bibliography: p.
    Includes index.
    1. Heat–Radiation and absorption. 2. Heat–
Transmission. 3. Materials–Thermal properties.
I. Howell, John R., joint author. II. Title.
QC331.S55 1980   536′.33   79-17242
ISBN 0-89116-506-1

CONTENTS

*Topics preceded by an asterisk can form the basis of a one-semester graduate course as discussed in the Preface to the First Edition.

8   Radiation Exchange in an Enclosure Composed of Diffuse-Gray Surfaces   233

9   Radiation in Enclosures Having Some Specularly Reflecting Surfaces   281

10   The Exchange of Thermal Radiation between Nondiffuse Nongray Surfaces   325

PREFACE

In this second edition of *Thermal Radiation Heat Transfer,* the authors have tried to retain the features that made the first edition unique as a radiation textbook and reference. These features include detailed derivation of the fundamentals of radiative transfer, careful exposition of the assumptions underlying the radiation laws, and sufficient explanatory material so that the student or practicing engineer can use the book as both a self-study text and a reference.

In the several years since the first edition, there has been considerable development in the field of solar-energy utilization. For this reason a new chapter was added on the transmission behavior of windows and multiple windows (as used in solar collectors), coated surfaces, and thin films. The material on glass properties has been expanded.

To improve the organization, two chapters were eliminated by consolidation with material in other chapters. The material on soot radiation was expanded and then combined with the information on radiation of gases in furnaces and combustion chambers. The treatment of scattering was combined with that on absorption to yield a more unified development of radiative behavior in attenuating media.

Many other additions have been made. The catalog of available configuration factors has been brought up to date. There has been considerable recent work on band radiation correlations, so this section was expanded. There is an updated section on the differential approximation, and a number of new homework problems and examples. The use of SI units has been increased but they are not used exclusively, as English units are also commonly used in engineering heat transfer calculations in the United States.

ROBERT SIEGEL
JOHN R. HOWELL

xiii

PREFACE TO THE FIRST EDITION

Several years ago it was realized that thermal radiation was becoming of increasing importance in aerospace research and design. This importance arose from several areas: high temperatures associated with increased engine efficiencies, high-velocity flight accompanied by elevated temperatures from frictional heating, and the operation of devices beyond the earth's atmosphere where convection vanishes and radiation becomes the only external mode of heat transfer. As a result, a course in thermal radiation at about a first-year graduate level was initiated at the NASA Lewis Research Center as part of an internal advanced study program.

The course was divided into three main sections. The first dealt with the radiation properties of opaque materials, including a discussion of the blackbody, electromagnetic theory, and measured properties. The second discussed radiation exchange in enclosures both with and without convection and wall conduction. The third section treated radiation in partially transmitting materials—chiefly gases.

When the course was originated, a single radiation textbook that covered the desired span was not available. As a result we began writing a set of notes; the present publication is an outgrowth of these notes. During the past few years, a few radiation textbooks have appeared in the literature; hence the need for a single reference has been partially satisfied.

The objectives of this volume are more extensive than providing the content of a standard textbook intended for a one-semester course. Many parts of the present discussion have been made quite detailed so that they will serve as a source of reference for some of the more subtle points in radiation theory. The detailed treatment has resulted in some rather long sections, but the intent in these instances was to be thorough rather than to try to conserve space. The sections have been subdivided so that specific portions can be located for easy reference.

This volume is divided into 21 chapters. The first five deal with some fundamentals of radiative transfer, the blackbody, electromagnetic theory, and the properties of solid materials. Chapters 6 through 12 treat energy exchange between surfaces and in enclosures when no attenuating medium is present. The final nine chapters concern radiative transfer in the presence of an attenuating medium.

Topics felt to be of primary interest for a one-semester course at a first-year graduate level have been marked with an asterisk in the contents. Experience in teaching the course has indicated that these topics can be covered provided that some of the subsections are considered in less detail according to the emphasis desired by the instructor. The authors have tried in most of the work to use a conversational style and, at the risk of being wordy, explain things in fundamental detail. As a teaching philosophy, the material usually proceeds from specific to more general (and hence usually more complex) situations. For these reasons the text is also intended for self-study by engineers having little previous knowledge of thermal radiation.

Each chapter contains numerical examples to acquaint the reader with the use of the analytical relations. It is hoped that these examples will help bridge the gap between theory and practical application. Homework exercises are also included at the end of each chapter.

As a closing note, let it be stated that this book comes as close as possible to being a 50:50 effort by the authors. This paragraph is written prior to flipping a coin to determine first authorship. Therefore, whoever appears as first author does not claim major credit for the work, but only a hot gambling hand.

The authors would like to acknowledge the help from many of their associates at NASA who reviewed the material and provided valuable detailed comments. Special thanks are extended to Curt H. Liebert who carefully reviewed the entire manuscript.

ROBERT SIEGEL
JOHN R. HOWELL

THERMAL RADIATION
HEAT TRANSFER

INTRODUCTION

All substances continuously emit electromagnetic radiation by virtue of the molecular and atomic agitation associated with the internal energy of the material. In the equilibrium state, this internal energy is proportional to the temperature of the substance. The emitted radiant energy can range from radio waves, which can have wavelengths of hundreds of meters, to cosmic rays with wavelengths of less than 10^{-14} meter (m). In this volume, only radiation that is detected as heat or light will be considered. Such radiation is termed *thermal radiation*, and it occupies an intermediate wavelength range (defined explicitly in Sec. 1-5).

Although radiant energy constantly surrounds us, we are not very aware of it because our bodies are able to detect only portions of it directly. Other portions require detection by use of instruments. Our eyes are sensitive detectors of light, being able to form images of objects, but are relatively insensitive to heat (infrared) radiation. Our skin is a direct detector for heat radiation but not a good one. The skin is not aware of images of warm or cool surfaces around us unless the heat radiation is large. We require indirect means, such as infrared-sensitive film in a camera, to form images using heat radiation.

Before discussing the nature of thermal radiation in detail, it is well to consider why thermal radiation is so important in modern technology.

1-1 IMPORTANCE OF THERMAL RADIATION

One of the factors that accounts for the importance of thermal radiation in some applications is the manner in which radiant emission depends on temperature. For conduction and convection the transfer of energy between two locations depends on

1

the temperature difference of the locations to approximately the first power.* The transfer of energy by thermal radiation between two bodies, however, depends on the differences of the individual absolute temperatures of the bodies each raised to a power in the range of about 4 or 5.

From this basic difference between radiation and the convection and conduction energy-exchange mechanisms, it is evident that the importance of radiation becomes intensified at high absolute-temperature levels. Consequently, radiation contributes substantially to heat transfer in furnaces and combustion chambers and in energy emission from a nuclear explosion. The laws of radiation govern the temperature distribution within the sun and the radiant emission from the sun or from a source duplicating the sun in a solar simulator. The nature of radiation from the sun is of obvious importance in the developing technology of solar-energy utilization. Some devices for space applications are designed to operate at high temperature levels to achieve high thermal efficiency. Hence radiation must often be considered in calculating thermal effects in devices such as a rocket nozzle, a nuclear power plant for space applications, or a gaseous-core nuclear rocket.

A second distinguishing feature of radiative transfer is that *no* medium need be present between two locations for radiant interchange to occur. Radiative energy will pass perfectly through a vacuum. This is in contrast to convection and conduction, where a physical medium must be present to carry the energy with the convective flow or to transport it by means of thermal conduction. When no medium is present, radiation becomes the only significant mode of heat transfer. Some common instances are heat leakage through the evacuated walls of a Dewar flask or thermos bottle, and heat dissipation from the filament of a vacuum tube. A more recent application is the radiation used to reject waste heat from a power plant operating in space.

Radiation can be of importance in some instances even though the temperature levels are not elevated and other modes of heat transfer are present. The following example is quoted from a Cleveland newspaper published in the spring of 1964. A florist "noted the recurrence of a phenomenon he has observed for two seasons since using plastic coverings over [flower] flats. Water collecting in the plastic has formed ice a quarter-inch thick [at night] when the official [temperature] reading was well above freezing. 'I'd like an answer to that, I supposed you couldn't get ice without freezing temperatures.'" The florist's oversight was in considering only the convection to the air and omitting the nighttime radiation loss occurring between the water-covered surface and the very cold heat sink of outer space.

A similar illustration is the discomfort that a person experiences in a room where cold surfaces are present. Cold window surfaces, for example, have a chilling effect as the body radiates directly to them without receiving compensating energy from them. Covering the windows with a shade or drape will greatly decrease the bodily discomfort.

An important application of thermal radiation is in the practical utilization of the sun's radiation as an energy source on earth. Solar energy is transferred to a solar

---

* For free convection, or when variable property effects are included, the power of the temperature difference may become larger than unity but usually in convection and conduction does not approach 2.

collector on earth through the vacuum of space and the earth's atmosphere. The collector operates by converting solar radiation into internal energy. If the arriving energy is not highly concentrated by means of a lens or curved mirror, the collector normally functions at temperatures near to, or at most a few hundred degrees Celsius above, ambient. The balance between the available solar energy, the useful energy transferred to a working fluid, and the various convective, conductive, and radiative losses is very sensitive, and it makes the design of an efficient collector quite challenging.

Finally, we note that the thermal radiation that we shall examine is in the wavelength region that gives mankind heat, light, photosynthesis, and all their attendant benefits. This in itself is strong justification for studying thermal radiation. Our existence depends on the solar radiant energy incident upon the earth. Understanding the interaction of this radiation with the atmosphere and surface of the earth can provide additional benefits in its utilization.

1-2 SYMBOLS

c	speed of electromagnetic radiation propagation in medium other than a vacuum
c_0	speed of electromagnetic radiation propagation in vacuum
k	thermal conductivity
n	index of refraction, c_0/c
q_c	energy per unit area per unit time resulting from heat conduction
q_r	radiant energy per unit area per unit time arriving at surface element
q_s	radiant energy per unit area per unit time arriving from unit surface element
q_v	radiant energy per unit area per unit time arriving from unit volume element
S	surface area
T	temperature
V	volume
x,y,z	coordinates in cartesian system
ζ	arbitrary direction
λ	wavelength in vacuum
ν	frequency

1-3 COMPLEXITIES INHERENT IN RADIATION PROBLEMS

First let us discuss some of the mathematical complexities that arise from the basic nature of radiation exchange. In conduction and convection heat transfer, energy is transported by means of a physical medium. The energy transferred into and from an infinitesimal volume element of solid or fluid depends on the temperature gradients and physical properties in the *immediate vicinity* of the element. For example, for the relatively simple case of heat conduction in a material (no convection) with temperature distribution $T(x,y,z)$ and constant thermal conductivity k, the heat conduction equation is derived by locally applying the Fourier conduction law:

$$q_c \big|_{\text{in } \zeta \text{ direction}} = -k \frac{\partial T}{\partial \zeta} \tag{1-1}$$

For an elemental cube within a solid as shown in Fig. 1-1a, consideration of the net heat flow in and out of all the faces, using the terms given in the figure, yields the Laplace equation governing the heat conduction within the material

$$\frac{\partial^2 T}{\partial x^2} + \frac{\partial^2 T}{\partial y^2} + \frac{\partial^2 T}{\partial z^2} = 0 \tag{1-2}$$

The terms in this energy-balance equation depend only on local temperature derivatives in the material.

A similar although more complex analysis can be made for the convection process, again demonstrating that the heat balance depends only on the conditions in the immediate vicinity of the location being considered.

In radiation, energy is transmitted between separated elements *without* the need of a medium between the elements. Consider a heated enclosure of surface S and volume V filled with radiating material (such as hot gas or glass) as shown in Fig. 1-1b. If $q_s \, dS$ is the radiant energy flux (energy per unit area and per unit time) arriving at dA from an element of the surface dS of the enclosure, and $q_v \, dV$ arrives at dA from

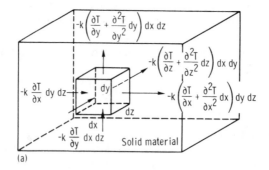

(a)

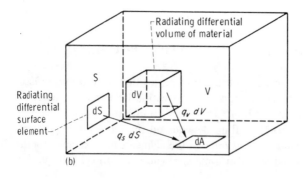

(b)

Figure 1-1 Comparison of types of terms for conduction and radiation heat balances. (*a*) Heat-conduction terms for volume element in solid; (*b*) radiation terms for enclosure filled with radiating material.

an element of the gas dV, then the total radiation arriving per unit area at dA is

$$q_r = \int_S q_s \, dS + \int_V q_v \, dV \qquad (1\text{-}3)$$

These types of terms lead to heat balances in the form of integral equations, which are generally not as familiar to the engineer as differential equations. When radiation is combined with conduction and/or convection, the presence of both integral and differential terms having different powers of temperature leads to nonlinear integro-differential equations. These are generally difficult to solve.

In addition to the mathematical difficulties, there is a second complexity inherent in radiation problems: accurately specifying the physical property values to be inserted into the equations. The difficulty in specifying accurate property values arises because the properties for solids depend on many variables, such as surface roughness and degree of polish, purity of material, thickness of a coating such as paint on a surface (for a thin coating the underlying material may have an effect), temperature, wavelength of radiation, and angle at which radiation leaves the surface. Unfortunately, many measurements have been reported where all the pertinent surface conditions have not been precisely defined.

1-4 WAVE AGAINST QUANTUM MODEL

The theory of radiant energy propagation can be considered from two viewpoints: classical electromagnetic wave theory and quantum mechanics. The classical view of the interaction of radiation and matter yields, in most cases, equations that are remarkably similar to the quantum-mechanical results. With a few exceptions, thermal radiation may therefore be viewed as a phenomenon based on the classical concept of the transport of energy by electromagnetic waves. These exceptions, however, include some of the most important effects common to radiative-transfer studies, such as the spectral distribution of the energy emitted from a body and the radiative properties of gases. These can only be explained and derived on the basis of quantum effects in which the energy is assumed to be carried by discrete particles (photons). The "true" nature of electromagnetic energy (that is, waves or quanta) is not known, nor is it generally important to the engineer. Throughout the present work, the wave theory will generally be applied because it has the greatest utility in engineering calculations and generally produces the same formal equations as the quantum theory. Occasional reference will be made to phenomena for which quantum arguments must be invoked.

1-5 ELECTROMAGNETIC SPECTRUM

Within the framework of the wave theory, electromagnetic radiation follows the laws governing transverse waves that oscillate in a direction perpendicular to the direction of travel. The speed of propagation for electromagnetic radiation in vacuum is the same as for light; light, after all, is simply the special case of electromagnetic radiation in a small

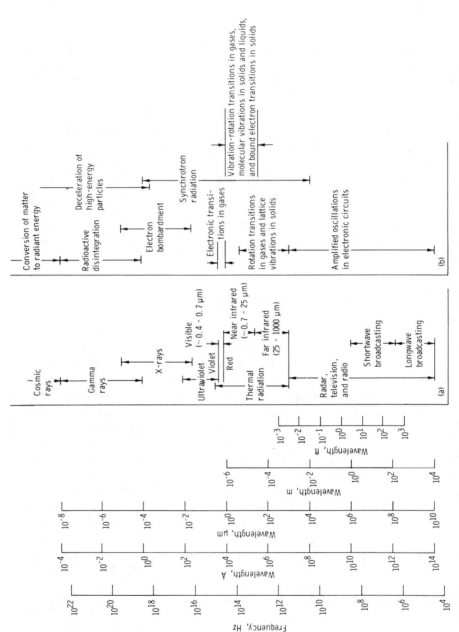

Figure 1-2 Spectrum of electromagnetic radiation. (*a*) Type of radiation; (*b*) production mechanism.

region of the spectrum. In vacuum the speed of propagation is $c_0 = 2.998 \times 10^8$ meters per second (m/s) or 186,000 miles per second (mi/s). The speed c in a medium is less than c_0 and is commonly given in terms of the index of refraction $n = c_0/c$, where n is greater than unity.* For glass, n is about 1.5, while for gases, n is very close to 1.

The types of electromagnetic radiation can be classified according to their wavelength λ in vacuum (or frequency ν where $c_0 = \lambda\nu$). Common units for wavelength measurement are the micrometer† (μm), where 1 μm $= 10^{-6}$ m or 10^{-4} cm, and the angstrom (Å) where 1 Å $= 10^{-10}$ m. Hence 10^4 Å $= 1$ μm. A chart of the radiation spectrum is shown in Fig. 1-2. A set of conversion factors for basic units in radiative transfer is given in Tables A-2 and A-3 in Appendix A.

The region of interest here includes a portion of the long-wave fringe of the ultraviolet, the visible-light region that extends from wavelengths of approximately 0.4 to 0.7 μm, and the infrared region that extends from beyond the red end of the visible spectrum to about $\lambda = 1000$ μm. The infrared region is sometimes divided into the near infrared, extending from the visible region to about $\lambda = 25$ μm, and the far infrared, composed of the longer-wavelength portion of the infrared spectrum.

The column at the far right in Fig. 1-2 indicates the various mechanisms by which electromagnetic radiation is produced. Some of the descriptions reflect a quantum-mechanical viewpoint in which electrons or molecules in a state of agitation undergo transitions from one energy state to another at lower energy. These transitions result in a radiative energy release. The transitions may occur spontaneously, or they may be initiated by the presence of a radiation field.

In this chapter we have discussed the importance of thermal radiation, the difficulties inherent in radiation problems, and the wavelength region occupied by thermal radiation within the electromagnetic spectrum. In the next chapter, the radiative behavior of the *ideal* radiating body, termed a *blackbody*, will be examined. With the behavior of this ideal as a standard for comparison, the behavior of radiative energy for conditions of interest to the engineer will be discussed in succeeding chapters.

---

* For attenuating media such as metals the index of refraction is a complex quantity of which n is only the real part. In some instances, such as in the region of anomalous dispersion, n can be less than unity which at first glance might convey the impression that the propagation speed can be greater than c_0. This is not the case; in this instance the waves propagating in the medium can be complicated in form. The c is then the phase velocity of the wave, which has no physical significance when it exceeds c_0. For further discussion see section 1.3 of Max Born and Emil Wolf, "Principles of Optics," 2nd rev. ed., The Macmillan Company, New York, 1964, and Sec. 4-5.2 of this text.

† The term *micron* (μ) has also been used to represent 10^{-6} m; micrometer (μm), however, is preferred.

TWO

RADIATION FROM A BLACKBODY

Before discussing the idealized concept of the blackbody, let us consider a few aspects of the interaction of incident radiant energy with matter. The idea we are concerned with is that the interaction at the surface of a body is not the result of only a surface property but depends as well on the bulk material beneath the surface.

When radiation is incident on a homogeneous body, some of the radiation is reflected, and the remainder penetrates into the body. The radiation may then be absorbed as it travels through the medium. If the material thickness required to substantially absorb the radiation is large compared with the thickness dimension of the body, then most of the radiation will be transmitted entirely through the body and will emerge with its nature unchanged. If, on the other hand, the material is a strong internal absorber, the radiation that is not reflected from the body will be converted into internal energy within a very thin layer adjacent to the surface. A distinction must be made between the ability of a material to let radiation pass through its surface and its ability to internally absorb the radiation after it has passed into the body. For example, a highly polished metal will generally reflect all but a small portion of the incident radiation, but the radiation passing into the body will be strongly absorbed and converted to internal energy within a very short distance within the material. Thus the metal has strong *internal* absorption ability, although it is a poor absorber for the incident beam since most of the incident beam is reflected. Nonmetals may exhibit the opposite tendency. Nonmetals may allow a substantial portion of the incident beam to pass into the material, but a larger thickness may be required than in the case of a metal to internally absorb the radiation and convert it into internal energy. A glass window readily allows radiation to pass through its surface, but is a poor absorber for visible radiation, and hence that radiation is transmitted. When *all* the radiation that passes into the body is absorbed internally, the body is called *opaque*.

To be a good absorber for incident energy, a material must have a low surface reflectivity and sufficiently high internal absorption to prevent the radiation from passing through. If metals in the form of very fine particles are deposited on a sub-surface, the result is a surface of low reflectivity. This effect, combined with the high internal absorption of the metal, causes this type of surface to be a good absorber. This is the basis for formation of the metallic "blacks," such as platinum or gold black [1]. A blackbody must have zero surface reflection and complete internal absorption.

2-1 SYMBOLS

A	surface area
C_1, C_2	constants in Planck's spectral energy distribution (see Table A-4 of Appendix A)
C_3	constant in Wien's displacement law (see Table A-4 of Appendix A)
c	speed of electromagnetic radiation propagation in medium other than a vacuum
c_0	speed of electromagnetic radiation propagation in vacuum
E	energy emitted per unit time
e	emissive power
$F_{0-\lambda}$	fraction of total blackbody intensity or emissive power lying in spectral region $0-\lambda$
h	Planck's constant
i	radiant intensity
k	Boltzmann constant
n	refractive index
Q	energy per unit time
R	radius
T	absolute temperature
ζ	the quantity $C_2/\lambda T$
η	wave number
θ	polar, or cone, angle (measured from normal of surface)
κ	extinction coefficient for electromagnetic radiation
λ	wavelength in vacuum
λ_m	wavelength in medium other than a vacuum
ν	frequency
φ	circumferential angle
σ	Stefan-Boltzmann constant [Eq. (2-22)]
ω	solid angle

Superscript

′	directional quantity

Subscripts

b	blackbody
max	corresponding to maximum energy
n	normal direction
p	projected
s	sphere
η	wave-number-dependent
λ	spectrally (wavelength) dependent
λ_1-λ_2	in wavelength span λ_1-λ_2
λT	evaluated at λT
ν	frequency-dependent

2-2 DEFINITION OF A BLACKBODY

A *blackbody* is defined as an ideal body that allows *all* the incident radiation to pass into it (no reflected energy) and absorbs internally *all* the incident radiation (no transmitted energy). This is true of radiation for all wavelengths and for all angles of incidence. Hence *the blackbody is a perfect absorber of incident radiation.* All other qualitative aspects of blackbody behavior can be derived from this definition.

The concept of a blackbody is basic to the study of radiative energy transfer. As a perfect absorber, it serves as a standard with which real absorbers can be compared. As will be seen, the blackbody also emits the maximum radiant energy and hence serves as an ideal standard of comparison for a body emitting radiation. The radiative properties of the ideal blackbody have been well established by quantum theory and have been verified by experiment.

Only a few surfaces, such as carbon black, carborundum, platinum black, and gold black, approach the blackbody in their ability to absorb radiant energy. The blackbody derives its name from the observation that good absorbers of incident visible light do indeed appear black to the eye. However, except for the visible region the eye is not a good indicator of absorbing ability in the wavelength range of thermal radiation. For example, a surface coated with white oil-base paint is a very good absorber for infrared radiation emitted at room temperature, although it is a poor absorber for the shorter-wavelength region characteristic of visible light.

2-3 PROPERTIES OF A BLACKBODY

Aside from being a perfect absorber of radiation, the blackbody has other important properties, now to be discussed.

2-3.1 Perfect Emitter

Consider a blackbody at a uniform temperature placed in vacuum within a perfectly insulated enclosure of arbitrary shape whose walls are also composed of blackbodies at some uniform temperature initially different from that of the enclosed blackbody

(Fig. 2-1). After a period of time, the blackbody and the enclosure will attain a common uniform equilibrium temperature. In this equilibrium condition, the blackbody must radiate exactly as much energy as it absorbs. To prove this, consider what would happen if the incoming and outgoing amounts of radiation were not equal with the system at a uniform temperature. Then the enclosed blackbody would either increase or decrease in temperature. This would involve a net amount of heat transferred between two bodies at the same temperature, which is in violation of the second law of thermodynamics. It follows then that because the blackbody is by definition absorbing the maximum possible radiation from the enclosure at each wavelength and from each direction, it must also be emitting the maximum total amount of radiation. This is made clear by considering any less-than-perfect absorber, which must emit less energy than the blackbody to remain in equilibrium. The fact that a body must continue to emit radiation even when in thermal equilibrium with its surroundings is called *Prevost's law.*

2-3.2 Radiation Isotropy in a Black Enclosure

Now consider the isothermal enclosure with black walls and arbitrary shape shown in Fig. 2-1, move the blackbody to another position, and rotate it to another orientation. The blackbody must still be at the same temperature because the whole enclosure remains isothermal. Consequently the blackbody must be emitting the same amount of radiation as before. To be in equilibrium, the body must still be receiving the same amount of radiation from the enclosure walls. Thus, the total radiation received by the blackbody is independent of body orientation or position throughout the enclosure; therefore, the radiation traveling through any point within the enclosure is independent

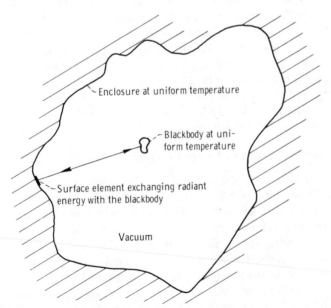

Figure 2-1 Enclosure geometry for derivation of blackbody properties.

of position or direction. This means that the black radiation filling the enclosure is *isotropic*.

In addition to emitting the maximum possible total radiation, the blackbody emits the maximum possible energy at each wavelength and in each direction. This is shown by the following arguments.

2-3.3 Perfect Emitter in Each Direction

Consider an area element on the surface of the black isothermal enclosure and an elemental blackbody within the enclosure. Some of the radiation from the surface element strikes the elemental body and is at an angle to the body surface. All this radiation, by definition, is absorbed. To maintain thermal equilibrium and isotropic radiation throughout the enclosure, the radiation emitted back into the incident direction must equal that received. Since the body is absorbing the maximum radiation from any direction, it must be emitting the maximum in any direction. Furthermore, since the black radiation filling the enclosure is isotropic, the radiation received or emitted in *any* direction by the enclosed black surface, per unit projected area normal to that direction, must be the same as that in any other direction.

2-3.4 Perfect Emitter at Every Wavelength

Consider a blackbody inside an evacuated enclosure with the whole system in thermal equilibrium. The enclosure boundary is specified as being of a very special type—it emits and absorbs radiation only in the small wavelength interval $d\lambda_1$ around λ_1. The blackbody, being a perfect absorber, absorbs all the incident radiation in this wavelength interval. To maintain the thermal equilibrium of the enclosure, the blackbody must reemit radiation in this same wavelength interval; the radiation can then be absorbed by the enclosure boundary which only absorbs in this particular wavelength interval. Since the blackbody is absorbing a maximum of the radiation in $d\lambda_1$, it must be emitting a maximum in $d\lambda_1$. A second enclosure can now be specified that only emits and absorbs in the interval $d\lambda_2$ around λ_2. The blackbody must then emit a maximum at the wavelength λ_2. In this manner the blackbody is shown to be a perfect emitter at each wavelength. The special nature of the enclosure assumed in this discussion is of no significance relative to the blackbody, because the emissive properties of a body depend only on the nature of the body and are independent of the enclosure.

2-3.5 Total Radiation into a Vacuum a Function Only of Temperature

If the enclosure temperature is altered, the enclosed blackbody temperature must adjust and become equal to the new enclosure temperature (that is, the complete isolated system must tend toward thermal equilibrium). The system is again isothermal and evacuated, and the absorbed and emitted energy of the blackbody will again be equal to each other, although the magnitude will differ from the value for the previous enclosure temperature. Since by definition the body absorbs (and hence emits) the maximum amount corresponding to this temperature, the characteristics of the surroundings do not affect the emissive behavior of the blackbody. Hence, *the total radiant energy emitted by a blackbody in vacuum is a function only of its temperature.*

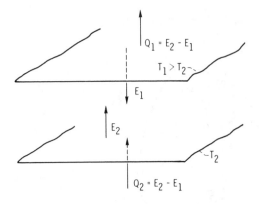

Figure 2-2 Device violating second law of thermodynamics.

Further, the second law of thermodynamics forbids net energy transfer from a cooler to a hotter surface without doing work on the system. If the radiant energy emitted by a blackbody increased with decreasing temperature, we could easily build a device to violate this law. Consider, for example, the infinite parallel black plates shown in Fig. 2-2. The upper plate is held at temperature T_1, which is higher than the temperature T_2 of the lower plate. If the emission of energy decreases with increasing temperature, then the energy E_2, emitted per unit time by plate 2, is larger than E_1, that emitted by plate 1. Because the plates are black, each absorbs all energy emitted by the other. To maintain the temperature of the plates, an amount of energy $Q_1 = E_2 - E_1$ must be extracted from plate 1 per unit time and an equal amount added to plate 2. Thus we are transferring net energy from the colder to the warmer plate without doing external work. Experience as embodied in the second law of thermodynamics says that this cannot be done. Therefore, the radiant energy emitted by a blackbody must *increase* with temperature.

From these arguments, the total radiant energy emitted by a blackbody is expected to be proportional only to a monotonically increasing function of temperature.

2-4 EMISSIVE CHARACTERISTICS OF A BLACKBODY

2-4.1 Definition of Blackbody Radiation Intensity*

Consider an elemental surface area dA surrounded by a hemisphere of radius R as shown in Fig. 2-3. A hemisphere has a surface area of $2\pi R^2$ and subtends a solid angle

*The system of units and definition of terms used here have been made as self-consistent as possible to avoid confusion. This is not true for all areas of radiation, where separate interests and needs have caused a great variety of inconsistent systems of units and definitions to be used. A good example of this was provided to the authors by Dr. Fred Nicodemus, who sent a data sheet used in the field of ophthalmology to define units of luminance. Enough comment is probably provided by the following equality taken from the data sheet:

1 nit (nt) = 3.142 apostilbs (asb) = 10^4 Bougie-Hectomètre-Carré = 0.2919 footlambert (fl)

Units for photometry, radiometry, and photon flux radiometry are defined for the various quantities in SI units in [2].

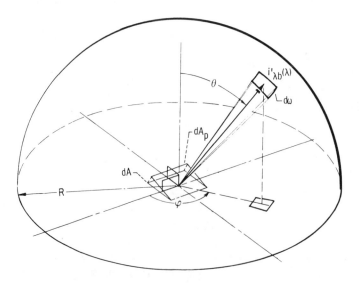

Figure 2-3 Spectral emission intensity from black surface.

of 2π steradians (sr) about a point at the center of its base. Hence by considering a hemisphere of unit radius, the solid angle about the center of the base can be regarded directly as the area on the unit hemisphere. Direction is measured by the angles θ and φ as shown in Fig. 2-3, where the angle θ is measured from the direction *normal* to the surface. The angular position for $\varphi = 0$ is arbitrary.

The radiation emitted in any direction will be defined in terms of the *intensity*. There are two types of intensities: the *spectral intensity* refers to radiation in an interval $d\lambda$ around a single wavelength, while the *total intensity* refers to the combined radiation including all wavelengths. The spectral intensity of a blackbody will be given by $i'_{\lambda b}(\lambda)$. The subscripts denote, respectively, that one wavelength is being considered and that the properties are for a blackbody. The prime denotes that radiation per unit solid angle in a single direction is being considered. The notation is explained in detail in Sec. 3-1.2. The emitted spectral intensity is defined as the energy emitted per unit time per unit small wavelength interval around the wavelength λ, per unit elemental *projected surface area* normal to the (θ,φ) direction and into a unit elemental solid angle centered around the direction (θ,φ). As will be shown in Sec. 2-4.2, the blackbody intensity defined in this way (that is, on the basis of projected area) is independent of direction; hence, the symbol for blackbody intensity is not modified by any (θ,φ) designation. The total intensity i'_b is defined analogously to $i'_{\lambda b}$ except that it includes the radiation for all wavelengths; hence, the subscript λ and the functional dependence (λ) do not appear. The spectral and total intensities are related by the integral over all wavelengths

$$i'_b = \int_{\lambda=0}^{\infty} i'_{\lambda b}(\lambda)\, d\lambda \tag{2-1}$$

2-4.2 Angular Independence of Intensity

The angular independence of the blackbody intensity can be shown by considering a spherical isothermal blackbody enclosure of radius R with a blackbody element dA at its center, as shown in Fig. 2-4a. Once again, the enclosure and the central elemental body are in thermal equilibrium. Thus all radiation in transit throughout the enclosure must be isotropic. Consider radiation in a wavelength interval $d\lambda$ about λ that is emitted by an element dA_s on the enclosure surface and travels toward the central element dA (Fig. 2-4b). The emitted energy in this direction per unit solid angle and time is $i'_{\lambda b,n}(\lambda)\,dA_s\,d\lambda$. The normal spectral intensity of a blackbody is used because the energy is emitted normal to the black wall element dA_s of the spherical enclosure. The amount of energy per unit time that impinges upon dA depends on the solid angle that dA occupies when viewed from the location of dA_s. This solid angle is the projected area of dA normal to the (θ,φ) direction divided by R^2. The projected area of dA is

$$dA_p = dA \cos\theta \tag{2-2}$$

Then the energy absorbed by dA is

$$d^3 Q'_{\lambda b}(\lambda,\theta,\varphi) = i'_{\lambda b,n}(\lambda)\,dA_s\,d\lambda\frac{dA\cos\theta}{R^2} \tag{2-3}$$

The energy emitted by dA in the (θ,φ) direction and incident on dA_s (Fig. 2-4c) must be equal to that absorbed from dA_s, or equilibrium would be disturbed; hence,

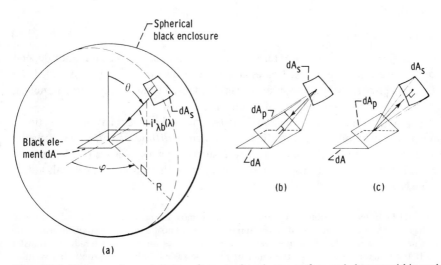

(a)

Figure 2-4 Energy exchange between element of enclosure surface and element within enclosure. (a) Black element dA within black spherical enclosure; (b) energy transfer from dA_s to dA_p; (c) energy transfer from dA_p to dA_s.

$$i'_{\lambda b}(\lambda,\theta,\varphi)\,dA_p\,\frac{dA_s}{R^2}\,d\lambda = d^3Q'_{\lambda b}(\lambda,\theta,\varphi) = i'_{\lambda b,n}(\lambda)\,dA_s\,\frac{dA\cos\theta}{R^2}\,d\lambda \tag{2-4}$$

Then, by virtue of (2-2),

$$i'_{\lambda b}(\lambda,\theta,\varphi) = i'_{\lambda b,n}(\lambda) \neq \text{function of } \theta,\varphi \tag{2-5}$$

Equation (2-5) shows that the *intensity of radiation from a blackbody*, as defined here on the basis of projected area, *is independent of the direction of emission*. Neither the subscript n nor the (θ,φ) notation is really needed for complete description of the black intensity. Since the blackbody is always a perfect absorber and emitter, these properties of the blackbody are independent of its surroundings. Hence, these results are independent of both the assumptions used in the derivation of a spherical enclosure and thermodynamic equilibrium with the surroundings.*

2-4.3 Blackbody Emissive Power—Definition and Cosine-Law Dependence

The intensity has been defined on the basis of projected area. It is useful also to define a quantity that gives the energy emitted in a given direction per unit of actual (unprojected) surface area. This is defined as $e'_{\lambda b}(\lambda,\theta,\varphi)$ which is the energy emitted by a black surface per unit time within a unit small wavelength interval centered around the wavelength λ, per unit elemental surface area and into a unit elemental solid angle centered around the direction (θ,φ). The energy in the wavelength interval $d\lambda$ centered about λ emitted per unit time in any direction $d^3Q'_{\lambda b}(\lambda,\theta,\varphi)$ can then be expressed in the two forms

$$d^3Q'_{\lambda b}(\lambda,\theta,\varphi) = e'_{\lambda b}(\lambda,\theta,\varphi)\,dA\,d\omega\,d\lambda = i'_{\lambda b}(\lambda)\,dA\cos\theta\,d\omega\,d\lambda$$

Consequently, there exists the relation

$$e'_{\lambda b}(\lambda,\theta,\varphi) = i'_{\lambda b}(\lambda)\cos\theta = e'_{\lambda b}(\lambda,\theta) \tag{2-6}$$

It is evident from the $i'_{\lambda b}(\lambda)\cos\theta$ term in Eq. (2-6) that $e'_{\lambda b}(\lambda,\theta,\varphi)$ does not depend on φ and hence can be expressed as $e'_{\lambda b}(\lambda,\theta)$. The quantity $e'_{\lambda b}(\lambda,\theta)$ is called the *directional spectral emissive power* for a black surface. In the case of some nonblack surfaces, there will be a dependence of e'_λ on angle φ.

Equation (2-6) is known as *Lambert's cosine law*, and surfaces having a directional emissive power that follows this relation are known as *diffuse* or *cosine-law* surfaces. A blackbody, because it is always a diffuse surface, serves as a standard for comparison with the directional properties of real surfaces that do not, in general, follow the cosine law.

---

*It should be noted that some exceptions do exist for most of the blackbody "laws" presented in this chapter. The exceptions are of minor importance in almost any practical engineering situation but need to be considered when extremely rapid transients are present in a radiative-transfer process. If the transient period is of the order of the time scale of whatever process is governing the emission of radiation from the body in question, then the emission properties of the body may lag the absorption properties. In such a case, the concepts of temperature used in the derivation of the blackbody laws no longer hold rigorously. The treatment of such problems is outside the scope of this work (see Sec. 13-9 for further discussion).

2-4.4 Hemispherical Spectral Emissive Power of a Blackbody

In calculations of total radiant energy rejection by a surface, one needs the spectral emissive power integrated over all solid angles of a hemispherical envelope placed over a black surface. This quantity is called the *hemispherical spectral emissive power* of a black surface $e_{\lambda b}(\lambda)$. It is the energy leaving a black surface per unit time per unit area and per unit wavelength interval around λ. Figure 2-5 shows the elemental area dA at the center of the base of a unit hemisphere. By definition a solid angle anywhere above dA is equal to the intercepted area on the unit hemisphere. An element of this hemispherical area is given by

$$d\omega = \sin \theta \, d\theta \, d\varphi$$

Hence, the spectral emission from dA per unit time and unit surface area passing through the element on the hemisphere is given by

$$e'_{\lambda b}(\lambda, \theta) \sin \theta \, d\theta \, d\varphi$$

By virtue of (2-6), this is equal to

$$e'_{\lambda b}(\lambda, \theta) \, d\omega = i'_{\lambda b}(\lambda) \cos \theta \sin \theta \, d\theta \, d\varphi \qquad (2\text{-}7)$$

To obtain the blackbody emission passing through the entire hemisphere, Eq. (2-7) is integrated over all solid angles to give

$$e_{\lambda b}(\lambda) = i'_{\lambda b}(\lambda) \int_{\varphi=0}^{2\pi} \int_{\theta=0}^{\pi/2} \cos \theta \sin \theta \, d\theta \, d\varphi \qquad (2\text{-}8a)$$

or

$$e_{\lambda b}(\lambda) = 2\pi i'_{\lambda b}(\lambda) \int_{0}^{1} \sin \theta \, d(\sin \theta) = \pi i'_{\lambda b}(\lambda) \qquad (2\text{-}8b)$$

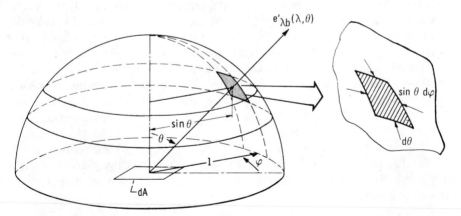

Figure 2-5 Unit hemisphere used to obtain relation between blackbody intensity and hemispherical emissive power.

where the prime notation is absent in the designation of the hemispherical quantity. Also from (2-6), when the emission is normal to the surface ($\theta = 0$) so that $\cos \theta = 1$,

$$e'_{\lambda b, n}(\lambda) = i'_{\lambda b}(\lambda)$$

and, from (2-8b),

$$e_{\lambda b}(\lambda) = \pi e'_{\lambda b, n}(\lambda) \tag{2-9}$$

Hence, purely from the geometry involved, this simple relation is found: *The blackbody hemispherical emissive power is π times the intensity or π times the directional emissive power normal to the surface.* This relation for a blackbody will prove to be very useful in relating directional and hemispherical quantities in following chapters.

2-4.5 Spectral Emissive Power through a Finite Solid Angle

Sometimes the emission through only part of the hemispherical solid angle enclosing an area element may be desired. The emission through the solid angle extending from θ_1 to θ_2 and φ_1 to φ_2 is found by modifying the limits of integration in Eq. (2-8a):

$$e_{\lambda b}(\lambda, \theta_1 - \theta_2, \varphi_1 - \varphi_2) = i'_{\lambda b}(\lambda) \int_{\varphi_1}^{\varphi_2} \int_{\theta_1}^{\theta_2} \cos \theta \sin \theta \, d\theta \, d\varphi$$

$$= i'_{\lambda b}(\lambda) \frac{\sin^2 \theta_2 - \sin^2 \theta_1}{2} (\varphi_2 - \varphi_1) \tag{2-10}$$

2-4.6 Spectral Distribution of Emissive Power; Planck's Law

Some of the blackbody characteristics that have been discussed are: the blackbody is defined as a perfect absorber and is also a perfect emitter; its total intensity and total emissive power in vacuum are functions only of the temperature of the blackbody; the emitted blackbody energy follows Lambert's cosine law.

All these blackbody properties have been demonstrated by thermodynamic arguments. However, a very important fundamental property of the blackbody remains to be presented. This is the formula that gives the magnitude of the emitted intensity at each of the wavelengths that, taken together, constitute the radiation spectrum. This relation cannot be obtained from purely thermodynamic arguments. Indeed, the search for this formula led Planck to an investigation and hypothesis that became the foundation of quantum theory. The derivation of the spectral distribution is beyond the scope of the present discussion; therefore, the results will be presented here without derivation. The interested reader may consult various standard physics texts [3–5] for the complete development.

It has been shown by the quantum arguments of Planck [6] and verified experimentally that for a blackbody the spectral distributions of hemispherical emissive power and radiant intensity in a *vacuum* are given as a function of absolute temperature and wavelength by

$$e_{\lambda b}(\lambda) = \pi i'_{\lambda b}(\lambda) = \frac{2\pi C_1}{\lambda^5 (e^{C_2/\lambda T} - 1)} \tag{2-11a}$$

This is known as *Planck's spectral distribution of emissive power.* As will be shown, for radiation into a medium in which the speed of light is not close to c_0, Eq. (2-11a) must be modified by including index-of-refraction multiplying factors (see Sec. 2-4.12). For most engineering work the radiant emission is into air or other gases with an index of refraction $n = c_0/c$, so close to unity that (2-11a) is applicable. The values of the constants C_1 and C_2 are given in Table A-4 in two common systems of units. These constants are equal to $C_1 = hc_0^2$ and $C_2 = hc_0/k$, where h is Planck's constant and k is the Boltzmann constant.* Equation (2-11a) is of great importance, as it provides quantitative results for the radiation from a blackbody.

EXAMPLE 2-1 A plane black surface is radiating at a temperature of 1500°F. What is the directional spectral emissive power of the blackbody at an angle of 60° from the normal and at a wavelength of 6 μm?
From Eq. (2-11a),

$$i'_{\lambda b}(6\ \mu m) = \frac{2 \times 0.1889 \times 10^8\ \text{Btu} \cdot \mu m^4/(\text{h} \cdot \text{ft}^2)}{6^5\ \mu m^5 (e^{25,898/(6 \times 1960)} - 1)\ \text{sr}} = 604\ \text{Btu}/(\text{h} \cdot \text{ft}^2 \cdot \mu m \cdot \text{sr})$$

From Eq. (2-6) the directional emissive power is

$$e'_{\lambda b}(6\ \mu m, 60°) = 604 \cos 60° = 302\ \text{Btu}/(\text{h} \cdot \text{ft}^2 \cdot \mu m \cdot \text{sr})$$

EXAMPLE 2-2 The sun emits like a blackbody at 5780 K. What is the sun's intensity at the center of the visible spectrum?
From Fig. 1-2 the wavelength of interest is 0.55 μm. Then, from (2-11a),

$$i'_{\lambda b}(0.55\ \mu m) = \frac{2 \times 0.59544 \times 10^8\ \text{W} \cdot \mu m^4/m^2}{0.55^5\ \mu m^5 (e^{14388/(0.55 \times 5780)} - 1)\ \text{sr}}$$

$$= 0.259 \times 10^8\ \text{W}/(m^2 \cdot \mu m \cdot \text{sr})$$

Alternative forms of Eq. (2-11a) are employed where frequency or wave number is used rather than wavelength. The use of frequency has an advantage when radiation travels from one medium into another, since in that situation the frequency remains constant while the wavelength changes because of the change in propagation velocity. To make the transformation of (2-11a) to frequency, note that in vacuum $\lambda = c_0/\nu$, and hence $d\lambda = -(c_0/\nu^2)\ d\nu$. Then the hemispherical emissive power in the wavelength interval $d\lambda$ becomes

$$e_{\lambda b}(\lambda)\ d\lambda = \frac{2\pi C_1\ d\lambda}{\lambda^5 (e^{C_2/\lambda T} - 1)} = \frac{-2\pi C_1 \nu^3\ d\nu}{c_0^4 (e^{C_2 \nu/c_0 T} - 1)} = -e_{\nu b}(\nu)\ d\nu \tag{2-11b}$$

The quantity $e_{\nu b}(\nu)$ is the emissive power in vacuum *per unit frequency interval* about ν. The intensity is $i'_{\nu b}(\nu) = e_{\nu b}(\nu)/\pi$.

---

*$h = 6.626 \times 10^{-27}\ \text{erg} \cdot \text{s} = 6.626 \times 10^{-34}\ \text{J} \cdot \text{s}$ and $k = 1.381 \times 10^{-16}\ \text{erg/K} = 1.381 \times 10^{-23}\ \text{J/K}$. In some literature, the constant C_1 is defined as $2\pi hc_0^2$.

The wave number $\eta = 1/\lambda$ is the number of waves per unit length. Then

$$d\lambda = -\frac{1}{\eta^2} d\eta$$

and

$$e_{\lambda b}(\lambda) \, d\lambda = -\frac{2\pi C_1 \eta^3 \, d\eta}{e^{C_2 \eta/T} - 1} = -e_{\eta b}(\eta) \, d\eta \tag{2-11c}$$

The quantity $e_{\eta b}(\eta)$ is the emissive power *per unit wave number interval* about η. The intensity is $i'_{\eta b}(\eta) = e_{\eta b}(\eta)/\pi$.

For better understanding of the implications of Eq. (2-11a), it has been plotted in Fig. 2-6. Here the hemispherical spectral emissive power is given as a function of wavelength for several different values of the absolute temperature. One characteristic that is quite evident is that the energy emitted at all wavelengths increases as the temperature increases. It was shown in Sec. 2-3.5, and is known from common experience, that the total (that is, including all wavelengths) radiated energy must increase with temperature; the curves show that this is also true for the energy at each wavelength. Another characteristic is that the peak spectral emissive power shifts toward a smaller wavelength as the temperature is increased. A cross plot of Fig. 2-6 giving energy as a function of temperature for fixed wavelengths shows that the energy emitted at the shorter-wavelength end of the spectrum increases more rapidly with temperature than the energy at the long wavelengths.

The position of the range of wavelengths included in the visible spectrum is included in Fig. 2-6. For a body at $1000°R$ (555 K) only a very small amount of energy would be in the visible region and would not be sufficient to be detected by eye. Since the curves at the lower temperatures slope downward from the red toward the violet end of the spectrum, the red light becomes visible first as the temperature is raised.* Higher temperatures make visible additional wavelengths of the visible light range, and at a sufficiently high temperature the light emitted becomes white, representing radiation composed of a mixture of all the visible wavelengths.

For the filament of an incandescent lamp to operate efficiently, the temperature must be high; otherwise too much of the electrical energy would be dissipated as radiation in the infrared region rather than in the visible range. Most tungsten filament lamps operate at about $5400°R$ (3000 K), and thus do give off a large fraction of their energy in the infrared, but their filament vaporization rate limits the temperature to near this value. The sun emits a spectrum quite similar to that of a blackbody at a temperature of about $10,400°R$ (5780 K), and an appreciable amount of energy release is in what we sense as the visible region. This may be because evolution has caused the human eye to be most sensitive in the spectral region of greatest energy received.† If the eye were sensitive in other regions (for example, the infrared, so that we could see thermal images in the "dark"), then our definition of the "visible region

---

*This occurs at the so-called *Draper point* of $977°F$ (798 K) [7], at which red light first becomes visible from a heated object in darkened surroundings.

†Although human vision is well tuned to the sun, this is not true of all beings on earth. Turtles have eyes sensitive to infrared but not blue; bees have eyes sensitive to ultraviolet but not red.

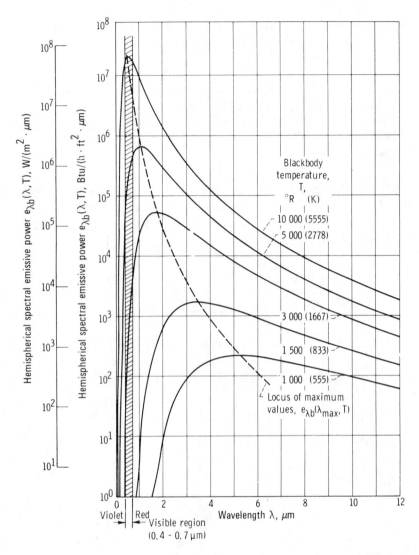

Figure 2-6 Hemispherical spectral emissive power of blackbody for several different temperatures.

of the spectrum" would change. If we find life in other solar systems having a sun with an effective temperature different from ours, it will be interesting to discover what wavelength range encompasses the "visible spectrum" if the beings there possess sight.

Equation (2-11a) can be placed in a more convenient form that eliminates the need for providing a separate curve for each value of T. This is done by dividing by the fifth power of temperature to obtain

$$\frac{e_{\lambda b}(\lambda,T)}{T^5} = \frac{\pi i'_{\lambda b}(\lambda,T)}{T^5} = \frac{2\pi C_1}{(\lambda T)^5 (e^{C_2/\lambda T} - 1)} \tag{2-12}$$

This equation gives the quantity $e_{\lambda b}(\lambda, T)/T^5$ in terms of the single variable λT. A plot of this relation is given in Fig. 2-7 and replaces the multiple curves in Fig. 2-6. A compilation of values is presented in Table A-5.

EXAMPLE 2-3 For a blackbody at $1500°R$, what is the spectral hemispherical emissive power at a wavelength of 2 μm? Use Table A-5.

The value of λT is 3000 $\mu m \cdot °R$. From Table A-5, at this λT, $e_{\lambda b}/T^5 = 87.047 \times 10^{-15}$ Btu/(h·ft²·μm·°R⁵). Then $e_{\lambda b}(2 \ \mu m) = 87.047 \times 10^{-15} (1500)^5 = 661$ Btu/(h·ft²·μm).

EXAMPLE 2-4 What is the blackbody intensity at $\lambda = 2$ μm for $T = 2500$ K?

From Table A-5 at $\lambda T = 5000$ $\mu m \cdot K$, $e_{\lambda b}/T^5 = 0.71383 \times 10^{-11}$ W/(m²·μm ·K⁵). Then since $i'_{\lambda b} = e_{\lambda b}/\pi$, $i'_{\lambda b}(2 \ \mu m) = 0.71383 \times 10^{-11}(1/\pi)(2500^5) = 22.2 \times 10^4$ W/(m²·μm·sr).

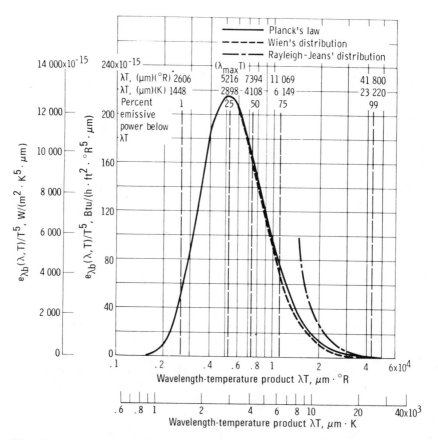

Figure 2-7 Spectral distribution of blackbody hemispherical emissive power.

2-4.7 Approximations for Spectral Distribution

Planck's spectral distribution gives the maximum (blackbody) intensity of radiation that any body can emit at a given wavelength for a given temperature. This intensity serves as an optimum standard with which real surface performance can be compared. In Chap. 3, the methods of comparison will be defined. Planck's distribution also provides a means to evaluate the maximum radiative performance that can be attained for any radiating device.

Some approximate forms of Planck's distribution are occasionally useful because of their simplicity. Care must be taken to use them only in the range in which their accuracy is acceptable.

Wien's formula. If the term $e^{C_2/\lambda T}$ is much larger than 1, Eq. (2-12) reduces to

$$\frac{i'_{\lambda b}(\lambda,T)}{T^5} = \frac{2C_1}{(\lambda T)^5 e^{C_2/\lambda T}} \tag{2-13}$$

which is known as *Wien's formula*. It is accurate to within 1% for λT less than 5400 μm · °R (3000 μm · K).

Rayleigh-Jeans formula. Another approximation is found by taking the denominator of (2-12) and expanding it in a series to obtain

$$e^{C_2/\lambda T} - 1 = 1 + \frac{C_2}{\lambda T} + \frac{1}{2!}\left(\frac{C_2}{\lambda T}\right)^2 + \frac{1}{3!}\left(\frac{C_2}{\lambda T}\right)^3 + \cdots - 1 \tag{2-14}$$

For λT much larger than C_2, this series can be approximated by the single term $C_2/\lambda T$, and Eq. (2-12) becomes

$$\frac{i'_{\lambda b}(\lambda,T)}{T^5} = \frac{2C_1}{C_2}\frac{1}{(\lambda T)^4} \tag{2-15}$$

This is known as the *Rayleigh-Jeans formula* and is accurate to within 1% for λT greater than 14×10^5 μm · °R (7.78×10^5 μm · K). This is well outside the range generally encountered in thermal radiation problems, since a blackbody emits over 99.9% of its energy at λT values below this. The formula has utility for long-wave radiation of other classifications, such as radio waves.

A comparison of these approximate formulas with the Planck distribution is shown in Fig. 2-7.

2-4.8 Wien's Displacement Law

Another quantity of interest with regard to the blackbody emissive spectrum is the wavelength λ_{max} at which the emissive power $e_{\lambda b}(\lambda)$ is a maximum for a given temperature. This maximum shifts toward shorter wavelengths as the temperature is increased, as shown by the dashed line in Fig. 2-6. The value of $\lambda_{max}T$ can be found at the peak of the distribution curve given in Fig. 2-7. Alternatively, it can be found analytically by differentiating Planck's distribution from Eq. (2-12) and setting the left side equal to zero. This gives the transcendental equation

$$\lambda_{max} T = \frac{C_2}{5} \frac{1}{1 - e^{-C_2/\lambda_{max}T}} \tag{2-16}$$

The solution to this equation is of the form

$$\lambda_{max} T = C_3 \tag{2-17}$$

which is one form of *Wien's displacement law*. Values of the constant C_3 are given in Table A-4. Equation (2-17) indicates that the peak emissive power and intensity shift to a shorter wavelength at a higher temperature in inverse proportion to T.

EXAMPLE 2-5 For a blackbody to radiate its maximum emissive power $e_{\lambda b}$ at the center of the visible spectrum, what would its temperature have to be?

Figure 1-2 shows that the visible spectrum spans the range 0.4–0.7 μm, and the center of the range is at 0.55 μm. From Eq. (2-17)

$$T = \frac{C_3}{\lambda_{max}} = \frac{5216 \ \mu m \cdot °R}{0.55 \ \mu m} = 9480°R$$

or

$$T = \frac{2898 \ \mu m \cdot K}{0.55 \ \mu m} = 5270 \ K$$

This is close to the effective radiating surface temperature of the sun, which is 10,400°R (5780 K).

2-4.9 Total Intensity and Emissive Power

The previous discussion has provided the energy per unit wavelength interval that a blackbody radiates into vacuum at each wavelength. It will now be shown how the total intensity can be determined, which includes the radiation for all wavelengths. The result is a surprisingly simple relation.

The intensity emitted over the small wavelength interval $d\lambda$ is given by $i'_{\lambda b}(\lambda) \, d\lambda$. Integrating over all wavelengths $\lambda = 0-\infty$ gives the *total intensity*

$$i'_b = \int_0^\infty i'_{\lambda b}(\lambda) \, d\lambda \tag{2-18}$$

This integral may be evaluated by substitution of Planck's distribution from (2-12) and a transformation of variables using $\zeta = C_2/\lambda T$. Equation (2-18) then becomes

$$
\begin{aligned}
i'_b &= \int_0^\infty \frac{2C_1}{\lambda^5 (e^{C_2/\lambda T} - 1)} \, d\lambda \\
&= \int_\infty^0 \left(\frac{C_2}{\lambda T}\right)^5 \left(\frac{T}{C_2}\right)^5 \frac{2C_1}{e^{C_2/\lambda T} - 1} \left(\frac{\lambda T}{C_2}\right)^2 \left(\frac{-C_2}{T}\right) d\left(\frac{C_2}{\lambda T}\right) \\
&= \frac{2C_1 T^4}{C_2^4} \int_0^\infty \frac{\zeta^3}{e^\zeta - 1} \, d\zeta \tag{2-19}
\end{aligned}
$$

From a table of integrals [8] this can be evaluated as

$$i'_b = \frac{2C_1 T^4}{C_2{}^4} \frac{\pi^4}{15} \tag{2-20}$$

Defining a new constant results in

$$i'_b = \frac{\sigma}{\pi} T^4 \tag{2-21}$$

where the constant is

$$\sigma = \frac{2C_1 \pi^5}{15 C_2{}^4} = 0.1712 \times 10^{-8} \text{ Btu}/(\text{h} \cdot \text{ft}^2 \cdot {}^\circ \text{R}^4) = 5.6696 \times 10^{-8} \text{ W}/(\text{m}^2 \cdot \text{K}^4) \tag{2-22}$$

The *hemispherical total emissive power* of a black surface into vacuum is then

$$e_b = \int_0^\infty e_{\lambda b}(\lambda)\, d\lambda = \int_0^\infty \pi i'_{\lambda b}(\lambda)\, d\lambda = \sigma T^4 \tag{2-23}$$

which is known as the *Stefan-Boltzmann law*, where σ is the *Stefan-Boltzmann constant*. The value of σ as determined experimentally differs slightly from that calculated by Eq. (2-22) (see Table A-4).

EXAMPLE 2-6 The beam emitted normal to a blackbody surface is found to have a total radiation per unit solid angle and per unit surface area of 10,000 W/$(\text{m}^2 \cdot \text{sr})$. What is the surface temperature?

The hemispherical total emissive power is related to the total emissive power in the normal direction by $e_b = \pi e'_{b,n}$. Hence, from Eq. (2-23) $T = (\pi e'_{b,n}/\sigma)^{1/4} = (10,000\pi/5.729 \times 10^{-8})^{1/4} = 861$ K. The experimental value of the Stefan-Boltzmann constant has been used.

EXAMPLE 2-7 A black surface is radiating with a hemispherical total emissive power of 2000 Btu/$(\text{h} \cdot \text{ft}^2)$. What is the surface temperature? At what wavelength is its maximum spectral emissive power?

From the Stefan-Boltzmann law, the temperature of the blackbody is $T = (e_b/\sigma)^{1/4} = (2000/0.173 \times 10^{-8})^{1/4} = 1037^\circ$R. Then from Wien's displacement law, $\lambda_{max} = C_3/T = 5216/1037 = 5.03$ μm.

EXAMPLE 2-8 An electric flat plate heater that is square with a 0.1 m edge length, is radiating 10^2 W from each side. If the heater can be considered black, what is its temperature?

Using the Stefan-Boltzmann law,

$$T = (Q/A\sigma)^{1/4} = \left[\frac{10^4 \text{ W}/\text{m}^2}{5.729 \times 10^{-8} \text{ W}/(\text{m}^2 \cdot \text{K}^4)} \right]^{1/4} = 646 \text{ K}$$

The spectral intensity of a black surface $i'_{\lambda b}(\lambda)$ was shown in Eq. (2-5) to be independent of the angle of emission. Integrating over all wavelengths of course did not change this angular independence. The intensity of a surface is what the eye interprets as "brightness." A black surface will exhibit the same brightness when viewed from any angle.

2-4.10 Behavior of Maximum Intensity with Temperature

The intensity at a given wavelength is found from Planck's spectral distribution. It is interesting to note that substitution of Wien's displacement law [Eq. (2-17)] into Eq. (2-12) gives

$$i'_{\lambda \max b} = T^5 \frac{2C_1}{C_3^5 (e^{C_2/C_3} - 1)} = C_4 T^5 \tag{2-24}$$

where C_4 is in Table A-4. This shows that the maximum intensity increases as temperature to the fifth power. Indeed, because $i'_{\lambda b}/T^5$ is a function only of λT as shown by Eq. (2-12), it is evident that if the blackbody temperature is changed from T_1 to T_2 and at the same time the wavelengths λ_1 and λ_2 are chosen such that $\lambda_1 T_1 = \lambda_2 T_2$, the value of $i'_{\lambda b}/T^5$ remains unchanged. Therefore, the intensity at λ_2 for temperature T_2 changes as $(T_2/T_1)^5$ from the intensity at λ_1 for temperature T_1. This is the general statement of Wien's law.

2-4.11 Blackbody Radiation in a Wavelength Interval

The Stefan-Boltzmann law shows that the hemispherical total emissive power of a blackbody radiating into vacuum is given by

$$e_b = \int_0^\infty e_{\lambda b}(\lambda)\, d\lambda = \sigma T^4$$

It is often desirable in calculations of radiative exchange to determine the fraction of the total emissive power that is emitted in a given wavelength band as illustrated by Fig. 2-8. This fraction is designated by $F_{\lambda_1 - \lambda_2}$ and is given by the ratio

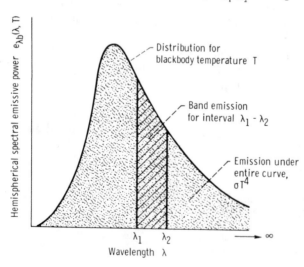

Figure 2-8 Emitted energy in wavelength band.

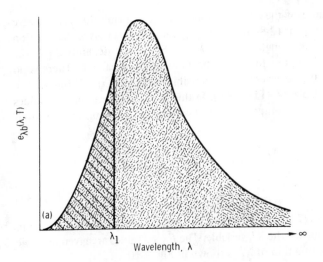

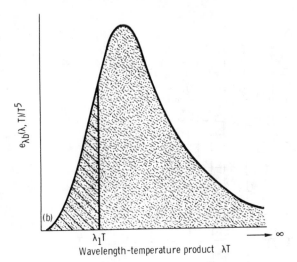

Figure 2-9 Physical representation of F factor, where $F_{0-\lambda_1}$ or $F_{0-\lambda_1 T}$ is ratio of crosshatched to shaded area. (*a*) In terms of curve for specific temperature. Entire area under curve, σT^4. (*b*) In terms of universal curve. Entire area under curve, σ.

$$F_{\lambda_1 - \lambda_2} = \frac{\int_{\lambda_1}^{\lambda_2} e_{\lambda b}(\lambda)\, d\lambda}{\int_0^\infty e_{\lambda b}(\lambda)\, d\lambda} = \frac{1}{\sigma T^4} \int_{\lambda_1}^{\lambda_2} e_{\lambda b}(\lambda)\, d\lambda \tag{2-25}$$

The last integral in Eq. (2-25) can be expressed by two integrals each with a lower limit at $\lambda = 0$:

$$F_{\lambda_1 - \lambda_2} = \frac{1}{\sigma T^4} \left[\int_0^{\lambda_2} e_{\lambda b}(\lambda)\, d\lambda - \int_0^{\lambda_1} e_{\lambda b}(\lambda)\, d\lambda \right] = F_{0-\lambda_2} - F_{0-\lambda_1} \tag{2-26}$$

The fraction of the emissive power for any wavelength band can therefore be found by having available values of $F_{0-\lambda}$ as a function λ. The $F_{0-\lambda_1}$ function is illustrated by Fig. 2-9*a*, where it equals the crosshatched area divided by the total area (shaded) under the curve.

For a blackbody, because of the simple manner in which hemispherical emissive power is related to intensity [Eq. (2-8b)], the $F_{\lambda_1-\lambda_2}$ function also gives the fraction of the intensity within the wavelength interval $\lambda_1-\lambda_2$. Since $e_{\lambda b}$ depends on T, the application of (2-26) would require that $F_{0-\lambda}$ be tabulated for each T. There is no need to have this complexity, however, as it is possible to arrange the F function in terms of only the single variable λT (Fig. 2-9b). In this way a universal set of F values is obtained that can apply for all temperatures and wavelengths. The universal form is found by rewriting (2-26) as

$$F_{\lambda_1-\lambda_2} = F_{\lambda_1 T - \lambda_2 T} = \frac{1}{\sigma}\left[\int_0^{\lambda_2 T} \frac{e_{\lambda b}(\lambda)}{T^5}\,d(\lambda T) - \int_0^{\lambda_1 T} \frac{e_{\lambda b}(\lambda)}{T^5}\,d(\lambda T)\right]$$

$$= F_{0-\lambda_2 T} - F_{0-\lambda_1 T} \tag{2-27}$$

As shown by Eq. (2-12), $e_{\lambda b}/T^5$ is only a function of λT so that the integrands in Eq. (2-27) are only dependent on the λT variable. The $F_{0-\lambda T}$ values are given in Table A-5, and a plot of $F_{0-\lambda T}$ as a function of λT is shown in Fig. 2-10.

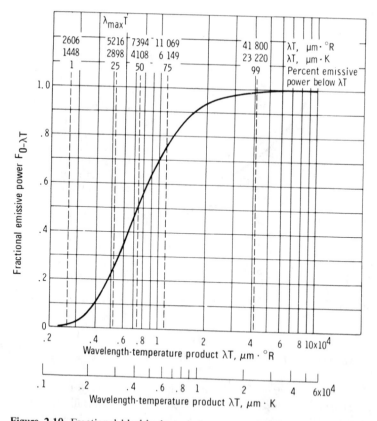

Figure 2-10 Fractional blackbody emissive power in range 0 to λT.

The tabulated $F_{0-\lambda T}$ values are also available in expanded form for use where even greater accuracy is required. The tables of Pivovonsky and Nagel [9], for example, tabulate values for every λT interval of 10 μm·K over a very wide range of λT. Other sources of blackbody functions are [10, 11]. Polynomial-approximation expressions for $F_{0-\lambda T}$ are also available and are included in Appendix A.

The compilation of values of $F_{0-\lambda T}$ has a number of uses, as illustrated in the following examples.

EXAMPLE 2-9 A blackbody is radiating at a temperature of 5000°R (2778 K). An experimenter wishes to measure the total radiant emission by use of a radiation detector. This detector absorbs all radiation in the λ range 0.8–5 μm, but detects no energy outside that range. What percentage correction will the experimenter have to apply to the energy measurement? If the sensitivity of the detector could be extended in range by 0.5 μm at only one end of the sensitive range, which end should be extended?

Taking $\lambda_1 T = 0.8 \times 5000 = 4000$ μm·°R (2222 μm·K) and $\lambda_2 T = 5 \times 5000 = 25,000$ μm·°R (13,890 μm·K) results in the fraction of energy outside the sensitive range being $F_{0-\lambda_1 T} + F_{\lambda_2 T-\infty} = F_{0-\lambda_1 T} + (1 - F_{0-\lambda_2 T}) = (0.1050 + 1 - 0.9621) = 0.1429$ or a correction of 14.3% of the total incident energy. Extending the sensitive range to the longer-wavelength side of the measurement interval adds little accuracy because of the small slope of the curve of F against λT in that region, so extending to shorter wavelengths would provide the greatest increase in detected energy.

EXAMPLE 2-10 The experimenter of the previous example has designed a radiant energy detector that can only be made sensitive over any 1-μm range of wavelength. The experimenter wants to measure the total emissive power of two blackbodies, one at 5000° and the other at 10,000°R, and plans to adjust the 1-μm interval to give a 0.5-μm sensitive band on each side of the peak blackbody emissive power. For which blackbody would you expect to detect the greatest percentage of the total emissive power? What will the percentage be in each case?

Wien's displacement law tells us that the peak emissive power will occur at $\lambda_{max} = 5216/T$ μm in each case. For the higher temperature a wavelength interval of 1 μm will give a wider spread of λT values around the peak of $\lambda_{max}T$ on the normalized blackbody curve (Fig. 2-7), so the measurement should be more accurate for the 10,000°R case. For the 10,000°R blackbody, $\lambda_{max} = 0.5216$ μm, and $\lambda_1 T = (0.5216 - 0.5000) \times 10,000 = 216$ μm·°R. Similarly, $\lambda_2 T = 1.0216 \times 10,000 = 10,216$ μm·°R. The percentage of detected emissive power is then

$$100(F_{0-10,216} - F_{0-216}) = 100 \times (0.708 - 0) = 70.8\%$$

A similar calculation for the 5000°R blackbody shows that 51.7% of the emissive power is detected.

EXAMPLE 2-11 A light-bulb filament is at 3000 K. Assuming the filament radiates a spectrum like a blackbody, what fraction of its energy is emitted in the visible region?

The visible region is within $\lambda = 0.4$–0.7 μm. The desired fraction is then

Table 2-1 Fraction of blackbody emission contained in the range 0-λT

λT		
$\mu m \cdot °R$	$\mu m \cdot K$	$F_{0-\lambda T}$
2,606	1,448	0.01
5,216 = $\lambda_{max}T$	2,898	0.25
7,394	4,108	0.50
11,069	6,149	0.75
41,800	23,220	0.99

$$F_{0-2100\,\mu m \cdot K} - F_{0-1200\,\mu m \cdot K} = 0.0831 - 0.0022 = 0.081$$

Some commonly used values of $F_{0-\lambda T}$ are given in Table 2-1. It is interesting to note that exactly one-fourth of the total emissive power lies in the wavelength range below the peak of the Planck spectral distribution at any temperature. This relation appears to have no simple physical explanation and must be put down alongside those other phenomena, such as gravitational attraction and the Stefan-Boltzmann fourth-power law, in which nature provides us with a simple law to describe an apparently complex event.

2-4.12 Blackbody Emission into a Medium Other than a Vacuum

The previous expressions for blackbody emission have been for emission into a vacuum or medium where $n \approx 1$. When emission is considered from a blackbody within a large volume of medium other than a vacuum, the quantities C_1 and C_2 appearing in Planck's energy distribution equation [Eq. (2-11a)] are replaced by

$$C_1' = hc^2 \tag{2-28a}$$

$$C_2' = \frac{hc}{k} \tag{2-28b}$$

so that

$$e_{\lambda mb}(\lambda_m)\, d\lambda_m = \frac{2\pi C_1'}{\lambda_m{}^5(e^{C_2'/\lambda_m T} - 1)}\, d\lambda_m \tag{2-29}$$

where k is the Boltzmann constant, h is Planck's constant, c is the speed of propagation of light *in the medium considered*, and λ_m is the wavelength *in the medium*.

Since the speed c depends on the medium, it is better to define C_1 and C_2 in terms of c_0, the speed of light in vacuum, so that C_1 and C_2 are then truly constants. The speed in the medium is given by $c = c_0/n$, where n is the index of refraction. Planck's distribution for the energy in a wavelength interval $d\lambda_m$ becomes (note that λ_m is the wavelength in the *medium*)

$$e_{\lambda mb}(\lambda_m)\,d\lambda_m = \frac{2\pi c^2 h}{\lambda_m{}^5(e^{ch/k\lambda_m T}-1)}\,d\lambda_m = \frac{2\pi c_0{}^2 h}{n^2\lambda_m{}^5(e^{c_0 h/nk\lambda_m T}-1)}\,d\lambda_m$$

$$e_{\lambda mb}(\lambda_m) = \frac{2\pi C_1}{n^2\lambda_m{}^5(e^{C_2/n\lambda_m T}-1)} \tag{2-30a}$$

In Eq. (2-30a), $C_1 = hc_0{}^2$ and $C_2 = hc_0/k$, which are the values of C_1 and C_2 presented in Table A-4. Equation (2-30a) can be placed in generalized form as in Eq. (2-12):

$$\frac{e_{\lambda mb}}{n^3 T^5} = \frac{2\pi C_1}{(n\lambda_m T)^5(e^{C_2/n\lambda_m T}-1)} \tag{2-30b}$$

Thus, Fig. 2-7 applies if the ordinate is replaced by $e_{\lambda mb}/n^3 T^5$, and the abscissa by $n\lambda_m T$. In frequency form, Eq. (2-30a) becomes

$$e_{\nu b} = \frac{2\pi n^2 C_1 \nu^3}{c_0{}^4(e^{C_2\nu/c_0 T}-1)} \tag{2-31}$$

The ν in the medium is the same as in vacuum. The $\lambda_m = \lambda/n$ where λ is the value in vacuum.

The integration of (2-30a) over all wavelengths follows as in (2-19) when n is constant. This yields the Stefan-Boltzmann law for hemispherical total emissive power into a medium of refractive index n:

$$e_{b,m} = n^2\sigma T^4 \tag{2-32}$$

The emission from a blackbody within glass ($n \approx 1.5$) is thus 2.25 times that leaving a blackbody in air (see Sec. 19-6.2).

Finally, by similar arguments Wien's displacement law becomes

$$n\lambda_{\max,m} T = C_3 \tag{2-33}$$

where $\lambda_{\max,m}$ is the wavelength at peak emission *into a medium*.

These refinements are only carried in a few succeeding sections because their applicability to engineering radiation problems is small. A notable exception is the work of Gardon and others [12–14] dealing with radiation effects in molten glass. These types of problems are discussed in Sec. 19-6, and in more detail by Viskanta and Anderson [15].

2-5 EXPERIMENTAL PRODUCTION OF A BLACKBODY

When making experimental measurements of the radiative properties of real materials, it is desirable to have a black surface for reference so that a direct comparison can be made between the real surface and the ideal (black) surface. Since perfectly black surfaces do not exist in nature, a special technique is utilized to provide a very close approximation to a black area. Figure 2-11 shows a metal cylinder that has been hollowed out to form a cavity with a small opening. If an incident beam passes into the cavity as shown, it strikes the cavity wall, and part is absorbed with the remainder

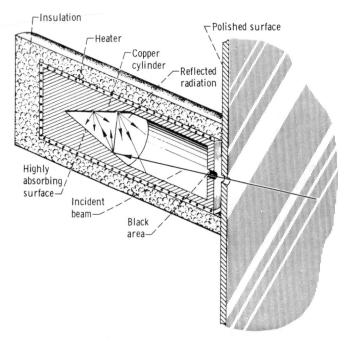

Figure 2-11 Cavity used to produce blackbody area.

being reflected. The reflected portion strikes other parts of the wall and is again partially absorbed. It is evident that, if the opening to the cavity is very small, very little of the original incident beam will manage to escape back out through the opening. Thus, if the opening is made sufficiently small, the opening area approaches the behavior of a black surface because essentially all the radiation passing in through it is absorbed. To help keep the cavity at a uniform temperature so that the internal radiation will all be in thermal equilibrium, the cavity shown in Fig. 2-11 is machined from a copper cylinder and surrounded by insulation. By heating the cavity, a source of black radiation is obtained at the opening since, as previously discussed in Sec. 2-3.1, a perfectly absorbing surface is also perfectly emitting. The polished surface at the front of the cavity aids in shielding the opening from stray radiation from the surroundings.

The attainment of isothermal conditions in such a cavity (often referred to as a *hohlraum*) is a difficult but necessary condition in the accurate experimental determination of radiative properties. Another difficulty in the use of a blackbody cavity is the upper temperature limit imposed by the heater and wall materials. This limits the energy output, especially in the short-wavelength portion of the spectrum.

If the cavity is assumed perfectly isothermal and perfectly insulated, and if the exit hole is infinitesimally small so that it does not disturb the radiative equilibrium in the enclosure, then the radiation from the hole is blackbody radiation and hence the cavity is completely filled with blackbody radiation. An interesting feature is that under these conditions, the radiation leaving the cavity wall will be blackbody radiation even though the wall is not a perfect emitter. When the blackbody radiation within the cavity strikes the wall, part of it will be absorbed and the remainder will be reflected. Since the wall is perfectly insulated on the outside, the absorbed energy must be reemitted on the inside. The combination of reflected and emitted energy from the wall must equal the incident blackbody radiation.

If a body that is small, so that it does not disturb conditions in the cavity, is placed in the cavity it will come to equilibrium at the cavity temperature. Then, since the net energy exchange with the body must be zero, the radiation leaving the body by combined reflection and emission must be blackbody radiation. Hence the radiation from the body will be the same as that from all the surroundings, and the body will not be visible inside the cavity.

2-6 SUMMARY OF BLACKBODY PROPERTIES

It has been shown in this chapter that the ideal blackbody possesses certain fundamental properties that make it a standard with which real radiating bodies can be compared. These properties, listed here for convenience, are the following:

1. The blackbody is the best possible absorber and emitter of radiant energy at any wavelength and in any direction.
2. The total radiant intensity and hemispherical total emissive power of a blackbody into a medium with index of refraction n are given by the Stefan-Boltzmann law:

$$\pi i_b' = e_b = n^2 \sigma T^4$$

3. The blackbody directional spectral and total emissive power follow Lambert's cosine law:

$$e_{\lambda b}'(\lambda, \theta) = e_{\lambda b, n}'(\lambda) \cos \theta$$

$$e_b'(\theta) = e_{b, n}' \cos \theta$$

4. The spectral distribution of intensity of a blackbody is given by Planck's distribution:

$$i_{\lambda b}'(\lambda) = \frac{2C_1}{\lambda^5 (e^{C_2/\lambda T} - 1)} \qquad \text{emission into vacuum}$$

$$i_{\lambda m b}'(\lambda_m) = \frac{2C_1}{n^2 \lambda_m^5 (e^{C_2/n\lambda_m T} - 1)} \qquad \text{emission into a medium}$$

5. The wavelength at which the maximum spectral intensity of radiation for a blackbody occurs is given by Wien's displacement law:

$$\lambda_{max} = \frac{C_3}{T} \qquad \text{emission into vacuum}$$

$$\lambda_{max, m} = \frac{C_3}{nT} \qquad \text{emission into a medium}$$

Because of the many definitions introduced in this chapter, it is convenient to summarize the quantities in tabular form. This has been done in Table 2-2. The formulas for the quantities are given in terms of either the spectral intensity $i'_{\lambda b}(\lambda)$, which is computed from Planck's law, or the surface temperature T.

2-7 HISTORICAL DEVELOPMENT

The derivation of the approximate spectral distributions of Wien and of Rayleigh and Jeans, the Stefan-Boltzmann law, and Wien's displacement law are all seen to be logical consequences of the spectral distribution of intensity as derived by Max Planck. However, it is interesting to note that all these relations were formulated *prior* to publication of Planck's work in 1901 and were originally derived through fairly complex thermodynamic arguments.

Joseph Stefan [16] proposed in 1879, after study of some experimental results, that emissive power was related to the fourth power of the absolute temperature of a radiating body. Ludwig Edward Boltzmann [17] was able to derive the same relation in 1884 by analyzing a Carnot cycle in which radiation pressure was assumed to act as the pressure of the working fluid.

Wilhelm Carl Werner Otto Fritz Franz (Willy) Wien [18] derived the displacement law in 1891 by consideration of a piston moving within a mirrored cylinder. He found that the spectral energy density in an isothermal enclosure and the spectral emissive power of a blackbody are both directly proportional to the fifth power of the absolute temperature when "corresponding wavelengths" are chosen. The relation presented in Sec. 2-4.8 [Eq. (2-17)] is more often cited as Wien's displacement law, but is actually a consequence of the previous sentence.

Wien [19] also derived his spectral distribution of intensity through thermodynamic argument plus assumptions concerning the absorption and emission processes.

Lord Rayleigh (1900) and Sir James Jeans (1905) based their spectral distribution on the assumption that the classical idea of equipartition of energy was valid [20, 21].

The fact that measurements and some theoretical considerations* indicated Wien's expression for the spectral distribution to be invalid at high temperatures and/or large wavelengths led Planck to an investigation of harmonic oscillators that were assumed to be the emitters and absorbers of radiant energy. Various further assumptions as to the average energy of the oscillators led Planck to derive both the Wien and the Rayleigh-Jeans distributions. Planck finally found an empirical equation that fit the measured energy distributions over the entire spectrum. In determining what modifications to the theory would allow derivation of this empirical equation, he was led to the assumptions that form the basis of the quantum theory. As we have seen, his equation

---

*It was felt that as temperature approaches large values, the intensity of a blackbody should not approach a finite limit. Examination of Wien's formula [Eq. (2-13)] shows that this condition is not met. Planck's distribution law [Eq. (2-11)], however, does satisfy the condition.

leads directly to all the results derived previously by Wien, Stefan, Boltzmann, Rayleigh, and Jeans.

For an interesting and informative comprehensive review of the history of the field of thermal radiation, the article by Barr [22] is recommended. Lewis [23] discusses the derivation of Planck's law.

REFERENCES

1. O'Neill, P., A. Ignatiev, and C. Doland: The Dependence of Optical Properties on the Structural Composition of Solar Absorbers: Gold Black, *Sol. Energy*, vol. 21, no. 6, pp. 465–468, 1978.
2. Nicodemus, Fred E. (ed.): "Self-Study Manual on Optical Radiation Measurements," part 1, "Concepts," National Bureau of Standards Technical Note 910-2, 1978.
3. Richtmyer, F. K., and E. H. Kennard: "Introduction to Modern Physics," 4th ed., McGraw-Hill Book Company, New York, 1947.
4. Ter Haar, D.: "Elements of Statistical Mechanics," 2nd ed., Holt, Rinehart and Winston Inc., New York, 1960.
5. Tribus, Myron: "Thermostatics and Thermodynamics: An Introduction to Energy, Information and States of Matter, with Engineering Applications," D. Van Nostrand Company, Inc., Princeton, N.J., 1961.
6. Planck, Max: Distribution of Energy in the Spectrum, *Ann. Phys.*, vol. 4, no. 3, pp. 553–563, 1901.
7. Draper, John W.: On the Production of Light by Heat, *Phil. Mag.*, ser. 3, vol. 30, pp. 345–360, 1847.
8. Dwight, Herbert B.: "Tables of Integrals and Other Mathematical Data," 4th ed., p. 231, The Macmillan Company, New York, 1961.
9. Pivovonsky, Mark, and Max R. Nagel: "Tables of Blackbody Radiation Functions," The Macmillan Company, New York, 1961.
10. Pisa, F. J.: Tables of Black-Body Radiation Functions and Their Derivatives, *NAVWEPS Rep.* 8646, NOTS TP 3687, U.S. Naval Ordnance Test Station, China Lake, Calif., December, 1964.
11. Gebel, Radames K. H.: The Normalized Cumulative Blackbody Functions, Their Applications in Thermal Radiation Calculations, and Related Subjects, ARL-69-0004, Aerospace Research Laboratories, January, 1969.
12. Gardon, Robert: The Emissivity of Transparent Materials, *J. Am. Ceram. Soc.*, vol. 39, no. 8, pp. 278–287, 1956.
13. Gardon, Robert: A Review of Radiant Heat Transfer in Glass, *J. Am. Ceram. Soc.*, vol. 44, no. 7, pp. 305–312, 1961.
14. Kellett, B. S.: The Steady Flow of Heat through Hot Glass, *J. Opt. Soc. Am.*, vol. 42, no. 5, pp. 339–343, 1952.
15. Viskanta, R., and E. E. Anderson: Heat Transfer in Semitransparent Solids, in J. P. Hartnett and T. Irvine Jr. (eds.) "Advances in Heat Transfer," vol. 11, pp. 317–458, Academic Press, N.Y., 1975.
16. Stefan, Joseph: Über die Beziehung zwischen der Wärmestrahlung und der Temperatur, *Sitzber. Akad. Wiss. Wien*, vol. 79, pt. 2, pp. 391–428, 1879.
17. Boltzmann, Ludwig: Ableitung des Stefan'schen Gesetzes, betreffend die Abhängigkeit der Wärmestrahlung von der Temperatur aus der electromagnetischen Lichttheorie, *Ann. Phys.*, ser. 2, vol. 22, pp. 291–294, 1884.
18. Wien, Willy: Temperatur und Entropie der Strahlung, *Ann. Phys.*, ser. 2, vol. 52, pp. 132–165, 1894.

Table 2-2 Blackbody radiation quantities ($n \approx 1$)

Symbol	Name	Definition	Geometry	Formula
$i'_{\lambda b}(\lambda, T)$	Spectral intensity	Emission in any direction per unit of projected area normal to that direction, and per unit time, wavelength interval about λ, and solid angle		$\dfrac{2C_1}{\lambda^5(e^{C_2/\lambda T} - 1)}$
$i'_b(T)$	Total intensity	Emission, including all wavelengths, in any direction per unit of projected area normal to that direction, and per unit time and solid angle		$\dfrac{\sigma T^4}{\pi}$
$e'_{\lambda b}(\lambda, \theta, T)$	Directional spectral emissive power	Emission per unit solid angle in direction θ per unit surface area, wavelength interval, and time		$i'_{\lambda b} \cos \theta$
$e'_b(\theta, T)$	Directional total emissive power	Emission, including all wavelengths, in direction θ per unit surface area, solid angle, and time		$\dfrac{\sigma T^4}{\pi} \cos \theta$
$e_{\lambda b}(\lambda, \theta_1 - \theta_2, \varphi_1 - \varphi_2, T)$	Finite solid-angle spectral emissive power	Emission in solid angle $\theta_1 \le \theta \le \theta_2$, $\varphi_1 \le \varphi \le \varphi_2$ per unit surface area, wavelength interval, and time		$i'_{\lambda b} \dfrac{\sin^2 \theta_2 - \sin^2 \theta_1}{2}(\varphi_2 - \varphi_1)$

$e(\theta_1 - \theta_2, \varphi_1 - \varphi_2, T)$	Finite solid-angle total emissive power	Emission, including all wavelengths, in solid angle $\theta_1 \leqslant \theta \leqslant \theta_2, \varphi_1 \leqslant \varphi \leqslant \varphi_2$ per unit surface area and time		$\dfrac{\sigma T^4}{\pi}(\varphi_2 - \varphi_1)\dfrac{\sin^2\theta_2 - \sin^2\theta_1}{2}$
$e_{\lambda b}(\lambda_1 - \lambda_2, \theta_1 - \theta_2, \varphi_1 - \varphi_2, T)$	Finite solid-angle band emissive power	Emission in solid angle $\theta_1 \leqslant \theta \leqslant \theta_2, \varphi_1 \leqslant \varphi \leqslant \varphi_2$ and wavelength band $\lambda_1 - \lambda_2$ per unit surface area and time		$\dfrac{\sigma T^4}{\pi}(\varphi_2 - \varphi_2)\dfrac{\sin^2\theta_2 - \sin^2\theta_1}{2}$ $\times (F_{0-\lambda_2} - F_{0-\lambda_1})$
$e_{\lambda b}(\lambda, T)$	Hemispherical spectral emissive power	Emission into hemispherical solid angle per unit surface area, wavelength interval, and time		$\pi i'_{\lambda b}$
$e_{\lambda b}(\lambda_1 - \lambda_2, T)$	Hemispherical band emissive power	Emission in wavelength band $\lambda_1 - \lambda_2$ into hemispherical solid angle per unit surface area and time		$(F_{0-\lambda_2} - F_{0-\lambda_1})\sigma T^4$
$e_b(T)$	Hemispherical total emissive power	Emission, including all wavelengths, into hemispherical solid angle per unit surface area and time		σT^4

19. Wien, Willy: Über die Energievertheilung im Emissionsspectrum eines schwarzen Körpers, *Ann. Phys.*, ser. 3, vol. 58, pp. 662–669, 1896.
20. Lord Rayleigh: The Law of Complete Radiation, *Phil. Mag.*, vol. 49, pp. 539–540, 1900.
21. Jeans, Sir James: On the Partition of Energy between Matter and the Ether, *Phil. Mag.*, vol. 10, pp. 91–97, 1905.
22. Barr, E. Scott: Historical Survey of the Early Development of the Infrared Spectral Region, *Am. J. Phys.*, vol. 28, no. 1, pp. 42–54, 1960.
23. Lewis, Henry R.: Einstein's Derivation of Planck's Radiation Law, *Am. J. Phys.*, vol. 41, no. 1, pp. 38–44, 1973.

PROBLEMS

1 A blackbody in air is at a temperature of 1000 K.

  (*a*) What is the spectral intensity in a direction normal to the black surface at $\lambda = 3\ \mu$m?

  (*b*) What is the spectral intensity at $\theta = 60°$ away from the normal of the black surface at $\lambda = 3\ \mu$m?

  (*c*) What is the directional spectral emissive power from the surface at $\theta = 60°$ and $\lambda = 3\ \mu$m?

  (*d*) At what λ is the maximum spectral intensity emitted from this blackbody, and what is this intensity?

  (*e*) What is the hemispherical total emissive power of the blackbody?

  Answer: (*a*) 0.408 W/(cm² ·μm·sr); (*b*) 0.408 W/cm² ·μm·sr);
         (*c*) 0.204 W/(cm² ·μm·sr); (*d*) 2.898 μm, 0.409 W/cm² ·μm·sr);
         (*e*) 5.729 W/cm² .

2 Plot the hemispherical spectral emissive power $e_{\lambda b}$ for a blackbody in air [(W/(m²·μm)] as a function of wavelength (μm) for surface temperatures of 1000 and 5000 K.

3 A blackbody at 2000°R (1111 K) is radiating to space.

  (*a*) What is the ratio of the spectral intensity of the blackbody at $\lambda = 1\ \mu$m to the spectral intensity of the blackbody at $\lambda = 5\ \mu$m?

  (*b*) What fraction of the blackbody emissive power lies between the wavelengths of $\lambda = 1$ and $\lambda = 5\ \mu$m?

  (*c*) At what wavelength does the peak in the blackbody spectrum occur for this blackbody?

  (*d*) How much energy is emitted by the blackbody in the range $1 \leqslant \lambda \leqslant 5\ \mu$m?

  Answer: (*a*) 0.0916; (*b*) 0.696; (*c*) 2.61 μm;
         (*d*) 19,100 Btu/(h·ft²), 60,300 W/m².

4 The surface of the sun has an effective blackbody temperature of 5780 K. What percentage of the radiant emission of the sun lies in the visible range (0.4–0.7 μm)? What percentage is in the ultraviolet? At what wavelength and frequency is the maximum energy per unit wavelength emitted? What is the maximum value of the hemispherical spectral emissive power?

  Answer: 36.7%;   12.2%; 0.502 μm, 5.98 × 10¹⁴ Hz; 8.30 × 10⁷ W/(m² ·μm).

5 A blackbody radiates such that the wavelength at maximum emissive power is 1.5 μm. What fraction of the total emissive power from this blackbody is in the range $\lambda = 1$ to $\lambda = 4\ \mu$m?

  Answer: 0.788.

6 How high must the temperature of a blackbody be in order that one quarter of the energy emitted be in the visible region?

  Answer: 4345 K.

7 Show that $i_{\lambda b}'$ increases with T at any fixed value of λ.

8 Blackbody radiation is leaving a small hole in a furnace at 1400 K. What fraction of the radiation is intercepted by the annular disk? What fraction passes through the hole in the disk?

  Answer: 0.301;   0.0588.

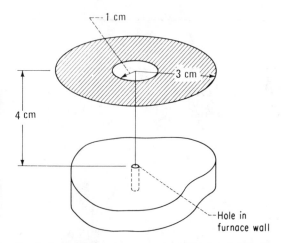

9   A sheet of silica glass transmits 92% of the incident radiation in the wavelength range between 0.35 and 2.7 μm, and is essentially opaque to radiation at longer and shorter wavelengths. Estimate the percent of solar radiation that the glass will transmit. (Consider the sun as a blackbody at 10,400°R (5780 K).)

   If the garden in a greenhouse radiates like a black surface and is at 100°F (38 C), what percent of this radiation will be transmitted through the glass?

   Answer: 83%;   0.003%.

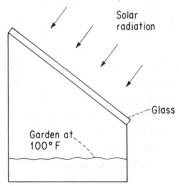

10   (*a*)   Derive Wien's displacement law by differentiation of Planck's spectral distribution in terms of wave number, and show that $T/\eta_{max} = 9160\ (\mu m \cdot °R)$.

   (*b*)   A student notes that the peak emission of the sun according to Wien's displacement law is at a wavelength of about $\lambda_{max} = C_3/10,400°R = 5216/10,400 = 0.501\ \mu m$. Using $\eta_{max} = 1/0.501\ \mu m$, the student solves again for the solar temperature using the result derived in part *a*. Does this computed temperature agree with the solar temperature? Why? (This is not trivial—put some thought into *why*.)

THREE

DEFINITIONS OF PROPERTIES FOR NONBLACK SURFACES

3-1 INTRODUCTION

In Chap. 2 the radiative behavior of a blackbody was presented in detail. The ideal behavior of the blackbody serves as a standard with which the performance of real radiating bodies can be compared. The radiative behavior of a real body depends on many factors such as composition, surface finish, temperature, wavelength of the radiation, angle at which radiation is either being emitted or intercepted by the surface, and the spectral distribution of the radiation incident on the surface. Various emissive, absorptive, and reflective properties, both unaveraged and averaged, are used to describe the radiative behavior of real materials relative to blackbody behavior.

The definitions of radiative properties of opaque materials are given in this chapter. To make them of greatest value, the definitions are presented rigorously and in detail. Since the definitions are numerous, the reader should not expect to read the present chapter in as complete detail as the discussion on the blackbody in Chap. 2. Rather, some of the alternative ways of defining the same quantity can be briefly scanned to obtain an overall view of what information is available, and the chapter then used as a reference source. The sections have been subdivided and made fairly independent to facilitate use for reference purposes.

The rigorous examination of radiative-property definitions arises from the need to properly interpret available property data for use in heat transfer computations. Limited amounts of data in the literature provide detailed directional and spectral measurements. Because of the difficulties in making these detailed measurements, most of the tabulated property values are *averaged* quantities. An averaged radiative performance has been measured for all directions, all wavelengths, or both. A clear understanding of the averages involved can be obtained from the definitions in this chapter. The definitions also reveal relations among various averaged properties in the form of equalities or reciprocity relations. This enables the researcher or engineer to make maximum use of the available property information. Thus, for example, absorptivity data can be obtained from measured emissivity data *if certain restrictions are observed*. These restrictions have often been misunderstood, resulting in confusion or inaccuracy in applying measured properties.

By detailed examination of the derivation of property definitions, the restrictions on the property relations are demonstrated. As an aid to understanding these definitions, Fig. 3-1 provides a schematic representation of the types of directional properties. The various parts of this figure will be referred to as each definition is introduced, to help provide a physical interpretation of the quantities being discussed. Further, Table 3-1 lists each of the properties, its symbolic notation, and the equation number of its definition. The notation is described in Sec. 3-1.2.

Measured properties of real materials are given in Chap. 5 to demonstrate the practical use of the relations derived here.

Quantities such as emissivity and absorptivity are usually thought of as surface properties. As discussed at the beginning of Chap. 2, for an opaque material the portion of incident radiation that is not reflected is absorbed within a layer that extends below the surface. This layer can be quite thin for a material with high internal absorptance such as a metal, or can be a millimeter or more for a less strongly internally absorbing dielectric material. Emission also takes place within the material extending below the surface, and the amount of energy leaving the surface depends on how much of the energy emitted by each internal volume element can reach and then penetrate the surface. As will be discussed in Sec. 19-6, some of the energy reaching the surface can be internally reflected from the surface back into the body, depending on the refractive index of the body relative to that of the medium outside the body. Thus, the emissive ability of a body depends on the refractive index of the surrounding medium. With perhaps rare exceptions, emissivity measurements are made for emission into air or vacuum so that tabulated values of ϵ_λ or ϵ are for an external medium with $n \approx 1$. It should be noted, however, that if the medium adjacent to the surface has $n > 1$, then ϵ_λ and α_λ, and hence ϵ and α, can be different from the tabulated values.

3-1.1 Nomenclature

A number of suggestions have been made in an effort to standardize the nomenclature of radiation. One controversy centers around the ending *ivity* for the various radiative properties of materials. The National Bureau of Standards (NBS) is attempting to standardize nomenclature, and in its publications reserves this ending for the properties of an optically smooth substance with an uncontaminated surface (emissivity, reflectivity, etc.), while assigning the *-ance* ending (emittance, reflectance, etc.) to measured properties where there is a need to specify surface conditions. In addition, NBS has published nomenclature for reflectance useful in the fields of illumination and measurement [1]. The nomenclature is quite close to that adopted here.

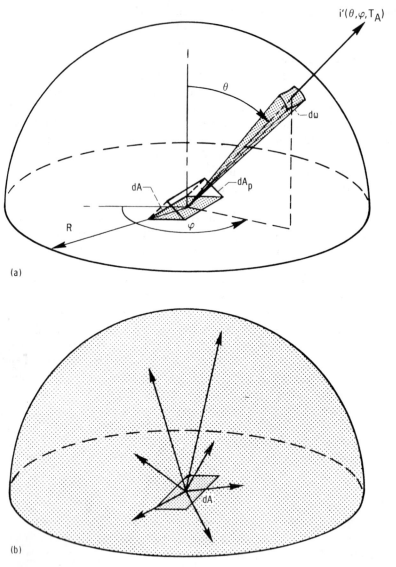

Figure 3-1 Pictorial description of directional and hemispherical radiation properties. (*a*) Directional emissivity $\epsilon'(\theta,\varphi,T_A)$; (*b*) hemispherical emissivity $\epsilon(T_A)$.

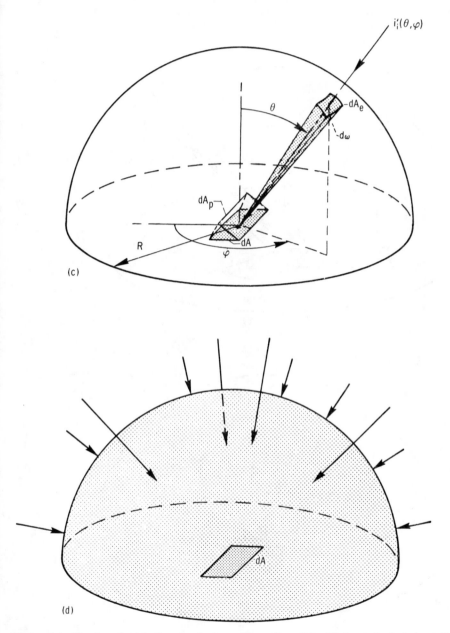

Figure 3-1 (*Continued*) (*c*) directional absorptivity $\alpha'(\theta,\varphi,T_A)$; (*d*) hemispherical absorptivity $\alpha(T_A)$.

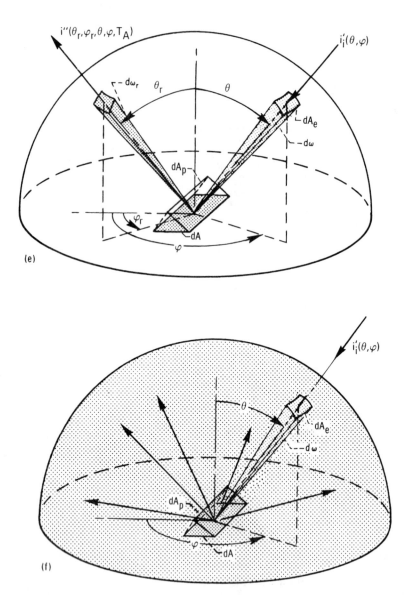

Figure 3-1 (*Continued*) (*e*) bidirectional reflectivity $\rho''(\theta_r,\varphi_r,\theta,\varphi,T_A)$; (*f*) directional-hemispherical reflectivity $\rho'(\theta,\varphi,T_A)$.

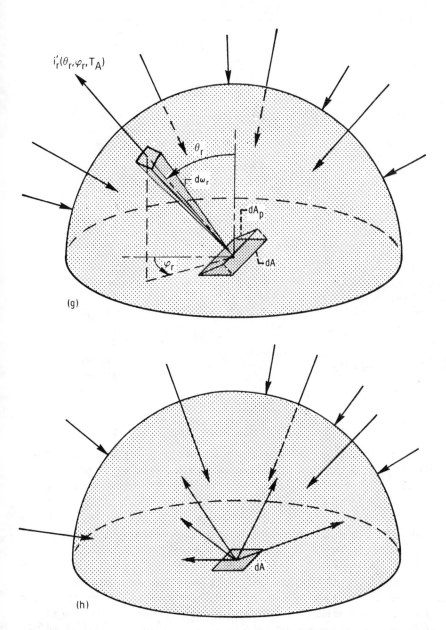

Figure 3-1 (*Continued*) (*g*) hemispherical-directional reflectivity $\rho'(\theta_r,\varphi_r,T_A)$; (*h*) hemispherical reflectivity $\rho(T_A)$.

Table 3-1 Summary of surface-property definitions

Quantity	Symbol	Defining equation	Descriptive figure
	Emissivity		
Directional spectral	ϵ'_λ	(3-2)	3-1a
Directional total	ϵ'	(3-3)	3-1a
Hemispherical spectral	ϵ_λ	(3-5)	3-1b
Hemispherical total	ϵ	(3-6)	3-1b
	Absorptivity		
Directional spectral	α'_λ	(3-10a)	3-1c
Directional total	α'	(3-14)	3-1c
Hemispherical spectral	α_λ	(3-16)	3-1d
Hemispherical total	α	(3-18)	3-1d
	Reflectivity		
Bidirectional spectral	ρ''_λ	(3-20)	3-1e
Directional-hemispherical spectral	$\rho'_\lambda(\theta,\varphi)$	(3-24)	3-1f
Hemispherical-directional spectral	$\rho'_\lambda(\theta_r,\varphi_r)$	(3-26)	3-1g
Hemispherical spectral	ρ_λ	(3-29)	3-1h
Bidirectional total	ρ''	(3-39)	3-1e
Directional-hemispherical total	$\rho'(\theta,\varphi)$	(3-41a)	3-1f
Hemispherical-directional total	$\rho'(\theta_r,\varphi_r)$	(3-41b)	3-1g
Hemispherical total	ρ	(3-43)	3-1h

It is the practice in most fields of science to assign the *-ivity* ending to intensive properties of materials, such as in the case of electrical resistivity, thermal conductivity, or diffusivity. The *-ance* ending is reserved, however, for extensive properties of materials as in electrical resistance or conductance. Use of the term *emittance* as defined in the previous paragraph does not follow this convention, since emittance would still be an intensive property as long as opaque materials are considered. Further, it seems cumbersome to define two terms for the same concept, using one term to differentiate the one very special case of the perfectly prepared pure substance.

For these reasons, the *-ivity* ending will be used throughout this book for the radiative properties of opaque materials, whether for ideal uncontaminated surfaces or for properties with some given surface condition. The *-ance* ending can then be reserved for an extensive property such as the emittance of a partially transmitting layer of water or glass where the emittance would vary with layer thickness. The derived relations of course apply regardless of the nomenclature adopted.

It must be noted that the *-ance* ending is often found in the literature dealing with the experimental determination of surface properties. The term emittance is also used in some references to describe what we have called emissive power.

3-1.2 Notation

Because of the many independent variables that must be specified for radiative properties, a concise but accurate notation is necessary. The notation to be used here is an
extension of that introduced in the preceding chapter. A functional notation is used to
explicitly give the variables upon which a quantity depends. For example, $\epsilon_\lambda'(\lambda,\theta,\varphi,T_A)$
shows that ϵ_λ' depends on the four variables noted. The prime denotes a directional
quantity, and the λ subscript specifies that the quantity is spectral. Certain quantities
depend upon *two* directions (four angles); these quantities will be given a double
prime. A hemispherical quantity will not have a prime, and a total quantity will not
have a λ subscript. A hemispherical-directional or directional-hemispherical quantity
will have a single prime. A quantity that is directional in nature, that is, it is evaluated
on a per-unit solid-angle basis, will always have a prime even if in a specific case its
numerical value is independent of direction; the independence of direction is denoted
by the absence of (θ,φ) in the functional notation. Similarly, a spectral quantity will
always have a λ subscript even when in specific cases the numerical value does not
vary with wavelength; such a specific case would not have a λ in the functional
notation.*

Additional notation is needed for Q, the energy rate for a finite area, to keep
consistent mathematical forms for energy balances. Thus, d^2Q_λ' denotes as before
a directional-spectral quantity, but the second differential is needed to denote that the
energy is of differential order in both wavelength and solid angle. Thus, dQ' and dQ_λ
are differential quantities with respect to solid angle and wavelength, respectively. If
a differential area is involved, the order of the derivative is correspondingly increased.

This notation may appear somewhat redundant, but the usefulness will become
clear in this chapter in dealing with certain special cases, such as gray and diffuse
bodies. In addition, the shorthand value of referring to ϵ_λ' in the text rather than
writing out the term *directional spectral emissivity* should be apparent. A study of
Table 3-1 will help clarify the notation system being used.

The three main sections of this chapter each deal with a different property, that is,
emissivity, absorptivity, and reflectivity. In each of these sections the most basic
unaveraged property is presented first; for example, in the first section, the directional
spectral emissivity is presented. Then the averaged quantities are obtained by integration. The section on absorptivity also contains forms of Kirchhoff's law relating
absorptivity to emissivity. The section on reflectivity includes the reciprocity relations.

3-2 SYMBOLS

A	surface area
C	a coefficient
e	radiative emissive power
F	fraction of blackbody total emissive power

---

*In later chapters, the functional notation will sometimes be abbreviated or omitted to
simplify the equations.

i	radiation intensity
Q	energy rate; energy per unit time
q	energy flux; energy per unit area per unit time
S	distance between emitting and absorbing elements
T	absolute temperature
α	absorptivity
θ	polar angle measured from normal of surface
ϵ	emissivity
λ	wavelength
φ	circumferential angle
ρ	reflectivity
σ	Stefan-Boltzmann constant, Table A-4
ω	solid angle
$\int_{\Omega}$	integration over solid angle of entire enclosing hemisphere

Subscripts

A	of surface A
a	absorbed
b	blackbody
d	diffuse
e	emitted or emitting
i	incident
p	projected
r	reflected
s	specular
λ	spectrally dependent

Superscipts

$'$	directional
$''$	bidirectional

3-3 EMISSIVITY

The *emissivity* is a measure of how well a body can radiate energy as compared with a blackbody. The emitting ability can depend on factors such as body temperature, particular wavelength being considered for the emitted energy, and angle at which the energy is being emitted. The emissivity is usually measured experimentally at a direction normal to the surface and as a function of wavelength. In calculating energy loss by a body the emission into all directions is required, and for such a calculation an emissivity is needed that is averaged over all directions and wavelengths. For radiant interchange between surfaces, emissivities averaged over wavelength but not direction might be needed; in other cases, when spectral effects become large, spectral values averaged only over direction are used. Thus, various averaged emissivities may be required by the analyst, and they must often be obtained from available measured values.

In this section, the basic definition of the directional spectral emissivity is given. This emissivity is then averaged with respect to wavelength, direction, and then wavelength and direction simultaneously.

Values averaged with respect to wavelength are termed *total* quantities; averages with respect to all directions are termed *hemispherical* quantities. This convention is adhered to throughout this book.

3-3.1 Directional Spectral Emissivity $\epsilon_\lambda'(\lambda,\theta,\varphi,T_A)$

Consider the geometry for emitted radiation shown in Fig. 3-1a. As discussed in Chap. 2, the radiation intensity is the energy per unit time emitted in the direction (θ,φ) per unit of the *projected* area dA_p *normal* to this direction, per unit solid angle and per unit wavelength band. In some texts the intensity is defined relative to the actual surface area rather than the projected value. By basing the intensity on the projected area as is done here, one gains the advantage that for a black surface the intensity has the same value for all directions. Unlike the intensity from a blackbody, the emission from a real body *does* depend on direction, and hence the (θ,φ) designation is included in the notation for intensity. The *energy* leaving a real surface dA of temperature T_A per unit time in the wavelength interval $d\lambda$ and within the solid angle $d\omega$ is then given by

$$d^3 Q_\lambda'(\lambda,\theta,\varphi,T_A) = i_\lambda'(\lambda,\theta,\varphi,T_A)\,dA\,\cos\theta\,d\lambda\,d\omega = e_\lambda'(\lambda,\theta,\varphi,T_A)\,dA\,d\lambda\,d\omega \qquad (3\text{-}1a)$$

For a blackbody the intensity is independent of direction and was designated in Chap. 2 by $i_{\lambda b}'(\lambda)$. The T_A notation is introduced here to clarify when properties are temperature-dependent so that the blackbody intensity is designated as $i_{\lambda b}'(\lambda,T_A)$. The energy leaving a black area element per unit time within $d\lambda$ and $d\omega$ is

$$d^3 Q_\lambda'(\lambda,\theta,T_A) = i_{\lambda b}'(\lambda,T_A)\,dA\,\cos\theta\,d\lambda\,d\omega = e_{\lambda b}'(\lambda,\theta,T_A)\,dA\,d\lambda\,d\omega \qquad (3\text{-}1b)$$

The emissivity is then defined as the ratio of the emissive ability of the real surface to that of a blackbody; this provides the definition

$$\textit{Directional spectral emissivity} \equiv \epsilon_\lambda'(\lambda,\theta,\varphi,T_A) = \frac{d^3 Q_\lambda'(\lambda,\theta,\varphi,T_A)}{d^3 Q_{\lambda b}'(\lambda,\theta,T_A)}$$

$$= \frac{i_\lambda'(\lambda,\theta,\varphi,T_A)}{i_{\lambda b}'(\lambda,T_A)} = \frac{e_\lambda'(\lambda,\theta,\varphi,T_A)}{e_{\lambda b}'(\lambda,\theta,T_A)} \qquad (3\text{-}2)$$

This is the most fundamental emissivity, because it includes the dependence on wavelength, direction, and surface temperature.

EXAMPLE 3-1 At $60°$ from the normal a surface heated to 1000 K has a directional spectral emissivity of 0.70 at a wavelength of 5 μm. The emissivity is isotropic with respect to the angle φ. What is the spectral intensity in this direction?

From Table A-5, for a blackbody at λT_A of 5000 $\mu m \cdot K$, $e_{\lambda b}(\lambda, T_A)/T_A^5 = 0.7138 \times 10^{-11}$ W/(m²·μm·K⁵). Then

$$i_\lambda'(5 \ \mu m, 60°, 1000 \ K) = \epsilon_\lambda'(5 \ \mu m, 60°, 1000 \ K) i_{\lambda b}'(5 \ \mu m, 1000 \ K)$$

$$= \epsilon_\lambda'(5 \ \mu m, 60°, 1000 \ K) \frac{e_{\lambda b}}{\pi} (5 \ \mu m, 1000 \ K)$$

$$= 0.70 \times \frac{7138 \times 10^{-15}}{\pi} (1000)^5 = 1590 \ W/(m^2 \cdot \mu m \cdot sr)$$

3-3.2 Averaged Emissivities

From the directional spectral emissivity as given in Eq. (3-2), an averaged emissivity can now be derived by proceeding along one of two approaches: averaging over all wavelengths or averaging over all directions.

Directional total emissivity $\epsilon'(\theta, \varphi, T_A)$. To obtain an average over all wavelengths, the radiation emitted into direction (θ, φ), including the contributions from all wavelengths, is found by integrating the directional spectral emissive power to give the *directional total emissive power* (as in Chap. 2 the term *total* denotes that radiation from all wavelengths is included)

$$e'(\theta, \varphi, T_A) = \int_0^\infty e_\lambda'(\lambda, \theta, \varphi, T_A) \, d\lambda$$

Similarly from Table 2-2 the directional total emissive power for a *blackbody* is given by

$$e_b'(\theta, T_A) = \int_0^\infty e_{\lambda b}'(\lambda, \theta, T_A) \, d\lambda = \frac{\sigma T_A^4 \cos \theta}{\pi}$$

The directional total emissivity is the ratio of $e'(\theta, \varphi, T_A)$ for the real surface to $e_b'(\theta, T_A)$ emitted by a blackbody at the same temperature; that is,

$$Directional \ total \ emissivity \equiv \epsilon'(\theta, \varphi, T_A) = \frac{e'(\theta, \varphi, T_A)}{e_b'(\theta, T_A)}$$

$$= \frac{\int_0^\infty e_\lambda'(\lambda, \theta, \varphi, T_A) \, d\lambda}{(\sigma T_A^4/\pi) \cos \theta} = \frac{\pi \int_0^\infty i_\lambda'(\lambda, \theta, \varphi, T_A) \, d\lambda}{\sigma T_A^4} \tag{3-3a}$$

The $e_\lambda'(\lambda, \theta, \varphi, T_A)$ or $i_\lambda'(\lambda, \theta, \varphi, T_A)$ in the numerator can be replaced in terms of $\epsilon_\lambda'(\lambda, \theta, \varphi, T_A)$ by using Eq. (3-2) to give

Directional total emissivity (in terms of directional spectral emissivity)

$$\equiv \epsilon'(\theta, \varphi, T_A) = \frac{\int_0^\infty \epsilon_\lambda'(\lambda, \theta, \varphi, T_A) \, e_{\lambda b}'(\lambda, \theta, T_A) \, d\lambda}{(\sigma T_A^4/\pi) \cos \theta}$$

$$= \frac{\pi \int_0^\infty \epsilon_\lambda'(\lambda, \theta, \varphi, T_A) \, i_{\lambda b}'(\lambda, T_A) \, d\lambda}{\sigma T_A^4} \tag{3-3b}$$

Thus if the wavelength dependence of $\epsilon_\lambda'(\lambda,\theta,\varphi,T_A)$ is known, the $\epsilon'(\theta,\varphi,T_A)$ is obtained as an integrated average weighted by the blackbody emissive power or intensity. The $\epsilon_\lambda'(\lambda,\theta,\varphi,T_A)$ must be known with good accuracy in the region where $i_{\lambda b}'(\lambda,T_A)$ is large, so that the integrand of Eq. (3-3b) will be accurate where it has large values.

EXAMPLE 3-2 At $1000°R$ the $\epsilon_\lambda'(\lambda,\theta,\varphi,T_A)$ can be approximated by 0.8 in the range $\lambda = 0-5\ \mu m$, and 0.4 for $\lambda > 5\ \mu m$. What is the value of $\epsilon'(\theta,\varphi,T_A)$?
From Eq. (3-3b),

$$\epsilon'(\theta,\varphi,T_A) = \frac{\pi \int_0^\infty \epsilon_\lambda'(\lambda,\theta,\varphi,T_A)\, i_{\lambda b}'(\lambda,T_A)\, d\lambda}{\sigma T_A^4}$$

Apply the following relation obtained from Table 2-2:

$$i_{\lambda b}'(\lambda,T_A) = \frac{e_{\lambda b}(\lambda,T_A)}{\pi}$$

This yields

$$\epsilon'(\theta,\varphi,T_A) = \int_0^{5T_A} \frac{0.8}{\sigma} \frac{e_{\lambda b}(\lambda,T_A)}{T_A^5}\, d(\lambda T_A) + \int_{5T_A}^\infty \frac{0.4}{\sigma} \frac{e_{\lambda b}(\lambda,T_A)}{T_A^5}\, d(\lambda T_A)$$

From Eq. (2-27),

$$\epsilon'(\theta,\varphi,T_A) = 0.8F_{0-5000} + 0.4F_{5000-\infty} = 0.8(0.223) + 0.4(0.777) = 0.489$$

Since 77.7% of the emitted blackbody energy at $1000°R$ is in the region for $\lambda > 5\ \mu m$, the result is weighted heavily toward the 0.4 emissivity value.

Hemispherical spectral emissivity $\epsilon_\lambda(\lambda,T_A)$ Now return to Eq. (3-2) and consider the average obtained by integrating the directional spectral quantities over all directions of a hemispherical envelope covering the surface (Fig. 3-1b). The spectral radiation emitted by a unit surface area into all directions of the hemisphere is termed the *hemispherical spectral emissive power* and is found by integrating the spectral energy per unit solid angle over all solid angles. This is analogous to Eq. (2-8a) for a blackbody and is given by

$$e_\lambda(\lambda,T_A) = \int_\cap i_\lambda'(\lambda,\theta,\varphi,T_A) \cos\theta\, d\omega$$

The notation $\int_\cap d\omega$ signifies integration over the hemispherical solid angle and $d\omega = \sin\theta\, d\theta\, d\varphi$. Here, $i_\lambda'(\lambda,\theta,\varphi,T_A)$ cannot in general be removed from under the integral sign as was done for a blackbody. By using Eq. (3-2) this can be written as

$$e_\lambda(\lambda,T_A) = i_{\lambda b}'(\lambda,T_A) \int_\cap \epsilon_\lambda'(\lambda,\theta,\varphi,T_A) \cos\theta\, d\omega \qquad (3\text{-}4a)$$

For a blackbody the hemispherical spectral emissive power is, from (2-8b),

$$e_{\lambda b}(\lambda,T_A) = \pi i_{\lambda b}'(\lambda,T_A) \qquad (3\text{-}4b)$$

The ratio of actual to blackbody emission from the surface [Eq. (3-4a) divided by (3-4b)] provides the following definition:

Hemispherical spectral emissivity (in terms of directional spectral emissivity)

$$\equiv \epsilon_\lambda(\lambda,T_A) = \frac{e_\lambda(\lambda,T_A)}{e_{\lambda b}(\lambda,T_A)} = \frac{1}{\pi} \int_\cap \epsilon_\lambda'(\lambda,\theta,\varphi,T_A) \cos\theta \, d\omega \tag{3-5}$$

Hemispherical total emissivity $\epsilon(T_A)$ To derive the hemispherical total emissivity, consider that from a unit area the spectral emissive power in any direction is derived from Eq. (3-2) as $\epsilon_\lambda'(\lambda,\theta,\varphi,T_A) i_{\lambda b}'(\lambda,T_A) \cos\theta$. This is integrated over all λ and ω to give the *hemispherical total emissive power*. Dividing by σT_A^4, which is the hemispherical total emissive power for a blackbody, results in the following emissivity:

Hemispherical total emissivity (in terms of directional spectral emissivity)

$$\equiv \epsilon(T_A) = \frac{e(T_A)}{e_b(T_A)} = \frac{\int_\cap [\int_0^\infty e_\lambda'(\lambda,\theta,\varphi,T_A) \, d\lambda] \, d\omega}{\sigma T_A^4}$$

$$= \frac{\int_\cap [\int_0^\infty \epsilon_\lambda'(\lambda,\theta,\varphi,T_A) i_{\lambda b}'(\lambda,T_A) \, d\lambda] \cos\theta \, d\omega}{\sigma T_A^4} \tag{3-6a}$$

By using Eq. (3-3b) this can be placed in a second form

Hemispherical total emissivity (in terms of directional total emissivity)

$$\equiv \epsilon(T_A) = \frac{1}{\pi} \int_\cap \epsilon'(\theta,\varphi,T_A) \cos\theta \, d\omega \tag{3-6b}$$

If the order of the integrations is interchanged in Eq. (3-6a), there results

$$\epsilon(T_A) = \frac{\int_0^\infty i_{\lambda b}'(\lambda,T_A) [\int_\cap \epsilon_\lambda'(\lambda,\theta,\varphi,T_A) \cos\theta \, d\omega] \, d\lambda}{\sigma T_A^4}$$

Equation (3-5) is then utilized to obtain a third form:

Hemispherical total emissivity (in terms of hemispherical spectral emissivity)

$$\equiv \epsilon(T_A) = \frac{\pi \int_0^\infty \epsilon_\lambda(\lambda,T_A) i_{\lambda b}'(\lambda,T_A) \, d\lambda}{\sigma T_A^4} \tag{3-6c}$$

Substituting Eq. (3-4b) gives

$$\epsilon(T_A) = \frac{\int_0^\infty \epsilon_\lambda(\lambda,T_A) e_{\lambda b}(\lambda,T_A) \, d\lambda}{\sigma T_A^4} \tag{3-6d}$$

To interpret Eq. (3-6d) physically, look at Fig. 3-2. In Fig. 3-2a is shown the emissivity ϵ_λ for a surface temperature T_A. The solid curve in Fig. 3-2b is the hemispherical spectral emissive power for a blackbody at T_A. The area under the solid curve is σT_A^4, which is the denominator of Eq. (3-6d) and is equal to the radiation emitted per unit

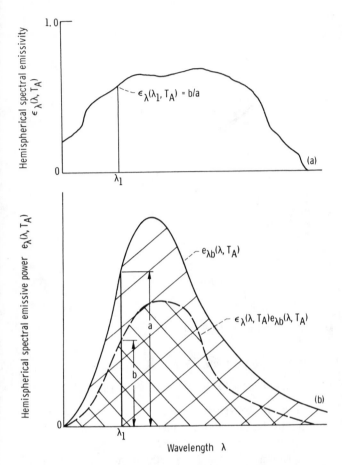

Figure 3-2 Physical interpretation of hemispherical spectral and total emissivities. (*a*) Measured emissivity values; (*b*) interpretation of emissivity as ratio of actual emissive power to blackbody emissive power.

area by a black surface including all wavelengths and directions. The dashed curve in Fig. 3-2*b* is the product $\epsilon_\lambda(\lambda,T_A) e_{\lambda b}(\lambda,T_A)$, and the area under this curve is the integral in the numerator of Eq. (3-6*d*), which is the emission from the real surface. Hence $\epsilon(T_A)$ is the ratio of the area under the dashed curve to that under the solid curve. From a slightly different viewpoint, at each λ the quantity ϵ_λ is the ordinate of the dashed curve divided by the ordinate of the solid curve. As shown in Fig. 3-2, for λ_1 the hemispherical spectral emissivity is $\epsilon_\lambda(\lambda_1,T_A) = b/a$.

EXAMPLE 3-3 A surface at 1000 K is isotropic in the sense that ϵ' is independent of φ, but depends on θ as shown in Fig. 3-3. What are the hemispherical total emissivity and the hemispherical total emissive power?

The $\epsilon'(\theta,1000\text{ K})$ can be approximated in this case quite well by the function $0.85 \cos\theta$ (dashed line). Then, from Eq. (3-6*b*), the hemispherical total emissivity

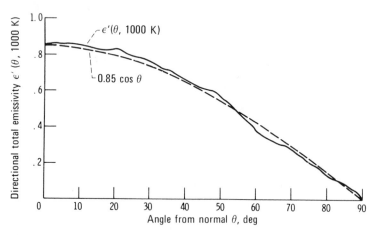

Figure 3-3 Directional total emissivity at 1000 K for Example 3-3.

is

$$\epsilon(T_A) = \frac{1}{\pi} \int_{\varphi=0}^{2\pi} \int_{\theta=0}^{\pi/2} 0.85 \sin\theta \cos^2\theta \, d\theta \, d\varphi = -1.70 \left.\frac{\cos^3\theta}{3}\right|_0^{\pi/2} = 0.57$$

The hemispherical total emissive power is then

$$e(T_A) = \epsilon(T_A) \, \sigma T_A^4 = 0.57 \times 5.729 \times 10^{-8} \times 1000^4 = 32{,}660 \text{ W/m}^2$$

Generally the $\epsilon'(\theta, T_A)$ will not be well approximated by a convenient analytical function, and the integration must be carried out numerically.

EXAMPLE 3-4 The $\epsilon_\lambda(\lambda, T_A)$ for a surface at $T_A = 2000°\text{R}$ can be approximated as shown in Fig. 3-4. What are the hemispherical total emissivity and the hemispherical total emissive power of the surface?
From Eq. (3-6d),

$$\epsilon(T_A) = \frac{1}{\sigma T_A^4} \int_0^\infty \epsilon_\lambda(\lambda, T_A) \, e_{\lambda b}(\lambda, T_A) \, d\lambda = \frac{1}{\sigma} \int_0^2 0.1 \frac{e_{\lambda b}(\lambda, T_A)}{T_A^5} T_A \, d\lambda$$

$$+ \frac{1}{\sigma} \int_2^6 0.4 \frac{e_{\lambda b}(\lambda, T_A)}{T_A^5} T_A \, d\lambda + \frac{1}{\sigma} \int_6^\infty 0.2 \frac{e_{\lambda b}(\lambda, T_A)}{T_A^5} T_A \, d\lambda$$

This yields

$$\epsilon(T_A) = \frac{0.1}{\sigma} \int_0^{4000} \frac{e_{\lambda b}}{T_A^5} d(\lambda T_A) + \frac{0.4}{\sigma} \int_{4000}^{12{,}000} \frac{e_{\lambda b}}{T_A^5} d(\lambda T_A)$$

$$+ \frac{0.2}{\sigma} \int_{12{,}000}^\infty \frac{e_{\lambda b}}{T_A^5} d(\lambda T_A)$$

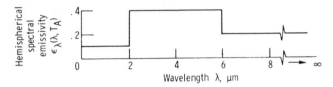

Figure 3-4 Hemispherical spectral emissivity for Example 3-4. Surface temperature T_A, 2000°R.

where the quantity $e_{\lambda b}/T_A^5$ is a function of λT_A. From Eq. (2-27) this can be written as

$$\epsilon(T_A) = 0.1F_{0-4000} + 0.4(F_{0-12,000} - F_{0-4000}) + 0.2(1 - F_{0-12,000})$$

$$= -0.3F_{0-4000} + 0.2F_{0-12,000} + 0.2$$

$$= -0.3(0.1051) + 0.2(0.7877) + 0.2 = 0.3260$$

The hemispherical total emissive power is

$$e(T_A) = \epsilon(T_A)\, \sigma T_A^4 = 0.326 \times 0.173 \times 10^{-8}(2000)^4 = 9020 \text{ Btu/(h·ft}^2)$$

3-4 ABSORPTIVITY

The absorptivity is defined as the fraction of the energy incident on a body that is absorbed by the body. The incident radiation is the result of the radiative conditions at the *source* of the incident energy. The spectral distribution of the incident radiation is independent of the temperature or physical nature of the absorbing surface (unless the radiation emitted from the surface is partially reflected back to the surface). Compared with emissivity, additional complexities are introduced into the absorptivity because the directional and spectral characteristics of the incident radiation must now be accounted for.

Experimentally it is often easier to measure the emissivity than the absorptivity; hence, it is desirable to have relations between these two quantities so that measured values of one will allow the other to be calculated. Such relations are developed in this section along with the definitions of the absorptivity quantities.

3-4.1 Directional Spectral Absorptivity $\alpha_\lambda'(\lambda,\theta,\varphi,T_A)$

Figure 3-5a illustrates the energy incident on a surface element dA from the (θ,φ) direction. The line from dA in the direction (θ,φ) passes normally through an area element dA_e on the surface of a hemisphere of radius R placed over dA. The incident spectral intensity passing through dA_e is $i_{\lambda,i}'(\lambda,\theta,\varphi)$. This is the energy per unit area of the hemisphere, per unit solid angle $d\omega_e$, per unit time, and per unit wavelength interval. The energy within the incident solid angle $d\omega_e$ strikes the area dA of the absorbing surface. Hence, the energy per unit time incident from the direction (θ,φ)

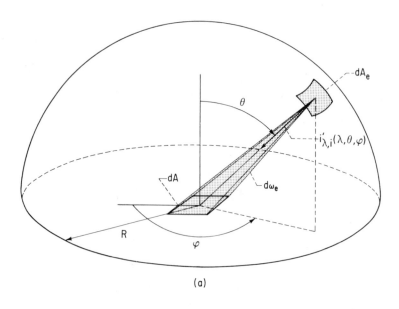

(a)

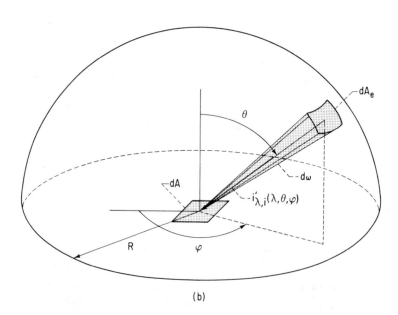

(b)

Figure 3-5 Equivalent ways of showing energy from dA_e that is incident upon dA. (*a*) Incidence within solid angle $d\omega_e$ having origin at dA_e; (*b*) incidence within solid angle $d\omega$ having origin at dA.

in the wavelength interval $d\lambda$ is

$$d^3Q'_{\lambda,i}(\lambda,\theta,\varphi) = i'_{\lambda,i}(\lambda,\theta,\varphi)\,dA_e\,d\omega_e\,d\lambda = i'_{\lambda,i}(\lambda,\theta,\varphi)\,dA_e\,\frac{dA\cos\theta}{R^2}\,d\lambda \qquad (3\text{-}7)$$

where $dA\cos\theta/R^2$ is the solid angle $d\omega_e$ subtended by dA when viewed from dA_e.

Equation (3-7) will be expressed in terms of the solid angle $d\omega$ shown in Fig. 3-5b. This is the solid angle subtended by dA_e when viewed from dA. The $d\omega$ has its vertex at the area dA and hence is the convenient solid angle to use when integrating to obtain energy incident from more than one direction. The use of this solid angle is also consistent with that used in an absorbing, emitting, and scattering medium as discussed in Chap. 13. For a nonabsorbing medium in the region above the surface, as is being considered here, the incident intensity does not change along the path from dA_e to dA (this is proved in Sec. 13-4). For these reasons, in the figures that follow, the energy incident on dA from dA_e will be pictured as that arriving in $d\omega$ as shown in Fig. 3-5b rather than as the energy leaving dA_e in $d\omega_e$ as in Fig. 3-5a.

To place (3-7) in terms of $d\omega$, note that

$$\frac{dA\cos\theta}{R^2}\,dA_e = \frac{dA_e}{R^2}\cos\theta\,dA = d\omega\cos\theta\,dA \qquad (3\text{-}8)$$

Equation (3-7) can then be written as

$$d^3Q'_{\lambda,i}(\lambda,\theta,\varphi) = i'_{\lambda,i}(\lambda,\theta,\varphi)\,d\omega\cos\theta\,dA\,d\lambda \qquad (3\text{-}9)$$

The fraction of the incident energy $d^3Q'_{\lambda,i}$ that is absorbed is defined as the *directional spectral absorptivity* $\alpha'_\lambda(\lambda,\theta,\varphi,T_A)$. In addition to depending on the wavelength and direction of the incident radiation, the spectral absorptivity is a function of the absorbing surface temperature. The amount of the incident energy that is absorbed is designated as $d^3Q'_{\lambda,a}$. Then the ratio is formed

$$\textit{Directional spectral absorptivity} \equiv \alpha'_\lambda(\lambda,\theta,\varphi,T_A) = \frac{d^3Q'_{\lambda,a}(\lambda,\theta,\varphi,T_A)}{d^3Q'_{\lambda,i}(\lambda,\theta,\varphi)}$$

$$= \frac{d^3Q'_{\lambda,a}(\lambda,\theta,\varphi,T_A)}{i'_{\lambda,i}(\lambda,\theta,\varphi)\,dA\cos\theta\,d\omega\,d\lambda} \qquad (3\text{-}10a)$$

If the incident energy is from black surroundings at uniform temperature T_b, then there is the special case

$$\alpha'_\lambda(\lambda,\theta,\varphi,T_A) = \frac{d^3Q'_{\lambda,a}(\lambda,\theta,\varphi,T_A)}{i'_{\lambda b,i}(\lambda,T_b)\,dA\cos\theta\,d\omega\,d\lambda} \qquad (3\text{-}10b)$$

3-4.2 Kirchhoff's Law

This law is concerned with the relation between the emitting and absorbing abilities of a body. The law can have various conditions imposed on it, depending on whether spectral, total, directional, or hemispherical quantities are being considered. From

Eqs. (3-1) and (3-2) the energy emitted per unit time by an element dA in a wavelength interval $d\lambda$ and solid angle $d\omega$ is

$$d^3 Q'_{\lambda,e} = i'_\lambda(\lambda,\theta,\varphi,T_A)\, dA \cos\theta\, d\omega\, d\lambda$$

$$= \epsilon'_\lambda(\lambda,\theta,\varphi,T_A)\, i'_{\lambda b}(\lambda,T_A)\, dA \cos\theta\, d\omega\, d\lambda \tag{3-11}$$

If the element dA at temperature T_A is assumed to be placed in an isothermal black enclosure also at temperature T_A, then the intensity of the energy incident on dA from the direction (θ,φ) (recalling the isotropy of intensity in a black enclosure) will be $i'_{\lambda b}(\lambda,T_A)$. To maintain the isotropy of the radiation within the black enclosure, the absorbed and emitted energies given by Eqs. (3-10b) and (3-11) must be equal. Equating these gives

$$\epsilon'_\lambda(\lambda,\theta,\varphi,T_A) = \alpha'_\lambda(\lambda,\theta,\varphi,T_A) \tag{3-12}$$

This equality is a fixed relation between the properties of the material and holds without restriction. This is the most *general form of Kirchhoff's law.* *

3-4.3 Directional Total Absorptivity $\alpha'(\theta,\varphi,T_A)$

The directional total absorptivity is the ratio of the energy including all wavelengths that is absorbed from a given direction to the energy incident from that direction. The total energy incident from the given direction is obtained by integrating the spectral incident energy [Eq. (3-9)] over all wavelengths to obtain

$$d^2 Q'_i(\theta,\varphi) = \cos\theta\, dA\, d\omega \int_0^\infty i'_{\lambda,i}(\lambda,\theta,\varphi)\, d\lambda \tag{3-13a}$$

The radiation absorbed is determined by integrating Eq. (3-10a) over all wavelengths,

---

*As will be discussed in Chap. 4 in connection with radiation properties from electromagnetic theory, radiation is polarized in the sense of having two wave components vibrating at right angles to each other and to the propagation direction. For the special case of black radiation the two components of polarization are equal. To be strictly accurate, Eq. (3-12) holds only for each component of polarization, and for (3-12) to be valid as written for all incident energy, the incident radiation must have equal components of polarization.

Kirchhoff's law was proved for thermodynamic equilibrium in an isothermal enclosure and hence is strictly true only when there is no *net* heat transfer to or from the surface. In an actual application there is usually a net heat transfer so that it is an approximation that Eq. (3-12) will still apply. The validity of this approximation is based on experimental evidence that in most applications α'_λ and ϵ'_λ are not significantly influenced by the surrounding radiation field. Another way of stating this is that the material is able to maintain itself in a local thermodynamic equilibrium in which the populations of energy states that take part in the absorption and emission processes are given to a very close approximation by their equilibrium distributions. Thus the extension of Kirchhoff's law to nonequilibrium systems is not a result of simple thermodynamic considerations. Rather it results from the physics of materials which allows them in most instances to maintain themselves in local thermodynamic equilibrium and thus have their properties not depend on the surrounding radiation field.

that is,

$$d^2Q_a'(\theta,\varphi,T_A) = \cos\theta \; dA \; d\omega \int_0^\infty \alpha_\lambda'(\lambda,\theta,\varphi,T_A) i_{\lambda,i}'(\lambda,\theta,\varphi) \, d\lambda \qquad (3\text{-}13b)$$

The following ratio is then formed:

$$\textit{Directional total absorptivity} \equiv \alpha'(\theta,\varphi,T_A) = \frac{d^2Q_a'(\theta,\varphi,T_A)}{d^2Q_i'(\theta,\varphi)}$$

$$= \frac{\int_0^\infty \alpha_\lambda'(\lambda,\theta,\varphi,T_A) i_{\lambda,i}'(\lambda,\theta,\varphi) \, d\lambda}{\int_0^\infty i_{\lambda,i}'(\lambda,\theta,\varphi) \, d\lambda} \qquad (3\text{-}14a)$$

By use of Kirchhoff's law (3-12), an alternative form of (3-14a) is

$$\alpha'(\theta,\varphi,T_A) = \frac{\int_0^\infty \epsilon_\lambda'(\lambda,\theta,\varphi,T_A) i_{\lambda,i}'(\lambda,\theta,\varphi) \, d\lambda}{\int_0^\infty i_{\lambda,i}'(\lambda,\theta,\varphi) \, d\lambda} \qquad (3\text{-}14b)$$

3-4.4 Kirchhoff's Law for Directional Total Properties

The general form of Kirchhoff's law [Eq. (3-12)] shows that ϵ_λ' and α_λ' are equal. It is now of interest to examine this equality for the directional total quantities. This can be accomplished by comparing a special case of Eq. (3-14b) with (3-3b). If in Eq. (3-14b) the incident radiation has a spectral distribution proportional to that of a blackbody at T_A, then $i_{\lambda,i}'(\lambda,\theta,\varphi) = C(\theta,\varphi) i_{\lambda b}'(\lambda,T_A)$, and Eq. (3-14b) becomes

$$\alpha'(\theta,\varphi,T_A) = \frac{\int_0^\infty \epsilon_\lambda'(\lambda,\theta,\varphi,T_A) i_{\lambda b}'(\lambda,T_A) \, d\lambda}{\int_0^\infty i_{\lambda b}'(\lambda,T_A) \, d\lambda(= \sigma T_A^4/\pi)} = \epsilon'(\theta,\varphi,T_A)$$

Hence when ϵ_λ' and α_λ' are dependent on wavelength, $\alpha'(\theta,\varphi,T_A) = \epsilon'(\theta,\varphi,T_A)$ only when the incident radiation meets the restriction $i_{\lambda,i}'(\lambda,\theta,\varphi) = C(\theta,\varphi) i_{\lambda b}'(\lambda,T_A)$, where C is independent of wavelength.

There is another important case when the relation $\alpha'(\theta,\varphi,T_A) = \epsilon'(\theta,\varphi,T_A)$ is valid. If the directional emission from a surface has the same wavelength dependence as a blackbody, $i_\lambda'(\lambda,\theta,\varphi,T_A) = C(\theta,\varphi) i_{\lambda b}'(\lambda,T_A)$, then the ϵ_λ' is independent of λ. From Eqs. (3-3b) and (3-14b), if $\epsilon_\lambda'(\theta,\varphi,T_A)$ and hence $\alpha_\lambda'(\theta,\varphi,T_A)$ do not depend on λ, then, for the direction (θ,φ), ϵ_λ', α_λ', ϵ', and α' are all equal. A surface exhibiting such behaviour is termed a *directional gray* surface.

3-4.5 Hemispherical Spectral Absorptivity $\alpha_\lambda(\lambda,T_A)$

The hemispherical spectral absorptivity is the fraction of the spectral energy that is absorbed from the spectral energy incident from all directions over a surrounding hemisphere (Fig. 3-1d). The spectral energy from an element dA_e on the hemisphere that is intercepted by a surface element dA is given by Eq. (3-9). The incident energy on dA from all directions of the hemisphere is then given by the integral

$$d^2Q_{\lambda,i} = dA \, d\lambda \int_\omega i'_{\lambda,i}(\lambda,\theta,\varphi) \cos\theta \, d\omega \qquad (3\text{-}15a)$$

The amount absorbed is found by integrating Eq. (3-10a) over the hemisphere:

$$d^2Q_{\lambda,a} = dA \, d\lambda \int_\omega \alpha'_\lambda(\lambda,\theta,\varphi,T_A) i'_{\lambda,i}(\lambda,\theta,\varphi) \cos\theta \, d\omega \qquad (3\text{-}15b)$$

The ratio of these quantities gives

$$\textit{Hemispherical spectral absorptivity} \equiv \alpha_\lambda(\lambda,T_A) = \frac{d^2Q_{\lambda,a}}{d^2Q_{\lambda,i}}$$

$$= \frac{\int_\omega \alpha'_\lambda(\lambda,\theta,\varphi,T_A) i'_{\lambda,i}(\lambda,\theta,\varphi) \cos\theta \, d\omega}{\int_\omega i'_{\lambda,i}(\lambda,\theta,\varphi) \cos\theta \, d\omega} \qquad (3\text{-}16a)$$

or by using Kirchhoff's law

$$\alpha_\lambda(\lambda,T_A) = \frac{\int_\omega \epsilon'_\lambda(\lambda,\theta,\varphi,T_A) i'_{\lambda,i}(\lambda,\theta,\varphi) \cos\theta \, d\omega}{\int_\omega i'_{\lambda,i}(\lambda,\theta,\varphi) \cos\theta \, d\omega} \qquad (3\text{-}16b)$$

The hemispherical spectral absorptivity and emissivity can now be compared by looking at Eqs. (3-16b) and (3-5). It is found that for the general case, where α'_λ and ϵ'_λ are functions of λ, θ, φ, and T_A, $\alpha_\lambda(\lambda,T_A) = \epsilon_\lambda(\lambda,T_A)$ only if $i'_{\lambda,i}(\lambda)$ is independent of θ and φ, that is, if the incident spectral intensity is uniform over all directions. If this is so, the $i'_{\lambda,i}$ can be canceled in (3-16b) and the denomonator becomes π, which then compares with (3-5).

For the case $\alpha'_\lambda(\lambda,T_A) = \epsilon'_\lambda(\lambda,T_A)$, that is, the directional spectral properties are independent of angle, the hemispherical spectral properties are related by $\alpha_\lambda(\lambda,T_A) = \epsilon_\lambda(\lambda,T_A)$ for any angular variation of incident intensity. Such a surface is termed a *diffuse spectral* surface.

3-4.6 Hemispherical Total Absorptivity $\alpha(T_A)$

The hemispherical total absorptivity represents the fraction of energy absorbed that is incident from all directions of the enclosing hemisphere and for all wavelengths, as shown in Fig. 3-1d. The total incident energy that is intercepted by a surface element dA is determined by integrating Eq. (3-9) over all λ and all (θ,φ) of the hemisphere, which results in

$$dQ_i = dA \int_\omega \left[\int_0^\infty i'_{\lambda,i}(\lambda,\theta,\varphi) \, d\lambda \right] \cos\theta \, d\omega \qquad (3\text{-}17a)$$

Similarly, by integrating (3-10a), the total amount of energy absorbed is found equal to

$$dQ_a(T_A) = dA \int_\omega \left[\int_0^\infty \alpha'_\lambda(\lambda,\theta,\varphi,T_A) i'_{\lambda,i}(\lambda,\theta,\varphi) \, d\lambda \right] \cos\theta \, d\omega \qquad (3\text{-}17b)$$

The ratio of absorbed to incident energy provides the definition

Hemispherical total absorptivity (in terms of directional spectral absorptivity or emissivity)

$$\equiv \alpha(T_A) = \frac{dQ_a(T_A)}{dQ_i} = \frac{\int_\omega [\int_0^\infty \alpha'_\lambda(\lambda,\theta,\varphi,T_A) i'_{\lambda,i}(\lambda,\theta,\varphi)\,d\lambda]\cos\theta\,d\omega}{\int_\omega [\int_0^\infty i'_{\lambda,i}(\lambda,\theta,\varphi)\,d\lambda]\cos\theta\,d\omega} \quad (3\text{-}18a)$$

or from Kirchhoff's law

$$\alpha(T_A) = \frac{\int_\omega [\int_0^\infty \epsilon'_\lambda(\lambda,\theta,\varphi,T_A) i'_{\lambda,i}(\lambda,\theta,\varphi)\,d\lambda]\cos\theta\,d\omega}{\int_\omega [\int_0^\infty i'_{\lambda,i}(\lambda,\theta,\varphi)\,d\lambda]\cos\theta\,d\omega} \quad (3\text{-}18b)$$

Equation (3-18b) can be compared with (3-6a) to determine under what conditions the hemispherical total absorptivity and emissivity are equal. It is recalled in (3-6a) that

$$\sigma T_A^4 = \int_\omega \left[\int_0^\infty i'_{\lambda b}(\lambda,T_A)\,d\lambda\right]\cos\theta\,d\omega$$

The comparison reveals that for the general case when ϵ'_λ and α'_λ vary with both wavelength and angle, $\alpha(T_A) = \epsilon(T_A)$ *only when the incident intensity is independent of the incident angle and has the same spectral form as that emitted by a blackbody with temperature equal to the surface temperature* T_A, that is, only when

$$i'_{\lambda,i}(\lambda,\theta,\varphi) = C i'_{\lambda b}(\lambda,T_A)$$

where C is a constant. Some more restrictive cases are listed in Table 3-2.

Substituting Eq. (3-14a) into (3-18a) gives the following alternative forms:

Hemispherical total absorptivity (in terms of directional total absorptivity)

$$\equiv \alpha(T_A) = \frac{\int_\omega [\int_0^\infty i'_{\lambda,i}(\lambda,\theta,\varphi)\,d\lambda]\alpha'(\theta,\varphi,T_A)\cos\theta\,d\omega}{\int_\omega [\int_0^\infty i'_{\lambda,i}(\lambda,\theta,\varphi)\,d\lambda]\cos\theta\,d\omega} \quad (3\text{-}18c)$$

or

$$\alpha(T_A) = \frac{\int_\omega \alpha'(\theta,\varphi,T_A) i'_i(\theta,\varphi)\cos\theta\,d\omega}{\int_\omega i'_i(\theta,\varphi)\cos\theta\,d\omega} \quad (3\text{-}18d)$$

where $i'_i(\theta,\varphi)$ is the incident total intensity from direction (θ,φ).

Changing the order of integration in Eq. (3-18a) and then substituting Eq. (3-16a) give

Hemispherical total absorptivity (in terms of hemispherical spectral absorptivity)

$$\equiv \alpha(T_A) = \frac{\int_0^\infty [\alpha_\lambda(\lambda,T_A)\int_\omega i'_{\lambda,i}(\lambda,\theta,\varphi)\cos\theta\,d\omega]\,d\lambda}{\int_0^\infty [\int_\omega i'_{\lambda,i}(\lambda,\theta,\varphi)\cos\theta\,d\omega]\,d\lambda} \quad (3\text{-}18e)$$

or

$$\alpha(T_A) = \frac{\int_0^\infty \alpha_\lambda(\lambda,T_A)\,d^2Q_{\lambda,i}}{\int_0^\infty d^2Q_{\lambda,i}} \quad (3\text{-}18f)$$

where $d^2Q_{\lambda,i}$ is the spectral energy incident from all directions that is intercepted by the surface element dA.

An interesting special case is in Prob. 1 of this chapter. If there is a uniform incident intensity from a gray source at T_i, and if $\epsilon_\lambda(\lambda,T_A)$ is independent of T_A, then the hemispherical total absorptivity for the incident radiation is equal to the hemispherical total emissivity of the material evaluated at the source temperature T_i, $\alpha(T_A) = \epsilon(T_i)$.

EXAMPLE 3-5 The hemispherical spectral emissivity of a surface at 300 K is 0.8 for $0 \leqslant \lambda \leqslant 3$ μm, and 0.2 for $\lambda > 3$ μm. What is the hemispherical total absorptivity for diffuse incident radiation from a black source at $T_i = 1000$ K? What is it for diffuse incident solar radiation?

For diffuse incident radiation $\alpha_\lambda(\lambda,T_A) = \epsilon_\lambda(\lambda,T_A)$, and for a black source $i'_{\lambda,i}(\lambda,\theta,\varphi) = i'_{\lambda b,i}(\lambda)$. Then (3-18e) becomes

$$\alpha(T_A) = \frac{\int_0^\infty \epsilon_\lambda(\lambda,T_A) i'_{\lambda b,i}(\lambda,T_i)\, d\lambda}{\int_0^\infty i'_{\lambda b,i}(\lambda,T_i)\, d\lambda} = \frac{\int_0^\infty \epsilon_\lambda(\lambda,T_A) e_{\lambda b,i}(\lambda,T_i)\, d\lambda}{\sigma T_i^4}$$

Expressing $\alpha(T_A)$ in terms of the two wavelength regions over which $\epsilon_\lambda(\lambda,T_A)$ is constant gives

$$\alpha(T_A) = 0.8 F_{0-3T_i} + 0.2(1 - F_{0-3T_i})$$

Using Table A-5, for a source at 1000 K (λT given in μm·K)

$$\alpha(T_A) = 0.8 F_{0-3000} + 0.2(1 - F_{0-3000})$$

$$= 0.8(0.27322) + 0.2(1 - 0.27322) = 0.364$$

For incident solar radiation, $T_i = 5780$ K:

$$\alpha(T_A) = 0.8 F_{0-17,340} + 0.2(1 - F_{0-17,340})$$

$$= 0.8(0.97879) + 0.2(1 - 0.97879) = 0.787$$

Hence, there is a considerable change in $\alpha(T_A)$ as a result of a shift with wavelength of the incident-energy spectrum as the source temperature is raised.

3-4.7 Diffuse-Gray Surface

As will be discussed in Chap. 8, it is often assumed when performing enclosure calculations that the surfaces are diffuse-gray. *Diffuse* signifies that the directional emissivity and directional absorptivity do not depend on direction. Hence in the case of emission, the emitted intensity will be uniform over all directions as in the case of a blackbody. The term *gray* signifies that the spectral emissivity and absorptivity do not depend on wavelength. They can, however, depend on temperature. Thus at each surface temperature the emitted radiation will be the same fraction of blackbody radiation for all wavelengths.

The diffuse-gray surface therefore *absorbs a fixed fraction of incident radiation from any direction and at any wavelength.* It emits radiation that is a *fixed fraction of blackbody radiation for all directions and wavelengths* (this is the motivation for the term gray). The directional-spectral absorptivity and emissivity then become

$$\alpha_\lambda'(\lambda,\theta,\varphi,T_A) = \alpha_\lambda'(T_A) \qquad \text{and} \qquad \epsilon_\lambda'(\lambda,\theta,\varphi,T_A) = \epsilon_\lambda'(T_A)$$

From Kirchhoff's law, Eq. (3-12), it follows that $\alpha_\lambda'(T_A) = \epsilon_\lambda'(T_A)$.

From Eqs. (3-18a), (3-18b), and (3-6a), since the $\alpha_\lambda'(T_A)$ and $\epsilon_\lambda'(T_A)$ are not functions of either direction or wavelength, they can be taken out of the integrals and the equations reduce to $\alpha(T_A) = \alpha_\lambda'(T_A) = \epsilon_\lambda'(T_A) = \epsilon(T_A)$. Thus for a diffuse-gray surface the directional-spectral and the hemispherical-total values of absorptivity and emissivity are all equal. The hemispherical total absorptivity is completely independent of the nature of the incident radiation.

3-4.8 Summary of Kirchhoff's-Law Relations

The restrictions on application of Kirchhoff's law are summarized in Table 3-2.

Table 3-2 Summary of Kirchhoff's-law relations between absorptivity and emissivity

Type of quantity	Equality	Restrictions
Directional spectral	$\alpha_\lambda'(\lambda,\theta,\varphi,T_A) = \epsilon_\lambda'(\lambda,\theta,\varphi,T_A)$	None
Directional total	$\alpha'(\theta,\varphi,T_A) = \epsilon'(\theta,\varphi,T_A)$	Incident radiation must have a spectral distribution proportional to that of a blackbody at T_A, $i_{\lambda,i}'(\lambda,\theta,\varphi) = C(\theta,\varphi)i_{\lambda b}'(\lambda,T_A)$; or $\alpha_\lambda'(\theta,\varphi,T_A) = \epsilon_\lambda'(\theta,\varphi,T_A)$ are independent of wavelength (directional-gray surface)
Hemispherical spectral	$\alpha_\lambda(\lambda,T_A) = \epsilon_\lambda(\lambda,T_A)$	Incident radiation must be independent of angle, $i_{\lambda,i}'(\lambda) = C(\lambda)$; or $\alpha_\lambda'(\lambda,T_A) = \epsilon_\lambda'(\lambda,T_A)$ do not depend on angle (diffuse-spectral surface)
Hemispherical total	$\alpha(T_A) = \epsilon(T_A)$	Incident radiation must be independent of angle and have a spectral distribution proportional to that of a blackbody at T_A, $i_{\lambda,i}'(\lambda) = Ci_{\lambda b}'(\lambda,T_A)$; or incident radiation independent of angle and $\alpha_\lambda'(\theta,\varphi,T_A) = \epsilon_\lambda'(\theta,\varphi,T_A)$ are independent of λ (directional-gray surface); or incident radiation from each direction has spectral distribution proportional to that of a blackbody at T_A and $\alpha_\lambda'(\lambda,T_A) = \epsilon_\lambda'(\lambda,T_A)$ are independent of angle (diffuse-spectral surface); or $\alpha_\lambda'(T_A) = \epsilon_\lambda'(T_A)$ are independent of wavelength and angle (diffuse-gray surface)

3-5 REFLECTIVITY

The reflective properties of a surface are more complicated to specify than either the emissivity or absorptivity. This is because the reflected energy depends not only on the angle at which the incident energy impinges on the surface, but additionally on the direction being considered for the reflected energy. Some of the pertinent reflectivity quantities will now be defined.

3-5.1 Spectral Reflectivities

Bidirectional spectral reflectivity $\rho_\lambda''(\lambda,\theta_r,\varphi_r,\theta,\varphi)$. Consider incident spectral radiation on a surface from the direction (θ,φ) as shown in Fig. 3-1e. Part of this energy is reflected into the (θ_r,φ_r) direction and provides a part of the reflected intensity in the (θ_r,φ_r) direction. The subscript r will denote quantities evaluated at the reflection angle. The entire magnitude of the $i_{\lambda,r}'(\lambda,\theta_r,\varphi_r)$ is the result of summing the reflected intensities produced by the incident intensities $i_{\lambda,i}'(\lambda,\theta,\varphi)$ from all incident directions (θ,φ) of the hemisphere surrounding the surface element. The contribution to $i_{\lambda,r}'(\lambda,\theta_r,\varphi_r)$ produced by the incident energy from only one (θ,φ) will be designated as $i_{\lambda,r}''(\lambda,\theta_r,\varphi_r,\theta,\varphi)$ and it depends on both the incidence and reflection angles.

The energy from direction (θ,φ) intercepted by dA per unit area and wavelength is, from Eq. (3-9),

$$\frac{d^3 Q_{\lambda,i}'(\lambda,\theta,\varphi)}{dA\, d\lambda} = i_{\lambda,i}'(\lambda,\theta,\varphi)\cos\theta\, d\omega \tag{3-19}$$

The *bidirectional spectral reflectivity* is a ratio expressing the contribution that $i_{\lambda,i}'(\lambda,\theta,\varphi)\cos\theta\, d\omega$ makes to the reflected spectral intensity in the (θ_r,φ_r) direction:

$$\text{Bidirectional spectral reflectivity} \equiv \rho_\lambda''(\lambda,\theta_r,\varphi_r,\theta,\varphi) = \frac{i_{\lambda,r}''(\lambda,\theta_r,\varphi_r,\theta,\varphi)}{i_{\lambda,i}'(\lambda,\theta,\varphi)\cos\theta\, d\omega} \tag{3-20}$$

Although the reflectivity is a function of surface temperature, the T_A notation modifying ρ will be omitted at present for simplicity. The ratio in Eq. (3-20) is a reflected intensity divided by the intercepted energy arriving within solid angle $d\omega$. Having $\cos\theta\, d\omega$ in the denominator means that when $\rho_\lambda''(\lambda,\theta_r,\varphi_r,\theta,\varphi)\, i_{\lambda,i}'(\lambda,\theta,\varphi)\cos\theta\, d\omega$ is integrated over all incidence angles to provide the reflected intensity $i_{\lambda,r}'(\lambda,\theta_r,\varphi_r)$, this reflected intensity will be properly weighted by the amount of energy intercepted from each direction. Since $i_{\lambda,r}''$ is generally one differential order smaller than $i_{\lambda,i}'$,* the $d\omega$ in the denominator prevents $\rho_\lambda''(\lambda,\theta_r,\varphi_r,\theta,\varphi)$ from being a differential quantity. For a diffuse reflection the incident energy from (θ,φ) contributes equally to the reflected intensity for all (θ_r,φ_r). It will also be shown that the form of equation (3-20) leads to some convenient reciprocity relations.

---

*For a mirrorlike (specular) reflection, the $i_{\lambda,r}''$ is of the same order as $i_{\lambda,i}'$ and ρ_λ'' can become quite large. Thus, in contrast to the other radiative properties, ρ_λ'' can be larger than unity. In the notation for ρ'' the outgoing angles are given first, followed by the incoming angles.

Reciprocity for bidirectional spectral reflectivity. It is generally true that $\rho_\lambda''(\lambda,\theta_r,\varphi_r,\theta,\varphi)$ is symmetric with regard to reflection and incidence angles; that is, ρ_λ'' for energy incident at (θ,φ) and reflected at (θ_r,φ_r) is equal to ρ_λ'' for energy incident at (θ_r,φ_r) and reflected at (θ,φ).

This is demonstrated by considering a nonblack element dA_2 located within an isothermal black enclosure as shown in Fig. 3-6. For the isothermal condition, the net energy exchange between black elements dA_1 and dA_3 must be zero. This energy is exchanged by two possible paths. The first is the direct exchange along the dashed line. This direct exchange between black elements is uninfluenced by the presence of dA_2 and hence is zero as it would be in a black isothermal enclosure without dA_2. If the net exchange along this path is zero and net exchange including all paths between dA_1 and dA_3 is zero, then net exchange along the remaining path having reflection from dA_2 must also be zero. We can now write the following for the energy traveling along the reflected path:

$$d^4Q_{\lambda,\,1-2-3}'' = d^4Q_{\lambda,\,3-2-1}'' \tag{3-21a}$$

The energy reflected from dA_2 that reaches dA_3 is

$$d^4Q_{\lambda,\,1-2-3}'' = i_{\lambda,\,r}''(\lambda,\theta_r,\varphi_r,\theta,\varphi)\cos\theta_r\,dA_2\,\frac{dA_3\cos\theta_3}{S_2^{\,2}}\,d\lambda$$

or, using Eq. (3-20),

$$d^4Q_{\lambda,\,1-2-3}'' = \rho_\lambda''(\lambda,\theta_r,\varphi_r,\theta,\varphi)\,i_{\lambda,\,1}'(\lambda,T)\cos\theta\,\frac{dA_1\cos\theta_1}{S_1^{\,2}}\cos\theta_r\,dA_2\,\frac{dA_3\cos\theta_3}{S_2^{\,2}}\,d\lambda \tag{3-21b}$$

Similarly,

$$d^4Q_{\lambda,\,3-2-1}'' = \rho_\lambda''(\lambda,\theta,\varphi,\theta_r,\varphi_r)\,i_{\lambda,\,3}'(\lambda,T)\cos\theta_r\,\frac{dA_3\cos\theta_3}{S_2^{\,2}}\cos\theta\,dA_2\,\frac{dA_1\cos\theta_1}{S_1^{\,2}}\,d\lambda \tag{3-21c}$$

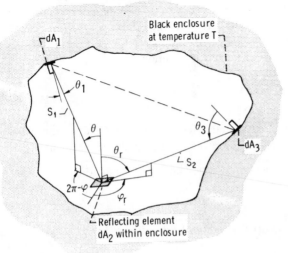

Black enclosure at temperature T

dA$_1$

θ_1

S_1

θ

θ_r

$2\pi-\varphi$

φ_r

θ_3

dA$_3$

S_2

← Reflecting element dA$_2$ within enclosure

Figure 3-6 Enclosure used to prove reciprocity of bidirectional spectral reflectivity.

Substituting Eqs. (3-21*b*) and (3-21*c*) into (3-21*a*) gives

$$\rho_\lambda''(\lambda,\theta_r,\varphi_r,\theta,\varphi)\,i_{\lambda,1}'(\lambda,T) = \rho_\lambda''(\lambda,\theta,\varphi,\theta_r,\varphi_r)\,i_{\lambda,3}'(\lambda,T)$$

or, because $i_{\lambda,1}'(\lambda,T) = i_{\lambda,3}'(\lambda,T) = i_{\lambda b}'(\lambda,T)$, we find the following *reciprocity relation for ρ_λ''*:

$$\rho_\lambda''(\lambda,\theta_r,\varphi_r,\theta,\varphi) = \rho_\lambda''(\lambda,\theta,\varphi,\theta_r,\varphi_r) \tag{3-22}$$

Directional spectral reflectivities. If $i_{\lambda,r}''$ is multiplied by $d\lambda \cos\theta_r\,dA\,d\omega_r$ and integrated over the hemisphere for all θ_r and φ_r, the energy per unit time is obtained that is reflected into the entire hemisphere as the result of an incident intensity from one direction:

$$d^3 Q_{\lambda,r}'(\lambda,\theta,\varphi) = d\lambda\,dA \int_\circ i_{\lambda,r}''(\lambda,\theta_r,\varphi_r,\theta,\varphi) \cos\theta_r\,d\omega_r$$

By use of Eq. (3-20) this is equal to

$$d^3 Q_{\lambda,r}'(\lambda,\theta,\varphi) = i_{\lambda,i}'(\lambda,\theta,\varphi) \cos\theta\,d\omega\,d\lambda\,dA \int_\circ \rho_\lambda''(\lambda,\theta_r,\varphi_r,\theta,\varphi) \cos\theta_r\,d\omega_r \tag{3-23}$$

The directional-hemispherical spectral reflectivity is then defined as the energy reflected into all solid angles divided by the incident energy from one direction (Fig. 3-1*f*). This gives Eq. (3-23) divided by the incident energy from Eq. (3-19):

Directional-hemispherical spectral reflectivity (in terms of bidirectional spectral reflectivity)

$$\equiv \rho_\lambda'(\lambda,\theta,\varphi) = \frac{d^3 Q_{\lambda,r}'(\lambda,\theta,\varphi)}{d^3 Q_{\lambda,i}'(\lambda,\theta,\varphi)} = \int_\circ \rho_\lambda''(\lambda,\theta_r,\varphi_r,\theta,\varphi) \cos\theta_r\,d\omega_r \tag{3-24}$$

Equation (3-24) defines how much of the radiant energy incident from one direction will be reflected into all directions. Another directional reflectivity is useful when one is concerned with the reflected intensity into one direction resulting from incident radiation coming from all directions. It is called the *hemispherical-directional spectral reflectivity* (Fig. 3-1*g*). The reflected intensity into the (θ_r,φ_r) direction is found by integrating Eq. (3-20) over all incident directions:

$$i_{\lambda,r}'(\lambda,\theta_r,\varphi_r) = \int_\circ \rho_\lambda''(\lambda,\theta_r,\varphi_r,\theta,\varphi)\,i_{\lambda,i}'(\lambda,\theta,\varphi) \cos\theta\,d\omega \tag{3-25}$$

The hemispherical-directional spectral reflectivity is then defined as the reflected intensity in the (θ_r,φ_r) direction divided by the integrated average incident intensity:

Hemispherical-directional spectral reflectivity (in terms of bidirectional spectral reflectivity)

$$\equiv \rho_\lambda'(\lambda,\theta_r,\varphi_r) = \frac{\int_\circ \rho_\lambda''(\lambda,\theta_r,\varphi_r,\theta,\varphi)\,i_{\lambda,i}'(\lambda,\theta,\varphi) \cos\theta\,d\omega}{(1/\pi) \int_\circ i_{\lambda,i}'(\lambda,\theta,\varphi) \cos\theta\,d\omega} \tag{3-26}$$

Reciprocity for directional spectral reflectivity. A reciprocity relation can also be found for ρ_λ' in the following manner: When the incident intensity is uniform over all incident directions, (3-26) reduces to

Hemispherical-directional spectral reflectivity (for uniform incident intensity)

$$\equiv \rho_\lambda'(\lambda,\theta_r,\varphi_r) = \int_{\,\cap} \rho_\lambda''(\lambda,\theta_r,\varphi_r,\theta,\varphi) \cos\theta \; d\omega \qquad (3\text{-}27)$$

By comparing Eqs. (3-24) and (3-27) and noting (3-22), one has the *reciprocal relation for* ρ_λ' resulting in (restricted to uniform incident intensity)

$$\rho_\lambda'(\lambda,\theta,\varphi) = \rho_\lambda'(\lambda,\theta_r,\varphi_r) \qquad (3\text{-}28)$$

where (θ_r,φ_r) and (θ,φ) are the same angles. This means that the reflectivity of a material irradiated at a given angle of incidence (θ,φ) as measured by the energy collected over the entire hemisphere of reflection is equal to the reflectivity for *uniform* irradiation from the hemisphere as measured by collecting the energy at a single angle of reflection (θ_r,φ_r) when (θ_r,φ_r) is the same angle as (θ,φ). This relation is employed in the design of "hemispherical reflectometers" for measuring radiative properties [2].

Hemispherical spectral reflectivity $\rho_\lambda(\lambda)$. If the incident spectral radiation arrives from all angles over the hemisphere (Fig. 3-1*h*), then all the radiation intercepted by the area element dA of the surface is given by Eq. (3-15*a*) as

$$d^2Q_{\lambda,i}(\lambda) = d\lambda \, dA \int_{\,\cap} i_{\lambda,i}'(\lambda,\theta,\varphi) \cos\theta \; d\omega$$

The amount of $d^2Q_{\lambda,i}$ that is reflected is, by integration of Eq. (3-24),

$$d^2Q_{\lambda,r}(\lambda) = \int_{\,\cap} \rho_\lambda'(\lambda,\theta,\varphi) \, d^3Q_{\lambda,i}'(\lambda,\theta,\varphi)$$

$$= d\lambda \, dA \int_{\,\cap} \rho_\lambda'(\lambda,\theta,\varphi) \, i_{\lambda,i}'(\lambda,\theta,\varphi) \cos\theta \; d\omega$$

The fraction of $d^2Q_{\lambda,i}(\lambda)$ that is reflected provides the definition

Hermispherical spectral reflectivity (in terms of directional-hemispherical spectral reflectivity)

$$\equiv \rho_\lambda(\lambda) = \frac{d^2Q_{\lambda,r}(\lambda)}{d^2Q_{\lambda,i}(\lambda)} = \frac{d\lambda \, dA}{d^2Q_{\lambda,i}(\lambda)} \int_{\,\cap} \rho_\lambda'(\lambda,\theta,\varphi) \, i_{\lambda,i}'(\lambda,\theta,\varphi) \cos\theta \; d\omega \qquad (3\text{-}29)$$

Limiting cases for spectral surfaces. Two important limiting cases of spectrally reflecting surfaces will be discussed in this section.

Diffusely reflecting surfaces For a *diffuse surface* the incident energy from the direction (θ,φ) that is reflected produces a reflected intensity that is uniform over all (θ_r,φ_r) directions, but the amount of energy reflected may vary as a function of *incident* angle.* When a diffuse surface element irradiated by an incident beam is viewed, the element will appear equally bright from all viewing directions. The bidirectional spectral reflectivity is then independent of (θ_r,φ_r), and Eq. (3-24) simplifies to

$$\rho'_{\lambda,d}(\lambda,\theta,\varphi) = \rho''_{\lambda}(\lambda,\theta,\varphi) \int_{\circ} \cos\theta_r \, d\omega_r$$

Carrying out the integration gives for a *diffuse* surface

$$\rho'_{\lambda,d}(\lambda,\theta,\varphi) = \pi\rho''_{\lambda}(\lambda,\theta,\varphi) \tag{3-30}$$

so that for any incidence angle the directional-hemispherical spectral reflectivity is equal to π times the bidirectional spectral reflectivity. This is because $\rho'_{\lambda,d}$ accounts for the reflected energy into all (θ_r,φ_r) directions, while ρ''_{λ} accounts for the reflected intensity into only one direction. This is analogous to the relation between blackbody hemispherical emissive power and intensity, $e_{\lambda b}(\lambda) = \pi i'_{\lambda b}(\lambda)$.

Equation (3-25) provides the intensity in the (θ_r,φ_r) directions when the incident radiation is distributed over (θ,φ) values. If the surface is *diffuse*, and if the *bidirectional reflectivity is independent of incidence angle, and if the incident intensity is uniform for all incident angles*, then Eq. (3-25) reduces to

$$i'_{\lambda,r}(\lambda) = \rho''_{\lambda}(\lambda) i'_{\lambda,i}(\lambda) \int_{\circ} \cos\theta \, d\omega = \pi\rho''_{\lambda}(\lambda) i'_{\lambda,i}(\lambda) \tag{3-31a}$$

By using Eq. (3-30) which applies for the diffuse surface and using reciprocity and Eq. (3-29), one has

$$i'_{\lambda,r}(\lambda) = \rho'_{\lambda,d}(\lambda) i'_{\lambda,i}(\lambda) = \rho_{\lambda,d}(\lambda) i'_{\lambda,i}(\lambda) \tag{3-31b}$$

so that the reflected intensity in any direction for this case is simply the hemispherical-directional reflectivity or the hemispherical reflectivity times the incident intensity. For the assumed uniform irradiation, the spectral energy per unit time intercepted by the surface element dA from all angular directions in the hemisphere is

$$d^2Q_{\lambda,i}(\lambda) = \pi i'_{\lambda,i}(\lambda) \, dA \, d\lambda$$

so that

$$i'_{\lambda,r}(\lambda) = \rho_{\lambda,d}(\lambda) \frac{d^2Q_{\lambda,i}(\lambda)}{\pi \, dA \, d\lambda} \tag{3-31c}$$

Specularly reflecting surfaces Mirrorlike, or specular, surfaces obey well-known laws of reflection. The perfect specular reflector and the perfect diffuse surface provide

---

*It is often tacitly assumed that diffuse reflectivities are independent of angle of incidence (θ,φ), but this is not a necessary condition for the diffuse definition.

two relatively simple special cases that can be used for the calculation of heat exchange in enclosures. For an incident beam from a single direction, a specular reflector, by definition, obeys a definite relation between incident and reflected angles. The reflected beam is at the same angle from the surface normal as the incident beam and is in the same plane as that formed by the incident beam and normal. Hence,

$$\theta_r = \theta \qquad \varphi_r = \varphi + \pi \tag{3-32}$$

and at all other angles the bidirectional spectral reflectivity of a specular surface is zero. We can write

$$\rho_\lambda''(\lambda,\theta,\varphi,\theta_r,\varphi_r)_{\text{specular}} = \rho_\lambda''(\lambda,\theta,\varphi,\theta_r = \theta,\varphi_r = \varphi + \pi) \equiv \rho_{\lambda,s}''(\lambda,\theta,\varphi) \tag{3-33}$$

and the bidirectional spectral reflectivity of a specular surface is considered to be only a function of the incident direction.

For the intensity of radiation reflected from a specular surface into the solid angle around (θ_r,φ_r), Eq. (3-25) gives, for an *arbitrary* directional distribution of incident intensity,

$$i_{\lambda,r}'(\lambda,\theta_r,\varphi_r) = \int_\cap \rho_{\lambda,s}''(\lambda,\theta,\varphi)\, i_{\lambda,i}'(\lambda,\theta,\varphi) \cos\theta\, d\omega \tag{3-34a}$$

The integrand of Eq. (3-34a) has a nonzero value only in the small solid angle around the direction (θ,φ) because of the properties of $\rho_{\lambda,s}''(\lambda,\theta,\varphi)$. Equation (3-34a) can then be written as

$$i_{\lambda,r}'(\lambda,\theta_r,\varphi_r) = \rho_{\lambda,s}''(\lambda,\theta,\varphi)\, i_{\lambda,i}'(\lambda,\theta,\varphi) \cos\theta\, d\omega \tag{3-34b}$$

Let us now consider for a moment the general equation for bidirectional spectral reflectivity [Eq. (3-20)]. When written for a specular surface, it becomes

$$i_{\lambda,r}''(\lambda,\theta_r = \theta, \varphi_r = \varphi + \pi) = \rho_{\lambda,s}''(\lambda,\theta,\varphi)\, i_{\lambda,i}'(\lambda,\theta,\varphi) \cos\theta\, d\omega \tag{3-35}*$$

This result is the intensity reflected into a solid angle around (θ_r,φ_r) from a single beam incident at $(\theta = \theta_r, \varphi = \varphi_r - \pi)$. The right side of Eq. (3-35) is seen to be identical to the right side of Eq. (3-34b), which gives the intensity reflected into the solid angle around (θ_r,φ_r) from distributed incident radiation. The point of this line of reasoning is to demonstrate the following rather obvious fact: In examining the radiation reflected from a specular surface into a given direction, only that radiation incident at the (θ,φ) defined by (3-32) need be considered as contributing to the reflected intensity *regardless* of the directional distribution of incident energy.

From Eqs. (3-26) and (3-34a), the hemispherical-directional spectral reflectivity for *uniform irradiation* of a specular surface is given by

---

*For a specular reflection $i_{\lambda,r}''$ can be of the same order as $i_{\lambda,i}'$. The $\rho_{\lambda,s}''$ then becomes very large because of the $d\omega$ on right side of (3-35). Hence, if a surface tends to be specular, the use of the bidirectional reflectivity can be of less practical value than in a situation with more diffuse reflection characteristics.

$$\rho'_{\lambda,s}(\lambda,\theta_r,\varphi_r) = \frac{\int_{\Omega} \rho''_{\lambda,s}(\lambda,\theta,\varphi) i'_{\lambda,i}(\lambda) \cos\theta \, d\omega}{(1/\pi) \int_{\Omega} i'_{\lambda,i}(\lambda) \cos\theta \, d\omega} = \frac{i'_{\lambda,r}(\lambda,\theta_r,\varphi_r)}{i'_{\lambda,i}(\lambda)} \qquad (3\text{-}36a)$$

Comparison with Eq. (3-34b) gives the relation between bidirectional and hemi-spherical-directional spectral reflectivities for a *specular surface* with uniform incident intensity as

$$\rho'_{\lambda,s}(\lambda,\theta_r,\varphi_r) = \rho''_{\lambda,s}(\lambda,\theta,\varphi) \cos\theta \, d\omega \qquad (3\text{-}36b)$$

Use of the reciprocity relation (3-28) shows that the directional-hemispherical reflectivity $\rho'_{\lambda,s}(\lambda,\theta,\varphi)$ for a single incident beam is

$$\rho'_{\lambda,s}(\lambda,\theta,\varphi) = \rho'_{\lambda,s}(\lambda,\theta_r,\varphi_r) = \rho''_{\lambda,s}(\lambda,\theta,\varphi) \cos\theta_r \, d\omega_r \qquad (3\text{-}36c)$$

where the restrictions of Eq. (3-32) still apply and the incident intensity is uniform.

The hemispherical spectral reflectivity of a uniformly irradiated specular reflector is, from (3-29),

$$\rho_{\lambda,s}(\lambda) = \frac{1}{\pi} \int_{\Omega} \rho'_{\lambda,s}(\lambda,\theta,\varphi) \cos\theta \, d\omega \qquad (3\text{-}37)$$

If $\rho'_{\lambda,s}$ is independent of incident angle, then evaluation of the integral in Eq. (3-37) gives

$$\rho_{\lambda,s}(\lambda) = \rho'_{\lambda,s}(\lambda) \qquad (3\text{-}38)$$

3-5.2 Total Reflectivities

The previous reflectivity definitions have dealt only with spectral radiation, but the expressions can be generalized to include the contributions from all wavelengths.

Bidirectional total reflectivity $\rho''(\theta_r,\varphi_r,\theta,\varphi)$. The bidirectional total reflectivity gives the contribution made by the total energy incident from direction (θ,φ) to the reflected total intensity into the direction (θ_r,φ_r). By analogy with Eq. (3-20)

$$Bidirectional\ total\ reflectivity \equiv \rho''(\theta_r,\varphi_r,\theta,\varphi) = \frac{\int_0^\infty i''_{\lambda,r}(\lambda,\theta_r,\varphi_r,\theta,\varphi) \, d\lambda}{\cos\theta \, d\omega \int_0^\infty i'_{\lambda,i}(\lambda,\theta,\varphi) \, d\lambda}$$

$$= \frac{i''_r(\theta_r,\varphi_r,\theta,\varphi)}{i'_i(\theta,\varphi) \cos\theta \, d\omega} \qquad (3\text{-}39a)$$

As an alternative form, the reflected energy is obtained by integrating (3-20) over all wavelengths

$$i''_r(\theta_r,\varphi_r,\theta,\varphi) = \cos\theta \, d\omega \int_0^\infty \rho''_\lambda(\lambda,\theta_r,\varphi_r,\theta,\varphi) i'_{\lambda,i}(\lambda,\theta,\varphi) \, d\lambda$$

so that Eq. (3-39a) can also be written as

Bidirectional total reflectivity (in terms of bidirectional spectral reflectivity)

$$\equiv \rho''(\theta_r,\varphi_r,\theta,\varphi) = \frac{\int_0^\infty \rho_\lambda''(\lambda,\theta_r,\varphi_r,\theta,\varphi) i_{\lambda,i}'(\lambda,\theta,\varphi)\, d\lambda}{i_i'(\theta,\varphi)} \tag{3-39b}$$

where $i_i'(\theta,\varphi) = \int_0^\infty i_{\lambda,i}'(\lambda,\theta,\varphi)\, d\lambda$.

Reciprocity. Rewriting equation (3-39b) for the case of energy incident from direction (θ_r,φ_r) and reflected into direction (θ,φ) gives

$$\rho''(\theta,\varphi,\theta_r,\varphi_r) = \frac{\int_0^\infty \rho_\lambda''(\lambda,\theta,\varphi,\theta_r,\varphi_r) i_{\lambda,i}'(\lambda,\theta_r,\varphi_r)\, d\lambda}{i_i'(\theta_r,\varphi_r)} \tag{3-39c}$$

Comparison of Eqs. (3-39b) and (3-39c) shows that

$$\rho''(\theta,\varphi,\theta_r,\varphi_r) = \rho''(\theta_r,\varphi_r,\theta,\varphi) \tag{3-40}$$

if the *spectral distribution of incident intensity is the same for all directions or, in a less restrictive sense, if* $i_{\lambda,i}'(\lambda,\theta,\varphi) = C i_{\lambda,i}'(\lambda,\theta_r,\varphi_r)$.

Directional total reflectivity ρ'. The *directional-hemispherical total reflectivity* is the fraction of the total energy incident from a single direction that is reflected into all angular directions. The spectral energy from a given direction that is intercepted by the surface is $i_{\lambda,i}'(\lambda,\theta,\varphi) \cos\theta\, d\omega\, d\lambda\, dA$. The portion of this energy that is reflected is $\rho_\lambda'(\lambda,\theta,\varphi) i_{\lambda,i}'(\lambda,\theta,\varphi) \cos\theta\, d\omega\, d\lambda\, dA$. If these quantities are integrated over all wavelengths to provide total values, the following definition is formed:

Directional-hemispherical total reflectivity (in terms of directional-hemispherical spectral reflectivity)

$$\equiv \rho'(\theta,\varphi) = \frac{d^2 Q_r'(\theta,\varphi)}{d^2 Q_i'(\theta,\varphi)} = \frac{\int_0^\infty \rho_\lambda'(\lambda,\theta,\varphi) i_{\lambda,i}'(\lambda,\theta,\varphi)\, d\lambda}{\int_0^\infty i_{\lambda,i}'(\lambda,\theta,\varphi)\, d\lambda} \tag{3-41a}$$

Another directional total reflectivity specifies the fraction of radiation reflected into a given (θ_r,φ_r) direction when there is uniform irradiation. The total radiation intensity reflected into the (θ_r,φ_r) direction when the incident intensity is uniform for all directions is

$$i_r'(\theta_r,\varphi_r) = \int_0^\infty i_{\lambda,r}'(\lambda,\theta_r,\varphi_r)\, d\lambda = \int_0^\infty i_{\lambda,i}'(\lambda) \rho_\lambda'(\lambda,\theta_r,\varphi_r)\, d\lambda$$

where $\rho_\lambda'(\lambda,\theta_r,\varphi_r)$ was discussed in connection with Eq. (3-27). Then the reflectivity can be defined as the reflected intensity divided by the incident intensity

Hemispherical-directional total reflectivity (for uniform irradiation)

$$\equiv \rho'(\theta_r,\varphi_r) = \frac{\int_0^\infty \rho_\lambda'(\lambda,\theta_r,\varphi_r) i_{\lambda,i}'(\lambda)\, d\lambda}{\int_0^\infty i_{\lambda,i}'(\lambda)\, d\lambda} \tag{3-41b}$$

Reciprocity. Equations (3-41a) and (3-41b) are now compared bearing in mind that the latter is restricted to uniform incident intensity. With this restriction, from Eq. (3-28) it is found that

$$\rho'(\theta_r, \varphi_r) = \rho'(\theta, \varphi) \tag{3-42}$$

where (θ_r, φ_r) and (θ, φ) are the same angles *when there is a fixed spectral distribution of the incident radiation such that*

$$i'_{\lambda, i}(\lambda, \theta, \varphi) = C i'_{\lambda, i}(\lambda)$$

Hemispherical total reflectivity. If the incident total radiation arrives from all angles over the hemisphere, then the total radiation intercepted by a unit area at the surface is given by Eq. (3-17a). The amount of this radiation that is reflected is

$$dQ_r = dA \int_{\cap} \rho'(\theta, \varphi) \, i'_i(\theta, \varphi) \cos \theta \, d\omega$$

The ratio of these two quantities is then the hemispherical total reflectivity, which is the fraction of all the incident energy that is reflected including all directions of reflection; that is,

Hemispherical total reflectivity (in terms of directional-hemispherical total reflectivity)

$$\equiv \rho = \frac{dQ_r}{dQ_i} = \frac{dA}{dQ_i} \int_{\cap} \rho'(\theta, \varphi) \, i'_i(\theta, \varphi) \cos \theta \, d\omega \tag{3-43a}$$

Another form is found by using $d^2 Q_{\lambda, i}(\lambda)$, which is the incident hemispherical spectral energy intercepted by the surface. The amount of this that is reflected is $\rho_\lambda(\lambda) \, d^2 Q_{\lambda, i}$, where $\rho_\lambda(\lambda)$ is the hemispherical spectral reflectivity from Eq. (3-29). Then integrating yields

Hemispherical total reflectivity (in terms of hemispherical spectral reflectivity)

$$\equiv \rho = \frac{\int_0^\infty \rho_\lambda(\lambda) \, d^2 Q_{\lambda, i}(\lambda)}{dQ_i} \tag{3-43b}$$

3-5.3 Summary of Restrictions on Reciprocity Relations between Reflectivities

In Table 3-3, a summary is presented of the restrictive conditions necessary for application of the various reciprocity relations for reflectivities.

3-6 RELATIONS AMONG REFLECTIVITY, ABSORPTIVITY, AND EMISSIVITY

From the definitions of absorptivity and reflectivity as fractions of incident energy absorbed or reflected, it is evident that for an opaque body (no radiation transmitted through the body) some simple relations exist between these surface properties.

Table 3-3 Summary of reciprocity relations between reflectivities

Type of quantity	Equality	Restrictions
A. Bidirectional spectral [Eq. (3-22)]	$\rho_\lambda''(\lambda,\theta,\varphi,\theta_r,\varphi_r) = \rho_\lambda''(\lambda,\theta_r,\varphi_r,\theta,\varphi)$	None
B. Directional spectral [Eq. (3-28)]	$\rho_\lambda'(\lambda,\theta,\varphi) = \rho_\lambda'(\lambda,\theta_r,\varphi_r)$ where $\theta = \theta_r$ $\varphi = \varphi_r$	$\rho_\lambda'(\lambda,\theta_r,\varphi_r)$ is for uniform incident intensity or $\rho_\lambda''(\lambda)$ independent of θ, φ, θ_r, and φ_r
C. Bidirectional total [Eq. (3-40)]	$\rho''(\theta,\varphi,\theta_r,\varphi_r) = \rho''(\theta_r,\varphi_r,\theta,\varphi)$	$i_{\lambda,i}'(\lambda,\theta,\varphi) = C i_{\lambda,i}'(\lambda,\theta_r,\varphi_r)$ or $\rho_\lambda''(\theta,\varphi,\theta_r,\varphi_r)$ independent of wavelength
D. Directional total [Eq. (3-42)]	$\rho'(\theta,\varphi) = \rho'(\theta_r,\varphi_r)$ where $\theta = \theta_r$ $\varphi = \varphi_r$	One restriction from both B and C

By using Kirchhoff's law (see Sec. 3-4.8) and taking note of the restrictions involved, further relations between the emissivity and the reflectivity can be found in certain cases.

Because the spectral energy per unit time $d^3 Q_{\lambda,i}'$ incident upon dA of an opaque body from a solid angle $d\omega$ is either absorbed or reflected, it is evident that

$$d^3 Q_{\lambda,i}'(\lambda,\theta,\varphi) = d^3 Q_{\lambda,a}'(\lambda,\theta,\varphi,T_A) + d^3 Q_{\lambda,r}'(\lambda,\theta,\varphi,T_A)$$

or

$$\frac{d^3 Q_{\lambda,a}'(\lambda,\theta,\varphi,T_A)}{d^3 Q_{\lambda,i}'(\lambda,\theta,\varphi)} + \frac{d^3 Q_{\lambda,r}'(\lambda,\theta,\varphi,T_A)}{d^3 Q_{\lambda,i}'(\lambda,\theta,\varphi)} = 1 \qquad (3\text{-}44)$$

Since the energy is incident from the direction (θ,φ), the two energy ratios of Eq. (3-44) are the directional spectral absorptivity [Eq. (3-10a)] and the directional-hemispherical spectral reflectivity [Eq. (3-24)], respectively. Substituting gives

$$\alpha_\lambda'(\lambda,\theta,\varphi,T_A) + \rho_\lambda'(\lambda,\theta,\varphi,T_A) = 1 \qquad (3\text{-}45)$$

Kirchhoff's law [Eq. (3-12)] can then be applied without restriction to yield

$$\epsilon_\lambda'(\lambda,\theta,\varphi,T_A) + \rho_\lambda'(\lambda,\theta,\varphi,T_A) = 1 \qquad (3\text{-}46)$$

When the total energy arriving at dA from a given direction is considered, (3-44) becomes

$$\frac{d^2 Q_a'(\theta,\varphi,T_A)}{d^2 Q_i'(\theta,\varphi)} + \frac{d^2 Q_r'(\theta,\varphi,T_A)}{d^2 Q_i'(\theta,\varphi)} = 1 \qquad (3\text{-}47)$$

Substituting Eqs. (3-14a) and (3-41a) for the energy ratios results in

$$\alpha'(\theta,\varphi,T_A) + \rho'(\theta,\varphi,T_A) = 1 \qquad (3\text{-}48)$$

The absorptivity is the directional total value, and the reflectivity is the directional-hemispherical total value.

Kirchhoff's law for directional total properties (Sec. 3-4.4) can then be applied to give

$$\epsilon'(\theta,\varphi,T_A) + \rho'(\theta,\varphi,T_A) = 1 \tag{3-49}$$

under the restriction that the incident radiation obeys the relation $i'_{\lambda,i}(\lambda,\theta,\varphi) = C(\theta,\varphi)i'_{\lambda b}(\lambda,T_A)$ or the surface is directional-gray.

If the incident spectral energy is assumed to be arriving at dA from all directions over the hemisphere, then (3-44) gives

$$\frac{d^2Q_{\lambda,a}(\lambda,T_A)}{d^2Q_{\lambda,i}(\lambda)} + \frac{d^2Q_{\lambda,r}(\lambda,T_A)}{d^2Q_{\lambda,i}(\lambda)} = 1 \tag{3-50}$$

Equation (3-50) can then be written as

$$\alpha_\lambda(\lambda,T_A) + \rho_\lambda(\lambda,T_A) = 1 \tag{3-51}$$

where the radiative properties are hemispherical spectral values from Eqs. (3-16) and (3-29). Substitution of the hemispherical spectral emissivity $\epsilon_\lambda(\lambda,T_A)$ for $\alpha_\lambda(\lambda,T_A)$ in this relation is valid *only* if the intensity of incident radiation is independent of incident angle, that is, it is uniform over all incident directions, or if the α_λ and ϵ_λ do not depend on angle (see Sec. 3-4.8). Under these restrictions, Eq. (3-51) becomes

$$\epsilon_\lambda(\lambda,T_A) + \rho_\lambda(\lambda,T_A) = 1 \tag{3-52}$$

If the incident energy on dA is summed over all wavelengths and directions, Eq. (3-44) becomes

$$\frac{dQ_a(T_A)}{dQ_i} + \frac{dQ_r(T_A)}{dQ_i} = 1 \tag{3-53}$$

The energy ratios are now the hemispherical total values of absorptivity and reflectivity [Eqs. (3-18) and (3-43a), respectively], and Eq. (3-53) becomes

$$\alpha(T_A) + \rho(T_A) = 1 \tag{3-54}$$

Again, certain restrictions apply if $\epsilon(T_A)$ is substituted for $\alpha(T_A)$ to obtain

$$\epsilon(T_A) + \rho(T_A) = 1 \tag{3-55}$$

The principal restrictions on the validity of this relation are that the incident spectral intensity is proportional to the emitted spectral intensity of a blackbody at T_A *and* the incident intensity is uniform over all incident angles; that is, $i'_{\lambda,i}(\lambda) = Ci'_{\lambda b}(\lambda,T_A)$. Other special cases where the substitution $\alpha(T_A) = \epsilon(T_A)$ can be made are listed in Sec. 3-4.8.

When the body is not opaque so that some radiation is transmitted entirely through it, a transmitted fraction must be introduced. This topic is more properly discussed later in connection with radiation in absorbing and transmitting media.

EXAMPLE 3-6 Radiation from the sun is incident on a surface in orbit above the earth's atmosphere. The surface is at 1000 K, and the directional total emissivity is given in Fig. 3-3. If the incident energy is at an angle of 25° from the normal to the surface, what is the reflected energy flux?

From Fig. 3-3, $\epsilon'(25°, 1000\ K) = 0.8$. The spectrum of radiation from the sun is similar to that of a blackbody. Section 3-4.8 shows that $\alpha'(25°, 1000\ K) = \epsilon'(25°, 1000\ K) = 0.8$ only when the incident spectrum is proportional to that emitted by a blackbody at $T_A = 1000$ K. This is not the case here, since the sun acts like a blackbody at 5780 K. Hence $\alpha' \neq 0.8$, and without α' we cannot determine ρ'; the emissivity data given are insufficient to work the problem.

EXAMPLE 3-7 A surface at $T_A = 1000°$R has a spectral emissivity in the normal direction that can be approximated as shown in Fig. 3-7. This surface is maintained at 1000°R by cooling water and is then enclosed by a black hemisphere heated to $T_i = 3000°$R. What will the reflected intensity be into the direction normal to the surface?

From Eq. (3-46),

$$\rho'_\lambda(\lambda, \theta = 0°, T_A) = 1 - \epsilon'_\lambda(\lambda, \theta = 0°, T_A)$$

which is the reflectivity into the hemisphere for radiation arriving from the normal direction. From reciprocity, for uniform incident intensity over the hemisphere,

$$\rho'_\lambda(\lambda, \theta_r = 0°, T_A) = \rho'_\lambda(\lambda, \theta = 0°, T_A)$$

Hence the reflectivity into the normal direction resulting from the incident radiation from the hemisphere is (by use of Fig. 3-7)

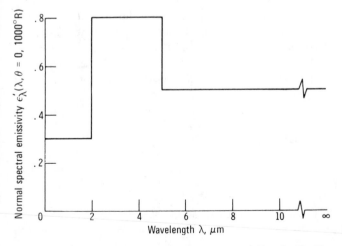

Figure 3-7 Directional spectral emissivity in normal direction for Example 3-7.

$\rho'_\lambda(0 \leqslant \lambda < 2, \theta_r = 0°, T_A) = 0.7$

$\rho'_\lambda(2 \leqslant \lambda < 5, \theta_r = 0°, T_A) = 0.2$

$\rho'_\lambda(5 \leqslant \lambda \leqslant \infty, \theta_r = 0°, T_A) = 0.5$

The incident intensity is $i'_{\lambda,i}(\lambda, T_i) = i'_{\lambda b}(\lambda, 3000°R)$. From the relation preceding Eq. (3-41b), the reflected intensity is

$$i'_r(\theta_r = 0°) = \int_0^\infty i'_{\lambda b}(\lambda, T_i) \rho'_\lambda(\lambda, \theta_r = 0°, T_A) \, d\lambda$$

$$= \frac{\sigma T_i^4}{\pi} \int_0^\infty \frac{e_{\lambda b}(\lambda, T_i)}{\sigma T_i^5} \rho'_\lambda(\lambda, \theta_r = 0°, T_A) \, d(\lambda T_i)$$

From Eq. (2-27) this becomes,

$$i'_r(\theta_r = 0°) = \frac{\sigma T_i^4}{\pi} (0.7 F_{0-2T_i} + 0.2 F_{2T_i-5T_i} + 0.5 F_{5T_i-\infty})$$

$$= \frac{0.173}{\pi} (30)^4 [0.7(0.347) + 0.2(0.869 - 0.347) + 0.5(1 - 0.869)]$$

$$= 18,400 \text{ Btu}/(\text{h·ft}^2\text{·sr})$$

3-7 CONCLUDING REMARKS

In this chapter, a precise system of nomenclature has been introduced, and careful definitions of the radiative properties have been given. The defining equations are summarized in Table 3-1 for convenience, along with the symbols used here.

By using these definitions it was possible to examine the restrictions on the various forms of Kirchhoff's law relating emissivity to absorptivity. These restrictions are sometimes a source of confusion, and it is hoped that the summary (Table 3-2, Sec. 3-4.8) will make clear the conditions governing when α can be set equal to ϵ. These restrictions are also invoked when deriving the relation $\epsilon + \rho = 1$ from the general relation $\alpha + \rho = 1$ for opaque bodies.

The detailed definitions made it possible to derive the reciprocal relations for reflectivities and examine the restrictions involved. These restrictions are listed in a convenient summary in Table 3-3 (Sec. 3-5.3).

REFERENCES

1. Nicodemus, F. E. et al.: Geometrical Considerations and Nomenclature for Reflectance, NBS monograph 160, National Bureau of Standards, United States Department of Commerce, 1977.
2. Brandenberg, W. M.: The Reflectivity of Solids at Grazing Angles. Measurement of Thermal Radiation Properties of Solids, Joseph C. Richmond (ed.), *NASA* SP-31, pp. 75–82, 1963.

PROBLEMS

1 A material has a hemispherical spectral emissivity that varies considerably with wavelength but is fairly independent of surface temperature (see, for example, the behavior of tungsten in Fig. 5-3). Radiation from a gray source at T_i is incident on the surface. Show that the total absorptivity for the incident radiation is equal to the total emissivity of the material evaluated at the source temperature T_i.

2 Using Fig. 5-3, estimate the hemispherical total emissivity of tungsten at 2800 K.

3 Suppose that ϵ_λ is independent of λ (gray-body radiation). Show that $F_{0-\lambda T}$ represents the fraction of the total radiant output of the gray body in the range from 0 to λT.

4 For a surface with hemispherical spectral emissivity ϵ_λ, does the maximum of the e_λ distribution occur at the same λ as the maximum of the $e_{\lambda b}$ distribution at the same temperature? (*Hint:* Examine the behavior of $de_\lambda/d\lambda$.) Plot the distributions of e_λ as a function of λ for the data of Fig. 3-7 at 1000°R and for the data in Prob. 6a at 600°R. At what λ is the maximum of e_λ? How does this compare with the maximum of $e_{\lambda b}$?

5 A white ceramic surface has a hemispherical spectral emissivity distribution at 3000°R (1667 K) as shown. What is the hemispherical total emissivity?

Answer: 0.28.

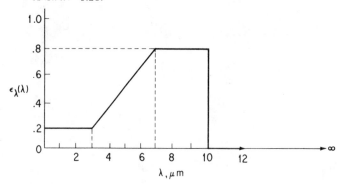

6 A surface has the following values of hemispherical spectral emissivity at a temperature of 140°F (600°R).

λ, μm	$\epsilon_\lambda(\lambda, 600°\ R)$
<1	0
1	0
1.5	0.2
2	0.4
2.5	0.6
3	0.8
3.5	0.8
4	0.8
4.5	0.7
5	0.6
6	0.4
7	0.2
8	0
>8	0

(a) What is the hemispherical total emissivity of the surface at 140°F?

(b) What is the hemispherical total absorptivity of the surface at 140°F if the incident radiation is from a gray source at 1540°F that has an emissivity of 0.8? The incident radiation is uniform over all incident angles.

Answer: (a) 0.064; (b) 0.50.

7 Using Fig. 5-22, estimate the absorptivity of typewriter paper for normally incident radiation from a blackbody source at 1178 K.

8 The spectral absorptivity of an SiO–Al selective surface can be approximated as shown below. The surface is in earth orbit around the sun and has incident on it in the normal direction the solar flux 442 Btu/(h·ft²) (1394 W/m²). What will be the equilibrium temperature of the surface?

Answer: 1197°R (665 K).

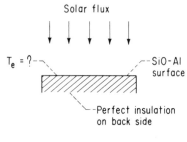

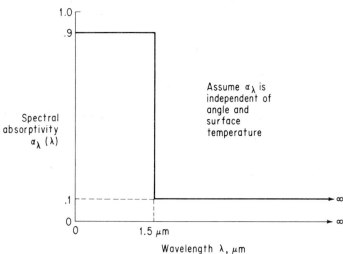

Wavelength λ, μm

9 A gray surface has directional emissivity as shown. The properties are isotropic with respect to circumferential angle φ.

(a) What is the value of the hemispherical emissivity of this surface?

(b) If energy from a blackbody source at 200°F (366 K) is incident uniformly from all directions, what fraction of the incident energy will be absorbed by this surface?

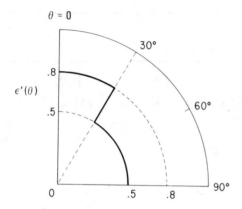

(c) If the surface is placed in an environment at absolute zero, at what rate must energy be added per unit area to maintain the surface at a temperature of 1000°R (550 K)?

Answer: (a) 0.575; (b) 0.575; (c) 995 Btu/(h·ft²) (3140 W/m²).

FOUR

PREDICTION OF RADIATIVE PROPERTIES BY CLASSICAL ELECTROMAGNETIC THEORY

4-1 INTRODUCTION

James Clerk Maxwell, in 1864, published an article defining what is generally conceded to be the crowning achievement of classical physics, the relation between electrical and magnetic fields, and the realization that electromagnetic waves propagate with the speed of light, indicating strongly that light itself is in the form of an electromagnetic wave [1]. Although quantum effects have since been shown to be the controlling phenomena in electromagnetic energy propagation, it is possible and indeed necessary to describe many of the properties of light and radiant heat by the classical wave approach.

It will be demonstrated in this chapter that the reflectivity, emissivity, and absorptivity of materials can in certain cases be calculated from the optical and electrical properties of the materials. The relations between the radiative properties of a material and its optical and electrical properties are found by considering the interaction that occurs when an electromagnetic wave traveling through one medium is incident on the surface of another medium.

The analysis will be based on the assumption that there is an ideal interaction between the incident waves and the surface. Physically this means that the results are for optically smooth, clean surfaces that reflect in a specular fashion. The wave propagation and surface interaction will be investigated here in a somewhat simplified fashion by using Maxwell's fundamental equations relating electric and magnetic fields. For ideal surface conditions it is possible to perform more accurate property computations by using theory that is more rigorous than the wave analysis presented here. However, the labor involved is generally not justified, because neither the simplified nor the more sophisticated approach can account for the effects of surface preparation. The departures of real materials from the ideal materials assumed in the theory are often responsible for introducing large variations of measured property values from theoretical predictions. These departures are caused by factors such as impurities, surface roughness, surface contamination, and crystal-structure modification by surface working.

Although in practice there can be large effects of surface condition, the theory presented here does serve a number of useful purposes. It provides an understanding of why there are basic differences in the radiative properties of insulators and electrical conductors, and reveals general trends that help unify the presentation of experimental data. These trends are also useful when it is required for engineering calculations to extrapolate limited experimental data into another range. The theory has utility in the theoretical understanding of the angular behavior of the directional reflectivity, absorptivity, and emissivity. Since the electromagnetic theory applies for pure substances with ideally smooth surfaces, it provides a means by which one limit of attainable properties can be computed; for example, the maximum reflectivity or minimum emissivity of a metallic surface can be determined.

The derivation of radiative property relations from classical theory is carried out in some detail in Secs. 4-3-4-5. The results are then applied to radiative property predictions in Sec. 4-6. Those readers interested only in the use of the results for property predictions are invited to pass over the derivation portions to Sec. 4-6.

4-2 SYMBOLS

C_1, C_2	constants in Planck spectral energy distribution
c	speed of electromagnetic wave in medium other than a vacuum
c_0	speed of electromagnetic wave in vacuum
E	amplitude of electric intensity wave
$\mathbf{E}$	electric intensity vector
e	emissive power
H	amplitude of magnetic intensity wave
$\mathbf{H}$	magnetic intensity vector
n	refractive index
$\bar{n}$	complex refractive index, $n - i\kappa$
r_e	electrical resistivity
S	instantaneous rate of energy transport per unit area
$\mathbf{S}$	Poynting vector, Eq. (4-24)
T	absolute temperature
t	time
x,y,z x',y',z'	coordinates in cartesian system
β	extinction coefficient in x direction

γ	permittivity
δ	propagation angle in medium
ϵ	emissivity
θ	angle measured from normal of surface, polar angle
κ	extinction coefficient
λ	wavelength
μ	magnetic permeability
ν	frequency
φ	circumferential angle
ρ	reflectivity
χ	angle of refraction
ω	angular frequency
$\int_{\Omega}$	integration over solid angle of entire enclosing hemisphere

Subscripts

A	property of body or surface A
b	black
i	incident
M	maximum value
n	normal
0	in a vacuum
r	reflected
s	specular
t	transmitted
x,y,z	components in x, y, z directions
x',y',z'	components in x', y', z' directions
λ	spectral
$1, 2$	medium 1 or 2
$\perp$	perpendicular component
$\parallel$	parallel component

Superscript

   $'$ directional quantity (except in x',y',z')

4-3 FUNDAMENTAL EQUATIONS OF ELECTROMAGNETIC THEORY

Maxwell's equations can be used to describe the interaction of electric and magnetic fields within any isotropic medium, including a vacuum, under the condition of no accumulation of static charge. With these restrictions the equations are, in SI (mks) units,

$$\nabla \times \mathbf{H} = \gamma \frac{\partial \mathbf{E}}{\partial t} + \frac{\mathbf{E}}{r_e} \qquad (4\text{-}1)$$

Table 4-1 Quantities for use in electromagnetic equations in SI units

Symbol	Quantity	Units	Value
c	Speed of electromagnetic wave propagation	m/s	
c_0	Speed of electromagnetic wave propagation in vacuum	m/s	2.9979×10^8
E	Electric intensity	N/C [newtons/coulomb]	
H	Magnetic intensity	C/(m·s)	
K	Dielectric constant, γ/γ_0		
r_e	Electrical resistivity	$\Omega \cdot$ m; N $\cdot$ m^2 $\cdot$ s/C^2	
S	Instantaneous rate of energy transport per unit area	N $\cdot$ m/(s $\cdot$ m^2); W/m^2	
x,y,z x',y',z'	Cartesian coordinate position	m	
γ	Electrical permittivity	C^2/(N $\cdot$ m^2)	
γ_0	Electrical permittivity of vacuum	C^2/(N $\cdot$ m^2)	$\dfrac{1}{4\pi \times 8.9875} \times 10^{-9}$
μ	Magnetic permeability	N $\cdot$ s^2/C^2	
μ_0	Magnetic permeability of vacuum	N $\cdot$ s^2/C^2	$4\pi \times 10^{-7}$

$$\nabla \times \mathbf{E} = -\mu \frac{\partial \mathbf{H}}{\partial t} \tag{4-2}$$

$$\nabla \cdot \mathbf{E} = 0 \tag{4-3}$$

$$\nabla \cdot \mathbf{H} = 0 \tag{4-4}$$

where **H** and **E** are the magnetic and electric intensities, respectively, γ is the permittivity, r_e is the electrical resistivity, and μ is the magnetic permeability of the medium. The SI units for these quantities are shown in Table 4-1. Zero subscripts denote quantities evaluated in a vacuum.

The solutions to these equations will reveal how radiation waves travel within a material and what the interaction is between the electric and magnetic fields. By knowing how the waves move in each of two adjacent media and applying coupling relations at the interface between the media, the relations governing reflection and absorption will be formulated.

4-4 RADIATIVE WAVE PROPAGATION WITHIN A MEDIUM

Propagation is first considered within an infinite, homogeneous, isotropic medium. The derivation of wave propagation within a perfect dielectric will be considered in Sec. 4-4.1; it is found that the wave is not attenuated in such a material. Media of finite electrical conductivity are then analyzed in Sec. 4-4.2; these media can be imperfect dielectrics (poor conductors) or metals (good conductors). The waves do attenuate in these materials because of absorption of energy within the material.

4-4.1 Propagation in Perfect Dielectric Media

For simplicity we first consider the situation in which the medium is a vacuum or other insulator having an electrical resistivity so large that the last term in Eq. (4-1), E/r_e, can be neglected. With this simplification Eqs. (4-1) and (4-2) can be written out in cartesian coordinates to provide two sets of three equations relating the x, y, and z components of the electric and magnetic intensities, that is,

$$\frac{\partial H_z}{\partial y} - \frac{\partial H_y}{\partial z} = \gamma \frac{\partial E_x}{\partial t} \tag{4-5a}$$

$$\frac{\partial H_x}{\partial z} - \frac{\partial H_z}{\partial x} = \gamma \frac{\partial E_y}{\partial t} \tag{4-5b}$$

$$\frac{\partial H_y}{\partial x} - \frac{\partial H_x}{\partial y} = \gamma \frac{\partial E_z}{\partial t} \tag{4-5c}$$

$$\frac{\partial E_z}{\partial y} - \frac{\partial E_y}{\partial z} = -\mu \frac{\partial H_x}{\partial t} \tag{4-6a}$$

$$\frac{\partial E_x}{\partial z} - \frac{\partial E_z}{\partial x} = -\mu \frac{\partial H_y}{\partial t} \tag{4-6b}$$

$$\frac{\partial E_y}{\partial x} - \frac{\partial E_x}{\partial y} = -\mu \frac{\partial H_z}{\partial t} \tag{4-6c}$$

From Eqs. (4-3) and (4-4), we get

$$\frac{\partial E_x}{\partial x} + \frac{\partial E_y}{\partial y} + \frac{\partial E_z}{\partial z} = 0 \tag{4-7}$$

and

$$\frac{\partial H_x}{\partial x} + \frac{\partial H_y}{\partial y} + \frac{\partial H_z}{\partial z} = 0 \tag{4-8}$$

The behavior of a wave of electromagnetic radiation within a material will be considered. The coordinate system x, y, z is fixed to the path of the wave that is propagating in the x direction (Fig. 4-1). For simplicity, a plane wave is considered; that is, all the quantities concerned with the wave are constant over any yz plane at any given time. Hence $\partial/\partial y = \partial/\partial z = 0$. For these conditions, (4-5)-(4-8) reduce to

$$0 = \gamma \frac{\partial E_x}{\partial t} \tag{4-9a}$$

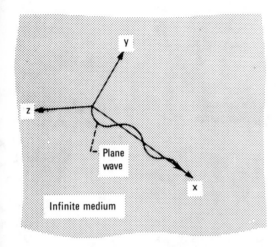

Figure 4-1 Propagation in homogeneous, isotropic material.

$$-\frac{\partial H_z}{\partial x} = \gamma \frac{\partial E_y}{\partial t} \qquad (4\text{-}9b)$$

$$\frac{\partial H_y}{\partial x} = \gamma \frac{\partial E_z}{\partial t} \qquad (4\text{-}9c)$$

$$0 = -\mu \frac{\partial H_x}{\partial t} \qquad (4\text{-}10a)$$

$$-\frac{\partial E_z}{\partial x} = -\mu \frac{\partial H_y}{\partial t} \qquad (4\text{-}10b)$$

$$\frac{\partial E_y}{\partial x} = -\mu \frac{\partial H_z}{\partial t} \qquad (4\text{-}10c)$$

$$\frac{\partial E_x}{\partial x} = 0 \qquad (4\text{-}11)$$

$$\frac{\partial H_x}{\partial x} = 0 \qquad (4\text{-}12)$$

The H components are then eliminated by differentiating Eqs. (4-9b) and (4-9c) with respect to t and Eqs. (4-10b) and (4-10c) with respect to x to obtain

$$-\frac{\partial^2 H_z}{\partial t\, \partial x} = \gamma \frac{\partial^2 E_y}{\partial t^2} \qquad (4\text{-}13a)$$

$$\frac{\partial^2 H_y}{\partial t\, \partial x} = \gamma \frac{\partial^2 E_z}{\partial t^2} \qquad (4\text{-}13b)$$

$$-\frac{\partial^2 E_z}{\partial x^2} = -\mu \frac{\partial^2 H_y}{\partial x \, \partial t} \tag{4-14a}$$

$$\frac{\partial^2 E_y}{\partial x^2} = -\mu \frac{\partial^2 H_z}{\partial x \, \partial t} \tag{4-14b}$$

Equations (4-13a) and (4-14b) are then combined to eliminate H_z, and similarly (4-13b) and (4-14a) to eliminate H_y. This provides the following two equations:

$$\mu\gamma \frac{\partial^2 E_y}{\partial t^2} = \frac{\partial^2 E_y}{\partial x^2} \tag{4-15a}$$

and

$$\mu\gamma \frac{\partial^2 E_z}{\partial t^2} = \frac{\partial^2 E_z}{\partial x^2} \tag{4-15b}$$

These wave equations govern the propagation of the y and z components of the electric intensity in the x direction. For simplicity in the remainder of the derivation, it will be assumed that the electromagnetic waves are polarized such that the vector **E** is contained only within the xy plane (see Fig. 4-2). Then E_z and its derivatives are zero and Eq. (4-15b) need not be considered. The vector **E** will have only x and y components.

With regard to the x components of **E** and **H**, from (4-9a), (4-10a), (4-11), and (4-12), $\partial E_x/\partial t = \partial E_x/\partial x = \partial H_x/\partial t = \partial H_x/\partial x = 0$. Hence, the electric and magnetic intensity components in the direction of propagation are both steady and independent of the propagation direction x. Consequently, the only time-varying component of **E** is E_y as governed by (4-15a). Since this component is normal to x, the direction of propagation, the wave is a transverse wave.

Equation (4-15a) is recognized as the *wave equation* that describes the propagation of the wave component E_y in the x direction. The general solution of this equation is

$$E_y = f\!\left(x - \frac{t}{\sqrt{\mu\gamma}}\right) + g\!\left(x + \frac{t}{\sqrt{\mu\gamma}}\right) \tag{4-16a}$$

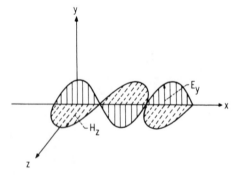

Figure 4-2 Electric field wave polarized in xy plane, traveling in x direction with companion magnetic field wave.

where f and g are *any* differentiable functions. The f function provides propagation in the positive x direction, while the g function accounts for propagation in the negative x direction. Since the present discussion deals with a wave moving in the positive direction, only the f function will be present in the analysis.

To obtain the wave propagation speed, consider an observer moving along with the wave; the observer will always be at a fixed value of E_y. The x location of the observer must then vary with time such that the argument of f, $x - t/\sqrt{\mu\gamma}$, is also fixed. Hence, $dx/dt = 1/\sqrt{\mu\gamma}$. The relation

$$E_y = f\left(x - \frac{t}{\sqrt{\mu\gamma}}\right) \tag{4-16b}$$

thus represents a wave with y component E_y, propagating in the positive x direction with speed $1/\sqrt{\mu\gamma}$. In free space, the propagation speed of the wave is c_0, *the speed of electromagnetic radiation in a vacuum*, so that there is the relation $c_0 = 1/\sqrt{\mu_0\gamma_0}$.[*]

Accompanying the E_y wave component is a companion wave component of the magnetic field. If (4-9b) is differentiated with respect to x, and (4-10c) with respect to t, the results can be combined to yield

$$\mu\gamma \frac{\partial^2 H_z}{\partial t^2} = \frac{\partial^2 H_z}{\partial x^2} \tag{4-17}$$

Equation (4-17) is the same wave equation as (4-15a). Hence, the H_z component of the magnetic field propagates along with E_y as shown in Fig. 4-2.

Any propagating waveform as designated by the f function in Eq. (4-16b) can be represented using Fourier series as a superposition of waves, each wave having a different fixed wavelength. Let us then consider only one such monochromatic wave, and note that any waveform could then be built up from a number of monochromatic components. For convenience in later portions of the analysis, the wave component will be given in complex form.

Suppose that at the origin ($x = 0$) the waveform variation with time is

$$E_y = E_{yM} \exp(i\omega t)$$

A position on the wave that leaves the origin ($x = 0$) at time t_1 arrives at location x after a time interval x/c, where c is the wave speed in the medium. Hence the time of arrival is $t = t_1 + x/c$, so that the time of leaving the origin was $t_1 = t - x/c$. A wave traveling in the positive x direction is then given by

$$E_y = E_{yM} \exp\left[i\omega\left(t - \frac{x}{c}\right)\right]$$

or

$$E_y = E_{yM} \exp[i\omega(t - \sqrt{\mu\gamma}\, x)] \tag{4-18a}$$

[*]Independent measurements of μ_0, γ_0, and c_0 validate the result. The fact that Maxwell's equations predict that all electromagnetic radiation propagates in a vacuum with speed c_0 was considered convincing evidence that light is a form of electromagnetic radiation, and was one of the early triumphs of the electromagnetic theory.

This is a solution to the governing wave equation (4-15a), as is shown by comparison with (4-16b). If desired, other forms of the solution can be obtained by using the relations $\omega = 2\pi\nu = 2\pi c/\lambda = 2\pi c_0/\lambda_0$, where λ and λ_0 are the wavelengths in the medium and in a vacuum, respectively.

The simple refractive index n is defined as the ratio of the wave speed in vacuum c_0 to the speed in the medium $c = 1/\sqrt{\mu\gamma}$. Hence,

$$n = \frac{c_0}{c} = c_0\sqrt{\mu\gamma} = \sqrt{\frac{\mu\gamma}{\mu_0\gamma_0}}$$

and (4-18a) can be written as

$$E_y = E_{yM} \exp\left[i\omega\left(t - \frac{n}{c_0}x\right)\right] \tag{4-18b}$$

$$= E_{yM}\left\{\cos\left[\omega\left(t - \frac{n}{c_0}x\right)\right] + i\sin\left[\omega\left(t - \frac{n}{c_0}x\right)\right]\right\}$$

As shown by (4-18), the wave propagates with undiminished amplitude through the medium. This is a consequence of the assumption that the medium can be regarded as a perfect dielectric, that is, one with zero electrical conductivity. In many real materials the conductivity is significant and the last term on the right in (4-1) cannot be neglected. As will now be shown, the inclusion of this term will lead to attenuation of the propagating wave.

4-4.2 Propagation in Isotropic Media of Finite Conductivity

This section includes consideration of imperfect dielectrics that have low electrical conductivity, and metals. For simplicity a single plane wave is again considered as described by Eqs. (4-18). If an exponential attenuation with distance is introduced [it will be shown by Eqs. (4-21)-(4-23) that this obeys Maxwell's equations], the wave takes the form

$$E_y = E_{yM} \exp\left[i\omega\left(t - \frac{n}{c_0}x\right)\right] \exp\left(-\frac{\omega}{c_0}\kappa x\right) \tag{4-19a}$$

where κ is termed the *extinction coefficient* for the medium. The attenuation term indicates an absorption of the energy of the wave as it travels through the medium. Such a wave with attenuation as a function of distance is called *evanescent*. The present form of the attenuation exponent was chosen so that the exponential terms could be combined into the relation

$$E_y = E_{yM} \exp\left\{i\omega\left[t - (n - i\kappa)\frac{x}{c_0}\right]\right\} \tag{4-19b}$$

$$= E_{yM}\left(\cos\left\{\omega\left[t - (n - i\kappa)\frac{x}{c_0}\right]\right\} + i\sin\left\{\omega\left[t - (n - i\kappa)\frac{x}{c_0}\right]\right\}\right)$$

A comparison of (4-19b) with (4-18b) shows that the simple refractive index n has been replaced by a complex term that will be termed the *complex refractive index* $\bar{n}$. Thus,

$$\bar{n} = n - i\kappa \tag{4-20}$$

It remains to be shown that (4-19b) constitutes a solution of the governing equations with the last term on the right of (4-1) included. With this term retained, (4-15a) takes the form

$$\mu\gamma \frac{\partial^2 E_y}{\partial t^2} = \frac{\partial^2 E_y}{\partial x^2} - \frac{\mu}{r_e} \frac{\partial E_y}{\partial t} \tag{4-21}$$

The wave given by (4-19b) is substituted into (4-21) and the following equality results:

$$c_0^2 \mu\gamma = (n - i\kappa)^2 + \frac{i\mu\lambda_0 c_0}{2\pi r_e} \tag{4-22a}$$

where λ_0 is the wavelength in a vacuum. Equation (4-22a) provides the relation between the wavelength and the properties of the medium necessary for the wave to satisfy Maxwell's equations. Equating the real and imaginary parts of (4-22a) yields

$$n^2 - \kappa^2 = \mu\gamma c_0^2 \tag{4-22b}$$

and

$$n\kappa = \frac{\mu\lambda_0 c_0}{4\pi r_e} \tag{4-22c}$$

These equations may be solved for the components of the complex refractive index, n and κ, in terms of μ, γ, λ_0, c_0, and r_e to yield

$$n^2 = \frac{\mu\gamma c_0^2}{2} \left\{ 1 + \left[1 + \left(\frac{\lambda_0}{2\pi c_0 r_e \gamma} \right)^2 \right]^{1/2} \right\} \tag{4-23a}$$

and

$$\kappa^2 = \frac{\mu\gamma c_0^2}{2} \left\{ -1 + \left[1 + \left(\frac{\lambda_0}{2\pi c_0 r_e \gamma} \right)^2 \right]^{1/2} \right\} \tag{4-23b}$$

In the solutions, positive signs were chosen in front of the square roots since n and κ are generally positive real quantities.

Comparison of (4-18b), the solution to the wave equation for perfect dielectric media, with (4-19b), the solution of the wave equation for conducting media, shows the solutions to be identical with one exception: the simple refractive index n appearing in the perfect dielectric solution is replaced for conductors by the complex refractive index $n - i\kappa$. This is a most important observation. It means that some of the relations that we derive for perfect dielectrics will also hold for conductors, providing we substitute the complex index $n - i\kappa$ for the simple refractive index n. Extensive use will be made of this analogy in succeeding sections, but there are some instances in which the analogy does not apply.

4-4.3 Energy of an Electromagnetic Wave

The instaneous energy carried per unit time and per unit area by an electromagnetic wave is given by the cross product of the electric and magnetic intensity vectors. This product is called the Poynting vector $\mathbf{S}$ where

$$\mathbf{S} = \mathbf{E} \times \mathbf{H}$$

and according to the properties of the cross product, $\mathbf{S}$ is a vector propagating at right angles to the $\mathbf{E}$ and $\mathbf{H}$ vectors in a direction defined by the right-hand rule. For the plane wave under consideration shown in Fig. 4-2, the propagation is in the positive x direction. The magnitude of $\mathbf{S}$ is given for the plane wave by

$$|\mathbf{S}| = E_y H_z \tag{4-24}$$

If E_y is given by (4-19b), then (4-10c), which holds for conductors as well as perfect dielectrics, can be used to find H_z as follows:

$$-\mu \frac{\partial H_z}{\partial t} = \frac{\partial E_y}{\partial x} = \frac{-i\omega}{c_0}(n - i\kappa) E_y = -\frac{i\omega \bar{n}}{c_0} E_y$$

Then noting the t dependence of E_y in (4-19b) and integrating yield the following relation between electric and magnetic intensities:

$$H_z = \frac{\bar{n}}{\mu c_0} E_y \tag{4-25}$$

The constant of integration has been taken to be zero. It would correspond to the presence of a steady magnetic intensity in addition to that induced by E_y and is zero for the conditions of the present discussion.

When H_z is substituted in (4-24), the magnitude of the Poynting vector becomes

$$|\mathbf{S}| = \frac{\bar{n}}{\mu c_0} E_y^2 \tag{4-26a}$$

Thus, the instantaneous energy per unit time and area carried by the wave is proportional to the square of the amplitude of the electric intensity.

Because $|\mathbf{S}|$ is a monochromatic property, it is seen by examination of its definition to be proportional to the quantity we have called spectral intensity. For radiation passing through a medium, the exponential decay factor in the spectral intensity must then be, by virtue of (4-26a), equal to the square of the decay term in E_y. Thus, from (4-19a), the intensity decay factor is $\exp(-2\omega\kappa x/c_0)$ or $\exp(-4\pi\kappa x/\lambda_0)$. The more general vector form of (4-26a) is

$$|\mathbf{S}| = \frac{\bar{n}}{\mu c_0} |\mathbf{E}|^2 \tag{4-26b}$$

4-5 LAWS OF REFLECTION AND REFRACTION

In the previous derivations the wave nature of the propagating radiation was revealed, and the characteristics of movement through infinite homogeneous isotropic media were found. Now the interaction of the electromagnetic wave with the interface between two media will be considered. This will provide laws of reflection and refraction in terms of the index of refraction and extinction coefficient, which are in turn related to the electric and magnetic properties of the media by means of Eq. (4-23).

4-5.1 Reflection and refraction at the interface between two perfect dielectrics ($\kappa \rightarrow 0$)

The interaction at the smooth interface between two nonattenuating materials will now be considered. For simplicity, throughout this discussion a simple cosine wave will be utilized as obtained by retaining only the cosine term in (4-19b). An x',y',z' coordinate system is fixed to the path of the incident wave, and the wave is moving in the x' direction. The wave strikes the interface between two media as shown in Fig. 4-3, where the interface is in the yz plane of the x,y,z coordinate system attached to the media. The plane containing both the normal to the interface and the incident direction x' is defined as the *plane of incidence*. In Fig. 4-3 the coordinate system has been drawn so that the y' direction is in the plane of incidence. The interaction of the wave with the interface depends on the wave orientation relative to the plane of incidence. For example, if the amplitude vector of the incident wave is in the plane of incidence (amplitude vector in y' direction), then the amplitude vector

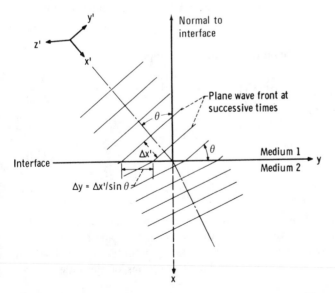

Figure 4-3 Plane wave incident upon interface between two media.

is at an angle to the interface. If the amplitude vector is normal to the plane of incidence (amplitude vector in z' direction), then the incident wave vector is parallel to the interface.

Figure 4-3 shows a plane, transverse wave front propagating in the x' direction. Although the wave will in general bend as it moves across the interface because of the difference in propagation velocity in the two media, the wave will be continuous, so that the velocity component tangent to the interface (y component) is the same in both media at the interface. This continuity relation will be used in deriving the laws of reflection.

Consider now an incident wave $E_{\parallel,i}$ polarized so that it has amplitude only in the $x'y'$ plane (Fig. 4-4) and hence is parallel to the plane of incidence. From Eq. (4-18b), retaining only the cosine term for simplicity, the wave propagating in the x' direction is characterized by

$$E_{\parallel,i} = E_{M\parallel,i} \cos \left[\omega \left(t - \frac{n_1 x'}{c_0} \right) \right] \tag{4-27}$$

From Fig. 4-4a, the components of the incident wave in the x,y,z coordinate system are (components are taken to be positive in the positive coordinate directions)

$$E_{x,i} = -E_{\parallel,i} \sin \theta \tag{4-28a}$$

$$E_{y,i} = E_{\parallel,i} \cos \theta \tag{4-28b}$$

$$E_z = 0 \tag{4-28c}$$

Substituting (4-27) into (4-28) and noting that x', the distance the wave front travels in a given time, is related to the y distance the front travels along the interface by

$$x' = y \sin \theta \tag{4-29}$$

as can be seen from Fig. 4-3, we obtain for the incident components

$$E_{x,i} = -E_{M\parallel,i} \sin \theta \cos \left[\omega \left(t - \frac{n_1 y \sin \theta}{c_0} \right) \right] \tag{4-30a}$$

$$E_{y,i} = E_{M\parallel,i} \cos \theta \cos \left[\omega \left(t - \frac{n_1 y \sin \theta}{c_0} \right) \right] \tag{4-30b}$$

$$E_{z,i} = 0 \tag{4-30c}$$

Upon striking the bounding yz plane between medium 1 and medium 2, the incident wave separates into a portion $E_{\parallel,r}$ reflected at angle θ_r and a portion $E_{\parallel,t}$ refracted at angle χ and transmitted into medium 2. From the geometry shown in Fig. 4-4 the components in the positive coordinate directions of the reflected ray evaluated at the interface are

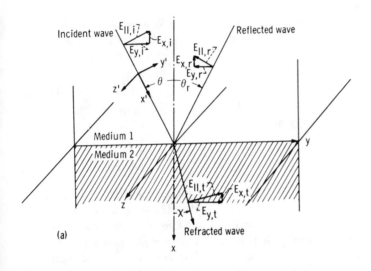

(a)

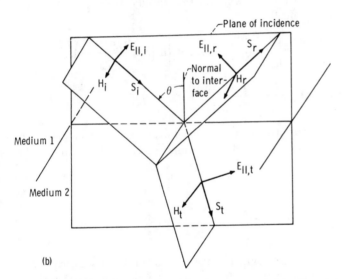

(b)

Figure 4-4 Interaction of electromagnetic wave with boundary between two media. (*a*) Plane electric field wave polarized in xy plane striking intersection of two media; (*b*) electric intensity, magnetic intensity, and Poynting vectors for incident wave polarized in plane of incidence.

$$E_{x,r} = -E_{M\|,r} \sin \theta_r \cos \left[\omega \left(t - \frac{n_1 y \sin \theta_r}{c_0} \right) \right] \qquad (4\text{-}31a)$$

$$E_{y,r} = -E_{M\|,r} \cos \theta_r \cos \left[\omega \left(t - \frac{n_1 y \sin \theta_r}{c_0} \right) \right] \qquad (4\text{-}31b)$$

$$E_{z,r} = 0 \qquad (4\text{-}31c)$$

The direction of $E_{\parallel,r}$ was drawn such that $E_{\parallel,r}$, H_r, and S_r would be consistent with the right-hand rule connecting the Poynting vector with the E and H fields. In a similar fashion, from Fig. 4-4, the components of the refracted portion of the wave are

$$E_{x,t} = -E_{M\parallel,t} \sin \chi \cos \left[\omega \left(t - \frac{n_2 y \sin \chi}{c_0} \right) \right] \tag{4-32a}$$

$$E_{y,t} = E_{M\parallel,t} \cos \chi \cos \left[\omega \left(t - \frac{n_2 y \sin \chi}{c_0} \right) \right] \tag{4-32b}$$

$$E_{z,t} = 0 \tag{4-32c}$$

Certain boundary conditions must be followed by the waves at the interface of the two media. The sum of the components, parallel to the interface, of the electric intensities of the reflected and incident waves must be equal to the intensity of the refracted wave in the same plane. This is because the intensity in medium 1 is the superposition of the incident and reflected intensities. For the polarized wave considered here, this condition gives the following for the equality of the y components (parallel to interface) in the two media:

$$\left\{ E_{M\parallel,i} \cos \theta \cos \left[\omega \left(t - \frac{n_1 y \sin \theta}{c_0} \right) \right] - E_{M\parallel,r} \cos \theta_r \cos \left[\omega \left(t - \frac{n_1 y \sin \theta_r}{c_0} \right) \right] \right.$$

$$\left. = E_{M\parallel,t} \cos \chi \cos \left[\omega \left(t - \frac{n_2 y \sin \chi}{c_0} \right) \right] \right\}_{x=0} \tag{4-33}$$

Since Eq. (4-33) must hold for arbitrary t and y, and the angles θ, θ_r, and χ are independent of t and y, the cosine terms involving time must be equal. This can only be true if

$$n_1 \sin \theta = n_1 \sin \theta_r = n_2 \sin \chi \tag{4-34}$$

which provides that

$$\theta = \theta_r \tag{4-35}$$

The angle of reflection of an electromagnetic wave is thus equal to its angle of incidence (rotated about the normal to the interface through a circumferential angle of $\theta = \pi$). These are the relations that define mirrorlike or specular reflections as discussed in Sec. 3-5.1 under the subheading Specularly Reflecting Surfaces.

Equation (4-34) also yields the following relation between θ and χ:

$$\frac{\sin \chi}{\sin \theta} = \frac{n_1}{n_2} \tag{4-36}$$

This relates the angle of refraction to the angle of incidence by means of the refractive indices. Equation (4-36) is known as *Snell's law*. For the often encountered case in which the incident wave is in air ($n_1 \approx 1$), then $n_2 = \sin \theta / \sin \chi$.

With the cosine terms involving time equal, and with the use of Eq. (4-35), there also follows from (4-33)

$$(E_{M\|,i}\cos\theta - E_{M\|,r}\cos\theta = E_{M\|,t}\cos\chi)_{x=0} \qquad (4\text{-}37)$$

This can be used to find how the reflected electric intensity is related to the incident value. The refracted component $E_{M\|,t}$ must be eliminated, and to accomplish this the magnetic intensities must be considered.

The magnetic intensity parallel to the boundary must be continuous at the boundary plane. The magnetic intensity vector is perpendicular to the electric intensity; since the electric intensity being considered is in the plane of incidence, the magnetic intensity will then be parallel to the boundary. Continuity at the boundary provides that

$$(H_i + H_r = H_t)_{x=0} \qquad (4\text{-}38)$$

The relation between electric and magnetic components was shown in (4-25). Although for simplicity this relation was derived for only the specific components H_z and E_y, it is true more generally so that the magnitudes of the **E** and **H** vectors are related by

$$|\mathbf{H}| = \frac{\bar{n}}{\mu c_0}|\mathbf{E}| \qquad (4\text{-}39)$$

For both dielectrics and metals the magnetic permeability is very close to that of a vacuum so that $\mu \approx \mu_0$. Then (4-38) can be written as

$$(\bar{n}_1 E_{M\|,i} + \bar{n}_1 E_{M\|,r} = \bar{n}_2 E_{M\|,t})_{x=0} \qquad (4\text{-}40)$$

Equations (4-37) and (4-40) are combined to eliminate $E_{M\|,t}$ and give the reflected electric intensity in terms of the incident intensity for nonattenuating materials $(\bar{n} \to n)$ as

$$\frac{E_{M\|,r}}{E_{M\|,i}} = \frac{\cos\theta/\cos\chi - n_1/n_2}{\cos\theta/\cos\chi + n_1/n_2} \qquad (4\text{-}41)$$

If the preceding derivation is repeated for an incident plane electric wave polarized perpendicular to the incident plane, the relation between reflected and incident components is

$$\frac{E_{M\perp,r}}{E_{M\perp,i}} = -\frac{\cos\chi/\cos\theta - n_1/n_2}{\cos\chi/\cos\theta + n_1/n_2} \qquad (4\text{-}42)$$

Equation (4-36) can be used in Eq. (4-41) to eliminate n_1/n_2 in terms of $\sin\chi/\sin\theta$. With some manipulation using trigonometric identities, the resulting expression can be cast into the form

$$\frac{E_{M\|,r}}{E_{M\|,i}} = \frac{\tan(\theta - \chi)}{\tan(\theta + \chi)} \qquad (4\text{-}43)$$

Similarly from Eq. (4-42)

$$\frac{E_{M\perp,r}}{E_{M\perp,i}} = -\frac{\sin(\theta - \chi)}{\sin(\theta + \chi)} \qquad (4\text{-}44)$$

The energy carried by a wave is proportional to the square of the amplitude of the wave as shown by (4-26). Squaring the ratio $E_{M,r}/E_{M,i}$ therefore gives the ratio of the energy reflected from a surface to the energy incident upon the surface from a given direction. This ratio was defined in Sec. 3-5 as the directional-hemispherical reflectivity. Because electromagnetic radiation for the ideal conditions examined here was shown by (4-35) to reflect specularly and because the electromagnetic theory relations are based on monochromatic waves, the energy ratio more exactly gives the directional-hemispherical spectral specular reflectivity (Sec. 3-5.1). The spectral dependence arises from the variation of the optical constants with wavelength.

The values of $\rho'_{\lambda,s}(\lambda,\theta,\varphi)$ for incident parallel and perpendicular polarized components are then obtained as

$$\rho'_{\lambda\|,s}(\lambda,\theta,\varphi) = \left(\frac{E_{M\|,r}}{E_{M\|,i}}\right)^2$$

$$\rho'_{\lambda\perp,s}(\lambda,\theta,\varphi) = \left(\frac{E_{M\perp,r}}{E_{M\perp,i}}\right)^2$$

$$(4\text{-}45)$$

The subscript s denotes a specular reflectivity. Because *all* reflectivities predicted by electromagnetic theory are specular, the subscript s will not be carried from this point on, to simplify an already complicated notation. Further, because of the assumption of isotropic behavior at the surface for the ideal surfaces considered, there is no dependence on the angle φ; hence, notation showing dependence on this variable will no longer be retained.

For unpolarized incident radiation the electric field has no definite orientation relative to the incident plane and can be resolved into parallel and perpendicular components that are equal. Then the directional-hemispherical spectral specular reflectivity is the average of $\rho'_{\lambda\|}(\lambda,\theta)$ and $\rho'_{\lambda\perp}(\lambda,\theta)$. Eqs. (4-43)–(4-45) result in

$$\rho'_{\lambda}(\lambda,\theta) = \frac{\rho'_{\lambda\|}(\lambda,\theta) + \rho'_{\lambda\perp}(\lambda,\theta)}{2} = \frac{1}{2}\left[\frac{\tan^2(\theta-\chi)}{\tan^2(\theta+\chi)} + \frac{\sin^2(\theta-\chi)}{\sin^2(\theta+\chi)}\right]$$

$$= \frac{1}{2}\frac{\sin^2(\theta-\chi)}{\sin^2(\theta+\chi)}\left[1 + \frac{\cos^2(\theta+\chi)}{\cos^2(\theta-\chi)}\right] \qquad (4\text{-}46)$$

Equation (4-46) is known as *Fresnel's equation*, and it gives the directional-hemispherical spectral reflectivity for an unpolarized ray incident upon a dielectric medium. The relation between χ and θ is given by (4-36).

In the special case when the incident radiation is normal to the interface between the two media, $\cos\theta = \cos\chi = 1$, and Eqs. (4-41) and (4-42) yield

$$\frac{E_{M\|,r}}{E_{M\|,i}} = \frac{E_{M\perp,r}}{E_{M\perp,i}} = \frac{1 - n_1/n_2}{1 + n_1/n_2} = \frac{n_2 - n_1}{n_2 + n_1} \qquad (4\text{-}47)$$

The normal directional-hemispherical spectral specular reflectivity is then

$$\rho'_{\lambda,n}(\lambda) = \rho'_{\lambda}(\lambda,\theta = \theta_r = 0) = \left(\frac{n_2 - n_1}{n_2 + n_1}\right)^2 \qquad (4\text{-}48)$$

For a wave entering the dielectric from air ($n_1 \approx 1$),

$$\rho'_{\lambda,n}(\lambda) = \left(\frac{n_2 - 1}{n_2 + 1}\right)^2 \tag{4-49}$$

The foregoing reflectivities are spectral quantities because n_1 and n_2 are functions of λ.

4-5.2 Incidence on an Absorbing Medium ($\kappa \neq 0$)

As was shown by Eq. (4-19b), the propagation of a wave in an infinite attenuating medium is governed by the same relations as in a nonattenuating medium if the index of refraction n for the latter case is replaced by the complex index $\bar{n} = n - i\kappa$. When the interaction of a wave with a boundary is considered, the theoretical expressions for reflected wave amplitudes as derived previously for $\kappa = 0$ also apply if $\bar{n}$ is used instead of n, but this leads to some complexities in interpretation. For example Snell's law becomes

$$\frac{\sin \chi}{\sin \theta} = \frac{\bar{n}_1}{\bar{n}_2} = \frac{n_1 - i\kappa_1}{n_2 - i\kappa_2} \tag{4-50}$$

Because this relation is complex, $\sin \chi$ is complex, and the angle χ can no longer be interpreted physically as a simple angle of refraction for propagation into the material. Except in the special case of normal incidence, n is no longer directly related to the propagation velocity. Some discussion will now be given by considering oblique incidence on an attenuating medium. This discussion is only of background interest and is not needed in the application of the reflection laws, which will ultimately be expressed in terms of the incidence angle and n and κ.

Figure 4-5 shows a plane wave incident from vacuum on an absorbing material with complex index of refraction $n - i\kappa$. After refraction, the planes of equal phase are still normal to the direction of propagation and these planes move with the *phase velocity*, which will be called c_0/α and is related to n and κ. In the case of a nonattenuating medium the phase velocity is simply c_0/n. The attenuation of the wave must depend on the distance traveled within the medium, and hence the planes of constant amplitude must be parallel to the interface. The wave in the material is called an *inhomogeneous wave* as the planes of constant amplitude and constant phase are not along the same direction. Only for normal incidence are the two sets of planes parallel, so the waves in the medium are homogeneous for normal incidence. The extinction coefficient in the x direction will be called β, and only for normal incidence will $\beta = \kappa$. By analogy with (4-19a) the wave in the medium can be described as

$$E_{y'} = E_{y'M} \exp\left[i\omega\left(t - \frac{\alpha}{c_0}x'\right)\right] \exp\left(-\frac{\omega}{c_0}\beta x\right) \tag{4-51}$$

where propagation is in the x' direction and extinction is in the x direction. The x coordinate can be written as $x = x' \cos \delta - y' \sin \delta$ where, from the phase velocity, $\delta = \sin^{-1}\left[(1/\alpha) \sin \theta\right]$. The propagation angle δ is not equal to the χ in (4-50) as χ is

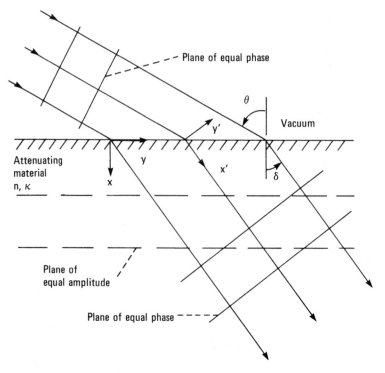

Figure 4-5 Planes of equal phase and amplitude for propagation into an attenuating material.

complex in this instance. Then $E_{y'}$ becomes

$$E_{y'} = E_{y'M} \exp(i\omega t) \exp\left[-x'\left(\frac{i\omega\alpha}{c_0} + \frac{\omega}{c_0}\beta\cos\delta\right)\right] \exp\left(y'\frac{\omega}{c_0}\beta\sin\delta\right) \qquad (4\text{-}52)$$

Since $E_{y'}$ is a function of t, x', and y', the wave equation (4-21) is written in two space dimensions as,

$$\mu\gamma\frac{\partial^2 E_{y'}}{\partial t^2} + \frac{\mu}{r_e}\frac{\partial E_{y'}}{\partial t} = \frac{\partial^2 E_{y'}}{\partial x'^2} + \frac{\partial^2 E_{y'}}{\partial y'^2}$$

Substituting (4-52) yields

$$-\mu\gamma\omega^2 + \frac{\mu}{r_e}i\omega = \left(\frac{i\omega\alpha}{c_0} + \frac{\omega}{c_0}\beta\cos\delta\right)^2 + \left(\frac{\omega}{c_0}\beta\sin\delta\right)^2$$

Equating real and imaginary parts and using Eqs. (4-22b) and (4-22c) yields the relations between α and β and n and κ:

$$\alpha^2 - \beta^2 = n^2 - \kappa^2$$

$$\alpha\beta\cos\delta = n\kappa$$

and, as given earlier, $\delta = \sin^{-1}[(1/\alpha) \sin \theta]$. This provides three simultaneous equations from which β, α, and δ can be calculated from θ, n, and κ. This yields the attenuation, propagation velocity (phase velocity), and direction of propagation within the material. It is evident that the propagation velocity c_0/α depends on δ; that is, the velocity depends on direction within the material even though the material is isotropic. In the case of normal incidence, $\delta = 0$, and then $\alpha = n$ and $\beta = \kappa$. Hence only in this instance is n directly related to the propagation velocity by $c = c_0/n$, and κ a direct measure of the rate of extinction with depth within the material.

Returning now to the reflection laws in terms of the complex index of refraction, we first consider the case of normal incidence. From (4-47) with $\bar{n}$ replacing n,

$$\frac{E_{M\|,r}}{E_{M\|,i}} = \frac{E_{M\perp,r}}{E_{M\perp,i}} = \frac{\bar{n}_2 - \bar{n}_1}{\bar{n}_2 + \bar{n}_1} = \frac{n_2 - i\kappa_2 - (n_1 - i\kappa_1)}{n_2 - i\kappa_2 + n_1 - i\kappa_1}$$

These are complex quantities, and it was noted in (4-26) that the energy in the wave depends on $|E|^2$. For a complex number z, $|z|^2 = zz^*$, where z^* is the complex conjugate. Then because the relations for $\|$ and $\perp$ polarization are the same for normal incidence, the reflectivity is

$$\rho'_{\lambda,n}(\lambda) = \left| \frac{E_{M\|,r}}{E_{M\|,i}} \right|^2 = \left[\frac{n_2 - i\kappa_2 - (n_1 - i\kappa_1)}{n_2 - i\kappa_2 + n_1 - i\kappa_1} \right]\left[\frac{n_2 + i\kappa_2 - (n_1 + i\kappa_1)}{n_2 + i\kappa_2 + n_1 + i\kappa_1} \right] \tag{4-53}$$

This simplifies to

$$\rho'_{\lambda,n}(\lambda) = \frac{(n_2 - n_1)^2 + (\kappa_2 - \kappa_1)^2}{(n_2 + n_1)^2 + (\kappa_2 + \kappa_1)^2} \tag{4-54}$$

For an incident ray in air ($n_1 = 1$, $\kappa_1 \approx 0$) striking an absorbing material (n_2, κ_2), Eq. (4-54) reduces to

$$\rho'_{\lambda,n}(\lambda) = \frac{(n_2 - 1)^2 + \kappa_2^2}{(n_2 + 1)^2 + \kappa_2^2} \tag{4-55}$$

When the material is transparent ($\kappa_2 \to 0$), Eq. (4-55) reduces to (4-49).

For oblique incidence the directional-hemispherical reflectivity can be derived by starting from (4-41) and (4-42) and using the complex index of refraction. For incident rays polarized parallel or perpendicular to the plane of incidence, (4-41) and (4-42) give the complex ratios

$$\frac{E_{M\|,r}}{E_{M\|,i}} = \frac{\cos \theta/\cos \chi - (n_1 - i\kappa_1)/(n_2 - i\kappa_2)}{\cos \theta/\cos \chi + (n_1 - i\kappa_1)/(n_2 - i\kappa_2)} \tag{4-56}$$

$$\frac{E_{M\perp,r}}{E_{M\perp,i}} = -\frac{\cos \chi/\cos \theta - (n_1 - i\kappa_1)/(n_2 - i\kappa_2)}{\cos \chi/\cos \theta + (n_1 - i\kappa_1)/(n_2 - i\kappa_2)} \tag{4-57}$$

The real and imaginary parts of (4-56) and (4-57) correspond respectively to the changes in amplitude and phase. The reflectivity for the parallel component is found by multiplying by the complex conjugate

$$\rho'_{\lambda,\parallel}(\lambda,\theta) = \frac{E_{M\parallel,r}}{E_{M\parallel,i}}\left(\frac{E_{M\parallel,r}}{E_{M\parallel,i}}\right)^*$$

and similarly for the perpendicular component. This requires considerable manipulation, as $\cos\chi$ is complex as shown by (4-50).

To provide some specific results, the important case of radiation incident in air or vacuum on a material with properties n and κ will now be considered. Then from (4-56), (4-57), and (4-50),

$$\frac{E_{M\parallel,r}}{E_{M\parallel,i}} = \frac{(n-i\kappa)\cos\theta - \cos\chi}{(n-i\kappa)\cos\theta + \cos\chi} = \frac{\bar{n}\cos\theta - \cos\chi}{\bar{n}\cos\theta + \cos\chi} \tag{4-58a}$$

$$\frac{E_{M\perp,r}}{E_{M\perp,i}} = -\frac{(n-i\kappa)\cos\chi - \cos\theta}{(n-i\kappa)\cos\chi + \cos\theta} = -\frac{\bar{n}\cos\chi - \cos\theta}{\bar{n}\cos\chi + \cos\theta} \tag{4-58b}$$

$$\frac{\sin\chi}{\sin\theta} = \frac{1}{n-i\kappa} = \frac{1}{\bar{n}}$$

The $\bar{n}\cos\chi$ in (4-58) can be found as

$$\bar{n}\cos\chi = \bar{n}(1-\sin^2\chi)^{1/2} = (\bar{n}^2 - \sin^2\theta)^{1/2} \tag{4-59}$$

The results can be presented more conveniently by letting

$$a - ib = (\bar{n}^2 - \sin^2\theta)^{1/2}$$

By squaring and equating real and imaginary parts, the resulting simultaneous equations can be solved for a and b to obtain

$$2a^2 = [(n^2 - \kappa^2 - \sin^2\theta)^2 + 4n^2\kappa^2]^{1/2} + n^2 - \kappa^2 - \sin^2\theta \tag{4-60a}$$

$$2b^2 = [(n^2 - \kappa^2 - \sin^2\theta)^2 + 4n^2\kappa^2]^{1/2} - (n^2 - \kappa^2 - \sin^2\theta) \tag{4-60b}$$

The quantity $a - ib$ is substituted for $\bar{n}\cos\chi$ in (4-58), and the resulting equations are multiplied through by their complex conjugates to yield the reflectivities [note that $\bar{n}^2 = (a-ib)^2 + \sin^2\theta$]:

$$\rho'_{\lambda\parallel}(\lambda,\theta) = \frac{a^2 + b^2 - 2a\sin\theta\tan\theta + \sin^2\theta\tan^2\theta}{a^2 + b^2 + 2a\sin\theta\tan\theta + \sin^2\theta\tan^2\theta}\,\rho'_{\lambda\perp}(\lambda,\theta) \tag{4-61a}$$

$$\rho'_{\lambda\perp}(\lambda,\theta) = \frac{a^2 + b^2 - 2a\cos\theta + \cos^2\theta}{a^2 + b^2 + 2a\cos\theta + \cos^2\theta} \tag{4-61b}$$

As before, if the incident beam has no specific polarization the reflectivity is an average of the parallel and perpendicular components, as in Eq. (4-46).

To this point in this chapter, the wave nature of radiation has been shown from a consideration of Maxwell's equations. Then the interaction of these waves with nonabsorbing and absorbing media has been discussed in terms of the refractive index n and the complex refractive index $\bar{n}$. Now the results will be applied more specifically to some actual radiative properties.

4-6 APPLICATION OF ELECTROMAGNETIC–THEORY RELATIONS TO RADIATIVE–PROPERTY PREDICTIONS

The electromagnetic theory, as applied here to radiative-property prediction, has a number of drawbacks that limit its usefulness for practical calculations. Aside from the many assumptions used in the derivations, the theory itself becomes invalid when the frequencies being considered become of the order of molecular vibrational frequencies. These qualifications restrict the equations used here to wavelengths longer than in the visible spectrum.

The theory neglects the effects of surface conditions on the radiative properties. This is its most serious limitation, since perfectly clean, optically smooth interfaces are not often encountered in practice. The greatest usefulness of the theory is probably in providing a means for intelligent extrapolation when only limited experimental data are available. In the following sections, the equations of electromagnetic theory that are useful for the prediction of properties will be examined, and the assumptions inherent in their derivation discussed.

4-6.1 Radiative Properties of Dielectrics ($\kappa \to 0$)

The equations to be examined in this section all contain these assumptions: (1) The medium is isotropic; that is, its electrical and internal optical properties are independent of direction. (2) The magnetic permeability of the medium is equal to that of a vacuum. (3) There is no accumulation of static electric charge. (4) No externally produced electrical conduction currents are present.

The measured index of refraction of the medium is in general a function of wavelength, and thus any calculated radiative properties will be wavelength-dependent. If, however, the refractive index is calculated from the permittivity γ or the *dielectric constant K* (where $K = \gamma/\gamma_0$), which are not generally given as functions of wavelength, the spectral dependency is lost. Because of these considerations, no notation is used in the following equations to signify spectral dependence, but the reader should be aware that such dependence can be included if the optical or electromagnetic properties are known as a function of wavelength.

The surfaces are assumed to be "optically smooth," that is, smooth in comparison with the wavelength of the incident radiation so that specular reflections result.

Reflectivity. Under the aforementioned restrictions, the directional-hemispherical specular reflectivity of a wave incident on a surface at angle θ and polarized parallel to the plane of incidence may be obtained from Eqs. (4-45) and (4-43) as

$$\rho'_{\parallel}(\theta) = \left[\frac{\tan(\theta - \chi)}{\tan(\theta + \chi)} \right]^2 \qquad (4\text{-}62a)$$

Similarly, from (4-45) and (4-44), for a wave polarized perpendicular to the incidence plane,

$$\rho'_{\perp}(\theta) = \left[\frac{\sin(\theta - \chi)}{\sin(\theta + \chi)} \right]^2 \qquad (4\text{-}62b)$$

where χ is the angle of refraction in the medium on which the ray impinges. For a given incident angle θ, the angle χ can be determined from (4-36) as

$$\frac{\sin \chi}{\sin \theta} = \frac{n_1}{n_2} = \frac{\sqrt{\gamma_1}}{\sqrt{\gamma_2}} = \frac{\sqrt{K_1}}{\sqrt{K_2}} \tag{4-63}$$

where γ is the permittivity, K is the dielectric constant, and n is the refractive index; the n, γ, and K are assumed not to have any angular dependence.

Alternative forms containing only θ can be obtained by eliminating χ in (4-62a) and (4-62b) by using (4-63):

$$\rho'_{\parallel}(\theta) = \left\{ \frac{(n_2/n_1)^2 \cos \theta - [(n_2/n_1)^2 - \sin^2 \theta]^{1/2}}{(n_2/n_1)^2 \cos \theta + [(n_2/n_1)^2 - \sin^2 \theta]^{1/2}} \right\}^2 \tag{4-64a}$$

$$\rho'_{\perp}(\theta) = \left\{ \frac{[(n_2/n_1)^2 - \sin^2 \theta]^{1/2} - \cos \theta}{[(n_2/n_1)^2 - \sin^2 \theta]^{1/2} + \cos \theta} \right\}^2 \tag{4-64b}$$

The $\rho'_{\parallel}(\theta) = 0$ when $\theta = \tan^{-1}(n_2/n_1)$; this θ is called *Brewster's angle*. Reflected radiation from this incidence angle will all be $\perp$ polarized.

The reflectivity for unpolarized radiation was shown in (4-46) (Fresnel's equation) to be given by

$$\rho'(\theta) = \frac{1}{2} \frac{\sin^2(\theta - \chi)}{\sin^2(\theta + \chi)} \left[1 + \frac{\cos^2(\theta + \chi)}{\cos^2(\theta - \chi)} \right] \tag{4-65a}$$

and for normal incidence

$$\rho'_n = \left(\frac{n_2 - n_1}{n_2 + n_1} \right)^2 \tag{4-65b}$$

EXAMPLE 4-1 An unpolarized beam of radiation is incident at angle $\theta = 30°$ from the normal on a dielectric surface (medium 2) in air (medium 1). The surface is of a material with $\kappa_2 \approx 0$ whose index of refraction is $n_2 = 3.0$. Find the directional-hemispherical reflectivity of the polarized components and the unpolarized beam.

Because the incident beam is in air, $n_1 - i\kappa_1 \approx 1$, and from Eq. (4-63), $n_1/n_2 = 1/3.0 = \sin \chi/\sin 30°$; therefore, $\chi = 9.6°$. The reflectivity of the parallel component is, from (4-62a), $\rho'_{\parallel}(\theta = 30°) = (\tan 20.4°/\tan 39.6°)^2 = 0.202$, and that of the perpendicular component is, from (4-62b), $\rho'_{\perp}(\theta = 30°) = (\sin 20.4°/\sin 39.6°)^2 = 0.301$. The reflectivity for the unpolarized beam obtained from (4-65a) or, more simply here, from the average of the components, is $\rho'(\theta = 30°) = (0.202 + 0.301)/2 = 0.252$.

EXAMPLE 4-2 What fraction of light is reflected for normal incidence from air on a glass surface? On the surface of water?

For glass $n \approx 1.55$, and for water $n \approx 1.33$. Then from (4-65b)

$$\rho_n'(\text{glass}) = \left(\frac{n-1}{n+1}\right)^2 = \left(\frac{0.55}{2.55}\right)^2 = 0.047$$

$$\rho_n'(\text{water}) = \left(\frac{0.33}{2.33}\right)^2 = 0.020$$

By performing the type of calculation shown in Example 4-1 or using (4-63) and (4-65) for various incidence angles and ratios of the indices of refraction, the reflectivity can be tabulated or presented graphically. This is a directional-hemispherical reflectivity in that it provides all the reflected energy resulting from an incident beam from one direction. It is a spectral quantity in the sense that the indices of refraction can correspond to a particular wavelength if the details of the wavelength dependency are available. Finally, it is a specular quantity in that it obeys the constraints of (4-35).

Emissivity. After the reflectivity has been evaluated, the directional spectral emissivity can be found from (3-46) as

$$\epsilon'(\theta) = 1 - \rho'(\theta)$$

when the body is opaque (if κ is very small, the body must be thick).

A graph of the directional emissivity is shown in Fig. 4-6 for various ratios n_2/n_1 where $n_2 \geqslant n_1$. When $n_2 < n_1$ there is a limiting angle such that radiation at larger incidence angles is totally reflected, giving $\epsilon'(\theta) = 0$ in that region. This is discussed in Sec. 19-6.2. When $\rho'(\theta)$ is computed for an incident beam in air ($n_1 \approx 1$), the ratio n_2/n_1 reduces to the refractive index for the material on which the beam is incident. Figure 4-6 can thus be regarded as giving the emissivity of a dielectric into air when the value of the parameter n_2/n_1 is set equal to the simple refractive index n of the dielectric material. In the following discussion, Fig. 4-6 will be interpreted in this sense.

For $n = 1$, the emissivity becomes unity (blackbody case), and the curve for this value on Fig. 4-6 is circular with a radius of unity. As n increases, the curves remain circular up to about $\theta = 70°$ and then begin to decrease rapidly to a zero value at $\theta = 90°$. Thus, dielectric materials emit poorly at large angles from the normal direction. For angles of less than $70°$, the emissivities are quite high so that, in a hemispherical sense, dielectrics are good emitters. It should be emphasized again that the assumptions used for the present interpretation of Maxwell's equations restrict these findings to wavelengths longer than the visible spectrum, as borne out by comparisons with experimental measurements.

From the directional spectral emissivity, the hemispherical spectral emissivity can be computed from Eq. (3-5) to be $\epsilon_\lambda(\lambda, T_A) = (1/\pi) \int_a \epsilon_\lambda'(\lambda, \theta, \varphi, T_A) \cos \theta \, d\omega$. Then an integration can be performed over all wavelengths to obtain the hemispherical total emissivity as given by (3-6a). Since the optical properties are generally not known in sufficient detail so that a wavelength integration of theoretical ϵ_λ can be made, in the theory spectral ϵ_λ values are used for total ϵ values for lack of anything better.

Angle of emission θ, deg

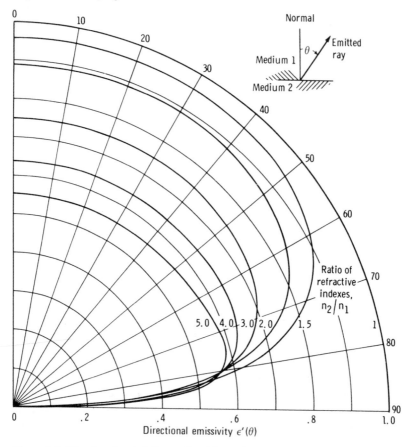

Figure 4-6 Directional emissivity predicted from electromagnetic theory.

The integration of $\epsilon'(\theta)$ to evaluate ϵ is complicated, as indicated by the forms of (4-64a) and (4-64b), but the integration has been carried out to yield

$$\epsilon = \frac{1}{2} - \frac{(3n+1)(n-1)}{6(n+1)^2} - \frac{n^2(n^2-1)^2}{(n^2+1)^3} \ln\left(\frac{n-1}{n+1}\right) + \frac{2n^3(n^2+2n-1)}{(n^2+1)(n^4-1)}$$
$$- \frac{8n^4(n^4+1)}{(n^2+1)(n^4-1)^2} \ln n \tag{4-66}$$

The normal emissivity provides a convenient value to which the hemispherical value may be referenced. The normal emissivity can be computed from Eq. (4-49) as

$$\epsilon_n' = 1 - \left(\frac{n-1}{n+1}\right)^2 = \frac{4n}{(n+1)^2} \tag{4-67}$$

for emission from a dielectric into air. The ϵ_n' is shown as a function of n in Fig. 4-7a.

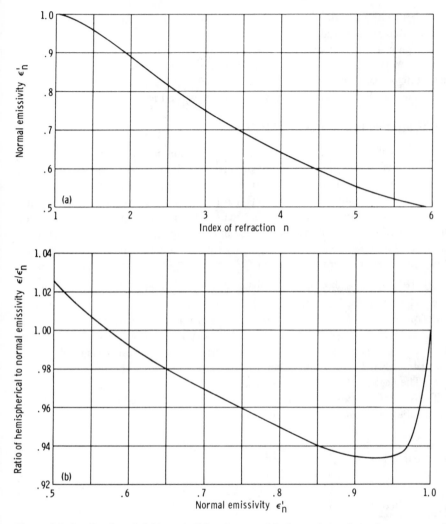

Figure 4-7 Predicted emissivities of dielectric materials for emission into air or vacuum from medium with refractive index n. (a) Normal emissivity as function of refractive index; (b) relation between hemispherical and normal emissivity.

Note that normal emissivities less than about 0.50 correspond to $n > 6$. Such large n values are not common for dielectrics, so that the curve is not extended to smaller ϵ_n'. The ratio of hemispherical to normal emissivity for dielectrics is provided as a function of normal emissivity in Fig. 4-7b.

> **EXAMPLE 4-3** A dielectric has a refractive index of 1.41. What is its hemispherical emissivity into air at the wavelength at which the refractive index was measured?
>
> From Eq. (4-67) the normal emissivity is $\epsilon_n' = 1 - (0.41/2.41)^2 = 0.97$. From Fig. 4-7$b$, $\epsilon/\epsilon_n' = 0.94$ and the hemispherical emissivity is $\epsilon = 0.97 \times 0.94 = 0.91$.

For large n the ϵ_n' values are relatively low, and with increasing n the curves shown in Fig. 4-6 depart more and more from the circular form of the curve corresponding to $n = 1$. Figure 4-7b reveals that the flattening of the curves of Fig. 4-6 in the region near the normal causes the hemispherical emissivity to exceed the normal value at large n. For n near unity (ϵ_n' near 1), the hemispherical value is lower than the normal value because of the poor emission at large θ, as shown in Fig. 4-6.

4-6.2 Radiative Properties of Metals

Metals are usually highly absorbing, and the extinction coefficient κ cannot be neglected. The use of the complex index of refraction leads to the expressions in Sec. 4-5.2 that are more complicated than those for a nonabsorbing dielectric. As will be shown, there are simplifying assumptions that lead to more convenient equations than the general results from the theory. The main difficulty in applying the theoretical results is that the optical properties for use in these equations are difficult to obtain; when measured values are available, they can be inaccurate because of the experimental problems involved in their measurement.

Reflectivity and emissivity relations using optical constants. For most metals, the simple index of refraction n and the extinction coefficient κ are quite large at wavelengths longer than those in the visible region. In this range of wavelengths κ is usually much larger than n (see Table 4-2). Hence in (4-60a) and (4-60b) $\sin^2 \theta$ can often be neglected as compared with the $n^2 - \kappa^2$ terms. This simplification can be used to reduce (4-61a) and (4-61b) to a simpler form; the a^2 and b^2 become $a^2 = n^2$ and $b^2 = \kappa^2$. The neglect of $\sin^2 \theta$ in (4-59) also implies that $\cos \chi \to 1$ for metals.

If $\cos \chi \to 1$ in Eqs. (4-41) and (4-42), these equations reduce to, for incidence through a dielectric, and with $\bar{n}_2$ replacing n_2,

$$\frac{E_{M\parallel,r}}{E_{M\parallel,i}} = \frac{\cos \theta - n_1/\bar{n}_2}{\cos \theta + n_1/\bar{n}_2} \tag{4-68a}$$

and

$$\frac{E_{M\perp,r}}{E_{M\perp,i}} = -\frac{1/\cos \theta - n_1/\bar{n}_2}{1/\cos \theta + n_1/\bar{n}_2} \tag{4-68b}$$

Multiplying by the complex conjugates yields the reflectivity components

$$\rho_\parallel'(\theta) = \frac{(n_2 - n_1/\cos \theta)^2 + \kappa_2^2}{(n_2 + n_1/\cos \theta)^2 + \kappa_2^2} \tag{4-69a}$$

and

$$\rho_\perp'(\theta) = \frac{(n_2 - n_1 \cos \theta)^2 + \kappa_2^2}{(n_2 + n_1 \cos \theta)^2 + \kappa_2^2} \tag{4-69b}$$

For a beam incident through air on a metal with complex refractive index $n_2 - i\kappa_2$, these equations reduce to (since the refractive index for air is $n_1 = 1$ as a very good approximation)

$$\rho'_{\parallel}(\theta) = \frac{(n_2 \cos \theta - 1)^2 + (\kappa_2 \cos \theta)^2}{(n_2 \cos \theta + 1)^2 + (\kappa_2 \cos \theta)^2} \tag{4-70a}$$

and

$$\rho'_{\perp}(\theta) = \frac{(n_2 - \cos \theta)^2 + \kappa_2{}^2}{(n_2 + \cos \theta)^2 + \kappa_2{}^2} \tag{4-70b}$$

These expressions can also be obtained from (4-61a) and (4-61b) by letting $a^2 = n^2$ and $b^2 = \kappa^2$. For an unpolarized beam,

$$\rho'(\theta) = \frac{\rho'_{\parallel}(\theta) + \rho'_{\perp}(\theta)}{2} \tag{4-71}$$

For the normal direction ($\theta = 0$),

$$\rho'_n = \frac{(n_2 - 1)^2 + \kappa_2{}^2}{(n_2 + 1)^2 + \kappa_2{}^2}$$

which is the same as the exact relation (4-55).

The corresponding emissivity values are found from $\epsilon'(\theta) = 1 - \rho'(\theta)$, and these simplify to

$$\epsilon'_{\parallel}(\theta) = \frac{4n_2 \cos \theta}{(n_2{}^2 + \kappa_2{}^2) \cos^2 \theta + 2n_2 \cos \theta + 1} \tag{4-72a}$$

$$\epsilon'_{\perp}(\theta) = \frac{4n_2 \cos \theta}{\cos^2 \theta + 2n_2 \cos \theta + n_2{}^2 + \kappa_2{}^2} \tag{4-72b}$$

For emission, which is unpolarized,

$$\epsilon'(\theta) = \frac{\epsilon'_{\parallel}(\theta) + \epsilon'_{\perp}(\theta)}{2} \tag{4-73}$$

In the normal direction ($\theta = 0$) this becomes

$$\epsilon'_n = \frac{4n_2}{(n_2 + 1)^2 + \kappa_2{}^2} \tag{4-74}$$

The use of these emissivity relations is demonstrated in Fig. 4-8 for a pure smooth platinum surface at a wavelength of 2 μm, and it is evident by comparison with the experimental data that, although the general shape of the curve predicted by (4-73) is correct, the magnitude is in error. The data for n and κ for platinum, taken from [2], are $n = 5.7$ and $\kappa = 9.7$.* A comment as to the difficulty of the measurement of

*The reader should be aware that the complex refractive index can be defined in other ways than $\bar{n} = n - i\kappa$ as used here. It is also commonly given as $\bar{n} = n - in\kappa$, and occasionally with a positive sign in front of the extinction factor. When consulting data references, care should be taken in determining what definition is used so that conversion to the system used in this text can be carried out if necessary.

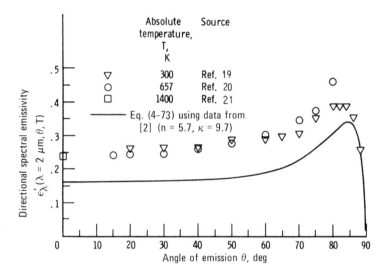

Figure 4-8 Directional spectral emissivity of platinum at wavelength $\lambda = 2\ \mu m$.

the optical properties of metals, perhaps because of the large influence of metal purity and the ease of contamination, is that the thirty-fifth edition of [2] lists values for platinum computed indirectly from an older measurement of angles of reflection giving circular polarization and a phase difference of $90°$ in the two reflected components of polarization. These measurements give $n = 0.70$ and $\kappa = 3.5$. The newer measurements thus differ by a factor of 8 in refractive index and 2.8 in the extinction coefficient.

Although the inaccuracy of the optical constants presents a difficulty in the precise evaluation of radiative property values, the theory does provide an understanding of the directional behavior of the properties. For metals, as illustrated by the results for platinum in Fig. 4-8, the emissivity is essentially constant for about $40°$ away from the normal, and then it increases to a maximum located within a few degrees of the tangent to the surface. This angular dependence for emission from metals is in contrast to the behavior for dielectrics for which the emission decreases substantially as the angle from the normal becomes larger than about $60°$.

In Table 4-2 the prediction of normal spectral emissivity by using Eq. (4-74) is compared with measured values. All data are taken from [2]. A wavelength of $\lambda = 0.589\ \mu m$ is used for some of the comparisons because of the wealth of data available. This is because of the ease with which a sodium-vapor lamp, which emits at this wavelength, can be employed as an intense monochromatic energy source in the laboratory. Since this wavelength is in the visible range, it is in the borderline short-wavelength region where the electromagnetic theory becomes inaccurate.

Comparison of the values in Table 4-2 shows the agreement between predicted and measured $\epsilon'_{\lambda,n}$ to be good, for example, for nickel and tungsten, but a factor of 4 in error for magnesium. For the cases of poor agreement, it is difficult to ascribe the error specifically to the optical constants, the measured emissivity, or the theory

Table 4-2 Comparison of spectral normal emissivity predictions from electromagnetic theory with experiment[a]

Metal	Wavelength λ, μm	Refractive index n	Extinction coefficient κ	Spectral normal emissivity $\epsilon'_{\lambda,n}(\lambda)$	
				Experimental	Calculated from (4-74)
Copper	0.650	0.44	3.26	0.20	0.140
	2.25	1.03	11.7	0.041	0.029
	4.00	1.87	21.3	0.027	0.014
Gold	0.589	0.47	2.83	0.176	0.184
	2.00	0.47	12.5	0.032	0.012
Iron	0.589	1.51	1.63	0.43	0.674
Magnesium	0.589	0.37	4.42	0.27	0.070
Nickel	0.589	1.79	3.33	0.355	0.381
	2.25	3.95	9.20	0.152	0.145
Silver	0.589	0.18	3.64	0.074	0.049
	2.25	0.77	15.4	0.021	0.013
	4.50	4.49	33.3	0.015	0.014
Tungsten	0.589	3.46	3.25	0.49	0.455

[a] Data from [2].

itself. Any or all could contribute to the discrepancy. Most probably the optical constants are somewhat in error, and the experimental samples do not meet the standards of perfection in surface preparation demanded by the theory.

Within the approximation of neglecting $\sin^2 \theta$ relative to $n^2 - \kappa^2$, the hemispherical emissivity for a metal (having complex refractive index $n - i\kappa$) in air or vacuum is found by substituting (4-73) into (3-5). After carrying out the integration, this yields

$$\epsilon = 4n - 4n^2 \ln \frac{1 + 2n + n^2 + \kappa^2}{n^2 + \kappa^2} + \frac{4n(n^2 - \kappa^2)}{\kappa} \tan^{-1} \frac{\kappa}{n + n^2 + \kappa^2}$$

$$+ \frac{4n}{n^2 + \kappa^2} - \frac{4n^2}{(n^2 + \kappa^2)^2} \ln (1 + 2n + n^2 + \kappa^2) - \frac{4n(\kappa^2 - n^2)}{\kappa(n^2 + \kappa^2)^2} \tan^{-1} \frac{\kappa}{1 + n} \quad (4\text{-}75)$$

Evaluation of (4-75) is difficult because it involves small differences of large numbers, and many significant figures must be carried in the calculations.

Without neglecting $\sin^2 \theta$, the ϵ can be calculated by a numerical integration of (3-5) using $\epsilon'(\theta) = [\epsilon'_\parallel(\theta) + \epsilon'_\perp(\theta)]/2 = [2 - \rho'_\parallel(\theta) - \rho'_\perp(\theta)]/2$, where the $\rho'(\theta)$ are given by the general expressions (4-64a) and (4-64b). This was done by Hering and Smith [3], and the results compared with those from (4-75) for various n and κ values. They found that for an accuracy of (4-75) within 1, 2, 5, or 10% the value of $n^2 + \kappa^2$

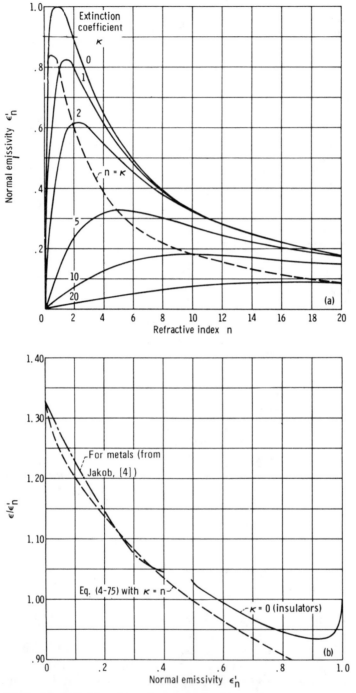

Figure 4-9 Emissivity of metals as computed from electromagnetic theory. (*a*) Normal emissivity of attenuating media emitting into air; (*b*) ratio of hemispherical to normal emissivity.

should be larger than 40, 3.25, 1.75, and 1.25, respectively. For most metals in view of the optical constants as indicated by Table 4-2, Eq. (4-75) should usually be accurate within a percent or two.

From (4-74) the normal emissivity from a metal into air can be computed as a function of n and κ, and this is shown in Fig. 4-9a. More complete results than those in Fig. 4-9a are given in [3].

It is of interest to compare the hemispherical emissivity with the normal value. The practical use for this comparison arises from the fact that it is often the normal emissivity that is measured experimentally because of the relative simplicity of placing a radiation detector in this one orientation. With regard to the total amount of heat dissipation, however, it is the hemispherical emissivity that is desired.

Figure 4-9b shows the ratio of hemispherical to normal emissivity as a function of the normal value. Equation (4-75) divided by ϵ'_n has been plotted for the case where $\kappa = n$. This is valid at large wavelengths for many metals, as shown in the next section. The curve is seen to be close to that presented by Jakob [4] for metals as derived from approximate equations and to lie somewhat below the curve for insulators (as taken from Fig. 4-7b) at high normal emissivities.

For polished metals when ϵ'_n is less than about 0.5, the hemispherical emissivity is larger than the normal value because of the increase in emissivity in the direction near tangency to the surface, as was pointed out in Fig. 4-8. Hence, in a table listing emissivity values for polished metals, if ϵ'_n is given, it should be multiplied by a factor larger than unity such as obtained from Fig. 4-9b to estimate the hemispherical value. Real surfaces that have roughness or may be slightly oxidized often tend to have a directional emissivity that is more diffuse than for polished specimens. For a practical case, therefore, the emissivity ratio may be closer to unity than indicated by Fig. 4-9b.

Relation between emissive and electrical properties. The wave solutions to Maxwell's equations provide a means for determining n and κ from the electric and magnetic properties of a material. The relations for n and κ are given by (4-23). For metals where r_e is small, and for relatively long wavelengths, say $\lambda_0 > \sim 5~\mu m$, the term $\lambda_0/2\pi c_0 r_e \gamma$ becomes dominating, and Eqs. (4-23) then reduce to (the magnetic permeability is taken equal to μ_0)

$$n = \kappa = \sqrt{\frac{\lambda_0 \mu_0 c_0}{4\pi r_e}} = \sqrt{\frac{30\lambda_0}{r_e}} \tag{4-76}$$

where λ_0 is in meters, and r_e in ohm-meters. If λ_0 is in micrometers and r_e in ohm-centimeters, Eq. (4-76) becomes

$$n = \kappa = \sqrt{\frac{0.003\lambda_0}{r_e}} \tag{4-77}$$

which is known as the *Hagen-Rubens equation* [5]. Predictions of n and κ from this equation can be greatly in error, as shown in Table 4-3. Nevertheless, some useful results will eventually be obtained.

With the simplification that $n = \kappa$, equation (4-55) reduces to the following expression for a material with refractive index n radiating in the normal direction into air or vacuum:

$$\epsilon'_{\lambda,n}(\lambda) = 1 - \rho'_{\lambda,n}(\lambda) = \frac{4n}{2n^2 + 2n + 1} \tag{4-78a}$$

where n can be inserted from (4-77). Although there is no difficulty in evaluating (4-78a), a further simplification is often made by expanding in a series to obtain

$$\epsilon'_{\lambda,n}(\lambda) = \frac{2}{n} - \frac{2}{n^2} + \frac{1}{n^3} - \frac{1}{2n^5} + \frac{1}{2n^6} - \cdots \tag{4-78b}$$

Because the index of refraction of metals as predicted from (4-77) is generally large at the long wavelengths being considered here, $\lambda_0 > \sim 5~\mu m$ (see Table 4-3, column 6), only the first term of the series is often retained, and the normal spectral emissivity is then given by substituting (4-77) to obtain the *Hagen-Rubens emissivity relation* (λ_0 in μm, r_e in ohm-centimeters),

$$\epsilon'_{\lambda,n}(\lambda) \approx \frac{2}{n} = \frac{2}{\sqrt{0.003\,\lambda_0/r_e}} \tag{4-79}$$

Data for polished nickel are shown in Fig. 4-10, and the extrapolation to long wavelengths by Eq. (4-79) appears reasonable. The predictions of normal spectral

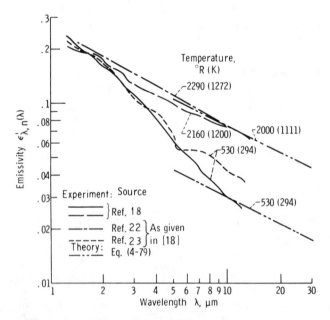

Figure 4-10 Comparison of measured values with theoretical predictions for normal spectral emissivity of polished nickel.

Table 4-3 Comparison of measured optical constants with electromagnetic theory predictions

| Metal | Wavelength λ_o, μm | Measured values | | | Calculated from (4-77) $n = \kappa$ | Spectral normal emissivity $\epsilon'_{\lambda,n}(\lambda)$ | |
		Electrical resistivity (at 20°C) r_e, Ω-cm[a]	Refractive index n	Extinction coefficient κ		Measured	Calculated from (4-79)
Aluminum	12	2.82×10^{-6}	33.6^b	76.4^b	113	0.02^a	0.018
Copper	4.20	1.72×10^{-6}	1.92^b	22.8^b	86	$0.027^{a,c}$	0.023
	4.20	1.72×10^{-6}	1.92^b	22.8^b	86	0.015^d	0.023
	5.50	1.72×10^{-6}	3.16^a	28.4^a	98	0.012^d	0.020
Gold	5.00	2.44×10^{-6}	1.81^a	32.8^a	78	$0.031^{a,c}$	0.026
Platinum	5.00	10×10^{-6}	11.5^a	15.7^a	39	0.050^d	0.051
Silver	4.50	1.63×10^{-6}	4.49^a	33.3^a	91	$0.015^{a,c}$	0.022
	4.37	1.63×10^{-6}	4.34^b	32.6^b	90	$0.015^{a,c}$	0.022

[a]Data from [2].
[b]Data from [17].
[c]Measured at μm.
[d]Data from [18].

emissivity at long wavelengths as presented in Table 4-3 are much better than the prediction of attendant optical constants.

The normal spectral emissivity given in (4-79) can be integrated with respect to wavelength to yield a normal total emissivity. The integration relating spectral and total quantities is given by (3-3b) (modified for a normal emissivity so that $\theta = 0$):

$$\epsilon'_n(T) = \frac{\pi \int_0^\infty \epsilon'_{\lambda,n}(\lambda,T) i'_{\lambda b}(\lambda,T) \, d\lambda}{\sigma T^4}$$

Equation (4-79) is only valid for $\lambda_0 > \sim 5$ μm, so in performing the integration starting from $\lambda = 0$ the condition is being imposed that the metal temperature is such that the energy radiated from $\lambda_0 = 0\text{-}5$ μm is small compared with that at wavelengths longer than 5 μm. Then substituting (4-79) and (2-11a) for $i'_{\lambda b}$ into the integral provides

$$\epsilon'_n(T) \approx \frac{\pi \int_0^\infty 2(r_e/0.003\lambda_0)^{1/2} 2C_1/[\lambda_0^5(e^{C_2/\lambda_0 T} - 1)] \, d\lambda_0}{\sigma T^4}$$

$$= \frac{4\pi C_1 (Tr_e)^{1/2}}{(0.003)^{1/2} \sigma C_2^{4.5}} \int_0^\infty \frac{\zeta^{3.5}}{e^\zeta - 1} \, d\zeta \tag{4-80}$$

where $\zeta = C_2/\lambda_0 T$ as was used in conjunction with (2-19). The integration is carried out by use of Γ functions to yield (T in K, r_e in ohm-cm)

$$\epsilon'_n(T) \approx \frac{4\pi C_1 (Tr_e)^{1/2}}{(0.003)^{1/2} \sigma C_2^{4.5}} (12.27) = 0.576(r_e T)^{1/2} \tag{4-81}$$

If additional terms in the series (4-78b) are retained, then

$$\epsilon'_n(T) = 0.576(r_e T)^{1/2} - 0.177 r_e T + 0.058(r_e T)^{3/2} - \cdots \tag{4-82a}$$

The recommended formula from [4] is

$$\epsilon'_n(T) = 0.576(r_e T)^{1/2} - 0.124 r_e T \tag{4-82b}$$

where T is in kelvins and r_e is in ohm-centimeters.

For pure metals, r_e is approximately described near room temperature by

$$r_e \approx r_{e,273} \frac{T}{273} \tag{4-83}$$

where $r_{e,273}$ is the electrical resistivity in ohm-centimeters, evaluated at 273 K (492°R). Substituting Eq. (4-83) into (4-81) gives the result

$$\epsilon'_n(T) \approx 0.0348 \sqrt{r_{e,273}} \, T \tag{4-84a}$$

where T is in kelvins, or for T in degrees Rankine,

$$\epsilon'_n(T) \approx 0.0194 \sqrt{r_{e,492}} \, T \tag{4-84b}$$

This indicates that, for long wavelengths ($\lambda_0 > \sim 5$ μm), the total emissivity of pure metals should be directly proportional to temperature. This result was originally derived by Aschkinass [6] in 1905. In some cases it holds to unexpectedly high tem-

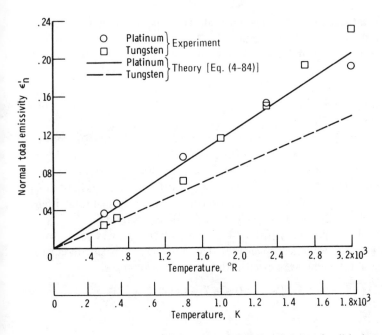

Figure 4-11 Temperature dependence of normal total emissivity of polished metals.

peratures where considerable radiation is in the short-wavelength region (for platinum, to near 1800 K (3200°R)), but in general applies only below about 550 K (1000°R). This is illustrated in Fig. 4-11 for platinum and tungsten (data from [2]).

In Fig. 4-12, a comparison is made at 100°C of the normal total emissivity from experiment and from (4-84) for a variety of polished surfaces of pure metals. Agree-

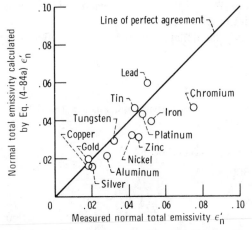

Figure 4-12 Comparison of data with calculated normal total emissivity for polished metals at 100°C.

ment is generally satisfactory. The experimental values are the minimum values of results available in three standard compilations [2, 7, 8].

By using the emissivity from Eq. (4-84), the total intensity in the normal direction emitted by a metal is obtained as

$$i'_{n,\text{metals}} = \epsilon'_{n,\text{metals}} \frac{\sigma T^4}{\pi} \propto T^5 \tag{4-85}$$

This indicates that the normal total intensity is proportional to the *fifth* power of absolute temperature rather than the fourth power as with a blackbody. Again, it must be emphasized that many assumptions were made to obtain this simplified result. If, for example, more than two terms had been retained from the series in (4-78b), it would have been found that the exact proportionality between normal total intensity and T^5 no longer holds, although the exponent would still be greater than 4.

Using the angular dependency from (4-73), an integration was made over all directions to provide hemispherical quantities. The following approximate equations for the hemispherical total emissive power fit the results in two ranges:

$$e(T) = \sigma T^4 (0.751 \sqrt{r_e T} - 0.396 r_e T) \qquad 0 < r_e T < 0.2 \tag{4-86a}$$

and

$$e(T) = \sigma T^4 (0.698 \sqrt{r_e T} - 0.266 r_e T) \qquad 0.2 < r_e T < 0.5 \tag{4-86b}$$

where the numerical factors in the parentheses and those used in specifying the ranges of validity apply for T in kelvins and r_e in ohm-centimeters. The resistivity r_e depends on T to the first power so that the first term inside the parentheses of each of these equations provides the T^5 dependency discussed earlier.

EXAMPLE 4-4 A polished platinum surface is maintained at temperature $T_A = 400°R$. Energy is incident upon the surface from a black enclosure, at temperature $T_i = 800°R$, that encloses the surface. What is the hemispherical-directional total reflectivity into the direction normal to the surface?

Equation (3-48) shows that the directional-hemispherical total reflectivity can be found from

$$\rho'_n(T_A = 400°R) = 1 - \alpha'_n(T_A = 400°R)$$

where $\alpha'_n(T_A = 400°R)$ is the normal total absorptivity of a surface at $400°R$ for incident black radiation at $800°R$; that is,

$$\alpha'_n(T_A = 400°R) = \frac{\int_0^\infty \alpha'_{\lambda,n}(\lambda, T_A = 400°R) i'_{\lambda b}(\lambda, 800°R) \, d\lambda}{\int_0^\infty i'_{\lambda b}(\lambda, 800°R) \, d\lambda}$$

For spectral quantities, $\alpha'_{\lambda,n}(\lambda, T_A = 400°R) = \epsilon'_{\lambda,n}(\lambda, T_A = 400°R)$. From Eq. (4-79) the variation of r_e with temperature provides the emissivity variation $\epsilon'_{\lambda,n}(\lambda, T_A) \propto T_A^{1/2}$. Then $\epsilon'_{\lambda,n}(\lambda, T_A = 400°R) = \epsilon'_{\lambda,n}(\lambda, T_A = 800°R)(400/800)^{1/2}$, and we obtain

$$\alpha_n'(T_A = 400°R) = \frac{\sqrt{\frac{1}{2}} \int_0^\infty \epsilon_{\lambda,n}'(\lambda, T_A = 800°R) i_{\lambda b}'(\lambda, 800°R)\, d\lambda}{\int_0^\infty i_{\lambda b}'(\lambda, 800°R)\, d\lambda}$$

$$= \frac{\epsilon_n'(T_A = 800°R)}{\sqrt{2}}$$

where the last equality is obtained by examination of the emissivity definition, Eq. (3-3b). The normal total emissivity of platinum at 800°R is given by (4-84b) as plotted in Fig. 4-11 as

$$\epsilon_n'(T_A = 800°R) = 0.0194\sqrt{r_{e,492}} \times 800 = 0.051$$

Note that (4-84) is only to be used when temperatures are such that most of the energy involved is at wavelengths greater than 5 μm. Examination of the black-body functions, Table A-5, shows that for a temperature of 800°R about 10% of the energy is at less than 5 μm so that possibly a small error is introduced.

The reciprocity relation of (3-28) for uniform incident intensity can now be employed to give the final result for the hemispherical-directional total reflectivity:

$$\rho_n'(T_A = 400°R) = 1 - \alpha_n'(T_A = 400°R) \approx 1 - \frac{\epsilon_n'}{\sqrt{2}}(T_A = 800°R)$$

$$= 1 - \frac{0.051}{\sqrt{2}} = 0.964$$

A summary of some of the property prediction equations for dielectrics and metals is given in Table 4-4.

4-7 EXTENSIONS OF THE THEORY OF RADIATIVE PROPERTIES

Much work has been expended in improving the theory of the radiative properties of materials, using both classical wave theory and quantum theory. A number of authors have successfully removed some restrictions that are present in the classical development presented here. Notable are the contributions of Davisson and Weeks [9], Foote [10], Schmidt and Eckert [11], and Parker and Abbott [12], all of whom extended the emissivity relations for metals to shorter wavelengths and higher temperatures, and of Mott and Zener [13], who derived predictions for metal emissivity at very short wavelengths on the basis of quantum relations. Edwards [14] reviews the advances in predictions of surface properties, and Sievers [15] gives some recent theory and results.

None of these treatments, however, accounts for surface effects. Because of the difficulty of specifying surface conditions and controlling surface preparation, it is found that comparison of the theory with experiment is not always adequate for even the refined theories. In fact, comparison to the less exact but simpler relations given here is often better. For even the purest materials given the most meticulous prepara-

Table 4-4 Summary of some equations for property prediction by electromagnetic theory

Property	Equation	Conditions
		Dielectrics ($\kappa = 0$)
Directional reflectivity	(4-64a)	Polarized in plane parallel to plane of incidence
Directional reflectivity	(4-64b)	Polarized in plane perpendicular to plane of incidence
Directional reflectivity	(4-65) (4-63)	Unpolarized
Normal reflectivity	(4-48)	Polarized or unpolarized
Hemispherical emissivity	(4-66)	Emission into medium having $n = 1$
Normal emissivity	(4-67)	Emission into medium having $n = 1$
		Metals (in contact with transparent medium of unity refractive index)
Directional reflectivity	(4-61a), (4-70a)	Parallel polarized component
Directional reflectivity	(4-61b), (4-70b)	Perpendicular polarized component
Directional reflectivity	(4-71)	Unpolarized
Directional emissivity	(4-73)	Unpolarized
Hemispherical emissivity	(4-75)	Unpolarized
Normal spectral emissivity	(4-74)	Unpolarized
	(4-78a) (4-79)	Unpolarized $\lambda > \sim 5 \ \mu m$
Normal total emissivity	(4-82), (4-84)	$T < \sim 1000°R$ (550 K)

tion, the elementary relations are often more accurate because the errors in the simpler theory are in the direction which causes compensation for surface working.

Polarization effects enter into the mathematical description of electromagnetic waves and wave reflections. A detailed discussion of these effects is beyond the intent of this book. A comprehensive discussion of the analytical methods and technology of polarization phenomena is given in [16].

REFERENCES

1. Maxwell, James Clerk: A Dynamical Theory of the Electromagnetic Field, in W. D. Niven (ed.), "The Scientific Papers of James Clerk Maxwell," vol. 1, Cambridge University Press, London, 1890.
2. Weast, Robert C. (ed.): "Handbook of Chemistry and Physics," 44th ed., Chemical Rubber Company, Cleveland, 1962.
3. Hering, R. G., and T. F. Smith: Surface Radiation Properties from Electromagnetic Theory, *Int. J. Heat Mass Transfer,* vol. 11, pp. 1567–1571, 1968.
4. Jakob, Max: "Heat Transfer," vol. 1, John Wiley and Sons, Inc., New York, 1949.

5. Hagen, E., and H. Rubens: Metallic Reflection, *Ann. Phys.*, vol. 1, no. 2, pp. 352–375, 1900. (See also E. Hagen and H. Rubens: Emissivity and Electrical Conductivity of Alloys, *Deutsch. Phys. Ges. Verhandl.*, vol. 6, no. 4, pp. 128–136, 1904.)
6. Aschkinass, E.: Heat Radiation of Metals, *Ann. Phys.*, vol. 17, no. 5, pp. 960–976, 1905.
7. Hottel, H. C.: Radiant Heat Transmission, in William H. McAdams (ed.), "Heat Transmission," 3d ed., pp. 55-125, McGraw-Hill Book Company, New York, 1954.
8. Eckert, E. R. G., and R. M. Drake, Jr.: "Heat and Mass Transfer," 2d ed., McGraw-Hill Book Company, New York, 1959.
9. Davisson, C., and J. R. Weeks, Jr.: The Relation between the Total Thermal Emissive Power of a Metal and Its Electrical Resistivity, *J. Opt. Soc. Am.*, vol. 8, no. 5, pp. 581–605, 1924.
10. Foote, Paul D.: The Emissivity of Metals and Oxides, III. The Total Emissivity of Platinum and the Relation between Total Emissivity and Resistivity, *NBS Bull.*, vol. 11, no. 4, pp. 607–612, 1915.
11. Schmidt, E., and E. R. G. Eckert: Über die Richtungsverteilung der Wärmestrahlung von Oberflächen, *Forsch. Geb. Ingenieurwes.*, vol. 6, no. 4, pp. 175–183, 1935.
12. Parker, W. J., and G. L. Abbott: Theoretical and Experimental Studies of the Total Emittance of Metals, *Symp. Therm. Radiat. Solids, NASA* SP-55, pp. 11–28, 1964.
13. Mott, N. F., and C. Zener: The Optical Properties of Metals, *Cambridge Phil. Soc. Proc.*, vol. 30, pt. 2, pp. 249–270, 1934.
14. Edwards, D. K.: Radiative Transfer Characteristics of Materials, *ASME J. Heat Transfer*, vol. 91, pp. 1–15, 1969.
15. Sievers, A. J.: Thermal Radiation from Metal Surfaces, *J. Opt. Soc. Am.*, vol. 68, no. 11, pp. 1505–1516, 1978.
16. Shurcliff, W. A.: "Polarized Light, Production and Use," Harvard University Press, Cambridge, Mass., 1962.
17. Garbuny, M.: "Optical Physics," Academic Press, Inc., New York, 1965.
18. Seban, R. A.: Thermal Radiation Properties of Materials, pt. III, WADD-TR-60-370, University of California, Berkeley, August, 1963.
19. Brandenberg, W. M.: The Reflectivity of Solids at Grazing Angles. Measurement of Thermal Radiation Properties of Solids, Joseph C. Richmond (ed.), *NASA* SP-31, pp. 75–82, 1963.
20. Brandenberg, W. M., and O. W. Clausen: The Directional Spectral Emittance of Surfaces between 200° and 600°C, *Symp. Therm. Radiat. Solids, NASA* SP-55, pp. 313–320, 1964.
21. Price, Derek J.: The Emissivity of Hot Metals in the Infra-Red, *Proc. Phys. Soc. London*, sec. A, vol. 59, pt. 1, pp. 118–131, 1947.
22. Hurst, C.: The Emission Constants of Metals in the Near Infra-Red, *Proc. R. Soc. London*, ser. A, vol. 142, no. 847, pp. 466–490, 1933.
23. Pepperhoff, W.: "Temperaturstrahlung," D. Steinkopf, Darmstadt, 1956.
24. Toscano, W. M., and E. G. Cravalho: Thermal Radiative Properties of the Noble Metals at Cryogenic Temperatures, *J. Heat Transfer*, vol. 98, no. 3, pp. 438–445, 1976.

PROBLEMS

1 An electrical insulator has a coefficient of refraction $n = 1.8$, and is radiating into air. What is the directional emissivity for the direction normal to the surface? What is it for the direction 85° from the normal?

   Answer: 0.919, 0.371.

2 At a temperature of 300 K these metals have the resistivities

   Silver 1.65×10^{-6} Ω-cm
   Platinum 11.0×10^{-6} Ω-cm
   Lead 20.8×10^{-6} Ω-cm

   What are the theoretical hemispherical total emissivities of these materials, and how do they compare with tabulated values for clean, unoxidized, polished surfaces?

   Answer: 0.017, 0.042, 0.057.

3 Evaluate the normal spectral reflectivity of aluminum at 200°F (367 K) when $\lambda_0 = 5$ μm, 10 μm, 20 μm.

   Answer: 0.969, 0.978, 0.984.

4 Polished gold at 80°F (300 K) is irradiated normally by a gray-body source at 1000°F (810 K). Evaluate the absorptivity α'_n. (Use the method of Example 4-4.)

   Answer: 0.026.

5 The hemispherical total emissive power emitted by a polished metallic surface is 1900 (W/m²) at some temperature T_s. What would you expect the emissive power to be if the temperature were doubled? What assumptions are involved in your answer?

   Answer: 60,800 W/m².

6 In Fig. 4-13 are given some experimental data for the hemispherical spectral reflectivity of polished aluminum at room temperature. Extrapolate the data to $\lambda = 12$ μm. Use whatever method you want, but *list your assumptions.* Discuss the probable accuracy of your extrapolation. (*Hint:* The electrical resistivity of pure aluminum is about 2.82×10^{-6} Ω-cm at 293 K. At 12 μm, $\bar{n} = 33.6 - 76.4i$. You may use any, all, or none of these data if you wish.)

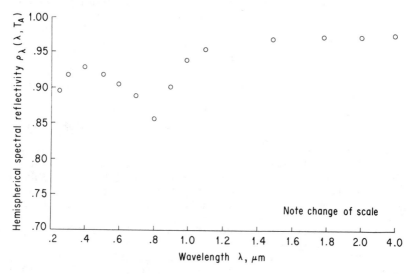

Figure 4-13 Spectral reflectivity of deposited aluminum. (*From G. Hass, Filmed Surfaces for Reflecting Optics, J.O.S.A., vol. 45, p. 945, 1955.*)

7 An unoxidized polished titanium sphere is heated until it is glowing red. From a distance it appears as a red disk. From electromagnetic theory how would you expect the brightness of the red color to vary across the disk? What would you expect after looking at Fig. 5-1?

8 Using the Hagen-Rubens emissivity relation, plot the normal spectral emissivity as a function of wavelength for a polished aluminum surface used in a cryogenic application at 60°R. What is the normal total emissivity? (*Note:* Do not use any relations valid only near room temperature.)

9 A sample of highly polished platinum has a value of normal spectral emissivity of 0.050 at a wavelength of 5.0 μm at 293 K. What value of normal spectral absorptivity do you expect the sample to have at

   (*a*) A wavelength of 10 μm and at 293 K?
   (*b*) A wavelength of 10 μm and at 570 K?

   Answer: (*a*) 0.035, (*b*) 0.049.

10 Metals cooled to very low temperatures approaching absolute zero become highly conducting; i.e., the value of $r_e(T \to 0) \to 0$. Based on electromagnetic theory predictions, what is your estimate of the values of the refractive index n and the extinction coefficient κ, and of the normal spectral and total emissivities of metals at such conditions? What assumptions are implicit in your estimates? (The results predicted by the Hagen-Rubens relation, and other results from classical electromagnetic theory, become inaccurate at $T < 100°$ K. Predictions of radiative properties at low absolute temperatures using more exact theoretical approaches are reviewed by Toscano and Cravalho [24].

11 A particular dielectric material has a refractive index near 2. Estimate:

(*a*) The hemispherical emissivity of the material for emission into air.

(*b*) The directional emissivity at $\theta = 70°$ into air.

(*c*) The directional hemispherical reflectivity for both components of polarized reflectivity. Plot both components on a graph similar to Fig. 4-6 for $n = 2$. Let θ be the angle of incidence.

Answer: (*a*) 0.84,   (*b*) 0.76.

FIVE

RADIATIVE PROPERTIES
OF REAL MATERIALS

5-1 INTRODUCTION

In this chapter, the general characteristics of the radiative properties of real materials will be examined. These properties can vary considerably from the idealized results of Chap. 4 for "optically smooth" materials as predicted by electromagnetic theory. The analytical predictions yield useful trends and provide a unifying basis to help explain various radiation phenomena. However, the analyses are inadequate in the sense that the engineer is generally dealing with surfaces coated in varying degrees with contaminants, oxide, paint, and so forth, and having a surface roughness that is difficult to specify completely. Examples of some typical variations of radiative properties as a function of these and other parameters will be presented in this chapter to illustrate the types of property variations that can occur. This should provide the reader with an appreciation of how sensitive the radiative performance is to the surface condition. In addition to the typical properties presented, a number of atypical examples will be given to demonstrate that a careful examination of individual properties must be made to properly select the property values to be used in radiative exchange calculations.

Most of the discussion in this chapter will be limited to opaque solids, where opaque is defined to mean that no transmission of radiant energy occurs through the entire thickness of the body. A composite body such as a thin coating on a substrate of a different material can have partial transmission through the coating, but for an opaque body, none of the transmitted radiation will pass entirely through the substrate. Information is also included on the transmission ability of glass and water.

This illustrates how transmission can differ considerably in short- and long-wavelength regions. In Chap. 19 the effects of thin films on various substrates are discussed in some detail.

No attempt to compile comprehensive property data will be made. Extensive tabulations and graphs of radiative properties have been gathered in [1–7]. A limited tabulation is given in Appendix D for the convenience of the reader.

As discussed in Chap. 4, basic differences in the radiative behavior of metals and dielectrics are evident from electromagnetic theory. For this reason the first two sections in this chapter will deal with these two classes of materials, with metals being discussed first. Then some special surfaces will be discussed that have specific desirable variations of properties with wavelength and direction.

5-2 SYMBOLS

A	area
c_0	speed of electromagnetic wave propagation in vacuum
E	overall emittance of plate
e	emissive power
$F_{0-\lambda}$	fraction of blackbody energy in spectral range $0-\lambda$
p	probability function
Q	energy rate, energy per unit time
q	energy flux, energy per unit time per unit area
R	overall reflectance of plate
r_e	electrical resistivity
T	absolute temperature; overall transmittance of plate
z	height of surface roughness
α	absorptivity
γ	electrical permittivity
ϵ	emissivity
θ	angle measured from normal of surface, polar angle
λ	wavelength
μ	magnetic permeability
φ	circumferential angle
ρ	reflectivity
σ	Stefan-Boltzmann constant (Table A-4)
σ_0	root-mean-square height of surface roughness

Subscripts

A	of surface A
a	absorber
b	blackbody condition
c	evaluated at cutoff wavelength
e	emitted
eq	at equilibrium
i	incident

max	maximum value
n	normal direction
R	radiator
r	reflected
s	specular
λ	spectrally dependent
$0-\lambda$	in wavelength range $0-\lambda$

Superscripts

$'$	directional
$''$	bidirectional

5-3 RADIATIVE PROPERTIES OF METALS

Pure, smooth metals are often characterized by low values of emissivity and absorptivity, and therefore comparatively high values of reflectivity. Figure 4-12 demonatrates that the emissivity in the direction normal to the surface is quite low for a variety of polished metals. However, low emissivity values are not an absolute rule for metals; in some of the examples that will be given the spectral emissivity rises to 0.5 or larger as the wavelength becomes short, or the total emissivity becomes large as the temperature is elevated.

5-3.1 Directional Variations

A behavior typical of polished metals is that except near $\theta = 90°$, the directional emissivity tends to increase with increasing angle of emission θ (θ is measured with respect to the surface normal). This is predicted by electromagnetic theory and was shown to be true for platinum in Fig. 4-8. At wavelengths shorter than the range for which the simple electromagnetic theory of Chap. 4 applies, a deviation from this behavior might be expected. To illustrate this deviation, the directional spectral emissivity of polished titanium is shown in Fig. 5-1. At wavelengths greater than about 1 μm, the directional spectral emissivity of titanium does indeed tend to increase with increasing θ over most of the θ range. The increase with θ becomes smaller as wavelength decreases; finally, at wavelengths less than about 1 μm, the directional spectral emissivity actually decreases with increasing θ over the entire range of θ. Hence, for polished metals the typical behavior of increased emission for directions nearly tangent to the surface can be violated at short wavelengths.

5-3.2 Effect of Wavelength

In the infrared region, it was shown in Chap. 4 that the spectral emissivity of metals tends to increase with decreasing wavelength. This trend remains true over a large span of wavelength as illustrated for several metals in Fig. 5-2, which gives the spectral emissivity in the normal direction. For other directions, the same effect is illustrated in Fig. 5-1 except at large angles from the normal, where curves for various wave-

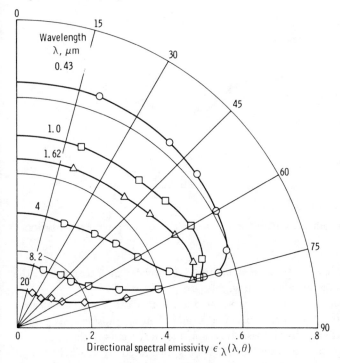

Figure 5-1 Effect of wavelength on directional spectral emissivity of pure titanium. Surface ground to 16 μin (0.4 μm) rms. *(Data from [6].)*

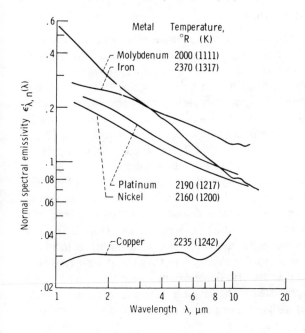

Figure 5-2 Variation with wavelength of normal spectral emissivity for polished metals. *(Data from [39].)*

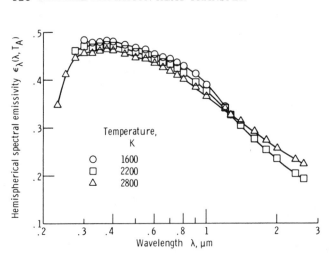

Figure 5-3 Effect of wavelength and surface temperature on hemispherical spectral emissivity of tungsten [40].

lengths may cross. The curve for copper in Fig. 5-2 provides an exception as the emissivity remains relatively constant with wavelength.

At very short wavelengths, the assumptions upon which the simplified electromagnetic theory of Chap. 4 are based become invalid. Indeed, most metals exhibit a peak emissivity somewhere near the visible region, and the emissivity then decreases rapidly with further decrease in wavelength. This is illustrated by the behavior of tungsten in Fig. 5.3.

5-3.3 Effect of Surface Temperature

The Hagen-Rubens relation [Eq. (4-79)] shows that, for wavelengths that are not too short ($\lambda > {\sim}5$ μm), the spectral emissivity of a metal is proportional to the resistivity of the metal to the one-half power. Hence, we can expect the spectral emissivity of pure metals to increase with temperature as does the resistivity, and this is found to be the case in most instances. Figure 5-3 is an example for the hemispherical spectral emissivity of tungsten. The expected trend is observed for $\lambda > 1.27$ μm. Figure 5-3 also illustrates a phenomenon characteristic of many metals as discussed in [8]. At short wavelengths (in the case of tungsten, $\lambda < 1.27$ μm), the temperature effect is reversed and the spectral emissivity decreases as temperature is increased.

The observed increase of spectral emissivity with decreasing wavelength for metals in the infrared radiation region (wavelengths longer than visible region), as discussed in Sec. 5-3.2, accounts for the increase in total emissivity with temperature. With increased temperature the peak of the blackbody radiation curve (Fig. 2-6) moves toward shorter wavelengths. Consequently, as the surface temperature is increased, proportionately more radiation is emitted in the region of higher spectral emissivity, which results in an increased total emissivity. Some examples are shown in Fig. 5-4. Here the behavior of metals is contrasted with that of a dielectric, magnesium oxide, for which the emissivity decreases with increasing temperatures.

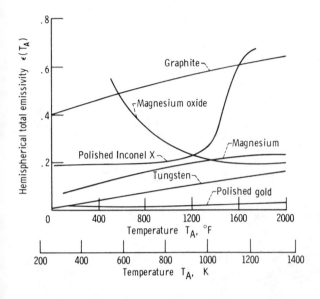

Figure 5-4 Effect of temperature on hemispherical total emissivity of several metals and one dielectric. *(Data from [1].)*

In evacuated insulation systems consisting of a series of radiation shields, an important heat loss is by radiation. At the low temperatures involved in the insulation of cryogenic systems, there is a lack of experimental property data especially at the long wavelengths characteristic of emission at these temperatures. The Hagen-Rubens result indicates that the normal spectral emissivity is proportional to $(r_e/\lambda)^{1/2}$, and the normal total emissivity is proportional to $(r_e T)^{1/2}$. If electrical resistivity is directly proportional to temperature, then $\epsilon'_{\lambda,n}(\lambda,T) \propto (T/\lambda)^{1/2}$ and $\epsilon'_n(T) \propto T$. This indicates that the emissivities should become quite small at low T and large λ. Experimental measurements as summarized in [9] for copper, silver, and gold indicate that the ϵ values do not decrease to such small values. As refinements in the theory, the Drude free-electron theory and anomalous skin-effect theory, as described in [9], include additional electron and quantum-mechanical interactions. The Drude theory, which reduces to the Hagen-Rubens relation at long wavelengths, predicts ϵ values that decrease much more rapidly with temperature than has been observed. The anomalous skin-effect model with diffuse electron reflections was found to predict the emissivity most accurately. Figure 5-5 shows the results of the two theoretical models. The limited data at low temperatures lie somewhat above the anomalous skin-effect model.

The next two factors to be discussed are surface roughness and surface impurities or coatings. These can cause major deviations from the electromagnetic-theory predictions of Chap. 4.

5-3.4 Effect of Surface Roughness

If the surface imperfections present on a material are much smaller than the wavelength of the radiation being considered, then the material is said to be *optically smooth*. A material that is optically smooth for long wavelengths may be comparatively quite rough at short wavelengths. The radiative properties of optically smooth

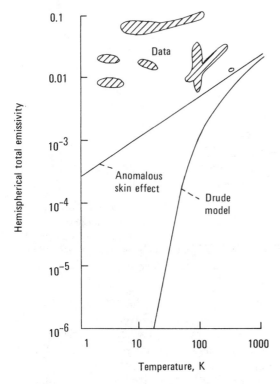

Figure 5-5 Effect of low temperature on hemispherical total emissivity of copper. *(Results from [9].)*

materials can be predicted within the limitations of electromagnetic theory as discussed in Chap. 4. An important parameter in characterizing roughness effects is the *optical roughness*, which is the ratio of a characteristic roughness height [usually the root-mean-square (rms) roughness σ_0] to the wavelength of the radiation.

When the optical roughness σ_0/λ is greater than about 1, there are multiple reflections in the cavities between the roughness elements. As a result the roughness increases the trapping of incident radiation, thereby increasing the hemispherical absorptivity and consequently the hemispherical emissivity of the surface.

The roughness also has a considerable effect on the directional emission and reflection characteristics when the roughness is large. For $\sigma_0/\lambda > 1$ the concepts of geometrical optics can be used to trace the radiation paths reflected within the cavities or from the roughness elements. If the roughness geometry is completely specified, it is possible in certain cases to predict the directional behavior. An example is the directional emissivity of a parallel grooved surface as discussed in Sec. 5-5.3. Ordinarily the roughness is very irregular, and a statistical model must be assumed. For example, the roughness can be represented as randomly oriented facets that each reflect in a specular manner.

When the optical roughness is small ($\sigma_0/\lambda < 1$), multiple reflection effects in the cavities usually become small, and the hemispherical properties approach those for optically smooth surfaces. However, as a result of diffraction effects, the directional properties (especially the bidirectional reflectivity) can be significantly influenced by the roughness.

Several analyses have been made with the intent of predicting the effect of surface roughness on the radiative properties of materials. The improvement and utilization of these analyses and the comparison with detailed property measurements is an area of active current research. The analyses have provided a basis for understanding many of the observed roughness effects.

A chief stumbling block in the prediction of radiative properties is in the precise definition of the surface characteristics for use in the analytical equations. Perhaps the most common way of characterizing surface roughness is by the method of preparation (lapping, grinding, etching, etc.) plus a specification of root-mean-square (rms) roughness. The latter is usually obtained by means of a profilometer, which is an instrument that traverses a sharp stylus over the surface and reads out the vertical perturbations of the stylus in terms of an rms value. It does not account for the horizontal spacing of the roughness and gives no indication of the *distribution* of the size of roughness around the rms value. It does not give any information on the average slope of the sides of the roughness peaks which influences the behavior of the cavities between them. At present, there is no generally accepted method of accurately specifying surface characteristics, and none of those mentioned in this paragraph is completely adequate for prediction of radiative properties.

Some of the analytical approaches utilized in the face of the aforementioned difficulties will now be briefly discussed, and some comparisons given with experimental data. Davies [10] has used diffraction theory to examine the reflecting properties of a surface with roughness that is assumed to be distributed according to a gaussian (normal) probability distribution, specified as a probability $p(z)$ of having a roughness of height z given by

$$p(z) = \frac{1}{\sigma_0 \sqrt{2\pi}} \exp\left(-\frac{z^2}{2\sigma_0^2}\right)$$

The individual surface irregularities were assumed of sufficiently small slope so that shadowing could be neglected, and σ_0 was assumed very much smaller than the wavelength of incident radiation λ. The material was assumed to be a perfect electrical conductor so that from (4-23b) the extinction coefficient is infinite. From (4-55) this provides perfect reflection, and consequently the theory is concerned with the directional distribution of the energy that is reflected rather than with the amount that is reflected. The reflected distribution was found to consist of a specular component and a component distributed about the specular peak.

A similar derivation, with σ_0 assumed much larger than λ, again yielded a distribution of reflected intensity about the specular peak, this time of larger angular spread than for the case of $\sigma_0 \ll \lambda$. This would be expected since the surface should behave increasingly like an ideal specular reflector as the roughness becomes very small compared with the wavelength of the incident radiation. Davies' treatment is found to be very inaccurate at near grazing angles because of the neglect of roughness shadowing.

Porteus [11] extended Davies' approach by removing the restrictions on the relation between σ_0 and λ and including more parameters for specification of the surface roughness characteristics. Some success in predicting the roughness characteristics of prepared samples from measured reflectivity data was obtained, but certain types of

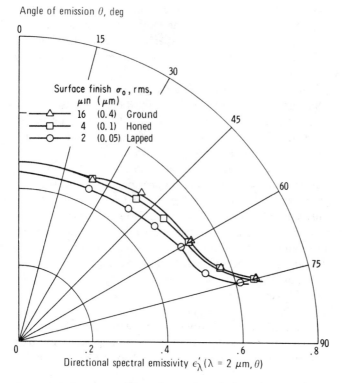

Angle of emission θ, deg

Surface finish σ_0, rms,
μin (μm)
—△— 16 (0.4) Ground
—□— 4 (0.1) Honed
—○— 2 (0.05) Lapped

Directional spectral emissivity $\epsilon'_\lambda (\lambda = 2\ \mu m, \theta)$

Figure 5-6 Roughness effects for small optical roughness, $\sigma_0/\lambda < 1$. (a) Effect of surface finish on directional spectral emissivity of pure titanium. Wavelength, 2 μm (78.7 μin). *(Data from [6].)*

surface roughness led to poor agreement. Measurements were mainly at normal incidence, and the neglect of shadowing makes the results of doubtful value at near grazing angles.

A more satisfactory treatment has been given by Beckmann and Spizzichino [12]. Their method includes the autocorrelation distance of the roughness in the prescription of the surface. This is a measure of the spacing of the characteristic roughness peaks on the surface and hence is related to the rms slope of the roughness elements. The method provides better data correlation than the earlier analyses. A critical evaluation and comparison of the Davies and Beckmann analyses is given by Houchens and Hering [13].

Some observed effects of surface roughness for small σ_0/λ are shown in Figs. 5-6a and b. The former shows the directional emissivity of titanium [6] at a wavelength of 2 μm for three surface roughnesses as obtained by grinding, honing, and lapping. The maximum roughness is 16 microinches (μin). Since 2 μm is equal to 78.7 μin, the wavelength of the radiation is significantly larger than the surface roughnesses. Hence, relative to this wavelength the specimens are smooth. As a result, the emissivity changes only a small amount as the roughness varies from 2 to 16 μin. Also there is very little effect on the directional variation of the emissivity. Reference [6] also gives data for sandblasted surfaces that produced larger increases in emissivity.

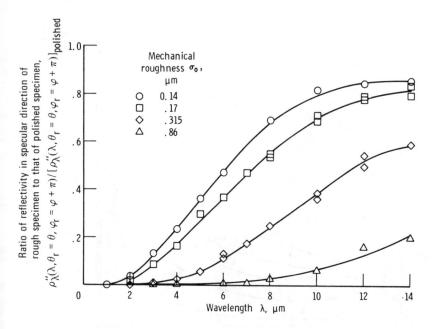

Figure 5-6 *(Continued)* *(b)* Effect of roughness on bidirectional reflectivity in specular direction for ground nickel specimens. Mechanical roughness for polished specimen, 0.015 μm. *(Data from [41].)*

Figure 5-6b provides the reflectivity of nickel for energy reflected into the specular direction from a beam incident at an angle 10° from the normal. In this figure, the reflectivities of the rough specimens are expressed as a ratio to the reflectivity of a polished surface to exhibit the effect of roughness on the directional characteristics rather than on the magnitude of the reflectivity. The polished surface used for comparison had a roughness about 10 times less than that of the rough specimens. A high value of the ordinate thus means that the specimen is behaving more like a polished surface. Data are shown for ground nickel specimens with four different roughnesses with $\sigma_0/\lambda < 1$. The reflectivity rises as wavelength is increased (that is, optical roughness is decreased) because for a given roughness the surface is smoother relative to the incident radiation. As expected, for a fixed wavelength the reflectivity for the specular direction decreases as the roughness is increased. Data exhibiting the same trends for aluminum have been well correlated in [14] by use of the Beckmann theory

For an optical roughness $\sigma_0/\lambda > 1$ some detailed experimental measurements of the bidirectional reflectivity, along with an analysis using geometrical optics, have been given by Torrance and Sparrow [15, 16]. The analysis substantiated the important trends of the data. Some typical results for the bidirectional reflectivity of aluminum are given in Fig. 5-7 for an optical roughness of 2.6. This gives the bidirectional reflectivity in the plane of incidence (plane formed by incident path and normal to surface) as a function of reflection angle θ_r. The reflectivity is ratioed to the value in the

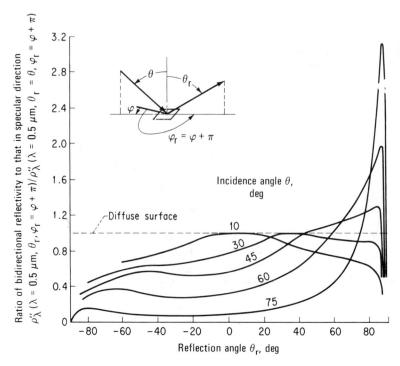

Figure 5-7 Bidirectional reflectivity in plane of incidence for various incidence angles; material, aluminum (2024-T4), aluminum coated; rms roughness, $\sigma_0 = 1.3$ μm; wavelength of incident radiation, $\lambda = 0.5$ μm. *(Data from [16].)*

specular direction. The various curves correspond to different angles of incidence. For a diffuse surface the reflected intensity is independent of θ_r so that the ratio given in the figure is unity as shown by the dashed line. A specular reflection would appear as a sharp high peak at $\theta_r = \theta$. For incidence at $30°$ the reflected intensity peaks in the specular direction ($\theta_r = 30°$). However, at larger incidence angles the maximum in ρ''_λ shifts to angles larger than the specular angle; for example, when $\theta = 60°$ the reflectivity peak is at $\theta_r = 85°$. This is in contrast to the behavior for a smooth surface where the peak would be at $\theta_r = \theta$. The theory indicates that this off-specular reflection, which occurs for large optical roughness and large incidence angles, is the result of shadowing effects of the roughness elements by each other.

References [17] and [18] analyze the angular distribution of the emissivity from a rough surface by considering the emission from regular V-grooves in a parallel or circular pattern. A more realistic model is the randomly roughened metal surface in [19, 20]. The distributions of heights and slopes of the roughnesses were assumed gaussian, and random blockage of radiation by adjacent roughness elements in the path of observation was included. Multiple reflections between surface elements were neglected. Calculations carried out for gold and chromium indicated that for directions less than about 60° away from the surface normal, the emissivity increases with surface roughness. At larger angles the emissivity was less than for a smooth surface; this was a result of both roughness effects and the behavior of smooth metallic surfaces where the emissivity becomes large at angles nearly tangent to the surface.

5-3.5 Effect of Surface Impurities

Impurities in this context include contaminants of any type that cause deviations of the surface properties from those of an optically smooth pure metal. The most common contaminants are thin layers of foreign materials deposited either by adsorption, such as in the case of water vapor, or by chemical reaction. The common example of the latter is the presence of a thin layer of an oxide on the metal. Because dielectrics, as will be discussed in Sec. 5-4, have generally high values of emissivity, an oxide or other nonmetallic contaminant layer will usually increase the emissivity of an otherwise ideal metallic body.

Figure 5-8 shows the directional spectral emissivity of titanium at an angle of 25° to the surface normal. The data points are for the unoxidized metal, and the solid line is the ideal emissivity predicted from electromagnetic theory. The dashed curve shown above the data points is the observed emissivity when an oxide layer only 0.06 μm in thickness is present. The emissivity is seen to be increased by a factor of almost 2 from that of the pure material over much of the wavelength range. Figure 5-9 shows a similar large increase in the normal spectral emissivity of Inconel X for an oxidized surface as compared with that for the polished metal.

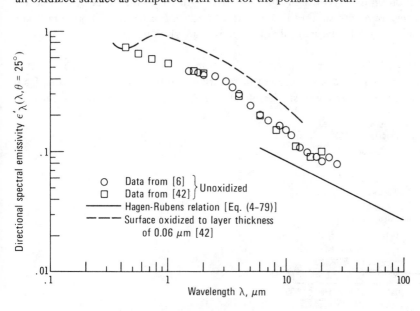

Figure 5-8 Effect of oxide layer on directional spectral emissivity of titanium. Emission angle, 25°; surface lapped to 2 μin (0.05 μm) rms; temperature 530°R (294 K).

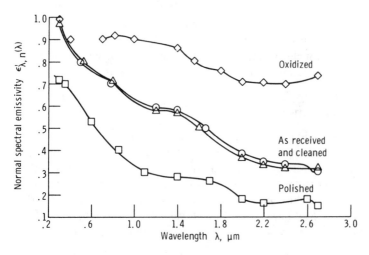

Figure 5-9 Effect of oxidation on normal spectral emissivity of Inconel X. *(Data from [5].)*

Figures 5-10 and 5-11a illustrate the effect of an oxide coating on the normal total emissivity of stainless steel and the hemispherical total emissivity of copper. The details of the oxide coatings are not specified, but the large effect of surface oxidation is apparent. More precise indications of oxide coating effect are shown in Figs. 5-11b and 5-12, where the normal total emissivity of copper and the hemispherical total emissivity of aluminum are given. An oxide thickness of even a few microns provides a very substantial emissivity increase. For oxidized aluminum some more recent results of a nature similar to Fig. 5-12 are given by Brannon and Goldstein [21].

Figure 5-13 shows approximately the directional total absorptivity of an anodized aluminum surface for radiation incident from various θ directions and originating from sources at various temperatures. The quantity $\rho_s'(\theta)$ is the fraction of the incident

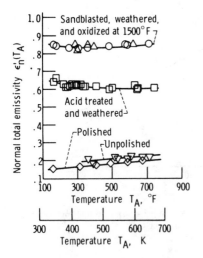

Figure 5-10 Effect of surface condition and oxidation on normal total emissivity of stainless steel type 18-8. *(Data from [5].)*

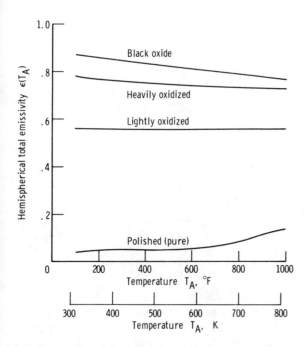

Figure 5-11 Effect of oxide coating on emissive properties of copper. (*a*) Effect of oxide coating on hemispherical total emissivity of copper. (*Data from [1].*)

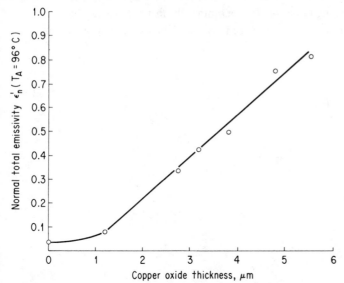

(*b*) Effect of oxide thickness on normal total emissivity of copper at 96°C. (*Data from [21].*)

energy that is reflected into the specular direction; hence, $1 - \rho'_s(\theta)$ is the fraction of the incident energy that is absorbed plus the fraction of the incident energy reflected into directions other than the specular direction. For the specimens tested, only a few percent of the energy was reflected into directions other than the specular direction.

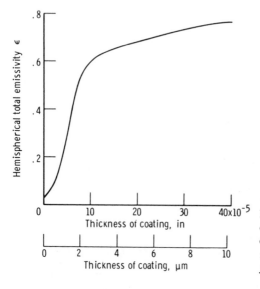

Figure 5-12 Typical curve illustrating effect of electrolytically produced oxide thickness on hemispherical total emissivity of aluminum. Temperature, 100°F (38°C). *(Data from [1].)*

Thus, in Fig. 5-13, the quantity $1 - \rho_s'(\theta)$ can be regarded as a good approximation to the directional total absorptivity. The curves have all been normalized to pass through unity at $\theta = 0$; hence, it is the shapes of the curves that are significant. At low source temperatures, the incident radiation is predominantly in the long-wavelength region. This incident radiation is barely influenced by the thin oxide film on the anodized

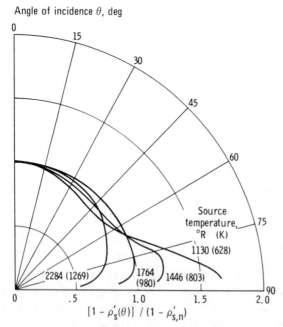

Figure 5-13 Approximate directional total absorptivity of anodized aluminum at room temperature relative to value for normal incidence. *(Redrawn from data of [43].)*

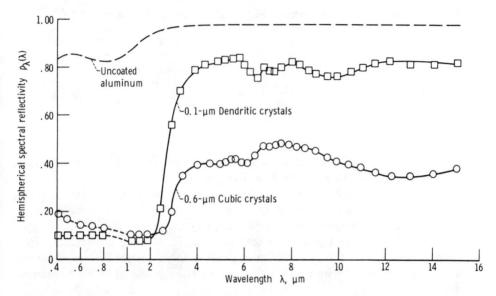

Figure 5-14 Hemispherical spectral reflectivity for normal incident beam on aluminum coated with lead sulfide. Coating mass per unit surface area, 0.68 mg/cm². *(Data from [44].)*

surface; consequently, the specimen acts like a bare metal and has large absorptivities at large angles from the normal. At high source temperatures where the incident radiation is predominantly at shorter wavelengths, the thin oxide film has a significant effect, and the surface behaves as a nonmetal where the absorptivity decreases with increasing θ.

The structure of the surface coating can also have a substantial effect on the radiative behavior. Figure 5-14 shows the hemispherical spectral reflectivity of aluminum coated with lead sulfide. The mass of the coating per unit area of surface is the same for both sets of data shown. The difference in crystal structure and size causes the reflectivity of the coated specimens to differ by a factor of 2 at wavelengths longer than about 3 μm.

5-4 RADIATIVE PROPERTIES OF OPAQUE NONMETALS

Roughly speaking, nonmetals are characterized by large values of total hemispherical emissivity and absorptivity at moderate temperatures and therefore generally small values of reflectivity in comparison with metals. For a clean, optically smooth surface, several results were arrived at in Chap. 4 by use of electromagnetic theory. These provide the following generalizations (bearing in mind the rather stringent assumptions of the theory): the directional emissivity will decrease with increasing angle from the surface normal; wavelength dependence is often weak as it enters the predicted properties through the refractive index, which varies slowly with wavelength for many nonmetals; finally, the temperature dependence of the properties of nonmetals will

Table 5-1 Normal total emissivity of nonmetals at room temperature (68°F)

Material	ϵ_n'	Material	ϵ_n'
Brick	0.94	Roofing paper	0.91
Lampsoot	0.95	Hard rubber	0.92
Oil paint	0.89–0.97	Wood	0.8–0.9

Source: Gubareff et al. [1].

also be small, since temperature also enters the prediction only through the refractive index, which is usually a weak function of temperature.

The difficulty with these generalizations is that most nonmetals cannot be polished to the degree necessary to allow their surfaces to be considered ideal, although some common exceptions exist, such as glass, large crystals of various types, gem stones, and some plastics (some of these are not opaque materials like those being discussed here). As a result of having such nonideal surface finishes, many nonmetals, in practice, deviate radically from the behavior predicted by electromagnetic theory.

Available property measurements for nonmetals are much less detailed than for metals. Specifications of the surface composition, texture, and so forth are often lacking. Table 5-1 illustrates this, as the type of wood, texture and color of the brick, and composition of the oil paint are unspecified. This table does reveal the large emissivity values typical of many nonmetal materials at room temperature (see also the table in Appendix D).

An effect that complicates the interpretation of the measured properties of nonmetals is that radiation passing into such a material may penetrate quite far (this is evident for visible wavelengths in glass) before being absorbed. To be opaque, a specimen must be of sufficient thickness to absorb essentially all the radiation that enters it; if it does not, transmitted radiation, which has not been included in the present discussion, must be accounted for. Often, samples of nonmetals such as paints are sprayed onto a metallic or other opaque base (substrate), and then the properties of the composite are measured. If, in such a case, it is desired to have the surface behave completely as the coating material, the thickness of the nonmetal coating must be sufficient to assure that no significant radiation is transmitted through the coating. Otherwise, in reflectivity measurements, some of the incident radiation can be reflected from the substrate and then transmitted again through the coating to reappear as energy measured by the instruments. The measured data will then be a function of both the coating material and the substrate.

In emissivity measurements of a coating material, the coating must be thick enough so that no emitted energy from the substrate penetrates the coating. A good illustration is given by Liebert [22]. He examined the spectral emissivity of zinc oxide on a variety of substrates, using various oxide thicknesses. The effect of coating thickness on the emissivity of the composite formed by the zinc oxide coating and a substrate of approximately constant normal spectral emissivity is shown in Fig. 5-15.

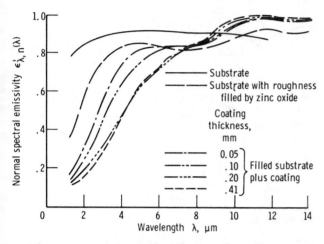

Figure 5-15 Emissivity of zinc oxide coatings on oxidized stainless steel substrate. Surface temperature, 880 ± 8 K. *(Data from [22].)*

The effect of increasing the coating thickness becomes small in the range of 0.2–0.4 mm thickness, indicating that the emissivity of the zinc oxide alone is being approached.

We shall now examine the effects of wavelength, temperature, and surface roughness on the radiative properties of dielectrics, and then briefly examine the radiative properties of semiconductors.

5-4.1 Spectral Measurements

There are fewer detailed spectral measurements for dielectrics than for metals. Figure 5-16 shows the hemispherical normal spectral reflectivity for three paint coatings on steel. From Kirchhoff's law and the reflectivity reciprocity relations, we can regard

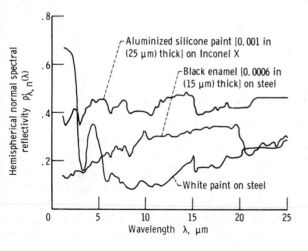

Figure 5-16 Spectral reflectivity of paint coatings. Specimens at room temperature. *(Data from [45].)*

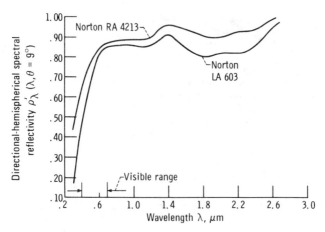

Figure 5-17 Directional-hemispherical spectral reflectivity of aluminum oxide. Incident angle, 9°; specimens at room temperature. *(Data from [5].)*

the difference between unity and these reflectivity values as the normal spectral emissivity. The three paints shown each exhibit somewhat different characteristics. White paint has a high reflectivity (low emissivity) at short wavelengths, and the reflectivity decreases at longer wavelengths. Black paint, on the other hand, has a relatively low reflectivity over the entire wavelength region shown. By using aluminum powder in a silicone base as a paint, the reflectivity is increased as would be expected for the more metallic coating. This particular specimen of aluminized paint acts approximately as a "gray" surface since the properties are reasonably independent of wavelength. Because of the large variation in spectral emissivity at short wavelengths, the gray approximation would be poor for the white paint unless very little of the participating radiation were at the shorter wavelengths.

Figure 5-17 illustrates that at the short wavelengths in the visible range the reflectivity for some nonmetals may decrease substantially. This behavior is very important in considering the suitability of a specific nonmetallic coating for reflecting radiation from a high-temperature source where much of the energy will be at short wavelengths.

5-4.2 Variation of Total Properties with Temperature

The effect of surface temperature on the total emissivity of several nonmetallic materials is shown in Figs. 5-18–5-20. Both increasing and decreasing emissivity trends with temperature are observed. Some of these effects may be caused by the fact that the dielectric coating is rather thin; hence, the properties are influenced by the temperature and spectral characteristics of the underlying material (substrate). For example, as shown in Fig. 5-18, magnesium oxide refractory exhibits a significant emissivity decrease with increasing temperature. For a silicon carbide coating on graphite, however, the emissivity increases with temperature; this may be partly caused

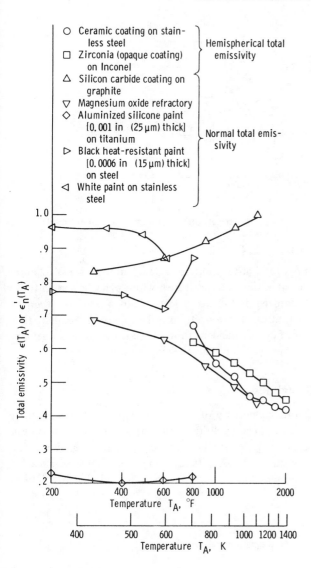

Figure 5-18 Effect of surface temperature on emissivity of dielectrics. *(Data from [1].)*

by the emissive behavior of the graphite substrate, which was shown in Fig. 5-4 to increase with temperature.

White and black paint both have high emissivities for the temperature range shown, as is typical for ordinary oil-base paint. Aluminized paint is considerably lower in emissive ability since it behaves partly like a metal. Note that the emissivity for aluminized paint in Fig. 5-18 is about one-half that in Fig. 5-16. This further emphasizes the wide variation in properties that can be found for samples having the same general description. For applications in which the property values are critical, it may often be necessary to make radiation measurements for the specific materials being used.

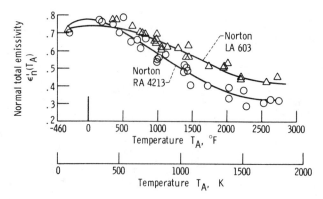

Figure 5-19 Effect of surface temperature on normal total emissivity of aluminum oxide. *(Data from [5].)*

Figure 5-21 gives the normal total absorptivity of a few materials for blackbody radiation incident from sources at various temperatures. White paper is shown to be a good absorber for radiation emitted at low temperatures but a poor absorber for the spectrum emitted at temperatures of several thousand degrees Fahrenheit. It is thus a reasonably good reflector of energy incident from the sun. An asphalt pavement or a gray slate roof, on the other hand, absorbs energy from the sun very well.

5-4.3 Effect of Surface Roughness

In Fig. 5-22 the bidirectional total reflectivity of typewriter paper is shown for three different angles of incidence. For an ideal (polished, smooth) surface, a specular peak would be expected with the angle of reflection and the angle of incidence symmetric about the normal; obviously, the surface finish of typewriter paper is not ideal, since the reflected intensity occupies a rather large angular envelope around the direction of

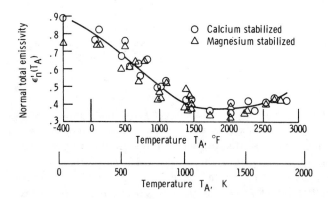

Figure 5-20 Effect of surface temperature on normal total emissivity of zirconium oxide. *(Data from [5].)*

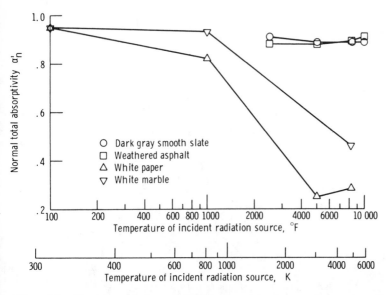

Figure 5-21 Normal total absorptivity of nonmetals at room temperature for incident black radiation from source at indicated temperatures. *(Data from [1].)*

specular reflection. Reference [15] provides detailed bidirectional reflectivity measurements for magnesium oxide ceramic with optical roughness σ_0/λ varying from 0.46 to 11.6. As the roughness and the incidence angle of the incoming radiation are increased, off-specular peaks are obtained as were discussed in connection with Fig. 5-7.

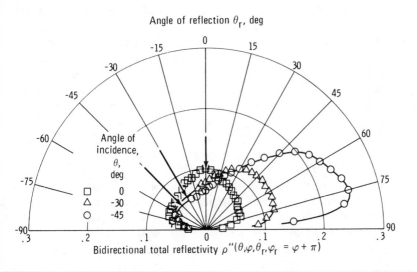

Figure 5-22 Bidirectional total reflectivity of typewriter paper in plane of incidence. Source temperature, 2120°R (1178 K). *(Replotted from data of [43].)*

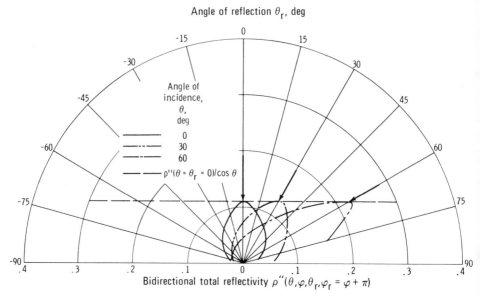

Figure 5-23 Bidirectional total reflectivity in plane of incidence $(\varphi_r = \varphi + \pi)$ for mountainous regions of lunar surface. *(After Orlova [46].)*

Dielectrics generally show a slight increase in emissivity with roughness. However, Cox [23] has shown that for materials with a ratio of cavity diameter to radiation mean penetration distance * in the dielectric of about 0.05, the emissivity may be *less* than for a smooth surface. This result was predicted by analysis and confirmed by experiment. Because of the high porosity of many dielectrics, it is difficult to study roughness parameters below a certain limiting smoothness.

The type of curves shown in Fig. 5-22 has suggested characterizing reflected energy as a combination of a purely diffuse plus a purely specular component. This type of approximation has merit in some cases and results in a simplification of radiant interchange calculations in comparison with the use of exact directional properties [24, 25]; in other cases, however, the approximation would fail completely. An example is shown in Fig. 5-23. This figure shows the observed bidirectional total reflectivity for visible light reflected from the surface of the moon. These particular curves are for the mountainous regions, but very similar curves are obtained for other areas. The interesting feature of these curves is that the peak of the reflected radiation is back into the direction of the incident radiation. This peak is located at a circumferential angle φ of 180° away from where a specular peak would occur.

A moment's thought will confirm that curves of this type must characterize lunar reflectivity. At full moon, which occurs when the sun, earth, and moon are almost (but not quite) in a straight line (Fig. 5-24), the moon appears equally bright across its face. For this to be true, it follows that an observer on earth sees equal intensities from all points on the moon. However, the solar energy incident upon a unit area of

*See Sec. 13-5.2 for definition.

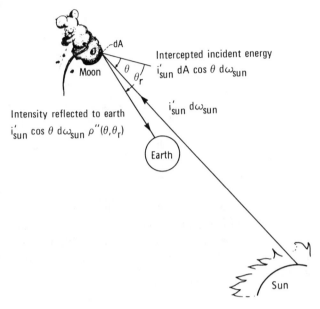

Moon

−dA

θ θ_r

Intercepted incident energy
$i'_{sun}\, dA \cos \theta\, d\omega_{sun}$

$i'_{sun}\, d\omega_{sun}$

Intensity reflected to earth
$i'_{sun} \cos \theta\, d\omega_{sun}\, \rho''(\theta,\theta_r)$

Earth

Sun

Figure 5-24 Reflected energy at full moon.

the lunar surface varies as the cosine of the angle θ between the sun and the normal to the lunar surface. The angle θ varies from 0 to 90°, as the position of the incident energy varies from the center to the edge of the lunar disk. To reflect a constant intensity to an observer on earth from all observable points on the lunar surface therefore requires that the product $\rho''(\theta,\theta_r) \cos \theta$ be constant. Consequently, the value of the bidirectional reflectivity in the direction of incidence must increase approximately in proportion to $1/\cos \theta$ (shown by the dashed line in Fig. 5-23) as the angle of incidence increases. This change in reflectivity with angle of incidence will compensate for the reduced energy incident per unit area on the moon at the large angles. The reflectivity behavior is confirmed by the curves in Fig. 5-23. Hence, the fact that the moon appears uniformly bright does not imply that it is a diffuse reflector. If the moon were diffuse, it would appear bright at the center and dark at the edges. The strong backscattering of the lunar surface causes the moon's brightness to peak strongly at full moon when the sun is almost directly behind the observer.

Further discussion of lunar infrared behavior can be found in [26–29].

5-4.4 Semiconductors

Semiconductors are arbitrarily considered here along with the nonmetals, but they behave partly as metals. Liebert [30] has shown that their radiative properties can be determined through electromagnetic theory by treating semiconductors as metals with high resistivity. In Fig. 5-25, the normal spectral emissivity of a silicon semiconductor is shown. The Hagen-Rubens relation shown for comparison is based on the dc resistivity measured for the same sample, one of the few cases where such comparable

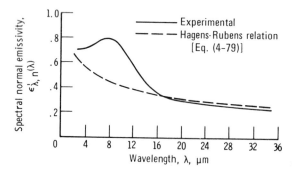

Figure 5-25 Normal spectral emissivity of a highly doped silicon semiconductor at room temperature. *(Data from [30].)*

emissive and electrical data are available. Agreement does not become good until wavelengths are reached that are much greater than those giving agreement for metals. This difference in range of agreement can be traced to the following assumption used in deriving the Hagen-Rubens equation (see Section 4-6.2, under Relation between Emissive and Electrical Properties):

$$\left(\frac{\lambda_0}{2\pi c_0 r_e \gamma}\right)^2 \gg 1$$

For semiconductors, in which the resistivity is larger than it is for metals, this inequality cannot hold until a range of wavelengths larger than those for metals is reached.

The shape of the curve measured for silicon (Fig. 5-25) resembles what would be expected for a polished metal (see, for example, the tungsten data in Fig. 5-3). The emissivity increases with decreasing wavelength over much of the measured spectrum, with a peak occurring at shorter wavelengths. However, most of the features of the semiconductor curve occur at longer wavelengths than for a metal; the peak emissivity, for example, is well outside the visible region.

Liebert [30] was also able to show excellent agreement between the measured emissivity and predictions from electromagnetic theory that included the effects of free electrons and was more sophisticated than that discussed in Chap. 4. The theoretical equations were evaluated by using required physical properties that were measured from the specific samples on which the emissivity measurements were made.

5-5 SPECIAL OPAQUE SURFACES AND SELECTIVE TRANSMISSION

For engineering purposes, it is often desirable to tailor the radiative properties of surfaces to increase or decrease their natural ability to absorb, emit, or reflect radiant energy. This can be done to provide two general types of behavior, a desired spectral performance or desired directional characteristics. A discussion is also given here on transmission through glass and water. These materials have wavelength-dependent behavior somewhat like the special surfaces being discussed.

5-5.1 Modification of Surface Spectral Characteristics

In applications of surfaces for use in the collection of radiant energy, such as in solar distillation units, solar furnaces, or solar collectors for energy conversion, it is desirable to maximize the energy absorbed by a surface while minimizing the amount lost by emission. In solar thermionic or thermoelectric devices, it is desirable to maintain the highest possible equlibrium temperature on the surface exposed to the sun. Here again a maximum-collection, minimum-loss performance is needed. Later in this section the condition will be discussed where it is desirable to keep a surface cool when it is exposed to the sun. In the latter case, it is desirable to have maximum solar reflection accompanied by maximum radiative emission from the surface.

For purposes of solar-energy collection, a black surface will of course maximize the absorption of incident solar energy; unhappily, it also maximizes the emissive losses. However, if a surface could be manufactured that had an absorptivity large in the spectral region of short wavelengths about the peak solar energy, yet small in the spectral region of longer wavelengths where the peak emission would occur, it might be possible to absorb nearly as well as a blackbody while emitting very little energy. Such surfaces are called "spectrally selective." One method of manufacture is to coat a thin, nonmetallic layer onto a metallic substrate. For radiation with large wavelengths, the thin coating is essentially transparent, and the surface behaves as a metal yielding low values for spectral absorptivity and emissivity. At short wavelengths, however, the radiation characteristics approach those of the nonmetallic coating so that the spectral emissivity and absorptivity are relatively large. Some examples of material behavior of this type are shown in Fig. 5-26.

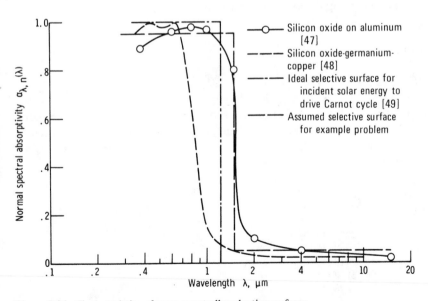

Figure 5-26 Characteristics of some spectrally selective surfaces.

An *ideal* solar selective surface would absorb a maximum of solar energy while emitting a minimum amount of energy. The surface would thus have an absorptivity of unity over the range of short wavelengths where the incident solar energy has a large intensity. At longer wavelengths, the absorptivity should drop sharply to zero. The wavelength λ_c at which this sharp drop occurs is termed the *cutoff wavelength*.

EXAMPLE 5-1 An *ideal* selective surface is exposed to a normally incident flux of radiation corresponding to the average solar constant q_i, where $q_i = 442$ Btu/ $(\mathrm{h \cdot ft^2})$.* The only means of heat transfer to or from the surface is by radiation. Determine the maximum equilibrium temperature T_{eq} corresponding to a cutoff wavelength of $\lambda_c = 1$ μm. (The energy arriving from the sun can be assumed to have a spectral distribution proportional to that of a blackbody at 10,400°R.)

Since the only means of heat transfer is by radiation, the radiant energy absorbed must be equal to that emitted. Since we have specified an ideal selective absorber, the hemispherical emissivity and absorptivity are given by

$$\epsilon_\lambda(\lambda) = \alpha_\lambda(\lambda) = 1 \qquad 0 \leqslant \lambda < \lambda_c$$

and

$$\epsilon_\lambda(\lambda) = \alpha_\lambda(\lambda) = 0 \qquad \lambda_c \leqslant \lambda < \infty$$

The energy absorbed by the surface per unit time is

$$Q_a = (1) F_{0-\lambda_c}(T_R) q_i A$$

where $F_{0-\lambda_c}(T_R)$ is the fraction of blackbody energy in the range of wavelengths between zero and the cutoff value, for a radiating source at temperature T_R. In this case, T_R is the effective solar radiating temperature of 10,400°R. Similarly, the energy emitted by the selective surface is

$$Q_e = (1) F_{0-\lambda_c}(T_{eq}) \sigma T_{eq}^4 A$$

Equating Q_e and Q_a, we obtain

$$T_{eq}^4 F_{0-\lambda_c}(T_{eq}) = \frac{q_i F_{0-\lambda_c}(T_R)}{\sigma}$$

For the chosen value for λ_c, all terms on the right are known, and we can solve for T_{eq} by trial and error. The equilibrium temperature for $\lambda_c = 1$ μm, as specified in the problem, is 2410°R. Values of T_{eq} corresponding to other values of λ_c are given in the following table:

Cutoff wavelength λ_c, μm	Equilibrium temperature, T_{eq}, °R (K)
0.6	3270 (1820)
0.8	2760 (1530)
1.0	2410 (1340)
1.2	2150 (1190)
1.5	1890 (1050)
$\rightarrow \infty$	713 (396)

*Recent measurements indicate that q_i may be about 3% lower, 429 Btu/$(\mathrm{h \cdot ft^2})$.

For a blackbody ($\lambda_c \rightarrow \infty$), the equilibrium temperature is 713°R (396 K); this is the equilibrium temperature of the surface of a black object in space near the earth's orbit when exposed to solar radiation and with all other surfaces of the object perfectly insulated. The same equilibrium temperature is reached by a gray body, since a gray emissivity would cancel out of the energy-balance equation.

As smaller values of λ_c are taken, T_{eq} continues to increase even though less energy is absorbed, because it becomes relatively more difficult to emit energy as λ_c is decreased.

A common measure of the performance of a given selective surface is the ratio of the directional total absorptivity $\alpha'(\theta, \varphi, T_A)$ of the surface for incident solar energy to the hemispherical total emissivity of the surface $\epsilon(T_A)$. The ratio α'/ϵ for the condition of incident solar energy is a measure of the theoretical maximum temperature that an otherwise insulated surface can attain when exposed to solar radiation. The significance of α'/ϵ is shown as follows.

The energy absorbed per unit time by any surface when exposed to a directional incident energy is given by

$$dQ_a'(\theta, \varphi, T_A) = \alpha'(\theta, \varphi, T_A) \, dQ_i'(\theta, \varphi) \tag{5-1}$$

For the case of solar energy with a flux of $q_i = 442$ Btu/(h·ft²) incident from the direction (θ, φ) on a surface element dA, this can be written as

$$dQ_a'(\theta, \varphi, T_A) = \alpha'(\theta, \varphi, T_A) \, q_i \, dA \, \cos\theta \tag{5-2}$$

The total energy emitted per unit time by the surface element is given by

$$dQ_e = e(T_A) \, dA = \epsilon(T_A) \, \sigma T_A^4 \, dA \tag{5-3}$$

If the only energy absorbed by the surface in question is that given by (5-2) and the surface only loses energy by radiation, then the emitted and absorbed energies as given by (5-3) and (5-2), respectively, may be equated to give

$$\frac{\alpha'(\theta, \varphi, T_{eq})}{\epsilon(T_{eq})} = \frac{\sigma T_{eq}^4}{q_i \cos\theta} \tag{5-4}$$

where T_{eq} is the equilibrium temperature that is achieved. Thus the ratio $\alpha'(\theta, \varphi, T_A)/\epsilon(T_A)$ is a measure of the equilibrium temperature of the element. Note also that the temperature at which the properties α' and ϵ are selected must be the equilibrium temperature that the body attains. In practice, temperature dependence of the properties is often assumed small so that this restriction can be somewhat relaxed.

The most common case considered is when the solar radiation is incident in the direction normal to the surface. Equation (5-4) becomes

$$\frac{\alpha_n'(T_{eq})}{\epsilon(T_{eq})} = \frac{\sigma T_{eq}^4}{q_i} \tag{5-5}$$

Equation (5-5) shows that the smaller the value of α_n'/ϵ that can be reached, the smaller will be the equilibrium temperature. For a cryogenic storage tank in space, α_n'/ϵ should be as small as possible. In practice, values of α_n'/ϵ in the range 0.20–0.25 can be obtained.

To attain high equilibrium temperatures, α_n'/ϵ should be as large as possible. Polished metals attain α_n'/ϵ values of 5–7, while specially manufactured surfaces have values of α_n'/ϵ approaching 20. Coatings with $\alpha_n'/\epsilon \approx 13$ and stability at temperatures up to about 900 K in air, were reported in [54].

The upper limit of α_n'/ϵ is established by the thermodynamic argument that the equilibrium temperature of the selective surface cannot exceed the effective solar temperature of 10,400°R. Substituting this solar temperature value into (5-5) gives

$$\frac{\alpha_n'(T_{eq})}{\epsilon(T_{eq})}\bigg|_{max} = \frac{\sigma(10,400)^4}{442} = 4.53 \times 10^4 \tag{5-6}$$

Attaining anything even close to this value of α_n'/ϵ is far beyond the present state of the art.

EXAMPLE 5-2 The properties of a real SiO–Al selective surface can be approximated by the long-dashed curve in Fig. 5-26 (it is assumed that the long-dashed curve can be extrapolated toward $\lambda = 0$ and $\lambda = \infty$). What is the equilibrium temperature of the surface for normally incident solar radiation when the only heat transfer is by radiation? What is α_n'/ϵ for the surface? Describe the spectra of the absorbed and emitted energy at the surface. (Assume normal and hemispherical emissivities are equal.)

As in the derivation of Eq. (5-5), we equate the absorbed and emitted energies. The emissivity has nonzero values on both sides of the cutoff wavelength, so that

$$Q_a = \epsilon_{0-\lambda_c} F_{0-\lambda_c}(T_R) q_i A + \epsilon_{\lambda_c-\infty} F_{\lambda_c-\infty}(T_R) q_i A = \alpha_n' q_i A$$

and

$$Q_e = \epsilon_{0-\lambda_c} F_{0-\lambda_c}(T_{eq}) \sigma T_{eq}^4 A + \epsilon_{\lambda_c-\infty} F_{\lambda_c-\infty}(T_{eq}) \sigma T_{eq}^4 A = \epsilon \sigma T_{eq}^4 A$$

Equating Q_e and Q_a gives

$$\{0.95 F_{0-\lambda_c}(T_R) + 0.05[1 - F_{0-\lambda_c}(T_R)]\} q_i$$
$$= \{0.95 F_{0-\lambda_c}(T_{eq}) + 0.05[1 - F_{0-\lambda_c}(T_{eq})]\} \sigma T_{eq}^4$$

Solving by trial and error as in Example 5-1, we obtain $\lambda_c = 1.5$ μm, $T_{eq} = 1430°R$. For $q_i = 442$ Btu/(h·ft²), Eq. (5-5) gives $\alpha_n'/\epsilon = \sigma(1430)^4/442 = 16.2$. The small difference in properties in this example from the properties of an ideal selective surface produces a significant change in T_{eq}, which in the previous example was 1890°R for an ideal selective surface having the same cutoff wavelength. The spectral curves of absorbed and emitted energy are shown in Fig. 5-27. The spectral curve of incident solar energy is given by

$$\epsilon_{\lambda, i}(\lambda, T_R) \propto e_{\lambda b}(\lambda, T_R)$$

It has the shape of the blackbody curve at the solar temperature, but it is reduced in magnitude so that the integral of $e_{\lambda, i}$ over all λ is equal to q_i, the total incident solar energy per unit area. Multiplying this curve by the spectral absorptivity of the selective surface gives the spectrum of the absorbed energy. The spectrum of emitted energy is that of a blackbody at 1430°R multiplied by the spectral emis-

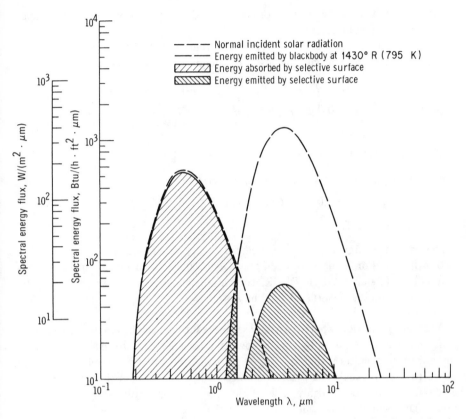

Figure 5-27 Spectral distribution of energy absorbed and emitted by example selective absorber.

sivity of the selective surface. The integrated energy under the spectral curves of absorbed and emitted energy are equal, although this is not obvious from the log-log plot.

It is noted that the energy equation solved in Example 5-2 is a two spectral band approximation to the following more general energy equation for a diffuse surface with a diffuse incident flux:

$$\int_{\lambda=0}^{\infty} \alpha_\lambda(\lambda, T_{eq}) \, dq_{\lambda,i}(\lambda) = \int_{\lambda=0}^{\infty} \epsilon_\lambda(\lambda, T_{eq}) \, e_{\lambda b}(\lambda, T_{eq}) \, d\lambda \qquad (5\text{-}7)$$

The $dq_{\lambda,i}$ can have any spectral distribution, and by Kirchhoff's law, $\alpha_\lambda(\lambda, T_{eq}) = \epsilon_\lambda(\lambda, T_{eq})$.

EXAMPLE 5-3 A selective surface having spectral characteristics as given in the previous example is to be used as a solar-energy absorber. The surface is to be maintained at a temperature of $T_A = 713°R$ by extracting energy to be used in a power generating cycle. If the absorber is placed in orbit around the sun at the same radius as the earth, how much net energy will a square foot of the surface provide? How does this energy compare with that supplied by a black surface at the same temperature?

The net energy extracted from the surface is the difference between that absorbed and that emitted. The absorbed energy flux is, as in Example 5-2,

$$q_a = \{0.95 F_{0-\lambda_c}(T_R) + 0.05[1 - F_{0-\lambda_c}(T_R)]\} q_i$$
$$= [0.95(0.880) + 0.05(1 - 0.880)] 442$$
$$= 372 \text{ Btu/(h·ft}^2)$$

The emitted flux is

$$q_e = \{0.95 F_{0-\lambda_c}(T_A) + 0.05[1 - F_{0-\lambda_c}(T_A)]\} \sigma T_A^4$$
$$= \{0.95 \times (\sim0) + 0.05[1 - (\sim0)]\} 0.1712 \times 10^{-8} \times (713)^4$$
$$= 22.1 \text{ Btu/(h·ft}^2)$$

and the net energy that can be used for power generation is $372 - 22 = 350$ Btu/(h·ft^2). For a blackbody or gray body, the equilibrium temperature was found in Example 5-1 as 713°R, so that the net useful energy that could be removed from such a surface would be zero.

Another application where spectrally selective surfaces can be employed to advantage is where it is desirable to cool an object that is exposed to incident radiation from a high-temperature source. The most common situation would be objects exposed to the sun, such as a gasoline storage tank, a cryogenic fuel tank in space, or the roof of a building. A highly reflecting coating such as a polished metal could be utilized. This would reflect much of the incident energy, but would be poor for radiating away any energy that was absorbed or generated within the enclosure (for example, an enclosure filled with electronic equipment). Also, some metals have a tendency toward lower reflectivity at the shorter wavelengths; this is shown, for example, for uncoated aluminum in Fig. 5-14. For some applications it may be advantageous to use a material that is spectrally selective; white paint as shown in Fig. 5-28 is an example. This will not only reflect the incident radiation predominant at short wavelengths but will also radiate well at the longer wavelengths characteristic of the relatively low temperature of the body.

5-5.2 Selective Transmission through Glass and Water

The previous section discussed wavelength-selective opaque surfaces. It is also of interest to point out the selective transmission and absorption behavior of two common nonopaque materials, glass and water.

Figure 5-29 shows the overall spectral transmittance of glass plates for normally incident radiation. As shown in Sec. 19-3.1 the overall transmittance includes the effect of multiple surface reflections and absorption within the glass; it is given by (19-2) as

$$T_\lambda = \frac{\tau_\lambda (1 - \rho_\lambda)^2}{1 - \rho_\lambda^2 \tau_\lambda^2}$$

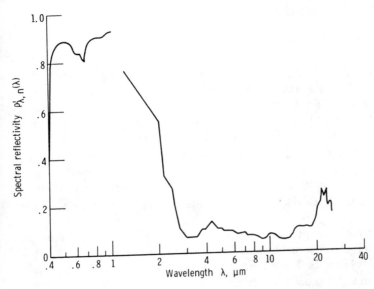

Figure 5-28 Reflectivity of white paint coating on aluminum. *(Replotted from [50].)*

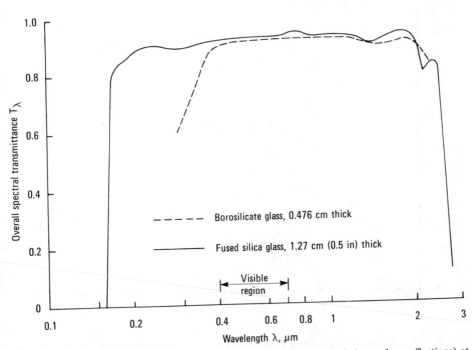

Figure 5-29 Normal overall spectral transmittance of glass plate (includes surface reflections) at 298 K. *(Replotted from [7].)*

where

$$\tau_\lambda = \exp\left(-a_\lambda d\right) \qquad \text{and} \qquad \rho_\lambda = \left(\frac{n-1}{n+1}\right)^2$$

For small $a_\lambda d$ this reduces to $T_\lambda = (1 - \rho_\lambda)/(1 + \rho_\lambda)$. Typically for glass n is about 1.5 so that $\rho_\lambda = (0.5/2.5)^2 = 0.04$. Then including only reflection losses gives $T_\lambda = (1 - 0.04)/(1 + 0.04) = 0.92$. In Fig. 5-29 the fused silica has very low absorption in the range $\lambda = 0.2$-2 μm, and $T_\lambda \approx 0.9$ in this region as a result of surface reflections. Ordinary glasses typically have two strong cutoff wavelengths beyond which the glass becomes highly absorbing and T_λ decreases rapidly to near zero except for very thin plates. The measured curve for fused silica in Fig. 5-29 shows this clearly. There is a strong cutoff in the far ultraviolet at $\lambda \approx 0.17$ μm and in the near infrared at $\lambda \approx 2.5$ μm. The glass will therefore be a strong absorber or emitter for $\lambda < 0.17$ μm and $\lambda > 2.5$ μm.

Figure 5-30 shows the transmittance for various thicknesses of soda-lime glass which is more absorbing than fused silica. The effect of absorption is illustrated quite well as the thickness increases. Typical optical constants for glass are given in [31].

For windows in high-temperature devices, such as furnaces or solar cavity receivers,

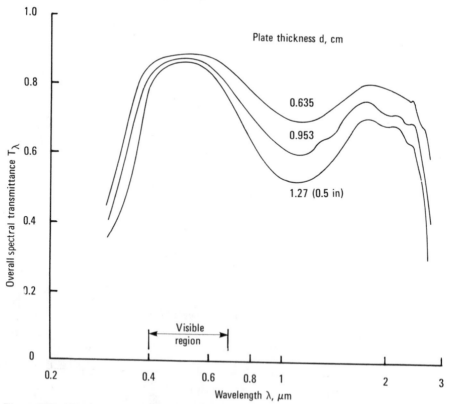

Figure 5-30 Effect of plate thickness on normal overall spectral transmittance of soda-lime glass (includes surface reflections) at 298 K. *(Replotted from [7].)*

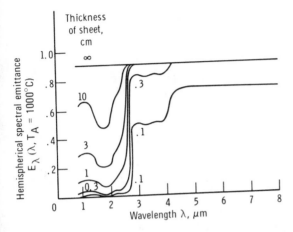

Figure 5-31 Emittance of sheets of window glass at 1000°C. *(Data from [51].)*

emission from the windows can be significant. From Kirchhoff's law the overall spectral emittance (including the effect of surface reflections) is equal to the spectral absorptance. Hence, from (19-4),

$$E_\lambda = \frac{(1-\rho_\lambda)(1-\tau_\lambda)}{(1-\rho_\lambda\tau_\lambda)}$$

For a thick plate, beyond the cutoff wavelength, $\tau_\lambda \to 0$ and $E_\lambda \approx 1 - \rho_\lambda$. In this instance the reflection from only one surface is significant as all the radiation is absorbed before it can be transmitted through to the second surface of the plate. If the n for glass is 1.5, then $\rho_\lambda = 0.04$ for incidence from the normal direction; hence $E_\lambda = 0.96$ in the normal direction for the highly absorbing wavelength regions of the glass. In a fashion similar to Example 4-3, the hemispherical value is found as $E_\lambda = 0.90$. This is the upper value in Fig. 5-31, which shows the emittance of window-glass sheets of various thicknesses.

The transmission behavior of glass as shown in Figs 5-29 and 5-30 provides glass windows with the important characteristic ability to trap solar energy. The sun radiates a spectral energy distribution very much like a blackbody at 5780 K (10,400°R). Considering the range $0.3 < \lambda < 2.7$ μm as being between the cutoff wavelengths in Fig. 5-30, the blackbody tables in Appendix A show that 95% of the solar energy will be in this range. This means that the glass has a low absorptance for solar radiation, which is primarily at short wavelengths, and consequently incident solar radiation passes readily through a glass window. The emission from objects at ambient temperature inside the enclosure is at long wavelengths and is trapped because of the high absorptance (poor transmission) of the glass in this long-wavelength spectral region ("greenhouse" effect).

An interesting application of a selectively transmitting layer is the transparent heat mirror. As mentioned in [32] the transparent heat mirror has been used to construct transparent metallurgical furnaces for observing the growth of crystals at temperatures up to 1000°C. The thermal insulation for this furnace is provided by a gold film about 0.02 μm thick deposited on the inside of a Pyrex tube that encloses the heated region. These films have a high reflectance in the infrared and are equivalent

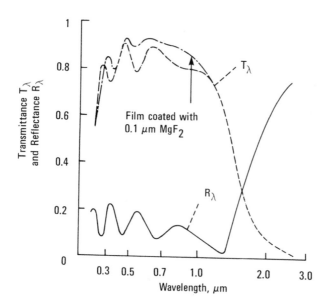

Figure 5-32 Transmittance and reflectance of 0.35-μm-thick film of Sn-doped In_2O_3 film on Corning 7059 glass. Also shown is the effect on T_λ of an antireflection coating of MgF_2. *(From [32].)*

to several inches of asbestos insulation in preventing radiation heat loss. The films, however, have a transmittance of about 0.2 in the visible region, which is adequate for high-temperature furnace observation.

The selectively transparent coating may also be useful in the collection of solar energy. It can allow the short-wavelength solar energy to pass into a solar collector, and prevent the escape of long-wavelength radiation reemitted by the receiver. Some possible coating materials are indium trioxide (In_2O_3), magnesium oxide (MgO), tin dioxide (SnO_2), and zinc oxide. Thin films of these materials are transparent in the solar part of the spectrum, and have a rapid increase of reflectance in the infrared. The measured transmittance and reflectance of a 0.35-μm-thick layer of Sn-doped In_2O_3 deposited on Corning 7059 glass is shown in Fig. 5-32. The application to a solar-energy receiver is discussed in [33].

Water is also a selective absorber for the transmission of radiation. Table 5-2 gives the spectral absorption coefficient of water as taken from [34] (see [35] and [36] for additional information). The absorption coefficient determines the exponential attenuation of the intensity as given by Eq. (13-21). In the visible region ($\lambda = 0.4$–0.7 μm) the absorption coefficient is quite small. In the vicinity of 1 μm, a_λ begins to increase, and at longer wavelengths in the near infrared region the absorption becomes quite large. In the visible region it is noted that a_λ is especially low in the blue-green region (0.5–0.55 μm). This accounts for the green appearance of sunlight penetrating underwater to large depths.

Application of the values of the absorption coefficient in Table 5-2 to the solar spectrum results in the energy penetrations given in Table 5-3 (from [37]). The second column shows the portions of the solar energy spectrum that are in various wavelength intervals. The successive columns demonstrate the high transparency for visible radi-

Table 5-2 Absorption coefficient of water (from [34])

λ, μm	a_λ, cm^{-1}	λ, μm	a_λ, cm^{-1}
0.20	0.0691	2.4	50.1
0.25	0.0168	2.6	153
0.30	0.0067	2.8	5160
0.35	0.0023	3.0	11,400
0.40	0.00058	3.2	3630
0.45	0.00029	3.4	721
0.50	0.00025	3.6	180
0.55	0.000045	3.8	112
0.60	0.0023	4.0	145
0.65	0.0032	4.2	206
0.70	0.0060	4.4	294
0.75	0.0261	4.6	402
0.80	0.0196	4.8	393
0.85	0.0433	5.0	312
0.90	0.0679	5.5	265
0.95	0.388	6.0	2240
1.0	0.363	6.5	758
1.2	1.04	7.0	574
1.4	12.4	7.5	546
1.6	6.72	8.0	539
1.8	8.03	8.5	543
2.0	69.1	9.0	557
2.2	16.5	9.5	587
		10.0	638

Table 5-3 Fractions of solar radiation spectrum transmitted through various thicknesses of water [37]

Spectral interval λ, μm	Incident solar-energy distribution	Transmitted energy distribution for water-layer thickness, cm							
		0.001	0.01	0.1	1	10	100	1000	10000
0.3–0.6	0.237	0.237	0.237	0.237	0.237	0.236	0.229	0.173	0.014
0.6–0.9	0.360	0.360	0.360	0.359	0.353	0.305	0.129	0.010	
0.9–1.2	0.179	0.179	0.178	0.172	0.123	0.008			
1.2–1.5	0.087	0.086	0.082	0.063	0.017				
1.5–1.8	0.080	0.078	0.064	0.027					
1.8–2.1	0.025	0.023	0.011						
2.1–2.4	0.025	0.025	0.019	0.001					
2.4–2.7	0.007	0.006	0.002						
Totals	1.000	0.994	0.953	0.859	0.730	0.549	0.358	0.183	0.014

ation as compared with very strong energy absorption in most of the near infrared region.

Ice also has a low absorption coefficient in the visible range, and the absorption coefficient increases by a factor of the order of 10^3 as the radiation wavelength increases from about 0.55 to 1.2 μm. Radiation in the visible and near visible range can therefore be passed through ice that is not cloudy due to impurities or air bubbles. If an ice layer is stuck to a surface, it may be possible to pass visible and near infrared radiation through the ice, thereby heating the surface and providing a means for ice removal [38].

5-5.3 Modification of Surface Directional Characteristics

As discussed in previous sections of this chapter, the roughness of a surface can have profound effects on the radiative properties and will indeed become a controlling factor when the roughness is large in comparison with the wavelength of the energy being considered. This leads to the concept of controlling the roughness to tailor the directional characteristics of a surface.

If the surface is used as an emitter, the surface might be roughened or designed in such a way as to emit strongly in preferred directions, while reducing emission in unwanted directions. Commercial radiant area-heating equipment would operate more efficiently by using such surfaces to direct energy to where it is most needed. The most common device for controlling the directional distribution of electromagnetic radiation in the visible region is called a *lamp shade*.

If the directional surface were primarily used as an absorber, then, using a solar absorber as an example, we might make it strongly absorbing in the direction of incident solar radiation but as close as possible to nonabsorbing in other directions. The surface would, because of Kirchhoff's law for directional properties, emit strongly toward the sun, but weakly in other directions. The surface would absorb the same energy as a nondirectional absorber since the incident energy is only from the direction of the sun but would emit less energy than a surface that emits well into all directions.

The characteristics of one such surface are shown in Fig. 5-33. The surface has very long grooves of angle 18.2° running parallel to each other. A highly reflecting specular coating is placed on the side walls of each groove, and a black surface is placed at the base of each groove. The solid line gives the behavior predicted by analysis of such an ideal surface, while the data points show experimental results at 8 μm for an actual surface. It is seen that the directional emissivity is very high for angles of emission less than about 30° from the surface normal. It then drops rapidly as the angle becomes larger. Many other such surface configurations exhibit similar characteristics.

EXAMPLE 5-4 Suppose that a directional surface has a directional total emissivity given for all φ by

$\epsilon(\theta) = 1 \qquad 0 \leqslant \theta \leqslant 30°$

$\epsilon(\theta) = 0 \qquad \theta > 30°$

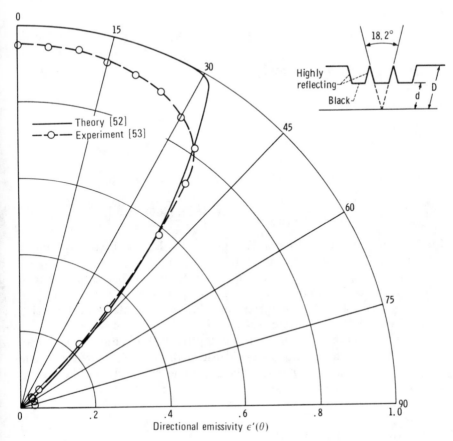

Figure 5-33 Directional emissivity of grooved surface with highly reflecting specular side walls and highly absorbing base; $d/D = 0.649$. Results in plane perpendicular to groove direction; data at 8 μm.

For solar radiation incident normally on such a surface in earth orbit with no other heat exchange except by radiation from the directional surface, what is the equilibrium temperature of the surface? How does this temperature compare with that achieved by a black surface?

The absorptivity of this surface for normal incident radiation is unity. Therefore, the absorbed energy per unit time is

$$Q_a = (1)q_i A$$

where q_i is the solar constant for an object at a distance equal to the mean radius of the earth's orbit from the sun ($q_i = 1394$ W/m²).*

*Recent reevaluations have indicated that the solar constant may be a few percent lower, 1353 ± 21 W/m² [429 Btu/(h·ft²)].

The energy emitted by the body when it is at thermal equilibrium is

$$Q_e = \epsilon \sigma T_{eq}^4 A$$

where ϵ is the hemispherical total emissivity given by Eq. (3-6b) as

$$\epsilon(T_{eq}) = \frac{1}{\pi} \int_\Omega \epsilon'(\theta,\varphi,T_{eq}) \cos \theta \, d\omega$$

For this problem ϵ becomes

$$\epsilon = \frac{2\pi}{\pi} \int_{\theta=0}^{30°} \sin \theta \cos \theta \, d\theta = 0.25$$

Equating Q_a and Q_e for radiative equilibrium gives

$$T_{eq} = \left(\frac{q_i}{\epsilon \sigma}\right)^{1/4} = \left(\frac{1394}{0.25 \times 5.670 \times 10^{-8}}\right)^{1/4} = 560 \text{ K } (1008°\text{R})$$

This is larger than the equilibrium temperature of a black or diffuse gray body of 396 K (713°R) as shown in Example 5-1.

Note that Eq. (5-5) can be used for the α_n'/ϵ of directional as well as spectrally selective surfaces. For the surface used in this example, $\alpha_n'/\epsilon = 4.0$. Combining selective and directional effects would be a way of obtaining considerably increased values of α_n'/ϵ for a given surface.

It should not be inferred that the directional distribution of emissivity assumed in this example corresponds to that of the parallel grooved surface in Fig. 5.33. In the case of Fig. 5-33, there is a strong dependence on the angle φ, which has been ignored in this example.

5-6 CONCLUDING REMARKS

The radiative-property examples discussed in this chapter have illustrated a number of the features that may be encountered when dealing with real surfaces. Certain broad generalizations could be attempted. For example, the total emissivities of dielectrics at moderate temperatures are larger than those for metals, and the spectral emissivity of metals increases with temperature over a broad range of wavelengths. However, these types of rules can be misleading because of the large property variations that can occur as a result of surface roughness, contamination, oxide coating, grain structure, and so forth. The presently available analytical procedures cannot account for all these factors so that it is not possible directly to predict radiative property values except for surfaces that approach ideal conditions of composition and finish. By coupling analytical trends with observations of experimental trends, it is possible to gain some insight into what classes of surfaces would be expected to be suitable for specific applications and how surfaces may be fabricated to obtain certain types of radiative behavior. The latter include spectrally selective surfaces that are of great value in a number of practical applications, such as the collection of solar energy.

Some other factors affecting radiative properties that evade prediction are outside the range of interest in this work, but they should be mentioned. For example, it is well known that exposure to ultraviolet radiation, cosmic rays, neutron, gamma and proton bombardment, and the solar wind can all cause significant changes in radiative properties. For the design of spacecraft, these effects are of major concern.

Finally, some comment on the measurement of radiative properties should be given. We note that relatively few precise measurements of directional spectral properties have been made. The reason for this lies in one of the many practical difficulties involved: For a directional measurement, the energy available for detection at a small solid angle centered about a given direction is itself small. If only the portion of this small energy that lies within a wavelength band is then measured to obtain directional spectral values, an even smaller energy is available for detection. Minor absolute errors in the measurement of the energy can then lead to large percentage errors in the directional properties being determined. Further, the sheer magnitude of the amount of data generated for such combined directional spectral properties precludes its gathering unless a very specific problem requires it. These and similar practical problems make the field of thermal-radiation property measurement a most exacting and difficult one.

REFERENCES

1. Gubareff, G. G., J. E. Janssen, and R. H. Torberg: "Thermal Radiation Properties Survey," 2d ed., Honeywell Research Center, Minneapolis, 1960.
2. Svet, Darii Ia.: "Thermal Radiation; Metals, Semiconductors, Ceramics, Partly Transparent Bodies, and Films," Consultants Bureau, Plenum Publishing Corporation, New York, 1965.
3. Goldsmith, Alexander, and Thomas E. Waterman: Thermophysical Properties of Solid Materials, WADC TR 58-476, Armour Research Foundation, January, 1959.
4. Hottel, H. C.: Radiant Heat Transmission, in William H. McAdams (ed.), "Heat Transmission," 3d ed., pp. 472-479, McGraw-Hill Book Company, New York, 1954.
5. Wood, W. D., H. W. Deem, and C. F. Lucks: "Thermal Radiative Properties," Plenum Press, Plenum Publishing Corporation, New York, 1964.
6. Edwards, D. K., and Ivan Catton: Radiation Characteristics of Rough and Oxidized Metals, *Adv. Thermophys. Properties Extreme Temp. Pressures* (Serge Gratch, ed.), ASME, 1965, pp. 189-199.
7. Touloukian, Y. S., et al.: *Thermal Radiative Properties*, vol. 7, "Metallic Elements and Alloys," vol. 8, "Nonmetallic Solids," vol. 9, "Coatings," Thermophysical Properties Research Center of Purdue University, Data Series, Plenum Publishing Corporation, New York, 1970.
8. Sadykov, B. S.: Temperature Dependence of the Radiating Power of Metals, *High Temp.*, vol. 3, no. 3, pp. 352-356, 1965.
9. Toscano, W. M., and E. G. Cravalho: Thermal Radiative Properties of the Noble Metals at Cryogenic Temperatures, *Trans. ASME, J. Heat Transfer*, vol. 98, no. 3, pp. 438-445, 1976.
10. Davies, H.: The Reflection of Electromagnetic Waves from a Rough Surface, *Proc. Inst. Elec. Eng. London*, vol. 101, pp. 209-214, 1954.
11. Porteus, J. O.: Relation between the Height Distribution of a Rough Surface and the Reflectance at Normal Incidence, *J. Opt. Soc. Am.*, vol. 53, no. 12, pp. 1394-1402, 1963.
12. Beckmann, P., and A. Spizzichino: "The Scattering of Electromagnetic Waves from Rough Surfaces," The Macmillan Company, New York, 1963.
13. Houchens, A. F., and R. G. Hering: Bidirectional Reflectance of Rough Metal Surfaces, *Progr. Astronautics and Aeronautics: Thermophys. Spacecraft Planetary Bodies*, vol. 20, pp. 65-89, 1967.

14. Smith, T. F., and R. G. Hering: Comparison of Bidirectional Measurements and Model for Rough Metallic Surfaces, *Fifth Symp. Thermophys. Properties ASME,* Boston, 1970.

15. Torrance, K. E., and E. M. Sparrow: Off-Specular Peaks in the Directional Distribution of Reflected Thermal Radiation, *J. Heat Transfer,* vol. 88, no. 2, pp. 223-230, 1966.

16. Torrance, K. E., and E. M. Sparrow: Theory for Off-Specular Reflection from Roughened Surfaces, *J. Opt. Soc. Am.,* vol. 57, no. 9, pp. 1105-1114, 1967.

17. Ody-Sacadura, J. F.: Influence de la rugosité sur le rayonnement thermique émis par les surfaces opaques: essai de modèle (Influence of Surface Roughness on the Radiative Heat Emitted by Opaque Surfaces: A Test Model), *Int. J. Heat Mass Transfer,* vol. 15, no. 8, pp. 1451-1465, 1972.

18. Kanayama, K.: Apparent Directional Emittance of V-Groove and Circular-Groove Rough Surfaces, *Heat Transfer Jpn. Res.,* vol. 1, no. 1, pp. 11-22, 1972.

19. Birkebak, R. C., and A. Abdulkadir: Random Rough Surface Model for Spectral Directional Emittance of Rough Metal Surfaces, *Int. J. Heat Mass Transfer*, vol. 19, no. 9, pp. 1039-1043, 1976.

20. Abdulkadir, A., and R. C. Birkebak: Spectral Directional Emittance of Rough Metal Surfaces: Comparison Between Semi-Random and Pyramidal Surface Approximations. AIAA paper 78-848, presented at 2d AIAA/ASME Thermophys. Heat Transfer Conf., Palo Alto, 1978.

21. Brannon, R. R., Jr., and R. J. Goldstein: Emittance of Oxide Layers on a Metal Substrate, *J. Heat Transfer,* vol. 92, no. 2, pp. 257-263, 1970.

22. Liebert, Curt H.: Spectral Emittance of Aluminum Oxide and Zinc Oxide on Opaque Substrates, *NASA TN D-3115,* 1965.

23. Cox, R. L.: Radiant Emission from Cavities in Scattering and Absorbing Media, *SMU Research Rep.* 68-2, Southern Methodist University Institute of Technology, Dallas, Tex., October, 1968.

24. Sarofim, A. F., and H. C. Hottel: Radiative Exchange among Non-Lambert Surfaces, *J. Heat Transfer,* vol. 88, no. 1, pp. 37-44, 1966.

25. Sparrow, E. M., and S. L. Lin: Radiation Heat Transfer at a Surface Having Both Specular and Diffuse Reflectance Components, *Int. J. Heat Mass Transfer,* vol. 8, pp. 769-779, 1965.

26. Saari, J. M., and R. W. Shorthill: Review of Lunar Infrared Observations, in S. F. Singer (ed.), "Physics of the Moon," vol. 13, AAS Science and Technology Series, 1967.

27. Harrison, James K.: Non-Diffuse Infrared Emission from the Lunar Surface, *Int. J. Heat Mass Transfer*, vol. 12, pp. 689-697, 1969.

28. Birkebak, Richard C.: Thermophysical Properties of Lunar Materials: Part I, Thermal Radiation Properties of Lunar Materials from the Apollo Missions, in J. P. Hartnett and T. F. Irvine, Jr. (eds.), "Advances in Heat Transfer," vol. 10, Academic Press, New York, 1974.

29. Birkebak, Richard C.: Spectral Emittance of Apollo-12 Lunar Fines, *ASME J. Heat Transfer,* vol. 94, no. 3, pp. 323-324, 1972.

30. Liebert, Curt H.: Spectral Emissivity of Highly Doped Silicon, paper no. 67-302, *AIAA,* April, 1967.

31. Hsieh, C. K., and K. C. Su: Thermal Radiative Properties of Glass From 0.32 to 206 μm, *Sol. Energy,* vol. 22, no. 1, pp. 37-43, 1979.

32. Fan, John C. C., and Frank J. Bachner: Transparent Heat Mirrors for Solar-Energy Applications, *Appl. Optics,* vol. 15, no. 4, pp. 1012-1017, 1976.

33. Jarvinen, P. O.: Heat Mirrored Solar Energy Receivers, paper no. 77-728, *AIAA 12th Thermophys. Conf.,* Albuquerque, 1977.

34. Hale, George M., and Marvin R. Querry: Optical Constants of Water in the 200-nm to 200-μm Wavelength Region, *Appl. Optics,* vol. 12, no. 3, pp. 555-563, 1973.

35. Irvine, William M., and James B. Pollack: Infrared Optical Properties of Water and Ice Spheres, *Icarus,* vol. 8, pp. 324-360, 1968.

36. Pinkley, L. W., P. P. Sethna, and D. Williams: Optical Constants of Water in the Infrared: Influence of Temperature, *J. Opt. Soc. Am.,* vol. 67, no. 4, pp. 494-499, 1977.

37. Kondratyev, Ya K.: "Radiation in the Atmosphere," Academic Press, Inc., New York, 1969.

38. Seki, N., M. Sugawara, and S. Fukusako: Back-Melting of a Horizontal Cloudy Ice Layer with Radiative Heating, *ASME J. Heat Transfer,* vol. 101, no. 1, pp. 90–95, 1979.

39. Seban, R. A.: Thermal Radiation Properties of Materials, pt. III, WADD TR-60-370, University of California, Berkeley, 1963.

40. De Vos, J. C.: A New Determination of the Emissivity of Tungsten Ribbon, *Physica,* vol. 20, pp. 690–714, 1954.

41. Birkebak, R. C., and E. R. G. Eckert: Effects of Roughness of Metal Surfaces on Angular Distribution of Monochromatic Reflected Radiation, *J. Heat Transfer,* vol. 87, no. 1, pp. 85–94, 1965.

42. Edwards, D. K., and N. Bayard de Volo: Useful Approximations for the Spectral and Total Emissivity of Smooth Bare Metals, *Adv. Thermophys. Properties Extreme Temp. Pressures* (Serge Gratch, ed.), ASME, 1965, pp. 174–188.

43. Munch, B.: "Directional Distribution in the Reflection of Heat Radiation and its Effect in Heat Transfer," Ph.D. thesis, Swiss Technical College of Zurich, 1955.

44. Williams, D. A., T. A. Lappin, and J. A. Duffie: Selective Radiation Properties of Particulate Coatings, *J. Eng. Power,* vol. 85, no. 3, pp. 213–220, 1963.

45. Ohlsen, P. E., and G. A. Etamad: Spectral and Total Radiation Data of Various Aircraft Materials, *Rep. NA57-330,* North American Aviation, July 23, 1957.

46. Orlova, N. S.: Photometric Relief of the Lunar Surface, *Astron. Z.,* vol. 33, no. 1, pp. 93–100, 1956.

47. Long, R. L.: A Review of Recent Air Force Research on Selective Solar Absorbers, *J. Eng. Power,* vol. 87, no. 3, pp. 277–280, 1965.

48. Hibbard, R. R.: Equilibrium Temperatures of Ideal Spectrally Selective Surfaces, *Sol. Energy,* vol. 5, no. 4, pp. 129–132, 1961.

49. Shaffer, L. H.: Wavelength-Dependent (Selective) Processes for the Utilization of Solar Energy, *J. Sol. Energy Sci. Eng.,* vol. 2, nos. 3, 4, pp. 21–26, 1958.

50. Dunkle, R. V.: Thermal Radiation Characteristics of Surfaces, in J. A. Clark (ed.), "Theory and Fundamental Research in Heat Transfer," pp. 1–31, Pergamon Press, New York, 1963.

51. Gardon, Robert: The Emissivity of Transparent Materials, *J. Am. Ceram. Soc.,* vol. 39, no. 8, pp. 278–287, 1956.

52. Perlmutter, Morris, and John R. Howell: A Strongly Directional Emitting and Absorbing Surface, *J. Heat Transfer,* vol. 85, no. 3, pp. 282–283, 1963.

53. Brandenberg, W. M., and O. W. Clausen: The Directional Spectral Emittance of Surfaces between 200° and 600°C, *Symp. Thermal Radiation Solids,* S. Katzoff (ed.), *NASA* SP-55 (AFML-TDR-64-159), 1965.

54. Craighead, H. G. et al.: Metal/Insulator Composite Selective Absorbers, *Sol. Energy Materials,* vol. 1, nos. 1 & 2, pp. 105–124, 1979.

PROBLEMS

1 The normal spectral absorptivity of a SiO–Al selective surface can be approximated as shown in Fig. 5-26. The surface receives from the normal direction a flux q. The equilibrium temperature of the surface is 2000°R. Assume $\epsilon_\lambda = \alpha'_\lambda(\theta = 0)$. What is the value of q if it comes from a gray-body source at 6000°R?

Answer: 3240 Btu/ (h·ft²) (10,200 W/m²).

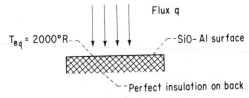

2   A directionally selective gray surface has properties as shown below. The α' is isotropic with respect to the angle φ.

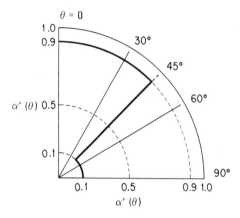

(a)   What is the ratio $\alpha'(\theta = 0)/\epsilon$ (the directional absorptivity over the hemispherical emissivity) for this surface?

(b)   If a thin plate with the above properties is in earth orbit around the sun with incident solar flux of 442 Btu/(h·ft²), what equilibrium temperature will it reach? Assume the plate is oriented normal to the sun's rays, and is perfectly insulated on the side away from the sun.

(c)   What is the equilibrium temperature if the plate is oriented at 60° to the sun's rays?

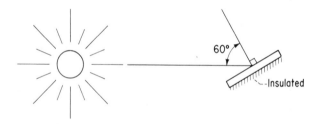

(d)   What is its equilibrium temperature if the plate is normal to the sun's rays, but is not insulated? Assume the plate is very thin, and has the same directional properties on both sides.

   Answer: (a) 1.8;   (b) 821°R;   (c) 399°R;   (d) 690°R.

3   A gray surface has directional total absorptivity given by $\alpha' = 0.75 \cos^3 \theta$. This flat surface is exposed to normally incident sunlight of flux 1250 W/m². A fluid flows past the back of the thin collector at $T_{\text{fluid}} = 300$ K at a velocity that gives a heat transfer coefficient of $h = 60$ W/m²·K. What is the equilibrium temperature of the collector?

   Answer: 313 K.

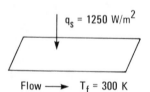

4   A plate made of material combining directional and spectral selectivity has a normal solar absorptivity of 0.97, and an infrared total hemispherical emissivity of 0.02. When placed in sunlight normal to the sun's rays, what temperature will the plate reach (neglecting conduction and convection and with no heat losses from the unexposed side of the plate)? What assumptions did you make in reaching your answer? Take the incident solar flux as 1150 W/m².

5   A hot-water heater consists of a sheet of glass 1 cm thick over a black surface that is assumed in perfect contact with the water below it. Estimate the water temperature for normally incident solar radiation. (Assume that Fig. 5-31 can be used for the glass properties and that the glass is perfectly transparent for wavelengths shorter than those shown. Take into account approximately the reflections at the glass surfaces; this will be treated in more detail in Chap. 19.)

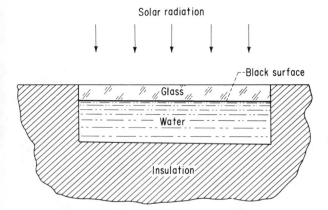

6   A flat gasoline storage tank is in the sun with the incident radiation essentially normal to the top of the tank. The tank is painted with white paint having the reflectivity in Fig. 5-28. Estimate the equilibrium temperature that the tank can achieve. (Neglect emitted and reflected radiation from the ground. Do not account for free or forced convection to the air although this will be appreciable.) What would the tank temperature be if the top is painted white as before but the sides are painted with a gray coating having an emissivity of 0.9? What is the temperature if the entire tank is painted with the gray coating?

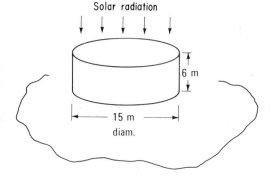

7   A spinning sphere 12 in (0.305 m) in diameter is in earth orbit and is receiving solar radiation. As a consequence of the sphere rotation, the surface temperature is assumed uniform. The sphere exterior is a SiO–Al selective surface as in Example 5-2, with a cutoff wavelength of

1.5 μm. The surface properties do not depend on angle. What is the equilibrium surface temperature for heat transfer only by radiation? How does this temperature depend on the sphere diameter?

It is desired to maintain the sphere surface at 1200°R (667 K). At what rate must electrical energy be supplied to the entire sphere to accomplish this?

Answer: 1015°R (564 K),   280 Btu/h   (82 W).

SIX

INTRODUCTION TO RADIATIVE EXCHANGE

The study of radiation interchange between individual surface elements in a system is required in a variety of engineering disciplines including applied optics, illumination engineering, and heat transfer. Indeed, such studies have been conducted for many years as evidenced by the publication dates of Refs. [1, 2]. More recently the study of radiant interchange has been given impetus by technological advances that have resulted in systems in which thermal radiation can be a very significant factor. Some examples are satellite temperature control, energy leakage into cryogenic vacuum systems, high-temperature phenomena in hypersonic flight, the heat transfer in nuclear propulsion systems, and devices for the collection and utilization of solar energy.

6-1 ENCLOSURE THEORY

In the succeeding six chapters a theory will be developed for computing thermal radiation exchanges within enclosures in vacuum or filled with nonabsorbing media. First it must be understood what is meant by an *enclosure*. Any surface can be considered as completely surrounded by an envelope of other solid surfaces or open areas. This envelope is the enclosure for the surface; thus an enclosure accounts for all directions surrounding the surface. By considering the radiation going from the surface to all parts of the enclosure, and the radiation arriving at the surface from all parts of the enclosure, it is assured that all the radiative contributions are accounted for. In working a problem, a convenient enclosure will usually be evident from the physical configuration. An opening can be considered as a plane of zero reflectivity. It will also act as a source of radiation when radiation is entering the enclosure from the environment.

For the enclosures considered in Chaps. 7-12 the medium in the region between the surfaces is assumed to be perfectly transparent and thus not to participate in the radiative interchange. The theory for enclosures filled with a radiating material such as a gas containing water vapor, carbon dioxide, or smoke will be treated in Chaps. 13-20. Enclosures with windows are treated in Chap. 19.

Chapters 1-5 discussed in detail the radiative properties of solid surfaces. It was demonstrated that for some materials there are substantial variations of properties with wavelength, surface temperature, and direction. For radiation computations within enclosures, the geometric effects governing how much radiation from one surface reaches another is a complication in addition to the variations of the surface properties. For simple geometries it may be possible to account in detail for property variations without the problem becoming unduly complex. As the geometry becomes more involved, it is often necessary to invoke more idealizations of the surface properties so that the problem can be solved with reasonable effort.

The treatment presented here could begin with the most general situation in which properties vary with wavelength, temperature, and direction, and where the radiation fluxes vary arbitrarily over the enclosure surfaces. All other situations would then be simplified special cases. However, this would entail the uninitiated reader plunging into the most complex treatment, which would be very difficult to understand. Hence the development presented here will begin with the most simple situation; successive complexities will then be added to build more comprehensive treatments.

6-1.1 Ideal Enclosures

The greatest simplification is to assume all the enclosure surfaces are black. In this instance there is no reflected radiation to be accounted for. Also, all the emitted energy is diffuse; that is, the intensity leaving a given isothermal surface is independent of direction. The exchange theory for a black enclosure is presented in Chap. 7. The heat balances involve the enclosure geometry that governs how much radiation leaving a surface will reach another surface. The geometric effects are expressed in terms of diffuse configuration factors; these factors are the fractions of radiation leaving a surface that reach another surface. The factors are derived on the basis that the distribution of radiation leaving a surface is both diffuse and uniformly distributed, and these restrictions should be kept in mind when the factors are applied in nonblack enclosures.

The computation of configuration factors involves integration over the solid angles by which the surfaces can view each other. Since these integrations are often tedious, it is desirable to use certain useful relations that exist between configuration factors. By using these relations, the desired factor can often be obtained from factors that are already known, and the integration will not have to be performed. These relations, along with various shortcut methods that can be used to obtain configuration factors, are presented in detail in Chap. 7. An appendix is also provided (Appendix B) giving references where configuration factors can be found for approximately 200 different geometrical configurations. Appendix C provides a catalog of some useful configuration factors.

After analyzing the black enclosure, the next step in complexity is an enclosure with gray surfaces that emit and reflect diffusely. It is also assumed that both the emitted and reflected energies are uniform over each surface. For these conditions the diffuse configuration factors found for black surfaces still apply for the radiation leaving a surface. For gray surfaces, reflections between surfaces must be accounted for. This is done in Chap. 8 by using a method developed by Poljak.

Another type of ideal surface is a perfect mirror reflector. The emission from this type of surface is approximated as being diffuse; hence, the emitted energy is treated by using the diffuse configuration factors. The reflected energy, however, is followed within the enclosure by using the characteristics of a mirror, where the angle of reflection is equal in magnitude to the angle of incidence. The method of tracing the reflected radiation paths and deriving the necessary heat balances is treated in Chap. 9.

6-1.2 Nonideal Enclosures

In some instances the black or diffuse-gray approximation is inadequate, and directional and/or spectral effects must be considered. The necessity of treating spectral effects was noticed quite early in the field of radiative transfer. In the remarkable paper [3] published in 1800 by Sir William Herschel entitled "Investigation of the Powers of the Prismatic Colours to Heat and Illuminate Objects; with Remarks, that prove the Different Refrangibility of Radiant Heat to which is added, an Inquiry into the Method of Viewing the Sun Advantageously, with Telescopes of Large Apertures and High Magnifying Powers," appears the following statement: "In a variety of experiments I have occasionally made, relating to the method of viewing the sun, with large telescopes, to the best advantage, I used various combinations of differently coloured darkening glasses. What appeared remarkable was, that when I used some of them, I felt a sensation of heat, though I had but little light; while others gave me much light, with scarce any sensation of heat. Now, as in these different combinations, the sun's image was also differently coloured, it occurred to me, that the prismatic rays might have the power of heating bodies very unequally distributed among them ..." This paper was the first in which what is now called the infrared region of the spectrum was defined and the energy radiated as "heat" shown to be of different wavelengths than those for "light."

The quotation shows an awareness that in some instances spectral effects must be included in the radiative analysis. The performance of spectrally selective surfaces such as are used in satellite temperature control and for solar collector surfaces can only be understood by considering the wavelength variations of the surface properties. Selectivity is also important in the transmitting characteristics of materials such as glass as used for an enclosure window.

A second nonideal surface property is that of strong directional dependence. In Chap. 5 a number of directionally dependent surface properties were examined, and some were shown to differ considerably from the diffuse or specular approximations. A good example is the lunar surface, which has a distribution of reflected energy strongly peaked back into the direction of incident radiation. This is in a sense the opposite of a specular reflector and can certainly not be considered diffuse.

Methods for treating surfaces that are nonideal in either spectral or directional properties, or both, are examined in Chaps. 10 and 11. Chapter 10 continues the enclosure-theory development of the previous chapters; Chap. 11 deals with an alternative approach—the Monte Carlo method. This is a general technique that involves following "bundles" of radiant energy along their paths within an enclosure. It can be applied to all types of radiation problems but is usually too detailed and costly in terms of computer time for use in simple situations. When directional and spectral effects must be considered, the Monte Carlo method is very valuable.

6-2 ENERGY TRANSFER BY COMBINED MODES

Chapter 12 deals with problems in which conduction and/or convection is combined with radiative heat transfer. Since opaque surfaces are dealt with in that chapter, the radiative interaction with a body is considered to occur only at the surface. Thus the radiation serves only as a boundary condition with regard to the conduction process within a body. This is analogous to the convective boundary condition at a surface. When a body is undergoing a transient temperature change, the radiative terms are applied at each instant when solving the energy balances governing the temperature distribution within the body.

The heat-conduction process is governed by local derivatives of the first power of the temperature. The convection process depends on local differences between the first power of the fluid and surface temperatures. Radiative exchange, however, depends approximately on differences of fourth powers of the surface temperatures, and also depends on the integral of the radiation incident from all the surroundings of the surface. As a result the energy balance for a combined convection, conduction, and radiation problem can result in an intergrodifferential equation. There are few standard mathematical methods for attacking these equations, and few closed-form analytical solutions are available. Numerical methods are usually employed for multimode problems.

6-3 NOTATION

The notation employed is now briefly reviewed. A prime denotes a directional quantity, while a λ subscript specifies that the quantity is spectral; for example, ϵ_λ' is the directional spectral emissivity. Certain quantities such as bidirectional reflectivities can depend on two directions, that is, the directions of the incoming and outgoing radiation. These bidirectional quantities are denoted by a double prime. A hemispherical quantity will not have a prime, and a total quantity will not have a λ subscript; thus ϵ is the hemispherical total emissivity. In addition, a notation such as $\epsilon_\lambda'(\lambda,\theta,\varphi,T)$ can be used to emphasize the functional dependencies or to state more specifically at what wavelength, angle, and surface temperature the quantity is being evaluated.

Additional notation is needed for the energy rate Q for a finite area to keep consistent mathematical forms for energy balances. The quantity d^2Q_λ' is directional-spectral, and the second derivative is used to indicate that the energy is of differential order in both solid angle and wavelength. The quantities dQ' and dQ_λ are of differential order with respect to solid angle and wavelength, respectively. If a differential area is involved, the order of the derivative is correspondingly increased.

6-4 CONCLUDING REMARKS

As mentioned previously, certain restrictions to ideal surfaces and nonparticipating media are present in each of the chapters that follow. In addition, some phenomena that are rather more specialized than is the intent of this work can be of importance in certain situations. For example, effects of polarization can lead to errors in energy transfer calculations if ignored under special conditions of geometry [4]. Interference effects [5], chemical and photochemical phenomena [6-9], and perhaps others can in some situations be the dominant mechanisms governing radiative transfer. The reader can only be referred to the specialized literature and warned to watch for such cases.

REFERENCES

1. Charle, M.: "Les Manuscripts de Léonard de Vinci, Manuscripts *C, E,* et *K* de la Bibliothèque de l'Institute Publiés en Facsimilés Phototypiques," Ravisson-Mollien, Paris, 1888. (Referenced in Middleton, W. E. Knowles: Note on the Invention of Photometry, *Am. J. Phys.,* vol. 31, no. 3, pp. 177–181, 1963.)
2. Francois d'Aguillon, S. J.: "Opticorum Libri Sex," Antwerp, 1613. (Referenced in Middleton, W. E. Knowles: Note on the Invention of Photometry, *Am. J. Phys.,* vol. 31, no. 3, pp. 177–181, 1963.)
3. Herschel, William: Investigation of the Powers of the Prismatic Colours to Heat and Illuminate Objects, *Trans. R. Soc. London,* vol. 90, pt. 2, pp. 255–283, 1800.
4. Edwards, D. K., and R. D. Tobin: Effect of Polarization on Radiant Heat Transfer through Long Passages, *J. Heat Transfer,* vol. 89, no. 2, pp. 132–138, 1967.
5. Cravalho, E. G., C. L. Tien, and R. P. Caren: Effect of Small Spacings on Radiative Transfer between Two Dielectrics, *J. Heat Transfer,* vol. 89, no. 4, pp. 351–358, 1967.
6. Garlick, G. F. J.: Luminescence in Solids, *Sci. Prog. Oxford,* vol. 52, pp. 3–25, 1964.
7. Pringsheim, Peter: "Fluorescence and Phosphorescence," Interscience Publishers, Inc., New York, 1949.
8. Curie, Daniel (G. F. J. Garlick, trans.): "Luminescence in Crystals," John Wiley & Sons, Inc., New York, 1963.
9. Bowen, E. J., and G. F. J. Garlick: Luminescence, *Int. Sci. Tech.,* no. 56, pp. 18–29, 1966.

SEVEN

EXCHANGE OF RADIANT ENERGY
BETWEEN BLACK ISOTHERMAL SURFACES

7-1 INTRODUCTION

This chapter begins the discussion of radiation exchange between surfaces and is concerned with the special situation in which all the surfaces involved are black. Black surfaces are dealt with first since they are perfect absorbers, and the energy exchange process is thus simplified because there is no reflected energy to be considered. Also, all black surfaces emit in a perfectly diffuse fashion, such that the radiation intensity leaving a surface is independent of the direction of emission. This simplifies the computation of how much of this radiation will reach another surface.

The fraction of the radiation leaving one surface that reaches another surface is defined as the geometric configuration factor between the two surfaces, because it depends on the geometric orientation of the surfaces with respect to each other. The geometric dependence is discussed here for black surfaces, but the results have a wider generality as they will apply for any uniform diffuse radiation leaving a surface. This geometric dependence leads to some algebraic relations between the factors, and these relations are demonstrated in this chapter for various surface configurations. In Appendix B, a tabulation is provided of references where known configuration factors can be found in the literature, and some selected configuration factors are given in Appendix C. Applications of these factors to example problems of engineering interest are then examined for radiative energy exchange between two surfaces.

After the relations for exchange between two surfaces have been developed, the relations can be applied to any number of surfaces arranged to form an enclosure of black surfaces each at a different temperature. The general set of equations governing the exchange within such an enclosure is developed, and some illustrative examples are provided.

In Chap. 8, the concepts developed in this chapter are extended for use in systems with diffuse-gray surfaces, and succeeding chapters introduce more and more complex systems. The concepts of the present chapter are discussed at some length because they are fundamental to the succeeding material dealing with less ideal surfaces.

7-2 SYMBOLS

A	area
e	emissive power
f	function defined by Eq. (7-56b)
F	configuration factor
i	intensity
l,m,n	direction cosines, Eq. (7-56a)
N	number of surfaces in an enclosure
P,Q,R	functions in contour integration used in Sec. 7-5.4
Q	energy per unit time
r	radius
S	distance between two differential elements
T	temperature
U	number of unknowns in equations describing an N-sided enclosure
x,y,z	cartesian coordinate positions
α,γ,δ	angles in direction cosines
β	angle in yz plane
λ	wavelength
θ	angle from normal
σ	Stefan-Boltzmann constant
ω	solid angle

Subscripts

b	blackbody
$d1,d2$	evaluated at differential element $d1$ or $d2$
i	inner
j,k	jth or kth surface
N	Nth surface
ring	ring area
s	sun
strip	elemental strip
λ	wavelength dependent
1, 2	at area 1 or 2

Superscript

$'$	denotes quantity is in one direction

7-3 RADIATIVE EXCHANGE BETWEEN TWO DIFFERENTIAL AREA ELEMENTS

The relations describing radiative exchange between differential elements are considered first, as they will be used in the succeeding sections to derive the relations for exchange between areas of finite size. Consider two differential black area elements as shown in Fig. 7-1. The elements dA_1 and dA_2 are isothermal at temperatures T_1 and T_2, respectively, are arbitrarily oriented, and have their normals at angles θ_1 and θ_2 to the line of length S joining them.

The total energy per unit time leaving dA_1 and incident upon dA_2 is

$$d^2Q'_{d1-d2} = i'_{b,1}\, dA_1 \cos\theta_1\, d\omega_1 \qquad (7\text{-}1)$$

where $d\omega_1$ is the solid angle subtended by dA_2 when viewed from dA_1. Equation (7-1) follows directly from the definition of $i'_{b,1}$, the total blackbody intensity of surface 1, as the total energy emitted by surface 1 per unit time, per unit of area dA_1 projected normal to S, and per unit of solid angle. As before, the prime indicates a quantity applied in a single direction. The quantity d^2Q' is a second differential to denote the dependence upon two differential quantities, dA_1 and $d\omega_1$.

Equation (7-1) can also be written for radiation at only one wavelength:

$$d^3Q'_{\lambda,d1-d2} = i'_{\lambda b,1}(\lambda)\, d\lambda\, dA_1 \cos\theta_1\, d\omega_1$$

The total radiation quantities are then found by integrating over all wavelengths:

$$d^2Q'_{d1-d2} = \int_{\lambda=0}^{\infty} d^3Q'_{\lambda,d1-d2} = dA_1 \cos\theta_1\, d\omega_1 \int_0^{\infty} i'_{\lambda b,1}(\lambda)\, d\lambda.$$

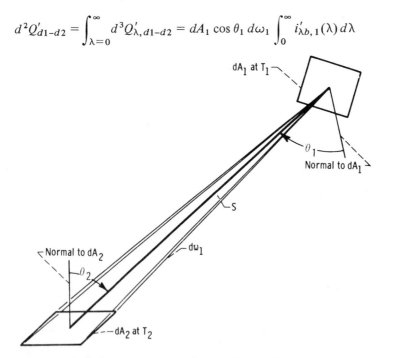

Figure 7-1 Radiative interchange between two black differential area elements.

For a black surface $i'_{\lambda b}(\lambda)$ does not depend on direction; hence all the geometric factors can be removed from under the integral sign, and the integration over wavelength is independent of any geometrical considerations. Thus the following results for geometric configuration factors apply for both spectral and total quantities. For simplicity in notation, the discussion will be carried out for total quantities.

The solid angle $d\omega_1$ is related to the projected area of dA_2 and the distance between the differential elements by the relation

$$d\omega_1 = \frac{dA_2 \cos \theta_2}{S^2} \tag{7-2}$$

Substituting this relation into Eq. (7-1) gives the following equation for the total energy per unit time leaving dA_1 that is incident upon dA_2:

$$d^2 Q'_{d1-d2} = \frac{i'_{b,1} \, dA_1 \cos \theta_1 \, dA_2 \cos \theta_2}{S^2} \tag{7-3}$$

An analogous derivation for the radiation leaving dA_2 that arrives at dA_1 results in

$$d^2 Q'_{d2-d1} = \frac{i'_{b,2} \, dA_2 \cos \theta_2 \, dA_1 \cos \theta_1}{S^2} \tag{7-4}$$

For later use, $d^2 Q'$ has been defined in Eqs. (7-3) and (7-4) as the energy emitted by one element that is *incident* upon the second element. For the special case of a black receiving element, all incident energy is absorbed so that Eqs. (7-3) and (7-4) in this case give the energy from one element that is *absorbed* by the second. As will be seen, the more general definition of $d^2 Q'$ allows the configuration factors derived here for black surfaces to be used in certain other cases. These will be examined in length in Chaps. 8 and 9.

The *net* energy per unit time $d^2 Q'_{d1 \rightleftharpoons d2}$ transferred from black element dA_1 to dA_2 along path S is then the difference of $d^2 Q'_{d1-d2}$ and $d^2 Q'_{d2-d1}$, or from (7-3) and (7-4)

$$d^2 Q'_{d1 \rightleftharpoons d2} \equiv d^2 Q'_{d1-d2} - d^2 Q'_{d2-d1} = (i'_{b,1} - i'_{b,2}) \frac{\cos \theta_1 \cos \theta_2}{S^2} dA_1 \, dA_2 \tag{7-5}$$

From (2-21), the blackbody total intensity is related to the blackbody total hemispherical emissive power by

$$i'_b = \frac{e_b}{\pi} = \frac{\sigma T^4}{\pi} \tag{7-6}$$

so that (7-5) can be written as

$$d^2 Q'_{d1 \rightleftharpoons d2} = \sigma (T_1^4 - T_2^4) \frac{\cos \theta_1 \cos \theta_2}{\pi S^2} dA_1 \, dA_2 \tag{7-7}$$

EXAMPLE 7-1 The sun emits energy at a rate that can be approximated by that of a blackbody at temperature 10,400°R. A blackbody area element in orbit around the sun at the mean radius of the earth's orbit (92.9×10^6 mi) is oriented normal to the line connecting the centers of the area element and sun. If the sun's radius is 4.32×10^5 mi, what energy flux is incident upon the element?

To the element in orbit, the sun appears as an isothermal disk element of area

$$dA_1 = \pi r_s{}^2 = \pi (4.32 \times 10^5)^2 = 5.86 \times 10^{11} \text{ mi}^2$$

From Eq. (7-3), the incident energy flux on the element in orbit is

$$\frac{d^2 Q'_{d1-d2}}{dA_2} = i'_{b,1} \, dA_1 \, \frac{\cos \theta_1 \cos \theta_2}{S^2} = \frac{\sigma T_s{}^4}{\pi} \frac{dA_1}{S^2}$$

$$= \frac{0.173 \times 10^{-8} (1.04 \times 10^4)^4}{\pi} \frac{5.86 \times 10^{11}}{(92.9 \times 10^6)^2} = 437 \text{ Btu/(h·ft}^2)$$

This value is consistent with the range of measured values of the mean solar constant, 422–436 Btu/(h·ft²) (1332–1374 W/m²).

An alternative procedure is to utilize the fact that the radiant energy will leave the sun in a spherically symmetric fashion. The energy radiated is $\sigma T_s{}^4 4\pi r_s{}^2$, and the area of a sphere surrounding the sun and having a radius equal to the earth's

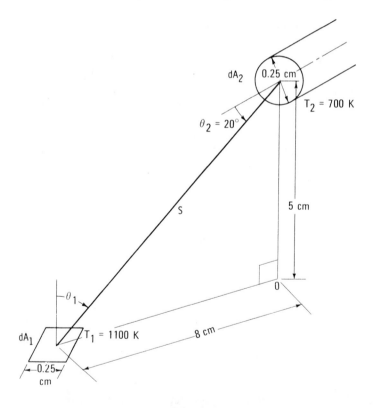

Figure 7-2 Radiative exchange between square element and circular-tube opening.

orbit is $4\pi S^2$. Hence the flux received at the earth's orbit is $\sigma T_s^4 4\pi r_s^2/4\pi S^2 = \sigma T_s^4 (r_s/S)^2$, giving the same result as before.

EXAMPLE 7-2 As shown in Fig. 7-2, a black square with side 0.25 cm is at temperature 1100 K and is near a tube 0.25 cm in diameter. The opening of the tube acts as a black surface, and the tube is at 700 K. What is the net radiation exchange along the connecting path S between the square and the tube opening?
From (7-7),

$$d^2 Q'_{d1 \rightleftharpoons d2} = \sigma(T_1^4 - T_2^4) \frac{\cos \theta_1 \cos \theta_2}{\pi S^2} dA_1 dA_2$$

The value of $\cos \theta_1$ is found from the known sides of the right triangle $dA_2\text{-}0\text{-}dA_1$ as $\cos \theta_1 = 5/(8^2 + 5^2)^{1/2} = 5/89^{1/2}$. The other factors in the energy exchange equation are given, and substituting them gives

$$d^2 Q'_{d1 \rightleftharpoons d2} = 5.729 \times 10^{-8}(1100^4 - 700^4) \frac{5}{89^{1/2}} \frac{\cos 20°}{\pi(89/10^4)} \frac{0.25^2}{10^4} \frac{\pi 0.25^2}{4 \times 10^4}$$

$$= 0.383 \times 10^{-4} \text{ W}$$

7-4 RADIATIVE GEOMETRIC CONFIGURATION FACTORS AND ENERGY EXCHANGE BETWEEN TWO SURFACES

One of the chief mathematical complexities in treating radiative transfer between surfaces is accounting for the geometric relations involved in how the surfaces view each other. These effects result mathematically in integrations of the radiative interchange over the finite areas involved in the exchange process. It would be helpful to have, as much as possible, handbook results to account for these geometric relations for often-encountered geometries. In this way repetitions of the tedious integrations could be avoided.

In this section, a method of accounting for the geometry is introduced in the form of a quantity called the *geometric configuration factor*. Such factors allow computation of radiative transfer in many systems by referring to formulas or tabulated data which have been previously obtained for the geometric relations between various surfaces. This removes what is often the most time-consuming and error-prone portion of the analysis.

7-4.1 Configuration Factor for Energy Exchange between Differential Elements

The *fraction* of energy leaving black surface element dA_1 that arrives at black element dA_2 is defined as the *geometric configuration factor* dF_{d1-d2}. [Either the total or spectral energy could be considered, as discussed with regard to Eq. (7-1), and the same results for dF would be obtained. The total energy is used here for convenience in not carrying the λ notation.] Using (7-3) and (7-6), the previous definition gives

$$dF_{d1-d2} = \frac{d^2 Q'_{d1-d2}}{\sigma T_1{}^4 dA_1} = \frac{\sigma T_1{}^4 (\cos \theta_1 \cos \theta_2 / \pi S^2) \, dA_1 \, dA_2}{\sigma T_1{}^4 \, dA_1}$$

$$= \frac{\cos \theta_1 \cos \theta_2}{\pi S^2} \, dA_2 \qquad (7\text{-}8)$$

where $\sigma T_1{}^4 dA_1$ is the total energy leaving dA_1 within the entire hemispherical solid angle over dA_1. Equation (7-8) shows that dF_{d1-d2} depends *only upon the size of dA_2 and its orientation with respect to dA_1*. By substituting (7-2), Eq. (7-8) can also be written in the form

$$dF_{d1-d2} = \frac{\cos \theta_1 \, d\omega_1}{\pi} \qquad (7\text{-}9)$$

Consequently, all elements dA_2 have the same configuration factor if they subtend the same solid angle $d\omega_1$ when viewed from dA_1 and are positioned along a path at angle θ_1 with respect to the normal of dA_1.

The factor dF_{d1-d2} has a variety of names, being called the view, angle, shape, interchange, exchange, or configuration factor. The last seems most specific, implying a dependence upon both orientation and shape, the latter variable entering when finite areas are involved.

The notation used here for configuration factors is based on subscript designation for the types of areas involved in the energy exchange and a derivative notation consistent with the mathematical meaning of the configuration factor. For the subscript notation, $d1, d2$, etc., will indicate differential area elements, while $1, 2$, and so forth indicate areas of finite size. Thus dF_{d1-d2} indicates a factor between two differential elements, as in Eq. (7-8). The notation dF_{1-d2} indicates a configuration factor from finite area A_1 to differential area dA_2.

The derivative notation dF indicates that the configuration factor is for energy transfer *to* a *differential* element, as in (7-8). This is redundant with the subscript notation, but keeps the mathematical form of equations [such as (7-8)] consistent in that a differential quantity appears on both sides (that is, the expression for dF contains a differential area). A configuration factor F denotes a factor *to* a *finite* area. Thus F_{d1-2} is the configuration factor from differential element dA_1 to finite area A_2.

Reciprocity for differential-element configuration factors. By a derivation similar to that used in obtaining Eq. (7-8), the configuration factor needed for calculating energy exchange from element dA_2 to dA_1 is

$$dF_{d2-d1} = \frac{\cos \theta_1 \cos \theta_2}{\pi S^2} \, dA_1 \qquad (7\text{-}10)$$

Multiplying (7-8) by dA_1 and (7-10) by dA_2 gives the *general reciprocity relation*

$$dF_{d1-d2} \, dA_1 = dF_{d2-d1} \, dA_2 = \frac{\cos \theta_1 \cos \theta_2}{\pi S^2} \, dA_1 \, dA_2 \qquad (7\text{-}11)$$

Finally, (7-7) for the net energy transferred from dA_1 to dA_2 along the path between the two black elements can be written by using (7-11); the result is

$$d^2Q'_{d1 \rightleftharpoons d2} = \sigma(T_1{}^4 - T_2{}^4)\, dF_{d1-d2}\, dA_1 = \sigma(T_1{}^4 - T_2{}^4)\, dF_{d2-d1}\, dA_2 \qquad (7\text{-}12)$$

Some sample configuration factors between differential elements. To this point, a series of algebraic manipulations has allowed a reduction of the equation for the net radiative transfer along the path between two black isothermal area elements to the apparently simple form of (7-12). This was done by introduction of the configuration factor dF, which encompasses the geometric complexities. The derivation of configuration factors will now be illustrated by considering some sample cases.

EXAMPLE 7-3 The two elemental areas shown in Fig. 7-3 are located on strips that have parallel generating lines. Derive an expression for the configuration factor between dA_1 and dA_2.

The distance S can be expressed as $S^2 = l^2 + x^2$, and $\cos \theta_1$ is then

$$\cos \theta_1 = \frac{l \cos \beta}{S} = \frac{l \cos \beta}{(l^2 + x^2)^{1/2}}$$

The solid angle subtended by dA_2, when viewed from dA_1, is

$$d\omega_1 = \frac{\text{projected area of } dA_2}{S^2} = \frac{(\text{projected width of } dA_2)(\text{projected length of } dA_2)}{S^2}$$

$$= \frac{(l\, d\beta)(dx \cos \psi)}{S^2} = \frac{l\, d\beta\, dx}{S^2}\frac{l}{S}$$

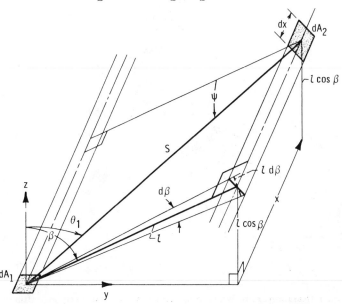

Figure 7-3 Geometry for configuration factor between elements on strips formed by parallel generating lines.

Substituting into (7-9) gives

$$dF_{d1-d2} = \frac{\cos \theta_1 \, d\omega_1}{\pi} = \frac{l \cos \beta}{(l^2 + x^2)^{1/2}} \frac{1}{\pi} \frac{l^2 \, d\beta \, dx}{(l^2 + x^2)^{3/2}} = \frac{l^3 \cos \beta \, d\beta \, dx}{\pi (l^2 + x^2)^2}$$

which is the desired configuration factor between dA_1 and dA_2.

EXAMPLE 7-4 Find the configuration factor between an elemental area and an infinitely long strip of differential width oriented as in Fig. 7-4, so that the generating lines of dA_1 and $dA_{\text{strip, 2}}$ are parallel.

Example 7-3 gave the configuration factor between differential element dA_1 and area element dA_2 of length dx as $dF_{d1-d2} = l^3 \cos \beta \, d\beta \, dx / \pi (l^2 + x^2)^2$. To find the factor when dA_2 becomes an infinite strip as in Fig. 7-4, integrate over all x to obtain

$$dF_{d1-\text{strip, 2}} = \frac{l^3 \cos \beta \, d\beta}{\pi} \int_{-\infty}^{\infty} \frac{dx}{(l^2 + x^2)^2}$$

$$= \frac{l^3 \cos \beta \, d\beta}{\pi} \left[\frac{x}{2l^2(l^2 + x^2)} + \frac{1}{2l^3} \tan^{-1} \frac{x}{l} \right]_{-\infty}^{\infty}$$

$$= \frac{\cos \beta \, d\beta}{2} = \tfrac{1}{2} d(\sin \beta)$$

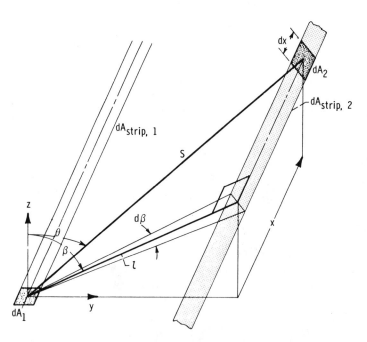

Figure 7-4 Geometry for configuration factor between elemental area and infinitely long strip of differential width; area and strip are on parallel generating lines.

where the angle β is in the yz plane. This useful configuration-factor relation will be used in later examples.

Figure 7-4 also shows that, if element dA_1 lies on an infinite strip $dA_{strip,1}$ with elements parallel to $dA_{strip,2}$, the configuration factor $dF_{d1-strip,2} = \frac{1}{2}d(\sin \beta)$ will be valid for dA_1 regardless of where dA_1 lies on $dA_{strip,1}$. Then, since any element dA_1 on $dA_{strip,1}$ has the same fraction of its energy reaching $dA_{strip,2}$, it follows that the fraction of energy from the *entire* $dA_{strip,1}$ that reaches $dA_{strip,2}$ is the same as the fraction for each element dA_1. Thus, the configuration factor between two infinitely long strips of differential width and having parallel generating lines must also be the same as for element dA_1 to $dA_{strip,2}$, or $\frac{1}{2}d(\sin \beta)$. The angle β is always in a plane normal to the generating lines of both strips.

EXAMPLE 7-5 Consider an infinitely long wedge-shaped groove as shown in cross section in Fig. 7-5. Determine the configuration factor between the differential strips dx and $d\xi$ in terms of x, ξ, and α.

As discussed in Example 7-4, the configuration factor is given by

$$dF_{dx-d\xi} = \tfrac{1}{2}d(\sin \beta) = \tfrac{1}{2}\cos \beta \, d\beta$$

From the construction in Fig. 7-5b, $\cos \beta = (\xi \sin \alpha)/L$. The quantity $d\beta$ is the angle subtended by the projection of $d\xi$ normal to L, that is,

$$d\beta = \frac{d\xi \cos (\alpha + \beta)}{L} = \frac{d\xi}{L}\frac{x \sin \alpha}{L}$$

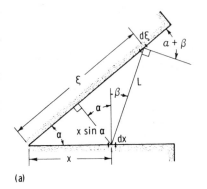

(a)

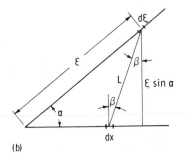

(b)

Figure 7-5 Configuration factor between two strips on sides of wedge groove. (*a*) Wedge-shaped groove geometry; (*b*) auxiliary construction.

From the law of cosines, $L^2 = x^2 + \xi^2 - 2x\xi \cos\alpha$. Then

$$dF_{dx-d\xi} = \frac{1}{2}\cos\beta\,d\beta = \frac{1}{2}\frac{x\xi\sin^2\alpha}{L^3}\,d\xi = \frac{1}{2}\frac{x\xi\sin^2\alpha}{(x^2 + \xi^2 - 2x\xi\cos\alpha)^{3/2}}\,d\xi$$

7-4.2 Configuration Factor between a Differential Element and a Finite Area

Consider now an isothermal black element dA_1 at temperature T_1 exchanging energy with a surface of finite area A_2 that is isothermal at temperature T_2 as shown in Fig. 7-6. The relations developed for exchange between differential elements must be extended to permit A_2 to be finite. Figure 7-6 shows (compare the solid and dashed cases) that the angle θ_2 will be different for different positions on A_2 and that θ_1 and S will also vary as different differential elements on A_2 are viewed from dA_1.

There are two configuration factors to be considered. The factor F_{d1-2} is from the differential area dA_1 to the finite area A_2, and dF_{2-d1} is from A_2 to dA_1. Each of these will be considered by using as the definition of configuration factor, the fraction of energy leaving one surface that reaches the second surface. To derive F_{d1-2}, note that the total energy radiated from the black surface element dA_1 is $dQ_1 = \sigma T_1^4\,dA_1$. The energy reaching dA_2 located on A_2 is

$$d^2Q'_{d1-d2} = \sigma T_1^4\frac{\cos\theta_1\cos\theta_2}{\pi S^2}\,dA_1\,dA_2$$

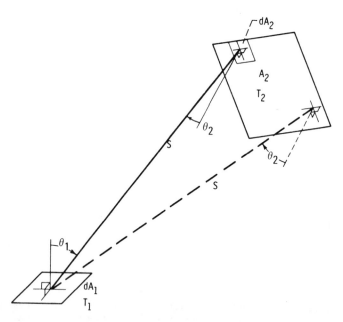

Figure 7-6 Radiant interchange between differential element and finite area.

Then, integrating over A_2 to obtain the energy reaching all of A_2 and dividing by the total energy leaving dA_1 result in

$$F_{d1-2} = \frac{\int_{A_2} d^2 Q'_{d1-d2}}{dQ_1} = \frac{\int_{A_2} \sigma T_1{}^4 (\cos\theta_1 \cos\theta_2 \, dA_1/\pi S^2) \, dA_2}{\sigma T_1{}^4 dA_1}$$

$$= \int_{A_2} \frac{\cos\theta_1 \cos\theta_2}{\pi S^2} \, dA_2 \qquad (7\text{-}13)$$

where the integration limits on A_2 extend over only the portion that can be viewed by dA_1. From Eq. (7-8) the quantity inside the integral of (7-13) is dF_{d1-d2}, so that F_{d1-2} can also be written as

$$F_{d1-2} = \int_{A_2} dF_{d1-d2} \qquad (7\text{-}14)$$

This merely expresses the fact that the fraction of the energy reaching A_2 is the sum of the fractions that reach all of the parts of A_2.

Now consider the configuration factor from the finite area A_2 to the elemental area dA_1. The energy reaching an elemental area dA_1 from a finite area A_2 is, by integrating (7-4) over A_2,

$$dQ_{2-d1} = dA_1 \int_{A_2} i'_{b,2} \frac{\cos\theta_1 \cos\theta_2}{S^2} \, dA_2 = dA_1 \int_{A_2} \sigma T_2{}^4 \frac{\cos\theta_1 \cos\theta_2}{\pi S^2} \, dA_2 \qquad (7\text{-}15)$$

The total hemispherical energy leaving A_2 is

$$Q_2 = \int_{A_2} \sigma T_2{}^4 \, dA_2 \qquad (7\text{-}16)$$

The configuration factor dF_{2-d1} is then the ratio of dQ_{2-d1} to Q_2 or

$$dF_{2-d1} = \frac{dA_1 \int_{A_2} \sigma T_2{}^4 (\cos\theta_1 \cos\theta_2/\pi S^2) \, dA_2}{\int_{A_2} \sigma T_2{}^4 \, dA_2}$$

$$= \frac{dA_1}{A_2} \int_{A_2} \frac{\cos\theta_1 \cos\theta_2}{\pi S^2} \, dA_2 \qquad (7\text{-}17)$$

The last integral on the right was obtained subject to the imposed condition that A_2 is isothermal. From (7-8) the quantity under the integral sign in Eq. (7-17) is dF_{d1-d2}, so the following alternative form is obtained:

$$dF_{2-d1} = \frac{dA_1}{A_2} \int_{A_2} dF_{d1-d2} \qquad (7\text{-}18)$$

Reciprocity for configuration factor between differential and finite areas. By use of (7-14) the factor dF_{2-d1} as given by (7-18) can be written as, $dF_{2-d1} = (dA_1/A_2)F_{d1-2}$ or

$$A_2 dF_{2-d1} = dA_1 F_{d1-2} \qquad (7\text{-}19)$$

which is a useful reciprocity relation.

Radiation interchange between differential and finite areas. The energy radiated from dA_1 that reaches A_2 is, from the definition of the configuration factor,

$$dQ_{d1-2} = \sigma T_1^4 dA_1 F_{d1-2}$$

Similarly, that radiated by A_2 and reaching dA_1 is

$$dQ_{2-d1} = \sigma T_2^4 A_2 dF_{2-d1}$$

The net transfer from dA_1 to A_2 is

$$dQ_{d1\rightleftharpoons2} = dQ_{d1-2} - dQ_{2-d1} = \sigma T_1^4 dA_1 F_{d1-2} - \sigma T_2^4 A_2 dF_{2-d1} \qquad (7\text{-}20)$$

By use of the reciprocity relation in Eq. (7-19) the net energy transfer can be expressed in the alternative forms

$$dQ_{d1\rightleftharpoons2} = \sigma(T_1^4 - T_2^4) dA_1 F_{d1-2} \qquad (7\text{-}21a)$$

$$dQ_{d1\rightleftharpoons2} = \sigma(T_1^4 - T_2^4) A_2 dF_{2-d1} \qquad (7\text{-}21b)$$

Some sample configuration factors involving a differential and a finite area. Certain geometries have configuration factors that can be represented by a simple closed-form algebraic solution (see Appendix C), while others require numerical integration of Eq. (7-13). Configuration factors can be tabulated for common geometries so that they need not be computed each time they are used. A list of references for available configuration factors is given in Appendix B.

Two geometries possessing closed-form configuration factors are given in the next examples, which illustrate how these factors are obtained.

EXAMPLE 7-6 An elemental area dA_1 is oriented perpendicular to a circular disk of finite area A_2 and outer radius r as shown in Fig. 7-7a. Find an equation describing the configuration factor F_{d1-2} in terms of the appropriate parameters h, l, and r.

The first step is to find expressions for the quantities inside the integral of (7-13) in terms of known quantities so that the integration can be carried out. The elemental area dA_2 is known in terms of the local radius on the disk and the angle φ as $dA_2 = \rho \, d\rho \, d\varphi$. Because the integral in (7-13) must be carried out over all ρ and φ, the other quantities in the integral must be put in terms of these two variables; this is done by using auxiliary constructions. Figure 7-7b is drawn to evaluate $\cos \theta_1$ and $\cos \theta_2$ which are seen to be

$$\cos \theta_1 = \frac{l + \rho \cos \varphi}{S} \quad \text{and} \quad \cos \theta_2 = \frac{h}{S}$$

Figure 7-7c allows evaluation of the remaining unknown S as $S^2 = h^2 + B^2$, where B^2 can be evaluated by using the geometric law of cosines on triangle aOb. This gives

$$B^2 = l^2 + \rho^2 - 2l\rho \cos(180 - \varphi) = l^2 + \rho^2 + 2l\rho \cos \varphi$$

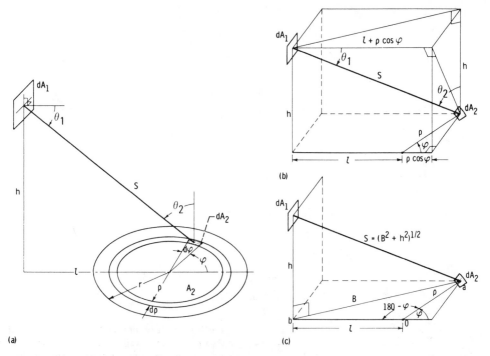

Figure 7-7 Geometry for radiative exchange between differential area and circular disk. (a) Geometry of problem; (b) auxiliary construction for determining cos θ_1 and cos θ_2; (c) auxiliary construction for determining S.

Substituting these relations into (7-13) results in

$$F_{d1-2} = \int_{A_2} \frac{\cos \theta_1 \cos \theta_2}{\pi S^2} \, dA_2 = \int_{A_2} \frac{h(l + \rho \cos \varphi)}{\pi S^4} \, \rho \, d\rho \, d\varphi$$

$$= \frac{h}{\pi} \int_{\rho=0}^{r} \int_{\varphi=0}^{2\pi} \frac{\rho(l + \rho \cos \varphi)}{(h^2 + l^2 + \rho^2 + 2\rho l \cos \varphi)^2} \, d\varphi \, d\rho$$

This integration is carried out using the symmetry of the configuration and is nondimensionalized to give, after considerable manipulation,

$$F_{d1-2} = \frac{2h}{\pi} \int_{\rho=0}^{r} \int_{\varphi=0}^{\pi} \frac{\rho(l + \rho \cos \varphi)}{(h^2 + \rho^2 + l^2 + 2\rho l \cos \varphi)^2} \, d\varphi \, d\rho$$

$$= \frac{2H}{\pi} \int_{\xi=0}^{R} \int_{\varphi=0}^{\pi} \frac{\xi(1 + \xi \cos \varphi)}{(H^2 + \xi^2 + 1 + 2\xi \cos \varphi)^2} \, d\varphi \, d\xi$$

$$= \frac{H}{2} \left\{ \frac{H^2 + R^2 + 1}{[(H^2 + R^2 + 1)^2 - 4R^2]^{1/2}} - 1 \right\}$$

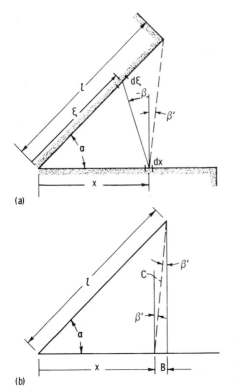

(a)

(b)

Figure 7-8 Configuration factor between one wall and strip on other wall of infinitely long wedge cavity. (*a*) Wedge-cavity geometry; (*b*) auxiliary construction to determine sin β'.

The nondimensionalization has been done by dividing numerator and denominator by l^4 and by letting $H = h/l$, $R = r/l$, and $\xi = \rho/l$. To find the net transfer of energy from dA_1 to A_2 in the configuration of Fig. 7-7, F_{d1-2} is evaluated by the previous expression and $dQ_{d1\rightleftharpoons2}$ is evaluated by using Eq. (7-21a). To avoid the complicated double integration for F_{d1-2}, the analysis can be carried out more conveniently by using the contour integration method of Sec. 7-5.4.

EXAMPLE 7-7 An infinitely long two-dimensional wedge cavity has an opening angle α. Derive an expression for the configuration factor from one wall of the wedge to a strip element of width dx on the other wall at a distance x from the wedge vertex as shown in Fig. 7-8a. (Such configurations approximate the geometries of long fins and ribs used in space radiators.)

From Example 7-4, the configuration factor between two infinitely long strip elements having parallel generating lines is $dF_{dx-d\xi} = \frac{1}{2}d(\sin\beta)$, where β is in a plane containing the normals of both strips. Note that β is measured clockwise from the normal of dx; Eq. (7-14) then gives

$$F_{dx-l} = \int_{\xi=0}^{l} dF_{dx-d\xi} = \int_{\beta=-\pi/2}^{0} \tfrac{1}{2}d(\sin\beta) + \int_{0}^{\beta'} \tfrac{1}{2}d(\sin\beta)$$

$$= \frac{\sin\beta}{2}\bigg|_{\beta=-\pi/2}^{0} + \frac{\sin\beta}{2}\bigg|_{\beta=0}^{\beta'} = \frac{1}{2} + \frac{\sin\beta'}{2}$$

The function $\sin \beta'$ can be found by the auxiliary construction of Fig. 7-8b to be

$$\sin \beta' = \frac{B}{C} = \frac{l \cos \alpha - x}{(x^2 + l^2 - 2xl \cos \alpha)^{1/2}}$$

Then,

$$F_{dx-l} = \frac{1}{2} + \frac{l \cos \alpha - x}{2(x^2 + l^2 - 2xl \cos \alpha)^{1/2}}$$

However, the problem requires dF_{l-dx}. Using the reciprocal relation of (7-19) gives

$$dF_{l-dx} = \frac{dx}{l} F_{dx-l} = dx \left[\frac{1}{2l} + \frac{\cos \alpha - x/l}{2(x^2 + l^2 - 2xl \cos \alpha)^{1/2}} \right]$$

By letting $X = x/l$, this can be placed in the dimensionless form

$$dF_{l-dx} = dX \left[\frac{1}{2} + \frac{\cos \alpha - X}{2(X^2 + 1 - 2X \cos \alpha)^{1/2}} \right]$$

The only parameters are the opening angle of the wedge and the dimensionless position from the vertex.

7-4.3 Configuration Factor for Two Finite Areas

Consider the configuration factor for radiation emitted from an isothermal surface A_1 shown in Fig. 7-9 and reaching A_2. By definition, F_{1-2} is the fraction of the energy leaving A_1 that arrives at A_2. The total energy leaving the black surface A_1 is $\sigma T_1^4 A_1$ since A_1 is isothermal at T_1. The radiation leaving an element dA_1 that reaches dA_2 was given previously as

$$d^2 Q'_{d1-d2} = \sigma T_1^4 \frac{\cos \theta_1 \cos \theta_2}{\pi S^2} dA_1 \, dA_2$$

If this is integrated over both A_1 and A_2, then the result will be the energy leaving A_1 that reaches A_2. The configuration factor is then found as

$$F_{1-2} = \frac{\int_{A_1} \int_{A_2} (\sigma T_1^4 \cos \theta_1 \cos \theta_2 / \pi S^2) \, dA_2 \, dA_1}{\sigma T_1^4 A_1}$$

$$F_{1-2} = \frac{1}{A_1} \int_{A_1} \int_{A_2} \frac{\cos \theta_1 \cos \theta_2}{\pi S^2} dA_2 \, dA_1 \tag{7-22}$$

This can be written in terms of the configuration factors involving differential areas as

$$F_{1-2} = \frac{1}{A_1} \int_{A_1} \int_{A_2} dF_{d1-d2} \, dA_1 = \frac{1}{A_1} \int_{A_1} F_{d1-2} \, dA_1 \tag{7-23}$$

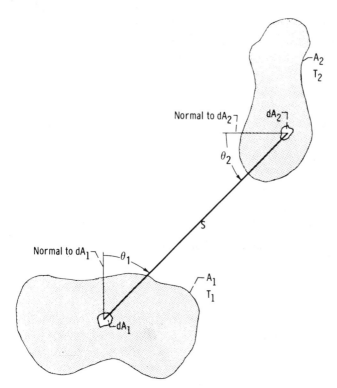

Figure 7-9 Geometry for energy exchange between finite areas.

In a manner similar to the derivation of (7-22), the configuration factor from A_2 to A_1 is found to be

$$F_{2-1} = \frac{1}{A_2} \int_{A_1} \int_{A_2} \frac{\cos\theta_1 \cos\theta_2}{\pi S^2} \, dA_2 \, dA_1 \qquad (7\text{-}24)$$

Reciprocity for configuration factor between finite areas. The double integrals in (7-22) and (7-24) are identical. Hence the reciprocity relation

$$A_1 F_{1-2} = A_2 F_{2-1} \qquad (7\text{-}25)$$

results. Further interrelations between configuration factors can be found by using (7-23) in conjunction with the reciprocity relations of (7-25) and (7-19), that is,

$$F_{2-1} = \frac{A_1}{A_2} F_{1-2} = \frac{A_1}{A_2} \frac{1}{A_1} \int_{A_1} F_{d1-2} \, dA_1 = \frac{1}{A_2} \int_{A_1} dF_{2-d1} \, A_2 = \int_{A_1} dF_{2-d1} \qquad (7\text{-}26)$$

Radiation exchange between finite areas. The energy radiated from A_1 that reaches A_2 is, from the definition of the configuration factor,

$$Q_{1-2} = \sigma T_1^4 A_1 F_{1-2}$$

Similarly, that radiated from A_2 and reaching A_1 is

$$Q_{2-1} = \sigma T_2^4 A_2 F_{2-1}$$

The net transfer from A_1 to A_2 is

$$Q_{1 \rightleftharpoons 2} = Q_{1-2} - Q_{2-1} = \sigma T_1^4 A_1 F_{1-2} - \sigma T_2^4 A_2 F_{2-1} \qquad (7\text{-}27)$$

By use of equation (7-25) this can be written in the two forms

$$Q_{1 \rightleftharpoons 2} = \sigma(T_1^4 - T_2^4) A_1 F_{1-2} \qquad (7\text{-}28a)$$

$$Q_{1 \rightleftharpoons 2} = \sigma(T_1^4 - T_2^4) A_2 F_{2-1} \qquad (7\text{-}28b)$$

EXAMPLE 7-8 Two isothermal plates of the same finite width and of infinite length are joined along one edge at angle α as shown in Fig. 7-8. Using the same nondimensional parameters as in Example 7-7, derive the configuration factor between the plates.

Example 7-7 gives the configuration factor between one plate and an infinite strip on the other plate as

$$dF_{l-dx} = \left[\frac{1}{2l} + \frac{\cos\alpha - x/l}{2(x^2 + l^2 - 2xl\cos\alpha)^{1/2}} \right] dx$$

Substituting into equation (7-26) gives

$$F_{l-l^*} = \int_{x=0}^{l^*} dF_{l-dx} = \int_0^{l^*} \left[\frac{1}{2l} + \frac{\cos\alpha - x/l}{2(x^2 + l^2 - 2xl\cos\alpha)^{1/2}} \right] dx$$

where, for convenience in labeling, the width of the side in Fig. 7-8 having element dx is specified as l^*. Using the dimensionless variable $X = x/l$ and the fact that $l^* = l$, this becomes

$$F_{l-l^*} = \int_0^1 \left[\frac{1}{2} + \frac{\cos\alpha - X}{2(X^2 + 1 - 2X\cos\alpha)^{1/2}} \right] dX$$

Carrying out the integration yields

$$F_{l-l^*} = 1 - \left(\frac{1 - \cos\alpha}{2} \right)^{1/2} = 1 - \sin\frac{\alpha}{2}$$

For the present case in which the two plate widths are equal, the only parameter is the angle α. Also, because the areas of the two sides are equal, the reciprocity relation [Eq. (7-25)] gives, as expected from symmetry,

$$F_{l-l^*} = F_{l^*-l}$$

7-4.4 Summary of Configuration-Factor and Energy-Exchange Relations

In Table 7-1 are summarized the net energy transfer equations, integral definitions of the configuration factors, and configuration-factor reciprocity relations.

Table 7-1 Summary of configuration-factor and energy-exchange relations

Geometry	Net energy transfer	Configuration factor	Reciprocity
Elemental area to elemental area	$d^2 Q'_{d1 \rightleftarrows d2}$ $= \sigma(T_1{}^4 - T_2{}^4)\, dA_1\, dF_{d1-d2}$	$dF_{d1-d2} = \dfrac{\cos\theta_1 \cos\theta_2}{\pi S^2}\, dA_2$	$dA_1\, dF_{d1-d2}$ $= dA_2\, dF_{d2-d1}$
Elemental area to finite area	$dQ_{d1 \rightleftarrows 2}$ $= \sigma(T_1{}^4 - T_2{}^4)\, dA_1 F_{d1-2}$	$F_{d1-2} = \displaystyle\int_{A_2} \dfrac{\cos\theta_1 \cos\theta_2}{\pi S^2}\, dA_2$	$dA_1 F_{d1-2}$ $= A_2\, dF_{2-d1}$
Finite area to finite area	$Q_{1 \rightleftarrows 2} = \sigma(T_1{}^4 - T_2{}^4) A_1 F_{1-2}$	F_{1-2} $= \dfrac{1}{A_1} \displaystyle\int_{A_1} \int_{A_2} \dfrac{\cos\theta_1 \cos\theta_2}{\pi S^2}\, dA_2\, dA_1$	$A_1 F_{1-2} = A_2 F_{2-1}$

7-5 METHODS FOR EVALUATING CONFIGURATION FACTORS

7-5.1 Configuration-Factor Algebra for Pairs of Surfaces

Because of the difficulty involved in directly computing configuration factors from the integral definitions in Table 7-1 for many geometries, it is desirable to utilize shortcut methods whenever possible. Such shortcuts can be obtained by using two concepts that have been developed in preceding sections: (1) the definition of configuration factor in terms of fractional intercepted energy and (2) the reciprocal relations. This section will show how these two concepts can be used to derive configuration factors for certain geometries from known configuration factors of other geometries. The interrelation between configuration factors is termed *configuration-factor algebra.*

Consider an arbitrary isothermal black area A_1 in Fig. 7-10 exchanging energy with a second area A_2. The configuration factor F_{1-2} is the fraction of the total energy emitted by A_1 that is incident upon A_2. If A_2 is divided into two parts A_3 and A_4, the fraction of the total energy leaving A_1 that is incident on A_3 and the fraction incident on A_4 must total to F_{1-2}. As a consequence, the following can be written:

$$F_{1-2} = F_{1-(3+4)} = F_{1-3} + F_{1-4} \tag{7-29}$$

Suppose then that F_{1-2} and F_{1-4} are known, but configuration factor F_{3-1} is desired. Then

$$F_{1-3} = F_{1-2} - F_{1-4} \tag{7-30}$$

The reciprocity relation, Eq. (7-25), gives

$$F_{3-1} = \frac{A_1}{A_3} F_{1-3} = \frac{A_1}{A_3} (F_{1-2} - F_{1-4}) \tag{7-31}$$

This is a powerful tool for obtaining new configuration factors from those previously computed. This method will be examined further by use of some examples.

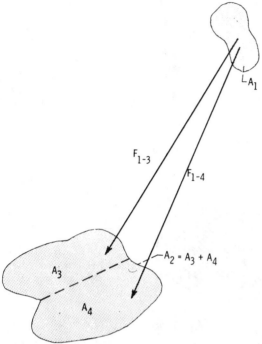

Figure 7-10 Energy exchange between finite areas with one area subdivided: $F_{1-3} + F_{1-4} = F_{1-2}$.

EXAMPLE 7-9 An elemental area dA_1 is oriented perpendicular to a ring of outer radius r_o and inner radius r_i as shown in Fig. 7-11. Derive an expression for the configuration factor $F_{d1-\text{ring}}$.

In Example 7-6, the configuration factor between element dA_1 and the entire disk of area A_2 and outer radius r_o was found to be

$$F_{d1-2} = \frac{H}{2} \left\{ \frac{H^2 + R_o^2 + 1}{[(H^2 + R_o^2 + 1)^2 - 4R_o^2]^{1/2}} - 1 \right\}$$

where $H = h/l$ and $R_o = r_o/l$. The configuration factor to the inner disk of area A_3 and radius r_i is similarly

$$F_{d1-3} = \frac{H}{2} \left\{ \frac{H^2 + R_i^2 + 1}{[(H^2 + R_i^2 + 1)^2 - 4R_i^2]^{1/2}} - 1 \right\}$$

where $R_i = r_i/l$. Using configuration-factor algebra, the desired configuration factor from dA_1 to the ring $A_2 - A_3$ is

$$F_{d1-\text{ring}} = F_{d1-2} - F_{d1-3}$$

$$= \frac{H}{2} \left\{ \frac{H^2 + R_o^2 + 1}{[(H^2 + R_o^2 + 1)^2 - 4R_o^2]^{1/2}} - \frac{H^2 + R_i^2 + 1}{[(H^2 + R_i^2 + 1)^2 - 4R_i^2]^{1/2}} \right\}$$

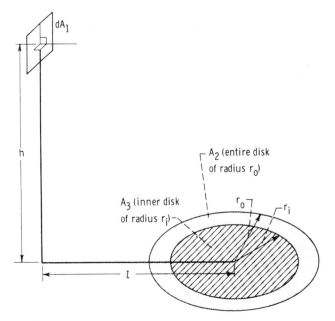

Figure 7-11 Interchange between elemental area and finite ring.

EXAMPLE 7-10 Suppose that the configuration factor is known between two parallel disks of arbitrary size whose centers lie on the same axis. From this, derive the configuration factor between the two rings A_2 and A_3 of Fig. 7-12. Give the answer in terms of known disk-to-disk factors from disk areas on the lower surface to disk areas on the upper surface.

The factor desired is F_{2-3}. From configuration-factor algebra, F_{2-3} is equal to $F_{2-3} = F_{2-(3+4)} - F_{2-4}$. The factor $F_{2-(3+4)}$ can be found from the reciprocal relation

$$A_2 F_{2-(3+4)} = (A_3 + A_4)F_{(3+4)-2}$$

Applying configuration-factor algebra to the right-hand side results in

$$A_2 F_{2-(3+4)} = (A_3 + A_4)(F_{(3+4)-(1+2)} - F_{(3+4)-1})$$
$$= (A_3 + A_4)F_{(3+4)-(1+2)} - (A_3 + A_4)F_{(3+4)-1}$$

Applying reciprocity to the right side gives

$$A_2 F_{2-(3+4)} = (A_1 + A_2)F_{(1+2)-(3+4)} - A_1 F_{1-(3+4)}$$

where the F factors on the right are both disk-to-disk factors from the lower surface to the upper.

Now the factor F_{2-4} remains to be determined. Again, apply the reciprocal relations and configuration-factor algebra to find

$$F_{2-4} = \frac{A_4}{A_2} F_{4-2} = \frac{A_4}{A_2}(F_{4-(1+2)} - F_{4-1}) = \frac{1}{A_2}[(A_1 + A_2)F_{(1+2)-4} - A_1 F_{1-4}]$$

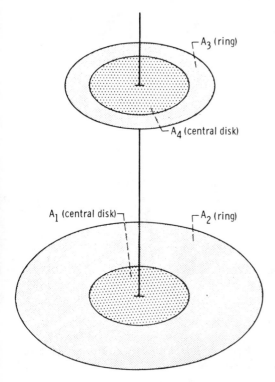

Figure 7-12 Interchange between parallel ring areas having common axis.

Substituting the relations for F_{2-4} and $F_{2-(3+4)}$ in the first equation gives

$$F_{2-3} = \frac{A_1 + A_2}{A_2} \left(F_{(1+2)-(3+4)} - F_{(1+2)-4} \right) - \frac{A_1}{A_2} \left(F_{1-(3+4)} - F_{1-4} \right)$$

and all configuration factors on the right-hand side of this equation are for exchange between two disks in the direction from disks on the lower surface to disks on the upper surface. The problem is now solved.

Because of the small differences of large numbers that can occur in obtaining an F factor by use of configuration-factor algebra (as might occur on the right side of the last equation of the preceding example), care must be taken that a sufficient number of significant figures are retained to ensure acceptable accuracy. Feingold [1] gives one example in which an error of 0.05% in a known factor causes an error of 57% in another factor computed from it by means of angle-factor algebra.

EXAMPLE 7-11 The internal surface of a hollow circular cylinder of radius R is radiating to a disk A_1 of radius r as shown in Fig. 7-13. Express the configuration factor from the cylindrical side A_3 to the disk in terms of disk-to-disk factors for the case of r less than R.

From any position on A_1 the solid angle subtended when viewing A_3 is the difference between the solid angle when viewing A_2, $d\omega_2$, and that viewing A_4,

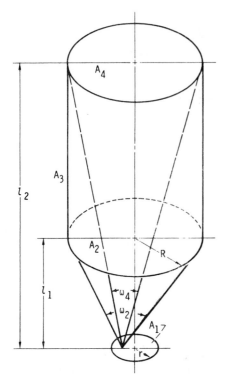

Figure 7-13 Internal surface of cylindrical cavity radiating to circular disk A_1 for $r < R$.

$d\omega_4$. This gives the F factor from an area element dA_1 on A_1 to area A_3 as $F_{d1-3} = F_{d1-2} - F_{d1-4}$. By integrating over A_1 and using Eq. (7-23), this can be written for the entire area A_1, as $F_{1-3} = F_{1-2} - F_{1-4}$. The factors on the right are between parallel disks. The final result for F from the cylindrical side A_3 to the disk A_1 is

$$F_{3-1} = \frac{A_1}{A_3} (F_{1-2} - F_{1-4})$$

From symmetry, the factor to any sector A_s of A_1 is $(A_s/A_1)F_{3-1}$.

There is a reciprocity relation that can be derived from the symmetry of a geometry. Consider the opposing areas in Fig. 7-14a. From the symmetry it is evident that $A_2 = A_4$ and $F_{2-3} = F_{4-1}$, so that $A_2 F_{2-3} = A_4 F_{4-1}$. From reciprocity, $A_4 F_{4-1} = A_1 F_{1-4}$. Hence, there is the derived relation

$$A_2 F_{2-3} = A_1 F_{1-4}$$

which relates the diagonal directions shown by the arrows on the figure. Similarly, the symmetry of Fig. 7-14b yields

$$A_2 F_{2-7} = A_3 F_{3-6}$$

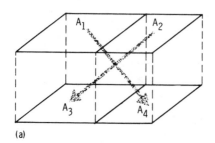

(a)

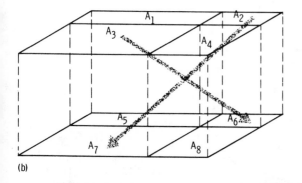

(b)

Figure 7-14 Geometry for reciprocity between opposing rectangles. (a) Two pairs of opposing rectangles: $A_1F_{1-4} = A_2F_{2-3}$; (b) four pairs of opposing rectangles: $A_2F_{2-7} = A_3F_{3-6}$.

Figure 7-15a shows four areas on two perpendicular rectangles having a common edge. Since these areas are of unequal sizes, there is no apparent symmetry relation. However, it will be shown that a valid relation is

$$A_1F_{1-2} = A_3F_{3-4} \tag{7-32}$$

To prove this, begin with the basic definition [Eq. (7-22)]; thus

$$A_1F_{1-2} = \frac{1}{\pi} \int_{A_1} \int_{A_2} \frac{\cos\theta_1 \cos\theta_2}{S^2} \, dA_2 \, dA_1$$

From Fig. 7-15b, $S^2 = (x_2 - x_1)^2 + y_1^2 + z_2^2$, $\cos\theta_1 = z_2/S$, and $\cos\theta_2 = y_1/S$. Then

$$A_1F_{1-2} = \frac{1}{\pi} \int_{x_1=0}^{c} \int_{y_1=0}^{a} \int_{x_2=c}^{c+d} \int_{z_2=0}^{b} \frac{y_1 z_2}{[(x_2-x_1)^2 + y_1^2 + z_2^2]^2} \, dz_2 \, dx_2 \, dy_1 \, dx_1 \tag{7-33a}$$

Similarly, reference to Fig. 7-15c reveals that

$$A_3F_{3-4} = \frac{1}{\pi} \int_{A_3} \int_{A_4} \frac{\cos\theta_3 \cos\theta_4}{S^2} \, dA_4 \, dA_3$$

$$= \frac{1}{\pi} \int_{x_3=c}^{c+d} \int_{y_3=0}^{a} \int_{x_4=0}^{c} \int_{z_4=0}^{b} \frac{y_3 z_4}{[(x_3-x_4)^2 + y_3^2 + z_4^2]^2} \, dz_4 \, dx_4 \, dy_3 \, dx_3 \tag{7-33b}$$

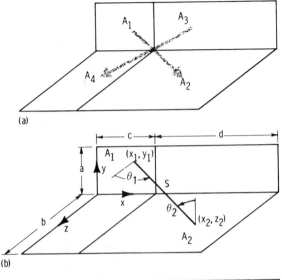

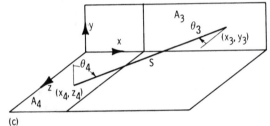

Figure 7-15 Reciprocity for diagonally opposite pairs of rectangles on two perpendicular planes having common edge. (*a*) Representation of reciprocity: $A_1 F_{1-2} = A_3 F_{3-4}$; (*b*) construction for F_{1-2}; (*c*) construction for F_{3-4}.

By interchanging the dummy integration variables x_1, y_1, x_2, and z_2 for x_4, y_3, x_3, and z_4, it is found that the integrals in Eqs. (7-33*a*) and (7-33*b*) are identical, thus proving (7-32).

EXAMPLE 7-12 If the configuration factor is known for two perpendicular rectangles with a common edge as shown in Fig. 7-16*a*, derive the configuration factor F_{1-6} for Fig. 7-16*b*.

First consider the geometry in Fig. 7-16*c* and derive the factor F_{7-6} as follows:

$$F_{(5+6)-(7+8)} = F_{(5+6)-7} + F_{(5+6)-8} = \frac{A_7}{A_5 + A_6} F_{7-(5+6)} + \frac{A_8}{A_5 + A_6} F_{8-(5+6)}$$

$$F_{(5+6)-(7+8)} = \frac{A_7}{A_5 + A_6} (F_{7-5} + F_{7-6}) + \frac{A_8}{A_5 + A_6} (F_{8-5} + F_{8-6})$$

Substitute $A_7 F_{7-6}$ for $A_8 F_{8-5}$ and solve the resulting relation for F_{7-6} to obtain

$$F_{7-6} = \frac{1}{2A_7} [(A_5 + A_6) F_{(5+6)-(7+8)} - A_7 F_{7-5} - A_8 F_{8-6}]$$

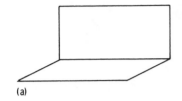

(a)

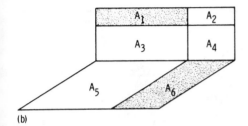

(b)

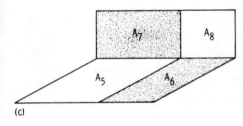

Figure 7-16 Orientation of areas for Example 7-12. (*a*) Perpendicular rectangles with one common edge; (*b*) geometry for F_{1-6}; (*c*) auxiliary geometry.

(c)

Now in Fig. 7-16*b*

$$F_{1-6} = \frac{A_6}{A_1} F_{6-1} = \frac{A_6}{A_1} F_{6-(1+3)} - \frac{A_6}{A_1} F_{6-3}$$

The factors $F_{6-(1+3)}$ and F_{6-3} are of the same type as F_{7-6} so that F_{1-6} can finally be written as

$$F_{1-6} = \frac{A_6}{A_1} \left\{ \frac{1}{2A_6} \left[(A_1 + A_2 + A_3 + A_4) F_{(1+2+3+4)-(5+6)} \right. \right.$$

$$\left. - A_6 F_{6-(2+4)} - A_5 F_{5-(1+3)} \right] - \frac{1}{2A_6} \left[(A_3 + A_4) F_{(3+4)-(5+6)} \right.$$

$$\left. \left. - A_6 F_{6-4} - A_5 F_{5-3} \right] \right\}$$

All the F factors on the right side are for two rectangles having one common edge as in Fig. 7-16*a*.

In formulating relations between configuration factors, it is sometimes useful to think in terms of energy quantities rather than fractions of energy leaving a surface that reach another surface. For example, in Fig. 7-10 the energy leaving A_2 that arrives

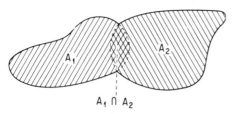

Figure 7-17 Union and intersection of finite areas.

at A_1 is proportional to $A_2 F_{2-1}$ and is equivalent to the sums of the energies from A_3 and A_4 that arrive at A_1. Thus,

$$(A_3 + A_4) F_{(3+4)-1} = A_3 F_{3-1} + A_4 F_{4-1} \tag{7-34}$$

This can also be proved by using reciprocity laws as follows:

$$(A_3 + A_4) F_{(3+4)-1} = A_1 F_{1-(3+4)} = A_1 F_{1-3} + A_1 F_{1-4} = A_3 F_{3-1} + A_4 F_{4-1}$$

7-5.2 Set-Theory Notation

In some instances, set-theory notation provides a convenient way of working with configuration factors. Consider two overlapping areas A_1 and A_2 as shown in Fig. 7-17. The double crosshatched area is defined in elementary set theory as the *intersection* of A_1 and A_2 and is denoted as $A_1 \cap A_2$. The net area enclosed by the continuous solid boundary around A_1 and A_2 is called the *union* of A_1 and A_2 and is denoted as $A_1 \cup A_2$.

An area A_E will have a configuration factor to $A_1 \cup A_2$ given by

$$F_{E-1\cup 2} = F_{E-1} + F_{E-2} - F_{E-1\cap 2} \tag{7-35}$$

This relation is made obvious by noting that the fraction of energy leaving A_E and incident upon $A_1 \cup A_2$ can be divided into, first, two portions: the fraction leaving A_E and incident upon A_1, and the fraction leaving A_E and incident upon A_2. However, these two fractions cover the portion $A_1 \cap A_2$ twice. To correct for this, we must subtract a fraction $F_{E-1\cap 2}$.

The configuration factor between an element dA_E and a rectangle in a parallel plane when the normal to the element passes through a corner of the rectangle (Fig. 7-18) is given in Appendix C. In Figs. 7-19a and b is shown a configuration made up of two such overlapping rectangles. The configuration factor between dA_E and the

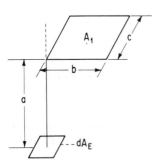

Figure 7-18 Geometry of known configuration factor.

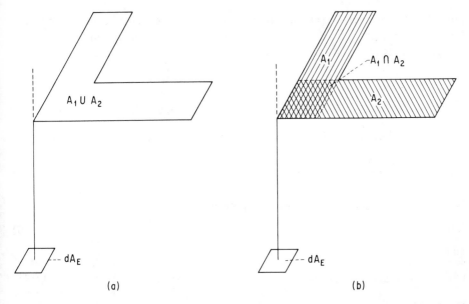

Figure 7-19 Geometry for derivation of configuration factor between element dA_E and L-shaped area $A_1 \cup A_2$. (a) Desired configuration; (b) configuration in terms of overlapping areas.

resulting L-shaped area is then, from Eq. (7-35),

$$F_{dE-1\cup2} = F_{dE-1} + F_{dE-2} - F_{dE-1\cap2} \qquad (7\text{-}36)$$

and all terms on the right are tabulated factors.

A somewhat more complex (and thus less obvious) geometry is shown in Fig. 7-20. Again, note that

$$F_{dE-1\cup2} = F_{dE-1} + F_{dE-2} - F_{dE-1\cap2} \qquad (7\text{-}37)$$

The factor F_{dE-1} between an element and a disk in a parallel plane with center on the

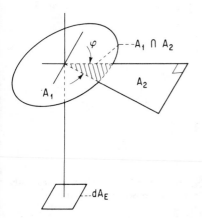

Figure 7-20 Geometry for derivation of configuration factor between element dA_E and overlapping circle and triangle.

normal to the element is given in Appendix C. F_{dE-2}, the factor between an element and a right triangle in a parallel plane with a vertex of the triangle on the normal to the element, is given in Example 7-18. Finally, $F_{dE-1\cap2}$ is observed to be, by symmetry,

$$F_{dE-1\cap2} = \frac{\varphi}{2\pi} F_{dE-1} \tag{7-38}$$

All terms on the right of (7-37) are known, and substituting (7-38) into (7-37) gives

$$F_{dE-1\cup2} = \frac{2\pi - \varphi}{2\pi} F_{dE-1} + F_{dE-2} \tag{7-39}$$

Note that the usual reciprocity relations apply in this notation:

$$(A_1 \cup A_2)F_{1\cup2-E} = A_E F_{E-1\cup2} \tag{7-40}$$

and

$$(A_1 \cap A_2)F_{1\cap2-E} = A_E F_{E-1\cap2} \tag{7-41}$$

7-5.3 Configuration Factors in Enclosures

To this point, only the radiation exchange between two black isothermal isolated surfaces has been considered, although subdivision of one or both of the surfaces into smaller portions has been examined. Consider the very useful class of problems in which the configuration factors are between black surfaces that form a complete enclosure. These configuration factors will later have a wider utility when nonblack diffuse enclosures are analyzed.

For an enclosure of N surfaces, such as shown in Fig. 7-21 (where $N = 8$ as an example), the entire energy leaving any surface inside the enclosure, for example surface A_k, must be incident on all the surfaces making up the enclosure. Thus all the

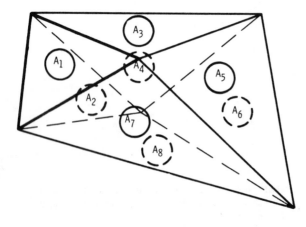

Figure 7-21 Isothermal enclosure composed of black surfaces.

fractions of energy leaving one surface and reaching the surfaces of the enclosure must total to unity; that is,

$$F_{k-1} + F_{k-2} + F_{k-3} + \cdots + F_{k-k} + \cdots + F_{k-N} = \sum_{j=1}^{N} F_{k-j} = 1 \qquad (7\text{-}42)$$

The factor F_{k-k} is included because when A_k is concave, it will intercept a portion of its own emitted energy.

EXAMPLE 7-13 Two black isothermal concentric spheres are exchanging energy. Find all the configuration factors for this geometry if the surface area of the inner sphere is A_1 and the area of the outer sphere is A_2.

All energy leaving A_1 is incident upon A_2, so immediately $F_{1-2} = 1$. Using the reciprocal relation reveals further that $F_{2-1} = A_1 F_{1-2}/A_2 = A_1/A_2$. Also, from (7-42), $F_{2-1} + F_{2-2} = 1$ or $F_{2-2} = 1 - F_{2-1} = (A_2 - A_1)/A_2$.

EXAMPLE 7-14 An isothermal cavity of internal area A_1 has a plane opening of area A_2. Derive an expression for the configuration factor of the internal surface of the cavity to itself.

Assume that a black plane surface A_2 replaces the cavity opening; this will have no effect on the F factor, which is a function only of geometry. Then $F_{2-1} = 1$ and $F_{1-2} = A_2 F_{2-1}/A_1 = A_2/A_1$, which is the configuration factor from the entire internal area to the opening. Since A_1 and A_2 form an enclosure, $F_{1-1} = 1 - F_{1-2} = (A_1 - A_2)/A_1$ which is the desired F factor.

EXAMPLE 7-15 An enclosure of triangular cross section is made up of three plane plates, each of finite width and infinite length (thus forming an infinitely long triangular prism). Derive an expression for the configuration factor between any two of the plates in terms of the plate widths L_1, L_2, and L_3.

For plate 1, $F_{1-2} + F_{1-3} = 1$. Using similar relations for each plate and multiplying through by the respective plate areas result in

$$A_1 F_{1-2} + A_1 F_{1-3} = A_1$$

$$A_2 F_{2-1} + A_2 F_{2-3} = A_2$$

$$A_3 F_{3-1} + A_3 F_{3-2} = A_3$$

By applying the reciprocal relations to some of the terms, these three equations become

$$A_1 F_{1-2} + A_1 F_{1-3} = A_1$$

$$A_1 F_{1-2} + A_2 F_{2-3} = A_2$$

$$A_1 F_{1-3} + A_2 F_{2-3} = A_3$$

thus giving three equations for the three unknown F factors. Subtracting the third from the second and adding the first give

$$F_{1-2} = \frac{A_1 + A_2 - A_3}{2A_1} = \frac{L_1 + L_2 - L_3}{2L_1}$$

For the special case of $L_1 = L_2$, this should reduce to the factor between infinitely long adjoint plates of equal width separated by an angle α as given in Example 7-8. For $L_1 = L_2$,

$$F_{1-2} = \frac{2L_1 - L_3}{2L_1} = 1 - \frac{L_3/2}{L_1} = 1 - \sin\frac{\alpha}{2}$$

which agrees with Example 7-8.

The set of three simultaneous equations from which the final result was derived in Example 7-15 will now be examined more closely. The first equation involves two unknowns F_{1-2} and F_{1-3}; the second equation has one additional unknown F_{2-3}; and the final equation has no additional unknowns. Generalizing the procedure from a three-surface enclosure to any N-sided enclosure made up of plane or convex surfaces shows that of N simultaneous equations, the first would involve $N - 1$ unknowns, the second $N - 2$ unknowns, and so forth. The total number of unknowns U is then

$$U = (N-1) + (N-2) + \cdots + 1 = N^2 - \sum_{j=1}^{N} j = N(N-1)/2 \qquad (7\text{-}43)$$

Thus, $N(N - 1)/2 - N = N(N - 3)/2$ factors must be provided. For a four-sided enclosure made up of planar or convex surfaces of known area, four equations relating $4(4 - 1)/2$ or six unknown configuration factors can be written. Specifying any two of these factors allows calculation of the rest by solving the set of four simultaneous equations.

If all the surfaces can view themselves, then the factor F_{k-k} must be included in each of the equations. Analyzing this situation, as previously done, shows that an N-sided enclosure allows the writing of N equations in $N(N + 1)/2$ unknowns. Thus $N(N + 1)/2 - N = N(N - 1)/2$ factors must be specified. For a four-sided enclosure, four equations involving 10 unknown F factors could be written. The specification of six factors would be required, and then the simultaneous relations could be solved to determine the remaining four factors. If only M of the surfaces can view themselves, then $N(N - 3)/2 + M$ factors must be specified.

Sowell and O'Brien [2] point out that in an enclosure with many surfaces, the configuration factors most easily specified or computed may not lend themselves to sequential calculation of the remaining factors. Reciprocity or the conservation relation (7-42) may be difficult to apply. They present a computer scheme using matrix algebra that allows calculation of all remaining factors in an N-surface planar or convex-surfaced enclosure once the needed configuration factors are specified.

7-5.4 Mathematical Techniques for the Evaluation of Configuration Factors

As shown by the summary of relations in Table 7-1, the evaluation of the configuration factors F_{d1-2} and F_{1-2} requires integration over the finite areas involved. A number of mathematical methods are useful in evaluating certain configuration factors when straightforward analytical integration methods become too cumbersome. These methods can encompass all techniques that are used in the evaluation of integrals, including numerical approaches.

A few methods that are especially valuable in dealing with configuration factors will be discussed here.

Hottel's crossed-string method. Consider the class of configurations such as long grooves in which all surfaces are assumed to extend infinitely far along one coordinate. Such surfaces can be generated by moving a line through space in such a way that it always remains parallel to its original position.

A typical configuration of this type is shown in cross section in Fig. 7-22. Suppose that the configuration factor is needed between A_1 and A_2 when some blockage of radiant transfer occurs because of the presence of other surfaces A_3 and A_4. To obtain F_{1-2}, first consider that A_1 may be concave. In this case draw the dashed line *agf* across A_1. Then draw in the dashed lines *cf* and *abc* to complete the enclosure *abcfga* which has three sides that are either convex or planar. The relation found in Example

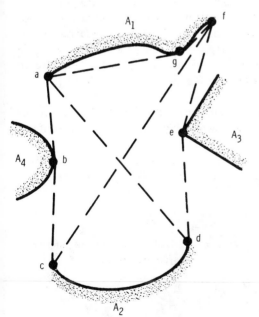

Figure 7-22 Hottel's crossed-string method for configuration-factor determination.

7-15 for enclosures of this type can be written as

$$A_{agf}F_{agf-abc} = \frac{A_{agf} + A_{abc} - A_{cf}}{2} \tag{7-44}$$

For the three-sided enclosure *adefga*, similar reasoning gives

$$A_{agf}F_{agf-def} = \frac{A_{agf} + A_{def} - A_{ad}}{2} \tag{7-45}$$

Further, note that

$$F_{agf-abc} + F_{agf-2} + F_{agf-def} = 1 \tag{7-46}$$

Substituting (7-44) and (7-45) into (7-46) results in

$$A_{agf}F_{agf-2} = A_{agf}(1 - F_{agf-abc} - F_{agf-def}) = \frac{A_{cf} + A_{ad} - A_{abc} - A_{def}}{2} \tag{7-47}$$

Now $F_{2-agf} = F_{2-1}$ since A_{agf} and A_1 subtend the same solid angle when viewed from A_2. Then, with the additional use of reciprocity, the left side of (7-47) can be written as

$$A_{agf}F_{agf-2} = A_2 F_{2-agf} = A_2 F_{2-1} = A_1 F_{1-2} \tag{7-48}$$

Substituting (7-48) into (7-47) results in

$$A_1 F_{1-2} = \frac{A_{cf} + A_{ad} - A_{abc} - A_{def}}{2} \tag{7-49}$$

If the dashed lines in Fig. 7-22 are imagined as being lengths of strings stretched tightly between the outer edges of the surfaces, then the term on the right of (7-49) is interpreted as one-half the total quantity formed by the sum of the lengths of the crossed strings connecting the outer edges of A_1 and A_2 minus the sum of the lengths of the uncrossed strings. This is a convenient way of determining configuration factors *in this type of two-dimensional geometry* and was first pointed out by Hottel [3].

EXAMPLE 7-16 Two infinitely long semicylindrical surfaces of radius R are separated by a minimum distance D as shown in Fig. 7-23a. Derive the configuration factor F_{1-2} for this case.

The length of crossed string *abcde* will be denoted as L_1, and of uncrossed string *ef* as L_2. From the symmetry of the problem, (7-49) may be written as

$$F_{1-2} = \frac{2L_1 - 2L_2}{2A_1} = \frac{L_1 - L_2}{\pi R}$$

The length L_2 is given by $L_2 = D + 2R$. The length of L_1 is twice the length *cde*. The segment of L_1 from c to d is found from right triangle $0cd$ to be

$$L_{1,c-d} = \left[\left(\frac{D}{2} + R\right)^2 - R^2\right]^{1/2} = \left[D\left(\frac{D}{4} + R\right)\right]^{1/2}$$

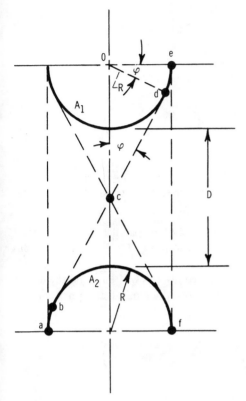

Figure 7-23a Example of crossed-string method. Configuration factor between infinitely long semicylindrical surfaces.

and the segment of L_1 from d to e is $L_{1,d-e} = R\varphi$. From triangle Ocd, the angle φ is given by $\varphi = \sin^{-1}[R/(D/2 + R)]$. Combining the known relations results in

$$F_{1-2} = \frac{L_1 - L_2}{\pi R} = \frac{2(L_{1,c-d} + L_{1,d-e}) - L_2}{\pi R}$$

$$= \frac{[4D(D/4 + R)]^{1/2} + 2R \sin^{-1}[R/(D/2 + R)] - D - 2R}{\pi R}$$

Letting $X = 1 + D/2R$ gives

$$F_{1-2} = \frac{2}{\pi}\left[(X^2 - 1)^{1/2} + \sin^{-1}\frac{1}{X} - X\right] \tag{7-50}$$

This can also be put in the form

$$F_{1-2} = \frac{2}{\pi}\left[(X^2 - 1)^{1/2} + \frac{\pi}{2} - \cos^{-1}\frac{1}{X} - X\right] \tag{7-51}$$

which agrees with the result in [4].

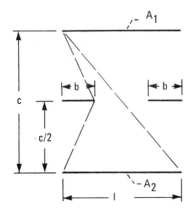

Figure 7-23b Example of crossed-string method. Partially blocked view between parallel strips.

EXAMPLE 7-17 The view between two infinitely long parallel strips of width l is partially blocked by strips of width b as shown in Fig. 7-23b. Obtain the configuration factor F_{1-2}.

The length of each crossed string (see dashed lines) is $\sqrt{l^2 + c^2}$, and the length of each uncrossed string is $2\sqrt{b^2 + (c/2)^2}$. From the crossed-string method the configuration factor is then,

$$F_{1-2} = \frac{\sqrt{l^2 + c^2} - 2\sqrt{b^2 + (c/2)^2}}{l} = \sqrt{1 + \left(\frac{c}{l}\right)^2} - \sqrt{\left(\frac{2b}{l}\right)^2 + \left(\frac{c}{l}\right)^2}$$

As expected, $F_{1-2} \to 0$ as $b \to l/2$.

Contour integration. Another tool that is useful in the evaluation of configuration factors is the application of Stokes' theorem for reduction of the multiple integration over a surface area to a single integration around the boundary of the area. This method is treated at some length by Moon [5], de Bastos [6], Sparrow and Cess [4], and Sparrow [7]. Consider a surface area A as shown in Fig. 7-24 with its boundary designated as C (where C is piecewise-continuous). An arbitrary point on the area is located at coordinate position x,y,z. At this point the normal to A is constructed, and the angles between this normal and the x, y, and z axes are designated as α, γ, and δ. Let the functions P, Q, and R be any twice-differentiable functions of x, y, and z. Stokes' theorem in three dimensions provides the following relation between an integral of P, Q, and R around the boundary C of the area and an integral over the surface A of the area:

$$\oint_C (P\,dx + Q\,dy + R\,dz)$$

$$= \int_A \left[\left(\frac{\partial R}{\partial y} - \frac{\partial Q}{\partial z}\right) \cos \alpha + \left(\frac{\partial P}{\partial z} - \frac{\partial R}{\partial x}\right) \cos \gamma + \left(\frac{\partial Q}{\partial x} - \frac{\partial P}{\partial y}\right) \cos \delta \right] dA \quad (7\text{-}52)$$

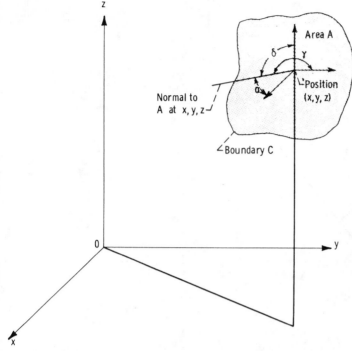

Figure 7-24 Geometry for quantities used in Stokes' theorem.

Now this relation will be applied to express area integrals in configuration-factor computations in terms of integrals around the boundaries of the areas.

Configuration factor between a differential and a finite area The integrand in the configuration factor F_{d1-2} is

$$\frac{\cos \theta_1 \cos \theta_2}{\pi S^2} dA_2$$

as shown in Table 7-1. In general, for the two cosines the following can be written (Fig. 7-25):

$$\cos \theta_1 = \frac{x_2 - x_1}{S} \cos \alpha_1 + \frac{y_2 - y_1}{S} \cos \gamma_1 + \frac{z_2 - z_1}{S} \cos \delta_1 \tag{7-53}$$

$$\cos \theta_2 = \frac{x_1 - x_2}{S} \cos \alpha_2 + \frac{y_1 - y_2}{S} \cos \gamma_2 + \frac{z_1 - z_2}{S} \cos \delta_2 \tag{7-54}$$

This follows from the relation that, for two vectors V_1 and V_2 having direction cosines (l_1, m_1, n_1) and (l_2, m_2, n_2), the cosine of the angle between the vectors is given by $l_1 l_2 + m_1 m_2 + n_1 n_2$.

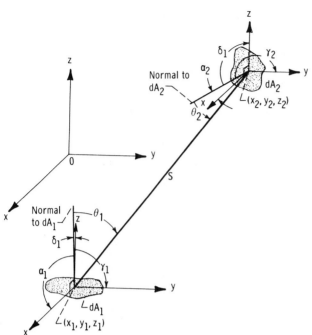

Figure 7-25 Geometry for contour integration.

Substituting Eqs. (7-53) and (7-54) into the integral relation for a configuration factor between a differential element and a finite area gives

$$F_{d1-2} = \int_{A_2} \frac{\cos \theta_1 \cos \theta_2}{\pi S^2} \, dA_2$$

$$= \frac{1}{\pi} \int_{A_2} \frac{(x_2 - x_1) \cos \alpha_1 + (y_2 - y_1) \cos \gamma_1 + (z_2 - z_1) \cos \delta_1}{S^4}$$

$$\times [(x_1 - x_2) \cos \alpha_2 + (y_1 - y_2) \cos \gamma_2 + (z_1 - z_2) \cos \delta_2] \, dA_2 \qquad (7\text{-}55)$$

Now let

$$l = \cos \alpha \qquad m = \cos \gamma \qquad n = \cos \delta \qquad (7\text{-}56a)$$

and

$$f = \frac{(x_2 - x_1) l_1 + (y_2 - y_1) m_1 + (z_2 - z_1) n_1}{\pi S^4} \qquad (7\text{-}56b)$$

Equation (7-55) can then be written in the abbreviated form

$$F_{d1-2} = \int_{A_2} [(x_1 - x_2) f l_2 + (y_1 - y_2) f m_2 + (z_1 - z_2) f n_2] \, dA_2 \qquad (7\text{-}57)$$

Comparison of (7-57) with the right side of (7-52) shows that Stokes' theorem can be applied if

$$\frac{\partial R}{\partial y_2} - \frac{\partial Q}{\partial z_2} = (x_1 - x_2)f \qquad (7\text{-}58a)$$

$$\frac{\partial P}{\partial z_2} - \frac{\partial R}{\partial x_2} = (y_1 - y_2)f \qquad (7\text{-}58b)$$

and

$$\frac{\partial Q}{\partial x_2} - \frac{\partial P}{\partial y_2} = (z_1 - z_2)f \qquad (7\text{-}58c)$$

Sparrow [7] indicates that useful solutions to these three equations are of the form

$$P = \frac{-m_1(z_2 - z_1) + n_1(y_2 - y_1)}{2\pi S^2} \qquad (7\text{-}59a)$$

$$Q = \frac{l_1(z_2 - z_1) - n_1(x_2 - x_1)}{2\pi S^2} \qquad (7\text{-}59b)$$

$$R = \frac{-l_1(y_2 - y_1) + m_1(x_2 - x_1)}{2\pi S^2} \qquad (7\text{-}59c)$$

Equation (7-52) is used to express F_{d1-2} in (7-57) as a contour integral; that is,

$$F_{d1-2} = \oint_{C_2} (P\,dx_2 + Q\,dy_2 + R\,dz_2) \qquad (7\text{-}60a)$$

Then P, Q, and R are substituted from (7-59), and the result is rearranged to obtain

$$\begin{aligned}
F_{d1-2} = \; & \frac{l_1}{2\pi} \oint_{C_2} \frac{(z_2 - z_1)\,dy_2 - (y_2 - y_1)\,dz_2}{S^2} \\
& + \frac{m_1}{2\pi} \oint_{C_2} \frac{(x_2 - x_1)\,dz_2 - (z_2 - z_1)\,dx_2}{S^2} \\
& + \frac{n_1}{2\pi} \oint_{C_2} \frac{(y_2 - y_1)\,dx_2 - (x_2 - x_1)\,dy_2}{S^2}
\end{aligned} \qquad (7\text{-}60b)$$

The double integration over area A_2 has been replaced by a set of three line integrals for the determination of F_{d1-2}. Sparrow [7] discusses the superposition properties of Eq. (7-57) that allows additions of the configuration factors of elements aligned parallel to the x, y, and z axes to obtain the factors for arbitrary orientation.

EXAMPLE 7-18 Determine the configuration factor F_{d1-2} from an element dA_1 to a right triangle as shown in Fig. 7-26.

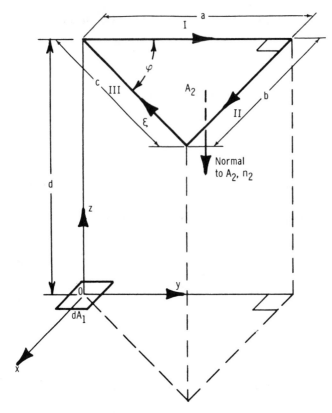

The normal to dA_1 is perpendicular to both the x and y axes and is thus parallel to z. The direction cosines for dA_1 are then $\cos\alpha_1 = l_1 = 0$, $\cos\gamma_1 = m_1 = 0$, and $\cos\delta_1 = n_1 = 1$, and (7-60b) becomes

$$F_{d1\text{-}2} = \frac{1}{2\pi}\oint_{C_2}\frac{(y_2-y_1)\,dx_2-(x_2-x_1)\,dy_2}{S^2}$$

Since dA_1 is situated at the origin of the coordinate system, $x_1 = y_1 = 0$ and $F_{d1\text{-}2}$ further reduces to

$$F_{d1\text{-}2} = \frac{1}{2\pi}\oint_{C_2}\frac{y_2\,dx_2-x_2\,dy_2}{S^2}$$

The distance S between dA_1 and any point (x_2,y_2,z_2) on A_2 is

$$S^2 = x_2{}^2+y_2{}^2+z_2{}^2 = x_2{}^2+y_2{}^2+d^2$$

The contour integration of the configuration-factor equation must now be carried out around the three sides of the right triangle. To keep the sign of $F_{d1\text{-}2}$ positive, the integration is performed by traveling around the boundary lines I, II, and III in a particular direction. The correct direction is that of a person walking around the boundary with his head in the direction of the normal n_2 and always keeping A_2 to his left. Along boundary line I, $x_2 = 0$, $dx_2 = 0$, and $0 \leqslant y_2 \leqslant a$.

On boundary II, $y_2 = a$, $dy_2 = 0$, and $0 \leqslant x_2 \leqslant b$. On boundary III, the integration is from $\xi = 0$ to c where ξ is a coordinate along the hypotenuse of the triangle so that

$$x_2 = (c - \xi) \sin \varphi \qquad\qquad dx_2 = -\sin \varphi \, d\xi$$
$$\text{and}$$
$$y_2 = (c - \xi) \cos \varphi \qquad\qquad dy_2 = -\cos \varphi \, d\xi$$

Substituting these quantities into the integral for F_{d1-2} gives

$$2\pi F_{d1-2} = \oint_{C_2} \frac{y_2 \, dx_2 - x_2 \, dy_2}{S^2} = \oint_{\text{I, II, III}} \frac{y_2 \, dx_2 - x_2 \, dy_2}{x_2^2 + y_2^2 + d^2}$$

$$2\pi F_{d1-2} = 0 + \int_{x_2=0}^{b} \frac{a \, dx_2}{x_2^2 + a^2 + d^2}$$

$$+ \int_{\xi=0}^{c} \frac{-(c - \xi) \cos \varphi \sin \varphi \, d\xi + (c - \xi) \sin \varphi \cos \varphi \, d\xi}{(c - \xi)^2 \sin^2 \varphi + (c - \xi)^2 \cos^2 \varphi + d^2}$$

or

$$2\pi F_{d1-2} = \int_0^b \frac{a \, dx_2}{x_2^2 + a^2 + d^2}$$

Use of the integral tables gives

$$F_{d1-2} = \frac{a}{2\pi(a^2 + d^2)^{1/2}} \tan^{-1} \frac{b}{(a^2 + d^2)^{1/2}}$$

or, in dimensionless variables,

$$F_{d1-2} = \frac{X}{2\pi(1 + X^2)^{1/2}} \tan^{-1} \frac{X \tan \varphi}{(1 + X^2)^{1/2}}$$

where $X = a/d$ and $\tan \varphi = b/a$.

Configuration factor between finite areas For configuration factors between two finite areas, substitution of (7-60b) into (7-23) gives

$$A_1 F_{1-2} = A_2 F_{2-1} = \int_{A_1} F_{d1-2} \, dA_1$$

$$= \frac{1}{2\pi} \oint_{C_2} \left[\int_{A_1} \frac{(y_2 - y_1)n_1 - (z_2 - z_1)m_1}{S^2} \, dA_1 \right] dx_2$$

$$+ \frac{1}{2\pi} \oint_{C_2} \left[\int_{A_1} \frac{(z_2 - z_1)l_1 - (x_2 - x_1)n_1}{S^2} \, dA_1 \right] dy_2$$

$$+ \frac{1}{2\pi} \oint_{C_2} \left[\int_{A_1} \frac{(x_2 - x_1)m_1 - (y_2 - y_1)l_1}{S^2} \, dA_1 \right] dz_2 \qquad (7\text{-}61)$$

where the integrals have been rearranged and dx_2, dy_2, and dz_2 factored out since these are independent of the area integration over A_1.

Stokes' theorem is applied in turn to each of the three area integrals. Consider the first of the integrals,

$$\int_{A_1} \frac{(y_2-y_1)n_1-(z_2-z_1)m_1}{S^2} dA_1$$

and compare it with the area integral in Stokes' theorem, Eq. (7-52). This gives

$$\frac{\partial R}{\partial y_1} - \frac{\partial Q}{\partial z_1} = 0$$

$$\frac{\partial P}{\partial z_1} - \frac{\partial R}{\partial x_1} = \frac{-(z_2-z_1)}{S^2}$$

$$\frac{\partial Q}{\partial x_1} - \frac{\partial P}{\partial y_1} = \frac{y_2-y_1}{S^2}$$

A solution to this set of partial differential equations [5] is $P = \ln S$, $Q = 0$, and $R = 0$; the area integral becomes, by use of Eq. (7-52) to convert it into a surface integral,

$$\int_{A_1} \frac{(y_2-y_1)n_1-(z_2-z_1)m_1}{S^2} dA_1 = \oint_{C_1} \ln S\, dx_1$$

By applying Stokes' theorem in a similar fashion to the other two integrals in (7-61), that equation can be written as

$$A_1 F_{1-2} = \frac{1}{2\pi} \oint_{C_2} \left(\oint_{C_1} \ln S\, dx_1 \right) dx_2 + \frac{1}{2\pi} \oint_{C_2} \left(\oint_{C_1} \ln S\, dy_1 \right) dy_2$$

$$+ \frac{1}{2\pi} \oint_{C_2} \left(\oint_{C_1} \ln S\, dz_1 \right) dz_2$$

or more compactly as

$$F_{1-2} = \frac{1}{2\pi A_1} \oint_{C_1} \oint_{C_2} (\ln S\, dx_2\, dx_1 + \ln S\, dy_2\, dy_1 + \ln S\, dz_2\, dz_1) \qquad (7\text{-}62)$$

Thus the integrations over two areas, which would involve integrating over four variables, have been replaced by integrations over the two surface boundaries. This allows considerable computational savings when numerical evaluations must be carried out, and can sometimes result in analytical integration being possible where it could not be carried out for the quadruple integral over the areas.

EXAMPLE 7-19 Using the contour-integration method, formulate the configuration factor for parallel rectangles as shown in Fig. 7-27.

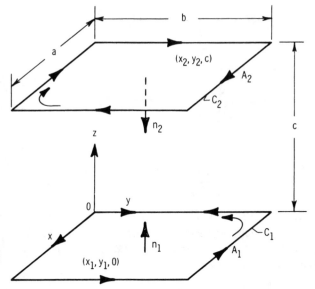

Figure 7-27 Contour integration to determine configuration between two parallel rectangles.

Note that on both surfaces dz will be zero. First integrate (7-62) around the boundary C_2. The value of S to be used in (7-62) is measured from an arbitrary point $(x_1, y_1, 0)$ on A_1 to a point on the portion of the boundary C_2 being considered. This gives

$$
F_{1-2} = \frac{1}{2\pi ab} \oint_{C_1} \left\{ \int_{y_2=0}^{b} \ln [x_1^2 + (y_2 - y_1)^2 + c^2]^{1/2} \, dy_2 \right.
$$

$$
\left. + \int_{y_2=b}^{0} \ln [(a - x_1)^2 + (y_2 - y_1)^2 + c^2]^{1/2} \, dy_2 \right\} dy_1
$$

$$
+ \frac{1}{2\pi ab} \oint_{C_1} \left\{ \int_{x_2=0}^{a} \ln [(x_2 - x_1)^2 + (b - y_1)^2 + c^2]^{1/2} \, dx_2 \right.
$$

$$
\left. + \int_{x_2=a}^{0} \ln [(x_2 - x_1)^2 + y_1^2 + c^2]^{1/2} \, dx_2 \right\} dx_1
$$

Then carrying the integration out over C_1 gives, in this case, eight integrals. The first four corresponding to the first two integrals of the previous equation are written out as

$$
2\pi ab F_{1-2} = \int_{y_1=0}^{b} \int_{y_2=0}^{b} \ln [a^2 + (y_2 - y_1)^2 + c^2]^{1/2} \, dy_2 \, dy_1
$$

$$
+ \int_{y_1=b}^{0} \int_{y_2=0}^{b} \ln [(y_2 - y_1)^2 + c^2]^{1/2} \, dy_2 \, dy_1
$$

$$+ \int_{y_1=0}^{b} \int_{y_2=b}^{0} \ln \left[(y_2 - y_1)^2 + c^2 \right]^{1/2} dy_2\, dy_1$$

$$+ \int_{y_1=b}^{0} \int_{y_2=b}^{0} \ln \left[a^2 + (y_2 - y_1)^2 + c^2 \right]^{1/2} dy_2\, dy_1$$

$$+ \text{(4 integral terms in } x)$$

$$= \int_{y_1=0}^{b} \int_{y_2=0}^{b} \ln \left[\frac{a^2 + (y_2 - y_1)^2 + c^2}{(y_2 - y_1)^2 + c^2} \right] dy_2\, dy_1$$

$$+ \int_{x_1=0}^{a} \int_{x_2=0}^{a} \ln \left[\frac{(x_2 - x_1)^2 + b^2 + c^2}{(x_2 - x_1)^2 + c^2} \right] dx_2\, dx_1$$

and the configuration factor is now given by the sum of two integrals. These can be integrated analytically by expressing each integrand as the difference of two log functions and letting $y_2 - y_1$ and $x_2 - x_1$ be new variables to reduce the integrals to standard forms. The final result is in Appendix C.

Differentiation of known factors. A further extension of configuration-factor algebra is the generation of configuration factors between differential elements by differentiating known factors between finite elements. This technique is very valuable in certain cases, and is best demonstrated by the use of an example.

EXAMPLE 7-20 As part of the determination of radiative exchange in a square channel whose temperature varies longitudinally, it is desired to find the configuration factor dF_{d1-d2} between an element dA_1 at one corner of the channel end and a differential length of wall section dA_2 as shown in Fig. 7-28 a.

Configuration-factor algebra plus differentiation can be used to find the required factor. Refer to Fig. 7-28 b. Since the fraction of energy leaving dA_1 that reaches dA_2 is the difference between the fractions reaching the squares A_3 and A_4, the factor dF_{d1-d2} is the difference between F_{d1-3} and F_{d1-4}. Then

$$dF_{d1-d2} = F_{d1-3} - F_{d1-4} = -\left. \frac{\Delta F_{d1-\square}}{\Delta x} \Delta x \right|_{\Delta x \to 0} = -\frac{\partial F_{d1-\square}}{\partial x} dx$$

Thus, if the configuration factor $F_{d1-\square}$ between a corner element and a square in a parallel plane were known, the derivative of this factor with respect to the separation distance could be used to give the required factor.

From Example 7-18, the configuration factor between a corner element and a parallel isosceles right triangle is given by setting $\tan \varphi = 1$ in the expression derived for a general right triangle. This yields (for the present case, where the distance $d = x$)

$$F_{d1-\triangle} = \frac{a}{2\pi(a^2 + x^2)^{1/2}} \tan^{-1} \frac{a}{(a^2 + x^2)^{1/2}}$$

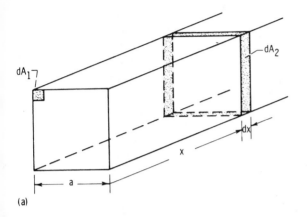

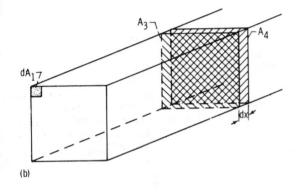

Figure 7-28 Derivation of configuration factor between differential length of square channel and element at corner of channel end. (*a*) Configuration factor between dA_1 and differential length of channel wall dA_2, (*b*) configuration factor between dA_1 and squares A_3 and A_4.

Inspection shows that, by symmetry, the factor between a corner element and a square is twice the factor $F_{d1-\triangle}$ The required factor dF_{d1-d2} is then

$$dF_{d1-d2} = -\frac{\partial F_{d1-\square}}{\partial x}\, dx = -\frac{a\, dx}{\pi}\frac{\partial}{\partial x}\left[\frac{1}{(a^2+x^2)^{1/2}}\tan^{-1}\frac{a}{(a^2+x^2)^{1/2}}\right]$$

$$= \frac{ax\, dx}{\pi(a^2+x^2)^{3/2}}\left[\tan^{-1}\frac{a}{(a^2+x^2)^{1/2}}+\frac{a(a^2+x^2)^{1/2}}{x^2+2a^2}\right]$$

$$= \frac{X\, dX}{\pi(1+X^2)^{3/2}}\left[\tan^{-1}\frac{1}{(1+X^2)^{1/2}}+\frac{(1+X^2)^{1/2}}{2+X^2}\right]$$

where $X = x/a$.

More generally, start with the configuration factor F_{1-2} for two parallel areas A_1 and A_2 that are cross sections of a cylindrical channel of arbitrary cross-sectional shape (Fig. 7-29*a*). This factor depends on the spacing $|x_2 - x_1|$ between the two areas and includes blockage due to the channel wall (that is, it is the factor by which A_2 is viewed from A_1 with the channel wall present). Note that for simple geometries such as a circular tube or rectangular channel the wall blockage is zero. The factor between

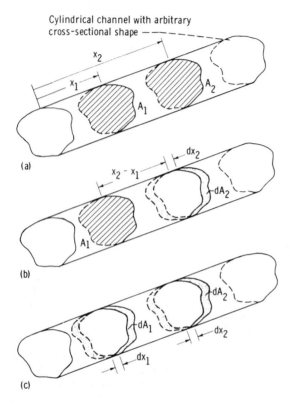

Figure 7-29 Configuration factors for differential areas as derived by differentiation of factor for finite areas. (a) Two finite areas, F_{1-2}, (b) finite to differential area, $dF_{1-d2} = -(\partial F_{1-2}/\partial x_2)\, dx_2$, ($c$) two differential areas, $dF_{d1-d2} = -(A_1/dA_1) (\partial^2 F_{1-2}/\partial x_1\, \partial x_2)\, dx_2\, dx_1$.

A_1 and dA_2 in Fig. 7-29b is then given by

$$dF_{1-d2} = -\frac{\partial F_{1-2}}{\partial x_2}\, dx_2 \tag{7-63}$$

as in Example 7-20. Equation (7-63) will now be used to obtain dF_{d1-d2}, the configuration factor between the two differential area elements in Fig. 7-29c.

By recriprocity,

$$F_{d2-1} = \frac{-A_1}{dA_2}\frac{\partial F_{1-2}}{\partial x_2}\, dx_2$$

Then in a fashion similar to the derivation of (7-63),

$$dF_{d2-d1} = \frac{\partial F_{d2-1}}{\partial x_1}\, dx_1$$

Substituting F_{d2-1} results in

$$dF_{d2-d1} = -\frac{A_1}{dA_2}\frac{\partial^2 F_{1-2}}{\partial x_1\, \partial x_2}\, dx_2\, dx_1 \tag{7-64a}$$

or after using reciprocity,

$$dF_{d1-d2} = -\frac{A_1}{dA_1} \frac{\partial^2 F_{1-2}}{\partial x_1 \partial x_2} dx_2 \, dx_1 \qquad (7\text{-}64b)$$

Hence, by two differentiations the factor dF_{d1-d2} can be found from F_{1-2} for the cylindrical configuration under consideration.

7-5.5 The Unit-Sphere Method

Experimental determination of configuration factors is possible by use of the *unit-sphere method* introduced by Wilhelm Nusselt [8].

If a hemisphere of unit radius is constructed over the area element dA_1 in Fig. 7-30, then the configuration factor from dA_1 to some area A_2 is, by Eq. (7-13),

$$F_{d1-2} = \frac{1}{\pi} \int_{A_2} \cos \theta_1 \frac{\cos \theta_2 \, dA_2}{S^2} = \frac{1}{\pi} \int_{A_2} \cos \theta_1 \, d\omega_1$$

Note that $d\omega_1$ is the projection of dA_2 onto the surface of the hemisphere, because

$$d\omega_1 = \frac{dA_s}{r^2} = dA_s = \frac{\cos \theta_2 \, dA_2}{S^2}$$

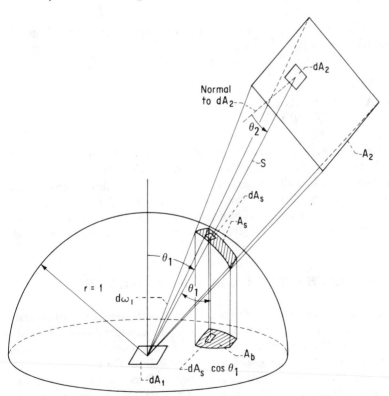

Figure 7-30 Geometry of unit-sphere method for obtaining configuration factors.

where r is the radius of the unit hemisphere. The configuration factor then becomes

$$F_{d1-2} = \frac{1}{\pi} \int_{A_s} \cos \theta_1 \, dA_s$$

However, $dA_s \cos \theta_1$ is the projection of dA_s onto the base of the hemisphere. It follows that integrating $\cos \theta_1 \, dA_s$ gives the projection A_b of A_s onto the base of the hemisphere, or

$$F_{d1-2} = \frac{1}{\pi} \int_{A_s} \cos \theta_1 \, dA_s = \frac{A_b}{\pi}$$

This relation forms the basis of several graphical and experimental methods of configuration-factor determination. In one such method, a hemisphere mirrored on the outside is placed over the area element dA_1. A photograph taken by a camera placed above the hemisphere and precisely normal to dA_1 then shows the projection of A_2, which we have called A_b. The measurement of A_b on the photograph then leads to the determination of F_{d1-2} by

$$F_{d1-2} = \frac{A_b}{\pi r_e^2}$$

where r_e is the radius of the experimental mirrored hemisphere. A means of optical projection is given in [9].

7-6 COMPILATION OF REFERENCES FOR KNOWN CONFIGURATION FACTORS

Many configuration factors have been given in analytical form or tabulated for specific geometries, and these formulas and tabulations are spread throughout the literature. Rather than attempt to gather all the factors here, a feat that would require a volume of considerable size, another course has been followed. In Appendix B, a list of geometries for which configuration factors are available and a reference list to aid in finding these factors are given. Some factors are given in Appendix C for convenient use.

7-7 RADIATION EXCHANGE IN A BLACK ENCLOSURE

In the preceding parts of this chapter, the net energy transfer between two separate surfaces or surface elements has been examined, and the configuration factor introduced. In this section, these ideas are generalized to consider the energy transfer *within* an enclosure composed of black surfaces that are individually isothermal.

In practice, the interior surfaces of an enclosure, such as a furnace, may not be isothermal. In such a case, the various nonisothermal surfaces are subdivided into smaller portions that can be considered individually isothermal. The theory for a black

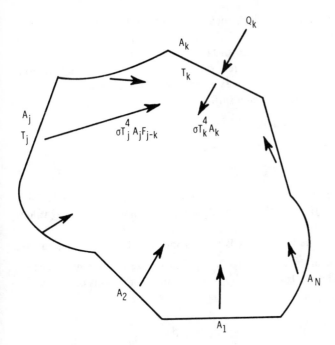

Figure 7-31 Enclosure composed of N black isothermal surfaces (shown in cross section for simplicity).

enclosure, which is an ideal case, will serve as an introduction to less restrictive theory in succeeding chapters.

Perform a heat balance on a typical surface A_k (Fig. 7-31). The combined energy supplied to A_k by all sources other than radiation, to maintain A_k at T_k, is Q_k. The Q_k could be composed of convection to the inside of the wall and/or conduction through the wall from a source outside. For wall cooling such as by channels within the wall, the contribution to Q_k is negative. The emission from A_k is $\sigma T_k{}^4 A_k$. The radiant energy received by A_k from another surface A_j is $\sigma T_j{}^4 A_j F_{j-k}$. The heat balance is then

$$Q_k = \sigma T_k{}^4 A_k - \sum_{j=1}^{N} \sigma T_j{}^4 A_j F_{j-k} \tag{7-65}$$

where the summation includes energy arriving from all surfaces of the enclosure including A_k if A_k is concave. Equation (7-65) can be written in some alternative forms. Applying reciprocity to the terms in the summation results in

$$Q_k = \sigma T_k{}^4 A_k - \sum_{j=1}^{N} \sigma T_j{}^4 A_k F_{k-j} \tag{7-66}$$

Also, for a complete enclosure, from (7-42), $\sum_{j=1}^{N} F_{k-j} = 1$ so that

$$Q_k = \sigma T_k{}^4 A_k \sum_{j=1}^{N} F_{k-j} - \sigma A_k \sum_{j=1}^{N} T_j{}^4 F_{k-j} = \sigma A_k \sum_{j=1}^{N} (T_k{}^4 - T_j{}^4) F_{k-j} \tag{7-67}$$

This is in the form of a sum of the net energy transferred from A_k to each surface.

EXAMPLE 7-21 The three-sided black enclosure of Example 7-15 has its surfaces maintained at temperatures T_1, T_2, and T_3, respectively. Determine the amount of energy that must be supplied to each surface per unit time by means other than radiation to maintain these temperatures—this is also the net radiative loss from each surface resulting from the radiative exchange within the enclosure.

Equation (7-67) is written for each surface as

$$Q_1 = A_1 F_{1-2} \sigma(T_1^4 - T_2^4) + A_1 F_{1-3} \sigma(T_1^4 - T_3^4)$$

$$Q_2 = A_2 F_{2-1} \sigma(T_2^4 - T_1^4) + A_2 F_{2-3} \sigma(T_2^4 - T_3^4)$$

$$Q_3 = A_3 F_{3-1} \sigma(T_3^4 - T_1^4) + A_3 F_{3-2} \sigma(T_3^4 - T_2^4)$$

The configuration factors have been found for this geometry in Example 7-15. Thus all factors on the right of this set of equations are known, and the Q values may be computed directly.

A check on the numerical results is that, from overall energy conservation, the net Q added to the entire enclosure, $\sum\limits_{k=1}^{N} Q_k$, must be zero to maintain steady-state temperatures. This is also shown by using reciprocal relations on the set of Q equations to obtain

$$\sum_{k=1}^{3} Q_k = A_1 F_{1-2} \sigma(T_1^4 - T_2^4) + A_1 F_{1-3} \sigma(T_1^4 - T_3^4)$$
$$+ A_1 F_{1-2} \sigma(T_2^4 - T_1^4) + A_2 F_{2-3} \sigma(T_2^4 - T_3^4)$$
$$+ A_1 F_{1-3} \sigma(T_3^4 - T_1^4) + A_2 F_{2-3} \sigma(T_3^4 - T_2^4)$$
$$= 0$$

EXAMPLE 7-22 The enclosure of Example 7-15 has two of its sides maintained at temperatures T_1 and T_2. The third side is an insulated (adiabatic) surface, $Q_3 = 0$. Determine Q_1, Q_2, and T_3.

Again Eq. (7-67) can be written for each surface as

$$Q_1 = A_1 F_{1-2} \sigma(T_1^4 - T_2^4) + A_1 F_{1-3} \sigma(T_1^4 - T_3^4)$$

$$Q_2 = A_2 F_{2-1} \sigma(T_2^4 - T_1^4) + A_2 F_{2-3} \sigma(T_2^4 - T_3^4)$$

$$0 = A_3 F_{3-1} \sigma(T_3^4 - T_1^4) + A_3 F_{3-2} \sigma(T_3^4 - T_2^4)$$

The final equation is solved for T_3, the only unknown in that equation. This T_3 is then inserted into the first two equations to obtain Q_1 and Q_2.

EXAMPLE 7-23 A very long black heated tube A_1 of length L is enclosed by a concentric black split cylinder as shown in Fig. 7-32. The diameter of the split cylinder is twice that of the heated tube, and one-half as much energy *flux* is to be removed from the upper area A_3 of the split cylinder as from the lower area A_2. If $T_1 = 1700$ K and a heat flux $Q_1/A_1 = 3 \times 10^5$ W/m² is supplied to the heated tube, what are the values of T_2, T_3, Q_2, and Q_3? Neglect the effect of the tube ends.

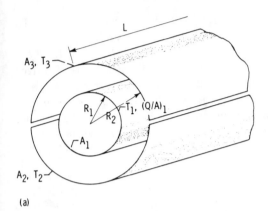

(a)

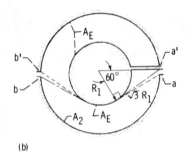

(b)

Figure 7-32 Radiant energy exchange in split-circular cylinder configuration, $L \gg R_2$. (*a*) Geometry of enclosure; (*b*) auxiliary construction to determine F_{2-2}.

Writing (7-67) for each surface gives

$$Q_1 = A_1 F_{1-2}\sigma(T_1{}^4 - T_2{}^4) + A_1 F_{1-3}\sigma(T_1{}^4 - T_3{}^4)$$

$$Q_2 = A_2 F_{2-1}\sigma(T_2{}^4 - T_1{}^4) + A_2 F_{2-3}\sigma(T_2{}^4 - T_3{}^4)$$

$$Q_3 = A_3 F_{3-1}\sigma(T_3{}^4 - T_1{}^4) + A_3 F_{3-2}\sigma(T_3{}^4 - T_2{}^4)$$

From the geometry,

$$\frac{A_1}{A_3} = \frac{A_1}{A_2} = \frac{\pi D_1 L}{\frac{1}{2}\pi D_2 L} = 1$$

since $D_2 = 2D_1$. From an energy balance, $Q_1 + Q_2 + Q_3 = 0$ and, since $A_1 = A_2 = A_3$, $Q_1/A_1 + Q_2/A_2 + Q_3/A_3 = 0$. From the statement of the problem, $Q_3/A_3 = Q_2/2A_2$, and this yields

$$\frac{Q_2}{A_2} = -\frac{2}{3}\frac{Q_1}{A_1} = -2.0 \times 10^5 \text{ W/m}^2$$

$$\frac{Q_3}{A_3} = -\frac{1}{3}\frac{Q_1}{A_1} = -1.0 \times 10^5 \text{ W/m}^2$$

From the symmetry of the geometry and configuration-factor algebra, it is known that $F_{1-2} = F_{1-3} = 1/2$, $F_{2-1} = F_{3-1} = A_1 F_{1-3}/A_3 = 1/2$, and $F_{2-3} = F_{3-2}$. To determine F_{2-3}, it is known that $F_{2-1} + F_{2-2} + F_{2-3} = 1$. Using $F_{2-1} = 1/2$ gives $F_{2-3} = 1/2 - F_{2-2}$. In the auxiliary construction of Fig. 7-32b, $F_{2-2} = 1 - F_{2-E}$. The effective area A_E has been drawn in to leave unchanged the view of surface 2 to itself and to simplify the geometry so that the crossed-string method can be used to determine F_{2-E}. The uncrossed strings extending from a to a' and b to b' have zero length. The crossed strings extend from a to b' and a' to b, and each has length $2\sqrt{3}\,R_1 + \pi R_1/3$. Then from Sec. 7-5.4 (Hottel's method) and the fact that $A_2 = A_1 = 2\pi R_1$,

$$F_{2-E} = \frac{2\sqrt{3}R_1 + \pi R_1/3}{2\pi R_1} = \frac{\sqrt{3}}{\pi} + \frac{1}{6}$$

It then follows

$$F_{2-3} = \frac{1}{2} - F_{2-2} = \frac{1}{2} - (1 - F_{2-E}) = \frac{\sqrt{3}}{\pi} - \frac{1}{3} = 0.218$$

By using this information, the energy exchange equations can now be written as

$$3 \times 10^5 = \frac{\sigma}{2}(1700^4 - T_2^4) + \frac{\sigma}{2}(1700^4 - T_3^4)$$

$$-2.0 \times 10^5 = \frac{\sigma}{2}(T_2^4 - 1700^4) + 0.218\sigma(T_2^4 - T_3^4)$$

$$-1.0 \times 10^5 = \frac{\sigma}{2}(T_3^4 - 1700^4) + 0.218\sigma(T_3^4 - T_2^4)$$

Adding the second and third equations results in the first, so only two of the equations are independent. Solving the first and second equations gives $T_2 = 1216$ K and $T_3 = 1418$ K.

7-8 HISTORICAL NOTE ON CONFIGURATION FACTORS

The first published statement that the geometrical factors in the radiant interchange equations could be separated from the energy factors has proved to be elusive. Certainly Nusselt realized that this was so in his derivation of the unit-sphere technique published in 1928 [8]. He refers to the "angle-factor" in the paper, and gives the modern interpretation of the factor as the fraction of radiant energy leaving one surface that is incident upon another.

One of the first calculations of the radiant exchange between two surfaces was carried out by Christiansen in 1883 [10]. He analyzed radiant exchange between concentric cylinders, treating the cylinders as both diffuse or both specular. By using a ray-tracing technique, he found the relation for diffuse cylinders given in most textbooks:

$$\frac{Q_1}{A_1} = \frac{\sigma(T_1^4 - T_2^4)}{1/\epsilon_1 + (1/\epsilon_2 - 1)A_1/A_2} \tag{7-68}$$

With such a derivation, the concept of the configuration factor is neither necessary nor obvious, and Christiansen makes no mention of it.

Sumpner, in 1894, discussed the validity of Lambert's cosine law in relation to some experiments in photometry [11]. He also came near to defining a configuration factor, but didn't quite make the final step. He did note a fact that remains valid today: "Terms in light are used vaguely, and it will not be deemed out of place to define those which will be here needed."

As late as 1907, Hyde discussed the geometrical theory of radiation [12]. He still did not separate the geometry terms and explicitly define a configuration factor, even though he went on to evaluate the integrals appearing in the interchange equations. He examined some rather complex geometries, looking, for example, at the radiant exchange from an ellipse to an area element.

Saunders extended Christiansen's work, and in so doing defined a factor he denoted as K [13]. This is the fraction of energy leaving a surface that returns to it by reflections from all other surfaces and is then reabsorbed by the surface. This is equivalent to Gebhart's B_{ii} factors [14]. Saunders applied this concept to simple geometrical arrangements of two bodies, and did not carry it further. It is in the period 1920–1930 that the concept of configuration factors appears in a great many references. Nusselt's paper is one example, but papers by Buckley and Yamauti make use of the idea [8, 15, 16]. On the other hand, the text by Schack does not, although a passing reference to Nusselt's paper is given [17, 18].

7-9 CONCLUDING REMARKS

In this chapter, methods have been introduced and developed for the computation of energy transfer between isothermal black surfaces and in enclosures consisting of individual isothermal black surfaces. The radiant interchange between individual isothermal black surfaces can be treated by reasonably straightforward techniques. The chief difficulties in such problems are not in the concepts involved, but rather in the geometrical and algebraic manipulations and the integrations that must be carried out to determine the configuration factors for specific geometries. These difficulties are minimized by the availability in the literature of fairly extensive formulas, graphs, and tabulations of configuration factors that have already been calculated. References to the sources of these factors are given in Appendix B, and a number of factors are given in Appendix C.

For practical radiation computations, the assumption of black surfaces is quite restrictive. Hence the results given here have limited direct application although there may be some instances, such as within certain furnaces, where a black computation will yield reasonable results. The black computation theory, in spite of its limitations, serves two important functions. First, it is a limiting case with which nonblack performance and computations can be compared. It provides a good numerical check for problems in which a parametric study is being made with radiation properties that vary over a range of values. The second function is that the black case provides a foundation for more general exchange and enclosure theories. In succeeding chapters, the approach used in this chapter will be adapted and extended for problems that deal with more complicated effects such as nonblack and nonisothermal surfaces.

REFERENCES

1. Feingold, A.: Radiation-Interchange Configuration Factors between Various Selected Plane Surfaces, *Proc. R. Soc. London*, ser. A, vol. 292, no. 1428, pp. 51–60, 1966.
2. Sowell, E. F., and P. F. O'Brien: Efficient Computation of Radiant-Interchange Configuration Factors within an Enclosure, *J. Heat Transfer*, vol. 94, no. 3, pp. 326–328, 1972.
3. Hottel, H.C.: Radiant Heat Transmission, in William H. McAdams (ed.), "Heat Transmission," 3rd ed., chap. 4, McGraw-Hill Book Company, New York, 1954.
4. Sparrow, E. M., and R. D. Cess: "Radiation Heat Transfer," augmented ed., Hemisphere Publishing Corp., Washington, D.C., 1978.
5. Moon, Parry: "The Scientific Basis of Illuminating Engineering," rev. ed., Dover Publications, Inc., New York, 1961.
6. de Bastos, R.: "Computation of Radiation Configuration Factors by Contour Integration," M.S. thesis, Oklahoma State University, Stillwater, 1961.
7. Sparrow, E. M.: A New and Simpler Formulation for Radiative Angle Factors, *J. Heat Transfer*, vol. 85, no. 2, pp. 81–88, 1963.
8. Nusselt, Wilhelm: Graphische Bestimmung des Winkelverhältnisses bei der Wärmestrahlung, *VDI Z.*, vol. 72, p. 673, 1928.
9. Farrell, R.: Determination of Configuration Factors of Irregular Shape, *J. Heat Transfer*, vol. 98, no. 2, pp. 311–313, 1976.
10. Christiansen, C.: II. Absolute Bestimmung des Emissions- und Absorptionsvermögens für Warmes, *Ann. Phys., Wied.* vol. 19, pp. 267–283, 1883.
11. Sumpner, W. E.: The Diffusion of Light, *Proc. Phys. Soc. London*, vol. 94, pp. 10–29, 1894.
12. Hyde, Edward P.: Geometrical Theory of Radiating Surfaces with Discussion of Light Tubes, *Nat. Bur. Stand. U.S. Bull.*, vol. 3, pp. 81–104, 1907.
13. Saunders, O. A.: Notes on Some Radiation Heat Transfer Formulae, *Proc. Phys. Soc. London*, vol. 41, pp. 569–575, 1928–1929.
14. Gebhart, B.: "Heat Transfer," 2nd ed., pp. 150–163, McGraw-Hill Book Company, New York, 1971.
15. Buckley, H.: Radiation from the Interior of a Reflecting Cylinder, *Phil. Mag.*, vol. 4, pp. 753–762, 1927.
16. Yamauti, Z.: Geometrical Calculation of Illumination, *Res. Electrotech. Lab. Tokyo*, vol. 148, 1924.
17. Schack, A.: "Industrielle Wärmeübergang," 1st ed., Verlag Stahleisen mbH, Düsseldorf, 1929.
18. Schack, Alfred: "Industrial Heat Transfer" (Hans Goldschmidt and Everett P. Partridge, trans.), John Wiley & Sons, Inc., New York, 1933.

PROBLEMS

1 Derive the configuration factor F_{d1-2} between a differential area centered above a disk and a finite disk of unit radius.

Answer: $1/(H^2 + 1)$.

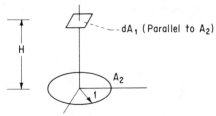

2   What is the net energy transfer from black surface dA_1 to black surface A_2?

    Answer: $dQ_{d1 \rightleftharpoons 2} = 14$ Btu/h.

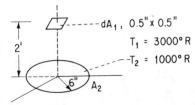

3   The configuration factor between two infinitely long directly opposed parallel plates of finite width L is F_{1-2}. The plates are separated by a distance D.

   (*a*) Derive an expression for F_{1-2} by integration of the configuration factor between differential strip elements.

   (*b*) Derive an expression for F_{1-2} by the crossed-string method.

    Answer: $\sqrt{1 + (D/L)^2} - D/L$.

4   The configuration factor between two parallel infinite plates of finite width L is F_{1-2} in the configuration shown below in cross section.

   (*a*) Derive an expression for F_{1-2} by the crossed-string method.

   (*b*) Derive an expression for F_{1-2} by using the results of Prob. 3 above and configuration-factor algebra.

    Answer: $\sqrt{1 + (D/2L)^2} + D/2L - \sqrt{1 + (D/L)^2}$.

5   Compute the configuration factor F_{1-2} between the infinitely long parallel plates shown below in cross section when the angle β is equal to, (*a*) 30° and (*b*) 90°.

    Answer: (*a*) 0.275.

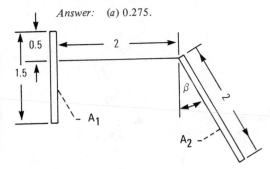

6 Derive the configuration factor F_{1-2} from a finite rectangle A_1 to an infinite plane A_2 where the rectangle is at an angle η relative to the plane.

$\quad$ *Answer:* $\quad F_{1-2} = \frac{1}{2}(1 - \cos \eta)$.

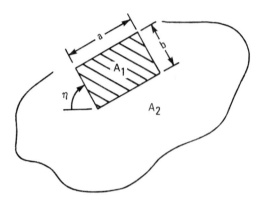

7 For the two-dimensional geometry shown, the view between A_1 and A_2 is partially blocked by an intervening structure. Determine the view factor F_{1-2}.

$\quad$ *Answer:* $\quad 0.2118$.

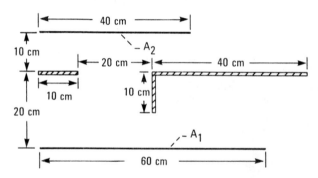

8 Using the crossed-string method, derive the configuration factor F_{1-2} between the infinitely long plate and cylinder shown below in cross section. Compare your result with that for configuration twenty-five in Appendix C.

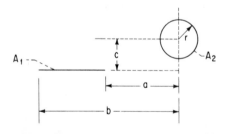

9 A long tube in a tube bundle is surrounded by six other identical equally spaced tubes, as shown in cross section below. What is the configuration factor from the central tube to each of the surrounding tubes?

$\quad$ *Answer:* $\quad 0.0844 \; (> \frac{1}{12})$.

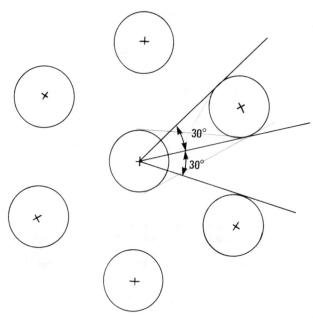

10 For the two-dimensional geometry shown, the view between A_1 and A_2 is partially blocked by two cylinders. Determine the view factor F_{1-2}.

Answer: 0.459.

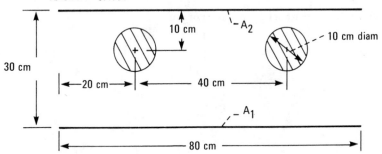

11 Given: F_{n-k}, the configuration factor between two perpendicular rectangles having a common edge. In terms of this factor, use configuration-factor relations to derive the F_{1-8} factor between the two areas A_1 and A_8.

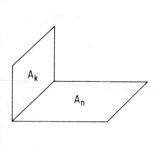

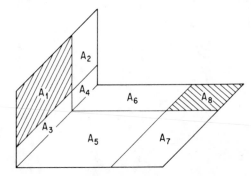

12 Determine the configuration factor F_{1-2} from the quarter disk to the plane ring.
    Answer: 0.19.

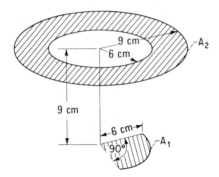

13 (*a*) Obtain the configuration factor between a sphere of radius R and a coaxial disk of radius r. (*Hint:* Think of the disk as being a cut through a spherical envelope concentric around the sphere.)

    (*b*) What is the F factor from a sphere to a sector of a disk?

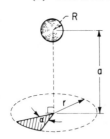

    (*c*) What is the dF factor from a sphere to a portion of a ring of infinitesimal width?

$$\text{Answer:} \quad (a) \frac{1}{2}\left(1 - \frac{a}{\sqrt{a^2 + r^2}}\right); \quad (b) \frac{\alpha}{4\pi}\left(1 - \frac{a}{\sqrt{a^2 + r^2}}\right); \quad (c) \frac{\alpha}{4\pi}\frac{ar}{(a^2 + r^2)^{3/2}}\, dr.$$

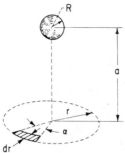

14 Do Example 7-1 by use of the sphere-to-disk configuration factor obtained in Prob. 13*a* above.

15 Use the crossed-string method to derive configuration factor two of Appendix C (infinitely long strip of differential width to infinitely long cylindrical surface).

16 Use the disk-to-disk configuration factor twenty-one of Appendix C to obtain the factor dF_{d1-d2} between two differential rings on the inside of a right circular cone.

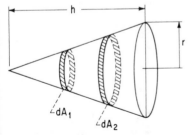

17 In terms of disk-to-disk configuration factors, derive the factor between the finite ring A_1 and the finite area A_2.

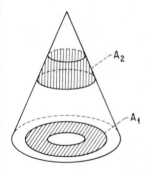

18 A closed right circular cylindrical shell with base diameter 1 m and height 1 m is located at the center of a spherical shell, 1 m in radius.

(*a*) Determine the diffuse configuration factor between the sphere and itself.

(*b*) If the top of the cylindrical shell were removed, determine the diffuse configuration factor between the sphere and the inside of the bottom of the cylindrical shell.

Answer: (*a*) $\frac{5}{8}$; (*b*) 0.01072.

19 Consider a black cubic enclosure. Determine the configuration factors between (*a*) two adjacent walls, and (*b*) two opposite walls. A sphere of diameter equal to one-half the length of a side of the cube is placed at the center of the cube. Determine the configuration factors between (*c*) the sphere and one wall of the enclosure, (*d*) one wall of the enclosure and the sphere, and (*e*) the enclosure and itself.

Answer: (*a*) 0.20004; (*b*) 0.19982; (*c*) $\frac{1}{6}$; (*d*) 0.1309; (*e*) 0.8691.

20 A person holds her open hand (approximated by a circular disk 12 cm in diameter) 10 cm directly above and parallel to a black heater element in the form of a circular disk 20 cm in diameter (such as on an electric stove). The heater element is at 700 K. How much radiant energy is incident on the hand.

Answer: 70.8 W.

21 A carbon steel billet 4 ft × 2 ft × 2 ft is initially at 2000°R and is supported in such a manner that it loses heat by radiation from all of its surface to surroundings at 70°F (assume surroundings act black). Neglect convective heat losses, and assume the billet radiates like a blackbody. Also assume for simplicity that the thermal conductivity of the steel is infinite. How long will it take for the billet to cool to $1000°R$?

Answer: 3.7 h [using $c_p = 0.11$ Btu/(lb·°R)].

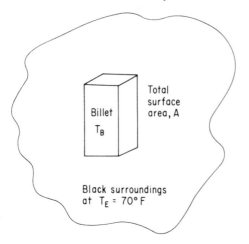

22 A black electrically heated rod is in a black vacuum jacket. The rod must dissipate 50 W without exceeding 800 K. Calculate the maximum allowable jacket temperature (neglect end effects).

Answer: 602.5 K.

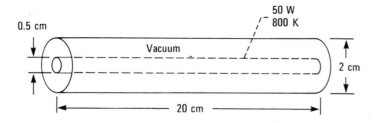

23 Two enclosures are identical in shape and size and have black surfaces. For one enclosure, the temperatures of the surfaces are $T_1, T_2, \cdots, T_N$. For the second, the surface temperatures are $(T_1^4 + k)^{1/4}, (T_2^4 + k)^{1/4}, \cdots, (T_N^4 + k)^{1/4}$, where k is a constant. How are the heat-transfer rates Q_j at any surface A_j related for the two enclosures?

Answer: The same.

24 An enclosure with black interior surfaces has one side open to an environment at temperature T_e. The sides of the enclosure are maintained at uniform temperatures $T_1, T_2, T_3, \cdots$. How are the heat inputs to the sides $Q_1, Q_2, Q_3, \cdots$ influenced by the value of T_e? How can the results for $T_e = 0$ be used to obtain solutions for other T_e?

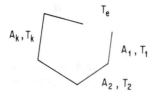

25 (*a*) A circular cylindrical enclosure has black interior surfaces, each maintained at uniform temperature as shown. The outside of the entire cylinder is insulated such that the outside does not radiate to the surroundings. How much Q (Btu/h) is supplied to each area as a result of the internal radiation exchange?

Answer: $Q_1 = 22,100;$ $Q_2 = -6300;$ $Q_3 = -15,800.$

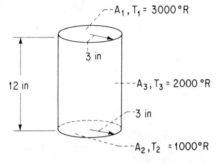

(*b*) For the same enclosure and the same surface temperatures, divide A_3 into two equal areas A_4 and A_5. What is the Q/A to each of these two areas, and how do they compare with Q_3/A_3 from part *a*?

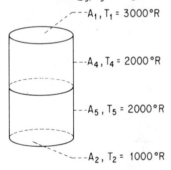

(*c*) What are Q_6/A_6 and Q_7/A_7 for the same enclosure and same surface temperatures?

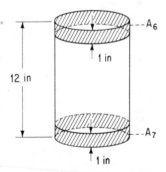

26 A hollow cylindrical heating element 15 cm long with 15 cm inside diameter with a black interior surface is to be maintained at 1100 K. The outside of the cylinder is insulated, and the surroundings are in vacuum at 800 K. Both ends of the cylinder are open. Estimate the energy that must be supplied to the element.

Answer: 1770 W.

27 By use of the contour-integration method of Sec. 7-5.4, obtain the final result in Example 7-6 for the configuration factor from an elemental area to a perpendicular circular disk.

28 A black 1-in-diameter sphere at a temperature of 1500°R is suspended in the center of a thin 2-in-diameter partial sphere having a black interior surface and an exterior surface with a hemispherical total emissivity of 0.4. The surroundings are at 1000°R. A 1.5-in-diameter hole is cut in the outer sphere. What is the temperature of the outer sphere? What is the Q being supplied to the inner sphere? (For simplicity, do not subdivide the surface areas into smaller zones.)

   Answer: 1233°R;   112 Btu/h.

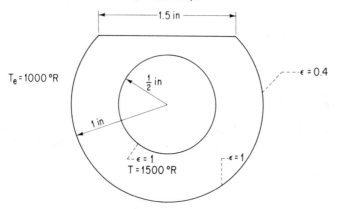

RADIATION EXCHANGE IN AN ENCLOSURE COMPOSED OF DIFFUSE-GRAY SURFACES

8-1 INTRODUCTION

8-1.1 Restrictions in the Analysis

In the previous chapter an enclosure composed of black surfaces was considered. As a next step in building toward more complex treatments that can account for the real property behavior of surfaces, the surfaces of the enclosure will now be taken as both *diffuse* and *gray*. In Chap. 3, the relations between emissivity and absorptivity are discussed. By definition, when a surface is diffuse-gray, the directional spectral emissivity and absorptivity do not depend on either angle or wavelength but can depend on surface temperature. As a result of this definition, at any surface temperature T_A the hemispherical total absorptivity and emissivity are equal and depend only on T_A; that is, $\alpha(T_A) = \epsilon(T_A)$. Even though this behavior is approached by only a limited number of real materials, the diffuse-gray approximation is often made to greatly simplify the enclosure theory.

Some comment is warranted as to what is meant by the individual "surfaces" or "areas" that comprise the total enclosure boundary. Usually, the geometry will tend to divide the enclosure into natural surface areas, such as the individual sides of a rectangular-prism enclosure. In addition, it may be necessary to specify surface areas on the basis of heating conditions; for example, if one side of an enclosure is partly at one temperature and partly at a second temperature, the side would be divided into two separate areas so that this difference in boundary condition could be accounted for. Hence, the surfaces or areas discussed in the radiation analysis are simply each separate portion of the enclosure boundary for which a heat balance is formed. These portions are selected on the basis of geometry and imposed heating conditions. A further consideration is the accuracy of the solution. If too few areas are designated, the accuracy will be poor; too many areas will require excessive computational time. Thus some engineering judgment is required in selecting both the shape of the surfaces and their number.

Surfaces of the enclosure can have various thermal boundary conditions imposed upon them. A given surface can be held at a specified temperature, have a specified imposed heat input, or be perfectly insulated from external heat addition or removal. It is a restriction in the present analysis that, whatever conditions are imposed, each separate surface of the enclosure must be at a uniform temperature. If the imposed heating conditions are such that the temperature would vary markedly over an area, the area should be subdivided into smaller, more nearly isothermal portions; these portions can be of differential size if necessary. As a consequence of this isothermal area requirement, the emitted energy is taken to be uniform over each surface of the enclosure.

Because a gray surface is not a perfect absorber (its absorptivity is less than unity rather than unity as for the black case), part of the energy incident on a surface is reflected. With regard to the reflected energy, two assumptions are made: (1) The reflected energy is diffuse, that is, the reflected intensity at each position on the boundary is uniform for all directions, and (2) the reflected energy is uniform over each surface of the enclosure. If the reflected energy is expected to vary over an area, the area should be subdivided into smaller areas over which the reflected energy will not vary too much. With these restrictions reasonably met, the reflected energy for each surface has the same diffuse and uniformly distributed character as the emitted energy. Hence, the reflected and emitted energy can be combined into a single energy quantity leaving the surface.

When a surface is both diffusely emitting and reflecting, the intensity of all the energy leaving the surface does not vary with angular direction. As a result, the geometric configuration factors (F factors) derived for black surfaces can be used for the present enclosure theory. It is well to emphasize that the derivation of the F factors in Chap. 7 was based on the condition of a diffuse uniform intensity leaving the surface; this diffuse-uniform condition must hold for both the emitted and reflected energies if one is to use the F factors for a nonblack surface.

Most of the problems encountered in practice are at steady state. However, the radiative heat balances considered here are not limited to steady-state conditions. The radiative balances can be directly applied to situations in which transient temperature changes are occurring. Instantaneously, the heat flux q that will be computed in the enclosure theory that follows can be considered as the net radiative loss from the location being considered on the enclosure boundary. For example, if a solid body is cooling by radiation, q provides the boundary condition for the transient heat conduction solution for the temperature distribution within the solid.

8-1.2 Summary of Restrictions

The assumptions for the present chapter are now summarized. The enclosure boundary is divided into areas so that over each of these areas the following restrictions are met:

1. The temperature is uniform.
2. The ϵ'_λ, α'_λ, and ρ'_λ are independent of wavelength and direction so that $\epsilon(T_A) = \alpha(T_A) = 1 - \rho(T_A)$, where ρ is the reflectivity.
3. All energy is emitted and reflected diffusely.
4. The incident and hence reflected energy flux is uniform over each individual area.

In some instances an analysis assuming diffuse-gray surfaces cannot yield good results. For example, if the temperatures of the individual surfaces of the enclosure differ considerably from each other, then a surface will be emitting predominantly in the range of wavelengths characteristic of its temperature while receiving energy predominantly in a different wavelength region. If the spectral emissivity varies with wavelength, the fact that the incident radiation has a different spectral distribution than the emitted energy will make the gray assumption invalid; that is, $\epsilon(T_A) \neq \alpha(T_A)$. When polished (specular) surfaces are present, the diffuse reflection assumption will be invalid, and the directional paths of the reflected energy must be considered. The treatments of specular and other more general surfaces are the subjects of Chaps. 9–11.

8-2 SYMBOLS

A	area
$\mathcal{A}$	inverse matrix coefficients, Eq. (8-29)
dA^*	differential element on same surface area as dA
a_{kj}	matrix elements defined by Eq. (8-25)
a^{-1}	inverse matrix
C_{kj}	matrix elements defined by Eq. (8-25)
D	diameter of tube or hole
F	configuration factor
G	function in integral equation, Eq. (8-57)
J	auxiliary variational function, Eq. (8-58)
J_1	Bessel function of the first kind
j,k	indices denoting individual surfaces
K	kernel of integral equation
L	length of surface
l	dimensionless length
m	coefficient in temperature distribution, Example 8-14
M_{kj}	minor of matrix element a_{kj}
N	number of surfaces in enclosure
Q	energy per unit time
q	energy flux, energy per unit area and time
R	radius of sphere
$\mathbf{r}$	direction vector
S	distance between areas
T	absolute temperature
x,y,z	coordinates

α	absorptivity
γ	polynomial coefficients, Eq. (8-59)
δ	Kronecker delta
ϵ	emissivity
ξ, η	dimensionless coordinates
ρ	reflectivity
θ	cone angle, angle from normal of surface
σ	Stefan-Boltzmann constant
φ	circumferential angle
Φ	dependent variable in integral equation

Subscripts

A	area
a	apparent value
black	blackbody property
e	external radiation entering through opening; environment
i	incoming
j,k	property of surface j or k
o	outgoing
s	sphere
λ	spectrally (wavelength) dependent
1, 2	surface 1 or 2

Superscript

| | quantity in one direction |

8-3 RADIATION BETWEEN FINITE AREAS

8-3.1 Net-radiation Method

Consider an enclosure composed of N discrete surface areas as shown in Fig. 8-1. The objectives of the analysis will be to analyze the radiation exchange between the surface areas for problems involving two types of boundary conditions: (1) The required energy supplied to a surface is to be determined when the surface temperature is specified, and (2) the temperature that a surface will achieve is to be found when a known heat input is imposed.

A complex radiative exchange occurs inside the enclosure as radiation leaves a surface, travels to the other surfaces, is partially reflected, and is then re-reflected many times within the enclosure with partial absorption at each contact with a surface. It would be very complicated to follow the beams of radiation as they undergo this process; fortunately, it is not necessary to do this. An analysis can be formulated in a convenient manner by using the *net-radiation method*. This method was first devised

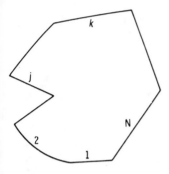

Figure 8-1 Enclosure composed of N discrete surface areas with typical surfaces j and k (shown in cross section for simplicity).

by Hottel [1] and later developed in a different manner by Poljak [2, 3]. An alternative approach was given by Gebhart [4]. All the methods are basically equivalent (as demonstrated in [5]); the Poljak approach, which the present authors generally prefer, will be given in this chapter. The Gebhart method is briefly presented in Appendix E.

Consider the kth inside surface area A_k of the enclosure shown in Figs. 8-1 and 8-2. The quantities q_i and q_o are the rates of incoming and outgoing *radiant* energy per unit inside area, respectively. The quantity q is the energy flux supplied *to the surface* by some means other than radiation to make up for the net radiative loss and thereby maintain the specified surface temperature. For example, if A_k is the inside surface of a wall of finite thickness, the q could be the heat conducted through the wall to A_k. A heat balance at the surface provides the relation

$$Q_k = q_k A_k = (q_{o,k} - q_{i,k}) A_k \tag{8-1}$$

A second equation results from the fact that the energy flux leaving the surface is composed of directly emitted plus reflected energy. This gives

$$q_{o,k} = \epsilon_k \sigma T_k{}^4 + \rho_k q_{i,k} = \epsilon_k \sigma T_k{}^4 + (1 - \epsilon_k) q_{i,k} \tag{8-2}$$

where the relations $\rho_k = 1 - \alpha_k = 1 - \epsilon_k$ have been used for opaque gray surfaces. The term *radiosity* is often used for the quantity q_o. The incident flux $q_{i,k}$ is derived from the portions of the energy leaving the surfaces in the enclosure that arrive at the kth surface. If the kth surface can view itself (is concave), a portion of its outgoing flux

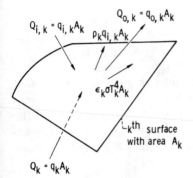

Figure 8-2 Energy quantities incident upon and leaving typical surface of enclosure.

will contribute directly to its incident flux. The incident energy is then equal to

$$A_k q_{i,k} = A_1 q_{o,1} F_{1-k} + A_2 q_{o,2} F_{2-k} + \cdots + A_j q_{o,j} F_{j-k} + \cdots$$
$$+ A_k q_{o,k} F_{k-k} + \cdots + A_N q_{o,N} F_{N-k} \tag{8-3}$$

From the configuration-factor reciprocity relation (7-25),

$$A_1 F_{1-k} = A_k F_{k-1}$$
$$A_2 F_{2-k} = A_k F_{k-2}$$
$$\cdots\cdots\cdots\cdots\cdots \tag{8-4}$$
$$A_N F_{N-k} = A_k F_{k-N}$$

Then Eq. (8-3) can be written so that the only area appearing is A_k:

$$A_k q_{i,k} = A_k F_{k-1} q_{o,1} + A_k F_{k-2} q_{o,2} + \cdots + A_k F_{k-j} q_{o,j} + \cdots$$
$$+ A_k F_{k-k} q_{o,k} + \cdots + A_k F_{k-N} q_{o,N} \tag{8-5a}$$

or

$$q_{i,k} = \sum_{j=1}^{N} F_{k-j} q_{o,j} \tag{8-5b}$$

Equations (8-2) and (8-5) provide two different expressions for $q_{i,k}$. These are each substituted into Eq. (8-1) to eliminate $q_{i,k}$ and provide these two basic heat-balance equations for Q_k in terms of $q_{o,k}$:

$$Q_k = A_k \frac{\epsilon_k}{1 - \epsilon_k} (\sigma T_k^4 - q_{o,k}) \tag{8-6}$$

$$Q_k = A_k \left(q_{o,k} - \sum_{j=1}^{N} F_{k-j} q_{o,j} \right) \tag{8-7}$$

The Q_k can be regarded as either the energy supplied to surface k by nonradiative means (convection or conduction to A_k), or as the net radiative loss from surface k resulting from radiation in the enclosure. Equation (8-6) or (8-7) is the balance between net radiative loss and energy supplied by means other than radiation.

As a first step in becoming familiar with this radiation analysis, consider that (8-6) and (8-7) can be written for each of the N surfaces in the enclosure. This provides $2N$ equations for $2N$ unknowns. The q_o's will be N of the unknowns. The remaining unknowns will consist of Q's and T's, depending on what boundary quanties are specified. As will be shown later, the q_o's can be eliminated giving N equations relating the N unknown Q's and T's.

Some examples will now be given to illustrate the use of Eqs. (8-6) and (8-7) as a system of simultaneous equations.

EXAMPLE 8-1 Derive the expression for the net radiative heat exchange between two infinite parallel flat plates in terms of their temperatures T_1 and T_2 ($T_1 > T_2$, Fig. 8-3).

Since all the radiation leaving one plate will arrive at the other plate, the configuration factors are $F_{1-2} = F_{2-1} = 1$. Equations (8-6) and (8-7) are then written for each plate:

$$\frac{Q_1}{A_1} = q_1 = \frac{\epsilon_1}{1 - \epsilon_1}(\sigma T_1^4 - q_{o,1}) \tag{8-8a}$$

$$\frac{Q_1}{A_1} = q_1 = q_{o,1} - q_{o,2} \tag{8-8b}$$

$$\frac{Q_2}{A_2} = q_2 = \frac{\epsilon_2}{1 - \epsilon_2}(\sigma T_2^4 - q_{o,2}) \tag{8-9a}$$

$$\frac{Q_2}{A_2} = q_2 = q_{o,2} - q_{o,1} \tag{8-9b}$$

By comparing Eqs. (8-8b) and (8-9b), it is evident that $q_1 = -q_2$ so that the heat added to surface 1 is removed from surface 2. The flux q_1 is thus the net heat transfer from 1 to 2 requested in the problem statement. Equation (8-8a) is solved for $q_{o,1}$:

$$q_{o,1} = \sigma T_1^4 - \frac{1 - \epsilon_1}{\epsilon_1} q_1$$

Similarly, from Eq. (8-9a),

$$q_{o,2} = \sigma T_2^4 - \frac{1 - \epsilon_2}{\epsilon_2} q_2 = \sigma T_2^4 + \frac{1 - \epsilon_2}{\epsilon_2} q_1$$

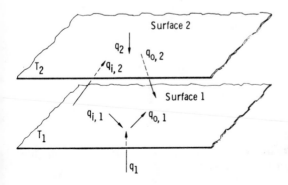

Figure 8-3 Heat fluxes for radiant interchange between infinite parallel flat plates.

These are substituted into (8-8b), and the result solved for q_1:

$$q_1 = -q_2 = \frac{\sigma(T_1^4 - T_2^4)}{1/\epsilon_1(T_1) + 1/\epsilon_2(T_2) - 1} \tag{8-10a}$$

The functional notation $\epsilon(T)$ has been introduced to emphasize that ϵ_1 and ϵ_2 can be functions of temperature. Since T_1 and T_2 are specified, ϵ_1 and ϵ_2 can be evaluated at their proper temperatures, and q_1 directly calculated.

EXAMPLE 8-2 For the parallel-plate geometry of the previous example, what temperature will surface 1 reach for a given heat input q_1 while T_2 is held at a specified value?

Equation (8-10a) still applies and when solved for T_1 gives

$$T_1 = \left\{ \frac{q_1}{\sigma} \left[\frac{1}{\epsilon_1(T_1)} + \frac{1}{\epsilon_2(T_2)} - 1 \right] + T_2^4 \right\}^{1/4} \tag{8-10b}$$

Since the emissivity $\epsilon_1(T_1)$ is a function of T_1 which is unknown, an iterative solution is necessary. A trial T_1 is selected, and then ϵ_1 is chosen at this value. Equation (8-10b) is then solved for T_1, and this value is used to select ϵ_1 for the next approximation. The process is continued until $\epsilon_1(T_1)$ and T_1 no longer change with further iterations.

EXAMPLE 8-3 Derive an expression for the net radiation exchange between two uniform temperature concentric diffuse-gray spheres as shown in Fig. 8-4.

This situation is more complicated than the parallel-plate geometry, as the two surfaces have unequal areas and surface 2 can partially view itself. The configuration factors were derived in Example 7-13 as $F_{1-2} = 1$, $F_{2-1} = A_1/A_2$, and $F_{2-2} = 1 - A_1/A_2$. The basic heat-balance equations [Eqs. (8-6) and (8-7)] are written for each of the two sphere surfaces as

$$Q_1 = A_1 \frac{\epsilon_1}{1 - \epsilon_1} (\sigma T_1^4 - q_{o,1}) \tag{8-11a}$$

$$Q_1 = A_1(q_{o,1} - q_{o,2}) \tag{8-11b}$$

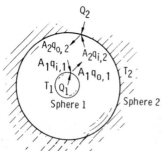

Figure 8-4 Energy quantities for radiant interchange between two concentric spheres.

$$Q_2 = A_2 \frac{\epsilon_2}{1-\epsilon_2}(\sigma T_2{}^4 - q_{o,2}) \tag{8-12a}$$

$$Q_2 = A_2 \left[q_{o,2} - \frac{A_1}{A_2} q_{o,1} - \left(1 - \frac{A_1}{A_2}\right) q_{o,2} \right] = A_1(-q_{o,1} + q_{o,2}) \tag{8-12b}$$

Comparing (8-11b) and (8-12b) reveals that $Q_1 = -Q_2$ as would be expected from an overall heat balance on the system. The four equations (8-11) and (8-12) can be solved for the four unknowns $q_{o,1}$, $q_{o,2}$, Q_1, and Q_2. This yields the net heat transfer (supplied to surface 1 and removed at surface 2):

$$Q_1 = \frac{A_1\sigma(T_1{}^4 - T_2{}^4)}{1/\epsilon_1(T_1) + (A_1/A_2)[1/\epsilon_2(T_2) - 1]} \tag{8-13}$$

When the spheres in Example 8-3 are not concentric, all the radiation leaving surface 1 is still incident on surface 2. The view factor F_{1-2} is again 1 and, with the use of the same assumptions, the analysis would follow as before, leading to (8-13). However, when sphere 1 is relatively small (say, one-half the diameter of sphere 2) and the eccentricity is large, the geometric appearance of the system is so different from the concentric case that using (8-13) would seem intuitively incorrect. The error in using (8-13) is that it was derived on the basis that q, q_i, and q_o are uniform over each of A_1 and A_2. These conditions are exactly met only for the concentric case.

EXAMPLE 8-4 A gray isothermal body with area A_1 and temperature T_1 is completely enclosed by a much larger gray isothermal enclosure having area A_2. How much energy is being transferred by radiation from A_1 to A_2? The area A_1 cannot see any part of itself, that is, $F_{1-1} = 0$, and A_1 is not near the boundary of A_2.

Since A_1 is completely enclosed and $F_{1-1} = 0$, the configuration factors and analysis are identical to that in Example 8-3, which results in the Q_1 given by Eq. (8-13). This is valid, as A_1 is specified as rather centrally located within A_2 and hence the heat fluxes tend to be uniform over A_1. For the present situation $A_1 \ll A_2$, and (8-13) reduces to

$$Q_1 = A_1\epsilon_1(T_1)\sigma(T_1{}^4 - T_2{}^4) \tag{8-14}$$

Note that this result is independent of the emissivity ϵ_2 of the enclosure (the enclosure acts like a black cavity).

EXAMPLE 8-5 Consider a long enclosure made up of three surfaces as shown in Fig. 8-5. The enclosure is long enough so that the ends can be neglected in the radiative heat balances. How much heat has to be supplied to each surface (equal to the net radiative heat loss from each surface resulting from exchange within the enclosure) to maintain the surfaces at temperatures T_1, T_2, and T_3?

To solve this problem, write (8-6) and (8-7) for each of the three surfaces:

$$\frac{Q_1}{A_1} = \frac{\epsilon_1}{1-\epsilon_1}(\sigma T_1{}^4 - q_{o,1}) \tag{8-15a}$$

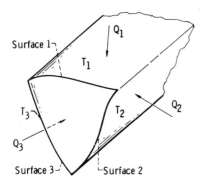

Figure 8-5 Long enclosure composed of three surfaces (ends neglected).

$$\frac{Q_1}{A_1} = q_{o,1} - F_{1-1}q_{o,1} - F_{1-2}q_{o,2} - F_{1-3}q_{o,3} \tag{8-15b}$$

$$\frac{Q_2}{A_2} = \frac{\epsilon_2}{1 - \epsilon_2}(\sigma T_2^{\,4} - q_{o,2}) \tag{8-16a}$$

$$\frac{Q_2}{A_2} = q_{o,2} - F_{2-1}q_{o,1} - F_{2-2}q_{o,2} - F_{2-3}q_{o,3} \tag{8-16b}$$

$$\frac{Q_3}{A_3} = \frac{\epsilon_3}{1 - \epsilon_3}(\sigma T_3^{\,4} - q_{o,3}) \tag{8-17a}$$

$$\frac{Q_3}{A_3} = q_{o,3} - F_{3-1}q_{o,1} - F_{3-2}q_{o,2} - F_{3-3}q_{o,3} \tag{8-17b}$$

The first equation of each of these three pairs of equations can be solved for q_o in terms of T and Q. These q_o's are then substituted into the second equation of each pair to obtain

$$\frac{Q_1}{A_1}\left(\frac{1}{\epsilon_1} - F_{1-1}^{\,0}\frac{1 - \epsilon_1}{\epsilon_1}\right) - \frac{Q_2}{A_2}F_{1-2}\frac{1 - \epsilon_2}{\epsilon_2} - \frac{Q_3}{A_3}F_{1-3}\frac{1 - \epsilon_3}{\epsilon_3}$$
$$= (1 - F_{1-1})\sigma T_1^{\,4} - F_{1-2}\sigma T_2^{\,4} - F_{1-3}\sigma T_3^{\,4} \tag{8-18a}$$

$$-\frac{Q_1}{A_1}F_{2-1}\frac{1 - \epsilon_1}{\epsilon_1} + \frac{Q_2}{A_2}\left(\frac{1}{\epsilon_2} - F_{2-2}^{\,0}\frac{1 - \epsilon_2}{\epsilon_2}\right) - \frac{Q_3}{A_3}F_{2-3}\frac{1 - \epsilon_3}{\epsilon_3}$$
$$= -F_{2-1}\sigma T_1^{\,4} + (1 - F_{2-2})\sigma T_2^{\,4} - F_{2-3}\sigma T_3^{\,4} \tag{8-18b}$$

$$-\frac{Q_1}{A_1}F_{3-1}\frac{1 - \epsilon_1}{\epsilon_1} - \frac{Q_2}{A_2}F_{3-2}\frac{1 - \epsilon_2}{\epsilon_2} + \frac{Q_3}{A_3}\left(\frac{1}{\epsilon_3} - F_{3-3}^{\,0}\frac{1 - \epsilon_3}{\epsilon_3}\right)$$
$$= -F_{3-1}\sigma T_1^{\,4} - F_{3-2}\sigma T_2^{\,4} + (1 - F_{3-3})\sigma T_3^{\,4} \tag{8-18c}$$

Since the T's are known, the ϵ's can be specified from surface-property data at their appropriate T values, and these three simultaneous equations solved for the desired Q values supplied to each surface. Note that the solutions are only first approximations, because the radiosity leaving each surface is not uniform as assumed by using (8-6) and (8-7). This is because the reflected flux is not uniform as a result of the enclosure geometry. Greater accuracy can be obtained by dividing each of the three sides into more surface elements.

Now that some familiarity with the radiant energy exchange equations has been achieved through a few simple examples, the system of equations will be written in a generalized form for an enclosure of N surfaces.

System of equations relating surface heating Q and surface temperature T. The form of (8-18) indicates that the Q's and T's for an enclosure of N surfaces can be related in a general system of N equations. Equation (8-6) is solved for $q_{o,k}$, and this is substituted into (8-7). (Note that $q_{o,j}$ is found by simply changing the subscript in the relation for $q_{o,k}$.) This results in the following form for the kth surface, a result that is also evident from (8-18):

$$-\frac{Q_1}{A_1}F_{k-1}\frac{1-\epsilon_1}{\epsilon_1} - \frac{Q_2}{A_2}F_{k-2}\frac{1-\epsilon_2}{\epsilon_2} - \cdots + \frac{Q_k}{A_k}\left(\frac{1}{\epsilon_k} - F_{k-k}\frac{1-\epsilon_k}{\epsilon_k}\right) - \cdots$$

$$-\frac{Q_N}{A_N}F_{k-N}\frac{1-\epsilon_N}{\epsilon_N} = -F_{k-1}\sigma T_1{}^4 - F_{k-2}\sigma T_2{}^4 - \cdots$$

$$+ (1 - F_{k-k})\sigma T_k{}^4 - \cdots - F_{k-N}\sigma T_N{}^4$$

A summation notation can be used to write this as

$$\sum_{j=1}^{N}\left(\frac{\delta_{kj}}{\epsilon_j} - F_{k-j}\frac{1-\epsilon_j}{\epsilon_j}\right)\frac{Q_j}{A_j} = \sum_{j=1}^{N}(\delta_{kj} - F_{k-j})\sigma T_j{}^4 \tag{8-19}$$

where corresponding to each surface, k takes on one of the values $1, 2, \ldots, N$, and δ_{kj} is the Kronecker delta defined as

$$\delta_{kj} = \begin{cases} 1 & \text{when } k = j \\ 0 & \text{when } k \neq j \end{cases}$$

When the surface temperatures are specified, the right side of Eq. (8-19) is known and there are N simultaneous equations for the unknown Q's.

In general the heat inputs to some of the surfaces may be specified and the temperatures of these surfaces are to be determined. There are still a total of N unknown Q's and T's, and Eq. (8-19) provides the necessary number of relations. Since the values of ϵ depend on temperature, it is necessary initially to guess the unknown T's. Then the ϵ values can be chosen and the system of equations solved. The resulting T values can be used to select new ϵ's, and the process repeated until the T and ϵ values no longer change upon further iteration. Again note that the results of this method will be approximate because the uniform radiosity assumption is not perfectly fulfilled over each finite area.

EXAMPLE 8-6 Consider an enclosure of three sides as shown in Fig. 8-5. Side 1 is held at T_1, side 2 is uniformly heated with a flux q_2, and the third side is insulated. What are the equations to determine Q_1, T_2, and T_3?

The conditions of the problem give $Q_2/A_2 = q_2$ and $Q_3 = 0$. Then (8-19) yields the following three equations, where the unknowns have been gathered on the left side:

$$\frac{Q_1}{A_1}\left(\frac{1}{\epsilon_1} - F_{1-1}\frac{1-\epsilon_1}{\epsilon_1}\right) + F_{1-2}\sigma T_2^{\ 4} + F_{1-3}\sigma T_3^{\ 4}$$

$$= (1 - F_{1-1})\sigma T_1^{\ 4} + q_2 F_{1-2}\frac{1-\epsilon_2}{\epsilon_2} \tag{8-20a}$$

$$-\frac{Q_1}{A_1}F_{2-1}\frac{1-\epsilon_1}{\epsilon_1} - (1 - F_{2-2})\sigma T_2^{\ 4} + F_{2-3}\sigma T_3^{\ 4}$$

$$= -F_{2-1}\sigma T_1^{\ 4} - q_2\left(\frac{1}{\epsilon_2} - F_{2-2}\frac{1-\epsilon_2}{\epsilon_2}\right) \tag{8-20b}$$

$$-\frac{Q_1}{A_1}F_{3-1}\frac{1-\epsilon_1}{\epsilon_1} + F_{3-2}\sigma T_2^{\ 4} - (1 - F_{3-3})\sigma T_3^{\ 4}$$

$$= -F_{3-1}\sigma T_1^{\ 4} + q_2 F_{3-2}\frac{1-\epsilon_2}{\epsilon_2} \tag{8-20c}$$

If ϵ_2 depends on temperature, an iterative procedure is needed wherein a T_2 is chosen, then $\epsilon_2(T_2)$ is specified, the equations are solved for T_2. and the iteration is continued until $\epsilon_2(T_2)$ and T_2 no longer change. (Note that for this particular simple geometry and conditions the solution can be simplified by using overall energy conservation to give $Q_1 = -Q_2$.)

Solution method in terms of outgoing radiative flux q_o An alternative approach for computing the radiative exchange within an enclosure involves first solving for q_o for each surface and then computing the Q's or T's. When a surface is sighted with a radiation detector, it is q_o that is intercepted, that is, the sum of both emitted and reflected radiation. For this reason, it is desirable in some instances to determine the q_o values as primary quantities. Of course, in the previous formulation the q_o's can be found from Q's and T's by using (8-6).

When the surface temperatures are all specified, the set of simultaneous equations for q_o's is obtained by eliminating Q_k's from (8-6) and (8-7). This yields the following equation for the kth surface:

$$q_{o,k} - (1 - \epsilon_k) \sum_{j=1}^{N} F_{k-j} q_{o,j} = \epsilon_k \sigma T_k^{\ 4} \tag{8-21}$$

To illustrate, for a system of two surfaces, (8-21) becomes

$$q_{o,1} - (1-\epsilon_1)F_{1-1}q_{o,1} - (1-\epsilon_1)F_{1-2}q_{o,2} = \epsilon_1\sigma T_1^4 \qquad (8\text{-}22a)$$

$$q_{o,2} - (1-\epsilon_2)F_{2-1}q_{o,1} - (1-\epsilon_2)F_{2-2}q_{o,2} = \epsilon_2\sigma T_2^4 \qquad (8\text{-}22b)$$

An alternative form of (8-21) is

$$\sum_{j=1}^{N} [\delta_{kj} - (1-\epsilon_k)F_{k-j}]q_{o,j} = \epsilon_k\sigma T_k^4 \qquad (8\text{-}23)$$

With the T's given, the q_o's can be found from (8-23). Then, if desired, (8-6) can be used to compute Q for each surface.

When Q is specified for some surfaces and T for others, (8-23) is used for the surfaces with known T in conjunction with (8-7) for the surfaces with known Q, to obtain the set of simultaneous equations for the unknown q_o's. Once q_o is obtained for a surface, it can be combined with the given Q (or T) and (8-6) used to determine the unknown T (or Q). In a general form, if an enclosure has surfaces $1, 2, \ldots, m$ with specified temperature and the remaining surfaces $m+1, m+2, \ldots, N$ with specified heat input, then the system of equations for the q_o's is, from (8-23) and (8-7),

$$\sum_{j=1}^{N} [\delta_{kj} - (1-\epsilon_k)F_{k-j}]q_{o,j} = \epsilon_k\sigma T_k^4 \qquad 1 \leqslant k \leqslant m \qquad (8\text{-}24a)$$

$$\sum_{j=1}^{N} (\delta_{kj} - F_{k-j})q_{o,j} = \frac{Q_k}{A_k} \qquad m+1 \leqslant k \leqslant N \qquad (8\text{-}24b)$$

Note that, for a black surface with T_k specified, (8-24a) gives $q_{o,k} = \sigma T_k^4$ so that the $q_{o,k}$ is known and the number of simultaneous equations can be immediately reduced by one.

EXAMPLE 8-7 A frustum of a cone has its base heated as shown in Fig. 8-6. The top is held at 550 K while the side is perfectly insulated. Surfaces 1 and 2 are assumed gray and diffuse, while surface 3 is black. What is the temperature of side 1? How important is the value of ϵ_2?

By using the configuration factor for two parallel disks (21 in Appendix C), it is found that $F_{3-1} = 0.33$. Then $F_{3-2} = 1 - F_{3-1} = 0.67$. From reciprocity, $A_1F_{1-3} = A_3F_{3-1}$ and $A_2F_{2-3} = A_3F_{3-2}$, it is found that $F_{1-3} = 0.147$ and $F_{2-3} = 0.13$. Then $F_{1-2} = 1 - F_{1-3} = 0.853$. From $A_1F_{1-2} = A_2F_{2-1}$, $F_{2-1} = 0.372$. Finally, $F_{2-2} = 1 - F_{2-1} - F_{2-3} = 0.498$. From Eq. (8-19) and by noting that $Q_2 = 0$ and $1 - \epsilon_3 = 0$, the three equations can be written as

$$\frac{3000}{0.6} = \sigma[T_1^4 - 0.853T_2^4 - 0.147(550)^4]$$

$$-3000(0.372)\frac{1-0.6}{0.6} = \sigma[-0.372T_1^4 + (1-0.498)T_2^4 - 0.13(550)^4]$$

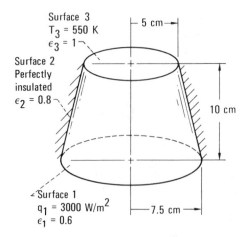

Surface 3
$T_3 = 550$ K
$\epsilon_3 = 1$

Surface 2
Perfectly insulated
$\epsilon_2 = 0.8$

5 cm

10 cm

Surface 1
$q_1 = 3000$ W/m^2
$\epsilon_1 = 0.6$

7.5 cm

Figure 8-6 Enclosure used in Example 8-7.

$$-3000(0.33)\frac{1-0.6}{0.6} + \frac{Q_3}{A_3} = \sigma[-0.33T_1^4 - 0.67T_2^4 + (550)^4]$$

These three equations can be solved for the unknowns T_1, T_2, and Q_3. (Note that for this particular example Q_3 can also be obtained from overall energy conservation; that is, $Q_3 = -Q_1$.) The result requested in the problem is $T_1 = 720$ K. Since $Q_2 = 0$, all the terms involving ϵ_2 were zero so that ϵ_2 does not appear in the simultaneous equations; hence, for this gray-diffuse analysis the emissivity of the insulated surface is of no importance. Physically, this results from the fact that for an insulated surface, all absorbed energy must be reemitted (no convection or conduction) and hence $q_o = q_i$ independently of the value of ϵ_2.

8-3.2 Matrix Inversion

When many surfaces are present in an enclosure, a large set of simultaneous equations such as (8-19) or (8-24) will result. These equations can be solved using a digital computer along with standard computer programs that can accommodate hundreds of simultaneous equations.

A set of equations such as (8-24) can be written in a shorter form. Let the known quantities on the right side be C_k, and the quantities in brackets on the left be a_{kj}. Then the k equations can be written as

$$\sum_{j=1}^{N} a_{kj}q_{o,j} = C_k \tag{8-25a}$$

where

$$a_{kj} = \begin{cases} \delta_{kj} - (1-\epsilon_k)F_{k-j} \\ \\ \delta_{kj} - F_{k-j} \end{cases} \qquad C_k = \begin{cases} \epsilon_k\sigma T_k^4 & 1 \leqslant k \leqslant m \\ \dfrac{Q_k}{A_k} & m+1 \leqslant k \leqslant N \end{cases} \tag{8-25b}$$

For an enclosure of N surfaces, the set of equations then has the form

$$a_{11}q_{o,1} + a_{12}q_{o,2} + \cdots + a_{1j}q_{o,j} + \cdots + a_{1N}q_{o,N} = C_1$$
$$a_{21}q_{o,1} + a_{22}q_{o,2} + \cdots + a_{2j}q_{o,j} + \cdots + a_{2N}q_{o,N} = C_2$$
$$\cdots\cdots\cdots\cdots\cdots\cdots\cdots\cdots\cdots\cdots\cdots\cdots\cdots\cdots\cdots$$
$$a_{k1}q_{o,1} + a_{k2}q_{o,2} + \cdots + a_{kj}q_{o,j} + \cdots + a_{kN}q_{o,N} = C_k$$
$$\cdots\cdots\cdots\cdots\cdots\cdots\cdots\cdots\cdots\cdots\cdots\cdots\cdots\cdots\cdots$$
$$a_{N1}q_{o,1} + a_{N2}q_{o,2} + \cdots + a_{Nj}q_{o,j} + \cdots + a_{NN}q_{o,N} = C_N$$

(8-26)

The array of a_{kj} coefficients is termed the *matrix of coefficients* and is often designated by a bracket notation:

$$\text{matrix } a \equiv [a_{kj}] \equiv
\begin{bmatrix}
a_{11} & a_{12} & \cdots & a_{1j} & \cdots & a_{1N} \\
a_{21} & a_{22} & \cdots & a_{2j} & \cdots & a_{2N} \\
\cdots & \cdots & \cdots & \cdots & \cdots & \cdots \\
a_{k1} & a_{k2} & \cdots & a_{kj} & \cdots & a_{kN} \\
\cdots & \cdots & \cdots & \cdots & \cdots & \cdots \\
a_{N1} & a_{N2} & \cdots & a_{Nj} & \cdots & a_{NN}
\end{bmatrix}$$

(8-27)

A method of solving a set of equations such as (8-26) is to obtain a second matrix a^{-1}, which is called the *inverse* of matrix a, that is,

$$a^{-1} \equiv [\mathscr{A}_{kj}] \equiv
\begin{bmatrix}
\mathscr{A}_{11} & \mathscr{A}_{12} & \cdots & \mathscr{A}_{1j} & \cdots & \mathscr{A}_{1N} \\
\mathscr{A}_{21} & \mathscr{A}_{22} & \cdots & \mathscr{A}_{2j} & \cdots & \mathscr{A}_{2N} \\
\cdots & \cdots & \cdots & \cdots & \cdots & \cdots \\
\mathscr{A}_{k1} & \mathscr{A}_{k2} & \cdots & \mathscr{A}_{kj} & \cdots & \mathscr{A}_{kN} \\
\cdots & \cdots & \cdots & \cdots & \cdots & \cdots \\
\mathscr{A}_{N1} & \mathscr{A}_{N2} & \cdots & \mathscr{A}_{Nj} & \cdots & \mathscr{A}_{NN}
\end{bmatrix}$$

(8-28)

In the inverse matrix there is a term $\mathscr{A}_{kj}$ corresponding to each a_{kj} in the original matrix. The $\mathscr{A}$'s are found by operating on the a's in a way briefly described as follows: If the kth row and jth column that contain element a_{kj} in a square matrix a are deleted, the determinant of the remaining square array is called the *minor* of element a_{kj} and is denoted by M_{kj}. The cofactor of a_{kj} is defined as $(-1)^{k+j}M_{kj}$. To obtain the inverse of a square matrix $[a_{kj}]$, each element a_{kj} is first replaced by its cofactor. The rows and columns of the resulting matrix are then interchanged. The elements of the matrix thus obtained are then each divided by the determinant $|a_{kj}|$ of the original matrix $[a_{kj}]$. The elements obtained in this fashion are the $\mathscr{A}_{kj}$. For more detailed information on matrix inversion, the reader should refer to a mathematics text such as [6]. There are standard digital-computer programs that will numerically obtain the inverse coefficients $\mathscr{A}_{kj}$ from a matrix of a_{kj} values.

After the inverse coefficients have been obtained, the unknown q_o values in (8-26) are found as the sum of products of $\mathscr{A}$'s and C's

$$q_{o,1} = \mathcal{A}_{11}C_1 + \mathcal{A}_{12}C_2 + \cdots + \mathcal{A}_{1j}C_j + \cdots + \mathcal{A}_{1N}C_N$$

$$q_{o,2} = \mathcal{A}_{21}C_1 + \mathcal{A}_{22}C_2 + \cdots + \mathcal{A}_{2j}C_j + \cdots + \mathcal{A}_{2N}C_N$$

$$\cdots\cdots\cdots\cdots\cdots\cdots\cdots\cdots\cdots\cdots\cdots\cdots\cdots\cdots\cdots$$

$$q_{o,k} = \mathcal{A}_{k1}C_1 + \mathcal{A}_{k2}C_2 + \cdots + \mathcal{A}_{kj}C_j + \cdots + \mathcal{A}_{kN}C_N$$

(8-29a)

or

$$q_{o,k} = \sum_{j=1}^{N} \mathcal{A}_{kj}C_j \tag{8-29b}$$

Thus the solution for each $q_{o,k}$ is in the form of a sum of $\epsilon\sigma T^4$ and Q/A that the C's represent, each weighted by an $\mathcal{A}$ coefficient.

For a given enclosure the configuration factors F_{k-j} in (8-25b) remain fixed. If in addition the ϵ_k's are constant, then the elements a_{kj}, and hence the inverse elements $\mathcal{A}_{kj}$, remain fixed for the enclosure. The fact that the $\mathcal{A}_{kj}$ remain fixed has utility when it is desired to compute the radiation quantities within an enclosure for many different values of the T's and Q's at the surfaces. The matrix need only be inverted once, and then (8-29b) can be applied for different values of the C's. These comments also apply to the system of equations (8-19). After the inverse is taken, the Q's can be found as a weighted sum of the T^4's.

8-4 RADIATION BETWEEN INFINITESIMAL AREAS

8-4.1 Generalized Net-Radiation Method for Infinitesimal Areas

In the previous section the enclosure was divided into finite areas. The accuracy of the results is limited by the assumptions in the analysis that the temperature and energy incident on and leaving each surface are uniform over that surface. If these quantities are nonuniform over part of the enclosure boundary, then the boundary surface must be subdivided until the variation over each area used in the analysis is not too large. It may be necessary to carry out several calculations in which successively smaller areas (and hence more simultaneous equations) are used until the solution no longer changes significantly when the area sizes are further diminished. In the limit, the enclosure boundary or a portion of it can be divided into infinitesimal parts; this will allow large variations in T, q, q_i, and q_o to be accounted for.

The formulation in terms of infinitesimal areas leads to heat balances in the form of a set of integral equations. Bu using both exact and approximate mathematical techniques that have been developed for integral equations, it is sometimes possible to obtain a closed-form analytical solution. When it is not possible to obtain an analytical solution, the integral equations can be solved numerically. In the case of a numerical solution, the solution method is similar to that used in the previous discussion dealing with finite areas.

Consider as before an enclosure composed of N finite areas. These areas would generally be the major geometric divisions of the enclosure or the areas on which a specified boundary condition is held constant. Each of these areas is further sub-divided into differential area elements as shown for two typical areas in Fig. 8-7. As before, throughout the following analysis *the surfaces will be considered diffuse-gray. The additional restriction is now also made that the radiative properties are indepen-dent of temperature.*

A heat balance on element dA_k located at position $\mathbf{r}_k$ gives

$$q_k(\mathbf{r}_k) = q_{o,k}(\mathbf{r}_k) - q_{i,k}(\mathbf{r}_k) \tag{8-30}$$

The outgoing flux is composed of emitted and reflected energy:

$$q_{o,k}(\mathbf{r}_k) = \epsilon_k \sigma T_k^4(\mathbf{r}_k) + (1 - \epsilon_k) q_{i,k}(\mathbf{r}_k) \tag{8-31}$$

The incoming flux in (8-31) is composed of portions of the outgoing fluxes from the other area elements of the enclosure. This is a generalization of (8-3) in the respect that over each finite surface an integration is performed to determine the total contri-bution that the local flux leaving that surface makes to the quantity $q_{i,k}$:

$$dA_k q_{i,k}(\mathbf{r}_k) = \int_{A_1} q_{o,1}(\mathbf{r}_1)\, dF_{d1-dk}(\mathbf{r}_1,\mathbf{r}_k)\, dA_1 + \cdots$$

$$+ \int_{A_k} q_{o,k}(\mathbf{r}_k^*)\, dF_{dk^*-dk}(\mathbf{r}_k^*,\mathbf{r}_k)\, dA_k^* + \cdots \tag{8-32}$$

$$+ \int_{A_N} q_{o,N}(\mathbf{r}_N)\, dF_{dN-dk}(\mathbf{r}_N,\mathbf{r}_k)\, dA_N$$

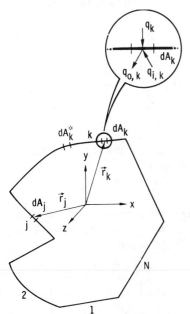

Figure 8-7 Enclosure composed of N discrete surface areas with areas subdivided into infinitesimal elements.

The second integral on the right is the contribution that other differential elements dA_k^* on surface A_k make to the incident energy at dA_k.

By using reciprocity, $dA_j\, dF_{dj-dk} = dA_k\, dF_{dk-dj}$, a typical integral in (8-32) can be transformed to obtain

$$\int_{A_j} q_{o,j}(\mathbf{r}_j)\, dF_{dj-dk}(\mathbf{r}_j,\mathbf{r}_k)\, dA_j = \int_{A_j} q_{o,j}(\mathbf{r}_j)\, dF_{dk-dj}(\mathbf{r}_j,\mathbf{r}_k)\, dA_k$$

By operating on all the integrals in (8-32) in this manner, the dA_k are divided out of the equation and the result becomes

$$q_{i,k}(\mathbf{r}_k) = \sum_{j=1}^{N} \int_{A_j} q_{o,j}(\mathbf{r}_j)\, dF_{dk-dj}(\mathbf{r}_j,\mathbf{r}_k) \tag{8-33}$$

Equations (8-31) and (8-33) provide two different expressions for $q_{i,k}(\mathbf{r}_k)$. These are each substituted into (8-30) to provide two expressions for $q_k(\mathbf{r}_k)$ comparable to (8-6) and (8-7):

$$q_k(\mathbf{r}_k) = \frac{\epsilon_k}{1 - \epsilon_k} [\sigma T_k^{\,4}(\mathbf{r}_k) - q_{o,k}(\mathbf{r}_k)] \tag{8-34}$$

$$q_k(\mathbf{r}_k) = q_{o,k}(\mathbf{r}_k) - \sum_{j=1}^{N} \int_{A_j} q_{o,j}(\mathbf{r}_j)\, dF_{dk-dj}(\mathbf{r}_j,\mathbf{r}_k) \tag{8-35}$$

As shown by Eq. (7-10), the differential configuration factor dF_{dk-dj} contains the differential area dA_j. To place (8-35) in a more standard form in which the variable of integration is explicitly shown, it is convenient to define a quantity $K(\mathbf{r}_j,\mathbf{r}_k)$ by

$$K(\mathbf{r}_j,\mathbf{r}_k) \equiv \frac{dF_{dk-dj}(\mathbf{r}_j,\mathbf{r}_k)}{dA_j} \tag{8-36}$$

Then (8-35) becomes the integral equation

$$q_k(\mathbf{r}_k) = q_{o,k}(\mathbf{r}_k) - \sum_{j=1}^{N} \int_{A_j} q_{o,j}(\mathbf{r}_j) K(\mathbf{r}_j,\mathbf{r}_k)\, dA_j \tag{8-37}$$

The quantity $K(\mathbf{r}_j,\mathbf{r}_k)$ that appears under the integral sign with the dependent variable such as in (8-37) is called the *kernel* of the integral equation.

As in the previous discussion for finite areas, there are two paths that can now be followed: (1) When the temperatures and imposed heat fluxes are important, (8-34) and (8-35) can be combined to eliminate the variables q_o. This gives a set of simultaneous relations directly relating the surface temperatures T and the imposed heat fluxes q. Along each surface area, either the T or the q will be specified by the boundary conditions. The remaining unknown T's and q's can then be found by solving the simultaneous relations. (2) Alternatively, when q_o is an important quantity, the unknown q's can be eliminated by combining (8-34) and (8-35) for each surface that does not have its q specified as a boundary condition. For a surface where q is known, (8-35) can be used to directly relate the q_o's to each other. This yields a set of simultaneous relations for the q_o's in terms of the known q's and T's that are specified by the boundary conditions. After solving for the q_o's, Eqs. (8-34) can be used, if desired, to relate the q's and T's, where either the q or T will be known at each surface from the boundary conditions. Each of these procedures will now be examined.

Relations between surface temperature T and surface heating q. To eliminate the q_o in the first method of solution, (8-34) is solved for $q_{o,k}(\mathbf{r}_k)$, giving

$$q_{o,k}(\mathbf{r}_k) = \sigma T_k^{4}(\mathbf{r}_k) - \frac{1 - \epsilon_k}{\epsilon_k} q_k(\mathbf{r}_k) \tag{8-38}$$

Equation (8-38) in the form shown and with k changed to j is then substituted into (8-35) to eliminate $q_{o,k}$ and $q_{o,j}$, which yields

$$\frac{q_k(\mathbf{r}_k)}{\epsilon_k} - \sum_{j=1}^{N} \frac{1 - \epsilon_j}{\epsilon_j} \int_{A_j} q_j(\mathbf{r}_j)\, dF_{dk-dj}(\mathbf{r}_j,\mathbf{r}_k)$$

$$= \sigma T_k^{4}(\mathbf{r}_k) - \sum_{j=1}^{N} \int_{A_j} \sigma T_j^{4}(\mathbf{r}_j)\, dF_{dk-dj}(\mathbf{r}_j,\mathbf{r}_k) \tag{8-39}$$

Equation (8-39) directly relates the surface temperatures to the heat fluxes supplied to the surfaces.

EXAMPLE 8-8 An enclosure of the general type in Fig. 8-5 is composed of three plane surfaces and for simplicity is infinitely long so that the heat transfer quantities do not vary with length. Surface 1 is heated uniformly, and surface 2 is at a uniform temperature. Surface 3 is black and at zero temperature. What are the governing equations needed to determine the temperature distribution over the perimeter of surface 1?

With $T_3 = 0$, $\epsilon_3 = 1$, and the self-view factors $dF_{dj-dj*} = 0$, Eq. (8-39) can be written for the two plane surfaces 1 and 2 having uniform q_1 and T_2 as

$$\frac{q_1}{\epsilon_1} - \frac{1 - \epsilon_2}{\epsilon_2} \int_{A_2} q_2(\mathbf{r}_2)\, dF_{d1-d2}(\mathbf{r}_2,\mathbf{r}_1) = \sigma T_1^{4}(\mathbf{r}_1) - \sigma T_2^{4} \int_{A_2} dF_{d1-d2}(\mathbf{r}_2,\mathbf{r}_1)$$

$$\tag{8-40a}$$

$$\frac{q_2(\mathbf{r}_2)}{\epsilon_2} - q_1 \frac{1 - \epsilon_1}{\epsilon_1} \int_{A_1} dF_{d2-d1}(\mathbf{r}_1,\mathbf{r}_2) = \sigma T_2^{4} - \int_{A_1} \sigma T_1^{4}(\mathbf{r}_1)\, dF_{d2-d1}(\mathbf{r}_1,\mathbf{r}_2)$$

$$\tag{8-40b}$$

A similar equation for surface 3 is not needed since Eqs. (8-40) do not involve the unknown $q_3(\mathbf{r}_3)$ as a consequence of $\epsilon_3 = 1$ and $T_3 = 0$. From the definitions of F factors.

$$\int_{A_2} dF_{d1-d2} = F_{d1-2} \qquad \text{and} \qquad \int_{A_1} dF_{d2-d1} = F_{d2-1}$$

Equations (8-40) simplify to the following relations where the unknowns have been placed on the left:

$$\sigma T_1^4(\mathbf{r}_1) + \frac{1-\epsilon_2}{\epsilon_2} \int_{A_2} q_2(\mathbf{r}_2) \, dF_{d1-d2}(\mathbf{r}_2,\mathbf{r}_1) = \sigma T_2^4 F_{d1-2}(\mathbf{r}_1) + \frac{q_1}{\epsilon_1} \qquad (8\text{-}41a)$$

$$\int_{A_1} \sigma T_1^4(\mathbf{r}_1) \, dF_{d2-d1}(\mathbf{r}_1,\mathbf{r}_2) + \frac{q_2(\mathbf{r}_2)}{\epsilon_2} = \sigma T_2^4 + q_1 \frac{1-\epsilon_1}{\epsilon_1} F_{d2-1}(\mathbf{r}_2) \qquad (8\text{-}41b)$$

Equations (8-41) can be solved simultaneously for the unknown distributions $T_1(\mathbf{r}_1)$ and $q_2(\mathbf{r}_2)$. Some methods for solving such a set of integral equations will be discussed in Sec. 8-4.2.

Solution method in terms of outgoing radiative flux q_o. A second method of solution results from eliminating the $q_k(\mathbf{r}_k)$ terms from (8-34) and (8-35) for the surfaces where $q_k(\mathbf{r}_k)$ is unknown. This provides a relation between q_o and the T variation specified along a surface:

$$q_{o,k}(\mathbf{r}_k) = \epsilon_k \sigma T_k^4(\mathbf{r}_k) + (1-\epsilon_k) \sum_{j=1}^{N} \int_{A_j} q_{o,j}(\mathbf{r}_j) \, dF_{dk-dj}(\mathbf{r}_j,\mathbf{r}_k) \qquad (8\text{-}42)$$

When $q_k(\mathbf{r}_k)$, the heat supplied to surface k, is known, (8-35) can be used directly to relate q_k and q_o. The combination of (8-42) and (8-35) thus provides a complete set of relations for the unknown q_o's in terms of known T's and q's.

This set of equations for the q_o's will now be formulated more explicitly. In general, an enclosure can have surfaces 1, 2, . . . , m with specified temperature distributions. For these surfaces, (8-42) is utilized. The remaining $N - m$ surfaces $m + 1, m + 2, \ldots, N$ have an imposed heat-flux distribution specified. For these surfaces (8-35) is applied. This results in a set of N equations for the unknown q_o distributions:

$$q_{o,k}(\mathbf{r}_k) - (1-\epsilon_k) \sum_{j=1}^{N} \int_{A_j} q_{o,j}(\mathbf{r}_j) \, dF_{dk-dj}(\mathbf{r}_j,\mathbf{r}_k) = \epsilon_k \sigma T_k^4(\mathbf{r}_k) \qquad 1 \leqslant k \leqslant m$$

$$(8\text{-}43a)$$

$$q_{o,k}(\mathbf{r}_k) - \sum_{j=1}^{N} \int_{A_j} q_{o,j}(\mathbf{r}_j) \, dF_{dk-dj}(\mathbf{r}_j,\mathbf{r}_k) = q_k(\mathbf{r}_k) \qquad m+1 \leqslant k \leqslant N \qquad (8\text{-}43b)$$

After the q_o's are found, (8-34) is applied to determine the unknown q or T distributions

$$q_k(\mathbf{r}_k) = \frac{\epsilon_k}{1-\epsilon_k} [\sigma T_k^4(\mathbf{r}_k) - q_{o,k}(\mathbf{r}_k)] \qquad 1 \leqslant k \leqslant m \qquad (8\text{-}44a)$$

$$\sigma T_k^{\,4}(\mathbf{r}_k) = \frac{1 - \epsilon_k}{\epsilon_k} q_k(\mathbf{r}_k) + q_{o,k}(\mathbf{r}_k) \qquad m + 1 \leqslant k \leqslant N \qquad (8\text{-}44b)$$

Special case when imposed heating q is specified for all surfaces. There is an interesting special case when the imposed energy flux q is specified for all surfaces of the enclosure and it is desired to determine the surface temperature distributions. For this case the use of the method of the previous section where the q_o's are first determined has an advantage over the method given by (8-39) where the T's are directly determined from the specified q's. This advantage arises from the fact that Eq. (8-43b) is independent of the radiative properties of the surfaces. This means that, for a given set of q's, the q_o's need be determined only once by writing (8-43b) for each of the surfaces. Then the temperature distributions are found from (8-44b), which introduces the emissivity dependence. This would have an advantage when it is desired to examine the temperature variations for various emissivity values when there is a fixed set of q's.

In the case in which the surfaces are all black, $\epsilon_k = 1$ and Eq. (8-44b) becomes

$$\sigma T_k^{\,4}(\mathbf{r}_k)_{\text{black}} = q_{o,k}(\mathbf{r}_k)$$

Since the $q_{o,k}$'s are independent of the emissivities, these $q_{o,k}$'s are also valid for surfaces where $\epsilon_k \neq 1$. The solution in (8-44b) can then be written as

$$\sigma T_k^{\,4}(\mathbf{r}_k) = \frac{1 - \epsilon_k}{\epsilon_k} q_k(\mathbf{r}_k) + \sigma T_k^{\,4}(\mathbf{r}_k)_{\text{black}} \qquad (8\text{-}45)$$

This relates the temperature distributions in an enclosure for $\epsilon_k \neq 1$ to the temperature distributions in a black enclosure having the same imposed heat fluxes. Thus, once the temperature distributions have been found for the black case, the fourth-power temperature distributions $\sigma T_k^{\,4}(\mathbf{r}_k)$ for gray surfaces are found by simply adding the term $[(1 - \epsilon_k)/\epsilon_k] q_k(\mathbf{r}_k)$.

To this point, a number of formulations of the governing equations of radiation interchange within an enclosure have been given. In Table 8-1, the relations that have been derived for finding quantities of interest, such as Q, T, and q_o on various surfaces in terms of given quantities, are summarized for convenience.

EXAMPLE 8-9 A relatively simple example of a heated enclosure is the circular tube shown in Fig. 8-8, open at both ends and insulated on the outside surface [7]. For a uniform heat addition along the tube wall and a surrounding environment temperature of 0 K, what is the temperature distribution along the tube? If the surroundings are at temperature T_e, how does this influence the temperature distribution?

Since the open ends of the tube are nonreflecting, they can be assumed to act as black disks at a specified temperature of 0 K. Then (8-44b) is used for these two disks to find their q_o. With $\epsilon_1 = \epsilon_3 = 1$, (8-44b) gives $q_{o,1} = q_{o,3} = \sigma T_1^{\,4} = \sigma T_3^{\,4} = 0$. Consequently, the summation in (8-43b) will provide only radiation from surface 2 to itself. Since the tube is axisymmetric, the two differential areas dA_k and dA_k^* can be taken as rings located respectively at x and y. For conveni-

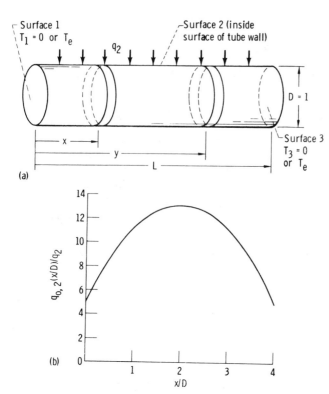

Figure 8-8 Uniformly heated tube insulated on outside and open to environment at both ends. (a) Geometry and coordinate system; (b) distribution of q_0 on inside of tube for $L/D = 4$.

ence, all lengths are nondimensionalized with respect to the tube diameter. Then Eq. (8-43b) yields

$$q_{o,2}(\xi) - \int_{\eta=0}^{\eta=l} q_{o,2}(\eta)\, dF_{d\xi-d\eta}(|\eta-\xi|) = q_2 \tag{8-46a}$$

where $\xi = x/D$, $\eta = y/D$, $l = L/D$, and $dF_{d\xi-d\eta}(|\eta-\xi|)$ is the configuration factor for two rings a distance $|\eta-\xi|$ apart and is given by 29 in Appendix C:

$$dF_{d\xi-d\eta}(|\eta-\xi|) = \left\{ 1 - \frac{|\eta-\xi|^3 + \frac{3}{2}|\eta-\xi|}{[(\eta-\xi)^2 + 1]^{3/2}} \right\} d\eta \tag{8-46b}$$

Absolute-value signs are used on $\eta - \xi$ because the configuration factor depends only on the magnitude of the distance between the rings. When $|\eta - \xi| = 0$, $dF = d\eta$, and this represents the view factor from a differential ring to itself. Equation (8-46a) can be divided by the constant q_2, and the solution for the dimensionless quantity $q_{o,2}(\xi)/q_2$ found by numerical or approximate methods for solving linear integral equations. A discussion of these methods will be given in Sec. 8-4.2. The resulting $q_{o,2}(\xi)/q_2$ distribution is shown in Fig. 8-8b for a tube four diameters in length. From Eq. (8-44b), the distribution of temperature

Table 8-1 Relations between energy flux and temperature in diffuse-gray enclosures

Type of areas	Boundary conditions	Desired quantities	Equation
Finite areas	T_k on all surfaces $1 \leq k \leq N$	Q_k	(8-19)
		$q_{o,k}$	(8-23)
	Q_k on all surfaces $1 \leq k \leq N$	T_k	(8-19)
		$q_{o,k}$	(8-7)
	T_k for $1 \leq k \leq m$ Q_k for $m + 1 \leq k \leq N$	Q_k for $1 \leq k \leq m$ T_k for $m + 1 \leq k \leq N$	(8-24) and (8-6), or (8-19)
		$q_{o,k}$	(8-24)
Infinitesimal areas	T_k on all surfaces $1 \leq k \leq N$	q_k	(8-39)
		$q_{o,k}$	(8-42)
	q_k on all surfaces $1 \leq k \leq N$	T_k	(8-39)
		$q_{o,k}$	(8-35)
	T_k for $1 \leq k \leq m$ q_k for $m + 1 \leq k \leq N$	q_k for $1 \leq k \leq m$ T_k for $m + 1 \leq k \leq N$	(8-39), or (8-43) and (8-44)
		$q_{o,k}$	(8-43a) and (8-43b)
	$q_{o,k}$ for $1 \leq k \leq N$ T_k for $1 \leq k \leq m$ q_k for $m + 1 \leq k \leq N$	q_k for $1 \leq k \leq m$ T_k for $m + 1 \leq k \leq N$	(8-44a) (8-44b)

to the fourth power along the tube is given by

$$\sigma T_2^4(\xi) = \frac{1 - \epsilon_2}{\epsilon_2} q_2 + q_{o,2}(\xi)$$

Since q_2 is a constant, the distribution $T_2^4(\xi)$ has the same shape as the distribution $q_{o,2}(\xi)$. The wall temperature is high in the central region of the tube and low near the end openings where heat can be radiated easily to the low-temperature environment.

Now consider the case in which the environment is at T_e rather than at zero. The open ends of the cylindrical enclosure can be regarded as perfectly absorbing disks at T_e. The integral equation (8-43b) now yields

$$q_{o,2}(\xi) - \int_{\eta=0}^{l} q_{o,2}(\eta) \, dF_{d\xi-d\eta}(|\eta - \xi|) - \sigma T_e^4 F_{d\xi-1}(\xi)$$

$$- \sigma T_e^4 F_{d\xi-3}(l - \xi) = q_2$$

where $F_{d\xi-1}(\xi)$ is the configuration factor from a ring element at ξ to disk 1 at $\xi = 0$, that is,

$$F_{d\xi-1}(\xi) = \frac{\xi^2 + \frac{1}{2}}{(\xi^2 + 1)^{1/2}} - \xi$$

Since the integral equation is linear in the variable $q_{o,2}(\xi)$, let a trial solution be in the form of a sum of two parts where, for each part, either $T_e = 0$ or $q_2 = 0$:

$$q_{o,2}(\xi) = q_{o,2}(\xi)|_{T_e=0} + q_{o,2}(\xi)|_{q_2=0}$$

Substitute the trial solution into the integral equation to get

$$q_{o,2}(\xi)|_{T_e=0} + q_{o,2}(\xi)|_{q_2=0} - \int_{\eta=0}^{l} q_{o,2}(\eta)|_{T_e=0}\, dF_{d\xi-d\eta}(|\eta - \xi|)$$

$$- \int_{\eta=0}^{l} q_{o,2}(\eta)|_{q_2=0}\, dF_{d\xi-d\eta}(|\eta - \xi|) - \sigma T_e{}^4 F_{d\xi-1}(\xi)$$

$$- \sigma T_e{}^4 F_{d\xi-3}(l - \xi) = q_2$$

For $T_e = 0$, (8-46a) applies; subtract this equation to give

$$q_{o,2}(\xi)|_{q_2=0} - \int_{\eta=0}^{l} q_{o,2}(\eta)|_{q_2=0}\, dF_{d\xi-d\eta}(|\eta - \xi|)$$

$$- \sigma T_e{}^4 F_{d\xi-1}(\xi) - \sigma T_e{}^4 F_{d\xi-3}(l - \xi) = 0$$

As can be verified by direct substitution and then integration, the solution is

$$q_{o,2}|_{q_2=0} = \sigma T_e{}^4$$

This would be expected physically for an unheated surface in a uniform temperature environment. The temperature distribution along the tube is found from Eq. (8-44b) as

$$\sigma T_2{}^4(\xi) = \frac{1 - \epsilon_2}{\epsilon_2} q_2 + q_{o,2}(\xi)|_{T_e=0} + q_{o,2}(\xi)|_{q_2=0}$$

$$\sigma T_2{}^4(\xi) = \frac{1 - \epsilon_2}{\epsilon_2} q_2 + q_{o,2}(\xi)|_{T_e=0} + \sigma T_e{}^4$$

where $q_{o,2}(\xi)|_{T_e=0}$ was found in the first part of this example. The superposition of an environment temperature has thus added a $\sigma T_e{}^4$ term to the solution for $\sigma T_2{}^4(\xi)$, found previously for $T_e = 0$.

The final result of Example 8-9 can be written as

$$\sigma[T_2{}^4(\xi) - T_e{}^4] = \frac{1 - \epsilon_2}{\epsilon_2} q_2 + q_{o,2}(\xi)|_{T_e=0}$$

Thus when T_e is nonzero, the quantity $T_2{}^4(\xi) - T_e{}^4$ is equal to $T_2{}^4(\xi)$ for the case when $T_e = 0$. This illustrates a general way of accounting for a finite environment temperature. For heat transfer only by radiation, the governing equations are linear in T^4. As a result, in a cavity type of enclosure the wall temperature $T_w|_{T_e=0}$ can be calculated for a zero environment temperature. Then by super-position the wall temperature for a finite T_e can be obtained as $T_w{}^4|_{T_e \neq 0} = T_w{}^4|_{T_e=0} + T_e{}^4$. Hence the thermal characteristics of a cavity having a wall temperature variation T_w and an external environment at T_e are the same as a cavity with wall temperature variation $(T_w{}^4 - T_e{}^4)^{1/4}$ and a zero environment temperature; it is easier to solve for the radiative behavior in this latter case.

EXAMPLE 8-10 This example concerns the emission from a long cylindrical hole drilled into a material that is all at uniform temperature T_w (Fig. 8-9a). The hole is assumed sufficiently long so that the surface at the bottom end of the hole can be neglected in the radiative heat balances. The environment outside the hole is at T_e. In accordance with the previous discussion, the solution is obtained by using the reduced temperature $T = (T_w{}^4 - T_e{}^4)^{1/4}$. If a position is viewed at x on the cylindrical side wall of the hole, the energy leaving the wall is composed of the

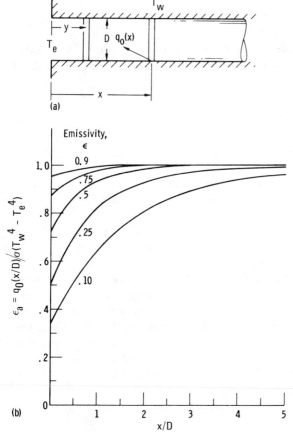

Figure 8-9 Radiant emission from cylindrical hole at uniform temperature. (a) Geometry and coordinate system; (b) apparent emissivity of cylinder wall.

direct emission plus the reflected energy, the total being the quantity $q_o(x)$. An apparent emissivity is defined as $\epsilon_a(x) = q_o(x)/\sigma T^4$. The objective of this analysis is to determine how $\epsilon_a(x)$ is related to the actual surface emissivity ϵ, where ϵ is constant over the side of the hole. The integral equation governing the radiation exchange within the hole was first derived by Buckley [8, 9] and later by Eckert [10]; both investigators obtained approximate analytical solutions. The results were later carried out numerically to greater accuracy by Sparrow and Albers with a digital computer [11].

By using the reduced temperature, the opening of the hole can be approximated by a perfectly absorbing (that is, black) disk at zero reduced temperature. Then from (8-44b) (because $\epsilon = 1$ and $T = 0$ for the opening area), the q_o for the opening disk is zero. Hence, the governing equation for the enclosure is Eq. (8-43a) written for the cylindrical side wall and including in the summation only the radiation from the cylindrical wall to itself. As in Example 8-9, the configuration factor is that for one ring of differential length on the cylindrical enclosure exchanging radiation with a second ring at a different axial location as given by (8-46b). Equation (8-43a) then yields

$$q_o(\xi) - (1 - \epsilon) \int_{\eta=0}^{\infty} q_o(\eta) \, dF_{d\xi-d\eta}(|\eta - \xi|) = \epsilon \sigma T^4 \tag{8-47}$$

where $\xi = x/D$, $\eta = y/D$, and $dF_{d\xi-d\eta}(|\eta - \xi|)$ is given by (8-46b). After division by σT^4 which is constant, the apparent emissivity is found to be governed by the following integral equation:

$$\epsilon_a(\xi) - (1 - \epsilon) \int_{\eta=0}^{\infty} \epsilon_a(\eta) \, dF_{d\xi-d\eta}(|\eta - \xi|) = \epsilon \tag{8-48}$$

The solution of (8-48) was carried out for various surface emissivities ϵ, and the results for ϵ_a as a function of location along the hole are shown in Fig. 8-9b. The radiation leaving the surface approaches that of a blackbody as the wall position is increased to greater depths into the hole. At the mouth of the hole, $\epsilon_a = \sqrt{\epsilon}$, as shown by Buckley [8, 9].

The radiation from a hole of finite depth was analyzed in [12], and results are given in Fig. 8-10. Approximate solutions are presented in [13]. The effective hemispherical emissivity ϵ_h shown is different from that in Fig. 8-9b, and it gives the total amount of energy leaving the mouth of the cavity ratioed to that emitted from a black-walled cavity. The latter is the same as the energy emitted from a black area across the mouth of the cavity. For each surface emissivity ϵ, the ϵ_h increases to a limiting value as the depth of the cavity is increased; the limiting values are less than unity. Unless ϵ is small, a cavity more than only a few diameters deep emits the same amount of radiation as an infinitely deep cavity.

To construct a cavity that will provide very close to blackbody emission for use in calibrating measuring equipment, the cavity opening can be partially closed as shown in Fig. 8-11. The apparent emissivity ϵ_a at the center of the bottom of

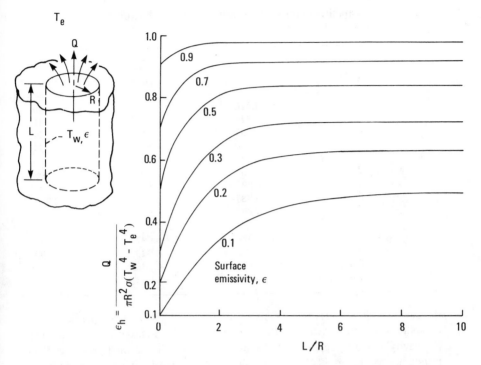

Figure 8-10 Apparent emissivity of cavity opening for a cylindrical cavity of finite length with diffuse reflecting walls at constant temperature [12].

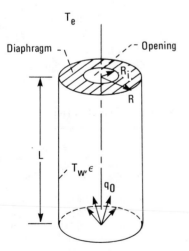

Figure 8-11 Cylindrical cavity with annular diaphragm partially covering opening [26, 27].

Table 8-2 Values of apparent emissivity at center of cavity bottom,
$\epsilon_a = q_0/\sigma(T_w{}^4 - T_e{}^4)$

ϵ	R_i/R	ϵ_a		
		$L/R = 2$	$L/R = 4$	$L/R = 8$
0.25	0.4	0.916	0.968	0.990
	0.6	0.829	0.931	0.981
	0.8	0.732	0.888	0.969
	1.0	0.640	0.844	0.965
0.50	0.4	0.968	0.990	0.998
	0.6	0.932	0.979	0.995
	0.8	0.887	0.964	0.992
	1.0	0.839	0.946	0.989
0.75	0.4	0.988	0.997	0.999
	0.6	0.975	0.993	0.998
	0.8	0.958	0.988	0.997
	1.0	0.939	0.982	0.996

the cavity is shown in Table 8-2 as a function of the diaphragm opening-to-outer-radius ratio and the cavity depth-to-radius ratio. The ϵ_a increases toward unity as the cavity is made deeper and the diaphragm opening is made smaller. Similar trends are shown in [14] for partially baffled conical cavities. In [15], cavities between radial fins are analyzed.

Some information on the directional nature of radiant emission from a cylindrical cavity is given in [16], where the emission normal to the cavity opening is calculated as would be received by a small detector facing the opening and located along the cavity centerline. The normal emissivity ϵ_n' of the cavity is defined as the energy received by the detector divided by the energy that would be received if the cavity walls were entirely black. The ϵ_n' depends on the detector distance from the cavity opening; the values given in Table 8-3 are for large distances where ϵ_n' no longer changes with distance. The table compares ϵ_n' with ϵ_h from Fig. 8-10 (ϵ is the emissivity of the cavity walls). The ϵ_n' is larger than ϵ_h except for small values of cavity L/R.

Table 8-3 Comparison of normal and hemispherical emissivity of cavity [16]

L/R	$\epsilon = 0.9$		$\epsilon = 0.7$		$\epsilon = 0.5$		$\epsilon = 0.3$	
	ϵ_h	ϵ_n'	ϵ_h	ϵ_n'	ϵ_h	ϵ_n'	ϵ_h	ϵ_n'
0.5	0.943	0.937	0.815	0.800	0.657	0.638	0.455	0.437
1	0.962	0.960	0.869	0.865	0.742	0.738	0.556	0.554
2	0.972	0.982	0.904	0.934	0.809	0.858	0.657	0.720
4	0.974	0.994	0.914	0.978	0.833	0.950	0.710	0.884
8	0.975	0.999	0.915	0.995	0.836	0.989	0.719	0.973

EXAMPLE 8-11 What are the integral equations governing the radiation exchange between two parallel opposed plates finite in one dimension and infinite in the other, as shown in Fig. 8-12? Each plate has a specified temperature variation that depends only on the x or y coordinate shown, and the environment is at T_e.

From the discussion in Example 7-4, the configuration factors between the infinitely long parallel strips dA_1 and dA_2 are

$$dF_{d1-d2} = \frac{1}{2} d(\sin \beta) = \frac{1}{2} \frac{a^2}{[(y-x)^2 + a^2]^{3/2}} dy$$

$$dF_{d2-d1} = \frac{1}{2} \frac{a^2}{[(y-x)^2 + a^2]^{3/2}} dx$$

The distribution of heat flux added to each plate can be found by applying Eq. (8-39) to each of the plates. As discussed prior to Example 8-10, reduced temperatures are used to account for T_e. With $T_1 = (T_{w1}{}^4 - T_e{}^4)^{1/4}$ and $T_2 = (T_{w2}{}^4 - T_e{}^4)^{1/4}$, the governing equations are

$$\frac{q_1(x)}{\epsilon_1} - \frac{1-\epsilon_2}{\epsilon_2} \int_{-L/2}^{L/2} q_2(y) \frac{1}{2} \frac{a^2}{[(y-x)^2 + a^2]^{3/2}} dy$$

$$= \sigma T_1{}^4(x) - \int_{-L/2}^{L/2} \sigma T_2{}^4(y) \frac{1}{2} \frac{a^2}{[(y-x)^2 + a^2]^{3/2}} dy \qquad (8\text{-}49a)$$

$$\frac{q_2(y)}{\epsilon_2} - \frac{1-\epsilon_1}{\epsilon_1} \int_{-L/2}^{L/2} q_1(x) \frac{1}{2} \frac{a^2}{[(y-x)^2 + a^2]^{3/2}} dx$$

$$= \sigma T_2{}^4(y) - \int_{-L/2}^{L/2} \sigma T_1{}^4(x) \frac{1}{2} \frac{a^2}{[(y-x)^2 + a^2]^{3/2}} dx \qquad (8\text{-}49b)$$

An alternative formulation can be obtained by applying (8-42). This yields the following two equations for $q_{o,1}(x)$ and $q_{o,2}(y)$:

$$q_{o,1}(x) - (1-\epsilon_1) \int_{-L/2}^{L/2} q_{o,2}(y) \frac{1}{2} \frac{a^2}{[(y-x)^2 + a^2]^{3/2}} dy = \epsilon_1 \sigma T_1{}^4(x) \qquad (8\text{-}50a)$$

$$q_{o,2}(y) - (1-\epsilon_2) \int_{-L/2}^{L/2} q_{o,1}(x) \frac{1}{2} \frac{a^2}{[(y-x)^2 + a^2]^{3/2}} dx = \epsilon_2 \sigma T_2{}^4(y) \qquad (8\text{-}50b)$$

After the q_o's are found, the desired $q_1(x)$ and $q_2(y)$ are obtained from (8-44a); they are

$$q_1(x) = \frac{\epsilon_1}{1-\epsilon_1} [\sigma T_1{}^4(x) - q_{o,1}(x)] \qquad (8\text{-}51a)$$

$$q_2(y) = \frac{\epsilon_2}{1-\epsilon_2} [\sigma T_2{}^4(y) - q_{o,2}(y)] \qquad (8\text{-}51b)$$

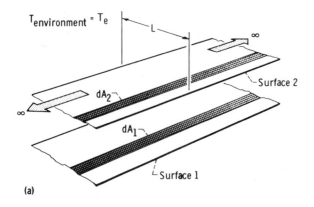

(a)

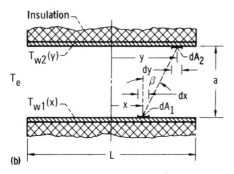

(b)

Figure 8-12 Geometry for radiation between two parallel plates infinitely long in one direction and of finite width. (*a*) Parallel plates of width L and infinite length; (*b*) coordinates in cross section of gap between parallel plates.

8-4.2 Methods for Solving Integral Equations

The previous examples have revealed that the unknown wall heat fluxes or temperatures along the surfaces of an enclosure are found from the solutions of single or simultaneous integral equations. The integral equations are linear; that is, the unknown q, q_o, or T^4 variables always appear to the first power (note that T^4 is considered as the variable rather than T). For linear integral equations there are a number of analytical and numerical solution methods that can be utilized. These are discussed in standard mathematics texts, for example, Chap. 4 of [6]. The use of some of these methods as applied to radiation problems will now be discussed, and some examples will be given.

Numerical integration yielding simultaneous equations. In most instances the functions inside the integrals of the integral equations are complicated algebraic quantities. This is because these functions involve a configuration factor that, for most geometries, is not of a simple form. There is generally little chance that an exact analytical solution can be found. A numerical solution must then be attempted in most cases. The integrals are expressed in finite-difference form by dividing each surface into a grid of small finite increments. The result is a set of simultaneous equations for the unknown quantities at each incremental position. This procedure is best illustrated by a specific example.

EXAMPLE 8-12 Referring to the integral equation (8-46a), derive a set of simultaneous algebraic equations to determine the $q_{o,2}$ distribution for a length $l = 4$.

For simplicity divide the length into four equal increments ($\Delta\eta = 1$), and use the trapezoidal rule for integration. When (8-46a) is applied at the end of the tube where $\xi = 0$, one obtains

$$q_{o,2}(0) - [\tfrac{1}{2}q_{o,2}(0)K(|0-0|) + q_{o,2}(1)K(|1-0|) + q_{o,2}(2)K(|2-0|)$$
$$+ q_{o,2}(3)K(|3-0|) + \tfrac{1}{2}q_{o,2}(4)K(|4-0|)](1) = q_2 \qquad (8\text{-}52)$$

The quantity included in brackets is the trapezoidal-rule approximation for the integral. The quantity $K(|\eta - \xi|) = dF(|\eta - \xi|)/d\eta$ is the algebraic expression within the braces of (8-46b). The $q_{o,2}(0)$ terms in (8-52) may be grouped together to provide the first of (8-53). The other four equations of the set are obtained by writing the finite-difference equation at the other incremental positions along the cylindrical enclosure:

$$q_{o,2}(0)[1 - \tfrac{1}{2}K(0)] - q_{o,2}(1)K(1) - q_{o,2}(2)K(2) - q_{o,2}(3)K(3)$$
$$- \tfrac{1}{2}q_{o,2}(4)K(4) = q_2$$

$$- \tfrac{1}{2}q_{o,2}(0)K(1) + q_{o,2}(1)[1 - K(0)] - q_{o,2}(2)K(1) - q_{o,2}(3)K(2)$$
$$- \tfrac{1}{2}q_{o,2}(4)K(3) = q_2$$

$$- \tfrac{1}{2}q_{o,2}(0)K(2) - q_{o,2}(1)K(1) + q_{o,2}(2)[1 - K(0)] - q_{o,2}(3)K(1)$$
$$- \tfrac{1}{2}q_{o,2}(4)K(2) = q_2$$

$$- \tfrac{1}{2}q_{o,2}(0)K(3) - q_{o,2}(1)K(2) - q_{o,2}(2)K(1) + q_{o,2}(3)[1 - K(0)]$$
$$- \tfrac{1}{2}q_{o,2}(4)K(1) = q_2$$

$$- \tfrac{1}{2}q_{o,2}(0)K(4) - q_{o,2}(1)K(3) - q_{o,2}(2)K(2) - q_{o,2}(3)K(1)$$
$$+ q_{o,2}(4)[1 - \tfrac{1}{2}K(0)] = q_2 \qquad (8\text{-}53)$$

These equations are solved simultaneously for the unknown $q_{o,2}$ values at the five surface locations. From the symmetry of the configuration and the fact that q_2 is uniform along the enclosure, it is possible in this instance to simplify the solution by using the equalities $q_{o,2}(0) = q_{o,2}(4)$ and $q_{o,2}(1) = q_{o,2}(3)$.

In practice a set of equations such as (8-53) are first solved for a moderate number of increments along the enclosure. Then the increment size is reduced, and the set of equations is solved again. This process is continued until sufficiently accurate q_o values are obtained. This procedure would generally be programmed for computer calculation in terms of an arbitrary increment size.

Equations (8-53) were derived using the trapezoidal rule as a simple numerical approximation to the integrals. Other more accurate numerical integration schemes can be used which may reduce the number of increments required to provide sufficient accuracy in a given problem.

One precaution should be noted. The quantity $q_{o,j} \, dF_{dk-dj}$ may in certain instances go through rapid changes in magnitude because of the geometry involved in the configuration factor; for example, dF_{dk-dj} may decrease very rapidly as the distance between dA_k and dA_j is increased. This may mean that an integration approximation such as Simpson's rule will not be very accurate since the shape of $q_{o,j} \, dF_{dk-dj}$ may not be approximated well by passing a parabola locally through the function. Care should be taken in selecting an integration scheme that can approximate well the general behavior of the functions involved.

Example 8-12 contained only one integral equation. The situation described by (8-49) involves two integral equations. Surfaces 1 and 2 can both be divided into increments, and the equations written in finite-difference form at each incremental location. This will yield a set of n simultaneous equations with n equal to the total number of chosen positions on both plates, and the equations can then be solved simultaneously for the $q_1(x)$ and $q_2(y)$ distributions.

Another way of solving the two integral equations numerically is by iteration. With $T_1(x)$ and $T_2(y)$ specified, the right sides of the equations are known as functions of x and y. Starting with (8-49a), a distribution for $q_2(y)$ is assumed as a first trial. Then the integration can be carried out numerically for various x values to yield $q_1(x)$ at these x locations. This $q_1(x)$ distribution is then inserted into Eq. (8-49b) and a $q_2(y)$ distribution is determined. This $q_2(y)$ is then used to compute a new $q_1(x)$, and the process is continued until $q_1(x)$ and $q_2(y)$ are no longer changing as the iterations proceed.

Use of approximate separable kernel. In an integral equation such as (8-46a) the solution can sometimes be simplified if the kernel is of a separable form, that is, equal to a product (or sum of products) of a function of $\mathbf{r}_j$ alone and a function of $\mathbf{r}_k$ alone. Recall from (8-36) that the kernel is $K(\mathbf{r}_j, \mathbf{r}_k) = dF_{dk-dj}(\mathbf{r}_j, \mathbf{r}_k)/dA_j$. For a separable kernel, the function of $\mathbf{r}_k$ can be taken out of the integral, thereby simplifying the integration. The general theory of integral equations with separable kernels is given in [6]. Generally for radiation problems K will not be in a separable form. However, it may be possible to find a separable function that closely approximates K and can thus be substituted into the integral equation to provide a simplification.

Buckley [8, 9] demonstrated that an especially useful form for a separable kernel is an exponential function or series of exponential functions. With this type of kernel it is possible to change the integral equation into a differential equation, and sometimes an analytical solution can be obtained. This will be demonstrated in Example 8-13. There is a mathematical point that should be mentioned here. The method of changing the integral equation into a differential equation requires taking derivatives of the approximate separable kernel. Even though the separable function may approximate the exact kernel fairly well, the approximation of the derivatives may become poor, especially when higher derivatives are taken. The use of the separable kernel will now be demonstrated with an example.

EXAMPLE 8-13 Determine $q_{o,2}/q_2$ from Eq. (8-46a) by use of an exponential approximate separable kernel [7].

The governing equation is

$$\frac{q_{o,2}(\xi)}{q_2} - \int_{\eta=0}^{1} \frac{q_{o,2}(\eta)}{q_2} K(|\eta - \xi|) \, d\eta = 1 \qquad (8\text{-}54a)$$

where

$$K(|\eta - \xi|) = 1 - \frac{|\eta - \xi|^3 + \frac{3}{2}|\eta - \xi|}{[(\eta - \xi)^2 + 1]^{3/2}} \qquad (8\text{-}54b)$$

The $K(|\eta - \xi|)$ is plotted in Fig. 8-13, and it is reasonably well approximated by the function $e^{-2|\eta-\xi|}$. When the approximate kernel is substituted into (8-54a), the part of the function depending on ξ can be taken out of the integral to give

$$\frac{q_{o,2}(\xi)}{q_2} - e^{-2\xi} \int_0^\xi \frac{q_{o,2}(\eta)}{q_2} e^{2\eta} \, d\eta - e^{2\xi} \int_\xi^1 \frac{q_{o,2}(\eta)}{q_2} e^{-2\eta} \, d\eta = 1 \qquad (8\text{-}55)$$

By differentiating (8-55) two times, the integrals can be removed and the following differential equation obtained:

$$\frac{d^2[q_{o,2}(\xi)/q_2]}{d\xi^2} = -4$$

This has the general solution, obtained by integrating twice,

$$\frac{q_{o,2}(\xi)}{q_2} = -2\xi^2 + C_1\xi + C_2 \qquad (8\text{-}56a)$$

To obtain C_1 and C_2, two boundary conditions are needed. From symmetry one

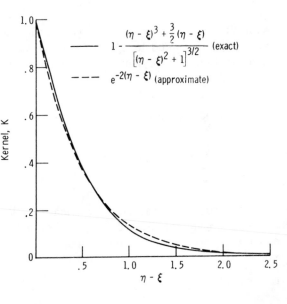

Figure 8-13 Exponential approximation to configuration-factor kernel for cylindrical enclosure.

boundary condition is

$$\frac{d(q_{o,2}/q_2)}{d\xi} = 0 \qquad \text{at } \xi = \frac{1}{2}$$

which yields $C_1 = 2l$. To determine C_2, a boundary condition can be obtained from (8-55) by evaluating it at $\xi = 0$ and $\xi = l$ and then utilizing the fact that $q_{o,2}(0) = q_{o,2}(l)$ to obtain the condition

$$\int_0^l \frac{q_{o,2}(\eta)}{q_2} e^{-2\eta} d\eta = e^{-2l} \int_0^l \frac{q_{o,2}(\eta)}{q_2} e^{2\eta} d\eta$$

Inserting $q_{o,2}/q_2 = -2\xi^2 + 2l\xi + C_2$ and integrating yield $C_2 = l + 1$. With C_1 and C_2 thus evaluated, the final result for $q_{o,2}/q_2$ by the separable-kernel method is the parabola

$$\frac{q_{o,2}(\xi)}{q_2} = l + 1 + 2(\xi l - \xi^2) \tag{8-56b}$$

More generally the boundary conditions to evaluate C_1 and C_2 could have been obtained even in an asymmetric case by evaluating the integral equation at both boundaries $\xi = 0$ and $\xi = l$. This yields, from (8-55),

$$\frac{q_{o,2}(0)}{q_2} - \int_0^l \frac{q_{o,2}(\eta)}{q_2} e^{-2\eta} d\eta = 1$$

$$\frac{q_{o,2}(l)}{q_2} - e^{-2l} \int_0^l \frac{q_{o,2}(\eta)}{q_2} e^{2\eta} d\eta = 1$$

Then $q_{o,2}/q_2$ from (8-56a) is substituted into these two boundary conditions. After integrating, two simultaneous equations result for C_1 and C_2, leading to the same solution as before. The advantage of using the symmetry condition was only algebraic simplicity.

Approximate solution by variational method. As mentioned in [6] (p. 495), an integral equation of the form

$$\Phi(\xi) = \int_a^b K(\xi,\eta)\Phi(\eta)\, d\eta + G(\xi) \tag{8-57}$$

can be solved by variational methods. A restriction is that $K(\xi,\eta)$ be symmetric, that is, K is not changed when the values of ξ and η are interchanged. The kernel of (8-54b) is an example of a symmetric kernel since, because of the absolute-value signs, it is evident that $K(|\eta - \xi|) = K(|\xi - \eta|)$.

The variational method depends on the use of an auxiliary function that is related in a particular way to Eq. (8-57). This auxiliary function is given by

$$J = \iint\limits_{a}^{b} K(\xi,\eta)\Phi(\xi)\Phi(\eta)\,d\xi\,d\eta - \int_{a}^{b}[\Phi(\xi)]^2\,d\xi + 2\int_{a}^{b}\Phi(\xi)G(\xi)\,d\xi \qquad (8\text{-}58)$$

The significance of the J function is that, when the correct solution for $\Phi(\xi)$ is found, J will have a minimum value.

The procedure for obtaining an approximate solution is to let $\Phi(\xi)$ be represented by a polynomial with unknown coefficients,

$$\Phi(\xi) = \gamma_0 + \gamma_1\xi + \gamma_2\xi^2 + \cdots + \gamma_n\xi^n \qquad (8\text{-}59)$$

This polynomial is substituted into (8-58), and the integration carried out. If K is so complicated algebraically that the integration cannot be performed analytically, the method is not practical. After the integration is carried out, the result is an analytical expression for J as a function of $\gamma_0, \gamma_1, \gamma_2, \cdots, \gamma_n$. These unknown coefficients are then determined by differentiating J with respect to each of the individual coefficients and setting each result equal to zero, that is, $\partial J/\partial\gamma_0 = 0$, $\partial J/\partial\gamma_1 = 0$, $\cdots$, $\partial J/\partial\gamma_n = 0$. This yields a set of $n + 1$ simultaneous equations for the $n + 1$ unknown coefficients. By differentiating J in this manner and setting the differentials equal to zero, the coefficients are found that make J a minimum value; thus the most accurate solution to the integral equation of the assumed form $\Phi(\xi) = \sum_{j=0}^{n}\gamma_j\xi^j$ is found.

This method has been applied to radiation in a cylindrical tube in [7] and radiation between parallel plates of finite width and infinite length in [17].

Approximate solution by Taylor-series expansion. The use of a Taylor-series expansion method for solving a radiation integral equation is demonstrated in [18, 19]. The physical idea that motivates this method of solution is that the geometric configuration factor can often decrease quite rapidly as the distance between the two elements exchanging radiation is increased. This means that the radiative heat balance at a given location may be significantly influenced only by the radiative fluxes leaving other surface elements in the immediate vicinity of that location.

As an example, consider the type of integral equation (8-54a). The function $K(|\eta - \xi|)$ decreases rapidly as $\eta - \xi$ is increased, as shown in Fig. 8-13. Then if it is assumed that the important values of η occur when η is close to the location ξ, the function $q_{o,2}(\eta)/q_2$ is expanded in a Taylor series about ξ:

$$\frac{q_{o,2}(\eta)}{q_2} = \frac{q_{o,2}(\xi)}{q_2} + (\eta - \xi)\left[\frac{d(q_{o,2}/q_2)}{d\xi}\right]_\xi + \frac{(\eta - \xi)^2}{2!}\left[\frac{d^2(q_{o,2}/q_2)}{d\xi^2}\right]_\xi + \cdots \qquad (8\text{-}60)$$

The derivatives in the Taylor expansion are evaluated at ξ and hence do not contain the variable η. This means that, when (8-60) is substituted into (8-54a), the derivatives can be taken out of the integrals to yield

$$\frac{q_{o,2}(\xi)}{q_2} - \frac{q_{o,2}(\xi)}{q_2} \int_{\eta=0}^{I} K(|\eta - \xi|)\, d\eta - \frac{d}{d\xi}\left[\frac{q_{o,2}(\xi)}{q_2}\right] \int_{\eta=0}^{I} (\eta - \xi) K(|\eta - \xi|)\, d\eta$$

$$-\frac{1}{2!}\frac{d^2}{d\xi^2}\left[\frac{q_{o,2}(\xi)}{q_2}\right] \int_{\eta=0}^{I} (\eta - \xi)^2 K(|\eta - \xi|)\, d\eta - \cdots = 1 \qquad (8\text{-}61)$$

The integrations are then carried out; if this cannot be done analytically, the method is not of practical utility because it is as easy to carry out a numerical solution of the exact integral equation as of (8-61). If the integrals can be carried out analytically, (8-61) becomes a differential equation for $q_{o,2}(\xi)/q_2$ that can be solved analytically or numerically if the boundary conditions can be specified. The boundary conditions can be derived, as illustrated in [19], from the physical constraints in the system, for example, symmetry or an overall heat balance. This method is probably of little value for enclosures involving more than one or two surfaces.

Solution by the method of Ambarzumian. Crosbie and Sawheny [20, 21] have applied the method of Ambarzumian [22, 23] to semi-infinite geometries. They studied the radiative exchange within two-dimensional cavities in a semi-infinite solid (Fig. 8-14) when the temperature or heat-flux distribution along the cavity walls can be described by an exponential variation or a summation of exponential terms. In the Ambarzumian method, the governing Fredholm integral equation of the second kind that describes the outgoing radiative flux $q_o(x)$ or $q_o(y)$ is transformed into an integrodifferential equation of the initial-value type.

EXAMPLE 8-14 Consider the semi-infinite rectangular cavity of Fig. 8-14 as analyzed in [20]. The walls along x and y stretch to infinity, and both have the same temperature distribution, described by

$$T(x) = T_0 \exp\left(\frac{-mx}{4a}\right) \qquad \text{and} \qquad T(y) = T_0 \exp\left(\frac{-my}{4a}\right)$$

Both walls are diffuse and gray with emissivity $\epsilon = 1 - \rho$. Equation (8-43a) for $q_o(x)$ gives

$$q_o(x,m) = \epsilon \sigma T^4(x,m) + \rho \int_0^\infty q_o(y,m)\, dF_{dx-dy}(x,y) \qquad (8\text{-}62)$$

where the functional dependence of q_o on the temperature variation parameter m is written explicitly. Since both walls have the same temperature distribution and properties, only one equation is required.

In this case, letting $\xi = x/a$ and $\eta = y/a$, one obtains

$$dF_{d\xi-d\eta}(\xi,\eta) = \frac{d\eta}{2[(\xi - \eta)^2 + 1]^{3/2}} = \frac{1}{2}K(|\xi - \eta|)\, d\eta$$

The kernel $K(|\xi - \eta|)$ is symmetric. Equation (8-62) becomes

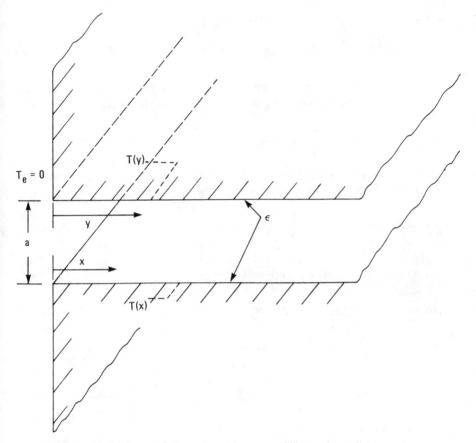

Figure 8-14 Semi-infinite rectangular cavity with exponentially varying wall temperature.

$$\Phi(\xi,m) = e^{-m\xi} + \frac{\rho}{2} \int_0^\infty \Phi(\eta,m)K(|\xi - \eta|)\, d\eta \qquad (8\text{-}63)$$

where

$$\Phi(\xi,m) = \frac{q_0(\xi,m)}{\epsilon\sigma T_0^4}$$

In the standard solutions, (8-63) would be solved by successive approximations or other techniques. In Ambarzumian's method, we proceed to transform (8-63) into an integrodifferential form. Equation (8-63) can be rewritten as

$$\Phi(\xi,m) = e^{-m\xi} + \frac{\rho}{2} \int_0^\xi \Phi(\eta,m)K(\xi - \eta)\, d\eta + \frac{\rho}{2} \int_\xi^\infty \Phi(\eta,m)K(\eta - \xi)\, d\eta \qquad (8\text{-}64)$$

Transforming variables in the first integral of (8-64) by substituting $z = \xi - \eta$, and in the second by $z = \eta - \xi$, one obtains

$$\Phi(\xi,m) = e^{-m\xi} + \frac{\rho}{2} \int_0^\xi \Phi(\xi - z,m)K(z)\,dz + \frac{\rho}{2} \int_0^\infty \Phi(\xi + z,m)K(z)\,dz \qquad (8\text{-}65)$$

Differentiating (8-65) with respect to ξ, again substituting to eliminate z, and gathering terms give the following integral equation for the derivative of $\Phi(\xi,m)$:

$$\frac{\partial \Phi(\xi,m)}{\partial \xi} = -me^{-m\xi} + \frac{\rho}{2}\Phi(0,m)K(\xi) + \frac{\rho}{2}\int_0^\infty \frac{\partial \Phi(\eta,m)}{\partial \eta}K(|\xi - \eta|)\,d\eta \qquad (8\text{-}66)$$

If Eq. (8-63) is multiplied by $mJ_1(m)\,dm/2$, where $J_1(m)$ is the Bessel function of the first kind, and then integrated from 0 to ∞, we find

$$\frac{1}{2}\int_0^\infty m\Phi(\xi,m)J_1(m)\,dm = \frac{1}{2}\int_0^\infty me^{-m\xi}J_1(m)\,dm$$

$$+ \frac{\rho}{4}\int_0^\infty mJ_1(m)\int_0^\infty \Phi(\eta,m)K(|\xi - \eta|)\,d\eta\,dm \qquad (8\text{-}67)$$

Multiplying (8-63) by $-m$ and (8-67) by $\rho\Phi(0,m)$ and adding give

$$-m\Phi(\xi,m) + \frac{\rho}{2}\Phi(0,m)\int_0^\infty t\Phi(\xi,t)J_1(t)\,dt$$

$$= -me^{-m\xi} + \frac{\rho}{2}\Phi(0,m)K(\xi) + \frac{\rho}{2}\int_0^\infty \frac{\partial \Phi(\eta,m)}{\partial \eta}K(|\xi - \eta|)\,d\eta \qquad (8\text{-}68)$$

after using the relation

$$\int_0^\infty me^{-xm}J_1(m)\,dm = \frac{1}{(x^2 + 1)^{3/2}} = K(x)$$

and noting that the result of the addition gives the same integral equation as (8-66).

Note that the right-hand sides of (8-68) and (8-66) are identical. Thus, (8-66) can be written as

$$\frac{\partial \Phi(\xi,m)}{\partial \xi} = -m\Phi(\xi,m) + \frac{\rho}{2}\Phi(0,m)\int_0^\infty t\Phi(\xi,t)J_1(t)\,dt \qquad (8\text{-}69)$$

The equation has thus been transformed into an initial-value problem for $\Phi(\xi,m)$, and only the initial condition $\Phi(0,m)$, the outgoing dimensionless wall flux at the cavity entrance $\xi = 0$, remains to be determined. References [20] and [21] derive the required relations, yielding the following equation, which is solved numerically and tabulated for the case of a prescribed exponential temperature distribution:

$$\frac{1}{\Phi(0,m)} = (1 - \rho)^{1/2} + \frac{\rho}{2}\int_0^\infty \frac{mJ_1(t)}{m + t}\Phi(0,t)\,dt$$

For $m = 0$ (isothermal cavity), this gives the well-known result (see Example 8-10)

$$\Phi(0,0) = \frac{1}{\epsilon^{1/2}}$$

After $\Phi(\xi,m)$ and thus $q_o(x,m)$ is obtained, the wall heat-flux distribution can be directly obtained from (8-44a).

In the past five sections, methods have been discussed for solving single equations or sets of integral equations by numerical methods and by some approximate analytical methods. The analytical methods are probably only of value when the integral equations are relatively simple. In almost all practical cases a numerical method would be resorted to. There are a few instances in which approximate or numerical solutions are not required since the radiation exchange integral equation has an exact analytical solution. One of these cases will now be discussed.

Exact solution of integral equation for radiation from a spherical cavity. The radiation from a spherical cavity as shown in Fig. 8-15a was analyzed by Jensen [24], discussed by Jakob [3], and further treated by Sparrow and Jonsson [25].

The spherical shape leads to a relatively simple integral-equation solution because there is an especially simple geometrical configuration factor between elements on the inside of the spherical cavity. The configuration factor between the two differential elements dA_j and dA_k shown in Fig. 8-15b is

$$dF_{dj-dk} = \frac{\cos \theta_j \cos \theta_k}{\pi S^2} dA_k \tag{8-70}$$

Since the sphere radius is normal to both elements dA_j and dA_k, the distance between these elements is given by $S = 2R \cos \theta_j = 2R \cos \theta_k$. Then (8-70) becomes

$$dF_{dj-dk} = \frac{dA_k}{4\pi R^2} \tag{8-71}$$

If, instead of an infinitesimal area dA_k, the element dA_j exchanges with the finite area A_k, then (8-71) becomes

$$F_{dj-k} = \frac{1}{4\pi R^2} \int_{A_k} dA_k = \frac{A_k}{4\pi R^2} \tag{8-72}$$

Equation (8-72) is independent of the area element dA_j; hence, dA_j could be replaced by a finite area A_j so that

$$F_{j-k} = \frac{A_k}{4\pi R^2} = \frac{A_k}{A_s} \tag{8-73}$$

where A_s is the surface area of the entire sphere.

Consider the spherical cavity shown in Fig. 8-15a. The cavity surface has a temperature distribution $T_1(dA_1)$ and has a total surface area A_1. The spherical cap that

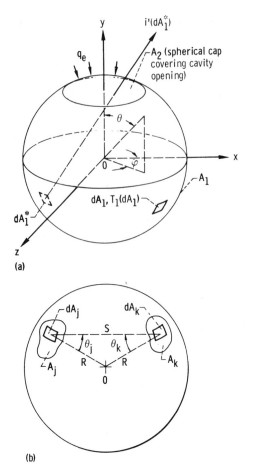

Figure 8-15 Geometry involved in radiation within spherical cavity. (a) Spherical cavity with diffuse entering radiation q_e and with surface at variable temperature T_1; (b) area elements on spherical surface.

would cover the cavity opening has area A_2. Assume there is diffuse radiative flux q_e (per unit area of A_2) entering from the environment through the cavity opening. The q_e can be variable over A_2. It is desired to compute the radiation intensity $i'(dA_1^*)$ leaving the cavity opening at a specified location and in a specified direction, as shown by the arrow in Fig. 8-15a. The figure shows that the desired intensity will result from the flux leaving the element dA_1^* and will equal $q_{o,1}(dA_1^*)/\pi$, where the factor π arises from the relation between hemispherical flux q_o and intensity i'. The flux $q_{o,1}(dA_1^*)$ can be found by applying (8-43a):

$$q_{o,1}(dA_1^*) - (1 - \epsilon_1) \int_{A_1} q_{o,1}(dA_1) \, dF_{d1*-d1}$$

$$- (1 - \epsilon_1) \int_{A_2} q_e(dA_2) \, dF_{d1*-d2} = \epsilon_1 \sigma T_1^4(dA_1^*) \qquad (8\text{-}74)$$

The F factors from (8-71) are then substituted to give

$$q_{o,1}(dA_1^*) - \frac{1-\epsilon_1}{4\pi R^2} \int_{A_1} q_{o,1}(dA_1) \, dA_1 = \frac{1-\epsilon_1}{4\pi R^2} \int_{A_2} q_e(dA_2) \, dA_2 + \epsilon_1 \sigma T_1{}^4(dA_1^*)$$

$$(8\text{-}75)$$

where the known quantities are grouped on the right side of the equation.

To solve (8-75), a trial solution of the form

$$q_{o,1}(dA_1^*) = f(dA_1^*) + C$$

is assumed, where f is an unknown function of the location of dA_1^*, and C is a constant. Substituting into (8-75) gives

$$f(dA_1^*) + C - \frac{1-\epsilon_1}{4\pi R^2} \int_{A_1} f(dA_1) \, dA_1 - \frac{1-\epsilon_1}{4\pi R^2} CA_1$$

$$= \frac{1-\epsilon_1}{4\pi R^2} \int_{A_2} q_e(dA_2) \, dA_2 + \epsilon_1 \sigma T_1{}^4(dA_1^*)$$

The only two terms that are functions of local position within the cavity are the first and last, which gives $f(dA_1^*) = \epsilon_1 \sigma T_1{}^4(dA_1^*)$. The remaining terms are then equated to determine C. This gives the result for $q_{o,1}(dA_1^*)$:

$$q_{o,1}(dA_1^*) = \epsilon_1 \sigma T_1{}^4(dA_1^*)$$

$$+ \frac{[(1-\epsilon_1)/4\pi R^2]\left[\int_{A_1} \epsilon_1 \sigma T_1{}^4(dA_1) \, dA_1 + \int_{A_2} q_e(dA_2) \, dA_2\right]}{1-(1-\epsilon_1)A_1/4\pi R^2} \qquad (8\text{-}76)$$

The desired solution is

$$i'(dA_1^*) = \frac{q_{o,1}(dA_1^*)}{\pi} \qquad (8\text{-}77)$$

8-5 CONCLUDING REMARKS

In this chapter, methods were developed for treating the energy exchange within enclosures having diffuse-gray surfaces; the surfaces can be of finite or infinitesimal size. The surfaces can have a specified net energy flux added to them by some external means such as conduction or convection, or can have a specified surface temperature, or can be subjected to some combination of these conditions. A number of methods were presented for solving the integral equations that resulted from the general formulation of these interchange problems. It was pointed out that most practical problems become so complex that numerical techniques are required for the solution of the governing equations.

In succeeding chapters, extensions of the present procedures to nonidealized surfaces are made, and methods for incorporating coupled conduction and convection of energy are introduced.

REFERENCES

1. Hottel, Hoyt C.: Radiant-heat Transmission, in William H. McAdams (ed.), "Heat Transmission," 3rd ed., chap. 4, McGraw-Hill Book Company, New York, 1954.

2. Poljak, G.: Analysis of Heat Interchange by Radiation between Diffuse Surfaces, *Tech. Phys. USSR*, vol. 1, nos. 5, 6, pp. 555–590, 1935.

3. Jakob, Max: "Heat Transfer," vol. II, John Wiley & Sons, Inc., New York, 1957.

4. Gebhart, B.: Unified Treatment for Thermal Radiation Transfer Processes – Gray, Diffuse Radiators and Absorbers, paper no. 57-A-34, *ASME*, December, 1957.

5. Sparrow, E. M.: On the Calculation of Radiant Interchange between Surfaces, in Warren Ibele (ed.), "Modern Developments in Heat Transfer," pp. 181–212, Academic Press, Inc., New York, 1963.

6. Hildebrand, Francis B.: "Methods of Applied Mathematics," Prentice-Hall, Inc., Englewood Cliffs, N.J., 1952.

7. Usiskin, C. M., and R. Siegel: Thermal Radiation from a Cylindrical Enclosure with Specified Wall Heat Flux, *J. Heat Transfer*, vol. 82, no. 4, pp. 369–374, 1960.

8. Buckley, H.: Radiation from the Interior of a Reflecting Cylinder, *Phil. Mag.*, vol. 4, pp. 753–762, 1927.

9. Buckley, H.: Radiation from Inside a Circular Cylinder, *Phil. Mag.*, vol. 6, pp. 447–457, 1928.

10. Eckert, E.: Das Strahlungsverhaltnis von Flachen mit Einbuchtungen und von zylindrischen Bohrungen, *Arch. Warmewirtsch.*, vol. 16, no. 5, pp. 135–138, 1935.

11. Sparrow, E. M., and L. U. Albers: Apparent Emissivity and Heat Transfer in a Long Cylindrical Hole, *J. Heat Transfer*, vol. 82, no. 3, pp. 253–255, 1960.

12. Lin, S. H., and E. M. Sparrow: Radiant Interchange among Curved Specularly Reflecting Surfaces – Application to Cylindrical and Conical Cavities," *Trans. ASME, J. Heat Transfer*, vol. 87, no. 2, pp. 299–307, 1965.

13. Kholopov, G. K.: Radiation of Diffuse Isothermal Cavities, *Inzh. Fiz. Zh.*, vol. 25, no. 6, pp. 1112–1120, 1973.

14. Heinisch, R. P., E. M. Sparrow, and N. Shamsundar: Radiant Emission from Baffled Conical Cavities, *J. Opt. Soc. Am.*, vol. 63, no. 2, pp. 152–158, 1973.

15. Masuda, H.: Radiant Heat Transfer on Circular-Finned Cylinders, *Rep. Inst. High Speed Mech. Tohoku Univ.*, vol. 27, no. 255, pp. 67–89, 1973. (See also *Trans. Jpn. Soc. Mech. Eng.*, vol. 38, pp. 3229–3234, 1972.)

16. Sparrow, E. M., and R. P. Heinisch: The Normal Emittance of Circular Cylindrical Cavities, *Appl. Opt.*, vol. 9, no. 11, pp. 2569–2572, 1970.

17. Sparrow, E. M.: Application of Variational Methods to Radiation Heat-transfer Calculations, *J. Heat Transfer*, vol. 82, no. 4, pp. 375–380, 1960.

18. Krishnan, K. S., and R. Sundaram: The Distribution of Temperature along Electrically Heated Tubes and Coils, I. Theoretical, *Proc. R. Soc. London*, ser. A, vol. 257, no. 1290, pp. 302–315, 1960.

19. Perlmutter, M., and R. Siegel: Effect of Specularly Reflecting Gray Surface on Thermal Radiation through a Tube and from Its Heated Wall, *J. Heat Transfer*, vol. 85, no. 1, pp. 55–62, 1963.

20. Crosbie, A. L., and T. R. Sawheny: Radiant Interchange in a Nonisothermal Rectangular Cavity, *AIAA J.*, vol. 13, no. 4, pp. 425–431, 1975.

21. Crosbie, A. L., and T. R. Sawheny: Application of Ambarzumian's Method to Radiant Interchange in a Rectangular Cavity, *ASME J. Heat Transfer*, vol. 96, pp. 191–196, 1974.

22. Ambarzumian, V. A.: Diffusion of Light by Planetary Atmospheres, *Astron. Zh.*, vol. 19, pp. 30–41, 1942.

23. Kourganoff, V.: "Basic Methods in Transfer Problems," Dover Publications, Inc., New York, 1963.

24. Jensen, H. H.: Some Notes on Heat Transfer by Radiation, *Kgl. Danske Videnskab. Selskab. Mat.-Fys. Medd.*, vol. 24, no. 8, pp. 1–26, 1948.

25. Sparrow, E. M., and V. K. Jonsson: Absorption and Emission Characteristics of Diffuse Spherical Enclosures, *NASA* TN D-1289, 1962.

26. Alfano, Gaetano: Apparent Thermal Emittance of Cylindrical Enclosures With and Without Diaphragms, *Int. J. Heat Mass Transfer*, vol. 15, no. 12, pp. 2671-2674, 1972.

27. Alfano, G., and A. Sarno: Normal and Hemispherical Thermal Emittances of Cylindrical Cavities, *ASME J. Heat Transfer*, vol. 97, no. 3, pp. 387-390, 1975.

PROBLEMS

1   Two infinite gray parallel plates are separated by a thin gray radiation shield. What is T_s, the temperature of the shield? What flux is transferred from plate 2 to plate 1? What is the ratio of the heat transferred from 2 to 1 with the shield to that transferred without the shield?

    Answer: 1655°R, 2200 Btu/(h · ft²), 0.176.

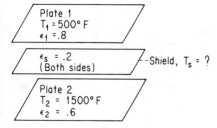

2   A heat flux q_0 is transferred across the gap between two gray parallel plates, having the same emissivity ϵ, that are at temperatures T_1 and T_2 (ϵ is independent of temperature). A single thin radiation shield also having emissivity ϵ on both sides is placed between the plates. Show that the resulting heat flux is $q_0/2$. Show that adding a second shield reduces the flux to $q_0/3$. It can be shown that for n shields the heat flux is $q_0/(n+1)$ when all surface emissivities are the same.

3   What is the effect of a single thin radiation shield on the flow of heat between two concentric spheres? Assume the sphere and shield surfaces are diffuse-gray with emissivities independent of temperature. Both sides of the shield have the same emissivity ϵ_s, and the inner and outer spheres have respective emissivities ϵ_1 and ϵ_2.

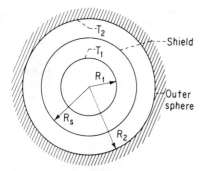

    Answer: $\dfrac{Q_{1,\text{ with shield}}}{Q_{1,\text{ without shield}}} = \dfrac{G_{12}}{(R_1/R_s)^2 G_{s2} + G_{1s}}$

    where $G_{ab} = \dfrac{1}{\epsilon_a} + \left(\dfrac{R_a}{R_b}\right)^2 \left(\dfrac{1}{\epsilon_b} - 1\right).$

4 Two infinitely long diffuse-gray concentric circular cylinders are separated by two concentric thin diffuse-gray radiation shields. The shields have identical emissivities on each side.

(a) Derive an expression for the heat transferred between the inner and outer cylinders in terms of their temperatures and the necessary radiative and geometric quantities. (Number the surfaces from the inside out; i.e., the inner surface is number 1, the outer surface is number 4.)

(b) Check this result by showing that in the proper limit it reduces to the correct result for four parallel plates with identical emissivities.

(c) Find the percent reduction in heat transfer when the shields are added if the radii for the surfaces are in the ratio 1:3:5:7, and if $\epsilon_1 = \epsilon_4 = 0.5$ and for the shields $\epsilon_2 = \epsilon_3 = 0.1$.

Answer: (c) $Q_{with}/Q_{without} = 0.175$

5 Consider a diffuse-gray right circular cylindrical enclosure. The diameter of the base and top are the same as the height of the cylinder, 1 m. The top is removed. If the remaining walls are maintained at 1000 K and have an emissivity of 0.5, determine the radiative energy escaping through the open end. Assume uniform irradiation on each surface so that subdivision of the base and curved wall is not necessary. The outside environment is at $T_e = 0$. How do the results compare with Fig. 8-10? Explain any difference. If $T_e = 600$ K, what percent reduction occurs in radiative heat loss?

6 Consider the gray cylindrical enclosure described in Prob. 5 above with the top in place. A hole 15 cm in diameter is drilled in the top. Determine the configuration factors between (a) the base and the hole, and (b) the curved wall and the hole. Estimate the radiant energy escaping through the hole. The outside environment is at $T_e = 0$.

7 A two-dimensional diffuse-gray enclosure (infinitely long into the page) has each surface at a uniform temperature. Compute the heat added per foot of enclosure length to each surface to account for the radiative exchange within the enclosure, Q_1, Q_2, Q_3. (Assume for simplicity that it is not necessary to subdivide the three areas.) The conditions are
$T_1 = 1500°F$ $\epsilon_1 = 0.6$ $T_2 = 500°F$ $\epsilon_2 = 0.9$ $T_3 = 800°F$ $\epsilon_3 = 0.5$

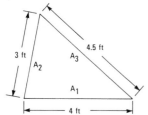

Answer: 45,900 Btu/h, −28,900, −17,000.

8 A thin gray disk with emissivity 0.8 on both sides is in earth orbit. It is exposed to normally incident solar radiation (neglect radiation from the earth). What is the equilibrium temperature of the disk? A single thin radiation shield having emissivity 0.1 on both sides is placed as shown. What is the disk temperature? What is the effect in both these calculations of reducing the disk emissivity to 0.5? (Assume the surroundings are at zero temperature. For simplicity do not subdivide the surface areas.)

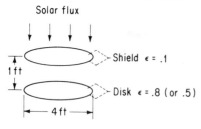

9 A frustum of a cone has its base heated as shown. The top is held at 550 K while the side is perfectly insulated. All the surfaces are diffuse-gray. What is the temperature achieved by surface 1 as a result of radiation exchange within the enclosure? (For simplicity do not subdivide the surface areas.)

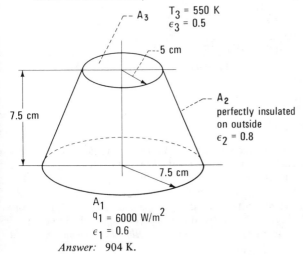

Answer: 904 K.

10 In a metal-processing operation, a metal sphere at uniform temperature is heated to high temperature by radiative exchange with a circular heating element in vacuum. The surrounding temperatures are low enough to be neglected in the heat flow calculations. The surfaces are diffuse and gray. Derive an expression for the net rate of energy absorption by the sphere. The expression should be given in terms of the quantities shown. For simplicity do not subdivide the surface areas. Discuss whether this is a reasonable approximation for this geometry with regard to the distribution of reflected energy from the sphere.

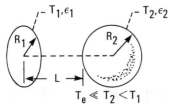

11 Consider two parallel plates of finite extent in one direction. Both plates are perfectly insulated on the outside. Plate 1 is uniformly heated electrically with heat flux q_e. Plate 2 has no external heat input. The environment is at zero temperature.

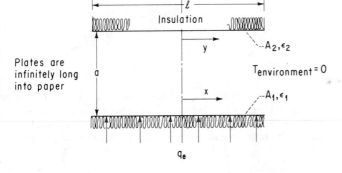

(a) For both surfaces *black*, show that the integral equations for the surface temperatures are

$$\theta_{b,1}(X) = 1 + \frac{1}{2}\int_{-L/2}^{L/2}\theta_{b,2}(Y)\frac{dY}{[(Y-X)^2+1]^{3/2}}$$

$$\theta_{b,2}(Y) = \frac{1}{2}\int_{-L/2}^{L/2}\theta_{b,1}(X)\frac{dX}{[(X-Y)^2+1]^{3/2}}$$

where $X = x/a$, $Y = y/a$, $\theta = \sigma T_e^4/q_e$, and $L = l/a$.

(b) If both plates are *gray*, show that

$$\theta_1(X) = \theta_{b,1}(X) + \frac{1-\epsilon_1}{\epsilon_1}$$

$$\theta_2(Y) = \theta_{b,2}(Y)$$

12 A 5-cm-diameter hole extends through the wall of a furnace having an interior temperature of 1400 K. The wall is 15-cm-thick refractory brick. Divide the wall thickness into two zones of equal length, and compute the radiation out of the hole into a room at 295 K. (Neglect heat conduction in the wall.)

13 A hemispherical cavity is in a block of lightly oxidized copper maintained at 550 K in vacuum (see Fig. 5-11a). The surroundings are at 295 K. Use the integral-equation method to compute the heat loss from the cavity in terms of cavity diameter.   (*Hint:* See Example 8-10 and Prob. 15.)

14 A cavity having a gray interior surface S is uniformly heated electrically and achieves a surface temperature distribution $T_{w,0}(S)$ while being exposed to a zero environment temperature $T_e = 0$. If the environment temperature is raised to T_e and the heating kept the same, what will the surface temperature distribution be?

   Answer: $T_w(S) = [T_{w,0}{}^4(S) + T_e{}^4]^{1/4}$.

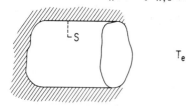

15 A gray circular tube insulated on the outside is exposed to an environment at $T_e = 0$ at both ends. The $q(x, T_e = 0)$ has been calculated to maintain the wall temperature at any constant value. Now let $T_e \neq 0$, and let the wall temperature be uniform at T_w. Show that the $q(x, T_e \neq 0)$ can be obtained as the $q(x, T_e = 0)$ corresponding to the wall temperature $(T_w{}^4 - T_e{}^4)^{1/4}$.

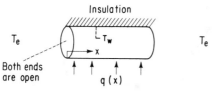

16 A rod $\frac{1}{2}$ in in diameter and 5 in long is at temperature $T_1 = 1800$ K and has a total emissivity of $\epsilon_1 = 0.22$. It is within a thin-walled concentric cylinder of the same length having a diameter of 2 in. The emissivity on the inside of the cylinder is $\epsilon_2 = 0.50$, and on the outside is $\epsilon_0 = 0.17$. All surfaces are gray. The entire assembly is suspended in a large vacuum chamber at $T_\infty = 300$ K. What is the temperature T_2 of the cylindrical shell? (For simplicity do not subdivide the surface areas. *Hint:* $F_{2-1} = 0.225$, $F_{2-2} = 0.617$.)

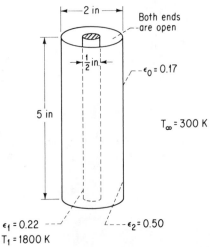

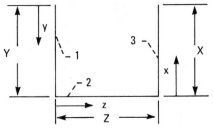

$\epsilon_1 = 0.22$

$T_1 = 1800$ K

Answer: 1048 K.

17 For the geometry and conditions shown below, set up the required integral equations for finding $q_1(y)$, $T_2(z)$, and $T_3(x)$. Place the equations in dimensionless form, and discuss how you would go about solving the equations. Which method of Chap. 8 appears most useful?

$T_1(y) = T_1 = $ constant
$q_2(z) = 0$ (insulated)
$q_3(x) = q_3 = $ constant
$\epsilon_1(y) = 1.0$
$\epsilon_2(z) = \epsilon_3(x) = 0.5$

18 A cube of side 3 m has a small sphere placed at its center. The sphere has emissivity $\epsilon = 0.5$, and is maintained electrically at $T_S = 1500$ K. The interior sides of the cube have the following properties:

Side	Temperature, K	Emissivity
1	1500	1.0
2	1000	0.5
3	900	0.5
4	500	1.0
5	200	0.5
6	0	1.0

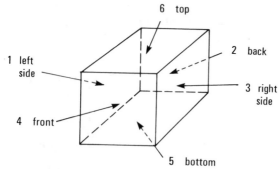

Determine the net Q added or removed from each side of the cube and the Q added to the sphere. Find the results in W/m². Tabulate all the configuration factors required. (Assume that incident radiation is uniform on all surfaces.)

RADIATION IN ENCLOSURES HAVING SOME SPECULARLY REFLECTING SURFACES

9-1 INTRODUCTION

In Chap. 8, all the surfaces considered were assumed to be diffuse emitters *and* diffuse reflectors. In this chapter, the characteristics of some of the surfaces will be changed. All the surfaces are still assumed to *emit* in a diffuse fashion. Some of the surfaces in an enclosure will be assumed to reflect diffusely as before; the remaining surfaces will be assumed specular, that is, to reflect in a mirrorlike manner. Recall from the discussion of roughness in Chap. 5 that an important parameter is the ratio of the root-mean-square roughness height to the wavelength of the radiation (optical roughness). For long wavelengths a surface tends toward being optically smooth, and the reflections tend to become more specular. Thus although a surface may not appear mirror-like to the eye (that is, for the short wavelengths of the visible spectrum), it may be specular for longer wavelengths in the infrared.

When reflection is diffuse, the directional history of the incident radiation is lost upon reflection: the reflected energy has the same directional distribution as if it had been absorbed and then diffusely reemitted. With a specular reflection, the reflection angle relative to the surface normal is equal in magnitude to the angle of incidence. Hence, in contrast to diffuse behavior, the directional history of the incident radiation is not lost upon reflection. Consequently, when dealing with specular surfaces, it will be necessary to account for the directional paths that the reflected radiation follows between surfaces.

The specular reflectivities used in this chapter are assumed *independent of incident angle* of radiation; that is, the same fraction of the incident energy is reflected, regardless of the angle of incidence of the energy. In addition, all the surfaces are assumed to have *gray properties*; that is, the properties do not depend on wavelength.

9-2 SYMBOLS

A	area
c_p	specific heat
D	tube diameter
d	number of diffuse surfaces
F	configuration factor
f	focal length
L	length of enclosure side
N	total number of surfaces
Q	energy rate, energy per unit time
q	energy flux, energy per unit area and per unit time
T	absolute temperature
V	volume
X,x	position coordinates
α	absorptivity; angle subtended by sun
γ	distance from mirror to image
ϵ	emissivity
θ	cone angle
ρ	reflectivity
ρ_M	density of material
ρ_m	reflectivity of mirror
σ	Stefan-Boltzmann constant
τ	time

Subscripts

e	emitted
F	final
I	initial
i	incoming
j,k	jth or kth surface
o	outgoing
s	specular; solar
$1,2$	surface 1 or 2

Superscripts

s	total specular exchange factor including all paths for specular inter-reflections plus direct exchange
$''$	bidirectional value
$*$	denotes a second portion of area on same surface

9-3 RADIATION BETWEEN PAIRS OF SURFACES WITH SPECULAR REFLECTIONS

9-3.1 Some Simple Cases

As an introduction, consider radiation exchange for some simple geometries: infinite parallel plates, concentric cylinders, and concentric spheres, as shown in Fig. 9-1. Specular radiation exchange in these cases is well understood, having been discussed by Christiansen [1] and Saunders [2] some years ago. Because the radiative exchange process is easy to grasp for these cases and is of practical importance in predicting the heat transfer performance of radiation shields, Dewar vessels, and cryogenic insulation, let us examine the exchange at some length.

Consider radiation between two infinite, grey, parallel specular plates as shown in Fig. 9-1a. All emitted and reflected radiation leaving surface 1 will directly reach surface 2; similarly, all emitted and reflected radiation leaving surface 2 will directly reach surface 1. This will be true whether the surfaces are specular or diffuse. Hence, for the specular case Eq. (8-10a) also applies and the net heat transfer from surface 1

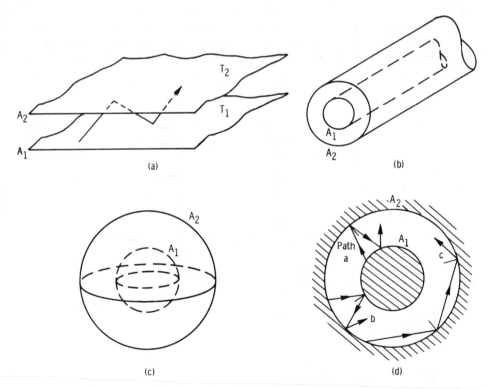

Figure 9-1 Radiation exchange for specular surfaces having simple geometries. (a) Infinite parallel plates; (b) gap between infinitely long concentric cylinders; (c) gap between concentric spheres; (d) paths for specular radiation in gap between concentric cylinders or spheres.

and surface 2 is

$$Q_1 = -Q_2 = \frac{A_1 \sigma (T_1{}^4 - T_2{}^4)}{1/\epsilon_1(T_1) + 1/\epsilon_2(T_2) - 1} \tag{9-1}$$

Now consider radiation between the concentric cylinders or spheres shown in Figs. 9-1b and c. Typical radiation paths for specular exchange are shown in Fig. 9-1d. As shown by path a, all the radiation emitted by surface 1 will directly reach 2. A portion will be reflected from surface 2 back to 1, and a portion of this will be rereflected from surface 1. This sequence of reflections between the surfaces continues until an insignificant amount of energy remains because the radiation has become partially absorbed on each contact with a surface. From the symmetry of the concentric geometry and the equal magnitudes of incidence and reflection angles for specular reflections, none of the radiation following path a can ever be reflected directly from a position on surface 2 to another element on surface 2. Thus the radiation-exchange process for radiation emitted from surface 1 is the same as though the two concentric surfaces were infinite parallel plates. However, the radiation emitted from the outer surface 2 can travel along either of two types of paths, b or c as shown in Fig. 9-1d. The fraction F_{2-2} will follow paths of type c. From the geometry of specular reflections these rays will always be reflected along surface 2 with none ever reaching 1. The fraction F_{2-1} will be reflected back and forth between the surfaces along path b in the same fashion as radiation emitted from surface 1. The amount of radiation following this type of path is

$$A_2 \epsilon_2 F_{2-1} \sigma T_2{}^4 = A_2 \epsilon_2 \frac{A_1}{A_2} \sigma T_2{}^4 = A_1 \epsilon_2 \sigma T_2{}^4$$

(the configuration factor $F_{2-1} = A_1/A_2$ has been employed). The fraction of the radiation leaving 2 that impinges on 1 thus depends on area A_1 and not on A_2. Hence, for specular surfaces the exchange behaves as if both surfaces were equal portions of infinite parallel plates equal in size to the area of the inner body. The net heat transfer from surface 1 to surface 2 is then given by Eq. (9-1).

EXAMPLE 9-1 A spherical vacuum bottle consists of two silvered, concentric glass spheres, the inner being 6 in in diameter and the evacuated gap between the spheres being $\frac{1}{4}$ in. The emissitivity of the silver coating is 0.02. If hot coffee at 200°F is in the bottle and the outside temperature is 70°F, what is the radiative heat leakage out of the bottle?

Equation (9-1) will apply for concentric specular spheres. For the small rate of heat leakage expected, it is assumed that the surfaces will be close to 200 and 70°F. This gives

$$Q_1 = \frac{\pi (\frac{1}{2})^2 0.173 \times 10^{-8} (660^4 - 530^4)}{1/0.02 + 1/0.02 - 1} = 1.52 \text{ Btu/h}$$

(If, instead of using the specular formulation, one assumes both surfaces to be diffuse, then (8-13) would be applied. The denominator of the Q_1 equation would become

$$\frac{1}{\epsilon_1} + \frac{A_1}{A_2}\left(\frac{1}{\epsilon_2} - 1\right) = \frac{1}{0.02} + \left(\frac{6}{6.5}\right)^2 \left(\frac{1}{0.02} - 1\right) = 91.8$$

instead of 99 as in the specular case. For diffuse surfaces the heat loss would be 1.64 Btu/h.)

EXAMPLE 9-2 For the previous example, how long will it take for the coffee to cool from 200 to 120°F if the heat loss is only by radiation?

The heat capacity of the coffee is $\rho_M V c_p T_1$. Assuming the coffee is always well enough mixed so that it is at uniform temperature, the cooling rate will be equal to the *instantaneous* loss by radiation. The loss of energy by radiation at any time τ, given by Eq. (9-1), is related to the loss of internal energy of the coffee by

$$-\rho_M V c_p \frac{dT_1}{d\tau} = \frac{A_1 \sigma [T_1{}^4(\tau) - T_2{}^4]}{1/\epsilon_1 + 1/\epsilon_2 - 1}$$

where it is assumed that surface 1 is at the coffee temperature and surface 2 is at the outside environment temperature. Then

$$-\int_{T_1 = T_I}^{T_1 = T_F} \frac{dT_1}{T_1{}^4 - T_2{}^4} = \frac{A_1 \sigma}{\rho_M V c_p (1/\epsilon_1 + 1/\epsilon_2 - 1)} \int_0^\tau d\tau$$

where T_I and T_F are the initial and final temperatures of the coffee, and ϵ_1 and ϵ_2 are assumed independent of temperature. Carrying out the integragion gives

$$\left(\frac{1}{4T_2{}^3} \ln\left|\frac{T_1 + T_2}{T_1 - T_2}\right| + \frac{1}{2T_2{}^3} \tan^{-1}\frac{T_1}{T_2}\right)\Bigg|_{T_I}^{T_F} = \frac{A_1 \sigma \tau}{\rho_M V c_p (1/\epsilon_1 + 1/\epsilon_2 - 1)}$$

Then the cooling time from T_I to T_F is

$$\tau = \frac{\rho_M V c_p (1/\epsilon_1 + 1/\epsilon_2 - 1)}{A_1 \sigma} \left[\frac{1}{4T_2{}^3} \ln\left|\frac{(T_F + T_2)/(T_F - T_2)}{(T_I + T_2)/(T_I + T_2)}\right|\right.$$

$$\left. + \frac{1}{2T_2{}^3}\left(\tan^{-1}\frac{T_F}{T_2} - \tan^{-1}\frac{T_I}{T_2}\right)\right]$$

Substituting the values $\rho_M = 62.4$ lb/ft³, $V = \frac{1}{6}\pi(\frac{1}{2})^3$ ft³, $c_p = 1$ Btu/(lb·°F), $\epsilon_1 = \epsilon_2 = 0.02$, $A_1 = \pi(\frac{1}{2})^2$ ft², $\sigma = 0.173 \times 10^{-8}$ Btu/(h·ft²·°R⁴), $T_2 = 530°$R, $T_I = 660°$R, and $T_F = 580°$R gives the cooling time as $\tau = 379$ h.

The coffee will stay hot for about 16 days if heat losses occur only by radiation. Conduction losses through the glass wall of the bottleneck usually cause the cooling rate to be much higher.

Table 9-1 Radiant interchange between some simply arranged surfaces

Geometry	Configuration	Surface type	Energy rate Q_1
Infinite parallel plates		A_1 or A_2, either specular or diffuse	$\dfrac{A_1\sigma(T_1{}^4 - T_2{}^4)}{1/\epsilon_1 + 1/\epsilon_2 - 1}$
Infinitely long concentric cylinders		A_1, specular *or* diffuse; A_2, diffuse	$\dfrac{A_1\sigma(T_1{}^4 - T_2{}^4)}{1/\epsilon_1 + (A_1/A_2)(1/\epsilon_2 - 1)}$
		A_1, specular *or* diffuse; A_2, specular	$\dfrac{A_1\sigma(T_1{}^4 - T_2{}^4)}{1/\epsilon_1 + 1/\epsilon_2 - 1}$
Concentric spheres		A_1, specular *or* diffuse; A_2, diffuse	$\dfrac{A_1\sigma(T_1{}^4 - T_2{}^4)}{1/\epsilon_1 + (A_1/A_2)(1/\epsilon_2 - 1)}$
		A_1, specular *or* diffuse; A_2, specular	$\dfrac{A_1\sigma(T_1{}^4 - T_2{}^4)}{1/\epsilon_1 + 1/\epsilon_2 - 1}$

Equation (9-1) applies for infinite parallel plates, infinitely long concentric cylinders, and concentric spheres when both surfaces are specular. For infinite parallel plates it also applies when both surfaces are diffuse or when one surface is diffuse and the other specular. For cylinders and spheres, (9-1) still applies if the surface of the inner body (surface 1) is diffuse as long as the outer body (surface 2) remains specular. This is because all radiation leaving surface 1 will go directly to 2 regardless of whether 1 is specular or diffuse. When surface 2 is diffuse, (8-13) applies and may be used when surface 1 is either specular or diffuse. The relations are summarized in Table 9-1.

EXAMPLE 9-3 Two infinite parallel plates are specular and have emissivities that depend on the angle θ (nondiffuse surfaces). What is the equation of radiative exchange between the surfaces?

The radiation per unit area emitted in solid angle $d\omega$ from surface 1 in the direction θ is, $\epsilon_1'(\theta)(\sigma T_1{}^4/\pi) \cos \theta \, d\omega = 2\epsilon_1'(\theta) \, \sigma T_1{}^4 \cos \theta \sin \theta \, d\theta = \epsilon_1'(\theta) \, \sigma T_1{}^4 \, d(\sin^2 \theta)$. Similarly for surface 2. The same geometric relation will apply as in Eq. (9-1). Hence, integrating over the exchanges for all θ yields

$$\frac{Q_1}{A_1} = \sigma(T_1{}^4 - T_2{}^4) \int_0^1 \frac{d(\sin^2 \theta)}{1/\epsilon_1'(\theta) + 1/\epsilon_2'(\theta) - 1}$$

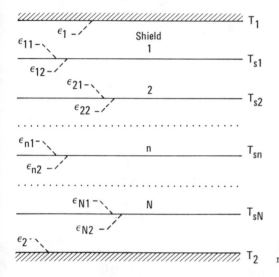

Figure 9-2 Parallel walls separated by N radiation shields.

The results in Table 9-1 can be used to obtain the performance of multiple radiation shields, as shown in Figs. 9-2 and 9-3. The shields are thin, parallel, highly reflecting sheets placed between surfaces. A highly effective insulation can be formed by using many layers separated by vacuum to provide a series of alternate radiation and conduction barriers. One construction is to deposit highly reflecting metallic films on both sides of thin sheets of plastic spaced apart by placing between them a cloth net having a large open area between the fibers. Typically a stacking of 50 radiation shields per inch of thickness can be obtained in this manner. An important use of multilayer insulation is in low-temperature applications such as insulation of cryogenic storage tanks.

For the results here, the spaces between the shields are evacuated so that heat transfer is only by radiation. To analyze shield performance, consider the situation of N different radiation shields between two surfaces at temperatures T_1 and T_2 with emissivities ϵ_1 and ϵ_2. As a general case let a typical shield n have an emissivity ϵ_{n1} on one side and ϵ_{n2} on the other, as shown in Fig. 9-2. As a result of the heat flow, the nth shield will be at temperature T_{sn}. Since the same q passes through the entire series of shields, Eq. (9-1) can be written for each set of adjacent shields as

$$q\left(\frac{1}{\epsilon_1} + \frac{1}{\epsilon_{11}} - 1\right) = \sigma(T_1^4 - T_{s1}^4)$$

$$q\left(\frac{1}{\epsilon_{12}} + \frac{1}{\epsilon_{21}} - 1\right) = \sigma(T_{s1}^4 - T_{s2}^4)$$

$$q\left(\frac{1}{\epsilon_{22}} + \frac{1}{\epsilon_{31}} - 1\right) = \sigma(T_{s2}^4 - T_{s3}^4)$$

. . . .

$$q\left(\frac{1}{\epsilon_{(N-1)2}} + \frac{1}{\epsilon_{N1}} - 1\right) = \sigma(T_{s(N-1)}^4 - T_{sN}^4)$$

$$q\left(\frac{1}{\epsilon_{N2}} + \frac{1}{\epsilon_2} - 1\right) = \sigma(T_{sN}^4 - T_2^4)$$

Adding these equations and dividing by the resulting factor multiplying q on the left-hand side give

$$q = \frac{\sigma(T_1^4 - T_2^4)}{1/\epsilon_1 + 1/\epsilon_{11} - 1 + \sum_{n=1}^{N-1}(1/\epsilon_{n2} + 1/\epsilon_{(n+1)1} - 1) + 1/\epsilon_{N2} + 1/\epsilon_2 - 1} \qquad (9\text{-}2a)$$

This can also be written

$$q = \frac{\sigma(T_1^4 - T_2^4)}{1/\epsilon_1 + 1/\epsilon_2 - 1 + \sum_{n=1}^{N}(1/\epsilon_{n1} + 1/\epsilon_{n2} - 1)} \qquad (9\text{-}2b)$$

The sum in (9-2a) is zero if $N = 1$. In most instances the ϵ is the same on both sides of each shield, and all the shields have the same ϵ. Let all the shield emissivities be ϵ_s; then the q becomes

$$q = \frac{\sigma(T_1^4 - T_2^4)}{1/\epsilon_1 + 1/\epsilon_2 - 1 + N(2/\epsilon_s - 1)} \qquad (9\text{-}3)$$

If the wall emissivities are the same as the shield emissivities, $\epsilon_1 = \epsilon_2 = \epsilon_s$, then (9-3) further reduces to

$$q = \frac{\sigma(T_1^4 - T_2^4)}{(N+1)(2/\epsilon_s - 1)} \qquad (9\text{-}4)$$

In this instance the q decreases as $1/(N+1)$ as the number of shields increases. When there are no shields, $N = 0$, and (9-3) reduces to $q = \sigma(T_1^4 - T_2^4)/(1/\epsilon_1 + 1/\epsilon_2 - 1)$ as given in Table 9-1. As an example to illustrate the performance of the shields, if in Eq. (9-3) $\epsilon_1 = \epsilon_2 = 0.8$ and $\epsilon_s = 0.05$, then $q = \sigma(T_1^4 - T_2^4)/(1.5 + 39N)$. The ratio $q(N \text{ shields})/q(\text{no shields})$ is $1.5/(1.5 + 39N)$. For various numbers of shields this yields

	$N = 0$	$N = 1$	$N = 10$	$N = 100$
$q(N)/q(N = 0)$	1	0.0370	0.00383	0.00038
$1/(N + 1)$	1	0.5	0.0909	0.0099

In the table, the factor $1/(N+1)$ is the result when the walls have the same ϵ as the shields. This illustrates that the fractional reduction in q is much larger when the wall ϵ's are large compared to the ϵ for the shields.

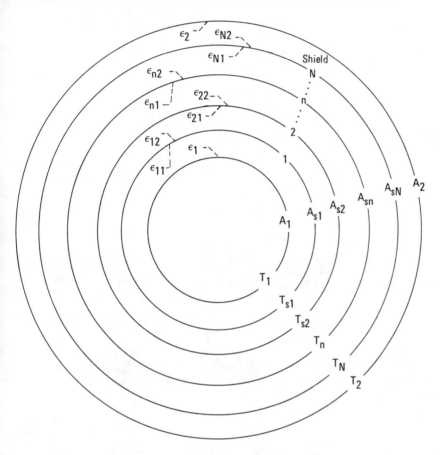

Figure 9-3 Radiation shields between concentric cylinders or spheres.

In a fashion similar to the previous derivation for flat plates, the expressions in Table 9-1 can be used to derive the heat flow through a series of concentric cylindrical or spherical radiation shields, as shown in Fig. 9-3. If the walls A_1 and A_2 and all the shields A_{sn} are diffuse, the heat flow is

$$Q = \frac{A_1 \sigma (T_1^4 - T_2^4)}{\{1/\epsilon_1 + (A_1/A_{s1})(1/\epsilon_{11} - 1) + \sum_{n=1}^{N-1} (A_1/A_{sn}) [1/\epsilon_{n2} + (A_{sn}/A_{s(n+1)})(1/\epsilon_{(n+1)1} - 1)]}$$

$$+ (A_1/A_{sN}) [1/\epsilon_{N2} + (A_{sN}/A_2)(1/\epsilon_2 - 1)] \}$$

(9-5a)

$$= \frac{A_1 \sigma (T_1^4 - T_2^4)}{1/\epsilon_1 + (A_1/A_2)(1/\epsilon_2 - 1) + \sum_{n=1}^{N} (A_1/A_{sn})(1/\epsilon_{n1} + 1/\epsilon_{n2} - 1)}$$

(9-5b)

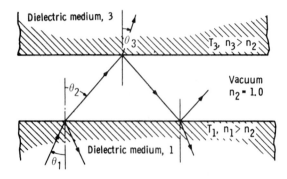

Figure 9-4 Reflection and transmission of electromagnetic wave in gap between two dielectrics; $T_3 < T_1$.

The sum in (9-5a) is zero if $N = 1$. If the walls are diffuse and all the shields are specular, then

$$Q = \frac{A_1 \sigma (T_1^4 - T_2^4)}{\{1/\epsilon_1 + 1/\epsilon_{11} - 1 + \sum_{n=1}^{N-1} (A_1/A_{sn})(1/\epsilon_{n2} + 1/\epsilon_{(n+1)1} - 1)}$$

$$+ (A_1/A_{sN}) [1/\epsilon_{N2} + (A_{sN}/A_2)(1/\epsilon_2 - 1)]\}$$

(9-6)

If all the surfaces are specular, then (9-6) applies if we replace A_{sN}/A_2 by unity in the last term in the denominator. In this instance, if all the shield emissivities are the same and equal to ϵ_s, then

$$Q = \frac{A_1 \sigma (T_1^4 - T_2^4)}{1/\epsilon_1 + 1/\epsilon_s - 1 + \sum_{n=1}^{N-1} (A_1/A_{sn})(2/\epsilon_s - 1) + (A_1/A_{sN})(1/\epsilon_s + 1/\epsilon_2 - 1)}$$

(9-7)

With regard to the radiative transfer between parallel surfaces, there is an additional effect when reflecting layers are spaced very close to each other. The question is whether very small spacings can have an influence on the radiative transfer. The transfer between closely spaced surfaces was examined by Cravalho et al. [3] who considered the geometry in Fig. 9-4, consisting of two semi-infinite dielectric media having refractive indices n_1 and n_3, separated by a vacuum gap. In the usual analysis for radiative transfer between two surfaces such as 1 and 3, the heat flux transferred across the gap is given by [see Eq. (10-11)]

$$q_1 = \int_0^\infty \frac{e_{\lambda b,1}(\lambda, T_1) - e_{\lambda b,3}(\lambda, T_3)}{1/\epsilon_{\lambda,1}(\lambda, T_1) + 1/\epsilon_{\lambda,3}(\lambda, T_3) - 1} d\lambda$$

and the spacing between the plates does not appear. When the spacing between the surfaces is very small, however, two effects enter that are a function of spacing. The first effect is wave interference, in which a wave reflecting back and forth in a gap between two dielectrics may undergo cancellation or reinforcement as described in Chap. 19 in connection with thin films.

The second effect is radiation tunneling. The evanescent fields present in thermal equilibrium at the outer surface of each body are able to reach to the opposite body and transfer energy if the distance is sufficiently small. Figure 9-4 reveals that for ordinary behavior at an interface, as discussed in Sec. 19-6.2, some of the radiation in medium 1 traveling toward region 2 can undergo total internal reflection at the interface when $n_1 > n_2$. For ordinary radiative behavior this would occur when the incidence angle θ_1 is equal to or larger than the angle for total reflection given by (19-46), that is, $\theta_1 \geq \sin^{-1}(n_2/n_1)$. When region 2 in Fig. 9-4 is sufficiently thin, however, electromagnetic theory predicts that, even for an intensity incident at θ_1 greater than $\sin^{-1}(n_2/n_1)$, total internal reflection will not occur. Rather, part of the incident intensity will propagate across the thin region 2 and enter medium 3. This effect is radiation tunneling as viewed classically.

As shown in [3], both tunneling and interference can become important only when the spacing between radiating bodies separated by vacuum is less than about $\lambda_{max,vac}(T_3)$ which is the wavelength in vacuum at maximum blackbody emissive power from a surface at the sink temperature T_3. The $\lambda_{max,vac}(T_3)$ is found from Wien's displacement law as C_3/T_3 [Eq. (2-17)]. The tunneling and interference effects also depend on temperature. Even for very small spacings on the order of $\lambda_{max,vac}(T_3)$, the effects become very small at normal temperatures, and hence are only important in certain cryogenic applications where temperatures of a few degrees absolute are encountered. Figure 9-5 shows some representative results under conditions giving maximum effects and illustrates the influence of temperature T_1.

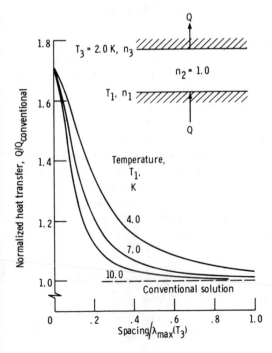

Figure 9-5 Effects of wave interference and radiation tunneling on radiative transfer between two dielectric surfaces; $n_1 = n_3 = 1.25$; $\lambda_{max}(T_3) = 0.28978/n_3 T_3$ cm. *(From [3].)*

Note that $\lambda_{max}(T_3)$ in Fig. 9-5 is the wavelength in medium 3 and hence is given by $C_3/n_3 T_3$ from (2-33). The conventional solution referred to in the figure is obtained when wave interference and radiation tunneling are neglected in the analysis. In [4] an analysis is carried out for closely spaced metallic surfaces. Good agreement was obtained with published data at $T \approx 315$ K and spacings in the range 1–10 μm.

9-3.2 Ray Tracing and the Construction of Images

When mirrorlike reflections occur in enclosures, the well-developed procedures of geometric optics can be applied to simplify both the concepts and the mathematics of the radiative exchange process. The basic ideas are outlined in this section. More advanced ideas may be found in [5, 6].

An incident ray striking a specular surface is reflected in a symmetric fashion about the surface normal so that the angle of reflection is equal in magnitude to the angle of incidence. This fact is used to formulate the concept of *images*. An image is simply an apparent point of origin for an observed ray. For example, in Fig. 9-6a, an observer views an object in a mirror. To the observer, the object appears to be *behind* the mirror in the position shown by the dotted object. This apparent object is called the image.

This concept is readily extended to cases in which a series of reflections occur, as shown in Fig. 9-6b. An interesting example of this is the "barber-chair" geometry, where mirrors are present on opposite walls of the barber shop. In this case, if the mirrors are parallel, an infinite number of reflections of a ray can occur, and a person receiving a haircut can view an infinite number of images of himself, or herself (if the mirrors are perfect—that is, if $\rho_s = 1$).

To this point, mirrors have only been discussed with regard to their ability to change the direction of the rays originating at the source. In the formulation of thermal-radiation problems, the specular surfaces will, in general, have a nonzero absorptivitiy. They will thus attenuate the energy of the rays from an object.

In addition to reflecting and absorbing energy, mirrors can emit energy. This emission can be conveniently analyzed with an image system rather than with the real mirror system. In the image system all radiation acts along straight lines without the complexity of directional changes at each reflecting surface. The attenuation at each surface is accounted for by multiplying the intensity of the ray by the specular reflectivity at each reflection. The emission from three surfaces is illustrated in Fig. 9-6c. For example, emitted energy reaching the viewer from surface 3 is considered to be coming directly from the image of 3, with attenuation due to reflections at 2 and 1 because of passage through these surfaces or their images.

9-3.3 Radiation Transfer by Means of Simple Specular Surfaces for Diffuse Energy Leaving a Surface

As an introduction to the analysis of radiation exchange in an enclosure having some specularly reflecting surfaces, a few examples will be considered for plane surfaces.

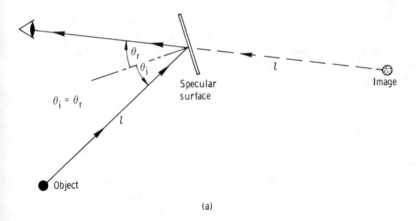

(a)

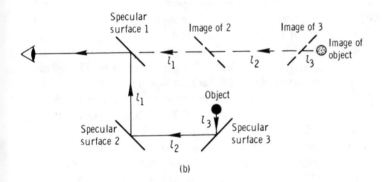

(b)

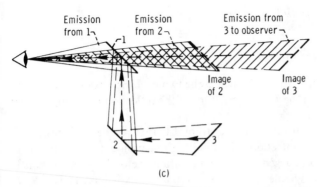

(c)

Figure 9-6 Ray tracing and images formed by specular reflections. (*a*) Image formed by single reflection; (*b*) image formed by multiple reflections; (*c*) contributions due to emission from specular surfaces.

The examples will further demonstrate the new features that enter when mirrorlike surfaces are present.

In the enclosure analysis the emission from all surfaces is assumed diffuse. This is a fairly good assumption in most cases, as can be shown by the electromagnetic-theory predictions of the emissivity of specular surfaces (Fig. 4-6). The reflected energy is assumed either diffuse or specular. In the enclosure theory that is developed later, the diffuse reflected energy is combined with the emitted energy, and it is then necessary to know how the transfer of this diffuse energy is influenced by the presence of specularly reflecting surfaces. The exchange factors F^s that will be obtained by con-sidering the presence of the specular surfaces will automatically account for the specularly reflected energy. Hence it is necessary to consider here *only* the transfer of *diffuse energy* leaving a surface.

Figure 9-7a shows a diffusely emitting and reflecting plane surface A_1 facing a specularly reflecting plane surface A_2. Surface 1 cannot directly view itself; the ordinary configuration factor from any part of A_1 to any other part of A_1 is thus zero. However, if A_2 is specular, then A_1 can view its image, and a path exists, by means of a reflection from the specular surface A_2, for diffuse radiation to travel from the differential area dA_1 to dA_1^*. From Fig. 9-7a, it is evident by ray-tracing that the radiation arriving at dA_1^* from dA_1 appears to come from the image $dA_{1(2)}$. Thus the geometric configuration factor between dA_1 and dA_1^* resulting from one reflec-tion can be obtained as $dF_{d1(2)-d1*}$ for diffuse radiation leaving dA_1. The subscript notation refers to a factor from the image of dA_1 (as seen in A_2) to dA_1^*.

There are points of similarity that should be noted when comparing the specular and diffuse reflecting cases. When A_1 and A_2 in Fig. 9-7a are both diffuse reflectors, the diffuse radiation leaving dA_1 is received at dA_1^* by means of diffuse reflection from A_2 and is governed by dF_{d1-d2} and then dF_{d2-d1} for each element dA_2 on A_2. The portion of the diffuse energy from dA_1 that reaches dA_1^* after one reflection from A_2 is the following, for the two cases of diffuse and specular A_2, respectively:

$$d^2 Q_{d1-d1*(2)} = dA_1\, q_{o,1}\, \rho_2 \int_{A_2} dF_{d1-d2}\, dF_{d2-d1*}$$

$$d^2 Q_{d1-d1*(2)} = dA_1\, q_{o,1}\, \rho_{s,2}\, dF_{d1(2)-d1*}$$

This reveals that, for $\rho_2 = \rho_{s,2}$, the difference in the two exchanges is incorporated in the configuration factors resulting from the nature of the reflection being considered; this is a purely geometric effect.

Figure 9-7b describes the diffuse radiation from dA_1 that reaches the entire area A_1 by means of one specular reflection. The reflected radiation appears to originate from the image $dA_{1(2)}$. Thus the geometric configuration factor involved from dA_1 to A_1 is $F_{d1(2)-1}$. From the symmetry revealed by the dot-dash lines in Fig. 9-7b, $F_{d1(2)-1} = F_{d1-1(2)}$. Thus the radiative transfer can also be expressed as a view factor from the first surface to the image of the second surface.

Figure 9-7c shows several typical rays leaving A_1 that are reflected back to A_1. These rays appear to originate from the image $A_{1(2)}$. The configuration factor from

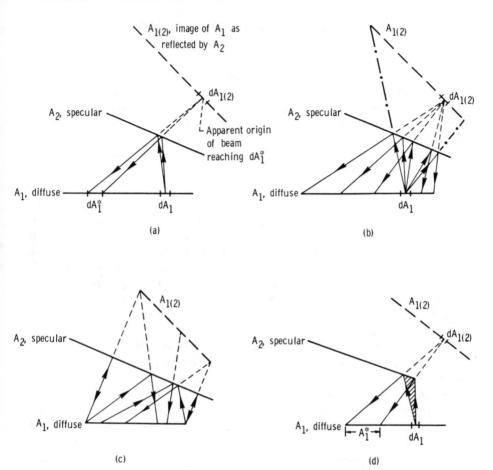

Figure 9-7 Radiation between a diffuse surface and itself by means of a specular surface. (*a*) Radiation between two differential areas with one intermediate specular reflection; (*b*) radiation from differential area to finite area by means of one intermediate specular reflection; (*c*) radiation from finite area that is reflected back to that area by means of one specular reflection; (*d*) radiation from dA_1 can reach only a portion of A_1 by means of specular reflection from A_2.

A_1 back to itself by means of one specular reflection is then $F_{1(2)-1}$. In this instance all of the image $A_{1(2)}$ is visible in A_2 from any position on A_1. In some instances this will not be true. An example is shown in Fig. 9-7*d*. The radiation from dA_1 has to be within the limited range of solid angle shown shaded in order for the radiation to be reflected back to A_1. The geometric configuration factor between dA_1 and A_1 is still $F_{d1(2)-1}$, but this factor is only evaluated over the portion of A_1 that receives reflected rays. $F_{d1(2)-1}$ is the factor by which $dA_{1(2)}$ views A_1, and it must be kept in mind that *the view may be a partial one*. This factor will have a different value as the location of dA_1 along A_1 is changed. The fact that the view between dA_1 and A_1 varies with the position of dA_1 along A_1 means that the energy from A_1 that is reflected back to A_1 will have a nonuniform distribution along A_1. The reflection of some of this energy from A_1 will provide a nonuniform q_o from A_1 that violates the assumption in the enclosure theory of uniform q_o from each surface. When partial images are present, caution should be exercised in subdividing the enclosure area into sufficiently small portions so that the accuracy of the solution is adequate.

Now consider the geometry when there are more than one specular surface involved in the radiation exchange. This will lead to multiple reflections and many different paths by which radiation can travel between surfaces. At each reflection, the radiation is modified by the ρ_s of the reflecting surface. At present in this discussion, only the geometry is being considered; the ρ_s factors will be included later when heat balances are formulated.

In Fig. 9-8 are shown two specular surfaces. Energy is being emitted diffusely from A_2 and is traveling to surface A_1. The fraction arriving at dA_1 is given by the geometric configuration factor dF_{2-d1}. This direct path is illustrated in Fig. 9-8a. A portion of the energy intercepted by A_1 will be reflected back to A_2 and then reflected back again to A_1. Hence, A_2 not only views dA_1 directly but also by means of an image formed by two reflections. This image is constructed in Fig. 9-8b. First the reflected image $A_{1(2)}$ of A_1 reflected in A_2 is drawn. Then A_2 is reflected into this image to form $A_{2(1-2)}$. The notation $A_{2(1-2)}$ is read as the image of area 2 formed by reflections in area 1 and area 2 (in that order). The radiation paths and the shaded area shown in Fig. 9-8b reveal that the solid angle within which radiation leaving A_2 will reach dA_1 by means of two reflections is the same as the solid angle by which dA_1 views the image $A_{2(1-2)}$. Thus the configuration factor involved for two reflections is $dF_{2(1-2)-d1}$. This is read as the factor from the image of surface 2 formed by reflections in surfaces 1 and 2 (in that order) to area element $d1$.

Consider the possibility of additional images. The geometric factor involved is always found by viewing dA_1 from the appropriate reflected image of A_2 as seen through the surface A_2 and all intermediate images. In the case of Fig. 9-8c, the image of A_2 after four reflections $A_{2(1-2-1-2)}$ cannot view dA_1 by looking through A_2. Hence, there is no radiation leaving A_2 that reaches dA_1 by means of four reflections, and no additional images need be considered.

EXAMPLE 9-4 An infinitely long groove as shown in Fig. 9-9 has specularly reflecting sides that emit diffusely. What fraction of the emitted energy from A_2 reaches the black receiver surface element dA_3? Express the result in terms of diffuse geometric configuration factors.

Consider first the energy that reaches dA_3 directly from A_2 and by means of an even number of reflections. The fraction of emitted radiation that reaches dA_3 directly from A_2 is dF_{2-d3} as illustrated in Fig. 9-9a. A second portion will be emitted from A_2 to A_1, reflected back to A_2, and then reflected to dA_3. From the diagram of images in Fig. 9-9b only part of the reflected image $A_{2(1-2)}$ can be viewed by dA_3 through A_2. The fraction of emitted energy reaching dA_3 by this path is the configuration factor *evaluated only over the part of $A_{2(1-2)}$ visible to dA_3* multiplied by the two specular reflectivities, $\rho_{s,1}\rho_{s,2}\,dF_{2(1-2)-d3}$. This is not an ordinary view factor but *takes into account the view through the image system*. In a similar fashion there will be a contribution after two reflections from each of

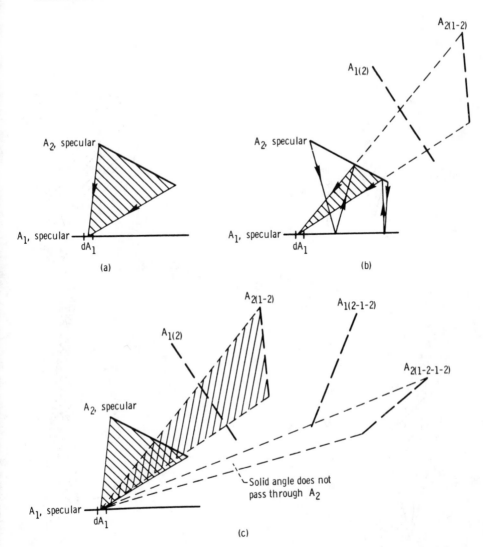

Figure 9-8 Radiant interchange between two specular reflecting surfaces. (*a*) Energy emitted from A_2 that directly reaches dA_1; (*b*) energy emitted from A_2 that reaches dA_1 after two reflections; (*c*) none of energy emitted by A_2 reaches dA_1 by means of four reflections.

A_1 and A_2. This is illustrated by the shaded solid angle in Fig. 9.9*c*. The third image of A_2, $A_{2(1-2-1-2-1-2)}$, cannot be viewed by dA_3 through A_2 and hence will not make a contribution. Also, the third image of A_2 cannot view A_1 through A_2 so there will be no additional images of A_2. The fraction of energy emitted by A_2 that reaches dA_3 both directly and by means of the images of A_2 resulting from an even number of reflections is then

$$dF_{2-d3} + \rho_{s,1}\rho_{s,2}\, dF_{2(1-2)-d3} + \rho_{s,1}^2\rho_{s,2}^2\, dF_{2(1-2-1-2)-d3}$$

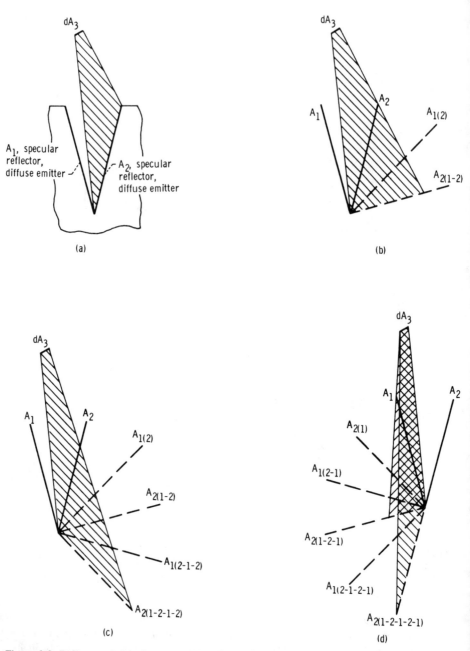

Figure 9-9 Diffuse emission from one side of specularly reflecting groove that reaches a differential-strip receiving area outside groove opening. (*a*) Geometry of direct exchange from A_2 to dA_3; (*b*) geometry of exchange for radiation from A_2 that reaches dA_3 by means of one intermediate reflection from each of A_1 and A_2; (*c*) geometry of exchange for radiation from A_2 that reaches dA_3 by means of two intermediate reflections from each of A_1 and A_2; (*d*) geometry of exchange for radiation from A_2 to A_3 by means of odd number of reflections.

Now consider the energy fraction that will reach dA_3 from A_2 by means of an odd number of reflections. Using Fig. 9-9d and arguments similar to those for an even number of reflections results in

$$\rho_{s,1}\, dF_{2(1)-d3} + \rho_{s,1}^2\rho_{s,2}\, dF_{2(1-2-1)-d3} + \rho_{s,1}^3\rho_{s,2}^2\, dF_{2(1-2-1-2-1)-d3}$$

The first two of the F factors are only evaluated over the portions of the images that can be viewed by dA_3.

The fraction of energy emitted by surface A_2 that reaches dA_3 directly and after all interreflections from both A_1 and A_2 is then

$$\frac{dQ_{2-d3}}{\epsilon_2\sigma T_2^4 A_2} = dF_{2-d3} + \rho_{s,1}\, dF_{2(1)-d3} + \rho_{s,1}\rho_{s,2}\, dF_{2(1-2)-d3}$$

$$+ \rho_{s,1}^2\rho_{s,2}\, dF_{2(1-2-1)-d3} + \rho_{s,1}^2\rho_{s,2}^2\, dF_{2(1-2-1-2)-d3}$$

$$+ \rho_{s,1}^3\rho_{s,2}^2\, dF_{2(1-2-1-2-1)-d3}$$

Additional information on the absorption and emission of radiation by specular grooves can be found in [7]. An application where such grooves may prove important is in the collection and concentration of solar energy. For this purpose an absorbing surface is placed at the bottom of each groove, which can have sides in the form of parabolic segments [8] as shown in Fig. 9-10a or in the form of a series of straight sections [9] as in Fig. 9-10b. These configurations provide concentration of the incident energy onto the absorber plane, which enables the production of elevated temperatures needed for efficient energy conversion. The groove shapes give good concentration even when the incident radiation is not well aligned with the center plane of the groove. Thus this type of solar collector will perform well as the angle of the sun changes throughout the day, even though the collector remains in a fixed position.

The notation adopted for the specular configuration factors allows a check on the form of the equations arising in the study of radiant interchange among specular surfaces. The numbers within the subscripted parentheses of the configuration factors designate the sequence in which reflections have occurred from specular surfaces. A reflectivity for each of these specular surfaces should multiply the configuration factor to account for attenuation of energy by absorption at these surfaces. For example, the factor $F_{A(B-C-D)-E}$ would be multiplied by $\rho_{s,B}\rho_{s,C}\rho_{s,D}$. Examination of the individual terms in the final equation of Example 9-4 will show this to be the case.

9-3.4 Configuration-Factor Reciprocity for Specular Surfaces

Reciprocity relations analogous to those for configuration factors between diffuse surfaces apply for the factors involving specular surfaces under certain conditions. Consider a three-sided isothermal enclosure at temperature T, made up of two black surfaces 1 and 2 and a specular surface 3 of reflectivity $\rho_{s,3}$(Fig. 9-11a). The energy

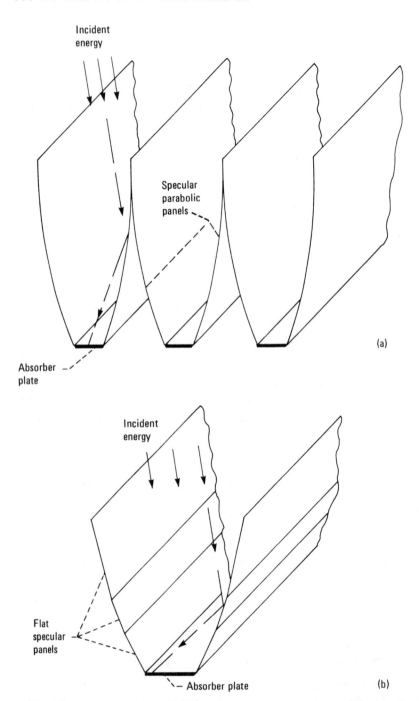

Figure 9-10 Concentrating solar collectors. (*a*) Parabolic trough concentrator; (*b*) concentrator made with flat segments.

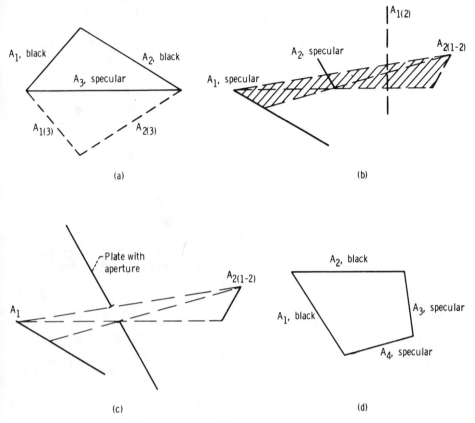

Figure 9-11 Reciprocity of configuration factors involving specular surfaces. (*a*) Three-sided enclosure with one specular reflecting surface; (*b*) system of images of surfaces 1 and 2; (*c*) energy-exchange analog of image system in (*b*); (*d*) enclosure with two specular and two black surfaces.

emitted by black surface 1 that reaches black surface 2 directly and by reflection from specular surface 3 is given by

$$Q_{1-2} = \sigma T^4 (A_1 F_{1-2} + A_1 \rho_{s,3} F_{1(3)-2})$$ (9-8)

The energy leaving surface 2 and reaching surface 1 directly and by specular reflection from surface 3 is

$$Q_{2-1} = \sigma T^4 (A_2 F_{2-1} + A_2 \rho_{s,3} F_{2(3)-1})$$ (9-9)

Noting that, for the isothermal enclosure, $Q_{2-1} = Q_{1-2}$ and $A_1 F_{1-2} = A_2 F_{2-1}$ results in the reciprocity relation for the case of one specular surface in the enclosure:

$$A_1 F_{1(3)-2} = A_2 F_{2(3)-1}$$ (9-10)

This relation can also be deduced from the symmetry about A_3 shown in Fig. 9-11*a* and the reciprocity relations for diffuse configuration factors.

A second type of reciprocity relation exists for configuration factors involving specular surfaces. To derive this relation, we examine the energy exchange between two surfaces A_1 and A_2 contained within an isothermal enclosure. If both surfaces are specular, then the image system shown in Fig. 9-11b can be constructed for the case of radiation from 2 to 1 by means of a reflection at 1 and at 2. For any such system, an analogous system can be constructed in which a plate with an aperture is substituted for the restraints on the ray paths that are present, as is done in Fig. 9-11c. The aperture is placed to allow passage of only those rays that pass through the image system by which A_1 can view at least a portion of $A_{2(1-2)}$ through A_2 and $A_{1(2)}$.

The emitted energy leaving specular surface A_2 in the analog system and absorbed by A_1 is

$$Q_{2(1-2)-1} = Q_{e,2}\rho_{s,1}\rho_{s,2}F_{2(1-2)-1}\alpha_1 = A_{2(1-2)}\epsilon_2\sigma T^4\rho_{s,1}\rho_{s,2}F_{2(1-2)-1}\epsilon_1 \tag{9-11}$$

The reflectivities account for the reduction in energy by the two intermediate specular reflections. $F_{2(1-2)-1}$ is the diffuse-surface configuration factor computed for the constrained paths passing through the aperture (see Example 9-5). But these paths are exactly those through the image system, so this is also the specular configuration factor. Similarly, the energy along the reverse path is

$$Q_{1-2(1-2)} = A_1\epsilon_1\sigma T^4\rho_{s,2}\rho_{s,1}F_{1-2(1-2)}\epsilon_2 \tag{9-12}$$

Equating the energy exchanges in either direction between A_1 and $A_{2(1-2)}$ for the isothermal enclosure results in the following reciprocity relation:

$$A_1 F_{1-2(1-2)} = A_{2(1-2)}F_{2(1-2)-1} = A_2 F_{2(1-2)-1} \tag{9-13}$$

By generalizing for many intermediate reflections from surfaces A, B, C, D, and so forth, Eq. (9-13) can be written as

$$A_1 F_{1-2(A-B-C-D \cdots)} = A_2 F_{2(A-B-C-D \cdots)-1} \tag{9-14}$$

For two-dimensional areas, the crossed-string method (Sec. 7-5.4) can be used to obtain the configuration factors. For example, in Fig. 9-11b the A_2 and $A_{1(2)}$ are regarded as apertures in the view between A_1 and $A_{2(1-2)}$. The $F_{1-2(1-2)}$ is found by having the crossed and uncrossed strings pass through these apertures.

EXAMPLE 9-5 A black surface A_1 faces a smaller parallel mirror A_2 as in Fig. 9-12. Compute the configuration factor $F_{1-1(2)}$ between A_1 and the image of A_1 formed by means of one specular reflection in A_2. The surfaces are infinitely long in the direction normal to the plane of the drawing.

The factor is computed from the integral $F_{1-1(2)} = (1/A_1) \int_{A_1} F_{d1-1(2)}\, dA_1$. Consider the element dA_1 at location x on A_1. The configuration factor for radiation from dA_1 to the portion of $A_{1(2)}$ in view through A_2 is (see Example 7-4)

$$F_{d1-1(2)} = \frac{1}{2}(\sin\beta' - \sin\beta'') = \frac{1}{2}\left[\frac{x+a}{\sqrt{(x+a)^2 + b^2}} - \frac{x-a}{\sqrt{(x-a)^2 + b^2}}\right]$$

This is valid until position $x = l - 2a$ is reached (Fig. 9-12b). For larger x values

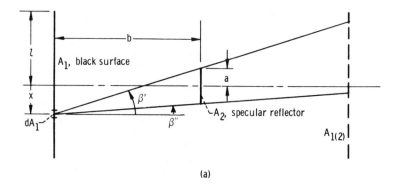

(a)

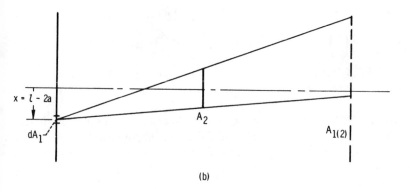

(b)

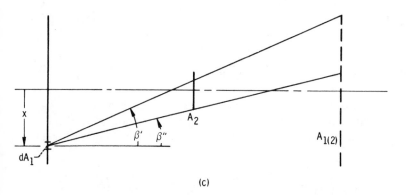

(c)

Figure 9-12 Configuration-factor computation involving partial views of surface and image (Example 9-5). (*a*) Portion of $A_{1(2)}$ in view from dA_1 through entire A_2; (*b*) limiting x for portion of $A_{1(2)}$ to be in view through entire A_2; (*c*) portion of $A_{1(2)}$ in view through part of A_2.

the geometry is as shown in Fig. 9-12c. Then

$$F_{d1-1(2)} = \frac{1}{2}(\sin\beta' - \sin\beta'') = \frac{1}{2}\left[\frac{x+l}{\sqrt{(x+l)^2+4b^2}} - \frac{x-a}{\sqrt{(x-a)^2+b^2}}\right]$$

The desired configuration factor is then

$$F_{1-1(2)} = \frac{1}{2l}2\int_0^l F_{d1-1(2)}\,dx$$

$$= \frac{1}{l}\left\{\frac{1}{2}\int_0^{l-2a}\left[\frac{x+a}{\sqrt{(x+a)^2+b^2}} - \frac{x-a}{\sqrt{(x-a)^2+b^2}}\right]dx\right.$$

$$\left. + \frac{1}{2}\int_{l-2a}^{l}\left[\frac{x+l}{\sqrt{(x+l)^2+4b^2}} - \frac{x-a}{\sqrt{(x-a)^2+b^2}}\right]dx\right\}$$

The integrations are carried out, and the results simplify to

$$F_{1-1(2)} = \sqrt{1+\left(\frac{b}{l}\right)^2} - \sqrt{\left(1-\frac{a}{l}\right)^2+\left(\frac{b}{l}\right)^2}$$

The derivation of this result by the crossed-string method is given as a homework problem.

Consider a case in which there are more than one specular surface in an isothermal enclosure at temperature T. For simplicity we discuss an enclosure such as Fig. 9-11d, where there are two specular and two black surfaces. If the heat exchange between the two black surfaces by direct exchange and all specular reflection paths is considered, the following relations result:

$$\frac{Q_{1-2}}{\sigma T^4} = A_1(F_{1-2} + \rho_{s,3}F_{1(3)-2} + \rho_{s,4}F_{1(4)-2}$$

$$+ \rho_{s,3}\rho_{s,4}F_{1(3-4)-2} + \cdots + \rho_{s,3}^m\rho_{s,4}^n F_{1(3^m-4^n)-2} + \cdots) \qquad (9\text{-}15a)$$

$$\frac{Q_{2-1}}{\sigma T^4} = A_2(F_{2-1} + \rho_{s,3}F_{2(3)-1} + \rho_{s,4}F_{2(4)-1}$$

$$+ \rho_{s,3}\rho_{s,4}F_{2(4-3)-1} + \cdots + \rho_{s,3}^m\rho_{s,4}^n F_{2(4^n-3^m)-1} + \cdots) \qquad (9\text{-}15b)$$

The shorthand notation $(3^m\text{-}4^n)$ means m reflections in 3 and n in 4; hence $F_{1(3^m-4^n)-2}$ is the configuration factor to area 2 from the image of 1 formed by these m and n reflections. Equation (9-15) can also be written as (for an isothermal enclosure $Q_{1-2} = Q_{2-1}$)

$$\frac{Q_{1-2}}{\sigma T^4} = \frac{Q_{2-1}}{\sigma T^4} = A_1 F^s_{1-2} = A_2 F^s_{2-1} \qquad (9\text{-}16)$$

where F^s is an *exchange factor* equal to the quantity in parenthesis in (9-15), that is,

$$F^s_{1-2} = F_{1-2} + \rho_{s,3}F_{1(3)-2} + \rho_{s,4}F_{1(4)-2} + \rho_{s,3}\rho_{s,4}F_{1(3-4)-2}$$
$$+ \cdots + \rho^m_{s,3}\rho^n_{s,4}F_{1(3^m-4^n)-2} + \cdots \tag{9-17}$$

The exchange factor is the fraction of diffuse energy leaving a surface that arrives at a second surface both directly and by all possible intermediate specular reflections. Equation (9-17) gives F^s_{1-2} in terms of *images of 1* to surface 2. An alternative form of F^s_{1-2} (or similarly of F^s_{2-1}) can be derived in terms of the radiation from 1 directly to 2 and by means of reflections to the *images of 2*. From (9-16),

$$F^s_{1-2} = \frac{A_2}{A_1} F^s_{2-1} = \frac{A_2}{A_1} (F_{2-1} + \rho_{s,3}F_{2(3)-1} + \rho_{s,4}F_{2(4)-1} + \rho_{s,3}\rho_{s,4}F_{2(4-3)-1} + \cdots)$$

Equation (9-14) is applied to each of the F factors in the series. The area ratio then cancels and gives the desired result,

$$F^s_{1-2} = F_{1-2} + \rho_{s,3}F_{1-2(3)} + \rho_{s,4}F_{1-2(4)} + \rho_{s,3}\rho_{s,4}F_{1-2(4-3)} + \cdots \tag{9-18}$$

Now look at (9-15) in more detail. Since $A_1F_{1-2} = A_2F_{2-1}$, and from (9-10) for one reflection $A_1F_{1(3)-2} = A_2F_{2(3)-1}$ and $A_1F_{1(4)-2} = A_2F_{2(4)-1}$, the equality in (9-16) reduces to

$$A_1(\rho_{s,3}\rho_{s,4}F_{1(3-4)-2} + \cdots + \rho^m_{s,3}\rho^n_{s,4}F_{1(3^m-4^n)-2} + \cdots)$$
$$= A_2(\rho_{s,3}\rho_{s,4}F_{2(4-3)-1} + \cdots + \rho^m_{s,3}\rho^n_{s,4}F_{2(4^n-3^m)-1} + \cdots) \tag{9-19}$$

Dividing by $\rho_{s,3}\rho_{s,4}$ results in

$$A_1(F_{1(3-4)-2} + \cdots + \rho^{m-1}_{s,3}\rho^{n-1}_{s,4}F_{1(3^m-4^n)-2} + \cdots)$$
$$= A_2[F_{2(4-3)-1} + \cdots + \rho^{m-1}_{s,3}\rho^{n-1}_{s,4}F_{2(4^n-3^m)-1} + \cdots) \tag{9-20}$$

This equality must hold in the limit as $\rho_{s,3}$ and $\rho_{s,4}$ approach zero so that

$$A_1F_{1(3-4)-2} = A_2F_{2(4-3)-1} \tag{9-21}$$

which is a geometric property of the system. A continuation of this reasoning leads to the general reciprocity relation

$$A_1F_{1(A-B-C-D\ldots)-2} = A_2F_{2(\ldots D-C-B-A)-1} \tag{9-22}$$

Note that combining (9-14) and (9-22) results in the identity

$$A_1F_{1(A-B-C-D\ldots)-2} = A_2F_{2(\ldots D-C-B-A)-1} = A_1F_{1-2(\ldots D-C-B-A)} \tag{9-23}$$

or

$$F_{1(A-B-C-D\ldots)-2} = F_{1-2(\ldots D-C-B-A)} \tag{9-24}$$

This latter relation can also be deduced directly from the fact that an image system can be constructed either starting with the real surface 1 and working toward image $2(\cdots D-C-B-A)$, or starting with image $1(A-B-C-D\cdots)$ and working toward real surface 2; the geometry of the construction will be identical in both systems. Thus, the configuration factors between the initial surface and the final surface must be the same.

9-4 NET-RADIATION METHOD IN ENCLOSURES HAVING SPECULAR AND DIFFUSE SURFACES

9-4.1 Enclosures with Plane Surfaces

In this section the radiation exchange theory in an enclosure composed of specularly and diffusely reflecting surfaces will be developed. The enclosure is composed of N surfaces, where d of the surfaces are diffuse reflectors and $N - d$ are specular. All the surfaces emit diffusely and are grey. Let the diffusely reflecting surfaces be numbered 1 through d, and the specularly reflecting surfaces $d + 1$ through N. If there are no external sources of unidirectional energy entering through a window and striking a specular surface, all the energy within the enclosure originates by surface emission that is diffuse. At each diffuse surface the reflected diffuse energy is combined with the emitted energy to form q_o, which is entirely diffuse. At each specular surface the only diffuse energy leaving is $\epsilon \sigma T^4$, since the reflected energy has a directional character. The transport between surfaces of the diffuse energies q_o and $\epsilon \sigma T^4$ is provided by the exchange factors. It is recalled that F_{A-B}^s gives the fraction of diffuse energy leaving surface A and arriving at surface B by the direct path and by all possible paths involving intermediate specular reflections. The reflected energy from the specular surfaces is already included in the F^s and hence does not have to be considered after the F^s have been obtained for the enclosure surfaces. Hence, by accounting for the fractions of q_o from the diffusely reflecting surfaces and fractions of $\epsilon \sigma T^4$ from the specularly reflecting surfaces that reach a particular surface, one accounts for all the incident energy, including the effect of both diffuse and specular reflections. Then at any surface, the incident energy is

$$q_{i,k} A_k = \sum_{j=1}^{d} q_{o,j} A_j F_{j-k}^s + \sigma \sum_{j=d+1}^{N} \epsilon_j T_j^4 A_j F_{j-k}^s \qquad 1 \leqslant k \leqslant N$$

After applying reciprocity (9-16), the areas cancel and $q_{i,k}$ becomes

$$q_{i,k} = \sum_{j=1}^{d} q_{o,j} F_{k-j}^s + \sigma \sum_{j=d+1}^{N} \epsilon_j T_j^4 F_{k-j}^s \qquad 1 \leqslant k \leqslant N \qquad (9\text{-}25)$$

For the diffuse surfaces it is convenient to use the heat-balance equations in the form of (8-1) and (8-2):

$$Q_k = q_k A_k = (q_{o,k} - q_{i,k}) A_k \qquad 1 \leqslant k \leqslant d \qquad (9\text{-}26)$$

$$q_{o,k} = \epsilon_k \sigma T_k^4 + (1 - \epsilon_k) q_{i,k} \qquad 1 \leqslant k \leqslant d \qquad (9\text{-}27)$$

For the specular reflecting surfaces, although these same heat balances apply, the $q_{o,k}$ is eliminated as this is composed of a combination of diffuse emission $\epsilon_k \sigma T_k^4$ and specular reflection $(1 - \epsilon_k) q_{i,k}$ and hence has a directional character that is not conveniently dealt with. Eliminating the $q_{o,k}$ from (9-26) and (9-27) gives

$$Q_k = A_k \epsilon_k (\sigma T_k^4 - q_{i,k}) \qquad d+1 \leqslant k \leqslant N \qquad\qquad (9\text{-}28)$$

Equations (9-25)-(9-28) can now be combined in various ways to obtain convenient equations to calculate the desired unknowns, depending on what quantities are specified. Consider now the case in which the temperatures are specified for all the surfaces and it is desired to obtain the net external energy Q_k added to each surface. Equation (9-25) is substituted into (9-27) to eliminate $q_{i,k}$ and obtain the following equation for each diffuse surface:

$$q_{o,k} - (1 - \epsilon_k) \sum_{j=1}^{d} q_{o,j} F_{k-j}^s = \epsilon_k \sigma T_k^4 + (1 - \epsilon_k)\sigma \sum_{j=d+1}^{N} \epsilon_j T_j^4 F_{k-j}^s \qquad 1 \leqslant k \leqslant d$$

$$(9\text{-}29)$$

This set of equations is solved for the q_o for the diffuse surfaces. This is somewhat simpler than for an enclosure having all diffuse surfaces, as there are now only d equations to solve simultaneously, rather than N equations. For each specular surface, the q_o's for the diffuse surfaces are used to obtain $q_{i,k}$ directly from (9-25):

$$q_{i,k} = \sum_{j=1}^{d} q_{o,j} F_{k-j}^s + \sigma \sum_{j=d+1}^{N} \epsilon_j T_j^4 F_{k-j}^s \qquad d+1 \leqslant k \leqslant N \qquad\qquad (9\text{-}30)$$

The net external energy added to each diffuse surface is obtained by eliminating $q_{i,k}$ from (9-26) and (9-27),

$$Q_k = A_k \frac{\epsilon_k}{1 - \epsilon_k} (\sigma T_k^4 - q_{o,k}) \qquad 1 \leqslant k \leqslant d \qquad\qquad (9\text{-}31)$$

and the Q_k to each specular surface is found from (9-28):

$$Q_k = A_k \epsilon_k (\sigma T_k^4 - q_{i,k}) \qquad d+1 \leqslant k \leqslant N \qquad\qquad (9\text{-}32)$$

Equations (9-28)-(9-32) are the general energy-interchange relations for enclosures made up of diffuse reflecting surfaces and specular reflecting surfaces.

If the kth diffuse surface is black, then $q_{o,k} = \sigma T_k^4$ and $1 - \epsilon_k = 0$, so that (9-31) is indeterminate. In this case, the following equation can be used:

$$Q_k = A_k (\sigma T_k^4 - q_{i,k}) \qquad\qquad (9\text{-}33)$$

where $q_{i,k}$ is found from (9-30) with $1 \leqslant k \leqslant d$.

If the heat input Q_k rather than T_k is specified for a diffuse surface $1 \leqslant k \leqslant d$, then T_k is unknown in (9-29). Equation (9-31) can be used to eliminate this unknown in terms of $q_{o,k}$ and the known Q_k.

If the heat input Q_k is specified for a specular surface, $d+1 \leqslant k \leqslant N$, then one of the T_j^4 in the last term of (9-29) will be unknown. Equation (9-32) is combined with (9-30) to eliminate $q_{i,k}$, which gives

$$\sigma T_k^4 - \frac{Q_k}{A_k \epsilon_k} = \sum_{j=1}^{d} q_{o,j} F_{k-j}^s + \sigma \sum_{j=d+1}^{N} \epsilon_j T_j^4 F_{k-j}^s \qquad d+1 \leqslant k \leqslant N \qquad (9\text{-}34)$$

Since Q_k is known, (9-34) can be combined with (9-29) to yield a simultaneous set of equations to determine the q_o of the diffusely reflecting surfaces and the T for the specularly reflecting surfaces having specified Q.

If all the surfaces are specular, then a simultaneous solution is not required. The $q_{i,k}$ are given by

$$q_{i,k} = \sigma \sum_{j=1}^{N} \epsilon_j T_j^4 F_{k-j}^s \qquad 1 \leqslant k \leqslant N \tag{9-35}$$

and the Q_k are then found from (9-32). By substituting (9-35) into (9-32) the Q_k can be found directly from the specified surface temperatures:

$$Q_k = \sigma A_k \epsilon_k (T_k^4 - \sum_{j=1}^{N} \epsilon_j T_j^4 F_{k-j}^s) \tag{9-36}$$

An alternative form of the enclosure equations can be found by using (9-31) and (9-32) to eliminate q_i and q_o from (9-29) and (9-30). This gives a set of equations, all of the same form, that directly relate the Q's and T's:

$$\frac{1}{\epsilon_k} \frac{Q_k}{A_k} - \sum_{j=1}^{d} \frac{Q_j}{A_j} \frac{1-\epsilon_j}{\epsilon_j} F_{k-j}^s = \sigma T_k^4 - \sum_{j=1}^{d} \sigma T_j^4 F_{k-j}^s - \sum_{j=d+1}^{N} \sigma \epsilon_j T_j^4 F_{k-j}^s \qquad 1 \leqslant k \leqslant N$$

$$\tag{9-37}$$

Equation (9-36) is the special case of (9-37) for $d = 0$.

Equation (9-37) can be used to obtain some relations between the F^s exchange factors analogous to the relation for diffuse surfaces $\sum_{j=1}^{N} F_{k-j} = 1$. Consider the situation when the entire enclosure is at uniform temperature. Then there is no net energy exchange and all the Q's are zero, so that (9-37) reduces to

$$\sum_{j=1}^{d} F_{k-j}^s + \sum_{j=d+1}^{N} \epsilon_j F_{k-j}^s = 1 \tag{9-38}$$

If all the surfaces in the enclosure are specular ($d = 0$), this further reduces to

$$\sum_{j=1}^{N} \epsilon_j F_{k-j}^s = \sum_{j=1}^{N} (1 - \rho_{s,j}) F_{k-j}^s = 1 \tag{9-39}$$

The set of enclosure equations is not difficult to solve after the exchange factors have been found; determining these factors, however may not be easy, depending on the complexity of the enclosure geometry. To illustrate the calculations, some specific enclosures will now be considered.

Figure 9-13 shows an enclosure composed of three plane surfaces at different uniform specified temperatures; two sides are diffuse reflectors, and the third is a specular reflector. In Fig. 9-13a, the energy arriving at surface 1 comes directly from the diffuse surfaces 2 and 3 without any intermediate specular reflections. Hence,

$$F_{2-1}^s = F_{2-1} \qquad \text{and} \qquad F_{3-1}^s = F_{3-1}$$

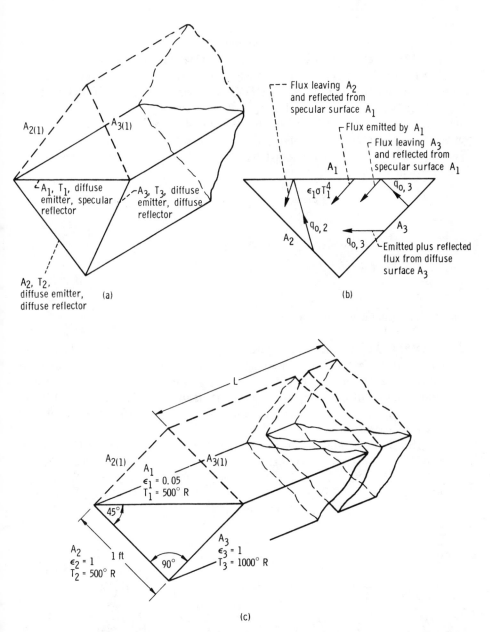

Figure 9-13 Enclosure having one specular reflecting surface and two surfaces that are diffuse reflectors. (*a*) General geometry; (*b*) energy fluxes that contribute to flux incident upon A_2; (*c*) enclosure for Example 9-6.

From reciprocity,

$$F_{1-2}^s = F_{1-2} \qquad \text{and} \qquad F_{1-3}^s = F_{1-3}$$

For surface 2 the incoming radiation is composed of four parts that originate as shown in Fig. 9-13b. The first is the diffuse energy originating from A_3 and going directly to A_2, which is $q_{o,3}A_3F_{3-2}$. The remaining three parts arrive by means of A_1 and consist of an emitted portion $\epsilon_1\sigma T_1^4 A_1 F_{1-2}$ and two specularly reflected portions. The latter arise from the energy leaving A_2 and A_3 that is specularly reflected to A_2 and will appear to come from the images $A_{2(1)}$ and $A_{3(1)}$ in Fig. 9-13a. The specularly reflected portions are $q_{o,2}\rho_{s,1}A_2F_{2(1)-2} + q_{o,3}\rho_{s,1}A_3F_{3(1)-2}$. Note that multiple specular reflections cannot occur when only one planar specular surface is present. The specular exchange factors are then

$$F_{1-2}^s = F_{1-2} \qquad F_{2-2}^s = \rho_{s,1}F_{2(1)-2} \qquad F_{3-2}^s = F_{3-2} + \rho_{s,1}F_{3(1)-2}$$

By using reciprocity, one obtains

$$F_{2-1}^s = F_{2-1} \qquad F_{2-2}^s = \rho_{s,1}F_{2-2(1)} \qquad F_{2-3}^s = F_{2-3} + \rho_{s,1}F_{2-3(1)}$$

Similarly, for surface 3,

$$F_{3-1}^s = F_{3-1} \qquad F_{3-2}^s = F_{3-2} + \rho_{s,1}F_{3-2(1)} \qquad F_{3-3}^s = \rho_{s,1}F_{3-3(1)}$$

A numerical example will now be given, using these factors.

EXAMPLE 9-6 An enclosure is made up of three sides as shown in Fig. 9-13c. The length L is sufficiently long so that the triangular ends can be neglected in the radiative heat balances. Two of the surfaces are black, and the third is a gray diffuse emitter of emissivity $\epsilon_1 = 0.05$. What is the heat added per foot of length to each surface for each of the two cases: (1) area 1 is a diffuse reflector and (2) area 1 is a specular reflector.

The configuration factors are computed first. From symmetry $F_{1-2} = F_{1-3}$. Also $F_{1-2} + F_{1-3} = 1$, so that $F_{1-2} = F_{1-3} = \frac{1}{2}$. From reciprocity $F_{2-1} = A_1F_{1-2}/A_2 = \sqrt{2}/2 = F_{3-1}$. Now $F_{2-1} + F_{2-3} = 1$. Hence $F_{2-3} = 1 - \sqrt{2}/2 = F_{3-2} = F_{2-2(1)} = F_{3-3(1)}$. Finally, $F_{3-2(1)} = F_{2-3(1)} = 1 - F_{3-2} - F_{3-3(1)} = \sqrt{2} - 1$.

For case 1, apply Eq. (8-18) to obtain

$$\frac{Q_1}{\sigma\sqrt{2}}\frac{1}{0.05} = 500^4 - \tfrac{1}{2}(500)^4 - \tfrac{1}{2}(1000)^4$$

$$-\frac{Q_1}{\sigma\sqrt{2}}\frac{\sqrt{2}}{2}\frac{1-0.05}{0.05} + \frac{Q_2}{\sigma(1)} = \frac{-\sqrt{2}}{2}500^4 + 500^4 - \left(1 - \frac{\sqrt{2}}{2}\right)1000^4$$

$$-\frac{Q_1}{\sigma\sqrt{2}}\frac{\sqrt{2}}{2}\frac{1-0.05}{0.05} + \frac{Q_3}{\sigma(1)} = \frac{-\sqrt{2}}{2}500^4 - \left(1 - \frac{\sqrt{2}}{2}\right)500^4 + 1000^4$$

The solution of these three equations yields the Q's per foot of enclosure length as $Q_1 = -57$ Btu/h, $Q_2 = -1018$ Btu/h, and $Q_3 = 1075$ Btu/h. The heat supplied to A_3 is removed from A_1 and A_2. The amount removed from A_1 is small because A_1 is a poor absorber.

For case 2 apply Eq. (9-29) to compute $q_{o,2}$ and $q_{o,3}$. Since $\epsilon_2 = \epsilon_3 = 1$, this yields simply $q_{o,2} = \sigma T_2{}^4$ and $q_{o,3} = \sigma T_3{}^4$, which would be expected for the outgoing fluxes from black surfaces. Then Eq. (9-25) yields the q_i for each surface as

$$\frac{q_{i,1}}{\sigma} = \tfrac{1}{2}(500)^4 + \tfrac{1}{2}(1000)^4 = 5313 \times 10^8 \; ^\circ R^4$$

$$\frac{q_{i,2}}{\sigma} = 0.05(500)^4 \frac{\sqrt{2}}{2} + 500^4(1 - 0.05)\left(1 - \frac{\sqrt{2}}{2}\right)$$

$$+ 1000^4 \left[1 - \frac{\sqrt{2}}{2} + (1 - 0.05)(\sqrt{2} - 1)\right] = 7060 \times 10^8 \; ^\circ R^4$$

$$\frac{q_{i,3}}{\sigma} = 0.05(500)^4 \frac{\sqrt{2}}{2} + 500^4 \left[1 - \frac{\sqrt{2}}{2} + (1 - 0.05)(\sqrt{2} - 1)\right]$$

$$+ 1000^4(1 - 0.05)\left(1 - \frac{\sqrt{2}}{2}\right) = 3234 \times 10^8 \; ^\circ R^4$$

With q_i known for each of the surfaces, Eqs. (9-32) and (9-33) are applied to find Q. This yields, per foot of enclosure length, $Q_1 = -57$ Btu/h, $Q_2 = -1113$ Btu/h, and $Q_3 = 1170$ Btu/h.

Comparing cases 1 and 2 reveals that, by making A_1 specular, the heat transferred from A_3 to A_2 is increased from 1018 to 1113 Btu/h, or an increase of 10%.

To further demonstrate the radiative analysis in an enclosure having some specular reflecting surfaces, a rectangular geometry is shown in Fig. 9-14. All the surfaces are diffuse emitters; two of the surfaces are diffuse reflectors, while the remaining two are specular. Shown dashed are the reflected images. The reflection process continues until all the outer perimeter enclosing the composite of original enclosure plus reflected images is made up of either diffuse (or nonreflecting, such as an opening) surfaces or images of diffuse surfaces.

For the enclosure in Fig. 9-14, the first step is to use (9-29) to obtain $q_{o,1}$ and $q_{o,2}$ for the two diffuse areas. To obtain the exchange factors, consider first F_{1-1}^s. Part of the energy leaving A_1 returns to A_1 by three paths: direct reflection from A_3, reflection from A_3 to A_4 and then to A_1, and reflection from A_4 to A_3 and then to A_1. Thus the fraction of the energy leaving A_1 that returns to A_1 is

$$F_{1-1}^s = \rho_{s,3}F_{1(3)-1} + \rho_{s,3}\rho_{s,4}F_{1(3-4)-1} + \rho_{s,4}\rho_{s,3}F_{1(4-3)-1}$$

The factor $F_{1(3-4)-1}$ is the view factor by which $A_{1(3-4)}$ is viewed from A_1 through A_4 and then $A_{3(4)}$, which are the reflection areas by means of which the $A_{1(3-4)}$ image was formed. Similarly, $F_{1(4-3)-1}$ is the view factor by which the same area $A_{1(3-4)}$ is viewed from A_1 through A_3 and then $A_{4(3)}$.

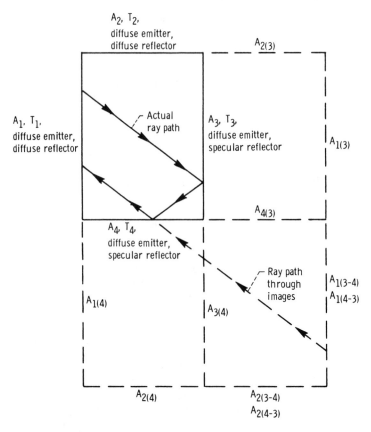

Figure 9-14 Rectangular enclosure and reflected images when two adjacent surfaces are specular reflectors and the other two are diffuse reflectors.

Radiation leaving A_2 reaches A_1 along four paths: direct exchange, reflection from A_3, reflection from A_4, and reflection from A_3 to A_4. No energy from A_2 reaches A_1 by means of reflections from A_4 and then A_3. This is because A_1 cannot view the image $A_{2(4-3)}$ through area A_3. This yields

$$F_{2-1}^s = F_{2-1} + \rho_{s,3}F_{2(3)-1} + \rho_{s,4}F_{2(4)-1} + \rho_{s,3}\rho_{s,4}F_{2(3-4)-1}$$

The diffuse energy leaving the specular surface A_3 (and similarly for A_4) consists only of the emitted energy $\epsilon_3 A_3 \sigma T_3^4$. There are two paths by which some of this will reach A_1: by direct exchange, and by means of specular reflection from A_4. This yields

$$F_{3-1}^s = F_{3-1} + \rho_{s,4}F_{3(4)-1} \qquad F_{4-1}^s = F_{4-1} + \rho_{s,3}F_{4(3)-1}$$

After applying view-factor reciprocity, the factors are substituted into (9-29) to yield

$$q_{o,1} - (1 - \epsilon_1)\{q_{o,1}[\rho_{s,3}F_{1-1(3)} + \rho_{s,3}\rho_{s,4}(F_{1-1(3-4)} + F_{1-1(4-3)})]$$
$$+ q_{o,2}(F_{1-2} + \rho_{s,3}F_{1-2(3)} + \rho_{s,4}F_{1-2(4)} + \rho_{s,3}\rho_{s,4}F_{1-2(3-4)})\} \qquad (9\text{-}40)$$
$$= \epsilon_1 \sigma T_1^4 + (1 - \epsilon_1)\sigma[\epsilon_3 T_3^4(F_{1-3} + \rho_{s,4}F_{1-3(4)}) + \epsilon_4 T_4^4(F_{1-4} + \rho_{s,3}F_{1-4(3)})]$$

In a similar fashion, considering $q_{i,2}$ for surface 2 yields

$$q_{o,2} - (1 - \epsilon_2)\{q_{o,1}(F_{2-1} + \rho_{s,3}F_{2-1(3)} + \rho_{s,4}F_{2-1(4)} + \rho_{s,3}\rho_{s,4}F_{2-1(4-3)})$$
$$+ q_{o,2}[\rho_{s,4}F_{2-2(4)} + \rho_{s,3}\rho_{s,4}(F_{2-2(4-3)} + F_{2-2(3-4)})]\}$$
$$= \epsilon_2 \sigma T_2^4 + (1 - \epsilon_2)\sigma[\epsilon_3 T_3^4(F_{2-3} + \rho_{s,4}F_{2-3(4)}) + \epsilon_4 T_4^4(F_{2-4} + \rho_{s,3}F_{2-4(3)})] \tag{9-41}$$

Equations (9-40) and (9-41) are solved simultaneously for $q_{o,1}$ and $q_{o,2}$.

For the two specular surfaces, the $q_{i,3}$ and $q_{i,4}$ can be found as soon the q_o's for the diffuse surfaces are known. From (9-30),

$$q_{i,3} = q_{o,1}(F_{3-1} + \rho_{s,4}F_{3-1(4)}) + q_{o,2}(F_{3-2} + \rho_{s,4}F_{3-2(4)}) + \epsilon_4 \sigma T_4^4 F_{3-4} \tag{9-42}$$

$$q_{i,4} = q_{o,1}(F_{4-1} + \rho_{s,3}F_{4-1(3)}) + q_{o,2}(F_{4-2} + \rho_{s,3}F_{4-2(3)}) + \epsilon_3 \sigma T_3^4 F_{4-3} \tag{9-43}$$

With the q_o for the diffuse surfaces and the q_i for the specular surfaces now known, the Q's to maintain the specified temperatures of the surfaces can be found from (9-31) and (9-32).

9-4.2 Curved Specular Reflecting Surfaces

In the previous discussion the specular surfaces were planar. Here curved specular reflecting surfaces will be considered, and in this instance the geometry of the reflected images can become quite complex. To demonstrate some of the basic ideas we consider the relatively simple case of radiation exchange within a specular tube [10], as shown in Fig. 9-15.

It is assumed that the imposed temperature or heating conditions depend only on axial position and are independent of the location around the tube circumference. To compute the radiative exchange within the tube for axisymmetric heating conditions, it is necessary to have the configuration factor between two ring elements on the tube wall. The direct exchange (Fig. 9-15a) is governed by the factor (see Example 8-9, and note that $|\eta - \xi|$ in that example is equal to X/D here):

$$dF_{dX_1 - dX} = \left\{1 - \frac{(X/D)^3 + 3X/2D}{[(X/D)^2 + 1]^{3/2}}\right\} dX$$

Figure 9-15b illustrates the configuration factor for one reflection. Because of the symmetry of the tube, all the radiation from dX_1 that reaches dX by one reflection will be reflected from a ring element halfway between dX_1 and dX. The ring at $X/2$ is only $dX/2$ wide so that the beam subtending it will spread to a width dX at the location X. The configuration factor for one reflection is then the factor between dX_1 and the dashed element $dX/2$:

$$dF_{dX_1 - dX/2} = \left\{1 - \frac{(X/2D)^3 + 3X/4D}{[(X/2D)^2 + 1]^{3/2}}\right\}\frac{dX}{2}$$

In a similar fashion, the geometric factor for exchange between dX_1 and dX by two

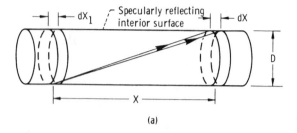

(a)

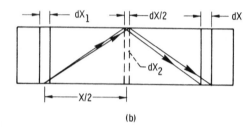

(b)

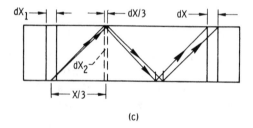

(c)

Figure 9-15 Radiation exchange within specularly reflecting cylindrical tube. (a) Direct exchange between two ring elements; (b) exchange by one reflection; (c) exchange by two reflections.

reflections is given by

$$dF_{dX_1-dX/3} = \left\{ 1 - \frac{(X/3D)^3 + 3X/6D}{[(X/3D)^2 + 1]^{3/2}} \right\} \frac{dX}{3}$$

and for n reflections

$$dF_{dX_1-dX/(n+1)} = \left(1 - \frac{[X/(n+1)D]^3 + 3X/2(n+1)D}{\{[X/(n+1)D]^2 + 1\}^{3/2}} \right) \frac{dX}{n+1}$$

In general, the geometric factor for *any* number of reflections can be found by considering the exchange between the originating element (dX_1 in this case) and the element (call it dX_2) from which the *first* reflection is made (the dashed element in Fig. 9-15b and c). This is because the fraction of energy leaving dX_1 in the solid angle subtended by dX_2 remains the same through the succeeding reflections along the path to dX.

At each reflection the energy must be multiplied by the specular reflectivity ρ_s. If all the contributions are summed, the fraction of energy leaving dX_1 that reaches dX

by direct exchange and all reflection paths provides the specular exchange factor

$$dF^s_{dX_1-dX} = \sum_{n=0}^{\infty} \rho_s^n \left(1 - \frac{[X/(n+1)D]^3 + 3X/2(n+1)D}{\{[X/(n+1)D]^2 + 1\}^{3/2}} \right) \frac{dX}{n+1} \quad (9\text{-}44)$$

When the geometry is even slightly more involved than the cylindrical geometry, the reflection patterns can become quite complex. Some further specific examples of radiation within a specular conical cavity and a specular cylindrical cavity with a specular end plane are given in [11]; a more generalized treatment of nonplanar reflections is given in [12]. An approximate analysis of the transfer through specular passages is given in [13]. It is based on estimating the average number of reflections that occur during transmission.

EXAMPLE 9-7 A cylindrical cavity has a specularly reflecting cylindrical wall and base (Fig. 9-16a). Determine the fraction of radiation from ring element dX_1 that reaches dX by means of one reflection from the base with reflectivity $\rho_{s,1}$, and one reflection from the cylindrical wall with reflectivity $\rho_{s,2}$.

As shown in Fig. 9-16b, for this geometry the reflected radiation from the base can be regarded as originating from an image of dX_1. The second reflection will occur from an element of width $dX/2$ located midway between the image dX_1 and dX. The desired radiation fraction is given by the view factor from the

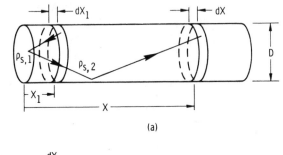

(a)

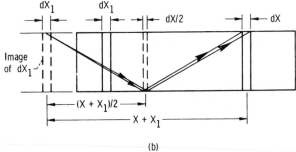

(b)

Figure 9-16 Reflection in cylindrical cavity with specular curved wall and base. (a) Cavity geometry; (b) image of dX_1 formed by reflection in cavity base.

image dX_1 to the dashed ring area $dX/2$ multiplied by the two reflectivities:

$$\rho_{s,1}\rho_{s,2}dF_{dX_1-dX} = \rho_{s,1}\rho_{s,2}\left(1 - \frac{[(X+X_1)/2D]^3 + 3(X+X_1)/4D}{\{[(X+X_1)/2D]^2 + 1\}^{3/2}}\right)\frac{dX}{2}$$

Another type of curved specular surface of practical importance is a paraboloidal mirror such as is used in a solar furnace. The mirror axis is aligned in the direction of the sun as shown in Fig. 9-17a, and a receiver is placed at the focal plane. It is desired to estimate the receiver temperature.

When *parallel* rays reflect from a perfect concentrator in the form of a paraboloid, the rays all go through the focal point. The sun's rays, however, are not quite parallel. At the earth-sun distance the solar diameter subtends about 32' of angle. The effective blackbody temperature varies somewhat across the surface of the solar disk. The radiant emission decreases somewhat at the outer edges of the disk (limb darkening), but this will not be accounted for here. Including this effect can increase the calcu-

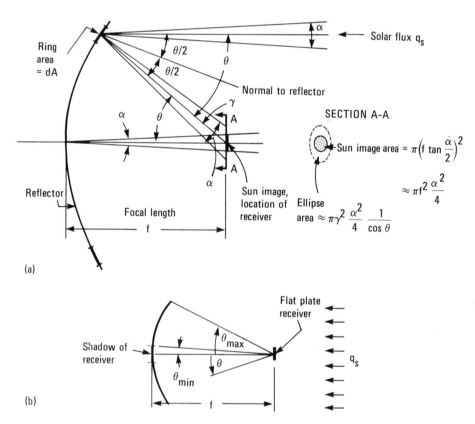

Figure 9-17 Paraboloidal-mirror solar concentrator. (a) Images formed at focal plane; (b) receiver at focal plane.

lated receiver temperature in a solar furnace by a few percent because of the higher temperatures in the central region of the solar disk.

To obtain a sharp image of the sun, the concentrator mirror should have a long focal length relative to its diameter. Only the portion of the radiation arriving at the mirror within a diameter that is small compared with the mirror focal length reflects in a direction sufficiently close to being normal to the focal plane to create a sharp image. A long focal length relative to the diameter may not, however, be a practical design. When the focal length is short, elliptical images are produced by reflections from all but a small region near the center of the mirror. For undistorted reflection from the center of the mirror, the diameter of the sun image at the focal plane is given by $2f \tan 16' = f/107.3$, where f is the focal length. It is desired to compute the flux received within this sun image, as this is where the highest receiver temperature can be obtained. Surrounding the sun image, there will be less flux received as a result of the elliptical images that are formed at the focal plane.

For the ellipitical image shown in Section A-A, Fig. 9-17a, the approximate fraction of incident energy q_s in angle α that lies within the sun image (assuming uniform illumination over the ellipse) is $(\pi f^2 \alpha^2/4)/(\pi \gamma^2 \alpha^2/4 \cos \theta) = (f/\gamma)^2 \cos \theta$. The energy received on a ring element dA of the reflector is $q_s dA \cos (\theta/2)$. If ρ_m is the mirror reflectivity, the resulting energy rate incident at the focal plane within the sun image is $dQ = q_s dA \cos (\theta/2) \rho_m (f/\gamma)^2 \cos \theta$. The area of a ring on the mirror is

$$dA = 2\pi\gamma \sin \theta \ [(d\gamma)^2 + (\gamma \, d\theta)^2]^{1/2} = 2\pi\gamma \sin \theta \ \gamma \, d\theta \left[\left(\frac{d\gamma}{\gamma \, d\theta} \right)^2 + 1 \right]^{1/2}$$

For a paraboloid γ, f and θ are related by $\gamma = 2f/(1 + \cos \theta)$. This yields $(1/\gamma)(d\gamma/d\theta) = \tan (\theta/2)$ so that dA becomes

$$dA = 2\pi\gamma^2 \sin \theta \, d\theta \left(\tan^2 \frac{\theta}{2} + 1 \right)^{1/2} = 2\pi\gamma^2 \sin \theta \, d\theta \, \sec \frac{\theta}{2}$$

Substitute into dQ to obtain

$$dQ = 2\pi f^2 \rho_m q_s \sin \theta \cos \theta \, d\theta$$

Now consider the heat balance on a flat-plate receiver covering the sun's image at the concentrator focal plane, as shown in Fig. 9-17b. Assume that the absorptivity in the range of θ from θ_{min} to θ_{max} can be approximated by a cosine function (in many instances a constant value would be adequate). Then the energy absorbed within the sun image is

$$\int dQ_{absorbed} = 2\pi f^2 \rho_m q_s \alpha_n \int_{\theta_{min}}^{\theta_{max}} \cos^2 \theta \sin \theta \, d\theta$$

$$= \frac{2}{3} \pi f^2 \rho_m q_s \alpha_n (\cos^3 \theta_{min} - \cos^3 \theta_{max}) \qquad (9\text{-}45)$$

Let the receiver surface be insulated on the back side, and have no heat loss except by radiation (no convection or conduction). Since the receiver will be at high temperature and hence will emit a spectrum with considerable energy at short wavelengths, wavelength-selective effects will be neglected here. Then $\epsilon(\theta) = \alpha(\theta) = \alpha_n \cos\theta$ and the emission loss is

$$2\alpha_n \sigma T_{eq}{}^4 \frac{\pi f^2 \alpha^2}{4} \int_0^{\pi/2} \cos^2\theta \sin\theta \, d\theta = \frac{2}{3}\alpha_n \sigma T_{eq}{}^4 \frac{\pi f^2 \alpha^2}{4} \qquad (9\text{-}46)$$

A heat balance is formed by equating the absorbed and emitted energies (9-45) and (9-46):

$$\frac{2}{3}\pi f^2 q_s \rho_m \alpha_n (\cos^3\theta_{min} - \cos^3\theta_{max}) = \frac{2}{3}\alpha_n \sigma T_{eq}{}^4 \frac{\pi f^2 \alpha^2}{4}$$

This is solved for the desired equilibrium temperature of the receiver,

$$T_{eq} = \rho_m{}^{1/4} \left(\frac{q_s}{\sigma}\right)^{1/4} \left(\frac{4}{\alpha^2}\right)^{1/4} (\cos^3\theta_{min} - \cos^3\theta_{max})^{1/4} \qquad (9\text{-}47)$$

Additional information is given [14, 15].

EXAMPLE 9-8 Estimate the receiver temperature if there is 30% attenuation of the solar flux by the atmosphere, the mirror reflectivity is 0.9, and $\theta_{min} \approx 0$, $\theta_{max} = 60°$.

If T_s, r_s, and d_s are the effective temperature, radius, and distance to the sun, the q_s with 30% atmospheric attenuation is given by

$$q_s = 0.7\sigma T_s{}^4 \left(\frac{r_s}{d_s}\right)^2 = 0.7\sigma T_s{}^4 \tan^2\frac{\alpha}{2} \approx 0.7\sigma T_s{}^4 \frac{\alpha^2}{4}$$

Inserting this into Eq. (9-47) gives

$$T_{eq} = (0.7\rho_m)^{1/4} T_s (1 - \cos^3\theta_{max})^{1/4}$$

$$\frac{T_{eq}}{T_s} = (0.7 \times 0.9)^{1/4} (1 - \cos^3 60)^{1/4} = 0.862$$

If $T_s = 5780$ K, then $T_{eq} = 4980$ K. This is somewhat higher than achieved in practice because the collector has been assumed to be of perfect shape. There are usually losses as a result of reflector distortions.

9-5 CONCLUDING REMARKS

This chapter presents treatments of radiative interchange between specularly reflecting surfaces and in enclosures containing both specularly and diffusely reflecting surfaces. In many instances, as in Example 9-6, the interchange of energy in enclosures is modified only a small amount by the consideration of specular in place of diffuse reflecting surfaces; however, in certain configurations, for example those found in the design of solar concentrators and furnaces, large effects of specular reflection are present. Bobco [16], Sparrow and Lin [17], Sarofim and Hottel [18], and Mahan [19] et al. have examined radiative exchange in enclosures involving surfaces with a reflectivity having both diffuse *and* specular components. Schornhorst and Viskanta [20] compared experimental and analytical results for radiant exchange between various types of surfaces and found that *regardless* of the presence of specular surfaces, the diffuse-surface analysis agreed best with experimental results.

Another remark may be apropos at this point. It has sometimes been implied that the actual energy transfer between two real surfaces can be bracketed by calculation of two limiting magnitudes: (1) interchange between diffuse surfaces of the same total hemispherical emissivities as the real surfaces and (2) interchange between specularly reflecting surfaces of the same total hemispherical emissivities as the real surfaces. This implication is not always true, however. Consider a surface that has a reflectivity as given by Fig. 9-18a (this is the type of reflectivity expected for the

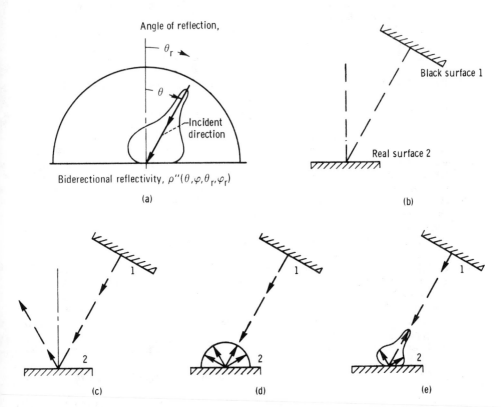

Figure 9-18 Radiant interchange with various idealizations of surface properties. (*a*) Bidirectional reflectivity of real surface; (*b*) geometry of interchange; (*c*) surface 2 reflects specularly; (*d*) surface 2 reflects diffusely; (*e*) surface 2 reflects with real properties.

surface of the moon; see Sec. 5-4.3). Now consider the radiant exchange between this real surface 2 and a black surface 1 as shown in Fig. 9-18*b*. If surface 2 is given specular properties, it will return no energy to the black surface by reflection (Fig. 9-18*c*). If given diffuse properties, it will return a portion of the incident energy by reflection (Fig. 9-18*d*). If allowed to take on its real directional properties, however, it will reflect more energy to the black surface than *either* of the so-called limiting ideal surfaces (Fig. 9-18*e*). Thus, the ideal directional surfaces do not constitute limiting cases for energy transfer in general. Figure 10-11 demonstrates another case in which diffuse and specular properties do not provide limiting solutions. At best, calculations based on specular and diffuse assumptions for the surface characteristics give some indication of the possible magnitude of directional effects. Within enclosures, these directional effects may be small because of the many reflections taking place between the surfaces.

REFERENCES

1. Christiansen, C.: Absolute Determination of the Heat Emission and Absorption Capacity, *Ann. Phys. Wied.,* vol. 19, pp. 267–283, 1883.
2. Saunders, O. A.: Notes on Some Radiation Heat Transfer Formulae, *Proc. Phys. Soc. London,* vol. 41, pp. 569–575, 1929.
3. Cravalho, E. G., C. L. Tien, and R. P. Caren: Effect of Small Spacings on Radiative Transfer between Two Dielectrics, *J. Heat Transfer,* vol. 89, no. 4, pp. 351–358, 1967.
4. Polder, D., and M. Van Hove: Theory of Radiative Heat Transfer between Closely Spaced Bodies, *Phys. Rev. B,* vol. 4, no. 10, pp. 3303–3314, 1971.
5. Born, Max, and Emil Wolf: "Principles of Optics," 2d ed., The Macmillan Company, New York, 1964.
6. Stone, John M.: "Radiation and Optics," McGraw-Hill Book Company, New York, 1963.
7. Howell, John R., and Morris Perlmutter: Directional Behavior of Emitted and Reflected Radiant Energy from a Specular, Gray, Asymmetric Groove, *NASA* TN D-1874, 1963.
8. Winston, Roland: Principles of Solar Concentrators of a Novel Design, *Sol. Energy,* vol. 16, no. 2, pp. 89–95, 1974.
9. Mannan, K. D., and L. S. Cheema: Compound-wedge Cylindrical Stationary Concentrator, *Sol. Energy,* vol. 19, no. 6, pp. 751–754, 1977.
10. Perlmutter, M., and R. Siegel: Effect of Specularly Reflecting Gray Surface on Thermal Radiation through a Tube and from Its Heated Wall, *J. Heat Transfer,* vol. 85, no. 1, pp. 55–62, 1963.
11. Lin, S. H., and E. M. Sparrow: Radiant Interchange among Curved Specularly Reflecting Surfaces–Application to Cylindrical and Conical Cavities, *J. Heat Transfer,* vol. 87, no. 2, pp. 299–307, 1965.
12. Plamondon, J. A., and T. E. Horton: On the Determination of the View Function to the Images of a Surface in a Nonplanar Specular Reflector, *Int. J. Heat Mass Transfer,* vol. 10, no. 5, pp. 665–679, 1967.
13. Rabl, Ari: Radiation Transfer through Specular Passages – A Simple Approximation, *Int. J. Heat Mass Transfer,* vol. 20, no. 4, pp. 323–330, 1977.
14. Cobble, M. H.: Theoretical Concentrations for Solar Furnaces, *Sol. Energy,* vol. 5, no. 2, pp. 61–72, 1961.
15. Kamada, O.: Theoretical Concentration and Attainable Temperature in Solar Furnaces, *Sol. Energy,* vol. 9, no. 1, pp. 39–47, 1965.
16. Bobco, R. P.: Radiation Heat Transfer in Semigray Enclosures with Specularly and Diffusely Reflecting Surfaces, *J. Heat Transfer,* vol. 86, no. 1, pp. 123–130, 1964.

17. Sparrow, E. M., and S. H. Lin: Radiation Heat Transfer at a Surface Having Both Specular and Diffuse Reflectance Components, *Int. J. Heat Mass Transfer*, vol. 8, no. 5, pp. 769–779, 1965.
18. Sarofim, A. F., and H. C. Hottel: Radiative Exchange among Non-Lambert Surfaces, *J. Heat Transfer*, vol. 88, no. 1, pp. 37–44, 1966.
19. Mahan, J. R., J. B. Kingsolver, and D. T. Mears: Analysis of Diffuse-Specular Axisymmetric Surfaces with Application to Parabolic Reflectors, *J. Heat Transfer*, vol. 101, no. 4, pp. 689–694, 1979.
20. Schornhorst, J. R., and R. Viskanta: An Experimental Examination of the Validity of the Commonly Used Methods of Radiant Heat Transfer Analysis, *J. Heat Transfer*, vol. 90, no. 4, pp. 429–436, 1968.

PROBLEMS

1 Derive Eqs. (9-5) and (9-6) for concentric radiation shields. Derive an equation for the temperature of the nth shield.

2 Obtain the result for $F_{1-1(2)}$ in Example 9-5 by use of the crossed-string method.

3 Two dielectric media, both having $n = 1.25$, are spaced 0.02 cm apart in vacuum and have temperatures as shown. To what extent is the heat flow across the gap influenced by wave

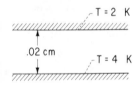

interference and radiation tunneling? How large must the spacing be to reduce these effects to 5% of the ordinary radiative transfer?

Answer: 38% increase, 0.093 cm.

4 Obtain an analytical expression for the view factor $F_{1-1(2)}$ between the cylinder A_1 and its image $A_{1(2)}$ for each of the two situations shown (the geometries are two-dimensional).

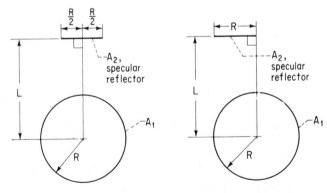

5 A long black cylinder is partially surrounded by two plane specular surfaces as shown. What is the rate of heat loss from the cylinder in terms of the quantities shown?

$$Answer: \quad \frac{Q}{Length} = 2\pi R\sigma T_1^{\,4} \left\{ 1 - \frac{2}{\pi} \left[\left(\frac{L^2}{R^2} - 1 \right)^{1/2} + \sin^{-1} \frac{R}{L} - \frac{L}{R} \right] \right\}.$$

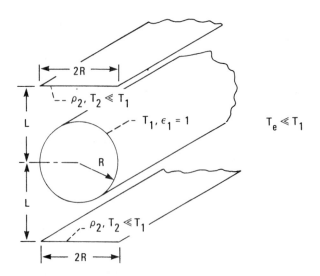

6   An enclosure is made up of two specular and two diffuse surfaces as shown. Draw a diagram of all the images that are needed to determine the energy-exchange process. Then write the equations for F_{1-2}^S and F_{1-3}^S in terms of the required specular configuration factors and reflectivities (that is, $F_{1-2}^S = F_{1-2} + \rho_{s,3}F_{1-2(3)} + \cdots$). Now write the set of energy-exchange equations for finding Q_1, Q_2, Q_3, Q_4.

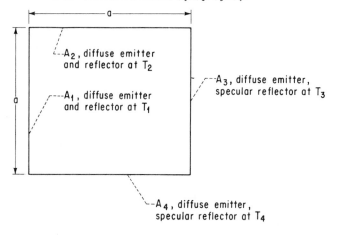

7   An equilateral triangular enclosure of infinite length has black surfaces 1 and 2 and a specularly reflecting surface 3 with reflectivity $\rho_{s,3} = 0.8$. Find the values of F_{1-1}^S, F_{1-2}^S, and F_{1-3}^S.
    Answer: 0.1072, 0.7928, 0.5000.

8   (a) What is the value of the summation $\sum_{j=1}^N F_{1-j}^S$ for Prob. 7?
    (b) What is the value of the summation $\sum_{j=1}^N (1 - \rho_{s,j}) F_{1-j}^S$ for Prob. 7?
    (c) Explain the results of parts a and b in terms of the definition of F_{1-j}^S. Is the result of part b a general relation for all specular enclosures?
       Answer: (a) 1.40;   (b) 1.00.

9 An enclosure is made up of three sides as shown. The length L is sufficiently long so that the triangular ends can be neglected in the radiation heat balances. Two of the surfaces are black, and the third is a diffuse gray emitter of emissivity $\epsilon_1 = 0.05$. What is the heat added per meter along the length L to each surface because of radiation exchange within the enclosure for each of the two cases: (a) area 1 is a diffuse reflector, and (b) area 1 is a specular reflector?

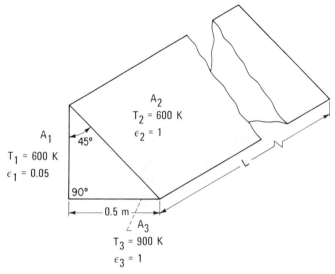

10 Compute the specular exchange factor F_{2-1}^s for the two-dimensional rectangular enclosure shown. All surfaces are gray and are diffuse emitters. Surfaces A_1 and A_2 are diffuse reflectors, while A_3 and A_4 are specular reflectors with $\rho_{s,3} = 0.8$ and $\rho_{s,4} = 0.9$.

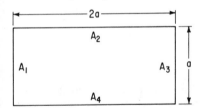

11 An infinitely long square bar, 1 m on a side, is enclosed by an infinitely long concentric cylinder of 1 m radius. The temperatures and emissivities of the bar and cylinder are, respectively, T_b, ϵ_b, and T_c, ϵ_c. Find the rate at which radiant energy is exchanged between the two surfaces per unit of length if

(a) Both are diffuse.
(b) Both are specular.
(c) The bar is diffuse, and the cylinder is specular.

    Answer: (a) $4\sigma(T_b{}^4 - T_c{}^4)/[1/\epsilon_b + (2/\pi)(1/\epsilon_c - 1)]$;
    (b), (c) $4\sigma(T_b{}^4 - T_c{}^4)/(1/\epsilon_b + 1/\epsilon_c - 1)$.

12 A two-dimensional rectangular enclosure has gray surfaces that are all diffuse emitters. Two opposing surfaces are specular reflectors, while the other two reflect diffusely. Write the set of energy equations to determine Q_1, Q_2, Q_3, and Q_4, including the expressions for the F^S factors in terms of the F factors. (*Note:* the F^S will each consist of an infinite sum.)

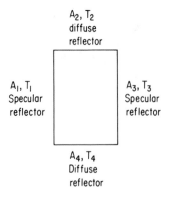

A₂, T₂
diffuse
reflector

A₁, T₁
Specular
reflector

A₃, T₃
Specular
reflector

A₄, T₄
Diffuse
reflector

13 An enclosure of equilateral triangular cross section and of infinite length has two diffuse reflecting surfaces and one specularly reflecting surface that is perfectly insulated on the outside (that is, $q_3 = 0$). All surfaces are gray and are diffuse emitters. Compute T_3 and the Q added to A_1 and A_2 as a result of radiation exchange within the enclosure for the conditions shown. (For simplicity do not subdivide the surface areas.)

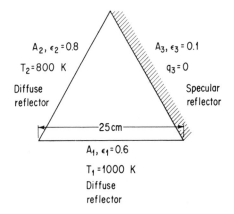

$A_2, \epsilon_2 = 0.8$

$T_2 = 800$ K

Diffuse
reflector

$A_3, \epsilon_3 = 0.1$

$q_3 = 0$

Specular
reflector

—25 cm—

$A_1, \epsilon_1 = 0.6$

$T_1 = 1000$ K

Diffuse
reflector

14 Two parallel plates as shown in Prob. 11 of Chap. 8 are of finite width (infinite length into the paper). Both plates are perfectly insulated on the outside. Plate 1 is uniformly heated electrically with heat flux q_e. Plate 2 has no externally supplied heat input ($q_2 = 0$). The environment is at zero temperature. Plate 1 is black, while plate 2 is a gray diffuse emitter and specular reflector with emissivity ϵ_2. Derive an integral equation formulation for the surface temperature distributions. Compare with the results in Prob. 11a of Chap. 8.

THE EXCHANGE OF THERMAL RADIATION BETWEEN NONDIFFUSE NONGRAY SURFACES

10-1 INTRODUCTION

The analysis of radiation exchange within enclosures in Chaps. 8 and 9 was restricted to either black or gray enclosure surfaces. If gray, the surfaces were assumed to both emit and reflect diffusely, or to be diffusely emitting and specularly reflecting. The additional restriction was sometimes made that the radiative properties were independent of temperature. As shown by the graphs of real properties in Chap. 5, most engineering materials deviate (in some instances radically) from the idealizations of being black, gray, diffuse, specular, or having temperature-independent radiative properties. In many practical engineering situations the assumption of idealized surfaces is made to simplify the computations. This is often the most reasonable approach for two reasons. First, the radiative properties are not known to high accuracy, especially with regard to their detailed dependence on wavelength and direction; hence, performing a refined computation would be fruitless when only crude property data are available. Second, in an enclosure, the many reflections and rereflections tend to average out radiative nonuniformities; for example, the radiation leaving (emitted plus reflected) a directionally emitting surface may be fairly diffuse if it consists mostly of reflected energy arising from radiation incident from all directions.

To gain some insight as to where simplifying assumptions are at all reasonable, it is necessary to carry out some exchange computations using as exact a solution procedure as possible. Then the results of those computations can be compared with results from simplified methods such as those in Chaps. 8 and 9. To provide the tools for making refined computations, some methods of treating the radiative interchange between nonideal surfaces will be examined in this chapter. Analysis of such problems is inherently more difficult than for ideal surfaces, and a complete treatment of real surfaces including all variations, while possible in principle when all radiative properties are known, is seldom attempted or justified. As stated earlier, the directional spectral properties are often not available. Property variations with wavelength for the normal direction are available for a number of materials; the data are usually sparse at the short- and long-wavelength ranges of the spectrum.* Directional variations for some materials with optically smooth surfaces can be computed using electromagnetic theory (Chap. 4).

Certain problems demand inclusion of the effects of spectral and directional property variations, and the methods presented here must be utilized. One example is the use of spectrally selective coatings for temperature control or energy collection in systems involving solar radiation.

10-2 SYMBOLS

A	area
a_0	autocorrelation distance of surface roughness
C_1, C_2	first and second constants, respectively, in Planck's spectral energy distribution
D	perpendicular distance between parallel areas
e	emissive power
F	configuration factor
$F_{0-\lambda}$	fraction of blackbody intensity in spectral range $0-\lambda$
G	function of emissivities in Example 10-5
h	height of cavity
i	intensity
L	width of infinitely long parallel plates
l	L/D, parameter in Example 10-6
Q	energy rate, energy per unit time
q	energy flux, energy per unit area and per unit time
R	radius of disk in Example 10-7
r	R/D, parameter in Example 10-7
S	distance between area elements
T	absolute temperature
w	width of cavity
x, y, z	cartesian coordinates
α	absorptivity
ϵ	emissivity
η	angle in plane perpendicular to surface
θ	cone angle
φ	circumferential angle
λ	wavelength

*The range for which data are available depends on the equipment used in taking the data, and of course on whether data have been gathered at all for the particular material. Typically, data are less available for many materials at wavelengths smaller than 0.3 or greater than 15 μm.

Ξ	absorption efficiency defined in Example 10-6
ξ	distance along width of plane surface having finite width and infinite length
ρ	reflectivity
σ	Stefan-Boltzmann constant
σ_0	rms amplitude of surface roughness
ω	solid angle
$\int_{\cap}$	integration over solid angle of entire enclosing hemisphere

Superscripts

$'$	directional quantity
$''$	bidirectional quantity

Subscripts

a	absorbed
b	blackbody
e	emitted
i	incident, incoming
k	quantity for kth surface
max	maximum
min	minimum
o	outgoing
r	reflected
s	specular
λ	spectrally dependent
$\Delta\lambda$	average over wavelength region $\Delta\lambda$
1, 2, 3	property of surface 1, 2, or 3

10-3 ENCLOSURE THEORY FOR DIFFUSE SURFACES WITH SPECTRALLY DEPENDENT PROPERTIES

By considering diffusely emitting and reflecting surfaces we eliminate directional effects, and it is possible to see more clearly how the spectral variations of properties can be accounted for. The surface emissivity, absorptivity, and reflectivity are independent of direction, but may depend on both wavelength λ and surface temperature T. These properties must be available as a function of T and λ to evaluate the radiative interchange between surfaces.

For diffusely emitting and reflecting spectral surfaces, the concept of configuration factor is still valid since these factors involve only geometric effects and were computed for diffuse radiation leaving a surface. In general, then, the energy-balance equations and methods developed in Chaps. 7-9 remain valid so long as they are written for the energy in each *wavelength interval $d\lambda$*. Often, however, the boundary conditions that are specified apply to the *total* (including all wavelengths) energy, and

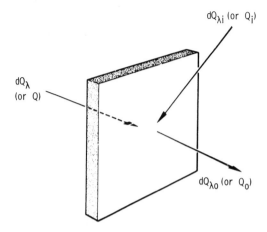

Figure 10-1 Spectral (or total) energy quantities at a surface.

care must be taken to apply the boundary conditions correctly. These total boundary conditions cannot generally be applied to the spectral energies. As an example, consider the surface of Fig. 10-1 having an incident total radiation Q_i and a radiation leaving by combined emission and reflection Q_0. If the surface is otherwise perfectly insulated and there is no heat Q being added externally (an *adiabatic* surface), then Q_i and Q_o will be equal to each other:

$$Q_o - Q_i = Q = 0 \tag{10-1}$$

However, at a given wavelength, the incident and outgoing dQ_λ are *not* necessarily equal, so that in general

$$dQ_{\lambda o} - dQ_{\lambda i} = dQ_\lambda \neq 0 \tag{10-2}$$

Rather, an *adiabatic* surface only has a *total* radiation gain or loss of zero, or, with Eq. (10-1) restated in terms of the quantities in (10-2),

$$Q = \int_{\lambda=0}^{\infty} dQ_\lambda = \int_{\lambda=0}^{\infty} (dQ_{\lambda o} - dQ_{\lambda i}) = 0 \tag{10-3}$$

The dQ_λ is net energy supplied at λ as a result of incident energy at other wavelengths. At a given wavelength, the dQ_λ can vary widely from zero for an adiabatic surface, depending on the property variations with wavelength and the spectral distribution of incident energy.

More generally, consider now a *diabatic** surface. The total energy added to the surface by some means other than radiation is equal to

$$Q = \int_{\lambda=0}^{\infty} dQ_\lambda = \int_{\lambda=0}^{\infty} (dQ_{\lambda o} - dQ_{\lambda i}) \tag{10-4}$$

---

*To quote Breene [1], "In certain circles 'nonadiabatic' is considered as an 'atrocious, pleonastic synonym,' and the author quit using the term as soon as he discovered how bad it was."

The Q may be either specified as an imposed condition or a quantity that is to be determined in order that a surface can be maintained at a specified temperature. In any small wavelength interval, the net energy $dQ_{\lambda o} - dQ_{\lambda i}$ may be positive or negative. The boundary condition only states that the integral of all such spectral energy values must be equal to Q. To build familarity with the use of these concepts, they are now applied to some example situations.

EXAMPLE 10-1 Two infinite parallel plates of tungsten at specified temperatures T_1 and T_2 $(T_1 > T_2)$ are exchanging radiant energy. Branstetter [2] has determined the hemispherical spectral temperature-dependent emissivity of tungsten by using the relations of electromagnetic theory to extrapolate limited experimental data, and a portion of his results are shown in Fig. 10-2. Using these data, compare the net energy exchange between the tungsten plates with that for the case of gray parallel plates.

The solution for gray plates has been given in Example 8-1. The present case follows in the same fashion except that the equations are written spectrally. From Eqs. (8-1) and (8-2) the energy quantities at surface 1 per unit area and time in a wavelength interval $d\lambda$ are related by

$$dq_{\lambda, 1} = dq_{\lambda o, 1} - dq_{\lambda i, 1} \tag{10-5}$$

$$dq_{\lambda o, 1} = \epsilon_{\lambda, 1}(\lambda, T_1) e_{\lambda b, 1}(\lambda, T_1) d\lambda + \rho_{\lambda, 1}(\lambda, T_1) dq_{\lambda i, 1} \tag{10-6}$$

For diffuse-opaque surfaces the hemispherical properties are related by $\rho_\lambda = 1 - \alpha_\lambda = 1 - \epsilon_\lambda$, and Eq. (10-6) becomes

$$dq_{\lambda o, 1} = \epsilon_{\lambda, 1}(\lambda, T_1) e_{\lambda b, 1}(\lambda, T_1) d\lambda + [1 - \epsilon_{\lambda, 1}(\lambda, T_1)] dq_{\lambda i, 1} \tag{10-7}$$

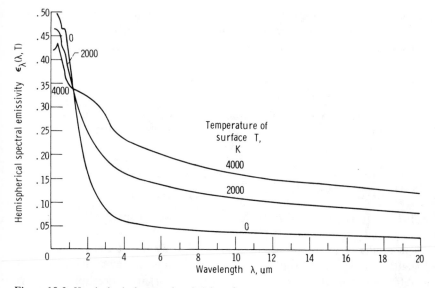

Figure 10-2 Hemispherical spectral emissivity of tungsten.

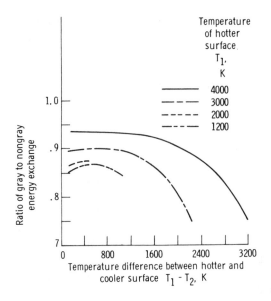

Figure 10-3 Comparison of effect of gray and nongray surfaces on computed energy exchange between infinite tungsten plates [2].

Eliminating $dq_{\lambda i, 1}$ from (10-5) and (10-7) gives

$$dq_{\lambda, 1} = \frac{\epsilon_{\lambda, 1}(\lambda, T_1)}{1 - \epsilon_{\lambda, 1}(\lambda, T_1)} \left[e_{\lambda b, 1}(\lambda, T_1) \, d\lambda - dq_{\lambda o, 1} \right] \tag{10-8}$$

For infinite parallel plates the configuration factor $F_{2-1} = 1$; then $q_{\lambda i, 1} = q_{\lambda o, 2}$ [see (8-5b)], and (10-5) becomes

$$dq_{\lambda, 1} = dq_{\lambda o, 1} - dq_{\lambda o, 2} \tag{10-9}$$

Equations (10-8) and (10-9) are analogous to (8-8a) and (8-8b) for the gray case. The equations for surface 2 are written in a similar fashion. Then the $dq_{\lambda o}$'s are eliminated, and the solution for one wavelength $d\lambda$ follows as in Eq. (8-10a):

$$dq_{\lambda, 1} = -dq_{\lambda, 2} = \frac{e_{\lambda b, 1}(\lambda, T_1) - e_{\lambda b, 2}(\lambda, T_2)}{1/\epsilon_{\lambda, 1}(\lambda, T_1) + 1/\epsilon_{\lambda, 2}(\lambda, T_2) - 1} \, d\lambda \tag{10-10}$$

The total heat flux exchanged (supplied to 1 and removed from 2) is found by substituting the property data of Fig. 10-2 into Eq. (10-10) and then integrating over all wavelengths:

$$q_1 = -q_2 = \int_{\lambda=0}^{\infty} dq_{\lambda, 1} = \int_{0}^{\infty} \frac{e_{\lambda b, 1}(\lambda, T_1) - e_{\lambda b, 2}(\lambda, T_2)}{1/\epsilon_{\lambda, 1}(\lambda, T_1) + 1/\epsilon_{\lambda, 2}(\lambda, T_2) - 1} \, d\lambda \tag{10-11}$$

The integration is performed numerically for each set of specified plate temperatures T_1 and T_2.

The results of such integrations as carried out by Branstetter [2] are shown in Fig. 10-3, where the ratio of diffuse-gray to diffuse-nongray exchange is given. The diffuse-gray results were obtained using (8-10a) with hemispherical total emissivities computed from the hemispherical spectral emissivities of Fig. 10-2.

(In the gray computation by Branstetter, the emissivity of the colder surface 2 was inserted at the mean temperature $\sqrt{T_1 T_2}$ rather than at T_2, which is a modification based on electromagnetic theory that is sometimes recommended for metals [3].) Over the range of surface temperatures shown, deviations of 25% below the nongray energy exchange are noted in the gray results.

EXAMPLE 10-2 Two infinite parallel plates and their spectral emissivities at their respective temperatures are shown in Fig. 10-4. What is the net total heat flux q transferred across the gap?

From Eq. (10-11),

$$q = \int_0^3 \frac{e_{\lambda b,1}(\lambda,T_1) - e_{\lambda b,2}(\lambda,T_2)}{1/0.4 + 1/0.7 - 1} \, d\lambda + \int_3^5 \frac{e_{\lambda b,1}(\lambda,T_1) - e_{\lambda b,2}(\lambda,T_2)}{1/0.8 + 1/0.7 - 1} \, d\lambda$$

$$+ \int_5^\infty \frac{e_{\lambda b,1}(\lambda,T_1) - e_{\lambda b,2}(\lambda,T_2)}{1/0.8 + 1/0.3 - 1} \, d\lambda$$

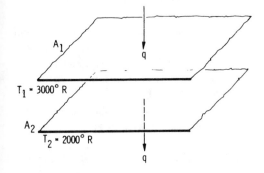

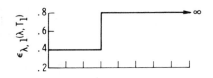

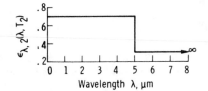

Figure 10-4 Example of heat transfer across space between infinite parallel plates having spectrally dependent emissivities.

which can be written as

$$q = \sigma T_1^4 \left[\frac{0.341}{\sigma T_1^4} \int_0^3 e_{\lambda b, 1}(\lambda, T_1)\, d\lambda + \frac{0.596}{\sigma T_1^4} \int_3^5 e_{\lambda b, 1}(\lambda, T_1)\, d\lambda \right.$$

$$+ \left. \frac{0.279}{\sigma T_1^4} \int_5^\infty e_{\lambda b, 1}(\lambda, T_1)\, d\lambda \right] - \sigma T_2^4 \left[\frac{0.341}{\sigma T_2^4} \int_0^3 e_{\lambda b, 2}(\lambda, T_2)\, d\lambda \right.$$

$$+ \frac{0.596}{\sigma T_2^4} \int_3^5 e_{\lambda b, 2}(\lambda, T_2)\, d\lambda + \left. \frac{0.279}{\sigma T_2^4} \int_5^\infty e_{\lambda b, 2}(\lambda, T_2)\, d\lambda \right]$$

An integral such as $(1/\sigma T_1^4) \int_3^5 e_{\lambda b, 1}(\lambda, T_1)\, d\lambda$ is the fraction of blackbody radiation at T_1 between $\lambda = 3$ and 5 μm, which is $F_{3T_1 - 5T_1} = F_{9000-15,000}$ and can be computed from the table of blackbody radiation functions (Table A-5). The $F_{\lambda T}$ should not be confused with the geometric configuration factor. Then

$$q = \sigma T_1^4 (0.341 F_{0-3T_1} + 0.596 F_{3T_1 - 5T_1} + 0.279 F_{5T_1 - \infty}) - \sigma T_2^4 (0.341 F_{0-3T_2}$$

$$+ 0.596 F_{3T_2 - 5T_2} + 0.279 F_{5T_2 - \infty}) = 43{,}600 \text{ Btu/(h·ft}^2)$$

EXAMPLE 10-3 An enclosure is made up of three plates of finite width and infinite length, as shown (in cross section) in Fig. 10-5. The radiative properties of each surface are dependent upon wavelength and temperature, and the temperatures of the plates are T_1, T_2, and T_3. Derive a set of equations governing the radiative energy exchange among the surfaces.

The configuration factors for such a geometry are derived in Example 7-15. The net spectral energy flux supplied to surface 1 can be written as

$$dq_{\lambda, 1} = \frac{dQ_{\lambda, 1}}{A_1} = \frac{\epsilon_{\lambda, 1}(\lambda, T_1)}{1 - \epsilon_{\lambda, 1}(\lambda, T_1)} \left[e_{\lambda b, 1}(\lambda, T_1)\, d\lambda - dq_{\lambda o, 1} \right] \qquad (10\text{-}12)$$

and

$$dq_{\lambda, 1} = \frac{dQ_{\lambda, 1}}{A_1} = dq_{\lambda o, 1} - F_{1-2}\, dq_{\lambda o, 2} - F_{1-3}\, dq_{\lambda o, 3} \qquad (10\text{-}13)$$

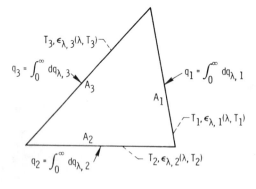

Figure 10-5 Radiant interchange in enclosure with surfaces having spectrally varying radiation properties.

These equations are derived in direct analogy with those for a gray surface, Eqs. (8-6) and (8-7), and by noting that $F_{1-1} = 0$. It should be emphasized that dq_λ is the energy supplied to the surface in wavelength interval $d\lambda$ as a result of external heat addition to the surface (e.g., conduction and/or convection) and energy transferred in from other wavelength regions.

Similar equations are written for surfaces 2 and 3. The result is a set of six simultaneous equations for the six unknowns $dq_{\lambda o,1}, dq_{\lambda o,2}, dq_{\lambda o,3}, dq_{\lambda,1}, dq_{\lambda,2}$, and $dq_{\lambda,3}$. The solution is carried out for the dq_λ in each wavelength interval $d\lambda$.

If the properties of the surfaces are invariant over some fairly large spectral interval $\Delta\lambda$, then the equations may be solved over this entire interval. In this instance the emissive power $e_{\lambda b,1}(\lambda,T_1)\,\Delta\lambda$ would be replaced by $\sigma T_1^4 F_{\lambda T_1 - (\lambda + \Delta\lambda)T_1}$, the amount of blackbody radiation at T_1 in the interval from λ to $\lambda + \Delta\lambda$. Finally, q at each surface is found by integrating dq_λ for that surface over all wavelengths:

$$q = \int_{\lambda=0}^{\infty} dq_\lambda \qquad (10\text{-}14)$$

This is the heat flux that must be externally supplied to the surface to maintain its specified surface temperature.

EXAMPLE 10-4 Consider the geometry of Fig. 10-5. Total energy flux is supplied to the three infinitely long plates at the rates q_1, q_2, and q_3. Determine the temperatures of the plates.

The equations are exactly the same as in Example 10-3. Now, however, the prescribed boundary conditions have made the problem much more difficult to solve. Because the surface temperatures are unknown, the emissivities are also unknown because of their temperature dependence. The solution is carried out as follows: A temperature is assumed for each surface, and $dq_\lambda(\lambda,T)$ for each surface is computed. The $dq_\lambda(\lambda,T)$ values are then integrated to find q_1, q_2, and q_3, which are compared to the specified boundary values. New temperatures are chosen, and the process is repeated until the computed q values agree with the specified values. The new temperatures for successive iterations must be guessed on the basis of the property variations and experience as to how changes in T are reflected in changes in q throughout the system.

10-4 THE BAND–ENERGY APPROXIMATION

The solution method presented and demonstrated in Sec. 10-3 for spectrally dependent surfaces required integrations over all wavelengths to compute the net total energy transfer. These integrations are the complication that makes surfaces with spectrally dependent properties so much more difficult to deal with than gray surfaces. Shortcut methods are desirable for circumventing the tedious numerical integrations that are required for rigorous solution of these problems. Some loss of accuracy in the integration may be acceptable in practical applications, because of the uncertainty already present in many of the spectral property values that are used.

10-4.1 Multiple Bands

One method of approximating the integrals is the *band-energy approximation*. This is the conceptually simple method of replacing the single integral extending over all wavelengths by a summation of smaller integrals, where each of the smaller integrals extends over a portion of the spectrum. An example will serve to illustrate the application of this method.

> **EXAMPLE 10-5** Two infinite parallel plates of tungsten are at temperatures of 4000 and 2000 K. Using the data of Fig. 10-2, compute the net energy exchange between the surfaces by using the band-energy approximation.
>
> In Example 10-1, the net energy transferred from plate 1 to plate 2 is given by the exact expression [Eq. (10-11)]
>
> $$q_1 = -q_2 = \int_0^\infty \frac{e_{\lambda b,1}(\lambda,T_1) - e_{\lambda b,2}(\lambda,T_2)}{1/\epsilon_{\lambda,1}(\lambda,T_1) + 1/\epsilon_{\lambda,2}(\lambda,T_2) - 1} \, d\lambda$$
>
> By using the substitution
>
> $$G_\lambda = \frac{1}{1/\epsilon_{\lambda,1} + 1/\epsilon_{\lambda,2} - 1}$$
>
> to shorten the notation, this can be written as
>
> $$q_1 = \int_0^\infty G_\lambda e_{\lambda b,1} \, d\lambda - \int_0^\infty G_\lambda e_{\lambda b,2} \, d\lambda$$
>
> The integrals are now written as approximate sums

$$q_1 \approx \sum_l (G_{\Delta\lambda} e_{\Delta\lambda,b,1} \Delta\lambda)_l - \sum_m (G_{\Delta\lambda} e_{\Delta\lambda,b,2} \Delta\lambda)_m \tag{10-15}$$

where $G_{\Delta\lambda}$ and $e_{\Delta\lambda,b}$ are average values applicable to the wavelength interval $\Delta\lambda$. Depending upon the way in which $G_{\Delta\lambda}$ and $e_{\Delta\lambda,b}$ are evaluated, (10-15) can have various degrees of accuracy. As a simple approximation, the terms $e_{\Delta\lambda,b}$ can be taken as an arithmetic mean of the blackbody emissive power over $\Delta\lambda$. To obtain a better degree of approximation for large $\Delta\lambda$ intervals, $e_{\Delta\lambda,b}$ can be evaluated using the blackbody functions, as

$$e_{\Delta\lambda,b} \Delta\lambda = \int_\lambda^{\lambda+\Delta\lambda} e_{\lambda b} \, d\lambda = (F_{0-(\lambda+\Delta\lambda)} - F_{0-\lambda}) \sigma T^4 \tag{10-16}$$

where λ and $\lambda + \Delta\lambda$ are the upper and lower limits of wavelength for the interval $\Delta\lambda$. Equation (10-16) will be used for the computations given here. It was previously used to evaluate the exact integrals for a simple problem in Example 10-2.

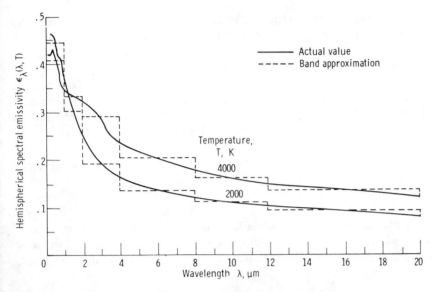

Figure 10-6 Band approximations to hemispherical spectral emissivity of tungsten.

The $G_{\Delta\lambda}$ terms in (10-15) are approximated most simply by

$$G_{\Delta\lambda} = \frac{1}{1/\epsilon_{\Delta\lambda,1} + 1/\epsilon_{\Delta\lambda,2} - 1} \tag{10-17}$$

where the $\epsilon_{\Delta\lambda}$ are appropriate mean emissivities over the wavelength interval $\Delta\lambda$.

In Fig. 10-6, the required emissivities of tungsten are plotted, and *arithmetic* mean values are shown for seven $\Delta\lambda$ intervals (the seventh interval being for $\lambda >$ 20 μm). For temperatures of 2000 and 4000 K, the peak in the $e_{\lambda b}$ function occurs at about 1.5 and 0.75 μm, respectively. For large values of λ, say $\lambda > 4$ μm in this example, $e_{\lambda b}$ is small and $G_\lambda e_{\lambda b}$ will contribute little to the integrals in this wavelength region. Thus the accuracy of the averages at large values of λ is not important in this example. The computations for q_1 are carried out using these seven intervals in the following tabulation:

$\Delta\lambda$, μm	$\epsilon_{\Delta\lambda,1}$	$\epsilon_{\Delta\lambda,2}$	$G_{\Delta\lambda}$	$e_{\Delta\lambda,b,1}\Delta\lambda$, W/m^2	$e_{\Delta\lambda,b,2}\Delta\lambda$, W/m^2	$G_{\Delta\lambda}e_{\Delta\lambda,b,1}\Delta\lambda$ W/m^2	$G_{\Delta\lambda}e_{\Delta\lambda,b,2}\Delta\lambda$ W/m^2
0–1	0.410	0.445	0.271	0.698×10^7	0.0061×10^7	1.89×10^6	0.017×10^6
1–2	0.335	0.300	0.188	0.545×10^7	0.0374×10^7	1.03×10^6	0.071×10^6
2–4	0.290	0.195	0.132	0.171×10^7	0.034×10^7	0.23×10^6	0.045×10^6
4–8	0.205	0.140	0.0907	0.032×10^7	0.011×10^7	0.03×10^6	0.010×10^6
8–12	0.160	0.115	0.0717	0.004×10^7	0.002×10^7	~ 0	0.002×10^6
12–20	0.140	0.095	0.0600	0.001×10^7	~ 0	~ 0	~ 0
>20	~ 0	~ 0	~ 0	~ 0	~ 0	~ 0	~ 0
Totals						3.18×10^6	0.145×10^6

Substituting the sums from the tabulation into the approximate band-energy exchange equation gives $q_1 = (3.18 - 0.15) \times 10^6 = 3030$ kW/m².

Branstetter [2], using numerical integration, found the exact result of $q_1 = 3000$ kW/m² for this case. The approximate band solution using seven intervals is thus in error by a very small amount. Examination of Fig. 10-3 shows that the gray-body assumption, which can be considered as a one-band approximation, yields answers that are in error by almost 10% (note that the gray results in Fig. 10-3 were modified from the usual gray analysis by inserting ϵ_2 at $\sqrt{T_1 T_2}$ rather than at T_2).

A close examination of the tabulation shows that most of the significant energy transfer for this example occurs in the wavelength range of 0–2 μm. If necessary, the accuracy of the band-energy approximation could be improved by dividing this range of most significant energy transfer into a larger number of increments and repeating the calculation. The errors in the band-energy approximation will arise in the regions where both $e_{\lambda b}$ and ϵ_λ are large; thus the wavelength range should be divided such that most of the bands lie within these regions.

The band-energy approximation is nothing more than a simple form of numerical integration carried out by using a relatively small number of wavelength intervals. If the number of intervals is increased, the exact results for energy transfer are approached. Dunkle and Bevans [4] give a calculation similar to Example 10-5 and shows errors from an exact numerical result of less than 2% for the band-energy solution, as compared to about 30% error for the gray-surface approximation. They give other examples of applications in enclosures with specified temperatures or net energy fluxes.

Some additional references providing analyses of energy exchange between spectrally dependent surfaces are those of Love and Gilbert [5], Goodman [6], and Rolling and Tien [7]. In [5] the analytical results compare well with experimental results for a geometry closely approximating infinite parallel plates.

10-4.2 The Semigray Approximations

In some practical situations there is a natural division of the energy within an enclosure into two well-defined spectral regions. This is the case for an enclosure with an opening through which solar energy is entering. The solar energy will have a spectral distribution concentrated in a short-wavelength region, while the energy originating by emission from the lower-temperature surfaces within the enclosure will be in a longer-wavelength region. A practical way of treating this situation is to define a hemispherical total absorptivity for incident solar radiation and a second hemispherical total absorptivity for incident energy originating by emission within the enclosure. This approach can be carried to the point of defining j different absorptivities for surface k, one absorptivity for incident energy from each enclosure surface j.

The assumption entering these analyses is that each absorptivity $\alpha_k(T_k, T_j)$ is based on an incident *blackbody* spectral distribution at the temperature of the originating surface T_j. Of course, the incident spectrum may actually be quite far removed from the Planckian form, and this is the weakness of the method. Often the dependence of α_k on T_k is small, so that the principal dependence is on T_j or, in other words, on the distribution of the incident spectrum. Because the absorptivity $\alpha_k(T_k, T_j)$ and emissivity $\epsilon_k(T_k)$ of surface k are not in general equal, this approach is often called the *semigray enclosure theory*. Reference [8] contains the formulation of a semigray analysis for a general enclosure.

Plamondon and Landram [9] have compared the semigray and exact solutions of the temperature profiles along the surface of a nongray wedge cavity exposed to incident solar radiation as shown in Fig. 10-7. The wedge cavity is assumed to be non-conducting, to be in a vacuum with an environment at zero degrees except for the solar radiation source, to have properties independent of surface temperature, and to have diffuse surfaces. Three solution techniques are given in [9]. The first is an exact solution of the complete integral equations and is called the *exact* solution. The first approximation to the exact solution, called *method I*, is the semigray analysis, which assigns an absorptivity α_{solar} for radiation (direct and reflected) that originated from the incident solar energy, and a second absorptivity α_{infrared} (equal to the surface emissivity) for radiation originating by emission from the wedge surfaces. Finally, *method II* is a poorer approximation that retains the same two absorptivities but applies α_{solar} only for the incident solar energy, and then uses α_{infrared} for *all* energy after reflection, regardless of its source. The results of these methods are shown in Fig. 10-7b for a polished aluminum surface in a 30° wedge. Method I, the full semi-gray analysis, is seen to give excellent agreement with the exact solution, while method II underestimates the exact temperatures by about 10%.

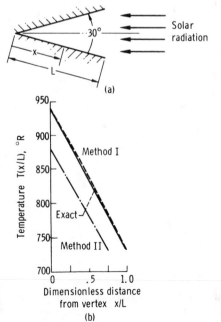

(a)

(b)

Figure 10-7 Effect of semigray approximations on computed temperature distribution in wedge cavity. (a) Geometry of wedge cavity; (b) temperature distribution along wedge, $\alpha_{\text{solar}} = 0.220$; $\alpha_{\text{infrared}} = 0.099$.

10-5 DIRECTIONAL–GRAY SURFACES

Some attention has been paid to the development of treatments of radiation interchange between surfaces or in enclosures where directionally dependent properties must be considered. The bulk of radiation analyses invoke the assumption of diffuse emitting and reflecting surfaces, although some treatments do include the effect of specular reflections as outlined in Chap. 9. The diffuse or specular surface conditions are convenient to treat analytically, and in most instances the detailed consideration of directional emission and reflection effects is unwarranted. There are nevertheless certain materials and certain geometric situations that require the consideration of directional effects. In this section, some methods of considering radiant interchange between surfaces with directional properties will be presented.

The difficulty in treating the general case of directionally dependent properties is perhaps best illustrated by performing an energy balance in a simple geometry. Let such a balance for the radiative interchange between two infinitely long parallel nondiffuse-gray surfaces of finite width L (Fig. 10-8) be examined. The intensity of radiation leaving element dA_1 in direction (θ_1,φ_1) is composed of an emitted intensity $i'_{1,e}(\theta_1,\varphi_1)$ and a reflected intensity $i'_{1,r}(\theta_1,\varphi_1)$, or

$$i'_1(\theta_1,\varphi_1) = i'_{1,e}(\theta_1,\varphi_1) + i'_{1,r}(\theta_1,\varphi_1) \tag{10-18}$$

These two components are given by modifications of (3-3a) and (3-25) as

$$i'_{1,e}(\theta_1,\varphi_1) = \epsilon'_1(\theta_1,\varphi_1) i'_{b,1}(T_1) \tag{10-19}$$

and

$$i'_{1,r}(\theta_1,\varphi_1) = \int_{A_2} \rho''_1(\theta_1,\varphi_1,\theta_2,\varphi_2) i'_2(\theta_2,\varphi_2) \frac{\cos\theta_2 \cos\theta_2}{S^2} dA_2 \tag{10-20}$$

In Eq. (10-20) the energy incident upon dA_1 from each element dA_2 is multiplied by

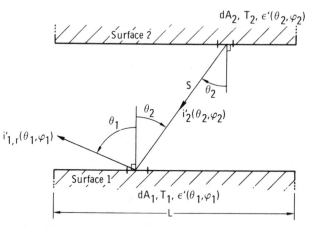

Figure 10-8 Radiant interchange between infinitely long parallel directional surfaces of finite width L.

the bidirectional total reflectivity ρ_1'' to give the contribution to the reflected intensity from dA_1 into direction (θ_1,φ_1). This is then integrated over all energy incident on dA_1 from A_2. The definition of ρ'' is given by (3-20) as

$$\rho''(\theta_r,\varphi_r,\theta,\varphi) = \frac{i_r''(\theta_r,\varphi_r,\theta,\varphi)}{i_i'(\theta,\varphi)\cos\theta\, d\omega} \qquad (10\text{-}21)$$

The $\rho''(\theta_r,\varphi_r,\theta,\varphi)$ is the ratio of reflected *intensity* in the (θ_r,φ_r) direction to the *energy flux* incident from the (θ,φ) direction.

Equation (10-18) for the intensity leaving the element dA_1 then becomes

$$i_1'(\theta_1,\varphi_1) = \epsilon_1'(\theta_1,\varphi_1)\, i_{b,1}'(T_1) + \int_{A_2} \rho_1''(\theta_1,\varphi_1,\theta_2,\varphi_2)\, i_2'(\theta_2,\varphi_2)\frac{\cos^2\theta_2}{S^2}\, dA_2 \qquad (10\text{-}22)$$

A similar equation may be written for an arbitrary element dA_2 on surface 2. This results in a very complicated coupled pair of integral equations that must be solved for $i'(\theta,\varphi)$ at each point and for each direction on the two surfaces. This set of integral equations is analogous to Eqs. (8-50) that were derived for diffuse-gray surfaces. Tabulated property data of $\epsilon'(\theta,\varphi)$ and $\rho''(\theta_r,\varphi_r,\theta,\varphi)$ for such a situation are seldom available. For the case when T_1 and T_2 are not known and the temperature dependence of the properties is considerable, the solution for the entire energy-exchange distribution becomes prohibitively tedious. To avoid the extreme amount of computation, a number of approximations can be invoked in the situations where they are justified. Usually such approximations involve analytically simulating the real properties with simple functions, omitting certain portions of energy that are deemed negligible, or ignoring all directional effects except those expected to provide significant changes from diffuse or specular analyses. Some of these methods are outlined in [10–14]. Rather than present all the possible approximations, an example will be given, and the reader will be left to use ingenuity in approximating the conditions of more realistic problems.

EXAMPLE 10-6 Two parallel isothermal plates of infinite length and finite width L are arranged as shown in Fig. 10-9a. The upper plate 2 is black, while the lower is composed of a highly reflective material with parallel deep grooves of open angle $1°$ cut into the surface and running in the infinite direction. Such a surface might be made by stacking polished razor blades. The surroundings are at zero temperature. Compute the net energy gain by the directional surface if $T_2 > T_1$, and compare the result to the net energy gain by a diffuse surface with emissivity equivalent to the hemispherical emissivity of the directional surface.

The directional emissivity for the grooved surface is obtained from [15], where the directional emissivity at the opening of an infinitely long groove with specularly reflecting walls of surface emissivity 0.01 is calculated. The directional emissivity for the grooved surface is given by the dot-dashed line in Fig. 10-9b. The angle η is measured from the normal of the base plane of the grooved surface and is in a plane perpendicular to the length of the groove as shown in Fig. 10-9a.

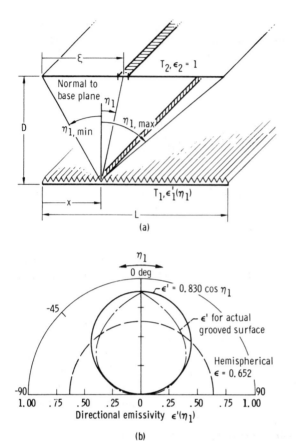

Figure 10-9 Interchange between grooved directional surface and black surface. (a) Geometry of problem (environment at zero temperature), (b) emissivity of directional surface.

The $\epsilon_1'(\eta_1)$ as given in [15] has already been averaged over all circumferential angles for a fixed η_1. Thus, it is an effective emissivity from a strip on the grooved surface to a parallel infinitely long strip element on an imaginary semicylinder over the groove and with its axis parallel to the grooves. The angle η_1 is different from the usual cone angle θ_1. The angle θ_1 would vary along the strip element of the semicylinder, while η_1 remains constant since it is the projected angle on a plane normal to the groove. The actual emissivity $\epsilon_1'(\eta_1)$ of Fig. 10-9b is approximated for convenience by the analytical expression $\epsilon_1'(\eta_1) \approx 0.830 \cos \eta_1$. By using cylindrical coordinates to perform the integration over all η_1, the corresponding hemispherical emissivity of this surface is found to be

$$\epsilon_1 = \frac{Q_1}{Q_b} = \frac{\int_{-\pi/2}^{\pi/2} \epsilon_1'(\eta_1) \cos \eta_1 \, d\eta_1}{\int_{-\pi/2}^{\pi/2} \cos \eta_1 \, d\eta_1} = 0.830 \int_0^{\pi/2} \cos^2 \eta_1 \, d\eta_1 = 0.652$$

and this result is shown in Fig. 10-9b as a dashed line.

The energy gained by surface 1 when 2 is a black surface and 1 is a diffuse surface with $\epsilon = 0.652$ will first be determined. The energy emitted by the diffuse surface per unit of the infinite length and per unit time is $Q_{e,1} = 0.652\sigma T_1^4 L$. Since surface 2 is black, none of this energy is reflected back to 1. The energy per unit length and time emitted by surface 2 that is absorbed by surface 1 is

$$Q_{a,1} = 0.652\sigma T_2^4 \int_{A_2} \int_{A_1} dF_{d2-d1} \, dA_2 = 0.652\sigma T_2^4 \int_{A_1} \int_{A_2} dF_{d1-d2} \, dA_1$$

The desired energy gained by 1 is $Q_{a,1} - Q_{e,1}$. For evaluating $Q_{a,1}$, the configuration factor between infinite parallel strips was found in Example 7-4 as $dF_{d1-d2} = d(\sin \eta_1)/2$. The double integral becomes, through integration over A_2,

$$\int_{A_1} \left(\int_{A_2} dF_{d1-d2} \right) dA_1 = \frac{1}{2} \int_{x=0}^{L} (\sin \eta_{1,\max} - \sin \eta_{1,\min}) \, dx$$

The value of $\sin \eta_1$ is found from Fig. 10-9a to be $\sin \eta_1 = (\xi - x)/[(\xi - x)^2 + D^2]^{1/2}$ and now solving for $Q_{a,1}$ gives

$$Q_{a,1} = 0.652\sigma T_2^4 \frac{1}{2} \int_{x=0}^{L} \left[\frac{L-x}{(x^2 - 2xL + L^2 + D^2)^{1/2}} + \frac{x}{(x^2 + D^2)^{1/2}} \right] dx$$

$$= 0.652\sigma T_2^4 [(L^2 + D^2)^{1/2} - D]$$

The net energy gained by surface 1, $Q_{a,1} - Q_{e,1}$, divided by the energy emitted by surface 2, is a measure of the efficiency of the surface as a directional absorber. For surface 1, being diffuse, this ratio is

$$\Xi_{\text{diffuse}} = \frac{Q_{a,1} - Q_{e,1}}{\sigma T_2^4 L} = \frac{0.652}{l} \left[(1 + l^2)^{1/2} - 1 - \frac{T_1^4}{T_2^4} l \right]$$

where $l = L/D$.

The analysis for surface 1 a directional (grooved) surface will now be carried out. The emitted energy from 1 is the same as for the diffuse surface since both have the same hemispherical emissivity. The energy absorbed by the grooved surface is

$$Q_{a,1} = \sigma T_2^4 \int_{A_2} \int_{A_1} \alpha_1'(\eta_1) \, dF_{d2-d1} \, dA_2 = \frac{0.830\sigma T_2^4}{2} \int_{x=0}^{L} \int_{\eta_{1,\min}}^{\eta_{1,\max}} \cos^2 \eta_1 \, d\eta_1 \, dx$$

$$= \frac{0.830\sigma T_2^4}{2} \int_{x=0}^{L} \frac{1}{2} (\sin \eta_1 \cos \eta_1 + \eta_1) \Big|_{\eta_{1,\min}}^{\eta_{1,\max}} dx$$

$$= \frac{0.830\sigma T_2^4}{4} \int_{x=0}^{L} \left[\frac{D(L-x)}{x^2 - 2xL + L^2 + D^2} + \tan^{-1}\left(\frac{L-x}{D} \right) \right.$$

$$\left. + \frac{xD}{x^2 + D^2} + \tan^{-1}\frac{x}{D} \right] dx$$

$$Q_{a,1} = \frac{0.830\sigma T_2^4 L}{2} \tan^{-1}\frac{L}{D}$$

The absorption efficiency Ξ of the directional surface is then

$$\Xi_{\text{directional}} = \frac{0.830}{2} \tan^{-1} l - 0.652 \left(\frac{T_1}{T_2}\right)^4$$

The absorption efficiencies Ξ of the grooved and diffuse surfaces are plotted in Fig. 10-10 as a function of l with $(T_1/T_2)^4$ as a parameter. It is seen that Ξ for the directional surface is higher than that for the diffuse surface for all values of l, indicating that the directional surface will always be a more efficient absorber in this configuration. As l approaches zero, the configuration approaches that of infinite elemental strips, and emission from surface 1 becomes much larger than absorption from surface 2. Thus, Ξ_{diffuse} and $\Xi_{\text{directional}}$ are nearly equal since the surfaces always emit the same amount. As l approaches infinity, the configuration approaches infinite parallel plates for which directional effects are lost. Again, the Ξ become equal for the two different surface conditions. At intermediate values of l, a 10% difference in absorption efficiency appears attainable.

The effects of directional properties on the local heat loss can be a considerable factor in many geometries. In Fig. 10-11, a number of assumed directional distributions of reflectivity are examined for their influence on local heat loss from the walls of an infinitely long groove cavity. The results are taken from [12], where for comparison the curves were gathered from original work and from various sources [13, 14, 16, 17]. The walls of the groove are at 90° to each other, and the surface emissivity distributions are all normalized to give a hemispherical emissivity of 0.1. Curves are presented for diffuse reflectivity ρ, specular reflectivity assumed independent of incident angle ρ_s', specular reflectivity dependent upon incident angle $\rho_s'(\theta)$ based upon electromagnetic theory, and three distributions of bidirectional reflectivity $\rho''(\theta_r, \theta)$. The bidirectional distributions are based upon the work of Beckmann and

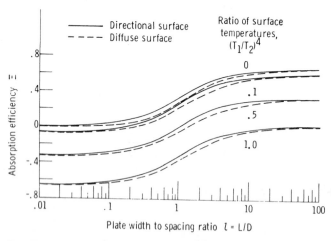

Figure 10-10 Effect of directional emissivity on absorption efficiency of surface.

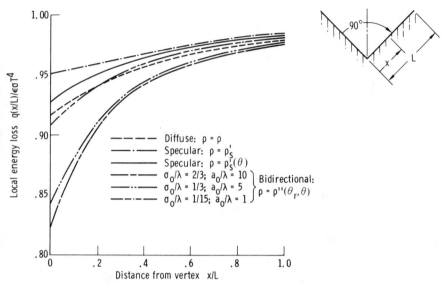

Figure 10-11 Local radiative energy loss from surface of isothermal groove cavity. Hemispherical emissivity of surface, 0.1.

Spizzichino [18] for rough surfaces having various combinations of rms optical-surface roughness amplitude to radiation wavelength ratio σ_0/λ and roughness autocorrelation distance to radiation wavelength ratio a_0/λ. Note that the results shown in Fig. 10-11 for the simple specular and diffuse models do *not* provide upper and lower limits to all the solutions as is sometimes claimed. Additional work on surface roughness as it affects the directional properties of surfaces is reported in [19–20]. Heat transfer was studied in a groove with two plane sides, each at uniform temperature, and having rough surfaces. The roughness was found to have a greater influence on the exchange of heat between the two sides than it has on the net energy radiated away from the groove.

Howell and Durkee [21] compared analysis and experiment for the case of a beam of collimated radiation entering a very long low-temperature three-sided cavity. Because of the low temperature the surface emission was not important. The cavity had two surfaces with reflectivities that were diffuse with a specular component, while the third surface was a honeycomb material with strong bidirectional reflectivity characteristics. It was found necessary to include all surface characteristics in the analysis to obtain agreement with experiment. This is characteristic of geometries involving point or collimated sources of incident relation.

Black [22] analyzed the optimization of two types of cavities that have directional emission characteristics. The cavities have some black and some specular sides, as shown in Fig. 10-12a and b. The V groove tends to emit in the normal direction, while the rectangular groove emits more in the grazing direction $\eta \to 90°$. Some emission results for the rectangular shape are shown in Fig. 10-12c, and they exhibit a strong directional characteristic for small values of h/w.

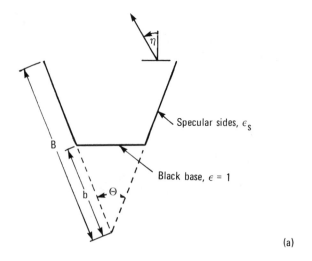

(a)

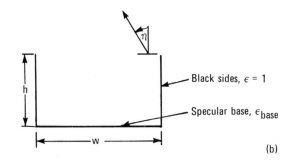

(b)

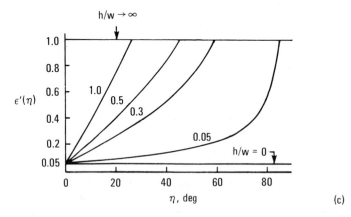

(c)

Figure 10-12 Directional emission from grooves. (a) V groove with specular sides; (b) rectangular groove with specular base; (c) directional emissivity for rectangular groove with specular base ($\epsilon_{base} = 0.05$) and black sides.

10-6 SURFACES WITH DIRECTIONALLY AND SPECTRALLY DEPENDENT PROPERTIES

The general case of radiative transfer in enclosures with surfaces having temperature-dependent radiative properties that depend on both wavelength and direction is a most complex and difficult one to treat fully. Closed-form solution of such problems is not possible unless many restrictive assumptions are introduced. When such problems must be treated, numerical techniques are necessary. The Monte Carlo method is a likely candidate, and example applications of the method to simple directional-spectral surfaces are made in Chap. 11. Toor [13] has studied radiation interchange by the Monte Carlo method for a variety of simply arranged surfaces with directional properties.

In this section, the general integral equations for radiation in such systems are formulated, and one considerably simplified example problem is solved. The procedure is a combination of the previously considered diffuse-spectral and directional-gray analyses. The equations will be formulated at one wavelength as in Sec. 10-3 and will also be formulated in terms of intensities as in Sec. 10-5. In this manner, both spectral and directional effects can be accounted for. For simplicity the interaction between only two plane surfaces will be treated. This treatment can then be generalized to a multisurface enclosure as has been done for gray surfaces in Chap. 8.

Consider an element dA_1 of surface A_1 in the xy plane as shown in Fig. 10-13. The surface is isothermal and has directional-spectral properties. Consider the spectral radiation intensity outgoing from dA_1 in the direction $(\theta_{r,1}, \varphi_{r,1})$ by means of both emission and reflection. The spectral intensity emitted by dA_1 in the direction

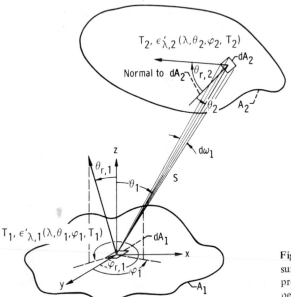

Figure 10-13 Interchange between surfaces having directional spectral properties (environment at zero temperature).

$(\theta_{r,1}, \varphi_{r,1})$ is

$$i'_{\lambda e,1}(\lambda, \theta_{r,1}, \varphi_{r,1}) = \epsilon'_{\lambda,1}(\lambda, \theta_{r,1}, \varphi_{r,1}) i'_{\lambda b,1}(\lambda) \tag{10-23}$$

These quantities are also functions of T_1, but this designation is omitted to simplify the notation somewhat. The intensity reflected from dA_1 into direction $(\theta_{r,1}, \varphi_{r,1})$ results from the intensity incident from A_2. It would be desirable to have an expression for the intensity incident within solid angle $d\omega_1$; then by integrating over all such $d\omega_1$, the incident radiation from all of A_2 would be accounted for. If the incident intensity within $d\omega_1$ is called $i'_{\lambda i,1}(\lambda, \theta_1, \varphi_1)$, then the energy reflected from dA_1 into direction $(\theta_{r,1}, \varphi_{r,1})$ is

$$i'_{\lambda r,1}(\lambda, \theta_{r,1}, \varphi_{r,1}) = \int_{A_2} \rho''_{\lambda,1}(\lambda, \theta_{r,1}, \varphi_{r,1}, \theta_1, \varphi_1) i'_{\lambda i,1}(\lambda, \theta_1, \varphi_1) \cos\theta_1 \, d\omega_1 \tag{10-24}$$

The surroundings are taken to be at zero temperature so that the only incident intensity is that from A_2. The spectral intensity outgoing from dA_1 in direction $(\theta_{r,1}, \varphi_{r,1})$ is then the sum of emitted and reflected quantities:

$$i'_{\lambda o,1}(\lambda, \theta_{r,1}, \varphi_{r,1}) = i'_{\lambda e,1}(\lambda, \theta_{r,1}, \varphi_{r,1}) + i'_{\lambda r,1}(\lambda, \theta_{r,1}, \varphi_{r,1})$$

$$= \epsilon'_{\lambda,1}(\lambda, \theta_{r,1}, \varphi_{r,1}) i'_{\lambda b,1}(\lambda)$$

$$+ \int_{A_2} \rho''_{\lambda,1}(\lambda, \theta_{r,1}, \varphi_{r,1}, \theta_1, \varphi_1) i'_{\lambda i,1}(\lambda, \theta_1, \varphi_1) \cos\theta_1 \, d\omega \tag{10-25}$$

In Eq. (10-25) the $i'_{\lambda i,1}(\lambda, \theta_1, \varphi_1)$ results from the outgoing intensity $i'_{\lambda o,2}(\lambda, \theta_2, \varphi_2)$ from surface 2. This outgoing intensity is composed of both emitted energy and energy incident from 1 that is reflected. The energy leaving dA_2 that reaches dA_1 is $i'_{\lambda o,2}(\lambda, \theta_2, \varphi_2) dA_2 \cos\theta_2 \, dA_1 \cos\theta_1 / S^2$. In terms of the incident intensity $i'_{\lambda i,1}(\lambda, \theta_1, \varphi_1)$, the incident energy in $d\omega_1$ is $i'_{\lambda i,1}(\lambda, \theta_1, \varphi_1) dA_1 \cos\theta_1 \, d\omega_1$ or $i'_{\lambda i,1}(\lambda, \theta_1, \varphi_1) dA_1 \cos\theta_1 \, dA_2 \cos\theta_2 / S^2$. Thus,

$$i'_{\lambda i,1}(\lambda, \theta_1, \varphi_1) = i'_{\lambda o,2}(\lambda, \theta_2, \varphi_2) \tag{10-26}$$

Substituting (10-26) into (10-25) gives

$$i'_{\lambda o,1}(\lambda, \theta_{r,1}, \varphi_{r,1}) = \epsilon'_{\lambda,1}(\lambda, \theta_{r,1}, \varphi_{r,1}) i'_{\lambda b,1}(\lambda)$$

$$+ \int_{A_2} \rho''_{\lambda,1}(\lambda, \theta_{r,1}, \varphi_{r,1}, \theta_1, \varphi_1) i'_{\lambda o,2}(\lambda, \theta_2, \varphi_2) \cos\theta_1 \, d\omega_1 \tag{10-27a}$$

Similarly, for surface 2,

$$i'_{\lambda o,2}(\lambda, \theta_{r,2}, \varphi_{r,2}) = \epsilon'_{\lambda,2}(\lambda, \theta_{r,2}, \varphi_{r,2}) i'_{\lambda b,2}(\lambda)$$

$$+ \int_{A_1} \rho''_{\lambda,2}(\lambda, \theta_{r,2}, \varphi_{r,2}, \theta_2, \varphi_2) i'_{\lambda o,1}(\lambda, \theta_1, \varphi_1) \cos\theta_2 \, d\omega_2 \tag{10-27b}$$

Equations (10-27) are both given in terms of outgoing intensities. Thus, they form a set of simultaneous integral equations for $i'_{\lambda o,1}$ and $i'_{\lambda o,2}$. An iterative numerical solution technique would generally be required. The radiative properties and temperature can in general vary across each surface.

When the $i'_{\lambda o,1}$ and $i'_{\lambda o,2}$ have been obtained, the total energy can be determined that must be supplied to each surface element to maintain the specified surface temperature. The total energy supplied is the difference between the total emitted energy Q_e and the total absorbed energy Q_a. For element dA_1,

$$dQ_1 = dQ_{e,1} - dQ_{a,1} = dA_1 \int_{\lambda=0}^{\infty} \int_{\Omega} \epsilon'_{\lambda,1}(\lambda,\theta_1,\varphi_1) i'_{\lambda b,1}(\lambda) \cos \theta_1 \, d\omega_1 \, d\lambda$$

$$- dA_1 \int_{\lambda=0}^{\infty} \int_{A_2} \alpha'_{\lambda,1}(\lambda,\theta_1,\varphi_1) i'_{\lambda i,1}(\lambda,\theta_1,\varphi_1) \cos \theta_1 \, d\omega_1 \, d\lambda$$

$$dQ_1 = \epsilon_1 \sigma T_1^4 \, dA_1 - dA_1 \int_{\lambda=0}^{\infty} \int_{A_2} \alpha'_{\lambda,1}(\lambda,\theta_1,\varphi_1) i'_{\lambda o,2}(\lambda,\theta_2,\varphi_2) \frac{\cos \theta_1 \cos \theta_2}{S^2} \, dA_2 \, d\lambda$$

$$(10\text{-}28)$$

where ϵ_1 is the hemispherical total emissivity of surface 1.

If $dQ_1(x,y)$ rather than $T_1(x,y)$ is specified, then $T_1(x,y)$ must be determined and the solutions can be quite tedious. A temperature distribution must be assumed for each surface, and the set of equations of the form of (10-27) solved to find $i'_{\lambda o}$ at each point. These outgoing intensities are substituted into (10-28), and the computed dQ_1 from Eq. (10-28) is compared to the given values. Adjustments are then made in the assumed distribution of temperatures, and this procedure is repeated until agreement between given and computed $dQ_1(x,y)$ is attained.

EXAMPLE 10-7 A small area element dA_1 is placed on the axis of and parallel to a black circular disk as shown in Fig. 10-14. The element is at temperature T_1,

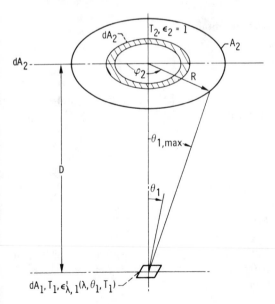

Figure 10-14 Energy exchange involving directional spectral surface element (environment at zero temperature).

and the disk is at T_2. The environment is at $T = 0$. The element has a directional spectral emissivity that is independent of φ and can be approximated by the expression

$$\epsilon'_{\lambda,1}(\lambda,\theta_1,T_1) = 0.8\cos\theta_1(1 - e^{-C_2/\lambda T_1})$$

where C_2 is one of the constants in Planck's spectral energy distribution. (As will be evident, this distribution was chosen to simplify this example.) Find the energy dQ_1 added to dA_1 to maintain it at T_1. Assume that T_1 is close to T_2.

Equation (10-28) can be employed immediately because $i'_{\lambda o, 2}$ is known from the specification of A_2 as a black surface. The emitted energy from dA_1 is given by

$$dQ_{e,1} = \epsilon_1 \sigma T_1{}^4 \, dA_1 = dA_1 \int_{\lambda=0}^{\infty} \int_{\cap} \epsilon'_{\lambda,1}(\lambda,\theta_1) i'_{\lambda b,1}(\lambda) \cos\theta_1 \, d\omega_1 \, d\lambda$$

Now insert the expressions for $\epsilon'_{\lambda,1}$, $i'_{\lambda b,1}$ [Eq. (2-11a)], and $d\omega_1 = \sin\theta_1 \, d\theta_1 \, d\varphi_1$ to obtain

$$dQ_{e,1} = 0.8 dA_1 \int_{\lambda=0}^{\infty} \int_{\cap} \cos\theta_1(1 - e^{-C_2/\lambda T_1}) \frac{2C_1}{\lambda^5(e^{C_2/\lambda T_1} - 1)}$$

$$\times \cos\theta_1 \sin\theta_1 \, d\theta_1 \, d\varphi_1 \, d\lambda$$

$$= 0.8 dA_1 \int_{\lambda=0}^{\infty} \left(\int_{\cap} \cos^2\theta_1 \sin\theta_1 \, d\theta_1 \, d\varphi_1\right) \frac{2C_1}{\lambda^5 e^{C_2/\lambda T_1}} \, d\lambda$$

Carrying out the integral over the hemisphere gives

$$dQ_{e,1} = 0.8 dA_1 \frac{2\pi}{3} \int_0^{\infty} \frac{2C_1}{\lambda^5 e^{C_2/\lambda T_1}} \, d\lambda$$

We use the transformation $\zeta = C_2/\lambda T_1$ to obtain

$$dQ_{e,1} = 0.8 dA_1 \frac{4C_1\pi}{3} \int_0^{\infty} \frac{T_1{}^4}{C_2{}^4} \frac{\zeta^3}{e^\zeta} \, d\zeta$$

Then using the relation $\int_0^{\infty} \zeta^3 e^{-\zeta} \, d\zeta = 3!$ from [23] gives $dQ_{e,1} = 6.4 dA_1 C_1\pi T_1{}^4/C_2{}^4$. But the Stefan-Boltzmann constant $\sigma = 2C_1\pi^5/15C_2{}^4$ so that

$$dQ_{e,1} = \frac{48}{\pi^4} \sigma T_1{}^4 \, dA_1$$

The energy absorbed by dA_1 is

$$dQ_{a,1} = dA_1 \int_{\lambda=0}^{\infty} \int_{A_2} \alpha'_{\lambda,1}(\lambda,\theta_1,\varphi_1) i'_{\lambda o, 2}(\lambda,\theta_2,\varphi_2) \frac{\cos\theta_1 \cos\theta_2}{S^2} \, dA_2 \, d\lambda$$

By using Kirchhoff's law, the directional spectral absorptivity and emissivity can be equated without restriction. Then, for dA_2 taken as a ring element the solid angle $\cos\theta_2 \, dA_2/S^2$ can be written as $2\pi \sin\theta_1 \, d\theta_1$. This is used to write the absorbed energy as

$$dQ_{a,1} = 2\pi(0.8)\,dA_1 \int_{\lambda=0}^{\infty} \int_{\theta_1=0}^{\theta_{1,\max}} (\cos^2\theta_1 \sin\theta_1\,d\theta_1) i'_{\lambda b,2} (1 - e^{-C_2/\lambda T_1})\,d\lambda$$

$$= -1.6\pi\,dA_1 \frac{\cos^3\theta_1}{3}\bigg|_0^{\theta_{1,\max}} \int_0^{\infty} \frac{2C_1(1 - e^{-C_2/\lambda T_1})}{\lambda^5(e^{C_2/\lambda T_2} - 1)}\,d\lambda$$

$$= \frac{3.2\pi C_1\,dA_1}{3}\left[1 - \frac{D^3}{(D^2 + R^2)^{3/2}}\right]\int_0^{\infty} \frac{1 - e^{-C_2/\lambda T_1}}{\lambda^5(e^{C_2/\lambda T_2} - 1)}\,d\lambda$$

If the approximation is invoked that T_1 is close in value to T_2, the integration over λ can be carried out with the following result:

$$dQ_{a,1} = \frac{48}{\pi^4}\left[1 - \frac{1}{(1 + r^2)^{3/2}}\right]\sigma T_2^4\,dA_1$$

where $r = R/D$. Finally, the heat added to dA_1 to maintain it at T_1 is given by

$$dQ_1 = dQ_{e,1} - dQ_{a,1} = \frac{48\sigma}{\pi^4}\left\{T_1^4 - T_2^4\left[1 - \frac{1}{(1 + r^2)^{3/2}}\right]\right\}\,dA_1$$

Even for this illustrative example, it was difficult to construct a realistic analytical function for ϵ'_λ that could be integrated in closed form over both angle and wavelength. Almost invariably it is necessary to use numerical methods to obtain solutions to problems of this type.

A surface with part specular and part diffuse reflectivity and a semigray analysis were used by Shimoji [24] to find the local temperatures in conical and V-groove cavities exposed to incident solar radiation parallel to the cone axis or V-groove bisector plane. Comparison with experiment of the various models using diffuse, specular, semigray, nongray, and combinations of these ideal behaviors was carried out by Toor and Viskanta [25]. They concluded from the particular geometries and materials studied that spectral effects were less important than directional effects and that the presence of one or more diffuse surfaces in an enclosure made the presence of specularly reflecting surfaces unimportant. Hering and Smith [19, 20] and Edwards and Bertak [26] applied various models of surface roughness in calculating radiant exchange between surfaces.

10-7 CONCLUDING REMARKS

Although the formulation of radiation-exchange problems involving directional and/or spectral property effects is not conceptually difficult, it is often very tedious to obtain solutions to the resulting integral equations. To simplify the equations, it is necessary to invoke many assumptions and approximations. The approximations that can be invoked with validity vary from case to case and are so numerous that they have not been discussed in any depth. Numerical techniques of many types can be used for directional spectral problems, since closed-form analytical solutions rarely can be obtained. The number and range of conditions and parameters in these problems precludes the specification of any one numerical technique as being the best. As more interchange problems of this type are investigated, perhaps the most valuable techniques will be evident. One technique is the Monte Carlo method, which is the subject of the next chapter.

REFERENCES

1. Breene, Robert G., Jr.: "The Shift and Shape of Spectral Lines," p. 52, Pergamon Press, New York, 1961.
2. Branstetter, J. Robert: Radiant Heat Transfer between Nongray Parallel Plates of Tungsten, *NASA* TN D-1088, 1961.
3. Eckert, E. R. G., and Robert M. Drake, Jr.: "Heat and Mass Transfer," 2d ed., p. 375, McGraw-Hill Book Company, New York, 1959.
4. Dunkle, R. V., and J. T. Bevans: Part 3, A Method for Solving Multinode Networks and a Comparison of the Band Energy and Gray Radiation Approximations, *J. Heat Transfer*, vol. 82, no. 1, pp. 14–19, 1960.
5. Love, Tom J., and Joel S. Gilbert: Experimental Study of Radiative Heat Transfer between Parallel Plates, ARL-66-0103, DDC no. AD-643307, Oklahoma University, June, 1966.
6. Goodman, Stanley: Radiant-heat Transfer between Nongray Parallel Plates, *J. Res. Nat. Bur. Stand.*, vol. 58, no. 1, pp. 37–40, 1957.
7. Rolling, R. E., and C. L. Tien: Radiant Heat Transfer for Nongray Metallic Surfaces at Low Temperatures, paper no. 67-335, *AIAA*, April, 1967.
8. Bobco, R. P., G. E. Allen, and P. W. Othmer: Local Radiation Equilibrium Temperatures in Semigray Enclosures, *J. Spacecr. Rockets*, vol. 4, no. 8, pp. 1076–1082, 1967.
9. Plamondon, J. A., and C. S. Landram: Radiant Heat Transfer from Nongray Surfaces with External Radiation. Thermophysics and Temperature Control of Spacecraft and Entry Vehicles, *Prog. Astronaut. Aeronaut.*, vol. 18, pp. 173–197, 1966.
10. Bevans, J. T., and D. K. Edwards: Radiation Exchange in an Enclosure with Directional Wall Properties, *J. Heat Transfer*, vol. 87, no. 3, pp. 388–396, 1965.
11. Hering, R. G.: Theoretical Study of Radiant Heat Exchange for Non-Gray Non-Diffuse Surfaces in a Space Environment, Rept. no. ME-TN-036-1, *NASA* CR-81653, Illinois University, September, 1966.
12. Viskanta, Raymond, James R. Schornhorst, and Jaswant S. Toor: Analysis and Experiment of Radiant Heat Exchange between Simply Arranged Surfaces, AFFDL-TR-67-94, DDC No. AD-655335, Purdue University, June 1967.
13. Toor, J. S.: "Radiant Heat Transfer Analysis among Surfaces Having Direction Dependent Properties by the Monte Carlo Method," M.S. thesis, Purdue University, 1967.
14. Hering, R. G.: Radiative Heat Exchange between Specularly Reflecting Surfaces with Direction-dependent Properties, *Proc. 3rd Int. Heat Transfer Conf.*, Chicago, Aug. 7–12, 1966. (*AIChE J.*, vol. 5, pp. 200–206, 1966.)
15. Howell, John R., and Morris Perlmutter: Directional Behavior of Emitted and Reflected Radiant Energy from a Specular, Gray, Asymmetric Groove, *NASA* TN D-1874, 1963.
16. Sparrow, E. M., J. L. Gregg, J. V. Szel, and P. Manos: Analysis, Results, and Interpretation for Radiation between Some Simply-arranged Gray Surfaces, *J. Heat Transfer*, vol. 83, no. 2, pp. 207–214, 1961.
17. Eckert, E. R. G., and E. M. Sparrow: Radiative Heat Exchange between Surfaces with Specular Reflection, *Int. J. Heat Mass Transfer*, vol. 3, no. 1, pp. 42–54, 1961.

18. Beckmann, Peter, and André Spizzichino: "The Scattering of Electromagnetic Waves from Rough Surfaces," The Macmillan Company, New York, 1963.
19. Hering, R. G., and T. F. Smith: Surface Roughness Effects on Radiant Transfer between Surfaces, *Int. J. Heat Mass Transfer*, vol. 13, no. 4, pp. 725–739, 1970.
20. Hering, R. G., and T. F. Smith: Surface Roughness Effects on Radiant Energy Interchange, *ASME J. Heat Transfer*, vol. 93, no. 1, pp. 88–96, 1971.
21. Howell, John R., and Ronald E. Durkee: Radiative Transfer between Surfaces in a Cavity with Collimated Incident Radiation: A Comparison of Analysis and Experiment, *J. Heat Transfer*, vol. 93, no. 2, pp. 129–135, 1971.
22. W. Z. Black: Optimization of the Directional Emission from V-Groove and Rectangular Cavities, *Trans. ASME, J. Heat Transfer*, vol. 95, no. 1, pp. 31–36, 1973.
23. Dwight, Herbert B.: "Tables of Integrals and Other Mathematical Data," 4th ed., p. 230, The Macmillan Company, 1961.
24. Shimoji, S.: Local Temperatures in Semigray Nondiffuse Cones and V-grooves, *AIAA J.*, vol. 15, no. 3, pp. 289–290, 1977.
25. Toor, J. S., and Viskanta, R.: A Critical Examination of the Validity of Simplified Models for Radiant Heat Transfer Analysis, *Int. J. Heat Mass Transfer*, vol. 15, pp. 1553–1567, 1972.
26. Edwards, D. K., and I. V. Bertak: Imperfect Reflections in Thermal Radiation Heat Transfer, *in* John W. Lucas (ed.), "Heat Transfer and Spacecraft Thermal Control," pp. 143–165, vol. 24 in *AIAA Progress in Astronautics and Aeronautics Series*, MIT Press, Cambridge, Mass., 1971.

PROBLEMS

1 Radiative heat flow is occurring across the space between two parallel plates with hemispherical emissivities as shown. What is the net heat flux q transferred from 1 to 2?

Answer: 56,100 Btu/(h·ft²) (177,000 W/m²)

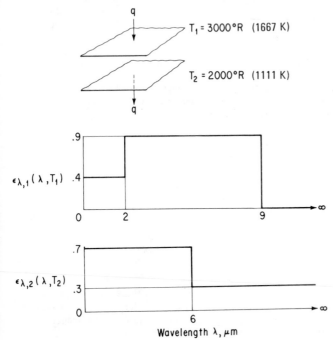

2 A polished aluminum tank in vacuum is surrounded by one thin aluminum radiation shield. The shield is painted on the outside with white paint (see Fig. 5-28). Treating the geometry as infinite parallel plates, what is the heat flux into the tank for normally incident solar radiation? Assume the tank is maintained at 170 K.

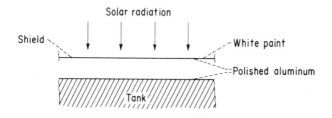

3 Consider the enclosure of Prob. 7 in Chap. 8. Compute the heat flows for the following values of hemispherical spectral emissivity. (For simplicity do not subdivide the three areas into smaller regions.)

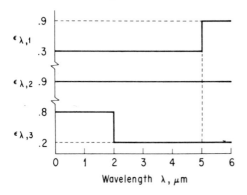

4 Estimate the heat flux leaking into a liquid-hydrogen container from an adjacent container of liquid nitrogen. The glass walls are coated with highly polished aluminum. Use electromagnetic theory to estimate the radiative properties.

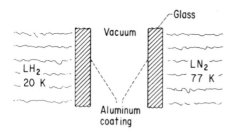

5 An area element dA at temperature T is radiating out through a circular opening of radius r in a plate parallel to dA. The directional total emissivity of dA is $\epsilon'(\theta) = 0.85 \cos \theta$ as in Fig. 3-3. Obtain a relation for the radiation ($Q_{\text{nondiffuse}}$) from dA through the opening. From Example 3-3 the hemispherical total emissivity of dA is $\epsilon = 0.57$. Using this ϵ and the diffuse

F factor, compute the radiation (Q_{diffuse}) through the opening. Plot $Q_{\text{nondiffuse}}/Q_{\text{diffuse}}$ as a function of h/r.

Answer: $\dfrac{Q_{nd}}{Q_d} = \dfrac{(H^2+1)^{3/2} - H^3}{(H^2+1)^{1/2}}$; $\quad H = \dfrac{h}{r}$.

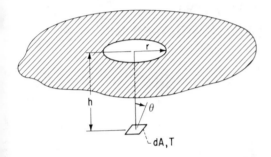

6 A sphere and area element are positioned as shown. The sphere is gray but has a nondiffuse emissivity $\epsilon_s = E\cos\theta$, where θ is the angle measured from the normal of the sphere surface and E is a constant. Set up the integral for the direct radiation from the sphere to the area element in terms of the quantities given.

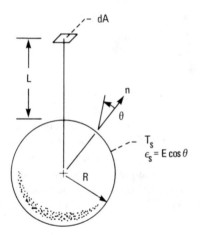

ELEVEN

THE MONTE CARLO APPROACH TO RADIANT-INTERCHANGE PROBLEMS

11-1 INTRODUCTION

In Chap. 10 it was found that the enclosure-theory analysis became very complex when directional- and spectral-surface property variations were accounted for. An alternative approach that can deal with these complexities is the Monte Carlo method. Since Monte Carlo is a statistical numerical method, it is first necessary to discuss some concepts of statistical theory. Then the basic procedure is outlined with regard to radiative exchange; to demonstrate the method, two example problems are formulated. Because the use of Monte Carlo requires a digital computer, complete example problem solutions are not given. Only the straightforward Monte Carlo approach will be presented. Some of the refinements that can shorten computation time by increasing accuracy will be briefly discussed.

A general view of the radiation heat transfer problems solved in the literature by Monte Carlo will be given. This survey will further demonstrate how the method can be utilized and will provide a source for available techniques that have been developed. Much of the material presented here is taken from [1].

11-1.1 Definition of Monte Carlo

Herman Kahn [2] has given the following definition of the Monte Carlo method which seems to incorporate the salient ideas: "The expected score of a player in any reasonable game of chance, however complicated, can in principle be estimated by averaging the results of a large number of plays of the game. Such estimation can be rendered more efficient by various devices which replace the game with another known to have the same expected score. The new game may lead to a more efficient estimate by being less erratic, that is, having a score of lower variance, or by being cheaper to play with the equipment on hand. There are obviously many problems about probability that can be viewed as problems of calculating the expected score of a game. Still more, there are problems that do not concern probability but are none the less equivalent for some purposes to the calculation of an expected score. The Monte Carlo method refers simply to the exploitation of these remarks."

This definition also provides a good outline for use of the method. Indeed, what must be done for a specific problem is to set up a game or model that has the same behavior, and hence is expected to produce the same outcome, as the physical problem that the model simulates; make the game as simple and fast to play as possible; use any available methods to reduce the variance of the average outcome of the game; then play the game many times and find the average outcome. After some remarks on the history of the method and on the approach being taken here to summarize and outline it, this formalism will be applied to problems in radiative heat transfer.

11-1.2 History

The history of "experimental mathematics" can be traced quite far into the past. Hammersley and Handscomb [3] give references to over 300 works dealing with Monte Carlo and closely related material published over the last six decades. They mention a determination of the value of π by a numerical experiment performed some thousands of years ago [4]. However, the great bulk of the literature has appeared since 1950.

Many early workers carried out numerical experiments by such means as throwing dice or playing card games many times over to determine the probability of a given outcome, but useful results from such methods awaited the unique abilities of high-speed digital computers. These machines could play simulations of the game at a high rate and thus compile accurate averages in a reasonable time.

Credit for development of Monte Carlo techniques as they are presently used in engineering and science goes to the extremely competent group of physicists and mathematicians who gathered at Los Alamos during the early work on nuclear weapons, including especially John von Neumann and Stanislaw Ulam.

11-1.3 General References

Referring to *the* Monte Carlo method is probably meaningless, although such terminology will be applied. Any specific problem more likely entails *a* Monte Carlo method, as the label has been placed on a large class of loosely related techniques. A number of general books and monographs are available that detail methods and/or review the literature. A valuable early outline is given in [5], which is the first work to use the term Monte Carlo for the approach being considered here. For clarity and usefulness, both [2] and [3] are valuable, as are the general texts by Cashwell and Everett [6], Schreider [7] (who gives 282 references, many to the foreign literature), Brown [8],

Halton [9], and the many excellent papers gathered in the symposium volume edited by Meyer [10]. The references cited give mathematical justification for some of the methods employed in Monte Carlo. The intention here is to give arguments based on physical foundations, with emphasis on why the mathematical forms evolve. No attempts to provide proofs of statistical laws will be made; the standard texts in statistics carry out these proofs in detail.

No definitive method of predicting computer running time exists for most Monte Carlo problems. The time used will depend, of course, on the machine used, and on the ability of the programmer to pick methods and shortcuts that will reduce the burden on the machine. An example of such a shortcut is the use of special sub-routines for computation of such functions as sine and cosine. These routines sacrifice some accuracy to a gain in speed. If problem answers accurate to a few percent are desired, then the use of eight-place functions from a relatively slow subroutine is a needless luxury, especially if the subroutine is to be used tens of thousands of times.

Finally, only this paragraph will be devoted to the fruitless argument as to whether Monte Carlo or some other method is a "better" way of attacking a given radiation problem. Suppose that a set of integral equations must be solved simultaneously to obtain an analytical solution to a given physical problem. A Monte Carlo solution of a physical analog may lead to a lengthy computer run. The question facing the programmer is then: Is it better to program the solution of the integral equations by finite-difference iterative techniques with the possibility that convergence to correct solutions will not be attained due to round-off errors or instabilities, or by Monte Carlo, which though long-running, will give the answer sooner or later? In general, there can be no definitive reply to this question. Only the background and intuition of the individual researcher can give some clue as to the most likely direction of attack. It is hoped that the following material will provide a basis for such decisions.

11-2 SYMBOLS

A	surface area
$A, B, C,$ $D, \cdots$	constants
E	exchange factor including direct exchange and all reflection paths
$F_{0-\lambda}$	fraction of total energy emitted by a blackbody in wavelength range $0-\lambda$
$f(\xi)$	frequency distribution of events occurring at ξ
I	total number of subsets used to compute mean
i	radiant intensity
l, m	lattice indices in square mesh corresponding to x, y positions, respectively
N	total number of sample bundles per unit time
n	individual sample index
P	probability density function
$\bar{P}$	mean of calculated values of P
Q	energy per unit time
R	number chosen at random from evenly distributed set of numbers in range $0-1$, random number

S	number of events occurring at some position
T	absolute temperature
w	energy carried by sample Monte Carlo bundle
x,y	positions in cartesian coordinate system
α	radiative surface absorptivity
γ	standard deviation defined by Eq. (11-15)
δ,δ',δ''	indices in computer program, Fig. 11-7
ϵ	radiative surface emissivity
η	function defined by Eq. (11-14)
θ	cone angle
λ	wavelength
μ	probable error
φ	circumferential angle
ξ	variable
σ	Stefan-Boltzmann constant

Subscripts

b	blackbody
e	emitted
λ	spectrally dependent
$1,2$	at surface 1 or 2

Superscripts

$'$	quantity in one direction
$''$	bidirectional quantity
$*$	denotes dummy variable

11-3 DETAILS OF THE METHOD

11-3.1 The Random Walk

Any reader looking into the background of the material to be presented here will soon encounter the term *Markov chain*. A Markov chain is simply a chain of events occurring in sequence with the condition that the probability of each succeeding event in the chain is uninfluenced by prior events. The usual example of this is a totally inebriated gentleman who begins a walk through a strange city. At each street corner that he reaches, he becomes confused. In continuing his walk, he chooses completely at random one of the streets leading from the intersection. In fact, he may walk up and down the same block several times before he chances to move off down a new street. Ths history of his walk is then a Markov chain, as his decision at any point is not influenced by where he has been.

Because of the randomness of his choice at each intersection, it might be possible to simulate a sample walk by constructing a "four-holer," that is, a roulette wheel with only four positions, each corresponding to a possible direction. The probability of the gentleman starting at his hotel bar and reaching any point in the city limits could then be found by simulating a large number of histories, using the four-holer to determine the direction of the walk at each decision point in each history.

It might be noted that the probability of the man reaching intersection (l,m) on a square grid representing the city street map is simply

$$P(l,m) = \tfrac{1}{4}[P(l+1,m) + P(l-1,m) + P(l,m+1) + P(l,m-1)] \tag{11-1}$$

where the factors in brackets are the probabilities of his being at each of the adjacent four intersections. This is because the probability of reaching $P(l,m)$ from a given adjacent intersection is one-fourth. This type of random walk is a convenient model for processes that are described by Laplace's equation; Eq. (11-1) is recognized as the finite-difference analog of the Laplace equation.

The probability of a certain occurrence for other processes is usually not as immediately obvious as Eq. (11-1). More often, the probability of an event must be determined from physical constraints, and then the decision as to what event will occur is made on the basis of this probability. Some of the basic methods of choosing an event from a known probability distribution of events will now be examined. Also, means of constructing these distributions will be discussed.

11-3.2 Choosing from Probability Distributions

Consider a very poor archer firing arrows at a target with an outer radius having 10 units of length. After firing many arrows, the number of arrows $F(\xi)$ that are found to have struck the target within a small radius increment $\Delta\xi$ about some radius ξ can be represented by a histogram of the frequency function $f(\xi) = F(\xi)/\Delta\xi$, where $0 \leqslant \xi \leqslant 10$. A smooth curve can then be passed through the histogram to give a continuous frequency distribution, similar to that of Fig. 11-1. What is now needed is a method

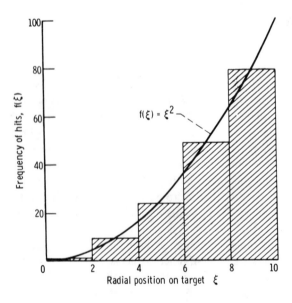

Figure 11-1 Frequency distribution of arrows at various target radii.

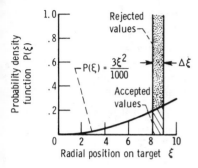

Figure 11-2 Probability density function of arrows on target.

for simulating further shots. This method should assign an expected radius ξ on the target to each of a group of succeeding arrows. In addition, the distribution of ξ values should correspond to the frequency distribution that the archer has previously fired as shown in Fig. 11-1. (It is assumed that all his arrows have hit somewhere on the target.)

This situation is analogous to that encountered in many Monte Carlo processes. The distribution of values that occurs in a given physical process is known, and a method of assigning values to individual samples is desired so that the distribution of values for all the samples will agree with the required distribution. In radiant heat transfer, for example, it is known that the distribution with wavelength of the spectral energy emitted by a blackbody must follow the Planck spectral emission curve. How are individual energy "bundles" of radiation each assigned a wavelength so that, after a large number of bundles are emitted from a blackbody, the distribution of emitted energy is indeed Planckian?

In addition, for a Markov process, the values at each step must be assigned in a random manner so that each decision in the chain is independent.

Following the archer's progress will show how this is done. The frequency curve given in Fig. 11-1 can in this example be approximated by the analytical expression

$$f(\xi) = \xi^2 \tag{11-2}$$

in the interval $0 \leqslant \xi \leqslant 10$, and $f(\xi) = 0$ elsewhere, because all the arrows struck the target. Equation (11-2) is normalized by dividing by the area under the frequency curve (that is, the total number of arrows) to obtain

$$P(\xi) = \frac{f(\xi)}{\int_0^{10} f(\xi)\, d\xi} = \frac{3\xi^2}{1000} \tag{11-3}$$

If the frequency with which arrows have struck the target radii is taken as the basis for estimating the locations the next set will strike, then the *probability density function* defined by (11-3) is the average distribution that must be satisfied by the ξ values determined by the simulation scheme. The probability density function is plotted in Fig. 11-2 and is interpreted physically as the proportion of values (arrows) that lie in the region $\Delta\xi$ around ξ.

To determine ξ values, the simulation scheme can proceed as follows: Choose two *random numbers* R_A and R_B from a large set of numbers evenly distributed in the range 0-1. (How these numbers are chosen in a practical calculation is discussed in Sec. 11-3.3.) The two random numbers are then used to select a point $(P(\xi), \xi)$ in Fig. 11-2 by setting $P(\xi) = R_A$ and $\xi = (\xi_{max} - \xi_{min})R_B = 10R_B$. This value of $P(\xi)$ is then compared to the value of $P(\xi)$ computed at ξ from (11-3). If the randomly selected value lies above the computed value of $P(\xi)$, then the randomly selected value of ξ is rejected and two new random numbers are selected. Otherwise, the value of ξ that has been found is listed as the location that the arrow will strike. Referring again to Fig. 11-2, it is seen that such a procedure assures that the correct fraction of ξ values selected for use will lie in each increment $\Delta\xi$ after a large number of completely random selections of $(P(\xi), \xi)$ is made.

The difficulty with such an event-choosing procedure is that in some cases a large portion of the ξ values may be rejected because they lie above the $P(\xi)$ curve. A more efficient method for choosing ξ is therefore desirable.

One such method is to integrate the probability density function $P(\xi)$ using the general relation

$$R(\xi) = \int_{-\infty}^{\xi} P(\xi^*) \, d\xi^* \tag{11-4}$$

where $R(\xi)$ can only take on values in the range 0-1 because the integral under the entire $P(\xi)$ curve is unity according to Eq. (11-3). Equation (11-4) is the general definition of the *cumulative distribution function*. A plot of R against ξ from (11-4) shows the probability of an event occuring in the range $-\infty$-ξ. For the method given here, the function R is taken to be a random number; each value of ξ is then obtained by choosing an R value at random and using the functional relation $R(\xi)$ to determine the corresponding value of ξ. To show that the probability density of ξ formed in this way corresponds to the required $P(\xi)$, the probability density function of Fig. 11-2 can be used as an illustrative example. Inserting the example $P(\xi)$ of (11-3) into (11-4) and noting that $P(\xi) = 0$ for $-\infty < \xi < 0$ give

$$R = \int_{0}^{\xi} P(\xi^*) \, d\xi^* = \frac{\xi^3}{1000} \qquad 0 \leqslant R \leqslant 1 \tag{11-5}$$

Equation (11-5) is shown plotted in Fig. 11-3.

Now it will be shown that choosing R at random and determining a corresponding value of ξ from (11-5) is equivalent to taking the derivative of the cumulative distribution function and that this derivative is, by examination of (11-5) and (11-3), simply $P(\xi)$. Divide the range of ξ into a number of equal increments $\Delta\xi$. Suppose that M values of R are now chosen in the range 0-1 and that these M values are chosen at equal intervals along R. There will be M values of ξ which correspond to these M values of R. The fraction of the M values of ξ which occurs per given increment $\Delta\xi$ is then $M_{\Delta\xi}/M = \Delta R$, which gives

$$\frac{M_{\Delta\xi}/M}{\Delta\xi} = \frac{\Delta R}{\Delta\xi} \tag{11-6}$$

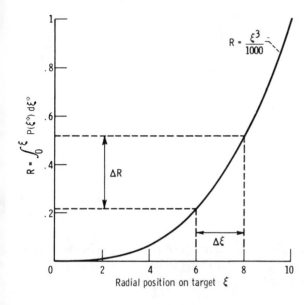

Figure 11-3 Cumulative distribution of arrows on target.

The quantity $\Delta R/\Delta \xi$ approaches $dR/d\xi$ if a large enough value is used for M and small increments $\Delta \xi$ are examined. But $dR/d\xi$ can be seen from (11-5) and (11-3) to be simply $P(\xi)$; therefore, by obtaining values of ξ as described preceding (11-5), the required probability distribution is indeed generated.

Often physical problems arise in which the frequency distribution depends on more than one variable. For example, if the archer discussed previously suffered from astigmatism, then a dependence on circumferential angle φ might appear in the distribution of arrows on the target in addition to the dependence on radius. If the interdependence of the variables is such that the frequency distribution can be factored into a product form, then the following can be written:

$$f(\xi,\varphi) = g(\xi)h(\varphi) \tag{11-7}$$

Values of $P(\xi)$ and $P(\varphi)$ can be found by integrating out each variable in turn to obtain

$$P(\xi) = \frac{\int_{\varphi_{min}}^{\varphi_{max}} f(\xi,\varphi)\,d\varphi}{\int_{\xi_{min}}^{\xi_{max}} \int_{\varphi_{min}}^{\varphi_{max}} f(\xi,\varphi)\,d\varphi\,d\xi} = \frac{g(\xi)\int_{\varphi_{min}}^{\varphi_{max}} h(\varphi)\,d\varphi}{\int_{\xi_{min}}^{\xi_{max}} g(\xi)\,d\xi \int_{\varphi_{min}}^{\varphi_{max}} h(\varphi)\,d\varphi} = \frac{g(\xi)}{\int_{\xi_{min}}^{\xi_{max}} g(\xi)\,d\xi}$$

and, similarly $\tag{11-8}$

$$P(\varphi) = \frac{\int_{\xi_{min}}^{\xi_{max}} f(\xi,\varphi)\,d\xi}{\int_{\varphi_{min}}^{\varphi_{max}} \int_{\xi_{min}}^{\xi_{max}} f(\xi,\varphi)\,d\xi\,d\varphi} = \frac{h(\varphi)}{\int_{\varphi_{min}}^{\varphi_{max}} h(\varphi)\,d\varphi} \tag{11-9}$$

The methods given previously in this section are used to evaluate ξ and φ independently of one another after two random numbers are chosen.

If $f(\xi,\varphi)$ cannot be placed in the form of Eq. (11-7) [that is, if there are no independent $g(\xi)$ and $h(\varphi)$], then it can be shown [2, 7] that ξ and φ values can be determined by choosing two random numbers R_ξ and R_φ. Note that

$$P(\xi,\varphi) = \frac{f(\xi,\varphi)}{\int_{\xi\,min}^{\xi\,max} \int_{\varphi\,min}^{\varphi\,max} f(\xi,\varphi)\,d\varphi\,d\xi}$$

Then ξ and φ are found from the equations

$$R_\xi = \int_{-\infty}^{\xi} \int_{\varphi_{min}}^{\varphi_{max}} P(\xi^*,\varphi)\,d\varphi\,d\xi^* \tag{11-10}$$

and

$$R_\varphi = \int_{-\infty}^{\varphi} P(\varphi^*, \xi = \text{fixed})\,d\varphi^* \tag{11-11}$$

where ξ in (11-11) is that value obtained from (11-10). This procedure may be extended to any number of variables. Equations (11-10) and (11-11) define the *marginal* and *conditional* distributions of $P(\xi,\varphi)$, respectively.

11-3.3 Random Numbers

What random numbers are. Formally, a random number is a number chosen without sequence from a large set of numbers spaced at equivalued intervals. For our purposes, the numbers are in the range 0-1. If the numbers $0, 0.01, 0.02, 0.03, \cdots, 0.99, 1.00$ are placed on slips of paper and then the jumbled slips are placed in a hat, there would be fair assurance that, if a few numbers are picked, they will be random numbers. If many choices are to be made, then perhaps smaller intervals (more slips) should be used; after it is drawn, each slip should be replaced and randomly mixed in the hat.

For a typical computer problem, random numbers might be needed for 10^5 or more decisions. It is desirable to have a rapid way of obtaining them; it is also desirable that the numbers chosen be truly random.

How random numbers are generated. On the modern digital computer, it is impractical to fit a mechanical arm and an optical scanner to choose and interpret slips pulled from a hat. To give truly random numbers, one possibility would be to sample a truly random process. Such phenomena as noise in an electronic circuit or radioactive-decay particle counts per unit time have been tried, but in the main they are found to be too slow for direct computer linkage.

A second means is to obtain or generate tables of random numbers [11, 12], perhaps by one of the processes mentioned previously, and then enter these tables in the computer memory. This allows rapid access to random numbers, but for complex problems requiring a large quantity of random numbers the required storage space becomes prohibitive. This method has been widely used, however, when a modest problem is to be solved.

The most widely practiced method at present for obtaining random numbers for a digital computer is a pseudorandom-number generator. This is simply a subroutine that exploits the apparent randomness of groups of digits in large numbers. One simple example of such a routine is to take an eight-digit number, square it, and then choose the middle eight digits of the resulting sixteen-digit number as the required random number. When a new random number is needed, square the previous random number, and the new random number is taken as the middle eight digits of the result. This process is said by Schreider [7] to degenerate after a few thousand cycles by propagating to an all-zero number.

A more satisfactory routine is based on suggestions in [13]. Here a random number is generated by taking the low-order 36 bits of the product $R_{n-1}K$, where $K = 5^{15}$ and R_{n-1} is the previously computed random number. The subroutine is started by taking $R_0 = 1$, or the programmer may give R_0 an arbitrary value. By always starting a given program with the same R_0, it is possible to check solutions through step-by-step tracing of a few histories. Many subroutines for generating random numbers that are available in software systems are based on this approach.

How the numbers are made sufficiently random. The fact that such subroutines generate *pseudo*random numbers immediately raises a danger flag. How can it be established that such pseudorandomness is sufficiently random for the problem being treated? Does the sequence repeat? If so, after how many numbers? Certain standard tests exist that give partial answers to these questions, and a full discussion of them is given in [3, 13, 14]. None of these tests is sufficient to establish randomness, although passing the tests is necessary. Kendall and Smith [14] describe four such tests. The names they ascribe give the flavor of the methods: the frequency test, serial test, poker test, and gap test. These tests are described as "... useful and searching. They are, however, not sufficient ...."

Perhaps the safest course to follow is to obtain a standard subroutine whose properties have been established by such tests and use it only within its proven limits. The applicability of a given pseudorandom-number generator can be checked to some extent by generating the mean of some known distributions appearing in the problem at hand and comparing the results with analytically determined mean values.

11-3.4 Evaluation of Error

Because the solutions obtained by Monte Carlo are averages over the results of a number of individual samples, they will in general contain fluctuations about a mean value. As in any process of this type, the mean can be more accurately determined by increasing the number of values used in determining the mean. Although it is not possible to ascribe a 100% confidence in the value obtained, such confidence can be approached as closely as desired so long as the budget for computer time can stand the strain. More generally some ad hoc rules of economy and an estimate of desired accuracy in a given problem can be applied, and solutions can be obtained by trading off within these limits.

To establish the accuracy of the solutions, one of several tests can be applied. For example, suppose we want to know the probability of the randomly staggering attendee of an engineering convention (discussed in Sec. 11-3.1) reaching a certain bar at the city limits. To exactly determine his success, an infinite number of hypothetical engineers have to be followed, and the probability $P(l,m)$ of reaching the boundary point (l,m) can be determined as

$$P(l,m) = \left[\frac{S(l,m)}{N}\right]_{N\to\infty} \tag{11-12}$$

where $S(l,m)/N$ is the number of samples $S(l,m)$ reaching the boundary point divided by the total number of samples N. Obviously, following an infinite number of samples would not be economical, and a probability would be computed based on some finite number of samples N, of order perhaps 10^2–10^6. Then an estimate is needed of the error μ involved in approximating infinity with these relatively small sample sizes.

For a sample size greater than about $N = 20$, it is found [3, 7], from application of the central limit theorem and the relations governing normal probability distributions, that the following relation holds whenever the samples S in question can be considered to leave a source and either reach a scoring position with probability P or not reach it with probability $1 - P$. The probability that the average $S(l,m)/N$ for finite N differs by less than some value μ from $[S(l,m)/N]_{N\to\infty}$ is given by

$$P\left[\left|\frac{S}{N} - \left(\frac{S}{N}\right)_{N\to\infty}\right| \leqslant \mu\right] = \frac{2}{\sqrt{\pi}} \int_0^{\eta/\sqrt{2}} e^{-\eta^{*2}} d\eta^* = \operatorname{erf}\frac{\eta}{\sqrt{2}} \tag{11-13}$$

where

$$\eta \approx \mu \left[\frac{N}{(S/N)(1 - S/N)}\right]^{1/2} \tag{11-14}$$

Compilations of the error function (erf) are given in many standard reference tables [15, 16].

In many problems, such an error-estimation procedure cannot be applied because the samples do not originate from a single source. For example, the radiative energy flux at a point on the boundary of an enclosure may depend on the energy arriving from many sources. For such situations, the most straightforward way of estimating the error in a result (such as the error in radiative heat flux at a point) is to subdivide the calculation of the desired statistical mean result into a group of I submeans. The *central limit theorem* then applies. This theorem states that the statistical fluctuations in the submeans are distributed in a normal or gaussian distribution about the overall mean. For such a distribution, a measure of the fluctuations in the means can be calculated. This measure is called the *variance*. For example, if 2000 samples are examined, a mean result $\bar{P}$ is calculated on the basis of the samples, and 20 submeans $P_1, P_2, \ldots, P_I$ of 100 samples each are calculated. Then the *variance* γ^2 of the mean solution $\bar{P}$ is given by

$$\gamma^2 = \frac{1}{I-1}\left[\sum_{i=1}^{I}(P_i - \bar{P})^2\right] = \frac{1}{I-1}\left[\sum_{i=1}^{I} P_i^2 - \frac{\left(\sum_{i=1}^{I} P_i\right)^2}{I}\right] \tag{11-15}$$

This variance is an estimate of the mean-square deviation of the sample mean $\bar{P}$ from the true mean, where the true mean would be obtained by using an infinite number of samples. From the properties of the normal frequency distribution, which the fluctuations in the results computed by Monte Carlo will in general follow, it is shown in most texts on statistics that the probability of the sample mean $\bar{P}$ lying within $\pm\gamma$ of the true mean is about 68%, of lying within $\pm 2\gamma$ is about 95%, and of lying within $\pm 3\gamma$ is 99.7%.

Another measure of the statistical fluctuations in the mean is γ, the *standard deviation*. Because γ is given by the square root of Eq. (11-15), it is evident that to reduce γ by half, the number of samples used in computing the results must be quadrupled (thereby quadrupling I for constant submean size). This probably means quadrupling the computer time involved unless the term in brackets can somehow be reduced by decreasing the variance (scatter) of the individual submeans. Much time and ingenuity have been expended in attempts at the latter, under such labels as stratified sampling, "splitting," importance sampling, and energy partitioning. These and other variance-reducing techniques are discussed in [3, 7, 17]. The savings in computer time available from application of these techniques is abundant reward for their study, and the reader who intends to use Monte Carlo for any problem of significant complexity is urged to apply them.

11-4 APPLICATION TO THERMAL RADIATIVE TRANSFER

11-4.1 Introduction

As discussed in Chaps. 8–10, the formulation of radiation-exchange heat balances in enclosures leads to integral equations for the unknown surface temperature or heat-flux distributions. Integral equations also result when radiation exchange is considered within a radiating medium such as a gas. These equations can be quite difficult to solve and are a consequence of using a "macroscopic" viewpoint when deriving the heat-flow quantities. By invoking a probabilistic model of the radiative-exchange process and applying Monte Carlo sampling techniques, it is possible to utilize a "semimacroscopic"* approach and avoid many of the difficulties inherent in the averaging processes of the usual integral-equation formulations. In this way, actions of small parts of the total energy can be examined on an individual basis, in place of an attempt to solve simultaneously for the entire behavior of all the energy involved. A microscopic type of model for the radiative exchange process will be examined; then the solution of two examples will be outlined.

11-4.2 Model of the Radiative Exchange Process

In engineering radiation calculations, the usual quantities of interest are the local temperatures and energy fluxes. It seems reasonable to model the radiative exchange process by following the progress of discrete amounts ("bundles") of energy, since

---

* Or, perhaps, "semimicroscopic."

local energy flux is then easily computed as the number of these energy "bundles" arriving per unit area and time at some position. The obvious bundle to visualize is the photon, but the photon has a disadvantage as a basis for a model; its energy depends on its wavelength, which would introduce a needless complication. Therefore, a model particle is devised that is more convenient. This is the *photon bundle*, which is a bundle carrying a given amount of energy w; it can be thought of as a group of photons bound together. For spectral problems where the wavelength of the bundle is specified, enough photons of that wavelength are grouped together to make the energy of the bundle equal to w.

By assigning equal energies to all photon bundles, local energy flux is computed by counting the number of bundles arriving at a position of interest per unit time and per unit area and multiplying by the energy of the bundle. The bundle paths and histories are computed by the Monte Carlo method as will now be demonstrated by an example problem.

11-4.3 Sample Problem

Let us look at a simple problem outlined in [18], and examine the energy radiated from element dA_1 at temperature T_1 that is absorbed by an infinite plane A_2 at temperature $T_2 = 0$ (see Fig. 11-4). Let element dA_1 have emissivity

$$\epsilon'_{\lambda,1} = \epsilon'_{\lambda,1}(\lambda, \theta_1, T_1) \tag{11-16}$$

and let area 2 have emissivity

$$\epsilon'_{\lambda,2} = \epsilon'_{\lambda,2}(\lambda, \theta_2, T_2) \tag{11-17}$$

and assume only that the emissivity of both surfaces is independent of circumferential angle φ.

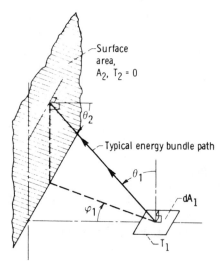

Surface area, A_2, $T_2 = 0$

θ_2

Typical energy bundle path

θ_1

dA_1

φ_1

T_1

Figure 11-4 Radiant interchange between two surfaces.

For surface element dA_1, the total emitted energy per unit time is

$$dQ_{e,1} = \epsilon_1(T_1)\sigma T_1^4 dA_1 \tag{11-18}$$

where $\epsilon_1(T_1)$ is the hemispherical total emissivity, given in this case by Eq. (3-6a):

$$\epsilon_1(T_1) = \frac{\int_0^\infty \int_\Omega \epsilon'_{\lambda,1}(\lambda,\theta_1,T_1)i'_{\lambda b,1}(\lambda,T_1)\cos\theta \, d\omega \, d\lambda}{\sigma T_1^4} \tag{11-19}$$

where $i'_{\lambda b,1}(\lambda,T_1)$ is the Planck spectral distribution of blackbody radiant intensity at T_1.

If it is assumed that $dQ_{e,1}$, the total energy emitted per unit time by dA_1, is composed of N energy bundles emitted per unit time, then the energy assigned to each bundle is

$$w = \frac{dQ_{e,1}}{N} \tag{11-20}$$

To determine the energy radiated from element dA_1 that is absorbed by surface A_2, follow N bundles of energy after their emission from dA_1 and determine the number S_2 absorbed at A_2. If the energy reflected from A_2 back to dA_1 and then rereflected to A_2 is neglected, the energy transferred per unit time from dA_1 to A_2 will be

$$dQ_{1\to\text{absorbed by 2}} = wS_2 = \frac{\epsilon_1(T_1)\sigma T_1^4 dA_1}{N}S_2 \tag{11-21}$$

The next question is how to determine the path direction and wavelength that are assigned to each bundle. This must be done in such a way that the directions and wavelengths of the N bundles conform to the constraints given by the emissivity of the surface and the laws governing radiative processes. For example, if wavelengths are assigned to N bundles, the spectral distribution of emitted energy generated by the Monte Carlo process (comprised of the energy $wN_\lambda \, \Delta\lambda$ for discrete intervals $\Delta\lambda$) must closely approximate the spectrum of the actual emitted energy (plotted as $\pi\epsilon_{\lambda,1}i'_{\lambda b,1} \, d\lambda$ against λ). To assure this, the methods of Sec. 11-3.2 are applied.

The energy emitted by element dA_1 per unit time in the wavelength interval $d\lambda$ about the wavelength λ, and in the angular interval $d\theta_1$ about θ_1 is

$$d^3Q'_{\lambda e,1}(\lambda,\theta_1) = 2\pi\epsilon'_{\lambda,1}(\lambda,\theta_1,T_1)i'_{\lambda b,1}(\lambda,T_1)\cos\theta_1 \, dA_1 \sin\theta_1 \, d\theta_1 \, d\lambda \tag{11-22}$$

The total energy emitted by dA_1 per unit time is given by Eq. (11-18). The probability $P(\lambda,\theta_1) \, d\theta_1 \, d\lambda$ of emission in a wavelength interval about λ and in an angular interval around θ_1 is then the energy in $d\theta_1 \, d\lambda$ [Eq. (11-22)] divided by the total emitted energy [Eq. (11-18)]:

$$P(\lambda,\theta_1) \, d\theta_1 \, d\lambda = \frac{d^3Q'_{\lambda e,1}(\lambda,\theta_1)}{dQ_{e,1}} = \frac{2\pi\epsilon'_{\lambda,1}(\lambda,\theta_1)i'_{\lambda b,1}(\lambda)\cos\theta_1 \sin\theta_1 \, d\theta_1 \, d\lambda}{\epsilon_1\sigma T_1^4} \tag{11-23}$$

(The T_1 in the functional notation has been dropped for simplicity.)

It is assumed here for simplicity that the directional spectral emissivity is a product function of the variables angle and wavelength; that is,

$$\epsilon'_{\lambda,1}(\lambda,\theta_1) = \Phi_1(\lambda)\Phi_2(\theta_1) \tag{11-24}$$

This assumption is probably not valid for many real surfaces since, in general, the angular distribution of emissivity depends on wavelength as shown, for example, by Fig. 5-1. For the assumed form in (11-24), it follows that the emissivity dependence on either variable may be found by integrating out the other variable [see Eq. (11-8)]. Then the normalized probability of emission occurring in the interval $d\lambda$ is

$$P(\lambda)\,d\lambda = d\lambda \int_0^{\pi/2} P(\lambda,\theta_1)\,d\theta_1 = \frac{2\pi\,d\lambda \int_0^{\pi/2} \epsilon'_{\lambda,1}(\lambda,\theta_1)i'_{\lambda b,1}(\lambda)\sin\theta_1\cos\theta_1\,d\theta_1}{\epsilon_1\sigma T_1^4}$$

$$\tag{11-25a}$$

Substituting into (11-4) and noting that $P(\lambda)\,d\lambda$ is zero in the range $-\infty < \lambda < 0$ give

$$R_\lambda = \frac{2\pi \int_0^\lambda \int_0^{\pi/2} \epsilon'_{\lambda,1}(\lambda^*,\theta_1)i'_{\lambda b,1}(\lambda^*)\sin\theta_1\cos\theta_1\,d\theta_1\,d\lambda^*}{\epsilon_1\sigma T_1^4} \tag{11-25b}$$

where the asterisk denotes a dummy variable of integration. If the number of bundles N is very large and this equation is solved for λ each time a random R_λ value is chosen, the computing time becomes too large for practical calculations. To circumvent this difficulty, equations like Eq. (11-25b) can be numerically integrated once over the range of λ values and a curve can be fitted to the result. A polynomial approximation

$$\lambda = A + BR_\lambda + CR_\lambda^2 + \cdots \tag{11-26}$$

is often adequate. Equation (11-26) rather than (11-25b) is used in the problem-solving program. Alternatively, a table of λ versus R_λ can be put into memory, and interpolated to obtain λ for chosen R_λ values.

Following a similar procedure for the variable cone angle of emission θ_1 gives the relation

$$R_{\theta_1} = \int_0^{\theta_1}\int_0^\infty P(\theta_1^*,\lambda)\,d\lambda\,d\theta_1^* = \frac{2\pi\int_0^{\theta_1}\int_0^\infty \epsilon'_{\lambda,1}(\lambda,\theta_1^*)i'_{\lambda b,1}(\lambda)\sin\theta_1^*\cos\theta_L^*\,d\lambda\,d\theta_1^*}{\epsilon_1\sigma T_1^4}$$

$$\tag{11-27}$$

which is curve-fit to give

$$\theta_1 = D + ER_{\theta_1} + FR_{\theta_1}^2 + \cdots \tag{11-28}$$

If dA_1 is a diffuse-gray surface, (11-25b) reduces to

$$R_{\lambda,\text{diffuse gray}} = \frac{\pi\int_0^\lambda i'_{\lambda b,1}(\lambda^*)\,d\lambda^*}{\sigma T_1^4} = F_{0-\lambda} \tag{11-29}$$

where $F_{0-\lambda}$ is the fraction of blackbody emission in the wavelength interval $0-\lambda$. Equation (11-27) for this case reduces to

$$R_{\theta_1,\text{diffuse gray}} = 2\int_0^{\theta_1}\sin\theta_1^*\cos\theta_1^*\,d\theta_1^* = \sin^2\theta_1 \tag{11-30a}$$

or

$$\sin \theta_1 = \sqrt{R_{\theta_1, \text{diffuse gray}}} \qquad (11\text{-}30b)$$

The point to be made here is that computational difficulty is not greatly different in obtaining λ from either (11-26) or (11-29), nor is it much different for obtaining θ_1 from either (11-28) or (11-30b). The difference between the nondiffuse-nongray case and the diffuse-gray case is mainly in the auxiliary numerical integrations of (11-25b) and (11-27). These integrations are performed once to obtain the curve fits; then as far as the main problem-solving program is concerned, the more difficult case might just as well be handled. Thus, increasing problem complexity leads to only gradual increases in the complexity of the Monte Carlo program and similar gradual increases in computer time.

For emission of an individual energy bundle from surface dA_1, a wavelength λ can be obtained from (11-26), and a cone angle of emission θ_1 can be obtained from (11-28) by choosing two random numbers R_λ and R_{θ_1}. To define the bundle path, there remains only specification of the circumferential angle φ_1. Because of the assumption made earlier that emission does not depend on φ_1, it is shown by the formalism outlined and is also fairly obvious from intuition that φ_1 can be determined by

$$\varphi_1 = 2\pi R_{\varphi_1} \qquad (11\text{-}31)$$

where R_{φ_1} is again a random number chosen from the range 0–1.

Because the position of plane A_2 with respect to dA_1 is known, it is a simple matter to determine whether a given energy bundle will strike A_2 after leaving dA_1 in direction (θ_1, φ_1). (It will hit A_2 whenever $\cos \varphi_1 \geqslant 0$ as shown in Fig. 11-4.) If it misses A_2 another bundle must be emitted from dA_1. If the bundle strikes A_2, it must be determined whether the bundle is absorbed or reflected. To do this, the geometry is used to find the angle of incidence θ_2 of the bundle onto A_2, that is,

$$\cos \theta_2 = \sin \theta_1 \cos \varphi_1 \qquad (11\text{-}32)$$

Knowing the absorptivity of A_2 from Kirchhoff's law,

$$\alpha'_{\lambda,2}(\lambda, \theta_2) = \epsilon'_{\lambda,2}(\lambda, \theta_2) \qquad (11\text{-}33)$$

and having determined the wavelength λ of the incident bundle from (11-26) and the incident angle θ_2 from (11-32), the probability of absorption of the bundle at A_2 can be determined. The probability of absorption is simply the absorptivity of A_2 evaluated at θ_2 and λ. This follows from the definition of directional spectral absorptivity $\alpha'_{\lambda,2}(\lambda, \theta_2)$ as the fraction of energy incident on A_2 in a given wavelength interval and within a given solid angle that is absorbed by the surface. This is also a precise definition of the probability of absorption of an individual bundle. The absorptivity is therefore the probability density function for the absorption of incident energy. It is now easy to determine whether a given incident energy bundle is absorbed by comparing the surface absorptivity $\alpha'_{\lambda,2}(\lambda, \theta_2)$ with a random number R_{α_2}. If

$$R_{\alpha_2} \leqslant \alpha'_{\lambda,2}(\lambda, \theta_2) \qquad (11\text{-}34)$$

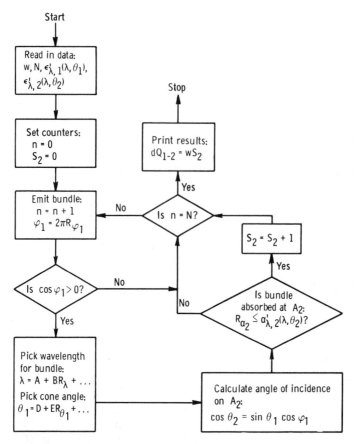

Figure 11-5 Computer flow diagram for example radiant-interchange problem.

the bundle of energy is absorbed and a counter S_2 in the computer memory is increased by 1 to keep account of the absorbed bundles. Otherwise, the bundle is assumed to be reflected and is not further accounted for. If the bundle path were followed further, rereflections from dA_1 would have to be considered. The neglect of rereflections is reasonable if the absorptivity of A_2 is large, or if the directional reflectivity is such that few bundles are reflected back along the direction of incidence. If such reflections cannot be neglected, angles of reflection must be chosen from known directional reflectivities, and the bundle is followed further along its path until it is absorbed by A_2 or lost from the system. For the purposes of this example, little is to be gained by following the bundle after reflection from surface A_2, because the derivation of the necessary relations is similar to that already presented.

A new bundle is now chosen at dA_1, and its history is followed. This procedure is continued until N bundles have been emitted from dA_1. The energy absorbed at A_2 is then calculated from (11-21).

The derivation of the equations needed for solution of the example is now complete. In putting together a flow chart to aid in formulating a computer program

(Fig. 11-5), some methods for shortening machine computing time can be invoked. For example, the angle φ_1 is computed first. If the bundle is not going to strike A_2 on the basis of the calculated φ_1, there is no point in computing λ and θ_1 for that bundle. Alternatively, because φ_1 values are isotropically distributed, it can be noted that exactly half the bundles must strike A_2. Therefore, the calculated φ_1 values can be constrained to the range $-\pi/2 < \varphi_1 < \pi/2$.

The formulation of this problem for a Monte Carlo solution is now complete. An astute observer will note that this example could be solved without much trouble by standard integral methods. A more astute observer might note further that extension to only slightly more difficult problems would cause serious consequences for the standard treatments. For example, consider introducing a third surface with directional properties into the problem and accounting for all interactions.

11-4.4 Useful Functions

A number of useful relations for choosing angles of emission and assigning a wavelength to bundles are given in the previous section. These and other functions from the literature dealing with radiative transfer are summarized in Table 11-1.

EXAMPLE 11-1 A wedge is made up of two very long parallel sides of equal width joined at an angle of $90°$, as shown in Fig. 11-6. The surface temperatures

Table 11-1 Convenient functions relating random numbers to variables for emission (assume no dependence on circumferential angle φ)

Variable	Type of emission	Relation
Cone angle θ	Diffuse	$\sin\theta = R_\theta^{1/2}$
	Directional-gray	$R_\theta = \dfrac{2\int_0^\theta \epsilon'(\theta^*)\sin\theta^*\cos\theta^*\,d\theta^*}{\epsilon}$
	Directional-nongray	$R_\theta = \dfrac{2\pi\int_0^\theta \int_0^\infty \epsilon_\lambda'(\lambda,\theta^*)i_{\lambda b}'(\lambda)\sin\theta^*\cos\theta^*\,d\lambda\,d\theta^*}{\epsilon\sigma T^4}$
Circumferential angle φ	Diffuse	$\varphi = 2\pi R_\varphi$
Wavelength λ	Black or gray	$F_{0-\lambda} = R_\lambda$
	Diffuse-nongray	$R_\lambda = \dfrac{\pi\int_0^\lambda \epsilon_\lambda(\lambda^*)i_{\lambda b}'(\lambda^*)\,d\lambda^*}{\epsilon\sigma T^4}$
	Directional-nongray	$R_\lambda = \dfrac{2\pi\int_0^\lambda \int_0^{\pi/2} \epsilon_\lambda'(\lambda^*,\theta)i_{\lambda b}'(\lambda^*)\sin\theta\cos\theta\,d\theta\,d\lambda^*}{\epsilon\sigma T^4}$

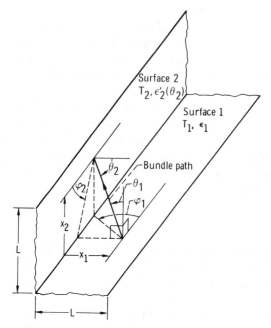

Figure 11-6 Geometry of Example 11-1.

are $T_1 = 1000$ K and $T_2 = 2000$ K. The effects of the ends may be neglected. Surface 1 is diffuse-gray with an emissivity of 0.5, while the properties of surface 2 are directional gray with directional total emissivity and absorptivity given by

$$\epsilon_2'(\theta_2) = \alpha_2'(\theta_2) = 0.5 \cos \theta_2 \qquad (11\text{-}35)$$

Assume for simplicity that surface 2 reflects diffusely. Set up a Monte Carlo flow sheet for determining the energy to be added to each surface to maintain its temperature. Assume that the environment is at $T = 0$ K.

The energy flux emitted by surface 1 is $q_{e,1} = \epsilon_1 \sigma T_1^4$. If N_1 emitted sample energy bundles are to be followed per unit time and area from surface 1, then the amount of energy per bundle will be

$$w = \frac{q_{e,1}}{N_1} = \frac{\epsilon_1 \sigma T_1^4}{N_1} \qquad (11\text{-}36)$$

The energy flux emitted from surface 2 is

$$q_{e,2} = 2\sigma T_2^4 \int_0^{\pi/2} \epsilon_2'(\theta) \cos \theta \sin \theta \, d\theta$$

$$= \sigma T_2^4 \int_0^{\pi/2} \cos^2 \theta \sin \theta \, d\theta = \frac{\sigma T_2^4}{3}$$

If the same amount of energy w is assigned to each bundle emitted by wall 2 as was used for wall 1, then $wN_2 = \sigma T_2^4/3$. Substituting (11-36), the value of ϵ_1, and the known surface temperatures gives

$$N_2 = \frac{\sigma T_2^4}{3} \frac{N_1}{\epsilon_1 \sigma T_1^4} = \frac{32}{3} N_1 \qquad (11\text{-}37)$$

Because all bundles have equal energy, and 32/3 as many bundles are emitted from surface 2 as from surface 1, it is obvious that surface 2 will make the major contribution to the energy transfer.

Now the distributions of directions for emitted bundles from the two surfaces will be derived. Surface 1 emits diffusely, so that (11-30b) applies. For surface 2, however, (11-27) must be used. Substituting (11-35) into (11-27) gives, for the directional-gray case,

$$R_{\theta_2} = \frac{2\pi i'_{b,2} \int_0^{\theta_2} (0.5 \cos \theta_2^*) \sin \theta_2^* \cos \theta_2^* \, d\theta_2^*}{\epsilon_2 \sigma T_2^4}$$

The hemispherical total emissivity is substituted from Eq. (11-19) to give

$$R_{\theta_2} = \frac{\int_0^{\theta_2} \cos^2 \theta_2^* \sin \theta_2^* \, d\theta_2^*}{\int_0^{\pi/2} \cos^2 \theta_2 \sin \theta_2 \, d\theta_2} = 1 - \cos^3 \theta_2$$

The fact that R and $1 - R$ are both uniform random distributions in the range $0 \leqslant R \leqslant 1$ can be used to write this as

$$\cos \theta_2 = R_{\theta_2}^{1/3}$$

Note that, by similar reasoning (11-30b) can be written

$$\cos \theta_1 = R_{\theta_1}^{1/2}$$

Since there is no dependence on angle φ for either surface, (11-31) applies for both surfaces.

The distribution of directions at which bundle emission will occur has now been determined. Next the position on each surface from which each bundle will be emitted must be determined. Because the wedge sides are isothermal, the emission from a given side will be uniform. In such a case, random positons x (Fig. 11-6) on a given side could be picked as points of emission. Such a procedure requires generation of a random number. The computer time required to generate a random number, can be saved by noting that the bundle emission is the initial process in each Monte Carlo history; hence, there is no prior history to be elimininated by using a random number. In this case, x positions along L can be sequentially chosen as $x = (n/N)L$, where n is the sample-history index for the history being begun, $1 \leqslant n \leqslant N$.

The point of emission and direction of emission for each bundle leaving either surface can now be determined. The remaining calculations involve determination of whether the emitted bundles will strike the adjacent wall or will leave the cavity. Examination of Fig. 11-6 shows that, for either surface, when

$\pi \leqslant \varphi \leqslant 2\pi$ the bundles will leave the cavity for any θ, and when $0 < \varphi < \pi$ they will leave if

$$\sin \theta < \frac{x/\sin \varphi}{[(x/\sin \varphi)^2 + L^2]^{1/2}} = \frac{1}{[1 + (L \sin \varphi/x)^2]^{1/2}}$$

The angle of incidence θ_i on a surface is given in terms of the angles θ_δ and φ_δ, at which the bundle leaves the other surface, by

$$\cos \theta_i = \sin \theta_\delta \sin \varphi_\delta$$

All the necessary relations are now at hand. Now a flow diagram is constructed to combine these relations in the correct sequence. Diffuse reflection is assumed from both surfaces. The resulting flow diagram is shown in Fig. 11-7. Study of this figure will show one way of constructing the flow of events for the problem at hand. The use of the indices δ, δ', and δ'' is an artifice to reduce the size of the chart. The index δ always refers to the wall from which the original emission of the bundle occurred, and δ' refers to the wall from which emission or reflection is presently occurring. The index δ'' is used to make the emitted distribution of θ angles correspond to either $R_{\theta_1}^{1/2}$ or $R_{\theta_2}^{1/3}$ and have all the reflected bundles correspond to a diffuse distribution.

11-4.5 Literature on Application to Radiation Exchange between Surfaces

The standard or conventional methods for solving problems of radiative transport between surfaces in the absence of absorbing media were formulated in Chaps. 7–10. The standard methods have advantages for certain types of problems and will outshine the Monte Carlo approach in speed and accuracy over some range of radiation calculations. This range is outlined roughly by the complexity of the problem, and the areas of usefulness of the Monte Carlo approach will now be discussed.

The chief usefulness of Monte Carlo to the thermal-radiation analyst lies in this fact: Monte Carlo program complexity increases roughly in proportion to problem complexity for radiative interchange problems, while the difficulty of carrying out conventional solutions increases roughly with the square of the complexity of the problem because of the matrix form into which the conventional formulations fall. However, because Monte Carlo is somewhat more difficult to apply to the simplest problems, it is most effective in problems in which complex geometries and variable properties must be considered. In complex geometries, Monte Carlo has the additional advantage that simple relations will specify the path of a given energy bundle, whereas most other methods require explicit or implicit integrations over surface areas. Such integrations become difficult when a variety of curved or skewed surfaces are present.

Configuration-factor computation. The calculation of radiative configuration factors by standard means usually involves certain assumptions that place restrictions on the application of these factors in exchange computations. The assumptions required when using the ordinary configuration factors as derived in Chap. 7 are that the surfaces

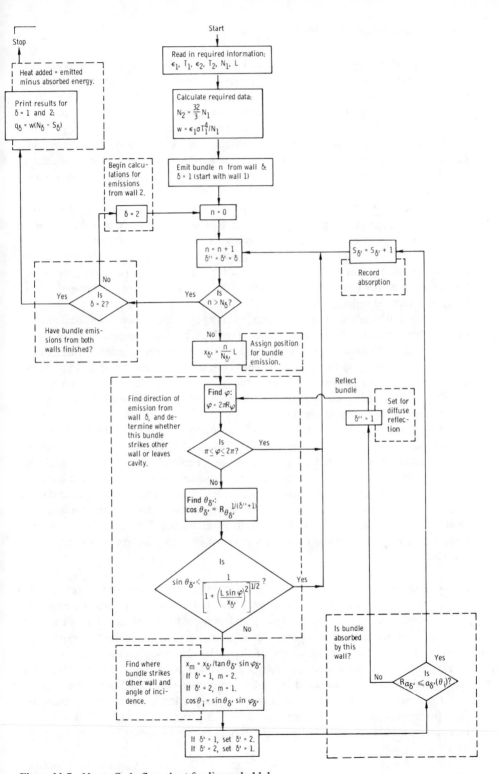

Figure 11-7 Monte Carlo flow chart for Example 11-1.

375

involved are diffuse-gray emitters and reflectors, that each surface is isothermal, and that the total flux arriving at and leaving each surface is evenly distributed across the surface. Any of these assumptions may be very poor; most surfaces are neither diffuse nor gray, and the distribution of reflected flux usually deviates from uniformity to some extent. Where deviations from the assumptions must be considered, calculation of the configuration factors becomes difficult; and if geometries with nonplanar surfaces are involved, Monte Carlo techniques may become invaluable. It should be noted, however, that unless a parametric study of the interchange of radiant energy within an enclosure with specified characteristics is being carried out, it may be easier to directly compute the entire radiative flux distribution by Monte Carlo. This would be simpler than computing configuration factors by Monte Carlo and then using an auxiliary program to calculate energy exchange by means of these factors.

As computed by Monte Carlo, configuration factors are identically equal to the fraction of the total energy bundles emitted from a surface that are incident upon a second surface. No restrictions are made to diffuse-gray surfaces with evenly distributed emitted and reflected flux.

Corlett [19] has computed exchange factors (as distinguished from configuration factors) for a variety of geometries, including louvers, and circular and square ducts with various combinations of diffusely and specularly reflecting interior surfaces and ends. These factors give the fraction of energy emitted by a given surface that reaches another surface by all paths, including intermediate reflections. One set of results, the exchange factors between the black ends of a cylinder with a diffusely reflecting internal surface, is shown in Fig. 11-8.

Weiner et al. [20] carried out the Monte Carlo evaluation of some simple configuration factors for comparison with analytical solutions. They then considered energy exchange within an enclosure with five specularly reflecting sides, each side being assumed to have a directional emissivity dependent upon cone angle of emission.

They also worked out the case of interchange within a simulated optical system. This system is constructed of a combination of spherical and conical surfaces that enclose a cylindrical specular reflector with two surfaces. This is obviously an interchange problem to cause many unhappy hours of analyzing integral limits in the usual formulations.

Cavity properties. At least one Monte Carlo solution exists in the literature for a surface interaction with a distant source. This is the case of a conical cavity with diffusely reflecting inner surface. Polgar and Howell [21] analyzed the bidirectional reflectivity of the cavity when exposed to a beam of parallel incident radiation and determined the directional emissivity of the cavity. Parameters varied were the angle of incidence, cone angle, and emissivity of the inner surface of the cone. One set of representative results is shown in Fig. 11-9. No results were found in the literature for direct comparison of the computed directional properties; however, the hemispherical absorptivity results were obtained by integrating the directional values and were compared in [22] to analytical results from [23]. The comparison is shown in Fig. 11-10.

The bidirectional reflectivity results computed by Monte Carlo in [21] illustrate

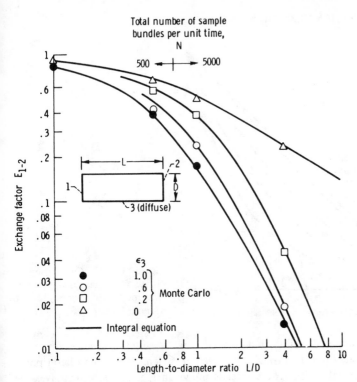

Figure 11-8 Radiation exchange factors between black ends of diffuse walled cylinder (from [19]). $\epsilon_1 = \epsilon_2 = 1$.

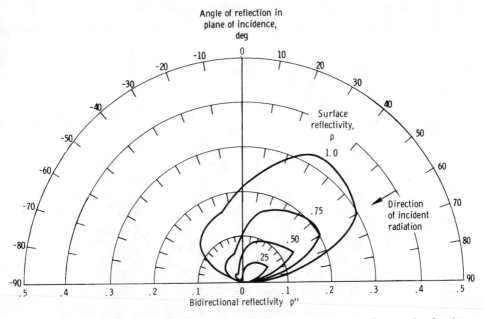

Figure 11-9 Bidirectional reflectivity of diffusely reflecting conical cavity. Cone angle of cavity, $30°$; angle of incidence of radiation, $60°$. (From [21].)

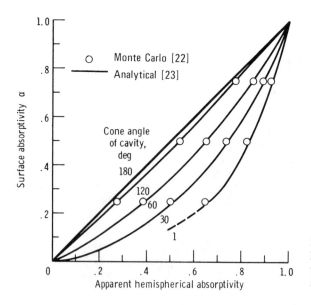

Figure 11-10 Comparison of Monte Carlo results for absorptivity of conical cavities [22] with analytical results [23].

the scatter of the computed points that depends on the number of energy bundles reflected from the cone interior through any given area element on a unit hemisphere imagined over the conical cavity. The scatter is shown in Fig. 11-11, which gives the standard deviation of the computed reflectivity at various angles of reflection. The solid angle subtended to the base by area elements of equal angular increment $\Delta\theta \ \Delta\varphi$ on the hemisphere varies with the sine of the angle of reflection, so the number of sample energy bundles per unit solid angle $d\omega = \sin\theta \ d\theta \ d\varphi$ near the cone axis

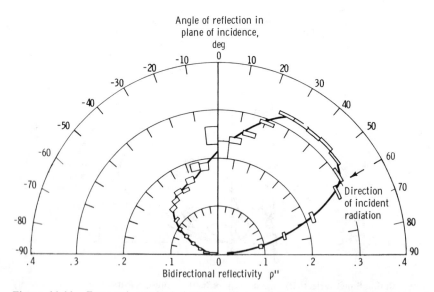

Figure 11-11 Expected standard deviation of results for bidirectional reflectivity of diffuse conical cavity. Cone angle of cavity, 30°; angle of incidence of radiation, 60°. (From [21].)

becomes very small. This leads to larger scatter at angles near the cone axis, where $\sin \theta \rightarrow 0$.

Shamsundar, Sparrow, and Heinisch applied Monte Carlo techniques to determine the emittance from baffled conical and cylindrical cavities [17, 24, 25]. The speed of computation was improved by application of "energy partitioning." In this approach, each bundle emitted from a given position on the interior surface of a cavity is divided into two parts. The first part is the fraction that leaves the cavity directly through the aperture without reflection from any surface. This fraction is obtained from geometry and is, in fact, simply the configuration factor from the surface element to the aperture. The remaining part of the bundle may undergo one or more reflections within the cavity. This part is traced to its next point of impingement. Determination of absorption is made following the procedures outlined in the examples in this chapter. If the bundle is not absorbed, it is again partitioned into a part leaving the aperture and a second part that strikes the cavity interior. The sequence is continued until the bundle is absorbed or its remaining energy is negligible. This procedure gives at least one statistical sample to the emitted energy for each original bundle emission. The results of [17, 24, 25] show more rapid convergence to a stable solution of low variance than by conventional Monte Carlo analysis.

Extension to directional and spectral surfaces. Few references exist that treat problems involving both directionally and spectrally dependent properties. The reasons for this omission seem twofold. First, accurate and complete directional spectral properties, especially the former, are not often found in the literature. An analyst desiring to include such effects might thus be unable to find the requisite data for a system. Second, when solutions are attained to such problems, they are often so specialized that little interest exists to warrant their wide dissemination in the open literature. As pointed out by Dunn et al. [26], when the radiative properties become available, the methods for handling such surface radiative-energy-exchange problems now exist, and Monte Carlo appears to be one of the better-suited techniques.

Toor and Viskanta et al. [27–30] have successfully applied Monte Carlo techniques to some interchange problems involving surfaces with directional and spectral property variations. Some of these results were discussed in Chap. 10. Howell and Durkee [31] compared experimental data and a Monte Carlo analysis of radiative exchange in an enclosure with directional surfaces, with good results.

Extension of Monte Carlo methods to thermal analysis problems with radiation and conduction, including property uncertainties, has been made in [32, 33]. In these treatments, mean values and standard deviations of both material properties and dimensional tolerances were used to choose sample systems. A large number of such systems were analyzed to predict the average expected performance of the thermal system and the variance in expected performance. The method has certain programming advantages over more conventional sensitivity analysis.

11-4.6 Statistical Difficulties of Monte Carlo Technique

Monte Carlo calculations give results that fluctuate around the "real" answer because the method is a repetitive experiment using a mathematical model in place of the actual physical situation. The uncertainty can be found by applying standard statistical tests; the uncertainty can be reduced in the same manner as experimental error, that is, by averaging over more tests (bundle histories), and/or by reducing the variance of individual tests.

No rigorous criteria exist to guarantee the convergence of Monte Carlo results to valid solutions; however, convergence has not as yet been a difficulty in thermal radiation problems. It would often be immediately evident that convergence to invalid solutions was occurring because of the limiting solutions and physical constraints that are known for most radiative problems. Shamsundar et al. [17] show that convergence for problems using energy partitioning is more rapid than for straightforward Monte Carlo solutions.

Most of the difficulties that do arise in Monte Carlo sampling techniques are concerned with obtaining an optimum sample size. Such difficulties have been sufficiently common in transport processes that are mathematically related to radiative transport so that special methods of weighting the free paths of bundles have been developed to obtain adequate samples. Using these methods saves computer time and increases accuracy; these gains, however, are at the expense of added complexity.

11-4.7 Closing Remarks

In this chapter, Monte Carlo has been discussed as a method suitable for the solution of complex radiative-exchange problems. Two sample problems were outlined to demonstrate its application, and some of the advantages and disadvantages of the technique were discussed along with pertinent literature references.

From this, certain conclusions emerge. First, Monte Carlo appears to have a definite advantage over other radiative-exchange calculation techniques when the difficulty of the problem being treated lies above some undefined level. This level usually cannot be defined since it depends not only on the specific problem but is probably a function of the experience, competence, and prejudice of the individual working the problem. However, problems with complexity above this nebulous benchmark can be treated by Monte Carlo with great flexibility, simplicity, and speed. The Monte Carlo approach does lack a kind of generality common to other approaches in that each problem may require an individual technique, and a dash of ingenuity often helps. This places a greater burden on the programmer's backlog of experience and intuition, whereas standard methods may allow programming through "cook-book" application of their formalism if they can be applied at all.

Second, for the thermal radiation problems carried out to date, the parameters and mathematical relations involved usually lie in ranges that allow straightforward Monte Carlo programming without the need of the more exotic schemes occasionally necessary in other Monte Carlo transport studies.

Third, with all its advantages, the method suffers from certain difficulties. The worst of these are the statistical nature of the results and the lack of guaranteed statistical convergence to the true mean value. It should be noted that the latter fault is common to many methods when complex problems are being treated, because rigorous mathematical criteria to guarantee convergence to a solution are available only in certain cases.

Finally, the person using Monte Carlo techniques often develops a physical grasp of the problems encountered because the model is simple, and the mathematics describing it is therefore less sophisticated. This is a contrast to the rather poor physical interpretations and predictions that can be made in working with, say, a matrix of integral equations.

REFERENCES

1. Howell, J. R.: Application of Monte Carlo to Heat Transfer Problems, *Adv. Heat Transfer* (J. P. Hartnett and T. F. Irvine, eds.), vol. 5, 1968.
2. Kahn, Herman: Applications of Monte Carlo, *Rept. No.* RM-1237-AEC (AEC No. AECU-3259), Rand Corp., April 27, 1956.
3. Hammersley, J. M., and D. C. Handscomb: Monte Carlo Methods, John Wiley & Sons, Inc., New York, 1964.
4. The Bible, 1 Kings 7:23; 2 Chron. 4:2.
5. Metropolis, Nicholas, and S. Ulam: The Monte Carlo Method, *J. Am. Stat. Assoc.*, vol. 44, no. 247, pp. 335-341, 1949.
6. Cashwell, E. D., and C. J. Everett: "A Practical Manual on the Monte Carlo Method for Random Walk Problems," Pergamon Press, New York, 1959.
7. Schreider, Yu. A (ed.): "Method of Statistical Testing – Monte Carlo Method," American Elsevier Publishing Company, Inc., New York, 1964.
8. Brown, G. W.: Monte Carlo Methods, in E. F. Bechenbach (ed.), "Modern Mathematics for the Engineer," pp. 279-307, McGraw-Hill Book Company, New York, 1956.
9. Halton, John H.: A Retrospective and Prospective Survey of the Monte Carlo Method, *SIAM Rev.*, vol. 12, no. 1, pp. 1-63, 1970.
10. Meyer, Herbert A. (ed.): "Symposium on Monte Carlo Methods," John Wiley & Sons, Inc., New York, 1956.
11. Rand Corp.: "A Million Random Digits with 100,000 Normal Deviates," The Free Press of Glencoe, Ill., Chicago, 1955.
12. Kendall, M. G., and B. Babington Smith: "Tables of Random Sampling Numbers," 2d ser., Cambridge University Press, New York, 1954.
13. Taussky, O., and J. Todd: Generating and Testing of Pseudo-Random Numbers, in Herbert A. Meyer (ed.), "Symposium on Monte Carlo Methods," pp. 15-28, John Wiley & Sons, Inc., New York, 1956.
14. Kendall, M. G., and B. B. Smith: Randomness and Random Sampling Numbers, *R. Stat. Soc. J.*, pt. 1, pp. 147-166, 1938.
15. Dwight, H. B.: "Mathematical Tables of Elementary and Some Higher Mathematical Functions," 2d ed., Dover Publications, Inc., New York, 1958.
16. Jahnke, Eugene, and Fritz Emde: "Tables of Functions with Formulae and Curves," 4th ed., Dover Publications, Inc., New York, 1945.
17. Shamsundar, N., E. M. Sparrow, and R. P. Heinisch: Monte Carlo Radiation Solutions – Effect of Energy Partioning and Number of Rays, *Int. J. Heat Mass Transfer*, vol. 16, no. 3, pp. 690-694, 1973.
18. Howell, John R.: Calculation of Radiant Heat Exchange by the Monte Carlo Method, *Paper No.* 65-WA/HT-54, *ASME*, November, 1965.
19. Corlett, R. C.: Direct Monte Carlo Calculation of Radiative Heat Transfer in Vacuum, *J. Heat Transfer*, vol. 88, no. 4, pp. 376-382, 1966.
20. Weiner, M. M., J. W. Tindall, and L. M. Candell: Radiative Interchange Factors by Monte Carlo, *Paper No.* 65-WA/HT-51, *ASME*, November, 1965.
21. Polgar, Leslie G., and John R. Howell: Directional Thermal-radiative Properties of Conical Cavities, *NASA* TN D-2904, 1965.

22. Polgar, Leslie G., and John R. Howell: Directional Radiative Characteristics of Conical Cavities and Their Relation to Lunar Phenomena, in G. B. Heller (ed.), "Thermophysics and Temperature Control of Spacecraft and Entry Vehicles," pp. 311–323, Academic Press, Inc., New York, 1966.

23. Sparrow, E. M., and V. K. Jonsson: Radiant Emission Characteristics of Diffuse Conical Cavities, *J. Am. Opt. Soc.*, vol. 53, no. 7, pp. 816–821, 1963.

24. Heinisch, R. P., E. M. Sparrow, and N. Shamsundar: Radiant Emission from Baffled Conical Cavities, *J. Opt. Soc. Am.*, vol. 63, no. 2, pp. 152–158, 1973.

25. Sparrow, E. M., R. P. Heinisch, and N. Shamsundar: Apparent Hemispherical Emittance of Baffled Cylindrical Cavities, *ASME J. Heat Transfer*, vol. 96, no. 1, pp. 112–114, 1974.

26. Dunn, S. Thomas, Joseph C. Richmond, and Jerome F. Parmer: Survey of Infrared Measurement Techniques and Computational Methods in Radiant Heat Transfer, *J. Spacecr. Rockets*, vol. 3, no. 7, pp. 961–975, 1966.

27. Toor, Jaswant S.: "Radiant Heat Transfer Analysis among Surfaces Having Direction Dependent Properties by the Monte Carlo Method," M.S. thesis, Purdue University, 1967.

28. Viskanta, Raymond, James R. Schornhorst, and Jaswant S. Toor: Analysis and Experiment of Radiant Heat Exchange between Simply Arranged Surfaces, Purdue University (AFFDL-TR-67-94, DDC No. AD-655335), June, 1967.

29. Toor, J. S., and R. Viskanta: A Numerical Experiment of Radiant Heat Exchange by the Monte Carlo Method, *Int. J. Heat Mass Transfer*, vol. 11, no. 5, pp. 883–897, 1968.

30. Toor, J. S., and R. Viskanta: Effect of Direction Dependent Properties on Radiant Interchange, *J. Spacecr. Rockets*, vol. 5, no. 6, pp. 742–743, 1968.

31. Howell, John R., and Ronald Durkee: Radiative Transfer between Surfaces with Collimated Incident Radiation: A Comparison of Analysis and Experiment, *ASME J. Heat Transfer*, vol. 93, no. 2, pp. 129–135, 1971.

32. Howell, J. R.: Monte Carlo Treatment of Data Uncertainties in Thermal Analysis, *J. Spacecr. Rockets*, vol. 10, no. 6, pp. 411–414, 1973.

33. Zigrang, Denis J.: Statistical Treatment of Data Uncertainties in Heat Transfer, *AIAA Paper 75-710*, May, 1975.

PROBLEMS

1 An area element dA_1 has directional emissivity given by $\epsilon_1'(\theta) = 0.8 \cos \theta$. Find, in the geometry pictured below, the fraction of energy leaving dA_1 that is absorbed by the black disk A_2. Use the Monte Carlo method. Compare your result with the analytical solution.

Answer: 0.35.

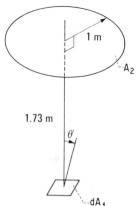

2 Construct a Monte Carlo flow diagram to compute the configuration factor F_{d1-2} from an area element to a perpendicular disk as shown in Example 7-6.

3 Construct a complete computer flow diagram for the Monte Carlo solution of the problem outlined in Prob. 11 of Chap. 8 for gray plates.

4 Draw a Monte Carlo computer flow diagram to obtain the specular exchange factor F^s_{2-1} in Prob. 10 of Chap. 9.

5 Program and solve prob. 3 above for $L = 1$, $\epsilon_1 = \epsilon_2 = 1$, and for $L = 1$, $\epsilon_1 = \epsilon_2 = 0.5$. By comparison of the two results, verify the result of Prob. 11b of Chap. 8.

6 Obtain a Monte Carlo solution for the nongray heat transfer between infinite parallel plates computed in Example 10-2.

7 Obtain a Monte Carlo solution for the second part of Example 9-6, that is, when surface A_1 is a specular reflector.

TWELVE

RADIATION IN THE PRESENCE OF OTHER MODES OF ENERGY TRANSFER

12-1 INTRODUCTION

In the preceding chapters, radiation exchange was considered without interaction with other modes of heat transfer. In many practical systems, however, a significant amount of heat conduction and/or convection may be occurring simultaneously, and the combined effect of all the heat transfer modes must be accounted for. The interaction of heat transfer modes may be simple in some cases; for example, the heat dissipation by radiation and convection may be essentially independent and hence may be computed separately and then added. In other instances the interaction can be quite complex.

The following are some examples of situations with combined heat transfer effects. For a vapor-cycle power plant operating in outer space the waste heat is rejected by radiation. In the space radiator, as shown in Fig. 12-1a, the vapor used as the working fluid in a thermodynamic cycle is condensed, thereby releasing its latent heat. The heat is then conducted through the condenser wall and into fins that radiate the energy into space. The temperature distribution in the fins and the fin efficiency depend on the combined radiation and conduction processes. The same fin-tube geometry is commonly used for the absorber plate in a flat-plate solar collector. Solar energy is incident on the absorber plate through one or two transparent cover plates. Water or other fluid is heated as it flows through the tubes.

In one type of steel-strip cooler in a steel mill (Fig. 12-1b), a sheet of hot metal moves past a bank of cold tubes and loses heat to them by radiation. At the same time cooling gas is blown over the sheet. A combined radiation and convection analysis must be performed to determine the temperature distribution along the steel strip moving through the cooler. Controlled radiative and convective cooling is also used in

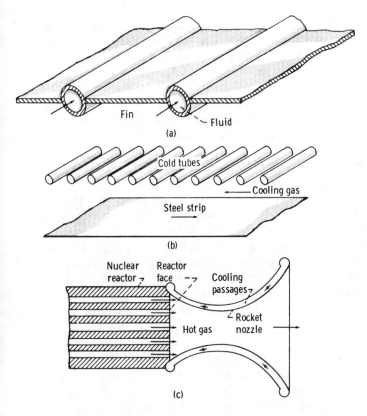

Figure 12-1 Heat transfer devices involving combined radiation, conduction, and convection. (*a*) Space radiator, or absorber plate of flat-plate solar collector; (*b*) steel-strip cooler; (*c*) nuclear rocket.

the tempering of sheets of high-strength plate glass such as those used for automobile windows.

In a nuclear rocket engine, illustrated by Fig. 12-1*c*, hydrogen gas is heated by flowing through a high-temperature nuclear-reactor core. The hot gas then passes out through the rocket nozzle. The interior surface of the rocket nozzle receives heat by radiation from the exit face of the reactor core and by convection from the flowing propellant stream. Both these energy quantities are conducted through the nozzle wall and removed by a flowing stream of coolant.

The examples cited all involve the transfer of heat by two or more heat transfer modes. Heat may flow first by one mode and then by a second, as is the case of conduction through a plate followed by radiation from the surface, and the modes are considered to be in series. Heat flow may also occur by parallel modes, such as by simultaneous conduction and radiation through a transparent medium as in glass tempering, or radiation and convection from a surface at elevated temperature. The modes can thus act in series, in parallel, or both.

In this chapter, combined radiation, conduction, and convection problems will be examined subject to an important restriction: The medium through which the radiation is passing does not absorb or emit radiation; that is, it is completely transparent. This restriction will be removed in Chaps. 13–20, which deal with media that absorb, emit, and scatter radiation.

The various heat transfer modes depend on temperature to different powers. When radiation exchange between black surfaces is considered, the energy fluxes depend upon surface temperatures to the fourth power. For nonblack surfaces the exponent on the temperature may differ somewhat from four because of the variation of emissivity with temperature. If conduction is present, the Fourier conduction law prescribes a dependence of heat flow upon local temperature gradient, thus introducing derivatives of the first power of the temperature (when the thermal conductivity does not depend on temperature). If convection enters the problem, it provides a heat flow that depends approximately on a difference of the first powers of the temperatures, the exact power depending on the type of flow; for example, free convection depends on temperature difference between the 1.25 and 1.4 power. Physical properties that vary with temperature will introduce additional temperature dependencies. The fact that such a wide variation in powers of temperature are involved in the energy transfer process means that the governing equations are highly nonlinear.

Because the radiation terms are usually in the form of integrals that give the amount of radiative energy incident from the surroundings, and the conduction terms involve derivatives, the energy-balance equations are in the form of nonlinear integrodifferential equations. Such equations are not easily solved using presently available mathematical techniques. Except in the simplest cases, it is usually necessary to resort to numerical evaluation of the solutions. Each problem requires its own most efficient method of attack, and for this reason no general discussion of numerical or other mathematical solution techniques will be given here. For such techniques the reader is referred to the extensive mathematical literature on numerical methods and the representative radiation papers referenced throughout this chapter. In this chapter we concentrate on the methods of setting up the energy-balance equations and on developing insight into the physical problems; the actual solution methods are left to the mathematical texts, except where specialized approaches are of value.

12-2 SYMBOLS

A	area
a	spacing between fins; coefficients in matrix
B	parameter in Example 12-4
b	thickness of conducting medium; fin thickness; tube wall thickness
c	correction factors
c_p	specific heat
D	tube diameter

F	configuration factor
f	coefficients in Eq. (12-12)
G	parameter in Eq. (12-30)
H	parameter defined in Example 12-5
h	heat transfer coefficient
k	thermal conductivity
L	length of tube
l	dimensionless tube length, L/D
M	parameter in Eq. (12-30)
N	parameter defined in connection with Eq. (12-25)
Nu	Nusselt number, hD/k
n	normal direction
P	perimeter
Pr	Prandtl number, $c_p \mu_f/k$
Q	energy rate, energy per unit time
q	energy flux, energy per unit area per unit time
R	dimensionless radius of Example 12-3
Re	Reynolds number, $Du_m \rho_f/\mu_f$
r	radius
S	parameter defined in Example 12-5
T	absolute temperature
t	dimensionless temperature
u_m	mean fluid velocity
W	width of fin in Example 12-4
X	distance from tube entrance to ring element
x,z	cartesian coordinate positions
α	absorptivity
γ,δ	dimensionless parameters of Example 12-3
ϵ	emissivity
η	fin efficiency, defined in Example 12-3
Θ	dimensionless temperatures in Examples 12-3 and 12-4
μ	dimensionless parameter defined in Example 12-4
μ_f	fluid viscosity
Ξ	distance from tube entrance
ξ	distance from fin base
ρ_f	density of fluid
ρ_m	density of solid material
σ	Stefan-Boltzmann constant
τ	time

Subscripts

a	base surface between fins
b	evaluated at base of fin
c	conduction
e	environment

f	fin or fluid
g	gas
i	in or inner
o	out or outer
R	radiation
r	reservoir
w	wall
x	at position x
ξ	at position ξ
1, 2	evaluated at surfaces 1, 2 or at inlet and exit ends of tube

12-3 PROBLEMS INVOLVING COMBINED RADIATION AND CONDUCTION

Situations involving only conduction and radiation are fairly common. Some examples are heat losses through the walls of a vacuum Dewar, heat transfer through "super-insulation" made up of many separated layers of highly reflective material, and heat losses and temperature distributions in satellite and spacecraft structures.

The sophistication of the radiative portion of the analysis can range from assuming finite black surfaces and using diffuse configuration factors to a complete treatment of local directional spectral effects via a Monte Carlo or integral-equation approach. The choice of radiative formulation depends on the accuracy required and the relative importance of the radiative mode in relation to the heat conduction. If conduction dominates, then fairly rough approximations can be invoked in the radiative portion of the analysis, and vice versa. Some simple examples of situations involving radiation and conduction are now examined; we then progress to more sophisticated treatments.

12-3.1 Uncoupled Problems

The most simple situation is where the radiation and conduction contributions to an unknown quantity, say heat flux, are independent; the contributions are then computed separately and the individual results added. The heat transfer modes are said to be *uncoupled* with regard to the desired quantity.

> **EXAMPLE 12-1** As an example of an uncoupled situation, consider two black infinite parallel plates separated by a medium of thickness b that has thermal conductivity k and is transparent to thermal radiation. If one plate is at temperature T_1 and the other is at temperature T_2, what is the net energy exchange between the plates?
>
> The net energy transferred is composed of the net radiative exchange and the transfer by conduction. The net energy transferred is also equal to the energy Q_1 that must be added to plate 1 to maintain it at its specified temperature. The energy transfer per unit time and area by radiation between two infinite

parallel black plates is simply

$$\frac{Q_R}{A} = \sigma(T_1{}^4 - T_2{}^4)$$

and that by conduction is

$$\frac{Q_c}{A} = \frac{k}{b}(T_1 - T_2)$$

The total energy transfer per unit time and area is then the sum of the separate contributions:

$$\frac{Q_1}{A} = \frac{Q_R}{A} + \frac{Q_c}{A} = \sigma(T_1{}^4 - T_2{}^4) + \frac{k}{b}(T_1 - T_2)$$

Example 12-1 demonstrates a situation in which the conductive and radiative components are *uncoupled* from one another; that is, the presence of one parallel heat transfer mode does not affect the other with regard to the computation of heat flow. The Q for each mode is computed independently, and the two contributions are added. In such problems all the methods of radiative computation developed heretofore can be applied without modification, since the radiation is computed independently.

12-3.2 Coupled Nonlinear Problems

Unhappily, the uncoupled problems described in the previous section are not as common as *coupled* problems. In coupled problems the *desired unknown quantity cannot be found by adding separate radiation and conduction solutions*, the governing energy equation must be solved with the two modes simultaneously included. In some situations it is possible to assume that the modes are uncoupled because of the weak coupling that occurs. This assumption, when valid, allows escape from some of the difficulties that will become manifest in succeeding sections of this chapter.

EXAMPLE 12-2 As a simple example of a coupled problem, consider another set of conditions for the geometry in the previous example, that is, two black infinite parallel plates separated by a transparent medium of thickness b and thermal conductivity k. Plate 2 is at temperature T_2, and a known amount of energy Q_1/A is added per unit area to plate 1 and removed at plate 2. What is the temperature T_1 of plate 1?

This is the same situation as in Example 12-1, except that Q_1 is now known and T_1 is to be found. The same energy equation applies as in Example 12-1 and is rewritten to place the unknown on the left:

$$\sigma T_1{}^4 + \frac{k}{b} T_1 = \sigma T_2{}^4 + \frac{k}{b} T_2 + \frac{Q_1}{A}$$

The problem is *coupled* with regard to the desired unknown T_1 in that T_1 must be found from an equation that simultaneously incorporates *both* heat transfer processes. The equation for T_1 is nonlinear and can be solved iteratively.

These first two examples demonstrate that the types of boundary conditions govern the possibility of uncoupling the radiative and conductive calculations. When all temperatures are specified, the heat fluxes can usually be uncoupled. If energy fluxes are specified, however, the entire problem must be treated simultaneously because of nonlinear coupling governing the unknown temperatures. The treatment can become more difficult if variations of physical properties as functions of temperature must be included.

In devices that operate in outer space, a means for dissipating energy is to employ radiating fins. The energy is conducted into the fin and radiated away from the fin surface. The determination of the unknown temperature distribution within the fin requires a coupled solution. The next example deals with an analysis of the performance of a single circular fin.

EXAMPLE 12-3 A thin annular fin in a vacuum is embedded in insulation so that it is insulated on one face and around its outside edge, as shown in Fig. 12-2a, b. The disk is of thickness b, inner radius r_i, outer radius r_o, and thermal conductivity k. Energy is supplied to the inner edge, say from a solid rod of radius r_i

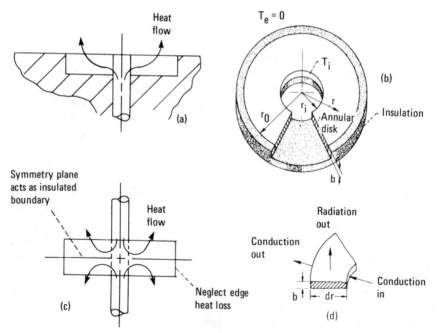

Figure 12-2 Geometry for finding temperature distribution in thin radiating annular plate insulated on one side and around outside edge. (a) Heat flow path through fin, (b) disk geometry; (c) application to annular fin; (d) portion of ring element on annular disk.

that fits the central hole, and this maintains the inner edge at T_i. The exposed annular surface, which is diffuse-gray with emissivity ϵ, radiates to the environment which is at temperature $T_e = 0$. Find the temperature distribution as a function of radial position along the annular disk. (The results also apply to the annular fin in Fig. 12-2c if the heat loss through the end edge of the fin is neglected. There is no heat flow across the symmetry plane of the fin, and hence this plane acts as an insulated boundary.)

Assume the disk is thin enough so that the local temperature can be taken as constant across the thickness b; then for any ring element of width dr as shown in Fig. 12-2d, an energy balance equates the conduction in and the radiation and conduction out:

$$-k2\pi rb \frac{dT}{dr} = \epsilon\sigma T^4 2\pi r\, dr - k2\pi rb \frac{dT}{dr} + \frac{d}{dr}\left(-k2\pi rb \frac{dT}{dr}\right) dr$$

Since the surroundings are at zero temperature, there is no incoming radiation.

If b and k are constant, the energy balance becomes

$$kb \frac{1}{r} \frac{d}{dr}\left(r \frac{dT}{dr}\right) - \epsilon\sigma T^4 = 0 \tag{12-1}$$

This equation is to be solved for the temperature distribution $T(r)$ subject to two boundary conditions: at the inner edge, $T = T_i$ at $r = r_i$, and, at the insulated outer edge where there is no heat flow, $dT/dr = 0$ at $r = r_o$. Using the dimensionless variables $\Theta = T/T_i$ and $R = (r - r_i)/(r_o - r_i)$ gives, for the energy equation,

$$\frac{d^2\Theta}{dR^2} + \frac{1}{R + r_i/(r_o - r_i)} \frac{d\Theta}{dR} - \frac{(r_o - r_i)^2\epsilon\sigma T_i^3}{kb} \Theta^4 = 0$$

Using the two parameters $\delta = r_o/r_i$ and $\gamma = (r_o - r_i)^2\epsilon\sigma T_i^3/kb$ results in

$$\frac{d^2\Theta}{dR^2} + \frac{1}{R + 1/(\delta - 1)} \frac{d\Theta}{dR} - \gamma\Theta^4 = 0 \tag{12-2}$$

with the following boundary conditions: $\Theta = 1$ at $R = 0$, and $d\Theta/dR = 0$ at $R = 1$. Equation (12-2) is a second-order differential equation that is nonlinear because it contains Θ raised to two different powers. The temperature distribution depends only on the two parameters δ and γ. A solution can be obtained by numerical methods.

A quantity of interest in the utilization of cooling fins is the *fin efficiency* η. This is defined as the energy actually radiated away by the fin divided by the energy that would be radiated if the entire fin were at the temperature T_i. The fin efficiency for the circular fin being studied here is then

$$\eta = \frac{2\pi\epsilon\sigma \int_{r_i}^{r_o} rT^4\, dr}{\pi(r_o^2 - r_i^2)\epsilon\sigma T_i^4} = \frac{2\int_0^1 [R(\delta - 1) + 1]\Theta^4\, dR}{\delta + 1}$$

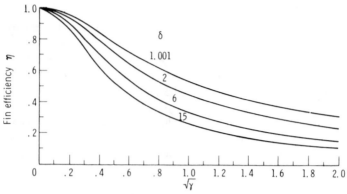

Figure 12-3 Radiation fin efficiency for fin of Example 12-3 [1].

This integral may be carried out after Θ has been determined from the differential equation. The fin efficiency for this type of annular fin has been obtained by Chambers and Somers [1] and is shown in Fig. 12-3. Keller and Holdredge [2] have extended the results to fins of radially varying thicknesses.

In a more general situation, if the environment is at T_e and the fin is nongray with a total absorptivity α for incoming radiation, the energy balance (12-1) becomes

$$kb\frac{1}{r}\frac{d}{dr}\left(r\frac{dT}{dr}\right) - \sigma(\epsilon T^4 - \alpha T_e^4) = 0$$

which can be written as

$$kb\frac{1}{r}\frac{d}{dr}\left(r\frac{dT}{dr}\right) - \epsilon\sigma\left\{T^4 - \left[\left(\frac{\alpha}{\epsilon}\right)^{1/4}T_e\right]^4\right\} = 0 \qquad (12\text{-}3a)$$

The $(\alpha/\epsilon)^{1/4}T_e$ is now a parameter in the solution, and results such as those in Fig. 12-3 can be determined for various values of the parameter. For a gray fin, $\alpha = \epsilon$; hence a nongray fin acts like a gray fin in an effective environment temperature $(\alpha/\epsilon)^{1/4}T_e$ and, by using the effective temperature, results for the gray case can be used for the nongray fin. Reference [3] contains design results for rectangular fins including incident radiation from the environment.

Because of the interest in radiator design for application in space power systems, many conducting-radiating systems have been analyzed. Typical are [1–23]; many other references are to be found in the literature.

For a transient situation where the temperature of the radiating fin is changing with time, a heat-storage term must be included in the energy balance. For the ring element in Example 12-3, this term is $\rho_m c_p b 2\pi r\, dr\, \partial T/\partial \tau$. With this term included, the energy balance (12-3a) becomes a partial differential equation in which temperature is a function of radius and time:

$$kb\frac{1}{r}\frac{\partial}{\partial r}\left(r\frac{\partial T}{\partial r}\right) - \epsilon\sigma\left\{T^4 - \left[\left(\frac{\alpha}{\epsilon}\right)^{1/4}T_e\right]^4\right\} = \rho_m c_p b\frac{\partial T}{\partial \tau} \qquad (12\text{-}3b)$$

Results for the transient behavior of a radiating fin are given in [5, 21].

For a *thin* radiating fin, the temperature within the fin is assumed uniform across the fin thickness, and hence the temperature variation is only in a direction parallel to the radiating surface. If the solid is thick, however, the temperature will also vary with distance normal to the radiating surface. The radiation acts as a boundary condition for the solid conduction problem; thus locally at the surface of a solid that is emitting but not receiving radiation, the boundary condition is

$$-k\frac{\partial T}{\partial n} = \epsilon\sigma T^4 \qquad (12\text{-}4a)$$

where n is the outward normal from the surface. More generally, when the surface is both receiving and losing radiant energy,

$$-k\frac{\partial T}{\partial n} = q_o - q_i \qquad (12\text{-}4b)$$

Time-dependent temperature distributions within solids having surface radiation are investigated in [9]. The transient heat conduction equation is solved with the boundary conditions (12-4).

Example 12-3 concerned only a single radiating fin. A complication that must usually be considered is the mutual interaction of radiation among the fins on a multi-finned surface. This introduces integral terms into the equations, as will be evident from the next example.

EXAMPLE 12-4 An infinite array of thin fins of thickness b, width W, and infinite length are attached to a black base that is held at a constant temperature T_b as pictured in Fig. 12-4. The fin surfaces radiate in a diffuse-gray manner, and the fins are in vacuum. Set up the equation describing the local fin temperature, assuming the environment to be at $T_e = 0$.

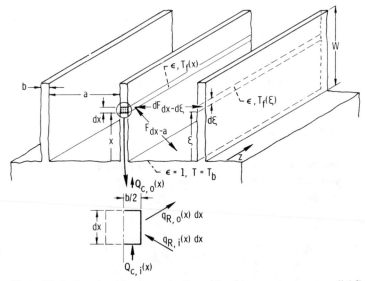

Figure 12-4 Geometry for determination of local temperatures on parallel fins.

Because the fins are thin, it will be assumed that the local temperature of the fin is constant across the thickness b. An energy balance will now be derived for the circled differential element of one fin shown in the inset of Fig. 12-4. Since there is an infinite row of fins, the surrounding environment is identical for each fin and is the same on both sides of each fin. Hence, from symmetry, only half the fin thickness need be considered. Also the problem is simplified because the temperature distribution $T_f(\xi)$ of the adjacent fin is the same as $T_f(x)$. Thus the energy balance need be considered for only one fin. The conduction terms for the energy into and out of the element dx per unit time and *per unit length of fin* in the z direction are

$$Q_{c,i}(x) = -k\frac{b}{2}\frac{dT_f}{dx}$$

$$Q_{c,o}(x) = Q_{c,i} + dQ_{c,i} = -k\frac{b}{2}\frac{dT_f}{dx} + \frac{d}{dx}\left(-k\frac{b}{2}\frac{dT_f}{dx}\right)dx$$

The radiation terms are formulated by using Poljak's net-radiation method from Sec. 8-4.1. The incoming radiation to the element originates from the adjacent fin and from the base surface (since the environment is at $T_e = 0$):

$$q_{R,i}(x)\,dx = \int_{\xi=0}^{W} q_{R,o}(\xi)\,dF_{d\xi-dx}\,d\xi + a\sigma T_b{}^4\,dF_{a-dx}$$

$$= dx\int_{\xi=0}^{W} q_{R,o}(\xi)\,dF_{dx-d\xi} + dx\sigma T_b{}^4 F_{dx-a} \tag{12-5}$$

The outgoing radiation is composed of emission plus reflected incident radiation:

$$q_{R,o}(x)\,dx = \epsilon\sigma T_f{}^4(x)\,dx + (1-\epsilon)q_{R,i}(x)\,dx \tag{12-6}$$

The energy balance on the element is composed of the conduction and radiation quantities:

$$q_{R,o}(x)\,dx + Q_{c,o}(x) = q_{R,i}(x)\,dx + Q_{c,i}(x)$$

By substituting the conduction terms and assuming constant thermal conductivity, the energy balance is given as

$$q_{R,i}(x)\,dx = q_{R,o}(x)\,dx - k\frac{b}{2}\frac{d^2T_f(x)}{dx^2}\,dx \tag{12-7}$$

Equation (12-7), along with (12-5) and (12-6) for $q_{R,i}(x)$ and $q_{R,o}(x)$, are three equations in the unknowns $q_{R,i}(x)$, $q_{R,o}(x)$, and $T_f(x)$. [Note that $q_{R,o}(\xi) = q_{R,o}(x)$.] Eliminating the two energy rates $q_{R,i}$ and $q_{R,o}$ from the three equations results in

$$-\mu\frac{d^2\Theta(X)}{dX^2} + \Theta^4(X) = F_{dX-B} + \int_{Z=0}^{1}\left[-\mu(1-\epsilon)\frac{d^2\Theta(Z)}{dZ^2} + \Theta^4(Z)\right]dF_{dX-dZ} \tag{12-8}$$

where $\Theta(X) = T_f(x)/T_b$, $B = a/W$, $\mu = kb/2\epsilon\sigma T_b^3 W^2$, $X = x/W$, and $Z = \xi/W$.

Equation (12-8) is a nonlinear integrodifferential equation and can be solved numerically. Since it is a second-order equation, two boundary conditions are needed. At the base of the fin $T_f(x = 0) = T_b$ so that

$$\Theta = 1 \qquad \text{at } X = 0 \tag{12-9a}$$

A second condition is obtained at the outer edge of the fin $x = W$. The conduction to this boundary must equal the heat radiated:

$$-k \left. \frac{\partial T_f}{\partial x} \right|_{x=W} = \epsilon\sigma T_f^4(W)$$

or, in terms of Θ,

$$-\frac{d\Theta}{dX} = \frac{\epsilon\sigma T_b^3 W}{k} \Theta^4 = \frac{1}{2\mu} \frac{b}{W} \Theta^4 \qquad \text{at } X = 1 \tag{12-9b}$$

and it is evident that the fin thickness-to-width ratio b/W now enters the problem as a new parameter. If $(b/W)/2\mu$ is very small, then $d\Theta/dX$ can be taken as zero.

The configuration factors in (12-8) are found by the methods of Examples 7-5 and 7-7 (for $\alpha = 90°$ in those equations), which give

$$F_{dX-B} = \frac{1}{2} \left(1 - \frac{x}{\sqrt{a^2 + x^2}} \right) = \frac{1}{2} \left(1 - \frac{X}{\sqrt{B^2 + X^2}} \right)$$

$$dF_{dX-dZ} = \frac{1}{2} \frac{a^2}{[a^2 + (\xi - x)^2]^{3/2}} d\xi = \frac{1}{2} \frac{B^2}{[B^2 + (Z - X)^2]^{3/2}} dZ$$

Note that if the array of fins is infinite, no benefit is gained by adding fins to the plane black base surface because no surface can emit more energy than a black surface. The directional characteristics of such a surface make it attractive in some applications, as discussed in Chap. 5.

Solutions to other fin problems involving mutual interactions are found in [10-23]. The optimization of a fin array with respect to minimum weight is carried out in [22]. Masuda [23] analyzed the radiant interchange between gray diffuse external circular fins on cylinders, and obtained the local heat-flux distribution on the fins and cylinders as well as the fin effectiveness.

The examples given in this section are simplified in that no property variations have been included. When properties are variable, the basic concepts are the same as demonstrated by the examples, although the inclusion of property variations does add some complexity to the functional form of the equations. The usual warnings concerning the inadequacy in some cases of the diffuse-gray assumptions carry over to multimode problems.

When finite-difference techniques are used in the solution of combined conduction-radiation problems, the energy equation is replaced by a set of simultaneous nonlinear algebraic equations. When the physical properties are constant, the conduction terms contain temperatures to the first power, while the radiation terms contain temperatures to the fourth power. To solve a set of nonlinear equations of this type, Ness [24] has presented a rapid-convergence iteration method for the digital computer based on the Newton-Raphson technique. Say that the set of finite-difference equations for the radiation-conduction problem has the form

$$(a_{11}t_1 + a'_{11}t_1{}^4) + (a_{12}t_2 + a'_{12}t_2{}^4) + \cdots + (a_{1n}t_n + a'_{1n}t_n{}^4) - b_1 = 0$$

$$\cdots\cdots\cdots\cdots\cdots\cdots\cdots\cdots\cdots\cdots\cdots\cdots\cdots\cdots\cdots$$

$$(a_{i1}t_1 + a'_{i1}t_1{}^4) + \cdots + (a_{ij}t_j + a'_{ij}t_j{}^4) + \cdots + (a_{in}t_n + a'_{in}t_n{}^4) - b_i = 0 \qquad (12\text{-}10)$$

$$\cdots\cdots\cdots\cdots\cdots\cdots\cdots\cdots\cdots\cdots\cdots\cdots\cdots\cdots\cdots$$

$$(a_{n1}t_1 + a'_{n1}t_1{}^4) + \cdots + (a_{nj}t_j + a'_{nj}t_j^4) + \cdots + (a_{nn}t_n + a'_{nn}t_n{}^4) - b_n = 0$$

The jth temperature is t_j, and the coefficients for the linear and nonlinear contributions of this temperature are a_{ij} and a'_{ij}, respectively.

In the Newton-Raphson procedure an approximate value for each temperature is assumed. Let t_{jo} be this approximation for the jth temperature. Then a correction factor c_j is computed so that $t_j = t_{jo} + c_j$. This corrected temperature is used to compute a new c_j, and the process continued until c_j becomes smaller than a specified value. The c_j are found from the following set of linear equations:

$$f_{11}c_1 + f_{12}c_2 + \cdots + f_{1n}c_n + f_1 = 0$$

$$\cdots\cdots\cdots\cdots\cdots\cdots\cdots\cdots\cdots\cdots\cdots$$

$$f_{i1}c_1 + \cdots + f_{ij}c_j + \cdots + f_{in}c_n + f_i = 0 \qquad (12\text{-}11)$$

$$\cdots\cdots\cdots\cdots\cdots\cdots\cdots\cdots\cdots\cdots\cdots$$

$$f_{n1}c_1 + f_{n2}c_2 + \cdots + f_{nn}c_n + f_n = 0$$

The coefficients f_i are given by

$$f_i = \sum_{j=1}^{n} (a_{ij}t_{jo} + a'_{ij}t_{jo}{}^4) - b_i \qquad (12\text{-}12)$$

and the f_{ij} are

$$f_{ij} = a_{ij} + 4a'_{ij}t_{jo}{}^3 \qquad (12\text{-}13)$$

12-4 RADIATION AND CONVECTION

The treatment of problems involving combined heat transfer by convection and radiation is quite similar to that for conduction-radiation problems. The temperature differences that govern convection appear in place of the derivatives governing conduction; otherwise, the governing energy equations remain of the same nature (that is, nonlinear and often almost intractable).

Radiation-convection interaction problems are found in consideration of the cooling of high temperature components; convection cells and their effect on radiation from stars; furnace design where heat transfer from surfaces occurs by parallel radiation and convection; the interaction of incident solar radiation with the earth's surface to produce complex free convection patterns and thus to complicate the art of weather forecasting; and marine environment studies for predicting free convection patterns in oceans and lakes. To illustrate the concepts involved in an engineering problem, an example will now be considered that involves gas flow through a heated tube. Representative solutions of this nature are found in [21, 25–30]. The discussion in this chapter is limited to convective media that do not absorb, scatter, or emit radiation.

The enclosure energy-balance equations, such as (8-6) and (8-7), can be used as before except that the Q_k will now specifically contain a convective heat addition to the wall. In Example 12-5 the Q_k/A_k [as revealed by (12-14)] consists of convection from the gas plus the electrical energy supplied by heat generation within the tube wall.

EXAMPLE 12-5 A transparent gas enters a black circular tube with geometry shown in Fig. 12-5. The wall of the tube is thin, and the outer surface of the tube is perfectly insulated. The tube wall is heated electrically to provide a uniform input of energy per unit area per unit time. The variation of local wall temperature along the tube length is to be determined. The convective heat transfer coefficient h between the gas and the inside of the tube is assumed constant. The gas has a mean velocity u_m, heat capacity c_p, and density ρ_f.

If radiation were not considered, the local heat addition to the gas would be equal to the local electrical heating (since the outside of the tube is insulated) and hence would be invariant with axial position X along the tube. As a consequence, both the gas temperature and wall temperature would rise linearly with X. On the other hand if convection were not considered, the only means for heat removal would be by radiation out the ends of the tube as discussed in Example 8-9. In this instance, for equal environment temperatures at both ends of the tube,

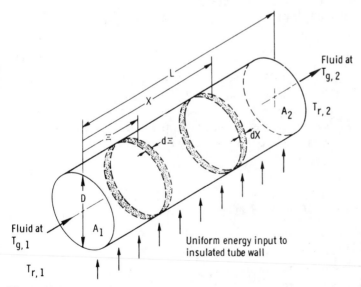

Figure 12-5 Flow through tube with uniform internal energy input to wall, and outer surface insulated.

the wall temperature has a maximum at the center of the tube and decreases continuously toward both ends. The solution of the combined radiation-convection problem is expected to partially exhibit the trends of both limiting solutions.

Consider an energy balance on a ring area element of length dX on the interior of the tube wall at position X as in Fig. 12-5. The energy supplied to the ring per unit time is

$$q_w \pi D \, dX + \int_{\Xi=0}^{L} \sigma T_w^4(\Xi) \, dF_{d\Xi-dX}(|\Xi - X|) \pi D \, d\Xi$$

$$+ \sigma T_{r,1}^4 \frac{\pi D^2}{4} dF_{1-dX}(X) + \sigma T_{r,2}^4 \frac{\pi D^2}{4} dF_{2-dX}(L - X)$$

The terms are, respectively, the energy supplied by electrical heating of the tube wall, the energy radiated to dA_X by other wall elements of the tube interior (see Example 8-9), and the energy radiated to dA_X from the inlet and the exit reservoirs. The reservoirs are assumed to act as black disks at the inlet and outlet reservoir temperatures, which would have to be specified. Usually the reservoirs are assumed to be at the inlet and outlet gas temperatures. The energy transferred away from the ring element at X by convection and radiation is

$$h \pi D \, dX [T_w(X) - T_g(X)] + \sigma T_w^4(X) \pi D \, dX$$

If axial heat conduction in the tube wall is neglected, the energy supplied to the ring element must equal that transferred away, and the energy quantities are equated to yield the following expression (reciprocity has been used on the F factors so that dX could be divided out of the equation):

$$h [T_w(X) - T_g(X)] + \sigma T_w^4(X) = q_w + \int_{\Xi=0}^{L} \sigma T_w^4(\Xi) \, dF_{dX-d\Xi}(|X - \Xi|)$$

$$+ \sigma T_{r,1}^4 F_{dX-1}(X) + \sigma T_{r,2}^4 F_{dX-2}(L - X) \qquad (12\text{-}14)$$

This is of the form of (7-66), where for (12-14), $Q_k/A_k = q_w + h[T_g(X) - T_w(X)]$. Equation (12-14) has two unknowns $T_w(X)$ and $T_g(X)$; hence, a second equation must be found before a solution can be obtained. This is done by forming an energy balance on the volume within the tube occupying the length dX. The energy that is carried into this volume by the flowing gas is

$$Q_{i,g} = u_m \rho_f c_p T_g(X) \frac{\pi D^2}{4}$$

An additional amount of energy is added to the volume by convection from the wall, namely,

$$dQ_{i,g} = \pi D h [T_w(X) - T_g(X)] \, dX$$

The energy carried out by the flowing gas is

$$Q_{o,g} = u_m \rho_f c_p \frac{\pi D^2}{4} \left[T_g(X) + \frac{dT_g(X)}{dX} dX \right]$$

Equating the outgoing and incoming energies gives the energy balance

$$u_m \rho_f c_p \frac{D}{4} \frac{dT_g(X)}{dX} = h[T_w(X) - T_g(X)] \tag{12-15}$$

By defining the dimensionless quantities

$$S = \frac{4h}{u_m \rho_f c_p} = \frac{4\,\text{Nu}}{\text{Re Pr}} \qquad H = \frac{h}{q_w}\left(\frac{q_w}{\sigma}\right)^{1/4} \qquad t = T\left(\frac{\sigma}{q_w}\right)^{1/4}$$

and $x = X/D$, $\xi = \Xi/D$, $l = L/D$, the energy balances on the wall and fluid elements can be written, respectively, as

$$t_w{}^4(x) + H[t_w(x) - t_g(x)] = 1 + \int_0^x t_w{}^4(\xi)\,dF_{dx-d\xi}(x - \xi)$$

$$+ \int_x^l t_w{}^4(\xi)\,dF_{dx-d\xi}(\xi - x) + t_{r,1}^4 F_{dx-1}(x) + t_{r,2}^4 F_{dx-2}(l - x) \tag{12-16}$$

$$\frac{dt_g(x)}{dx} = S[t_w(x) - t_g(x)] \tag{12-17}$$

giving two equations involving the unknowns $t_w(x)$ and $t_g(x)$, and having five parameters S, H, l, $t_{r,1}$, and $t_{r,2}$. The configuration factors can be obtained from known disk-to-disk factors by the technique of Example 7-20 and Eq. (7-64), and are given in Example 8-9.

To solve Eqs. (12-16) and (12-17), it is noted that (12-17) is a first-order linear differential equation that can be solved in general form by use of an integrating factor. The boundary condition is that at $x = 0$ the gas temperature has a specified value $t_{g,1}$. The general solution is then

$$t_g(x) = Se^{-Sx} \int_0^x e^{S\xi} t_w(\xi)\,d\xi + t_{g,1}e^{-Sx} \tag{12-18}$$

This can be substituted into (12-16) to eliminate $t_g(x)$ and yield the following integral equation for the desired variation in tube-wall temperature:

$$t_w{}^4 + Ht_w - HSe^{-Sx} \int_{\xi=0}^x e^{S\xi} t_w(\xi)\,d\xi - Ht_{g,1}e^{-Sx}$$

$$= 1 + \int_{\xi=0}^x t_w{}^4(\xi)\,dF_{dx-d\xi}(x - \xi) + \int_{\xi=x}^l t_w{}^4(\xi)\,dF_{dx-d\xi}(\xi - x)$$

$$+ t_{r,1}^4 F_{dx-1}(x) + t_{r,2}^4 F_{dx-2}(l - x) \tag{12-19}$$

Solutions to (12-19) have been obtained by Perlmutter and Siegel [25], and some representative results as calculated by numerical integration are shown in Fig. 12-6. Note that the predicted temperatures for combined radiation-convection fall below the temperatures predicted for either convection or radiation acting independently. For a short tube the radiation effects are significant

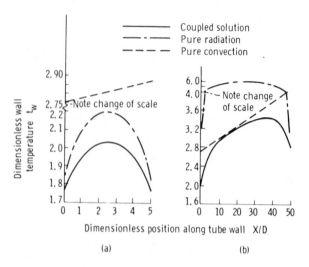

Figure 12-6 Tube wall temperatures resulting from combined radiation and convection for transparent gas flowing in uniformly heated black tube for $S = 0.02$, $H = 0.8$, $t_{r,1} = t_{g,1} = 1.5$, and $t_{r,2} = t_{g,2}$. (a) Tube length, $L/D = 5$; (b) tube length, $L/D = 50$.

over the entire tube length, and for the parameters shown the combined mode temperature distribution is similar to that for radiation alone. For a long tube, however, the combined mode distribution is very close to that for convection alone over the central portion of the tube. The heat transfer resulting from combined convection-radiation is more efficient than by either mode alone. This means that the wall temperature distribution predicted in the combined problem will always lie below both distributions predicted by using either mode alone.

EXAMPLE 12-6 If the tube in Example 12-5 had a diffuse-gray interior surface, rather than being black, what would be the governing energy equations?

Using the net-radiation method, a heat balance on an area element at X is

$$q_w(X) + q_i(X) = q_o(X) + h[T_w(X) - T_g(X)] \tag{12-20}$$

where q_i and q_o are the incoming and outgoing radiation fluxes. For the outgoing radiation flux one may write

$$q_o(X) = \epsilon \sigma T_w^4(X) + (1 - \epsilon) q_i(X) \tag{12-21}$$

Equations (12-20) and (12-21) are combined to eliminate q_i with the following result:

$$q_o(X) = \frac{1 - \epsilon}{\epsilon} \{h[T_w(X) - T_g(X)] - q_w\} + \sigma T_w^4(X) \tag{12-22}$$

The analysis leading to Eq. (12-14) applies for the gray case if the radiation leaving the surface $\sigma T_w{}^4$ is replaced by q_o. This gives

$$h[T_w(X) - T_g(X)] + q_o(X) = q_w + \int_{\Xi=0}^{L} q_o(\Xi)\, dF_{dX-d\Xi}(|X - \Xi|)$$

$$+ \sigma T_{r,1}^4 F_{dX-1}(X) + \sigma T_{r,2}^4 F_{dX-2}(L - X) \tag{12-23}$$

Equation (12-15) is unchanged by having the wall gray. Thus (12-22), (12-23), and (12-15) comprise a set of three equations in the unknowns $T_w(X)$, $q_o(X)$, and $T_g(X)$. Some numerical solutions for this system of equations are given in [26]. Note that (12-22) and (12-23) are the same as the general diffuse-gray enclosure equations (8-6) and (8-7). For the present situation the $Q_k/A_k = q_w + h[T_g(X) - T_w(X)]$.

12-5 RADIATION COMBINED WITH BOTH CONDUCTION AND CONVECTION

The basic elements of the derivations in Secs. 12-3 and 12-4 are combined when both conduction and convection are present in a radiating system. The energy equations now contain both temperature differences arising from convection, and temperature derivatives arising from conduction. There are also a greater number of independent parameters. These arise from such things as the convective heat transfer coefficients, thermal conductivity of the body, and body dimensions, in other words the quantities that govern both convection and conduction. As a result of these complexities there are no "classical" solutions or solution methods, and results must usually be obtained using numerical techniques.

The basic ideas involved will now be given by discussing a few specific problems. Additional information and results are given in [31–42].

EXAMPLE 12-7 Consider again the tube in Example 12-5. The tube is uniformly heated, perfectly insulated on the outside, and has a black interior surface. Gas flows through the tube, and the convective heat transfer coefficient h is assumed constant. Axial heat conduction within the tube wall will now be included. The tube wall has thermal conductivity k_w; its thickness is b; the tube inside diameter is D_i, and the outside diameter is D_o. The desired result is the temperature distribution along the tube length. The tube wall is assumed sufficiently thin that the temperature at each axial position is constant across the wall thickness.

The energy balance as given by Eq. (12-14) must be modified to include axial wall conduction. The heat conduction into an elemental length of the tube wall is

$$Q_{c,i} = -k_w \frac{\pi(D_o^2 - D_i^2)}{4} \frac{dT_w(X)}{dX}$$

while that conducted out of the element is

$$Q_{c,o} = -k_w \frac{\pi(D_o{}^2 - D_i{}^2)}{4} \left[\frac{dT_w(X)}{dX} + \frac{d^2T_w(X)}{dX^2} dX \right]$$

The net gain of energy by the element from conduction is then

$$k_w \pi \frac{D_o{}^2 - D_i{}^2}{4} \frac{d^2T_w(X)}{dX^2} dX$$

This term is divided by the internal area of the ring $\pi D_i \, dX$ and is then added to the right side of (12-14) to obtain the energy balance:

$$h[T_w(X) - T_g(X)] + \sigma T_w{}^4(X) = q_w + k_w \frac{D_o{}^2 - D_i{}^2}{4D_i} \frac{d^2T_w(X)}{dX^2}$$

$$+ \int_{\Xi=0}^{L} \sigma T_w{}^4(\Xi) \, dF_{dX-d\Xi}(|X - \Xi|) + \sigma T_{r,1}^4 F_{dX-1}(X)$$

$$+ \sigma T_{r,2}^4 F_{dX-2}(L - X) \tag{12-24}$$

This is of the same form as (7-66) with the nonradiative heat addition per unit area being

$$\frac{Q_k}{A_k} = q_w + h[T_g(X) - T_w(X)] + k_w \frac{D_o{}^2 - D_i{}^2}{4D_i} \frac{d^2T_w(X)}{dX^2}$$

As in connection with (12-16), all lengths are nondimensionalized by dividing by the internal tube diameter, and dimensionless parameters are introduced. The conduction term yields a new parameter

$$N = \frac{k_w}{4q_w D_i} \left[\left(\frac{D_o}{D_i} \right)^2 - 1 \right] \left(\frac{q_w}{\sigma} \right)^{1/4}$$

For thin walls where $(D_o - D_i)/2 = b \ll 1$ this reduces to

$$N = \frac{k_w b}{2 q_w D_i{}^2} \left(\frac{q_w}{\sigma} \right)^{1/4}$$

which is the parameter used in some of the references.

The dimensionless form of the energy equation is

$$t_w{}^4(x) + H[t_w(x) - t_g(x)] = 1 + N \frac{d^2 t_w(x)}{dx^2} + \int_{\xi=0}^{x} t_w{}^4(\xi) \, dF_{dx-d\xi}(x - \xi)$$

$$+ \int_{\xi=x}^{l} t_w{}^4(\xi) \, dF_{dx-d\xi}(\xi - x) + t_{r,1}{}^4 F_{dx-1}(x) + t_{r,2}{}^4 F_{dx-2}(l - x) \tag{12-25}$$

The energy equation for the fluid within the tube element remains

$$\frac{dt_g(x)}{dx} = S[t_w(x) - t_g(x)] \tag{12-26}$$

as in (12-17); these equations may be further combined as in (12-19).

Hottel [31] has discussed this problem in terms of slightly different parameters. He obtained a numerical solution before the common utilization of high-speed computers. For one set of parameters and for five ring-area intervals on the tube wall, the solution required 10 h of hand computation. The results are shown in Fig. 12-7 in terms of the parameters derived here.

Two additional factors that enter this problem are the conduction boundary conditions. The solution of (12-25) requires two boundary conditions because of the arbitrary constants introduced by integrating the d^2t_w/dx^2 term. These boundary conditions depend on the physical construction at the ends of the tube, which determine the amount of conduction present. In [32] some detailed results were obtained. It was assumed for simplicity that the end edges of the tube were insulated, that is,

$$\frac{dT_w}{dX}\bigg|_{X=0} = \frac{dT_w}{dX}\bigg|_{X=L} = 0$$

The extension was also made in [32] to let the convective heat transfer coefficient vary with position along the tube. This accounts for the variation of h in the thermal entrance region.

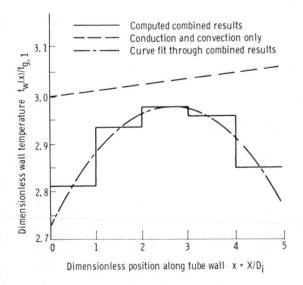

Figure 12-7 Wall temperature distribution for flow of transparent fluid through black tube with combined radiation, convection, and conduction for $l = 5$, $S = 0.005$, $N = 0.316$, $H = 1.58$, $t_{r,1} = t_{g,1} = 0.316$, and $t_{r,2} = t_{g,2}$.

EXAMPLE 12-8 As a second type of problem including combined conduction, convection, and radiation, consider a fin as shown in Fig. 12-8. A gas at T_e is flowing over the fin and removing heat by convection. The environment to which the fin radiates is assumed also to be at T_e. The cross section of the fin has area A and perimeter P. The fin is nongray with absorptivity α for incident radiation from the environment.

An energy balance on an element of length dX yields

$$kA \frac{d^2T}{dX^2} dX = \sigma(\epsilon T^4 - \alpha T_e{}^4) P \, dX + hP \, dX(T - T_e) \tag{12-27}$$

The term on the left is the net conduction into the element. The terms on the right are the radiative and convective losses. The radiative exchange between the fin and its base is being neglected. This equation is to be solved for T as a function of X. We multiply (12-27) by $[1/(kA \, dX)] \, dT/dX$ to obtain

$$\frac{d^2T}{dX^2} \frac{dT}{dX} = \frac{\epsilon\sigma P}{kA} \left(T^4 - \frac{\alpha}{\epsilon} T_e{}^4\right) \frac{dT}{dX} + \frac{hP}{kA} (T - T_e) \frac{dT}{dX}$$

This can be integrated once to obtain

$$\frac{1}{2}\left(\frac{dT}{dX}\right)^2 = \frac{\epsilon\sigma P}{kA} \left(\frac{T^5}{5} - \frac{\alpha}{\epsilon} TT_e{}^4\right) + \frac{hP}{kA} \left(\frac{T^2}{2} - TT_e\right) + C \tag{12-28}$$

where C is a constant of integration.

As a simplification let $T_e \approx 0$ and let the fin be very long. Then for large X, $T(X) \to 0$ and $dT/dX \to 0$, and from (12-28) the constant $C = 0$. Then solving for dT/dX results in

$$\frac{dT}{dX} = -\left(\frac{2}{5} \frac{P\epsilon\sigma}{kA} T^5 + \frac{hP}{kA} T^2\right)^{1/2} \tag{12-29}$$

The minus sign is used when taking the square root since T decreases as X increases. The variables in (12-29) can be separated and the equation integrated with the condition that $T(X) = T_b$ at $X = 0$:

$$\int_0^X dX = -\int_{T_b}^T \frac{dT}{T[\frac{2}{5}(P\epsilon\sigma/kA) T^3 + hP/kA]^{1/2}}$$

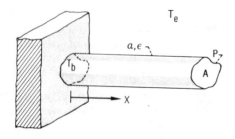

Figure 12-8 Fin of constant cross-sectional area transferring energy by radiation and convection. (Flowing gas and environment at T_e.)

Integration yields

$$X = \tfrac{1}{3}M^{-1/2}\left[\ln\frac{(GT_b^3 + M)^{1/2} - M^{1/2}}{(GT_b^3 + M)^{1/2} + M^{1/2}} - \ln\frac{(GT^3 + M)^{1/2} - M^{1/2}}{(GT^3 + M)^{1/2} + M^{1/2}}\right] \qquad (12\text{-}30)$$

where $G = \tfrac{2}{5}Pe\sigma/kA$ and $M = hP/kA$. Hence, for this simplified case a closed-form analytical solution for the temperature distribution is obtained. A detailed treatment of this type of fin problem with both convection and radiation from the surface is given in [38-40].

Variational methods have been applied by Campo [43] to longitudinal fins of this type for both steady-state and transient conditions. The Ritz method is applied in the steady-state case, while the Kantorovich method is applied to obtain the transient solution, assuming that the fin temperature profile is the product of a polynomial in position and an unknown function of time. The form of the position polynomial is evaluated from the steady-state solution. Comparison with numerical solutions shows good accuracy.

12-6 COMPUTER PROGRAMS FOR MULTIMODE ENERGY TRANSFER

Except in simple geometries, the solution of problems involving radiation transfer plus energy transfer by other modes becomes exceedingly difficult. Examination of the examples will show that for this reason only fairly simple cases have been solved here. Because of the mathematical difficulties involved, a number of generalized finite-difference computer programs have been developed for multimode problems, and some of these are outlined in [44-51]. Such programs allow "cookbook" solution of problems that fall within their limitations. Each program referenced allows consideration of combined conduction, radiation, and convection, and most of the programs also allow inclusion of the effects of internal energy generation, flow, transients, variable properties, mass transfer, changes of state, heat capacity in the media considered, and three-dimensional geometries. These programs are limited by the assumption of diffuse-gray surfaces, but are impressive in their generality.

12-7 CONCLUDING REMARKS

Multimode energy transfer problems involving radiant transfer through transparent media have been examined. Conceptually, the treatment involves only the careful construction of energy-balance equations over finite areas or on discrete elements. The chief difficulty then becomes the mathematical treatment of these energy-balance equations.

Many mathematical methods have been applied with some success to these multimode problems. When a problem of this type is encountered, the techniques that have been successful for similar problems in the literature should be examined. These range from brute-force finite-difference formulations through quite sophisticated analytical treatments. The reference list following gives representative problems and solution techniques, along with some expositions of specific mathematical techniques.

REFERENCES

1. Chambers, R. L., and E. V. Somers: Radiation Fin Efficiency for One-Dimensional Heat Flow in a Circular Fin, *J. Heat Transfer*, vol. 81, no. 4, pp. 327–329, 1959.
2. Keller, H. H., and E. S. Holdredge: Radiation Heat Transfer for Annular Fins of Trapezoid Profile, *J. Heat Transfer*, vol. 92, no. 6, pp. 113–116, 1970.
3. Mackay, Donald B.: "Design of Space Powerplants," Prentice-Hall, Inc., Englewood Cliffs, N.J., 1963.
4. Ungar, Eric E., and L. A. Mekler: Tube Metal Temperatures for Structural Design, *J. Eng. Ind.*, vol. 82, no. 3, pp. 270–276, 1960.
5. Hickman, R. S.: Transient Response and Steady-state Temperature Distribution in a Heated, Radiating, Circular Plate, *Tech. Rep.* 32-169, *Jet Propulsion Lab.*, California Institute of Technology, Nov. 22, 1961.
6. Wilkins, J. Ernest, Jr.: Minimum-Mass Thin Fins and Constant Temperature Gradients, *J. Soc. Ind. Appl. Math.*, vol. 10, no. 1, pp. 62–73, 1962.
7. Jaeger, J. C.: Conduction of Heat in a Solid with a Power Law of Heat Transfer at Its Surface, *Cambridge Phil. Soc. Proc.*, vol. 46, pt. 4, pp. 634–641, 1950.
8. Chambré, Paul L.: Nonlinear Heat Transfer Problem, *J. Appl. Phys.*, vol. 30, no. 11, pp. 1683–1688, 1959.
9. Abarbanel, Saul S.: Time Dependent Temperature Distribution in Radiating Solids, *J. Math. Phys.*, vol. 39, no. 4, pp. 246–257, 1960.
10. Stockman, Norbert O., and John L. Kramer: Effect of Variable Thermal Properties on One-Dimensional Heat Transfer in Radiating Fins, *NASA* TN D-1878, 1963.
11. Tatom, John W.: Shell Radiation, paper no. 60-WA-234, *ASME*, November, 1960.
12. Sparrow, E. M., and E. R. G. Eckert: Radiant Interaction between Fin and Base Surfaces, *J. Heat Transfer*, vol. 84, no. 1, pp. 12–18, 1962.
13. Hering, R. G.: Radiative Heat Exchange between Conducting Plates with Specular Reflection, paper no. 65-HT-28, *ASME*, August, 1965.
14. Sparrow, E. M., G. B. Miller, and V. K. Jonsson: Radiating Effectiveness of Annular-Finned Space Radiators, Including Mutual Irradiation between Radiator Elements, *J. Aerosp. Sci.*, vol. 29, no. 11, pp. 1291–1299, 1962.
15. Schreiber, L. H., R. P. Mitchell, G. D. Gillespie, and T. M. Olcott: Techniques for Optimization of a Finned-tube Radiator, paper no. 61-SA-44, *ASME*, June, 1961.
16. Heaslet, Max A., and Harvard Lomax: Numerical Predictions of Radiative Interchange between Conducting Fins with Mutual Irradiations, *NASA* TR R-116, 1961.
17. Nichols, Lester D.: Surface-Temperature Distribution on Thin-walled Bodies Subjected to Solar Radiation in Interplanetary Space, *NASA* TN D-584, 1961.
18. Kotan, K., and O. A. Arnas: On the Optimization of the Design Parameters of Parabolic Radiating Fins, paper no. 65-HT-42, *ASME*, August, 1965.
19. Russell, Lynn D., and Alan J. Chapman: Analytical Solution of the "Known-Heat-Load" Space Radiator Problem, *J. Spacecr. Rockets*, vol. 4, no. 3, pp. 311–315, 1967.
20. Donovan, R. C., and W. M. Rohrer: Radiative Conducting Fins on a Plane Wall, Including Mutual Irradiation, paper no. 69-WA/HT-22, *ASME*, November, 1969.
21. Eslinger, R., and B. Chung: Periodic Heat Transfer in Radiating and Convecting Fins or Fin Arrays, *AIAA J.*, vol. 17, no. 10, pp. 1134–1140, 1979.
22. Schnurr, N. M., A. B. Shapiro, and M. A. Townsend: Optimization of Radiating Fin Arrays with Respect to Weight, *ASME J. Heat Transfer*, vol. 98, no. 4, pp. 643–648, 1976.

23. Masuda, H.: Radiant Heat Transfer on Circular-Finned Cylinders, *Rep. Inst. High Speed Mech. Tohōku Univ.*, vol. 27, no. 255, pp. 67–89, 1973. (See also *Trans. Jpn. Soc. Mech. Eng.*, vol. 38, pp. 3229–3234, 1972.)

24. Ness, A. J.: Solution of Equations of a Thermal Network on a Digital Computer, *Sol. Energy*, vol. 3, no. 2, p. 37, 1959.

25. Perlmutter, M., and R. Siegel: Heat Transfer by Combined Forced Convection and Thermal Radiation in a Heated Tube, *J. Heat Transfer*, vol. 84, no. 4, pp. 301–311, 1962.

26. Siegel, R., and M. Perlmutter: Convective and Radiant Heat Transfer for Flow of a Transparent Gas in a Tube with a Gray Wall, *Int. J. Heat Mass Transfer*, vol. 5, pp. 639–660, 1962.

27. Cess, R. D.: The Effect of Radiation Upon Forced-Convection Heat Transfer, *Appl. Sci. Res.*, vol. 10, sec. A., pp. 430–438, 1961.

28. Keshock, E. G., and R. Siegel: Combined Radiation and Convection in an Asymmetrically Heated Parallel Plate Flow Channel, *J. Heat Transfer*, vol. 86, no. 3, pp. 341–350, 1964.

29. Love, T. J., and J. Francis: A Linearized Analysis for Longitudinal Fins with Radiative and Convective Exchange, AIAA paper 78-856, May, 1978.

30. Aziz, A., and J. Y. Benzies: Application of Perturbation Techniques to Heat Transfer Problems with Variable Thermal Properties, *Int. J. Heat Mass Transfer*, vol. 19, pp. 271–276, 1976.

31. Hottel, H. C.: Geometrical Problems in Radiant Heat Transfer, Heat Transfer Lectures (Don Cowen, ed.), *Comp. Rept.* NEPA-979-IER-13, Fairchild Engine and Airplane Corporation, vol. 2, pp. 76–95, 1949.

32. Siegel, Robert, and Edward G. Keshock: Wall Temperatures in a Tube with Forced Convection, Internal Radiation Exchange, and Axial Wall Heat Conduction, *NASA* TN D-2116, 1964.

33. Robbins, William H., and Carroll A. Todd: Analysis, Feasibility, and Wall-Temperature Distribution of a Radiation-Cooled Nuclear-Rocket Nozzle, *NASA* TN D-878, 1962.

34. Krebs, Richard P., Henry C. Haller, and Bruce M. Auer: Analysis and Design Procedures for a Flat, Direct-Condensing, Central Finned-Tube Radiator, *NASA* TN D-2474, 1964.

35. Okamoto, Yoshizo: Thermal Performance of Radiative and Convective Plate-Fins with Mutual Irradiation, *Bull. JSME*, vol. 9, no. 33, pp. 150–165, 1966.

36. Okamoto, Yoshizo: Temperature Distribution and Efficiency of a Single Sheet of Radiative and Convective Fin Accompanied by Internal Heat Source, *Bull. JSME*, vol. 7, no. 28, pp. 751–758, 1964.

37. Okamoto, Yoshizo: Temperature Distribution and Efficiency of a Plate and Annular Fin with Constant Thickness, *Bull. JSME*, vol. 9, no. 33, pp. 143–150, 1966.

38. Shouman, A. R.: An Exact Solution for the Temperature Distribution and Radiant Heat Transfer along a Constant Cross Sectional Area Fin with Finite Equivalent Surrounding Sink Temperature, *Proc. Ninth Midwestern Mech. Conf.*, Madison, Wis., Aug. 16–18, 1965, pp. 175–186.

39. Shouman, A. R.: Nonlinear Heat Transfer and Temperature Distribution through Fins and Electric Filaments of Arbitrary Geometry with Temperature-Dependent Properties and Heat Generation, *NASA* TN D-4257, 1968.

40. Shouman, A. R.: An Exact General Solution for the Temperature Distribution and the Radiation Heat Transfer along a Constant Cross-Sectional-Area Fin, paper no. 67-WA/HT-27, *ASME*, November, 1967.

41. Sohal, M., and John R. Howell: Thermal Modeling of a Plate with Coupled Heat Transfer Modes, in "Thermophysics and Spacecraft Thermal Control," vol. 35 of *Progress in Astronautics and Aeronautics*, MIT Press, Cambridge, Mass., 1974.

42. Sohal, M., and John R. Howell: Determination of Plate Temperature in Case of Combined Conduction, Convection and Radiation Heat Exchange, *Int. J. Heat Mass Transfer*, vol. 16, pp. 2055–2066, 1973.

43. Campo, Antonio: Variational Techniques Applied to Radiative-Convective Fins with Steady and Unsteady Conditions, *Wärme Stoffübertrag.*, vol. 9, pp. 139–144, 1976.

44. Sepetoski, W. K., C. H. Sox, and P. F. Strong: Description of a Transient Thermal Analysis

Program for Use with the Method of Zones, *Rept.* 2, C-65670, Arthur D. Little Company (*NASA* CR-56722), August, 1963.

45. Mintz, Michael D.: Finite Difference Representation and Solution of Practical Heat-transfer Problems, *Rept.* UCRL-7960 (rev. 1), University of California Lawrence Radiation Laboratory, 1965.

46. Smith, James P.: SINDA (Systems Improved Numerical Differencing Analyzer) Users Manual, *TRW Systems Group Rep.* 14690-H001-R0-00; *NASA* Contract 9-10435, April, 1971.

47. Oren, J. A., and D. R. Williams: Thermal and Flow Analysis Subroutines for the SINDA-Version 9 Computer Routine, *Rept.* 00.1582, Vought Systems Division, LTV Aerospace Corp. (*NASA-JSC* contract NAS 9-6807), Sept. 24, 1973.

48. Stephens, G. L., and D. J. Campbell: Program THTB for Analysis of General Transient Heat Transfer Systems, *Rept.* R60 FPD 647, General Electric Company, April, 1961.

49. Schultz, H. D.: Thermal Analyzer Computer Program for the Solution of General Heat Transfer Problems, *Rept.* LR-18902, Lockheed California Company (*NASA* CR-65581), July 16, 1965.

50. Strong, P. F., and A. G. Emslie: The Method of Zones for the Calculation of Temperature Distribution, *Rept.* 1, C-65670, Arthur D. Little Company (*NASA* CR-56800), July, 1963.

51. Mason, W. E., Jr.: TACO: A Finite Element Heat Transfer Code, UICD-17980, Lawrence Livermore Laboratory, California, Dec. 1978.

PROBLEMS

1 A thin two-dimensional fin in vacuum is radiating to outer space at temperature $T_e = 0$. The base of the fin is at T_b, and the heat loss from the end edge of the fin can be considered zero. The fin surface is gray with emissivity ϵ. What is the differential equation in dimensionless form for the temperature distribution along the fin (neglect any radiant interaction with the base surface)? What are the boundary conditions? Can you separate variables and indicate the integration to obtain the temperature distribution $T(x)$? [*Hint:* $\int (d^2\theta/dx^2) \, (d\theta/dx) = \frac{1}{2}(d\theta/dx)^2 + \text{constant}$.]

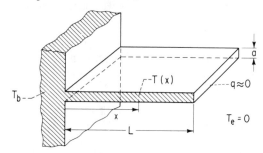

2 Consider the fin in Fig. 12-8 as analyzed in Example 12-8. The heat transfer coefficient at the tip of the fin is h_L, and the emissivity of the end area is ϵ as for the rest of the fin surface. Formulate the boundary condition at the end face of the fin, and apply this condition to the general solution of the fin energy equation. Formulate all the analytical relations, and describe how you would obtain the fin efficiency.

3 Assume the fin in Example 12-8 is on a black plane base surface (at T_b) that is very large compared with the fin length. How would the formulation be modified to account for the radiant interaction between the fin and base surface?

4 Consider Prob. 1 of Chap. 8. The radiation shield now consists of a layer of opaque plastic 0.05 in thick coated on each side with a thin layer of metal having the same emissivity, $\epsilon_s = 0.2$, as before. The thermal conductivity of the plastic is 0.12 Btu/(h·ft·°F). What is the

heat transfer from plate 2 to plate 1, and how does it compare with that for a very thin shield as in Prob. 1 of Chap. 8?

Answer: 2000 Btu/(h·ft²).

5   A space radiator is composed of a series of plane fins of thickness $2t$ between tubes at uniform temperature T_b. The tubes are black and the fins are gray with emissivity ϵ. The radiator is in a vacuum with an environment temperature $T_e = 0$. Formulate the differential equation (including the analytical expressions for the view factors) and boundary conditions to obtain the temperature distribution $T(x)$ along the fin. Include the interaction between the fin and tubes.

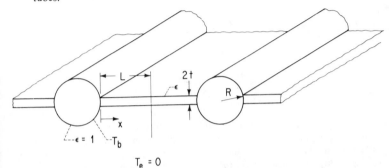

$T_e = 0$

6   A thin electrically heated wire is dissipating a total of Q_e W. The electrodes at the ends of the wire are water cooled, and both are maintained at temperature T_0. The surroundings are large in size and are at uniform temperature T_e. The wire surface is diffuse-gray with emissivity ϵ. Air at T_e is blowing across the wire, providing a uniform heat transfer coefficient h. Set up the differential equation and boundary conditions to obtain the wire temperature distribution.

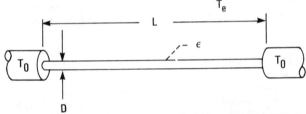

7   A copper-constantan thermocouple is in an inert-gas stream at 350 K adjacent to a blackbody surface at 900 K. The heat transfer coefficient from the gas to the thermocouple is 25 W/(m²·K). Estimate the temperature of the bare thermocouple. Estimate the thermocouple temperature if it is surrounded by one polished aluminum radiation shield in the form of a cylinder open at both ends. The heat transfer coefficient from the gas to both sides of the shield is 15 W/(m²·K).

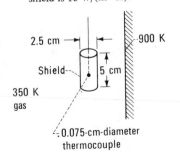

8   Thin wire is extruded at fixed velocity through a die at temperature T_0. The wire then passes through air at T_a until its temperature is reduced to T_L. The heat transfer coefficient to the air is h, and the wire emissivity is ϵ. It is desired to obtain the relation between T_L and T_0 as a function of wire velocity V and distance L. Derive the differential equation for wire temperature as a function of distance from the die; state the boundary conditions. (*Hint:* Compute the heat balance for flow in and out of a control volume fixed in space.)

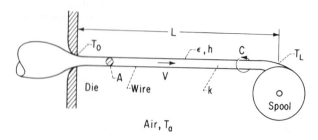

9   A single circular fin is to dissipate heat from both its sides in a vacuum at low temperature. The fin is on a tube with a $\frac{1}{2}$-in outer diamter. The tube wall is maintained at 1800°R by vapor condensing on the inside. The fin has a 4-in outer diameter and is $\frac{1}{16}$ in thick. Estimate the heat radiated away if the fin is made from: (*a*) copper with a polished surface (Fig. 5-11a); (*b*) copper with a lightly oxidized surface; (*c*) stainless steel [$k = 20$ Btu/(h·ft·°R)] with a clean surface. What is the effect on the heat dissipated of increasing the fin thickness to $\frac{1}{8}$ in?

   Answer: (*c*) 270 Btu/h (using $\epsilon = 0.3$), 390 Btu/h.

10 How would the analysis in Example 12-4 be modified to include the effect of a nonzero environment temperature?

11 The billet in Prob. 21 of Chap. 7 has air at 70°F blowing across it that provides an average convective heat transfer coefficient of 5 Btu/(h·ft²·°F). Estimate the cooling time with both radiation and convection included.

   Answer: 1.98 h.

12 Opaque liquid at temperature $T(0)$ and with mean velocity $\bar{u}$ enters a long tube that is surrounded by a vacuum jacket and a concentric electric heater that is kept at uniform temperature T_e along its length. The heater can be considered black, and the tube exterior is diffuse-gray with emissivity ϵ. The convective heat transfer coefficient between the liquid and the tube wall is h, and the tube-wall thermal conductivity is k_w. Derive the relations to determine the mean liquid temperature and the tube-wall outer surface temperature as a function of distance x along the tube (assume that liquid properties are constant).

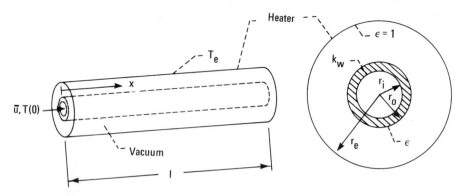

13 A solar collector is designed to fit onto the horizontal section of a roof as diagrammed at the left below. Flow is from right to left in the tubes of the collector. A white, diffuse, roof section helps reflect additional solar flux onto the collector. Set up equations for determining the local temperature of the tubes for two cases: (*a*) no flow in the tubes, and (*b*) flow in each tube of $\dot{m} = 5$ lb/min. Indicate a solution method for this problem. Assume the roof and collector are infinitely long into the page. Do not look up actual properties.

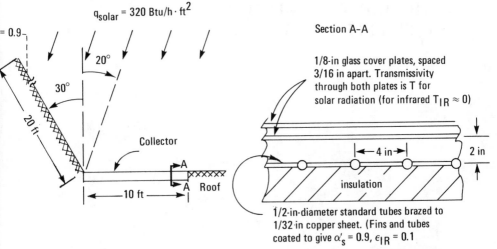

$q_{solar} = 320$ Btu/h · ft^2

$_s = 0.9$

20°

30°

20 ft

Collector

10 ft

A

A

Roof

Section A-A

1/8-in glass cover plates, spaced 3/16 in apart. Transmissivity through both plates is T for solar radiation (for infrared $T_{IR} \approx 0$)

4 in

2 in

insulation

1/2-in-diameter standard tubes brazed to 1/32 in copper sheet. (Fins and tubes coated to give $\alpha'_s = 0.9$, $\epsilon_{IR} = 0.1$

14 A highly polished copper tube carries condensing steam at temperature T_s. Heat losses are Q W, which is deemed excessive. Insulation of thermal conductivity k and thickness t cm is added. The infrared emissivity of the insulation surface is ϵ. When the system is put back in operation, the heat losses are found to be greater than before! Over what ranges of parameters (t, ϵ, k, T_s) is this possible? Put your results in terms of the most compact set of dimensionless groups that you can. A plot might be a good way to present your results. The surrounding air and environment can be taken to be at T_e. Neglect the temperature drop in the copper tube wall, and assume the heat transfer coefficient for condensing steam is very high. (Hint: For simplicity the radiation contribution can be linearized by assuming that $T_s/T_e = 1 + \Delta$ where $\Delta \ll 1$.)

THIRTEEN

FUNDAMENTALS OF RADIATION IN ABSORBING, EMITTING, AND SCATTERING MEDIA

13-1 INTRODUCTION

The study of energy transfer through media that can absorb, emit, and scatter radiation has received increased attention in the past two decades. This interest stems from the complicated and interesting phenomena associated with nuclear explosions, hypersonic shock layers, rocket propulsion, plasma generators for nuclear fusion, ablating systems, and combustion chambers at high pressure and temperature. Although some of these applications are fairly recent, the study of radiation in gases has been of continuing interest for over 100 years. One of the early considerations was the absorption and scattering of radiation in the earth's atmosphere. This has always plagued astronomers when observing on earth the light from the sun and more distant stars. The arriving form of the solar spectrum was recorded by Samuel Langley over a period of years beginning in 1880 [1]. Figure 13-1 shows some recent results [2]. The uppermost solid curve is the incident solar spectrum outside the earth's atmosphere, and a 6000-K blackbody spectrum is shown for comparison. The lowest solid curve, which has a number of sharp dips, shows the spectrum received at ground level after the solar radiation has passed through the atmosphere in a direction normal to the earth. The shaded regions show where radiation has been absorbed by various

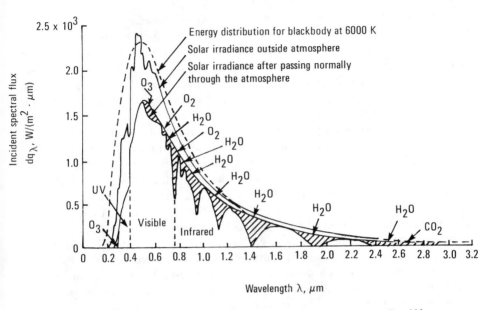

Figure 13-1 Attenuation by earth's atmosphere of incident solar spectral energy flux [2].

atmospheric constituents, mainly water vapor and carbon dioxide. The absorption occurs in specific wavelength regions, illustrating that gas-radiation properties vary considerably with wavelength. Extensive discussions of absorption in the atmosphere are given by Goody [3] and Kondratyev [4].

Gas radiation has also been of interest to astrophysicists with regard to the study of stellar structure. Models of stellar atmospheres, such as for the sun, and the energy transfer processes within them have been constructed; then emitted energy spectra calculated on the basis of the models are compared with observed stellar spectra.

In industry the importance of gas radiation was recognized in the 1920s in connection with heat transfer inside furnaces. The carbon dioxide and water vapor formed as products of combustion were found to be significant emitters and absorbers of radiant energy. Radiation can also be appreciable in engine combustion chambers, where peak temperatures can reach a few thousand kelvins. The energy emitted from flames depends not only on the gaseous emission but also arises from the heated carbon (soot) particles formed within the flame.

Another interesting example of radiation within an absorbing-emitting medium is in a glass-melting furnace. As described in [5], the temperature distribution measured within a deep tank of molten glass was found to be more uniform than expected from heat conduction alone. It was thought that convection might account for the discrepancy, but experimental investigations did not indicate that this was the contributing heat transfer mode. In the late 1940s it became evident that radiative transfer by absorption and reemission within the glass provided a significant means of energy transport.

When radiation interacts with a substance, part of the energy may be redirected by scattering. Scattering may occur by interaction with particles or objects of any size from electrons to planets. Common examples of scattering situations are radiant transfer through fog, dust suspensions, and some pigmented materials. The fundamentals of scattering will be introduced here. Information on the extent of scattering for various types of particles is given in Chap. 16. For *elastic* scattering the photon energy, and therefore its frequency, is unchanged by the scattering; for *inelastic* scattering the photon energy is changed. The analysis here will deal only with elastic scattering so that no frequency changes need be accounted for upon scattering.

Two difficulties encountered in studying radiation within absorbing, emitting, and scattering media make these studies, to say the least, challenging. First, with a participating medium present, absorption, emission, and scattering of energy occur not only at system boundaries, but at all locations within the medium. A complete solution of the energy exchange problem requires knowledge of the temperature, radiation intensity, and physical properties at every point within the medium. The mathematics describing the radiative situation is inherently complex. A second difficulty is that spectral effects are often much more pronounced in gases than for solid surfaces, and a detailed spectrally dependent analysis may be required. When approximations based on spectrally averaged properties are used, special care must be taken. Most of the simplifications introduced in gas-radiation problems are aimed at dealing with one or both of these complexities.

In this chapter the fundamental concepts are introduced for radiant intensity within a medium, and for the effects of absorption, emission, and scattering on radiant propagation. These are the quantities that are needed in Chap. 14 to formulate the equations of radiant energy transfer.

13-2 SYMBOLS

A	area
a	absorption coefficient
C_i	concentration of gas in mixture
C_2	second constant in Planck's spectral energy distribution, hc_0/k
c	speed of light in medium other than vacuum
c_0	speed of light in vacuum
E	energy
E_I	ionization potential
e	emissive power
$F_{0-\lambda T}$	fraction of blackbody emissive power in spectral region $0-\lambda T$
h	Planck's constant
i	radiation intensity
K	extinction coefficient, $a + \sigma_s$
k	Boltzmann constant

l_m	extinction mean free path
N	number density, particles per unit volume
n	simple refractive index
P	pressure
p	partial pressure of gas in mixture
Q	energy per unit time
q	energy flux, energy per unit area and time
R	radius of sphere
S	coordinate along path of radiation
s	scattering cross section
T	absolute temperature
V	volume
α	absorptance
ϵ	emittance
η	wave number
θ	cone angle, angle from normal of area; scattering angle measured relative to forward direction
κ	optical thickness [Eq. (13-17)]
λ	wavelength in medium
ν	frequency
φ	circumferential angle
ρ	density
σ	Stefan-Boltzmann constant
σ_s	scattering coefficient
τ	transmittance
Φ	phase function for scattering
ω	solid angle

Subscripts

a	absorbed
b	blackbody
e	emitted
g	gas
i	component i; incident
m	mass coefficient; mean value
p	projected
s	source or scatter
η	wave number-dependent
λ	wavelength-dependent
ν	frequency-dependent

Superscripts

$'$	directional quantity
$+$	true value, not modified by addition of induced emission
$*$	dummy variable of integration

13-3 PHYSICAL MECHANISMS OF ABSORPTION, EMISSION, AND SCATTERING

13-3.1 Absorption and Emission

Although the remaining chapters are devoted to radiation in absorbing, emitting, and scattering media in general, it will usually be gases that are used as examples. If the radiation properties of gases and opaque solids are compared, a difference in spectral behavior is quite evident. As shown by the plots of radiation properties in Chap. 5, the property variations with wavelength for opaque solids are fairly smooth, although in some instances the variation is somewhat irregular. Gas properties, however, exhibit very irregular wavelength dependencies. As a result the absorption or emission by gases is significant only in certain wavelength regions, especially at temperature levels below a few thousand kelvins. The absorptance of a gas layer as a function of wavelength typically looks like that shown for carbon dioxide (CO_2) in Fig. 13-2.

The radiation emitted from a solid actually originates within the solid (that is, not at the solid surface) so the solid can be considered an absorbing and emitting medium like a gas. The differences in emission spectra are caused by the various types of energy transitions that occur within the medium. A gas has different types of transitions, a fact which leads to a less continuous spectrum than for a solid. The energy transitions that account for radiation emission and absorption will now be discussed.

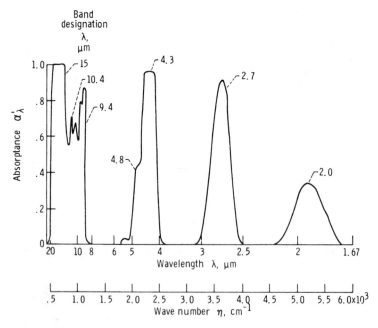

Figure 13-2 Low-resolution spectrum of absorption bands for CO_2 gas at 830 K, 10 atm, and for path length through gas of 38.8 cm.

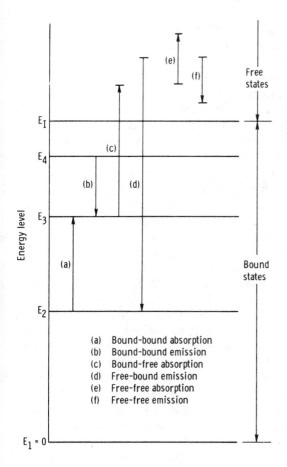

(a) Bound-bound absorption
(b) Bound-bound emission
(c) Bound-free absorption
(d) Free-bound emission
(e) Free-free absorption
(f) Free-free emission

Figure 13-3 Schematic diagram of energy states and transitions for atom, ion, or electron.

A radiating gas can be composed of molecules, atoms, ions, and free electrons. These particles can have various energy levels associated with them. In a molecule, for example, the atoms form a dynamic system that has certain vibrational and rotational modes. These modes have specific energy levels associated with them. A schematic diagram of the energy levels for an atom, ion, or electron is shown in Fig. 13-3. (The levels for a molecule are diagramed in Fig. 16-5.) The zero energy level is assigned to the ground state (lowest energy bound state), with the higher bound states being at positive energy levels. The energy E_I in Fig. 13-3 is the ionization potential, that is, the energy required to produce ionization from the ground state. Energies above E_I denote that ionization has taken place and free electrons have been produced.

It is convenient to discuss the radiation process by utilizing a photon or quantum point of view. The photon is the basic unit of radiative energy. Radiative emission consists of the release of photons of energy, and absorption is the capture of photons by a particle. When a photon is emitted or absorbed, the energy of the emitting or absorbing particle is correspondingly decreased or increased. Figure 13-3 is a diagram of the three types of transitions that can occur. These are bound-bound, bound-free, and free-free; they will be discussed a little further on in more detail. In addition to emission and absorption processes, it is possible for a photon to transfer part of its energy in certain inelastic scattering processes. These processes are of minor importance in engineering radiative transfer.

The magnitude of the energy transition is related to the frequency of the emitted or absorbed radiation. The energy of a photon is $h\nu$, where h is Planck's constant and ν is the frequency of the photon energy. For an energy transition, say from bound state E_3 down to bound state E_2 in Fig. 13-3, a photon is emitted with energy $E_3 - E_2 = h\nu$. The frequency of the emitted energy is then $\nu = (E_3 - E_2)/h$ so that a fixed frequency is associated with the transition from a specific energy level to another. Thus, in the absence of any other effects, the emitted radiation will be in the form of a spectral line. Conversely, in a transition between two bound states when a particle absorbs energy, the quantum nature of the process dictates that the absorption is such that the particle can only go to one of the discrete higher energy levels. Consequently, the frequency of the photon energy must have one of certain discrete values for the photon to be absorbed. For example, a particle in the ground state in Fig. 13-3 may absorb photons with frequencies $(E_2 - E_1)/h, (E_3 - E_1)/h,$ or $(E_4 - E_1)/h$ and undergo a transition to a higher bound energy level. Photons with other frequencies in the range $0 < \nu < E_I/h$ cannot be absorbed.

When a photon is absorbed or emitted by an atom or molecule and there is no ionization or recombination of ions and electrons, the process is termed a *bound-bound* absorption or emission (see processes *a* and *b* in Fig. 13-3). The atom or molecule moves from one quantized bound energy state to another. These states can be rotational, vibrational, or electronic in molecules, and electronic in atoms. Since bound-bound energy changes are associated with specific energy levels, the absorption and emission coefficients will be sharply peaked functions of frequency in the form of a series of spectral lines. These lines do have a finite width resulting from various broadening effects that will be discussed in Sec. 16-6.1.

The vibrational energy modes are always coupled with rotational modes. The rotational spectral lines superimposed on the vibrational line give a band of closely spaced spectral lines. If these are averaged together into one continuous region, it becomes a *vibration-rotation band* (see Sec. 16-6.4). Rotational transitions within a given vibrational state are associated with energies at long wavelengths, $\sim 8-1000~\mu m$ (see Fig. 1-2). Vibration-rotation transitions are at infrared energies of about 1.5–20 μm. Electronic transitions are at short wavelengths in the visible region 0.4–0.7 μm, and at portions of the ultraviolet and infrared near the visible region. At industrial temperatures the radiation is principally from vibrational and rotational transitions; at high temperatures (above several thousand degrees Rankine), it is the electronic transitions that are important.

Process *c* in Fig. 13-3 is a *bound-free* absorption (photoionization). An atom absorbs a photon with sufficient energy to cause ionization. The resulting ion and electron are free to take on any kinetic energy; hence, the bound-free absorption coefficient is a continuous function of photon energy frequency ν as long as the photon energy $h\nu$ is sufficiently large to cause ionization. The reverse (process *d* in Fig. 13-3) is free-bound emission (photorecombination). Here an ion and free electron combine, a photon of energy is released, and the energy of the resulting atom drops to that of a discrete bound state. The free-bound emission produces a continuous spectrum, as the combining particles can have any initial kinetic energy.

In an ionized gas a free electron can pass near an ion and interact with its electric field. This can produce a *free-free* transition (often called *bremsstrahlung*, meaning brake radiation). The electron can absorb a photon (process *e* in Fig. 13-3) thereby going to a higher kinetic energy, or it can emit a photon (process *f*) and drop to a lower free energy. Since the initial and final free energies can have any values, a continuous absorption or emission spectrum is produced. Bremsstrahlung can also be produced if an electron passes very close to a neutral atom, since there can be an electric field very close to an atom. This process is much less probable than electron-positive-ion interactions.

13-3.2 The Scattering of Energy

Scattering is taken here to be any encounter between a photon and one or more other particles during which the photon does not lose its *entire* energy. It may undergo a change in direction, and a partial loss or a gain of energy. In any of these cases, the photon is said to have been scattered.

The *scattering coefficient* $\sigma_{s\lambda}$ is the inverse of the mean free path that a photon of wavelength λ will travel before undergoing scattering (strictly true only when $\sigma_{s\lambda}$ does not vary along the path). The scattering can be characterized by four types of events: *elastic* scattering in which the energy (and, therefore, frequency and wavelength) of the photon is unchanged by the scattering, *inelastic* scattering in which the energy is changed, *isotropic* scattering in which scattering into any direction is equally likely, and *anisotropic* scattering in which there is a distribution of scattering directions. Elastic-isotropic scattering is most amenable to analysis without resort to sophisticated analytical or numerical techniques. Most scattering events of importance in engineering are elastic, or very nearly so, and the analysis given here will consider only the elastic case.

In theoretical developments, scattering is usually considered for a single particle. When a cloud of many particles is dealt with, the scattering intensities from the individual particles are usually added, under the assumption that each particle scatters independently. The criteria for independent scatter have been tentatively suggested [6] as clearances between particles exceeding 0.3 wavelength and clearance-diameter ratios exceeding 0.4. In most practical situations the assumption of independent scatter can be made, as the particles are separated by much larger distances.

Various phenomena may occur when incident radiation strikes a particle. Some of the incident radiation may be reflected from the particle surface. The remaining portion of the radiation will penetrate into the particle, where part of the radiation can be absorbed. If the particle is not a strong internal absorber, some of this radiation will pass back out. This may occur after travel along only a single path through the particle, or the radiation may undergo multiple internal reflections and travel about within the particle before escaping. When interacting with the particle boundary, the radiation will be refracted and will also have its direction changed by subsequent internal reflections. The redirection by these processes of the energy penetrating into the particle and then escaping is termed *scattering by refraction*. Additional scattering is caused by *diffraction*, which produces, for example, the interference patterns observed when light passes through an aperture in a screen. Diffraction is the result of slight bending of the radiation propagation paths near the edges of an obstruction.

For an elastic scattering process without emission or absorption, there is no exchange of energy between the radiation field and the medium. Therefore, the local thermodynamic conditions of the medium are not affected by the radiation field. Scattering calculations for this special case become more tractable than for situations in which the internal energy of the medium and radiation field can interact strongly because absorption and emission are present.

13-3.3 Luminescence

The term *luminescence* covers a broad range of mechanisms that result in radiant energy emission by the transition of electrons from an excited state to a lower energy state, where the original excitation took place by means *other than thermal agitation*. Common modes of excitation are by visible light, ultraviolet radiation, and electron bombardment. The absorption and reemission result in a change of wavelength as well as direction of the radiation. This is an example of a process that is not in local thermodynamic equilibrium (Sec. 13-9). Because the electronic transitions are between discrete energy states, the span of wavelengths over which the emission occurs is quite small. Luminescence, therefore, does not add significant energy to the spectrum of emission in engineering situations and can almost invariably be neglected in engineering heat transfer calculations. However, there are some situations in which effects other than total energy transport are of interest; hence a very brief mention of luminescence is included here.

Luminescence is categorized in various ways. A common classification is by its persistence. Luminescence that persists only during the influence of some external exciting agent, such as an ultraviolet lamp, is called *fluorescence*, a name arising from the strong luminescence shown by fluorspar when so irradiated. Luminescence that continues with slowly diminishing intensity after the excitation is removed is called *phosphorescence*, a word derived from the luminescence of white phosphorous.†

Another categorization is by excitation agent. Luminescence arising from a chemical reaction such as the oxidation of white phosphorous is called *chemiluminescence*; luminescence caused by a beam of incident electrons as on a color TV screen is *cathodoluminescence*; a biochemical reaction producing luminosity, as in fireflies and some marine animals, is called *bioluminescence*; luminous emission by the presence of an electric field, as in certain commercial panel lamps, is *electroluminescence*; and luminescence due to photon bombardment is often called *photoluminescence*. This last effect is the same mechanism that causes the laser to function. Other luminescence mechanisms are proton bombardment, believed to be responsible for the luminous red patches observed on the moon, and nuclear reactions causing luminous emission.

Because luminescence is common to materials at room temperature, it obviously

†Phosphorus itself is named from the Greek word *phosphoros*, meaning "light bearing."

cannot be predicted by the usual laws that govern thermal radiation, as these would predict no visible radiation at such temperatures. This is the origin of the term *cold light* for fluorescent-lamp emission. Rather, the quantum-mechanical properties of such luminescent materials must be examined to explain their behavior. Detailed material on luminescence is contained in [7,8]. The calculation of luminescence effects is not relevant to this work.

13-4 SOME FUNDAMENTAL PROPERTIES OF THE RADIATION INTENSITY

Radiation intensity is a convenient quantity for use in problems dealing with radiative transfer through absorbing, emitting, and scattering media. This convenience arises chiefly because of certain invariance properties. In Chap. 2 the radiation intensity leaving a surface in the direction (θ,φ) was defined as the energy leaving per unit time per unit of projected surface area normal to the (θ,φ) direction and per unit elemental solid angle centered around direction (θ,φ). As a result of this definition, the intensity of emission from a blackbody did not vary with the direction of emission. This invariance was a useful characteristic in comparing the directional intensity of emission from nonblack surfaces with that from a black surface. This led to a convenient measure of the difference between real-surface behavior and black-surface behavior; the ratio of the two directional emissions was defined as the surface directional emissivity.

In the case of a transmitting medium, the intensity has to be considered in terms of a local area within the medium. The intensity is then defined in a manner consistent with the solid surface (Sec. 2-4.1). It is as if the radiation traveling through an area within the medium originated at that area. *The intensity is then defined* (see Fig. 13-4a) *as the radiation energy passing through the area per unit time, per unit of the projected area, and per unit solid angle.* The projected area is formed by taking the area that the energy is passing through and projecting it *normal to the direction of travel.* The unit elemental solid angle is centered about the direction of travel and *has its origin at dA.* The spectral intensity is the intensity per unit small wavelength interval around a wavelength λ.

As stated previously, the emitted intensity from a blackbody is invariant with emission angle. Now a second invariant property of intensity will be examined. Consider radiation from a source dA_s, traveling in an ideal medium that is nonabsorbing, nonemitting, and nonscattering and has constant properties. Suppose that an imaginary area element dA_1 is considered at distance S_1 from dA_s and that dA_s and dA_1 are normal to S_1 as shown in Fig. 13-4b. From the definition of spectral intensity $i'_{\lambda,1}$ as the rate of energy passing through dA_1 per unit projected area of dA_1 per unit solid angle and per unit wavelength interval, the energy from dA_s passing through dA_1 in the direction of S_1 is

$$d^3Q'_{\lambda,1} = i'_{\lambda,1}\,dA_1\,d\omega_1\,d\lambda \qquad (13\text{-}1a)$$

where the third-derivative notation d^3 emphasizes that there are three differential

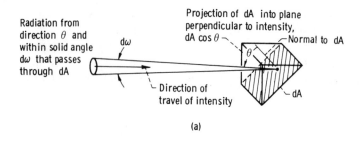

Radiation from direction θ and within solid angle $d\omega$ that passes through dA

Projection of dA into plane perpendicular to intensity, $dA \cos \theta$

Normal to dA

$d\omega$

Direction of travel of intensity

dA

(a)

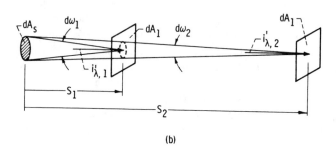

dA_s $d\omega_1$ dA_1 $d\omega_2$ dA_1

$i'_{\lambda, 1}$ $i'_{\lambda, 2}$

S_1

S_2

(b)

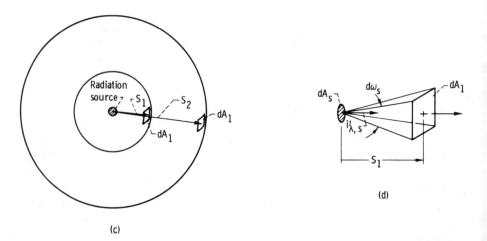

Radiation source S_1 S_2

dA_1 dA_1

dA_1

(c)

dA_s $d\omega_s$ dA_1

$i'_{\lambda, s}$

S_1

(d)

Figure 13-4 Derivations of intensity relations. (*a*) Geometry for definition of intensity in medium; (*b*) intensity from source to area element; (*c*) variation of energy flux with distance from source; (*d*) intensity of emitted radiation.

quantities on the right side of the equation. The solid angle $d\omega_1$ is equal to dA_s/S_1^2 so

$$d^3Q'_{\lambda,1} = i'_{\lambda,1}\, dA_1\, \frac{dA_s}{S_1^2}\, d\lambda \tag{13-1b}$$

Suppose that dA_1 is now placed a distance S_2 from the source along the same direction as the original position. The rate of energy passing through dA_1 in the new position is

$$d^3Q'_{\lambda,2} = i'_{\lambda,2}\, dA_1\, d\omega_2\, d\lambda = i'_{\lambda,2}\, dA_1\, \frac{dA_s}{S_2^2}\, d\lambda \tag{13-2}$$

Dividing (13-1b) by (13-2) gives

$$\frac{d^3Q'_{\lambda,1}}{d^3Q'_{\lambda,2}} = \frac{i'_{\lambda,1} S_2^2}{i'_{\lambda,2} S_1^2} \tag{13-3}$$

Now consider a differential source emitting energy equally in all directions, and draw two concentric spheres around it as in Fig. 13-4c. If $d^2Q_{\lambda,s}$ is the entire spectral energy leaving the source, then the energy *flux* crossing the inner sphere is $d^2Q_{\lambda,s}/4\pi S_1^2$, and that crossing the outer sphere is $d^2Q_{\lambda,s}/4\pi S_2^2$. The ratio of the energies passing through the two elements dA_1 is

$$\frac{d^3Q'_{\lambda,1}}{d^3Q'_{\lambda,2}} = \frac{(d^2Q_{\lambda,s}/4\pi S_1^2)\,dA_1}{(d^2Q_{\lambda,s}/4\pi S_2^2)\,dA_1} = \frac{S_2^2}{S_1^2} \tag{13-4}$$

Substituting (13-4) for the left side of (13-3) gives the following important result:

$$i'_{\lambda,1} = i'_{\lambda,2} \tag{13-5}$$

Thus, *the intensity in a given direction in a nonattenuating and nonemitting medium with constant properties is independent of position along that direction.* Note that these intensities are based on the solid angles subtended by the source as viewed from dA_1 as in Fig. 13-4b. As S is increased, the decrease in solid angle by which dA_1 views the source dA_s is accompanied by a comparable decrease in energy flux arriving at dA_1. Thus the flux per unit solid angle, used in forming the intensity, remains constant.

The radiant energy passing through dA_1 can also be written in terms of the intensity leaving the source. Using Fig. 13-4d results in

$$d^3Q'_{\lambda,1} = i'_{\lambda,s}\, dA_s\, d\omega_s\, d\lambda = i'_{\lambda,s}\, dA_s\, \frac{dA_1}{S_1^2}\, d\lambda \tag{13-6}$$

Equating this with the energy rate passing through dA_1 as given by (13-1b) results in

$$i'_{\lambda,1} = i'_{\lambda,s} \tag{13-7}$$

This relation again shows the invariance of intensity with position in a nonattenuating and nonemitting medium.

The invariance of intensity when no attenuation or emission is present provides a convenient way of specifying the magnitude of any attenuation or emission as these effects are given directly by the change of intensity with distance. By use of the foregoing intensity properties, the attenuation and emission of radiation within a medium can now be considered.

13-5 THE ATTENUATION OF ENERGY BY ABSORPTION AND SCATTERING

Consider spectral radiation of intensity i'_λ impinging normally on a layer of material of thickness dS as in Fig. 13-5. The medium in the layer absorbs and scatters radiation. Only these two effects will be considered for the present. In following sections emission will be introduced, along with the scattering into one direction of radiation from other directions. As the radiation passes through the layer, its intensity is reduced by absorption and scattering. The change in intensity has been found experimentally to depend on the magnitude of the local intensity. If a coefficient of proportionality K_λ which depends on the local properties of the medium is introduced, then the decrease is given by

$$di'_\lambda = -K_\lambda(S)i'_\lambda\,dS \tag{13-8}$$

The K_λ is the *extinction coefficient* of the material in the layer; it is a physical property of the material and has units of reciprocal length. It is a function of the temperature T, pressure P, composition of the material (specified here in terms of the concentrations C_i of the i components), and wavelength of the incident radiation:

$$K_\lambda = K_\lambda(\lambda, T, P, C_i) \tag{13-9}$$

As will be shown [see Eq. (13-16)], the K_λ is inversely related to the mean penetration distance of radiation in an absorbing and scattering medium.

Integrating (13-8) over path length S gives

$$\int_{i'_\lambda(0)}^{i'_\lambda(S)} \frac{di'_\lambda}{i'_\lambda} = -\int_0^S K_\lambda(S^*)\,dS^* \tag{13-10}$$

where $i'_\lambda(0)$ is the intensity entering the layer, and S^* is a dummy variable of integration. Carrying out the integration on the left side of (13-10) yields

$$\ln\frac{i'_\lambda(S)}{i'_\lambda(0)} = -\int_0^S K_\lambda(S^*)\,dS^* \tag{13-11}$$

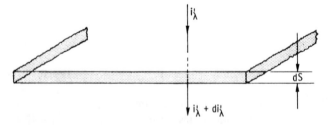

Figure 13-5 Intensity incident normally on absorbing and scattering layer of thickness dS.

or

$$i'_\lambda(S) = i'_\lambda(0) \exp\left[-\int_0^S K_\lambda(S^*)\, dS^*\right] \tag{13-12}$$

Equation (13-12) is known as Bouguer's law;[†] it shows that, as a consequence of the proportionality in (13-8), the intensity of monochromatic radiation along a path is attenuated exponentially while it passes through an absorbing-scattering medium (not including emission or scattering into direction S). The exponent is equal to the integral of the local extinction coefficient over the path length traversed by the radiation.

13-5.1 The Extinction Coefficient

The extinction coefficient is composed of two parts, an *absorption coefficient* $a_\lambda(\lambda, T, P)$ and a *scattering* coefficient $\sigma_{s\lambda}(\lambda, T, P)$:

$$K_\lambda(\lambda, T, P) = a_\lambda(\lambda, T, P) + \sigma_{s\lambda}(\lambda, T, P) \tag{13-13}$$

For simplicity the notation showing dependence upon the relative concentration of the constituents of the gas has been dropped. As noted previously, these coefficients have units of reciprocal length and are therefore called *linear coefficients* (they are also called *volumetric* coefficients). Some researchers prefer to work with mass coefficients,

$$K_{\lambda,m} = a_{\lambda,m} + \sigma_{s\lambda,m} = \frac{K_\lambda}{\rho} = \frac{a_\lambda}{\rho} + \frac{\sigma_{s\lambda}}{\rho} \tag{13-14}$$

where ρ is the local density of the absorbing-scattering species. The mass coefficients have units of area per unit mass and are directly related to the concept of a cross section in molecular physics (see Sec. 13-5.6 for a discussion of scattering cross sections). Since the extinction coefficient K_λ increases as the density of the absorbing or scattering species is increased, the use of $K_{\lambda,m} = K_\lambda/\rho$ has the advantage that it tends to remain more constant than K_λ. However, the K_λ, which will be used here, also has an advantage in that, when K_λ is constant, it can be interpreted as the reciprocal of the radiation mean penetration distance. This will now be shown.

13-5.2 Radiation Mean Penetration Distance

From (13-12) the fraction of the original radiation that penetrates through the path length S is

$$\frac{i'_\lambda(S)}{i'_\lambda(0)} = \exp\left[-\int_0^S K_\lambda(S^*)\, dS^*\right]$$

[†]Named after Pierre Bouguer (boo'gâr') (1698–1758) who first showed on a quantitative basis how light intensities could be compared. Equation (13-12) is sometimes called Lambert's law, the Bouguer-Lambert law, or Beer's law. Beer's law is more properly a restricted form of Eq. (13-9) stating that the absorption of radiation depends only on the concentration of the absorbing species along the path. To avoid confusion with Lambert's cosine law, (13-12) will be referred to herein as Bouguer's law.

The fraction absorbed in the layer from S to $S + dS$ is

$$\frac{i'_\lambda(S) - i'_\lambda(S + dS)}{i'_\lambda(0)} = \frac{-d[i'_\lambda(S)/i'_\lambda(0)]}{dS} dS = K_\lambda(S) \exp\left[-\int_0^S K_\lambda(S^*) dS^*\right] dS$$

The mean penetration distance of the radiation is obtained by multiplying the fraction absorbed at S by the distance S and then integrating over all path lengths from $S = 0$ to $S = \infty$:

$$l_m = \int_{S=0}^\infty SK_\lambda(S) \exp\left[-\int_0^S K_\lambda(S^*) dS^*\right] dS \tag{13-15}$$

When K_λ is constant, carrying out the integral gives

$$l_m = K_\lambda \int_0^\infty S \exp(-K_\lambda S) dS = \frac{1}{K_\lambda} \tag{13-16}$$

demonstrating that the average penetration distance before absorption or scattering is the reciprocal of K_λ when K_λ does not vary along the path. Equation (13-16) provides a simple way of gaining some insight as to whether or not an absorbing-scattering medium is very opaque with regard to radiation traveling through it. This will now be further discussed in connection with the definition of optical thickness.

13-5.3 Optical Thickness

The exponential factor in (13-12) is often written in an alternative form by defining the dimensionless quantity

$$\kappa_\lambda(S) \equiv \int_0^S K_\lambda(S^*) dS^* \tag{13-17}$$

so that (13-12) becomes

$$i'_\lambda(S) = i'_\lambda(0) \exp[-\kappa_\lambda(S)] \tag{13-18}$$

The quantity $\kappa_\lambda(S)$ is the *optical thickness* or *opacity* of the layer of thickness S and is a function of all the values of K_λ between 0 and S. Because K_λ is a function of the local parameters P, T, and C_i, the optical thickness becomes a function of all these conditions along the path between 0 and S.†

The optical thickness is a measure of the ability of a given path length of gas to attenuate radiation of a given wavelength. A large optical thickness means large attenuation. The quantity κ_λ is a convenient dimensionless parameter that will occur in the solutions of radiative transfer problems.

---

† The notation for the optical thickness κ_λ should not be confused with the extinction coefficient for electromagnetic radiation κ used in (13-22) and (13-23) that follow. It is regrettable but true that the notation possibilities of the English and Greek alphabets reach saturation when such interdisciplinary fields as gas radiation are discussed.

For a gas that is of uniform composition and is at uniform temperature and pressure (a *uniform* gas) or for a gas with K_λ independent of T, P, and C_i, Eq. (13-17) becomes

$$\kappa_\lambda(S) = K_\lambda S \tag{13-19}$$

The optical thickness then depends directly on the extinction coefficient and the thickness of the absorbing-scattering layer. By using (13-16), one obtains $\kappa_\lambda = S/l_m$, so that the optical thickness is the number of mean penetration distances. If $\kappa_\lambda \gg 1$, the medium is *optically thick*, that is, the mean penetration distance is quite small compared to the characteristic dimension of the medium. For this condition a volume element within the material is only influenced by the surrounding neighboring elements. If $\kappa_\lambda \ll 1$, the medium is *optically thin* and the mean penetration distance is much larger than the medium dimension. Radiation can pass entirely through the material without significant absorption, and each element within the medium interacts directly with the medium boundary. Radiation emitted within the material is not reabsorbed by the material; this condition is called *negligible self-absorption*.

13-5.4 The Absorption Coefficient

If scattering can be neglected ($\sigma_{s\lambda} \approx 0$), then $K_\lambda = a_\lambda$ and (13-12) becomes

$$i_\lambda'(S) = i_\lambda'(0) \exp\left[-\int_0^S a_\lambda(S^*)\,dS^*\right] \tag{13-20}$$

If, in addition, a_λ is not a function of position as is the case in a gas of uniform temperature, pressure, and composition, then

$$i_\lambda'(S) = i_\lambda'(0) \exp(-a_\lambda S) \tag{13-21}$$

In the electromagnetic theory of the propagation of radiant energy [see discussion following Eq. (4-26)], it is shown that the intensity of radiation is attenuated in conducting media according to the relation

$$\frac{i_\lambda'(S)}{i_\lambda'(0)} = \exp\frac{-4\pi\kappa S}{\lambda} \tag{13-22}$$

where κ is the extinction coefficient from electromagnetic theory and is related to the magnetic permeability, electrical resistivity, and electrical permittivity of the medium [Eq. (4-23b)]. Thus, a_λ is related to κ by

$$a_\lambda = \frac{4\pi\kappa}{\lambda} \tag{13-23}$$

Such a relation provides some theoretical basis for Bouguer's law, which was originally based on experimental observations.

The absorption coefficient $a_\lambda(\lambda, T, P)$ usually varies strongly with wavelength and often varies substantially with temperature and pressure. Considerable analytical and experimental effort has been expended in the determination of a_λ for various gases, liquids, and solids.

Analytical determinations of a_λ require detailed quantum-mechanical calculations beyond the scope of this volume, although some of the concepts are outlined in Chap. 16. Except for the simplest gases, such as atomic hydrogen, the calculations are very tedious and require many simplifying assumptions. For the methods used in the calculation of a_λ, [9–11] give detailed discussions.

The complexity of the calculations is presaged by examination of some measured solid and gas spectral absorption coefficients. In Fig. 13-6, a_λ is shown for pure diamond. Strong absorption peaks due to crystal-lattice vibrations at certain wavelengths are evident. Figure 13-7 shows the calculated emission spectrum of hydrogen gas at 40 atm and 11,300 K for a path length through the gas of 50 cm. The variations in emission are closely related to variations in the absorption coefficient. The presence of "spikes" or strong emission lines is the result of transitions between bound energy states. The continuous part of the emission spectrum is due to various photodissociations, photoionization, and free electron-atom-photon interactions of other types. The lines and the continuous regions are common features of both emission and absorption spectra. Figure 13-8 shows the absorption coefficient of air at 1 atm and 12,000 K. In this case, there is a merging of the contributions from the many closely spaced lines produced by vibrational and rotational transitions between energy states, and the absorption coefficient has the appearance of being continuous. Even when this merging is not complete, the resolution of experimental measurements causes the measured spectrum to appear continuous over these closely spaced lines.

Note that Figs. 13-6–13-8 each have a different abscissa (that is, wavelength, wave number, and frequency), emphasizing the lack of an accepted standard variable. When radiative properties of opaque surfaces were discussed in Chap. 5, it was found that the

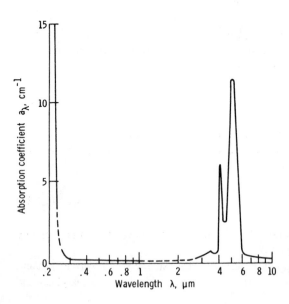

Figure 13-6 Spectral absorption coefficient of diamond *(from [17])*.

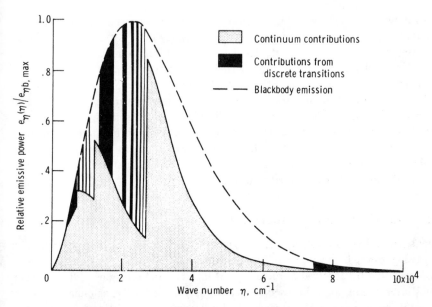

Figure 13-7 Normalized emission spectrum of hydrogen at 11,300 K and 40 atm, for a path length of 50 cm *(from [18])*.

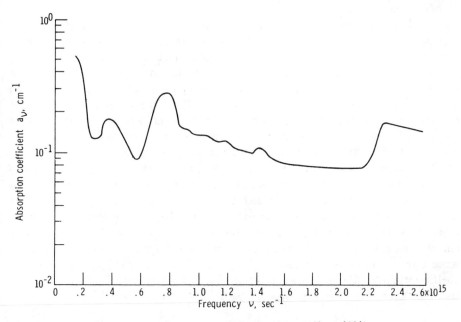

Figure 13-8 Absorption coefficient of air at 12,000 K and 1 atm *(from [19])*.

wavelength was generally used. In radiation from gases, however, the frequency is more common. It has the advantage that the frequency does not change when radiation passes from one medium into another with a different refractive index. The wavelength does change because of the change in propagation velocity.

Dealing with spectral line emission and absorption is one of the computational difficulties encountered in analyses of radiant energy transfer through gases. Incident radiation at wavelengths near a line center will be strongly absorbed, while radiation of only a slightly different wavelength may experience almost no attenuation. Integrating line absorption coefficients with respect to wavelength to obtain band or total absorption coefficients is generally tedious. These averaged coefficients are used in certain calculation methods of radiative transfer.

13-5.5 True Absorption Coefficient

Bouguer's law in the form of Eq. (13-20) gives the attenuation of a beam of radiation upon passing through a volume of nonemitting nonscattering gas along a path of length S as would be observed by detectors of incident and emerging radiation. Such observed information could be used in determining a_λ. Actually, as radiative energy passes through a gas, not only is it absorbed but there is an additional phenomenon in that its presence stimulates some of the gas atoms or molecules to emit energy. This is not the *ordinary* or *spontaneous* emission that will be discussed in Sec. 13-6. Spontaneous emission is the result of the excited state of the gas being unstable and decaying spontaneously to a state of lower energy. The emission resulting from the presence of the radiation field is termed *stimulated* or *induced* emission and is in a sense a negative absorption.

Physically, the induced-emission process can be pictured as follows: A photon of a certain frequency from the radiation field encounters a gas atom or molecule in an excited state, that is, an energy state above the ground state. There exists a certain probability that the incident photon will trigger a return of the gas particle to a lower energy state. If this occurs, the particle will emit a photon at the same frequency and in the same direction as the incident photon. Thus, the incident photon is not absorbed but is joined by a second identical photon. This process is often viewed as a negative absorption and is so treated in the equations of energy balance to be derived in Chap. 14. More discussion of the induced-emission process is given in Sec. 16-4.

The induced emission constitutes a portion of the intensity that is observed in the beam emerging from the gas volume. Consequently, the amount of energy that is actually absorbed by the gas is greater than that found by taking the difference between the entering and leaving intensities. This is because the observed emerging intensity is the result of the actual absorption modified by the addition of induced emission along the path of the beam. The actual absorbed energy should be calculated using a *true absorption coefficient* $a_\lambda{}^+(\lambda,T,P)$, which is larger than the absorption coefficient $a_\lambda(\lambda,T,P)$ calculated by using observed attenuation data and Bouguer's law. The "true" law for absorption along path S is then written as

$$i_\lambda'(S) = i_\lambda'(0) \exp\left[-\int_0^S a_\lambda{}^+(S^*)\,dS^*\right]$$ (13-24)

Statistical-mechanical considerations give the relation between $a_\lambda(\lambda,T,P)$ and $a_\lambda^+(\lambda,T,P)$ for a gas with refractive index $n = 1$ as

$$a_\lambda(\lambda,T,P) = \left[1 - \exp\left(-\frac{hc_0}{k\lambda T}\right)\right] a_\lambda^+(\lambda,T,P) = \left[1 - \exp\left(-\frac{C_2}{\lambda T}\right)\right] a_\lambda^+(\lambda,T,P) \quad (13\text{-}25)$$

Examination of Eq. (13-25) shows that, because of the negative exponential term, a_λ^+ will always be larger than a_λ (hence the use of the superscript +).

Because the induced emission depends on the incident radiation field, it is usually grouped together with the true absorption, thereby yielding the absorption coefficient a_λ. The emission term in the equation of radiative transfer then includes only the spontaneous emission and consequently depends only on the local conditions of the gas. As will be shown in Chap. 14, the grouping of induced emission into the absorption term simplifies the equations of radiative transfer.

The exponential term in (13-25) is small except at large values of λT. Thus a_λ and a_λ^+ are nearly equal except at large values of λT (long wavelengths and/or high temperatures). The values are within 1% for λT less than 3120 $\mu m \cdot K$, and within 5% for λT less than 4800 $\mu m \cdot K$.

When properties from the literature are used in calculations of radiative transfer in absorbing-emitting media, care must sometimes be exercised to determine whether the reported absorption coefficients include the effects of induced emission; usually it is a_λ that is given.

13-5.6 The Scattering Coefficient

The extent of scattering can be viewed in terms of the *scattering cross section s*. This is the apparent area that an object presents to an incident beam insofar as the ability of the object to deflect radiation from the beam is concerned. It is usually given in square centimeters for thermal radiation properties. This apparent area may be quite different from the physical cross-sectional area of the scatterers, as can be seen from some of the approximate cross sections in Table 13-1. In addition to depending on the particle size, the scattering cross section may depend upon the shape and material of the scattering body and the wavelength, polarization, and coherence of the incident radiation. The ratio of s to the actual geometric projected area of the particle normal to the incident beam is termed the *scattering efficiency factor*.

The scattering cross section can be determined experimentally by measuring the amount of radiation in a beam that is able to penetrate through a cloud of scattering particles. One experimental difficulty is in separating the radiation that is scattered into the forward direction from the radiation that is transmitted without any particle interaction. This difficulty can be diminished by using an incident beam with a very small divergence angle. Then the forward direction of the transmitted radiation will encompass only a small solid angle that includes only a small portion of incoming scattered radiation. The ratio of the scattered portion $di_{\lambda,s}'$ of the incident intensity to the intensity i_λ' of the incident beam is equal to the ratio of the apparent projected scattering area $d^2 A_{s\lambda}$ occupied by all scattering particles to the cross-sectional area of the incident beam dA. This gives the following for a beam traveling a differential

Table 13-1 Approximate scattering cross sections for various bodies exposed to incident photons

Body	Physical cross section, cm^2	Conditions	Type of scattering	Scattering cross section,[a] cm^2
Photon		Energy of incident photon small		$\sim 2 \times 10^{-56}$
Free electron		Energy of photon $\ll$ electron kinetic energy	Thomson	$\frac{8}{3}\pi r_0^2 = 6.65 \times 10^{-25} \equiv \sigma_T$
		Energy of photon $\gg$ electron kinetic energy	Compton	$\frac{3}{8}\sigma_T \frac{1}{\epsilon_p}(\frac{1}{2} + \ln 2\epsilon_p)$
Atom or molecule	0.88×10^{-16} (first Bohr electron orbit)	Elastic, $\lambda \gg$ size of molecule or atom		
		Energy of incident photon $\ll$ electronic binding energy	Rayleigh	Proportional to $\sim 6.65 \times 10^{-25}/\lambda^4$
		Energy of incident photon $\gg$ electronic binding energy	Rayleigh, approaches Thomson	$\sim 6.65 \times 10^{-25}$
		Inelastic		
		Energy of incident photon $\ll$ electronic binding energy	Raman	$\sim 6.65 \times 10^{-25}$
		Energy of incident photon $\gg$ electronic binding energy	Approaches Compton	$\sim \frac{3}{8}\sigma_T \frac{1}{\epsilon_p}(\frac{1}{2} + \ln 2\epsilon_p)$
Particles of diameter D	$\frac{\pi D^2}{4}$	$\lambda \gg D$, single scattering	Rayleigh	Proportional to V^2/λ^4
		$\lambda \approx D$	Mie	Varies widely
		$\lambda \ll D$	Fraunhofer and Fresnel diffraction plus reflection	$\sim 2\left(\frac{\pi D^2}{4}\right)$

[a] r_0 = classical electron radius 2.818×10^{-15} m; $\epsilon_p = h\nu/m_e c_0^2$, where h is Planck's constant; σ_T = cross section for Thomson scattering; m_e = electron mass, 9.1096×10^{-31} kg.

distance within a medium in which it encounters the scattering area $d^2A_{s\lambda}$:

$$\frac{di'_{\lambda,s}}{i'_\lambda} = \frac{d^2A_{s\lambda}}{dA} \tag{13-26}$$

Note that the apparent projected scattering area of the particles can and usually will depend on wavelength.

The apparent scattering area presented by a group of the scattering particles is related to the average scattering areas of the individual particles by

$$d^2A_{s\lambda} = s_\lambda N_s \, dV = s_\lambda N_s \, dA \, dS \tag{13-27}$$

where N_s is the number density of the particles, s_λ is the average scattering cross section of the particles, and dV is a differential volume of the particle containing cloud as shown in Fig. 13-9. Inserting (13-27) into (13-26) gives the change di'_λ of the intensity as a result of scattering from the incident beam:

$$-\frac{di'_\lambda}{i'_\lambda} = \frac{di'_{\lambda,s}}{i'_\lambda} = \frac{s_\lambda N_s \, dA \, dS}{dA} = s_\lambda N_s \, dS \tag{13-28}$$

There is also intensity scattered from all incident directions into the S direction, which will contribute to di'_λ, but this will be incorporated later.

By integrating (13-28) over a path from 0 to S, the intensity is found at S as a result of attenuation by scattering from the beam with original intensity $i'_\lambda(0)$:

$$i'_\lambda(S) = i'_\lambda(0) \exp\left(-\int_0^S s_\lambda N_s \, dS^*\right) \tag{13-29a}$$

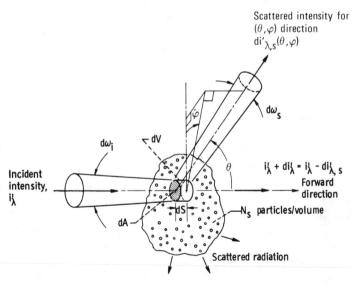

Figure 13-9 Scattering of intensity into direction (θ, φ) from incident radiation within solid angle $d\omega_i$.

The portion of the incident intensity that was scattered away along the path is thus

$$i'_\lambda(0) - i'_\lambda(S) = i'_\lambda(0) \left[1 - \exp\left(- \int_0^S s_\lambda N_s \, dS^* \right) \right] \tag{13-29b}$$

The *scattering coefficient* $\sigma_{s\lambda}$ is defined to be

$$\sigma_{s\lambda} \equiv s_\lambda N_s \tag{13-30}$$

so that (13-29a) becomes

$$i'_\lambda(S) = i'_\lambda(0) \exp\left[- \int_0^S \sigma_{s\lambda}(S^*) \, dS^* \right] \tag{13-31}$$

This is the pure-scattering form of Bouguer's law.

If a distribution of particle sizes is to be considered in detail, the preceding analysis can be generalized. Let $N_s(R) \, dR$ be the number of particles per unit volume in the radius range from R to $R + dR$, and let $s_\lambda(R)$ be the scattering cross section for a particle of radius R. Then by integrating over all the particles, the scattering coefficient is found as

$$\sigma_{s\lambda} = \int_{R=0}^\infty s_\lambda(R) N_s(R) \, dR \tag{13-32}$$

As in the interpretation of the extinction coefficient, the scattering coefficient $\sigma_{s\lambda}$ can be regarded as the reciprocal of the mean free path that the radiation will traverse before being scattered. The $\sigma_{s\lambda}$ is thus a reciprocal length and can be regarded as a scattering area per volume along the path, $\sigma_{s\lambda} = d^2 A_{s\lambda}/dV$ from (13-30) and (13-27). At particle densities near or below the molecular density of air at 1 atm ($N_s \approx 2.7 \times 10^{19}$ particles $/cm^3$), it can be seen that for most of the processes listed in Table 13-1 the scattering coefficient will be very small (and thus the scattering mean free path very long). This is especially true for photon-photon, Thomson, and Raman scattering, which may generally be ignored in engineering radiative transfer calculations.

13-6 THE EMISSION OF ENERGY

Having considered the various definitions connected with attenuation, we now turn to the emission of energy within the medium.

Consider an elemental volume dV of gas as shown in Fig. 13-10. The true absorption coefficient within dV is $a_\lambda{}^+(\lambda, T, P)$ and is considered constant over dV. Let dV be placed at the center of a large black hollow sphere of radius R at uniform temperature T. The space between dV and the sphere wall is filled with a nonparticipating material. The spectral intensity incident at the dA_s location on dV from an element dA on the surface of the enclosure is, by use of (13-7),

$$i'_\lambda(0) = i'_{\lambda b}(\lambda, T) \tag{13-33}$$

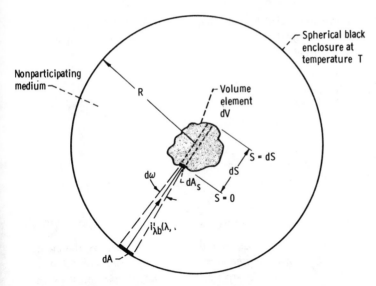

Figure 13-10 Geometry for derivation of emission from volume of gas.

The change of the intensity as a result of true absorption in dV is

$$di_\lambda' = -i_\lambda'(0)[1 - \exp(-a_\lambda{}^+ dS)] = -i_{\lambda b}'(\lambda, T)[1 - \exp(-a_\lambda{}^+ dS)] \qquad (13\text{-}34)$$

Since the true absorption coefficient has been used, this does not include induced emission. Let dV be small enough so that $a_\lambda{}^+ dS$ is sufficiently small to allow the approximation

$$di_\lambda' = -i_{\lambda b}'(\lambda, T)[1 - \exp(-a_\lambda{}^+ dS)] \approx -i_{\lambda b}'(\lambda, T) a_\lambda{}^+ dS \qquad (13\text{-}35)$$

Equation (13-35) is also evident from the differential form (13-8). The energy absorbed by the differential subvolume $dS\, dA_s$ from this incident radiation is

$$d^5 Q_{\lambda, a}' = -di_\lambda' \, dA_s \, d\lambda \, d\omega \qquad (13\text{-}36)$$

where $d\omega = dA/R^2$ and dA_s is a projected area normal to $i_\lambda'(0)$. Substituting (13-35) in (13-36) results in

$$d^5 Q_{\lambda, a}' = i_{\lambda b}'(\lambda, T) a_\lambda{}^+ dS \, dA_s \, d\lambda \, d\omega \qquad (13\text{-}37)$$

The energy emitted by dA and absorbed by all of dV is found by integration over dV; that is,

$$d^4 Q_{\lambda, a}' = \int_{dV} d^5 Q_{\lambda, a}' = i_{\lambda b}'(\lambda, T) a_\lambda{}^+ d\lambda \, d\omega \int_{\substack{\text{projected area of} \\ dV \text{ normal to} \\ \text{path from } dA}} dA_s \, dS$$

$$= a_\lambda{}^+ i_{\lambda b}'(\lambda, T) \, dV \, d\lambda \, d\omega \qquad (13\text{-}38)$$

where $d\omega$ is the solid angle subtended by dA when viewed from dV. To account for all energy incident upon dV from the entire spherical enclosure, integration is carried out over all such solid angles to give

$$d^3Q_{\lambda,a} = \int_\omega d^4Q'_{\lambda,a} = a_\lambda{}^+ i'_{\lambda b}(\lambda,T)\, dV\, d\lambda \int_{4\pi} d\omega$$

$$= 4\pi a_\lambda{}^+ i'_{\lambda b}(\lambda,T)\, dV\, d\lambda = 4a_\lambda{}^+ e_{\lambda b}(\lambda,T)\, dV\, d\lambda \qquad (13\text{-}39)\dagger$$

where $e_{\lambda b}$ is the blackbody spectral emissive power [Eq. (2-11a)]

To maintain equilibrium in the enclosure, dV must emit an amount of energy equal to that absorbed. Hence, *the energy emitted by an isothermal volume element in equilibrium with its surroundings* is

$$d^3Q_{\lambda,e} = d^3Q_{\lambda,a} = 4a_\lambda{}^+(\lambda,T,P)e_{\lambda b}(\lambda,T)\, dV\, d\lambda \qquad (13\text{-}40)$$

This result includes both spontaneous emission and emission induced by the incident equilibrium radiation field. *For only spontaneous emission, the coefficient a_λ would be used.* The shape of the element dV is arbitrary; however, its size must be small enough to justify the approximation (13-35) and small enough so that energy emitted within dV escapes before reabsorption within dV. Further, the gas must be in thermodynamic equilibrium with respect to its internal energy, a restriction discussed more fully in Sec. 13-9.

At this point an emission coefficient could be defined in a manner similar to the absorption coefficient. However, the radiation literature has other definitions of the emission coefficient‡ that do not follow an analogy to the absorption-coefficient definition, and there is no need to add confusion by defining a new coefficient here. Rather, (13-40) will be used directly as the relation for the emission of energy from an infinitesimal volume element of gas.

When the spontaneously emitted intensity is the same for all directions (isotropic spontaneous emission of energy), which is the condition for all cases discussed in this volume, the radiation intensity emitted spontaneously by a volume element into any direction is

$$di'_{\lambda,e}(\lambda,T) = \frac{d^3Q_{\lambda,e}}{4\pi\, dA_p\, d\lambda} = \frac{a_\lambda(\lambda,T,P)e_{\lambda b}(\lambda,T)\, dS}{\pi} = a_\lambda(\lambda,T,P)i'_{\lambda b}(\lambda,T)\, dS \qquad (13\text{-}41)$$

where dA_p is the projected area of dV normal to the direction of emission, and dS is the mean thickness of dV parallel to the direction of emission (that is, $dS = dV/dA_p$).

13-7 INCOMING SCATTERING

The scattering relations in Sec. 13-5.6 were concerned with the portion of the intensity in an incident beam that is lost as a result of being scattered away along a path. One

---

†Note that dV is regarded as a second-order differential, that is, $dA\, dS$.

‡In the astrophysical literature [12, 13], the emission coefficient is usually given the symbol j_λ defined by $j_\lambda = a_\lambda e_{\lambda b}$ and having units, therefore, of energy rate per unit volume per unit wavelength interval.

also needs to know how much the intensity can be increased by scattering into a path from other directions. To calculate this, the directional distribution of the scattered radiation is needed; this is given in terms of an angularly dependent phase function.

13-7.1 The Phase Function

Consider the radiation within solid angle $d\omega_i$ that is incident on area dA in Fig. 13-9. The portion of the incident intensity that is scattered away in dS is given by (13-28) and (13-30) as

$$di'_{\lambda,s} = \sigma_{s\lambda} i'_\lambda \, dS \tag{13-42}$$

The $di'_{\lambda,s}$ is the spectral energy scattered within path length dS per unit incident solid angle and area normal to the incident beam:

$$di'_{\lambda,s} = \frac{d^4 Q'_{\lambda,s}}{d\omega_i \, dA \, d\lambda} \tag{13-43}$$

As shown by Fig. 13-9, the scattered energy produces an intensity distribution as a function of the angle θ measured relative to the forward direction and the circumferential angle φ. A phase function $\Phi(\theta,\varphi)$ will be defined to describe the angular distribution of the scattered energy.

The scattered intensity in any direction (θ,φ) is defined as the energy scattered in that direction per unit solid angle of the scattered direction and per unit area and solid angle of the incident radiation:

$$di'_{\lambda,s}(\theta,\varphi) \equiv \frac{\text{spectral energy scattered in direction } (\theta,\varphi)}{d\omega_s \, dA \, d\omega_i \, d\lambda} = \frac{d^5 Q'_{\lambda,s}(\theta,\varphi)}{d\omega_s \, dA \, d\omega_i \, d\lambda} \tag{13-44}$$

The directional magnitude of $di'_{\lambda,s}(\theta,\varphi)$ is related to the entire intensity $di'_{\lambda,s}$ scattered away from the incident radiation by the phase function, such that

$$di'_{\lambda,s}(\theta,\varphi) = di'_{\lambda,s} \frac{\Phi(\theta,\varphi)}{4\pi} = \sigma_{\lambda s} i'_\lambda \, dS \frac{\Phi(\theta,\varphi)}{4\pi} \tag{13-45}$$

To better understand the phase function, note that the spectral energy per unit $d\lambda$, $d\omega_i$, and dA scattered into $d\omega_s$ is $di'_{\lambda,s}(\theta,\varphi) \, d\omega_s$, and hence that scattered into all $d\omega_s$ is $\int_{\omega_s=4\pi} di'_{\lambda,s}(\theta,\varphi) \, d\omega_s$. However, the scattered energy per unit $d\lambda$, $d\omega_i$, and dA, is $di'_{\lambda,s}$ from (13-43) so that

$$di'_{\lambda,s} = \int_{\omega_s=4\pi} di'_{\lambda,s}(\theta,\varphi) \, d\omega_s \tag{13-46}$$

Using (13-45) to eliminate $di'_{\lambda,s}$ gives the phase function as

$$\Phi(\theta,\varphi) = \frac{di'_{\lambda,s}(\theta,\varphi)}{(1/4\pi) \int_{\omega_s=4\pi} di'_{\lambda,s}(\theta,\varphi) \, d\omega_s} \tag{13-47}$$

Thus, $\Phi(\theta,\varphi)$ has the physical interpretation of being the scattered intensity in a direction, divided by the intensity that would be scattered in that direction if the scattering were isotropic. For isotropic scattering, then, $\Phi = 1$. By integrating (13-47) over all $d\omega_s$, it becomes evident that $\Phi(\theta,\varphi)$ is a function normalized such that

$$\frac{1}{4\pi} \int_{\omega_s=4\pi} \Phi(\theta,\varphi)\, d\omega_s = 1 \tag{13-48}$$

The phase function can be a complicated function of θ and φ, as will be shown in Sec. 16-7.6.

13-7.2 Augmentation of Intensity by Incoming Scattering

The local intensity along a path will be enhanced by radiation scattered into the direction being considered. To compute the scattering from all directions into the direction of $i_\lambda'(S)$, consider the radiation incident at angle (θ,φ) as shown in Fig. 13-11. This radiation has intensity $i_\lambda'(\theta,\varphi)$, and in the process of going through the volume element dV it will pass through a path length $dS/\cos\theta$. From Eq. (13-45) oriented with respect to the present coordinate system, the intensity scattered from $i_\lambda'(\theta,\varphi)$ into the direction of i_λ' is

$$di_{\lambda,s}' = \sigma_{s\lambda}\, i_\lambda'(\theta,\varphi) \frac{dS}{\cos\theta} \frac{\Phi(\theta,\varphi)}{4\pi} \tag{13-49}$$

However, from (13-44), $i_{\lambda,s}'$ is an intensity defined as energy in the scattered direction per unit $d\lambda$, per unit scattered solid angle, per unit incident solid angle $d\omega_i$, and per unit area normal to the incident intensity. This is the area normal to $i_\lambda'(\theta,\varphi)$, which is $dA\cos\theta$. Then the spectral energy scattered into the S direction as a result of $i_\lambda'(\theta,\varphi)$ is, by use of (13-49),

$$d^5Q_{\lambda,s}' = di_{\lambda,s}'\, d\omega\, d\omega_i\, d\lambda\, dA\cos\theta$$

$$= \sigma_{s\lambda} i_\lambda'(\theta,\varphi) \frac{dS}{\cos\theta} \frac{\Phi(\theta,\varphi)}{4\pi} d\omega\, d\omega_i\, d\lambda\, dA\cos\theta$$

$$= \sigma_{s\lambda} i_\lambda'(\theta,\varphi)\, dS \frac{\Phi(\theta,\varphi)}{4\pi} d\omega\, d\omega_i\, d\lambda\, dA$$

The contribution of this scattered energy to the spectral intensity in the S direction is then

$$\frac{d^5Q_{\lambda,s}'}{dA\, d\omega\, d\lambda} = \sigma_{s\lambda} i_\lambda'(\theta,\varphi) \frac{\Phi(\theta,\varphi)}{4\pi} d\omega_i\, dS \tag{13-50}$$

To account for the scattering contributions by the incident intensities from all directions, one integrates over all $d\omega_i$ to obtain

$$\int_{\omega_i=4\pi} \frac{d^5Q_{\lambda,s}'}{dA\, d\omega\, d\lambda} = \frac{dS}{4\pi} \sigma_{s\lambda} \int_{\omega_i=4\pi} i_\lambda'(\theta,\varphi)\, \Phi(\theta,\varphi)\, d\omega_i \tag{13-51}$$

The scattering particles have been assumed randomly oriented so that the scattering

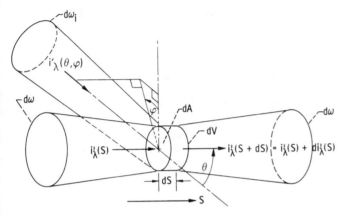

Figure 13-11 Scattering of energy into S direction.

cross section $\sigma_{s\lambda}$ is independent of the incidence direction. The augmentation of intensity in direction S by incoming scattering is then

$$\frac{di'_\lambda}{dS} = \frac{\sigma_{s\lambda}}{4\pi} \int_{\omega_i=4\pi} i'_\lambda(\theta,\varphi)\, \Phi(\theta,\varphi)\, d\omega_i \tag{13-52}$$

13-8 DEFINITIONS USED FOR ENGINEERING GAS PROPERTIES

It will be desirable, in analyzing systems with absorbing media, to make use of the extensive techniques developed for radiative interchange computations between surfaces without intervening absorbing media. With this objective, analogous concepts and terminology are developed for problems involving participating gases. This is done through the concept of the emittance and absorptance of a gas volume. These gas-property definitions are analogous to the emissivity and absorptivity of opaque bodies, and it is important to note that they are only for gases at *uniform temperature.* Because the energies emitted or absorbed by a volume of gas depend on the *size* and *shape* of the volume in addition to its physical properties and temperature, the absorptance and emittance are *extensive* properties. The nomenclature used here applies the *-ance* suffix to extensive properties.

13-8.1 Absorptance for Medium at Uniform Temperature

To be a reasonably simple engineering parameter, the absorptance should depend at most on the geometry, size, temperature, and physical properties of the volume for which it is evaluated. It is, therefore, defined for a volume with uniform conditions, so that no *gradients* in the physical conditions need be considered.

Consider energy of intensity $i'_\lambda(0)$ incident in the S direction in a uniform medium as shown in Fig. 13-12. The radiation passes through a thickness S and is incident on

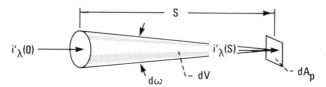

Figure 13-12 Geometry for absorptance along path length S.

a projected area dA_p normal to S. The energy arriving at dA_p in solid angle $d\omega$ is $i'_\lambda(S)\, dA_p\, d\omega\, d\lambda$. Without absorption in the medium the intensity arriving at dA_p would be $i'_\lambda(0)$, and the energy at dA_p would be $i'_\lambda(0)\, dA_p\, d\omega\, d\lambda$. Hence the energy in solid angle $d\omega$ absorbed by the medium is

$$d^3Q'_{\lambda,a} = [i'_\lambda(0) - i'_\lambda(S)]\, dA_p\, d\omega\, d\lambda \tag{13-53}$$

By substituting (13-21) for the attenuation of intensity by absorption in a uniform medium, this becomes

$$d^3Q'_{\lambda,a} = i'_\lambda(0)[1 - \exp(-a_\lambda S)]\, dA_p\, d\omega\, d\lambda \tag{13-54}$$

The energy incident upon dV in solid angle $d\omega$ is equal to the energy that would arrive at dA_p in the absence of absorption:

$$d^3Q'_{\lambda,i} = i'_\lambda(0)\, dA_p\, d\omega\, d\lambda \tag{13-55}$$

The absorptance for path length S in the volume is defined as the fraction of the incident energy in solid angle $d\omega$ that is absorbed while traversing S; then dividing (13-54) by (13-55) gives

Spectral absorptance for path length S in a uniform gas volume

$$\equiv \alpha'_\lambda(\lambda,T,P,S) = \frac{d^3Q'_{\lambda,a}}{d^3Q'_{\lambda,i}} = 1 - \exp(-a_\lambda S) \tag{13-56}$$

By substituting (13-56) into (13-21) the relation between intensities is found as

$$i'_\lambda(S) = i'_\lambda(0)[1 - \alpha'_\lambda(S)] \tag{13-57}$$

The $\alpha'_\lambda(S)$ is a directional spectral absorptance. The values of absorptance including energy of all wavelengths are also of use in engineering analyses. Integrating (13-54) and (13-55) to include all wavelengths and then taking the ratio of energies gives

Total absorptance for path length S in a uniform gas volume

$$\equiv \alpha'(T,P,S) = \frac{d^2Q'_a}{d^2Q'_i} = \frac{\int_0^\infty i'_\lambda(0)[1 - \exp(-a_\lambda S)]\, d\lambda}{\int_0^\infty i'_\lambda(0)\, d\lambda}$$

$$= \frac{\int_0^\infty \alpha'_\lambda(\lambda,T,P,S)\, i'_\lambda(0)\, d\lambda}{\int_0^\infty i'_\lambda(0)\, d\lambda} \tag{13-58}$$

Then, combining (13-58) with the integral of (13-21) over all λ, yields for total intensities, $i'(S) = i'(0)[1 - \alpha'(S)]$.

13-8.2 Emittance for Medium at Uniform Temperature

The directional emittance of the uniform gas volume in $d\omega$ of Fig. 13-12 is the ratio of the energy emitted by the volume in direction S to that emitted into that direction and $d\omega$ by a blackbody at the same temperature. Because Kirchhoff's law holds without restriction for directional spectral absorptance values, as discussed in relation to Table 3-2, it follows by use of Eq. (13-56) that

Directional spectral emittance for path length S in a uniform gas volume

$$\equiv \epsilon_\lambda'(\lambda,T,P,S) = 1 - \exp(-a_\lambda S) \tag{13-59}$$

For a blackbody radiating to dA_p, the energy arriving at dA_p within $d\omega$ is $i_{\lambda b}' dA_p d\omega d\lambda$. Then, from the definition of directional emittance, the energy in wavelength interval $d\lambda$ arriving at dA_p at location S as a result of emission by the medium in solid angle $d\omega$ is $\epsilon_\lambda' i_{\lambda b}' dA_p d\omega d\lambda = i_{\lambda b}' [1 - \exp(-a_\lambda S)] dA_p d\omega d\lambda$. The intensity incident on dA_p is $i_\lambda'(S) = \epsilon_\lambda' i_{\lambda b}'(T)$. It follows by analogy to Eq. (13-58) that for total quantities

Directional total emittance for path length S in a uniform gas volume

$$\equiv \epsilon'(T,P,S) = \frac{\int_0^\infty i_{\lambda b}'(\lambda,T)[1 - \exp(-a_\lambda S)] d\lambda}{\int_0^\infty i_{\lambda b}'(\lambda,T) d\lambda}$$

$$= \frac{\int_0^\infty i_{\lambda b}'(\lambda,T)[1 - \exp(-a_\lambda S)] d\lambda}{\sigma T^4/\pi}$$

$$= \frac{\int_0^\infty \epsilon_\lambda'(\lambda,T,P,S) i_{\lambda b}'(\lambda,T) d\lambda}{\sigma T^4/\pi} \tag{13-60}$$

where by using σT^4 for the total blackbody emission it is assumed that the index of refraction of the medium is $n = 1$. The total emitted intensity incident on dA_p is $i'(S) = \epsilon' i_b'(T) = \epsilon'(\sigma T^4/\pi)$.

In Fig. 13-13 the directional total emittance of carbon dioxide is shown as a function of temperature, partial pressure of the CO_2, and path length. This is an example of the extensive tabulations of such properties that are available for gases at conditions of importance in industrial design. The methods for using these properties in radiative exchange computations are developed in Chap. 17, where more detailed charts of the radiative properties are given.

Comparing (13-58) and (13-60) shows that Kirchhoff's law for directional *total* properties,

$$\alpha'(T,P,S) = \epsilon'(T,P,S) \tag{13-61}$$

holds only under the condition that the incident spectral radiation for absorption is proportional to a blackbody spectrum at the gas temperature T or the gas is gray, that is $\alpha_\lambda' = \epsilon_\lambda'$ are independent of wavelength. The same restrictions apply for opaque bodies as discussed in Chap. 3.

EXAMPLE 13-1 As a rough approximation, idealize the absorptance of CO_2 at $T_g = 830$ K ($1500°R$) and 10 atm as in Fig. 13-2 so that it consists of four bands

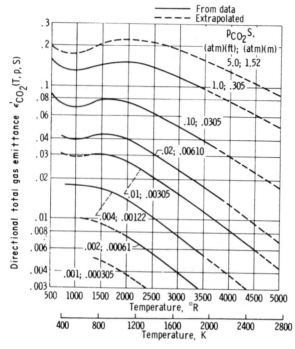

Figure 13-13 Emittance of carbon dioxide in mixture with non-absorbing gas at total pressure of 1 atm [20].

having vertical boundaries at the values 1.8 and 2.2, 2.6 and 2.8, 4.0 and 4.6, and 9 and 19 μm. What is the total emittance of a thick layer of gas at this temperature?

For a thick layer of gas, Eq. (13-59) indicates that ϵ'_λ will go to unity in the absorbing regions. Hence, the gas will emit like a blackbody in the four spectral absorption bands. In the nonabsorbing regions between the bands, ϵ'_λ is very small and is neglected in this simplified model. From (13-60) the total emittance becomes

$$\epsilon'(T_g,P,S) = \frac{\int_0^\infty \epsilon'_\lambda(\lambda,T_g,P,S)e_{\lambda b,g}\, d\lambda}{\sigma T_g^4} = \frac{\int_{\text{absorbing bands}} e_{\lambda b,g}\, d\lambda}{\sigma T_g^4}$$

The emittance is thus the fractional emission of a blackbody over the wavelength intervals of the absorbing bands, which can be obtained from the $F_{0-\lambda T_g}$ factors in Table A-5 in the appendix. The required values are as follows:

λ, μm	λT_g, μm · K	$F_{0-\lambda T_g}$
1.8	1,500	0.01285
2.2	1,830	0.04338
2.6	2,160	0.09478
2.8	2,320	0.12665
4.0	3,320	0.34734
4.6	3,820	0.44977
9	7,470	0.83435
19	15,800	0.97302

Then the emittance is

$$\epsilon'(T_g,P,S) = \sum_{\substack{\text{absorbing} \\ \text{bands}}} (F_{(\lambda T_g)_{\text{lower}} - (\lambda T_g)_{\text{upper}}})_{\text{band}}$$

$$= \sum_{\substack{\text{absorbing} \\ \text{bands}}} (F_{0-(\lambda T_g)_{\text{upper}}} - F_{0-(\lambda T_g)_{\text{lower}}})_{\text{band}}$$

The numerical values give

$$\epsilon' = 0.04338 - 0.01285 + 0.12665 - 0.09478 + 0.44977 - 0.34734$$

$$+ 0.97302 - 0.83435 = 0.304$$

EXAMPLE 13-2 What fraction of incident solar radiation will be absorbed by a thick layer of CO_2 at 10 atm and 830 K (1500°R)? Use the approximate absorption bands of Example 13-1.

The effective radiating temperature of the sun is about $T_s = 5780$ K (10,400°R). The desired result is the fraction of the solar spectrum that lies within the four CO_2 bands, as this is the only portion of the incident radiation that will be absorbed. Using the $F_{0-\lambda T_s}$ factors as in Example 13-1, but using the solar temperature, gives the following values (using Table A-5):

λ, μm	λT_s μm · K	$F_{0-\lambda T_s}$
1.8	10,400	0.92166
2.2	12,720	0.95255
2.6	15,030	0.96892
2.8	16,180	0.97446
4.0	23,120	0.99024
4.6	26,590	0.99336
9	52,020	0.99901
19	109,800	~1.00000

The fraction absorbed is then

$$\alpha' = \sum_{\substack{\text{absorbing} \\ \text{bands}}} (F_{0-(\lambda T_s)_{\text{upper}}} - F_{0-(\lambda T_s)_{\text{lower}}})_{\text{band}}$$

$$= 0.95255 - 0.92166 + 0.97446 - 0.96892 + 0.99336 - 0.99024$$

$$+ 1.00000 - 0.99901 = 0.041$$

Even though the gas layer is thick, only 4.1% of the incident energy is absorbed since the gas is essentially transparent in the region between the absorption bands.

13-8.3 Transmittance of a Medium at Uniform Temperature

The *transmittance* of a gas volume is the fraction of the incident energy that passes through the gas volume. If it is assumed that no reflection or scattering of the incident radiation occurs, then the energy transmitted along a path is the incident energy minus the energy absorbed along the path:

$$d^3 Q'_{\lambda,t} = d^3 Q'_{\lambda,i} - d^3 Q'_{\lambda,a} \tag{13-62}$$

Rearranging gives the transmittance as

$$\frac{d^3 Q'_{\lambda,t}}{d^3 Q'_{\lambda,i}} = 1 - \frac{d^3 Q'_{\lambda,a}}{d^3 Q'_{\lambda,i}} \tag{13-63}$$

Substituting Eq. (13-56) gives

Spectral transmittance for path length S in a uniform gas volume

$$\equiv \tau'_\lambda(\lambda,T,P,S) = \frac{d^3 Q'_{\lambda,t}}{d^3 Q'_{\lambda,i}} = 1 - \alpha'_\lambda(\lambda,T,P,S) = \exp(-a_\lambda S) \tag{13-64}$$

From Eq. (13-57) the intensities (as shown in Fig. 13-12) can then be related as

$$i'_\lambda(S) = i'_\lambda(0)\,\tau'_\lambda(S) \tag{13-65}$$

By analogous arguments, the directional total transmittance is given by the following (it is again assumed the reflectance of the gas volume is negligible):

Directional total transmittance for path length S in a uniform gas volume

$$\equiv \tau'(T,P,S) = 1 - \alpha'(T,P,S)$$

$$= \frac{\int_0^\infty \tau'_\lambda(\lambda,T,P,S)\,i'_\lambda(0)\,d\lambda}{\int_0^\infty i'_\lambda(0)\,d\lambda}$$

$$= \frac{\int_0^\infty i'_\lambda(0)\exp(-a_\lambda S)\,d\lambda}{\int_0^\infty i'_\lambda(0)\,d\lambda} \tag{13-66}$$

The total transmitted intensity at S is then $i'(S) = i'(0)\,\tau'(S)$.

EXAMPLE 13-3 Some types of nuclear explosions produce, at their peak, an emissive power spectrum like that of a blackbody at 6000 K. The sun also emits very close to this spectrum (see Fig. 13-1). Consequently, the transmissivity of the atmosphere for solar radiation can be used to determine the attenuation of energy from a nuclear explosion.

When the sun is directly overhead, the total transmittance of the atmosphere for solar radiation averages 35% throughout the fall and winter in the Great Lakes region. Assume that a 20-Mton weapon is detonated at a height of 10 km and dissipates its energy uniformly over a period of 4 s. Assume further that the fireball during this period is 1000 m in diameter, and that 50% of the total energy is dissipated as thermal radiation. Calculate the radiant energy flux directly below the burst at ground level.

The total energy expended by the fireball per unit time is (1 Mton $\approx 10^{12}$ kcal $= 4.187 \times 10^{12}$ kW·s) $Q = 20$ Mton/4 s $= 20.93 \times 10^{12}$ kW. For 50% of the energy going into thermal radiation, the emissive power of the fireball is $e = 0.5Q/A_{\text{fireball}} = 0.5Q/4\pi R^2_{\text{fireball}}$. The intensity of radiation leaving the diffuse fireball is $i'(0) = e/\pi$, and from (13-65) the intensity arriving at ground level is

$$i'_{\text{ground}} = \tau' i'(0) = 0.35 \frac{e}{\pi} = 0.35 \frac{0.5Q}{4\pi^2 R^2_{\text{fireball}}}$$

To compute the energy arriving at the ground, the fireball is treated approximately as a differential area with projected area $dA_p = \pi R^2_{\text{fireball}}$ seen from the ground. Then the energy reaching the ground directly below the fireball per unit time is

$$Q'_g = i'_{\text{ground}} A_{\text{ground}}\, d\omega = \frac{0.175Q}{4\pi^2 R^2_{\text{fireball}}} A_{\text{ground}} \frac{\pi R^2_{\text{fireball}}}{S^2}$$

The energy flux at ground level directly below the fireball is

$$q = \frac{Q'_g}{A_{\text{ground}}} = \frac{0.175Q}{4\pi S^2} = \frac{0.175 \times 20.93 \times 10^{12}}{4\pi \times 10^8} = 2915 \text{ kW/m}^2$$

Note that the result is independent of R_{fireball}.

An alternative method of solution is as follows:

For 50% of the energy going into thermal radiation, the radiant energy expended is $Q_{\text{rad}} = 10.47 \times 10^{12}$ kW. Imagine a concentric sphere of radius 10 km (the height above the ground) surrounding the fireball. Since the Q_{rad} leaves the fireball in a spherically symmetric fashion, the radiant energy flux arriving at the imaginary sphere in the absence of any attenuation is

$$\frac{Q_{\text{rad}}}{A_{\text{sphere}}} = \frac{10.47 \times 10^{12} \text{ kW}}{4\pi(10 \times 10^3)^2 \text{ m}^2} = \frac{10.47 \times 10^4}{4\pi} \text{ kW/m}^2$$

Since the transmittance is 35%, the energy flux received at ground level is

$$q = \frac{10.47 \times 10^4}{4\pi} 0.35 = 2915 \text{ kW/m}^2$$

The rather grim result is that this flux applied over a few seconds is more than four times that required to ignite newspapers. A more complete approach to problems of this type is given in [14], where the slant angle to the ground is included. Higher-altitude detonations than those of this example were studied. Figure 13-14 shows some results of [14] not recommended for the imaginative reader.

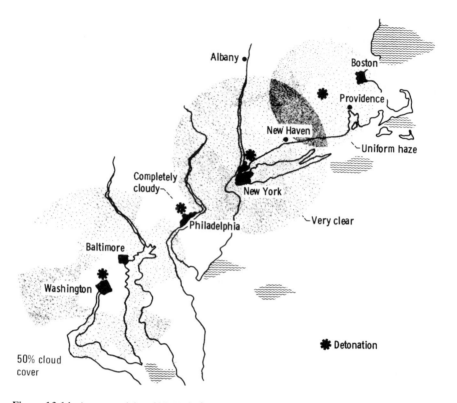

Figure 13-14 Areas receiving 628 kW/m² [15 cal/(s·cm²)] or more of radiative flux as function of weather conditions during four large-yield high-altitude weapon detonations [14].

13-9 THE CONCEPT OF LOCAL THERMODYNAMIC EQUILIBRIUM

It has been tacitly assumed in earlier chapters that opaque solids emit energy based solely on the temperature and physical properties of the body. The spectrum of emitted energy was assumed unaffected by the characteristics of any incident radiation. This is generally true because all the absorbed part of the energy incident on an opaque solid is quickly redistributed into internal energy states in an equilibrium distribution at the temperature of the solid.

In a gas, the redistribution of absorbed energy occurs by various types of collisions between the atoms, molecules, electrons, and ions that comprise the gas. Under most engineering conditions, this redistribution occurs quite rapidly, and the energy states of the gas *will* be populated in equilibrium distributions at any given locality. When this is true, the Planck spectral distribution correctly describes the emission from a blackbody, and (13-40) correctly describes the emission from a gas volume element.

The assumption that a gas will emit according to (13-40) regardless of the spectral distribution of intensity passing through and being absorbed by dV is a consequence of the assumption of "local thermodynamic equilibrium" or LTE. When the condition of LTE is not present, the calculation of radiant transfer becomes much more complex [15, 16].

Cases in which the LTE assumption breaks down are occasionally encountered. Examples are in very rarefied gases, where the rate and/or effectiveness of interparticle collisions in redistributing absorbed radiant energy is low; when rapid transients exist so that the populations of energy states of the particles cannot adjust to new conditions during the transient; where very sharp gradients occur so that local conditions depend on particles that arrive from adjacent localities at widely different conditions and may emit before reaching equilibrium; and where extremely large radiative fluxes exist, so that absorption of energy and therefore population of higher energy states occur so strongly that collisional processes cannot repopulate the lower states to an equilibrium density. Under any of these conditions, the spectral distribution of emitted radiation is not given by (13-40). Then the populations must be determined by detailed examination of the relation between the collision and radiation processes and their effect on the distribution of energy among the various possible states—a most formidable undertaking. It is, however, necessary in the examination of shock phenomena (sharp gradients), stellar atmospheres (extreme energy flux and low density), nuclear explosions (transients, sharp gradients, and extreme fluxes), and high-altitude and interplanetary gas dynamics (very low densities).

A gas with small optical thickness can have transmitted within it radiation from regions at widely different conditions. For this reason, a nearly transparent or "clear" gas is more likely to depart from LTE than is an optically thick gas of the same density.

A very prominent non-LTE effect is found in the laser. In this device, a material with a metastable energy state is excited by some external means. Because the excited state is metastable and is chosen so that no competing process is trying to depopulate it, its population can reach a value well above the equilibrium value. This condition is called a *population inversion*. The material is then exposed to radiation containing photons with the same frequency as the transition frequency from the excited to a lower state in the material. This radiation induces or stimulates the transition to the lower state. Consequently, a large number of photons with the transition frequency are emitted, thus amplifying the intensity of the incident radiation. This process leads to the acronym *l*ight *a*mplification by the *s*timulated *e*mission of *r*adiation, or *laser*.

Such non-LTE problems are beyond the scope of this work. It will be assumed here that LTE always exists and that although the flux arriving at a volume element dV may come from localities at widely different temperatures, the emission from dV will be governed by (13-40).

REFERENCES

1. Langley, S. P.: Experimental Determination of Wave-Lengths in the Invisible Prismatic Spectrum, *Mem. Natl. Acad. Sci.*, vol. 2, pp. 147–162, 1883.
2. Thekaekara, M. P.: Survey of the Literature on the Solar Constant and the Spectral Distribution of Solar Radiant Flux, *NASA* SP-74, 1965.

3. Goody, R. M.: "Atmospheric Radiation, Theoretical Basis," vol. 1, Clarendon Press, Oxford, 1964.
4. Kondratyev, K. Ya.: "Radiation in the Atmosphere," Academic Press, Inc., New York, 1969.
5. Gardon, Robert: A Review of Radiant Heat Transfer in Glass, *J. Am. Ceram. Soc.,* vol. 44, no. 7, pp. 305-312, 1961.
6 Hottel, H. C., A. F. Sarofim, I. A. Vasalos, and W. H. Dalzell: Multiple Scatter: Comparison of Theory with Experiment, *J. Heat Transfer,* vol. 92, no. 2, pp. 285-291, 1970.
7. Curie, Daniel (G. F. K. Garlick, trans.): "Luminescence in Crystals," John Wiley and Sons, Inc., New York, 1963.
8. Pringsheim, Peter: "Fluorescence and Phosphorescence," Interscience Publishers, Inc., New York, 1949.
9. Penner, S. S.: "Quantitative Molecular Spectroscopy and Gas Emissivities," Addison-Wesley Publishing Company, Inc., Reading, Mass., 1959.
10. Bond, John W., Jr., Kenneth M. Watson, and Jasper A. Welch, Jr.: "Atomic Theory of Gas Dynamics," Addison-Wesley Publishing Company, Inc., Reading, Mass., 1965.
11. Bates, David R. (ed.): "Atomic and Molecular Processes," Academic Press, Inc., New York, 1962.
12. Chandrasekhar, S.: "Radiative Transfer," Dover Publications, Inc., New York, 1960.
13. Kourganoff, Vladimir: "Basic Methods in Transfer Problems; Radiative Equilibrium and Neutron Diffusion," Dover Publications, Inc., New York, 1963.
14. Atlas, Reynold, and B. N. Charles: Atmospheric Attenuation of the Thermal Radiation from a High-altitude Nuclear Detonation, paper 64-318, *AIAA,* June, 1964.
15. Kulander, John L.: Non-Equilibrium Radiation, *Rept.* R64SD41, General Electric Company, June, 1965. (Available from DDC as AD-617383.)
16. Thomas, Richard N.: "Some Aspects of Nonequilibrium Thermodynamics in the Presence of a Radiation Field," University of Colorado Press, Boulder, 1965.
17. Garbuny, Max: "Optical Physics," Academic Press, Inc., New York, 1965.
18. Aroeste, Henry, and William C. Benton: Emissivity of Hydrogen Atoms at High Temperatures, *J. Appl. Phys.,* vol. 27, pp. 117-121, 1956.
19. Meyerott, R. E., J. Sokoloff, and R. A. Nicholls: Absorption Coefficients of Air, *Rept.* LMSD-288052, Lockheed Aircraft Corporation (AFCRC-TR-59-296), September, 1959.
20. Hottel, H. C.: Radiant-heat Transmission, in William H. McAdams (ed.), "Heat Transmission," 3d ed., chap. 4, McGraw-Hill Book Company, New York, 1954.

PROBLEMS

1 A monochromatic beam of radiation at $\lambda = 2.5$ μm and with intensity 3000 Btu/(h·ft²·μm) enters a gas layer 8 in thick. The gas is at 2000°R and has an absorption coefficient $a_{2.5\,\mu m} = 2$ ft⁻¹. What is the intensity of the beam emerging from the gas layer? Neglect scattering, but include emission.

   Answer: 2400 Btu/(h·ft²·μm·sr).

2 Radiation from a blackbody source at 3000 K is passing through a layer of air at 12,000 K and 1 atm. Considering only the transmitted radiation (that is, not accounting for the radiation by the air), what path length is required to attenuate by 25% the energy at the wavelength corresponding to the peak of the blackbody radiation?

   Answer: 2.2 cm.

3 Radiation with a wavelength of 1.5 μm is passing through a gas at a temperature of 10,000 K. What is the ratio of the true absorption coefficient to the absorption coefficient?

   Answer: 1.62.

4 A thin black plate 1 × 1 cm is at the center of a sphere of CO_2-air mixture at a uniform temperature of 2400 K and 1 atm total pressure. The partial pressure of the CO_2 is 0.6 atm,

and the sphere diameter is 1 m. How much energy is absorbed by the plate? What will the plate temperature be? (Assume the boundary of the sphere is black and kept cool so that it does not enter into the radiative exchange.)

Answer: 24.7 W, 1210 K.

5 Using Fig. 13-8 estimate the total absorptance for air at 12,000 K and 1 atm for solar radiation passing through a path length of 5 cm.

Answer: 0.58.

6 From the spectral absorptance α_λ' in Fig. 13-2 estimate the total emittance of CO_2 for the temperature, pressure, and path length given in the figure caption. Compare the result with the value obtained from Fig. 13-13 (or Fig. 17-11).

7 A gas layer at constant pressure P has a linearly decreasing temperature in it, and a constant mass absorption coefficient a_m (no scattering). For radiation passing in a normal direction through the layer, what is the ratio i_2'/i_1' as a function of T_1, T_2, and L? The temperature range T_2 to T_1 of the gas is low enough so that emission by the gas can be neglected (gas constant $= R$).

Answer: $\dfrac{i_2'}{i_1'} = \exp\left[-\dfrac{PLa_m}{R(T_1-T_2)}\ln\dfrac{T_1}{T_2}\right]$

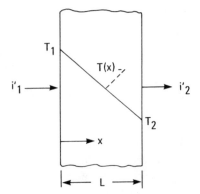

8 A thin cylinder of gas contains a uniform suspension of scattering particles giving a scattering coefficient $\sigma_{s\lambda}$. The phase function is independent of circumferential angle φ and depends only on the angle θ away from the path of the incident radiation, $\Phi(\theta) = \frac{3}{4}(1 + \cos^2 \theta)$. A beam of radiation with spectral intensity i_λ' is incident in a normal direction on the end of the cylinder. As a function of L, what fraction of energy exits in the forward direction ($0 \leqslant \theta < \pi/2$), and what fraction is back-scattered ($\pi/2 < \theta \leqslant \pi$)? There is no absorption or emission.

Answer: $\frac{1}{2}(1 + e^{-\sigma_{s\lambda}L})$, $\frac{1}{2}(1 - e^{-\sigma_{s\lambda}L})$.

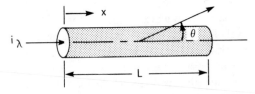

FOURTEEN

THE EQUATIONS OF ENERGY TRANSFER FOR ABSORBING, EMITTING, AND SCATTERING MEDIA

14-1 INTRODUCTION

In Chap. 13 some of the basic concepts and definitions were presented for intensity, emission, absorption, and scattering within a medium. Radiation traveling along a path is attenuated by absorption and scattering, and is enhanced by both spontaneous and induced emission and by radiation scattered in from other directions. The processes of absorption, emission, and scattering as discussed in Chap. 13 will be employed to develop a first-order integrodifferential equation governing the radiation intensity along a path through a medium. This is called the *equation of transfer*. In obtaining a solution to the equation of transfer, the constant of integration introduces the intensity at the origin of the radiation path being considered. Because the origin is usually at the boundary of the radiating medium, the radiation at the boundaries is coupled into the radiation distribution within the medium.

The intensity as given by the equation of transfer gives the radiation that is locally traveling in a single direction per unit solid angle and wavelength, and that is crossing a unit area normal to the direction of travel. To obtain the net *energy* crossing an area requires integrations that include the contributions of the intensities crossing in all directions and at all wavelengths. This results in an equation for local radiative flux that is used in forming energy balances within the medium. The energy balance is needed to obtain the temperature distribution within the medium.

Since the local intensities depend on scattering, if it is present, the energy balance will depend on the net energy scattered into a volume element. The scattering into a direction along a path is often combined with the local emission in that direction to form a quantity called the *source function*. When scattering is present, a source-function equation must be solved simultaneously with the energy equation to determine two unknowns: the temperature distribution and the source-function distribution, which contains the distribution of scattered energy. This is a consequence of the two coupled transfer processes absorption (and hence emission) and scattering. For a nonabsorbing medium, the scattering becomes decoupled from the energy equation, considerably simplifying the analysis.

In this chapter the equation of transfer for local intensities is first derived. After this, the radiative flux is considered, and the energy equation is presented in three dimensions. The formulation is then given in detail for a plane layer between infinite parallel plates. The most general equations include anisotropic scattering. Then simplified cases are presented for isotropic scattering, a gray medium, and a medium bounded by diffuse walls.

14-2 SYMBOLS

A	area
a	absorption coefficient
c	speed of light in a medium
D	spacing between parallel plates
E_n	exponential integral function, Eq. (14-59)
e	emissive power
f	photon distribution function
G	quantity defined in Eq. (14-81)
h	Planck's constant; enthalpy
I	source function, Eq. (14-9)
i	radiation intensity
i,j,k	unit vectors in x,y,z coordinate directions
K	extinction coefficient, $a + \sigma_s$
k	thermal conductivity
n	unit normal vector
P	pressure
Q	energy per unit time
$\mathbf{q}_r$	radiative flux vector
q	energy flux, energy per unit area and time
q'''	volume heat source
r	position vector
S	coordinate along path of radiation
s	unit vector in S direction
T	absolute temperature
U	radiant energy density
V	volume
x,y,z	coordinates in cartesian system
β	coefficient of volume expansion

ϵ	emissivity
θ	cone angle, angle from normal of area
κ	optical thickness
κ_D	optical thickness for path of length D
λ	wavelength
μ	the quantity $\cos\theta$
ν	frequency
ρ	density
σ	Stefan-Boltzmann constant
σ_s	scattering coefficient
Φ	phase function for scattering; viscous dissipation function
φ	circumferential angle
ϕ	dimensionless temperature ratio
ψ	dimensionless energy flux
τ	time
Ω_0	albedo for scattering
ω	solid angle

Subscripts

a	absorbed
b	blackbody
e	emitted
i	incident; incoming
o	outgoing
P	Planck mean value, Eq. (14-31)
s	scattering
λ,ν	spectrally dependent
$1,2$	surface 1 or 2

Superscripts

$+$	along directions having positive cosine θ
$-$	along directions having negative cosine θ
$*$	dummy variable of integration
$-$	(overbar) averaged over all incident solid angles
$'$	directional quantity

14-3 THE EQUATION OF TRANSFER

The equation of transfer in a medium will now be derived. This will describe the intensity of radiation at any position along its path through an absorbing, emitting, and scattering medium.

14-3.1 Derivation

Bouguer's law, Eq. (13-12), accounts for attenuation by absorption and scattering. The equation of transfer is an extension to include the contributions to the radiation intensity by emission and incoming scattering along the path.

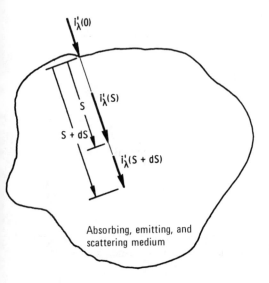

Figure 14-1 Geometry for derivation of equation of transfer.

Consider radiation of intensity $i'_\lambda(S)$ within a region of absorbing, emitting, and scattering medium (Fig. 14-1). Attention is directed to the change of intensity as the radiation passes through a distance dS. Using (13-8) gives the decrease due to absorption as

$$di'_{\lambda,a} = -a_\lambda(S)i'_\lambda(S)\,dS \tag{14-1}$$

Note that a_λ has been used in (14-1) rather than the "true" absorption coefficient a_λ^+. Thus the change in intensity is not only the result of true absorption, but also includes the contribution of induced emission as discussed in Sec. 13-5.5.

If the radiation along the path is in local thermodynamic equilibrium, the spontaneous emission contribution by the medium along the path length dS to the intensity in the S direction is given by Eq. (13-41) as

$$di'_{\lambda,e} = a_\lambda(S)i'_{\lambda b}(S)\,dS \tag{14-2}$$

From (13-28), (13-30), (13-52), and Fig. 13-11, the sum of the attenuation by scattering and the gain by incoming scattering in direction S can be written as

$$di'_{\lambda,s} = -\sigma_{s\lambda}(S)\,i'_\lambda(S)\,dS + \frac{dS\,\sigma_{s\lambda}}{4\pi}\int_{\omega_i=4\pi} i'_\lambda(S,\omega_i)\,\Phi(\lambda,\omega,\omega_i)\,d\omega_i \tag{14-3}$$

Adding (14-1)–(14-3) gives, for the change in intensity with S in the solid angle $d\omega$ about the direction of S,

$$\frac{di'_\lambda}{dS} = -a_\lambda i'_\lambda(S) + a_\lambda i'_{\lambda b}(S) - \sigma_{s\lambda} i'_\lambda(S) + \frac{\sigma_{s\lambda}}{4\pi}\int_{\omega_i=4\pi} i'_\lambda(S,\omega_i)\,\Phi(\lambda,\omega,\omega_i)\,d\omega_i \tag{14-4}$$

| Loss by absorption (including the contribution by induced emission) | Gain by emission (not including induced emission) | Loss by scattering | Gain by scattering into S direction |

The two terms representing decreases by absorption and scattering are combined, giving the equation of transfer for absorbing, emitting, and scattering media (for the case of *elastic* anisotropic scattering):

$$\frac{di_\lambda'}{dS} = -K_\lambda i_\lambda'(S) + a_\lambda i_{\lambda b}'(S) + \frac{\sigma_{s\lambda}}{4\pi} \int_{\omega_i=4\pi} i_\lambda'(S,\omega_i)\,\Phi(\lambda,\omega,\omega_i)\,d\omega_i \qquad (14\text{-}5)$$

where $K_\lambda = a_\lambda + \sigma_{s\lambda}$ is the extinction coefficient (Sec. 13-5.1) and in general is a function of position S.

The albedo for scattering Ω_0, defined as the ratio of the scattering coefficient to the extinction coefficient, is sometimes used:

$$\Omega_{0\lambda} \equiv \frac{\sigma_{s\lambda}}{K_\lambda} = \frac{\sigma_{s\lambda}}{a_\lambda + \sigma_{s\lambda}} \qquad (14\text{-}6)$$

For scattering alone $\Omega_{0\lambda} \rightarrow 1$, while for absorption alone $\Omega_{0\lambda} \rightarrow 0$. The optical depth or opacity when both scattering and absorption are present was defined in (13-17):

$$\kappa_\lambda(S) = \int_0^S K_\lambda(S^*)\,dS^* = \int_0^S (a_\lambda + \sigma_{s\lambda})\,dS^* \qquad (14\text{-}7)$$

where S^* is a dummy variable of integration. Equation (14-5) now becomes

$$\frac{di_\lambda'}{d\kappa_\lambda} = -i_\lambda'(\kappa_\lambda) + (1 - \Omega_{0\lambda}) i_{\lambda b}'(\kappa_\lambda) + \frac{\Omega_{0\lambda}}{4\pi} \int_{\omega_i=4\pi} i_\lambda'(\kappa_\lambda,\omega_i)\,\Phi(\lambda,\omega,\omega_i)\,d\omega_i \qquad (14\text{-}8)$$

where $d\kappa_\lambda = K_\lambda(S)\,dS$ is the *optical differential thickness*. The final two terms in (14-8) are combined into the source function $I_\lambda'(\kappa_\lambda,\omega)$ defined as

$$I_\lambda'(\kappa_\lambda,\omega) \equiv (1 - \Omega_{0\lambda}) i_{\lambda b}'(\kappa_\lambda) + \frac{\Omega_{0\lambda}}{4\pi} \int_{\omega_i=4\pi} i_\lambda'(\kappa_\lambda,\omega_i)\,\Phi(\lambda,\omega,\omega_i)\,d\omega_i \qquad (14\text{-}9)$$

This is the source of intensity along the optical path from both emission and incoming scattering. For anisotropic scattering the I_λ' is a function of ω (that is, the direction of S). The *equation of transfer* then becomes

$$\frac{di_\lambda'}{d\kappa_\lambda} + i_\lambda'(\kappa_\lambda) = I_\lambda'(\kappa_\lambda,\omega) \qquad (14\text{-}10)$$

This is an integrodifferential equation since i_λ' is within the integral of the source function.

There is a basic advantage to including the induced emission in the absorption coefficient. The induced emission as discussed in Sec. 13-5.5 is in the same direction as the transmitted radiation. The spontaneous emission, however, is uniform over all directions. Thus by combining the induced emission with the "true" absorption to form a_λ, the quantities depending on the direction of the incident radiation have been brought together. The resulting emission term in the equation of transfer contains only spontaneous emission and hence does not depend on direction.

14-3.2 Integration by Use of Integrating Factor

Equation (14-10) is a first-order linear integrodifferential equation, and an integrated form can be obtained by use of an integrating factor. Multiplying through by $\exp \kappa_\lambda$ gives

$$\exp \kappa_\lambda \frac{di_\lambda'}{d\kappa_\lambda} + i_\lambda'(\kappa_\lambda) \exp \kappa_\lambda = \frac{d}{d\kappa_\lambda} [i_\lambda'(\kappa_\lambda) \exp \kappa_\lambda] = I_\lambda'(\kappa_\lambda, \omega) \exp \kappa_\lambda \qquad (14\text{-}11)$$

Integrating over an optical thickness from $\kappa_\lambda = 0$ to $\kappa_\lambda(S)$ gives

$$i_\lambda'(\kappa_\lambda) \exp \kappa_\lambda - i_\lambda'(0) = \int_0^{\kappa_\lambda} I_\lambda'(\kappa_\lambda^*, \omega) \exp \kappa_\lambda^* \, d\kappa_\lambda^* \qquad (14\text{-}12)$$

or

$$i_\lambda'(\kappa_\lambda) = i_\lambda'(0) \exp(-\kappa_\lambda) + \int_0^{\kappa_\lambda} I_\lambda'(\kappa_\lambda^*, \omega) \exp[-(\kappa_\lambda - \kappa_\lambda^*)] \, d\kappa_\lambda^* \qquad (14\text{-}13)$$

where κ_λ^* is a dummy variable of integration.

Equation (14-13) is interpreted physically as the intensity being composed of two terms at optical depth κ_λ. The first is the attenuated incident radiation arriving at κ_λ (including, however, the contribution of induced emission along the path). The second is the intensity at κ_λ resulting from spontaneous emission and incoming scattering in the S direction by all thickness elements along the path, reduced by exponential attenuation between each point of emission and incoming scattering κ_λ^* and the location κ_λ. Equation (14-13) is the *integrated form of the equation of transfer*.

The equation of transfer in the form of (14-10) or (14-13) cannot be used by itself to obtain the local intensity because it contains the unknown temperature and scattering distributions that are in the source function [the temperature is in the $i_{\lambda b}'(\lambda, T)$ within I_λ']. The temperature distribution is also needed to determine the absorption and scattering coefficients so that the local optical depth $\kappa_\lambda(S)$ can be computed from (14-7), and the physical coordinate S thereby related to the optical coordinate κ_λ. The temperature distribution depends on conservation of energy within the medium, which in turn depends on the total absorbed radiation in each volume element along the path. This total energy will be obtained by utilizing the intensity passing through a location and integrating over all incident solid angles and all wavelengths. As will be shown, the energy equation is solved in conjunction with a source-function equation to yield compatible temperature and scattering distributions. The source-function equation is found by inserting (14-13) into (14-9) to eliminate i_λ' and obtain an equation for I_λ'.

The resulting equations are sufficiently complex so that numerical solutions are almost always required. In some instances advanced analytical techniques have been used to obtain closed-form solutions for a limited range of geometries and conditions. The application of singular eigenvalue expansions (also called *Case's method*) and the related method of Ambarzumian [1] to certain homogeneous-medium plane-geometry problems with simple source configurations can provide exact solutions for a gray

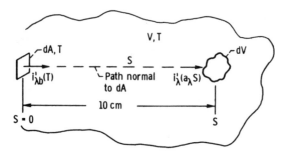

Figure 14-2 Geometry for Example 14-1.

medium. Singular eigenvalue expansions were used by Gritton and Leonard [2] for the gray case. Yener et al. [3] further expanded the method to a "picket-fence" nongray medium, including isotropic scattering and specularly reflecting boundaries; mathematically exact solutions are presented. The Wiener-Hopf technique [4] can be applied to certain multidimensional problems.

EXAMPLE 14-1 A black surface element dA is 10 cm from an element of gas dV (Fig. 14-2). The gas element is a part of a gas volume V that is isothermal and at the same temperature T as dA. If the gas has an absorption coefficient a_λ of 0.1 cm^{-1} at wavelength 1 μm and there is no scattering, what is the spectral intensity at $\lambda = 1$ μm that arrives at dV along the path S from dA to dV?

Because element dA is black and at temperature T, the intensity at $S = 0$ is $i'_\lambda(0) = i'_{\lambda b}(T)$. Since the gas is isothermal, the emitted blackbody intensity in the gas is $i'_{\lambda b}(\kappa_\lambda) = i'_{\lambda b}(T)$. Substituting into the integrated equation of transfer [Eq. (14-13)] with $\sigma_{s\lambda} = 0$ gives

$$i'_\lambda(\kappa_\lambda) = i'_{\lambda b}(T) \exp{(-\kappa_\lambda)} + i'_{\lambda b}(T) \exp{(-\kappa_\lambda)} \int_0^{\kappa_\lambda} \exp{(\kappa_\lambda^*)} \, d\kappa_\lambda^*$$

After integration, this reduces to

$$i'_\lambda(\kappa_\lambda) = i'_{\lambda b}(T)$$

The $i'_{\lambda b}(T)$ is given by (2-11a) for a gas with refractive index $n = 1$. The intensity arriving at dV along an isothermal path from a black surface element at the same temperature as the gas is thus equal to the blackbody intensity emitted by the wall and does not depend on a_λ or S. The attenuation by the gas of the intensity emitted by the wall was exactly compensated by emission from the gas along the path from dA to dV.

14-3.3 The Radiative Flux Vector

In Fig. 14-3, the intensity is the energy per unit solid angle crossing dA per unit area normal to i'. Hence the energy crossing dA as a result of i' is $i' \, dA \cos \theta \, d\omega$. The radiative flux crossing dA as a result of intensities incident from all directions is thus

$$q_{r,n} = \int_{\omega=4\pi} i' \cos \theta \, d\omega \tag{14-14}$$

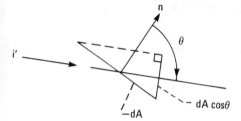

Figure 14-3 Quantities in derivation of radiative flux vector.

where θ is the angle from the normal of dA to the direction of i', and i' is a function of direction. Since θ is in the integral, the $q_{r,n}$ depends on the direction of n; hence $q_{r,n}$ is a vector. If the direction cosines of n are α', δ', γ' and those for i' are α, δ, γ, then $\cos\theta = \alpha\alpha' + \delta\delta' + \gamma\gamma'$ and

$$q_{r,n} = \int_{\omega=4\pi} i'(\alpha,\delta,\gamma)(\alpha\alpha' + \delta\delta' + \gamma\gamma')\, d\omega$$

Thus

$$q_{r,n} = \alpha' q_{r,x} + \beta' q_{r,y} + \gamma' q_{r,z} \tag{14-15}$$

The $q_{r,x}$, $q_{r,y}$, and $q_{r,z}$ are fluxes across areas normal to the x, y, and z directions and are the components of the *radiative flux vector*. Each component is obtained from the integral in (14-14) with n oriented in that coordinate direction. For example, $q_{r,x} = \int_{\omega=4\pi} i'(\alpha,\delta,\gamma)\alpha\, d\omega$, where α is the cosine of the angle between the x axis and the direction of i'.

For use in an energy balance on a volume element, the net radiative energy supplied to dV is desired. If a volume element $dx\, dy\, dz$ is considered, the radiative energies in and out of the $dy\, dz$ faces are shown in Fig. 14-4. The energies are similarly written for the other faces, the outgoing energies are subtracted from the incoming, and the result divided by $dx\, dy\, dz$. The result is the net radiative energy supplied per unit volume:

$$-\left[\frac{\partial q_{r,x}}{\partial x} + \frac{\partial q_{r,y}}{\partial y} + \frac{\partial q_{r,z}}{\partial z}\right] = -\nabla\cdot\mathbf{q}_r \tag{14-16}$$

which is the negative of the divergence of the radiative heat-flux vector

$$\mathbf{q}_r = i q_{r,x} + j q_{r,y} + k q_{r,z} \tag{14-17}$$

The vector form $-\nabla\cdot\mathbf{q}_r$ can be directly used in coordinate systems other than the rectangular coordinates considered here.

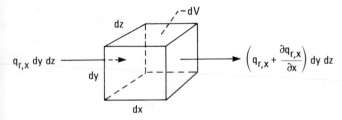

Figure 14-4 Radiative fluxes for volume element.

14-4 ENERGY CONSERVATION WITHIN THE MEDIUM

14-4.1 General Energy Conservation, Including Conduction and Convection

A general energy balance on a volume element includes conduction, convection, internal heat sources, compression work, viscous dissipation and energy storage due to transients, as well as the contribution by radiative heat transfer. Storage of radiant energy within an element is generally negligible; hence no modification of the usual transient terms will be considered here as a result of the radiation field. Radiation pressure is negligible relative to the fluid pressure and hence does not contribute to the compression work term. The net inflow of radiant energy per unit volume can be written as the negative of the divergence of a radiant flux vector $\mathbf{q}_r$ [Eq. (14-16)]. The net heat conduction into a volume element can also be written as the divergence of a vector $\nabla \cdot (k\, \nabla T)$. Thus the usual energy equation can be utilized by adding $-\mathbf{q}_r$ to $k\, \nabla T$ to yield

$$\rho c_p \frac{DT}{D\tau} = \beta T \frac{DP}{D\tau} + \nabla \cdot (k\, \nabla T - \mathbf{q}_r) + q''' + \Phi \tag{14-18}$$

The β is the thermal coefficient of volume expansion of the fluid, q''' is the local heat source per unit volume and time, and Φ is the heat production by viscous dissipation. An alternative form in terms of enthalpy is

$$\rho \frac{Dh}{d\tau} = \frac{DP}{d\tau} + \nabla \cdot (k\, \nabla T - \mathbf{q}_r) + q''' + \Phi \tag{14-19}$$

To obtain the temperature distribution in the medium by solving (14-18), an expression for $\nabla \cdot \mathbf{q}_r$ is needed. A derivation will now be given based on physical reasoning where scattering is absent. Then a more mathematical derivation will be given with scattering included.

14-4.2 Divergence of Radiative Flux for Absorption Alone (No Scattering)

In many instances scattering can be neglected, so that the relations obtained here will be quite useful. For zero scattering $\Omega_{0\lambda} \to 0$, and the source function reduces to $I'_\lambda = i'_{\lambda b}$. Equation (14-13) then becomes

$$i'_\lambda(\kappa_\lambda) = i'_\lambda(0)\exp(-\kappa_\lambda) + \int_0^{\kappa_\lambda} i'_{\lambda b}(\kappa^*_\lambda)\exp[-(\kappa_\lambda - \kappa^*_\lambda)]\,d\kappa^*_\lambda \tag{14-20}$$

Equation (14-20) describes the intensity at a single wavelength traveling in a single direction within the medium. To obtain the net radiative energy supplied to a volume element, consider the energy absorbed by a volume element dV within a medium as shown in Fig. 14-5. The energy absorbed from the incident intensity $i'_\lambda(\lambda,\omega,\kappa_\lambda)$ that arrives within the incremental solid angle $d\omega$ is, by analogy with (13-38),

$$d^4 Q'_{\lambda,a} = a_\lambda(dV) i'_\lambda(\lambda,\omega,\kappa_\lambda)\,dV\,d\lambda\,d\omega \tag{14-21}$$

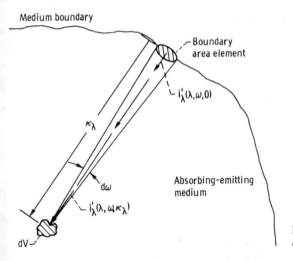

Medium boundary

Figure 14-5 Geometry for derivation of energy-conservation relation.

The incident intensity $i_\lambda'(\lambda,\omega,\kappa_\lambda)$ is given by (14-20) as

$$i_\lambda'(\lambda,\omega,\kappa_\lambda) = i_\lambda'(\lambda,\omega,0)\exp(-\kappa_\lambda) + \int_0^{\kappa_\lambda} i_{\lambda b}'(\lambda,\kappa_\lambda^*)\exp[-(\kappa_\lambda - \kappa_\lambda^*)]\,d\kappa_\lambda^* \quad (14\text{-}22)$$

where $i_\lambda'(\lambda,\omega,0)$ is the spectral intensity directed toward dV from the system boundary in the direction of $d\omega$.

The energy absorbed by dV from all incident directions is found by integrating (14-21) with respect to ω:

$$d^3Q_{\lambda,a} = \int_{\omega=0}^{4\pi} d^4Q_{\lambda,a}' = a_\lambda(dV)\,dV\,d\lambda \int_{\omega=0}^{4\pi} i_\lambda'(\lambda,\omega,\kappa_\lambda)\,d\omega \quad (14\text{-}23)$$

For convenience in writing the equations a little more compactly, a mean intensity $\bar{i}_\lambda(\lambda)$ at dV can be defined by

$$\bar{i}_\lambda(\lambda) \equiv \frac{1}{4\pi}\int_0^{4\pi} i_\lambda'(\lambda,\omega,\kappa_\lambda)\,d\omega \quad (14\text{-}24)$$

Then (14-23) becomes

$$d^3Q_{\lambda,a} = 4\pi a_\lambda(dV)\bar{i}_\lambda(\lambda)\,dV\,d\lambda \quad (14\text{-}25)$$

By integrating (14-25) over all wavelengths, the total energy absorbed by dV from the radiation field is obtained as

$$d^2Q_a = \int_{\lambda=0}^{\infty} d^3Q_{\lambda,a} = 4\pi\,dV\int_0^{\infty} a_\lambda(dV)\bar{i}_\lambda(\lambda)\,d\lambda \quad (14\text{-}26)\dagger$$

The total energy emitted spontaneously from dV is obtained by using a_λ in

† Note that the differential on the left is of second order since, in the nomenclature used, dV is of order $dA\,dS$.

(13-40) and integrating over all wavelengths:

$$d^2Q_e = \int_0^\infty d^3Q_{\lambda,e} = 4dV \int_0^\infty a_\lambda(dV)\, e_{\lambda b}(\lambda,T)\, d\lambda \tag{14-27}$$

This derivation uses the assumption that dV is very small so that all the energy emitted by dV escapes before any can be reabsorbed within dV. The net outflow of radiant energy per unit volume, which is the desired divergence of the radiant heat-flux vector, is then, from (14-26) and (14-27),

$$\nabla \cdot \mathbf{q}_r = 4 \int_0^\infty a_\lambda(\lambda,T,P)\, [e_{\lambda b}(\lambda,T) - \pi \bar{\imath}_\lambda(\lambda)]\, d\lambda \tag{14-28}$$

14-4.3 Radiative Equilibrium

For situations in which all other energy exchange mechanisms such as conduction and convection are negligible compared with radiation, the total emitted energy from dV is equal to the total absorbed energy. This is termed *radiative equilibrium* and is simply a statement of steady-state energy conservation in the absence of any other exchange mechanism but radiation. With only radiation present, Eq. (14-18) yields

$$\nabla \cdot \mathbf{q}_r = 0 \tag{14-29}$$

or, from (14-28),

$$\int_0^\infty a_\lambda(\lambda,T,P)\, e_{\lambda b}(\lambda,T)\, d\lambda = \pi \int_0^\infty a_\lambda(\lambda,T,P)\, \bar{\imath}_\lambda(\lambda)\, d\lambda \tag{14-30}$$

14-4.4 Some Mean Absorption Coefficients

As a result of the emission integral on the left of (14-30), it is convenient to define the *Planck mean absorption coefficient* $a_P(T,P)$ as

$$a_P(T,P) \equiv \frac{\int_0^\infty a_\lambda(\lambda,T,P)\, e_{\lambda b}(\lambda,T)\, d\lambda}{\int_0^\infty e_{\lambda b}(\lambda,T)\, d\lambda} = \frac{\int_0^\infty a_\lambda(\lambda,T,P)\, e_{\lambda b}(\lambda,T)\, d\lambda}{\sigma T^4} \tag{14-31}$$

The a_P is the mean of the spectral coefficient when weighted by the blackbody emission spectrum. It will prove useful in considering *emission* from a volume, and in certain limiting cases of radiative transfer.

Substituting (14-31) into Eq. (14-30) results in a relation for radiative equilibrium in the form

$$a_P(T,P)\, \sigma T^4 = \pi \int_0^\infty a_\lambda(\lambda,T,P)\, \bar{\imath}_\lambda(\lambda)\, d\lambda \tag{14-32}$$

Thus, if $\bar{\imath}_\lambda(\lambda)$ is known at a position within the gas, (14-32) can be solved for T at that location. The Planck mean a_P is convenient since it depends only on the properties at dV. It can be tabulated and is especially useful where the pressure is constant over the geometry of the system.

As a result of the absorption integral on the right side of (14-32), another mean absorption coefficient can be defined, the *incident mean* (or *modified Planck mean*) absorption coefficient $a_i(T,P)$:

$$a_i(T,P) = \frac{\int_0^\infty a_\lambda(\lambda,T,P)\,\bar{i}_\lambda(\lambda)\,d\lambda}{\int_0^\infty \bar{i}_\lambda(\lambda)\,d\lambda} \tag{14-33}$$

However, such a definition has little value for general use. A tabulation of a_i would have to be carried out for all combinations of incident spectral distributions and spectral variations of local absorption coefficients. Except in certain very limited special cases, the work involved in such a tabulation would not be warranted. Further discussion of the physical interpretation of various mean absorption coefficients is given in Sec. 15-6.

For the special case of a gray gas, $a_\lambda = a_P$ and (14-32) reduces to the convenient relation

$$\sigma T^4 = \pi \int_0^\infty \bar{i}_\lambda\,d\lambda = \frac{1}{4}\int_{\omega=0}^{4\pi}\int_0^\infty i'_\lambda\,d\lambda\,d\omega = \frac{1}{4}\int_{\omega=0}^{4\pi} i'\,d\omega \tag{14-34}$$

where i' is the total intensity $i' = \int_0^\infty i'_\lambda\,d\lambda$.

14-4.5 Divergence of Radiative Flux Including Scattering

This is an extension of Sec. 14-4.2, starting with the equation of transfer (14-4). The first term in that equation can be written as

$$\frac{di'_\lambda}{dS} = \frac{\partial i'_\lambda}{\partial x}\frac{dx}{dS} + \frac{\partial i'_\lambda}{\partial y}\frac{dy}{dS} + \frac{\partial i'_\lambda}{\partial z}\frac{dz}{dS}$$

The $dx/dS = \alpha$, $dy/dS = \delta$, and $dz/dS = \gamma$ where α, δ, γ are the direction cosines of i'_λ in the S direction in an x, y, z coordinate system.

Equation (14-4) is then integrated at location S over all solid angles $\omega = 0\text{-}4\pi$:

$$\int_{\omega=4\pi}\left(\frac{\partial i'_\lambda}{\partial x}\alpha + \frac{\partial i'_\lambda}{\partial y}\delta + \frac{\partial i'_\lambda}{\partial z}\gamma\right)d\omega = -(a_\lambda + \sigma_{s\lambda})\int_{\omega=4\pi} i'_\lambda(S,\omega)\,d\omega$$

$$+ a_\lambda\int_{\omega=4\pi} i'_{\lambda b}(S)\,d\omega + \frac{\sigma_{s\lambda}}{4\pi}\int_{\omega=4\pi}\int_{\omega_i=4\pi} i'_\lambda(S,\omega_i)\,\Phi(\lambda,\omega,\omega_i)\,d\omega_i \tag{14-35}$$

From Sec. 14-3.3, $\int_{\omega=4\pi} i'_\lambda(S,\omega)\alpha\,d\omega = q_{r\lambda,x}(S)$, and similarly in the y and z directions. Let

$$\bar{\Phi}(\lambda,\omega_i) = \frac{1}{4\pi}\int_{\omega=4\pi}\Phi(\lambda,\omega,\omega_i)\,d\omega$$

The $\bar{\Phi}(\lambda,\omega_i)$ is a measure of how much scattering occurs for radiation incident from the ω_i direction. If $\sigma_{s\lambda}(\omega_i)$ is the scattering coefficient for this direction of incidence, then $\bar{\Phi}(\lambda,\omega_i) = \sigma_{s\lambda}(\omega_i)/\sigma_{s\lambda}$, where $\sigma_{s\lambda} = (1/4\pi) \int_{\omega_i=4\pi} \sigma_{s\lambda}(\omega_i) \, d\omega_i$. Note that $i'_{\lambda b}$ is independent of angular direction ω. Then (14-35) becomes

$$\frac{\partial q_{r\lambda,x}}{\partial x} + \frac{\partial q_{r\lambda,y}}{\partial y} + \frac{\partial q_{r\lambda,z}}{\partial z} = \nabla \cdot \mathbf{q}_{r\lambda} = -(a_\lambda + \sigma_{s\lambda}) 4\pi \bar{i}_\lambda(S) + a_\lambda 4\pi i'_{\lambda b}(S)$$

$$+ \sigma_{s\lambda} \int_{\omega_i=4\pi} i'_\lambda(S,\omega_i) \bar{\Phi}(\lambda,\omega_i) \, d\omega_i \tag{14-36}$$

To obtain the local divergence of the total radiative flux, (14-36) is integrated over all λ to obtain, at any location S,

$$\nabla \cdot \mathbf{q}_r = 4 \int_0^\infty \left\{ a_\lambda(\lambda) e_{\lambda b}(\lambda) - \pi [a_\lambda(\lambda) + \sigma_{s\lambda}(\lambda)] \bar{i}_\lambda(\lambda) \right.$$

$$\left. + \frac{\sigma_{s\lambda}(\lambda)}{4} \int_{\omega_i=4\pi} i'_\lambda(\lambda,\omega_i) \bar{\Phi}(\lambda,\omega_i) \, d\omega_i \right\} d\lambda \tag{14-37}$$

This is the generalized form of (14-28) to include anisotropic scattering.

For isotropic scattering $\Phi(\lambda,\omega,\omega_i) = 1$ and hence $\bar{\Phi}(\lambda,\omega_i) = 1$. Then (14-37) reduces to

$$\nabla \cdot \mathbf{q}_r = 4 \int_0^\infty \left\{ a_\lambda(\lambda) e_{\lambda b}(\lambda) - \pi [a_\lambda(\lambda) + \sigma_{s\lambda}(\lambda)] \bar{i}_\lambda(\lambda) \right.$$

$$\left. + \frac{\sigma_{s\lambda}(\lambda)}{4} \int_{\omega_i=4\pi} i'_\lambda(\lambda,\omega_i) \, d\omega_i \right\} d\lambda$$

Since $\int_{\omega_i=4\pi} i'_\lambda(\lambda,\omega_i) \, d\omega_i = 4\pi \bar{i}_\lambda(\lambda)$, the two terms involving scattering cancel and the result is

$$\nabla \cdot \mathbf{q}_r = 4 \int_0^\infty a_\lambda(\lambda) [e_{\lambda b}(\lambda) - \pi \bar{i}_\lambda(\lambda)] \, d\lambda \tag{14-38}$$

which is the same as (14-28). Although there is no scattering coefficient in (14-38), the $\bar{i}_\lambda(\lambda)$ is a function of the scattering as it is obtained by integrating (14-13) over all ω and λ.

If scattering is anisotropic, it is often the case that Φ does not depend on ω_i. This would be the situation where scattering particles are randomly oriented in a medium. In this instance the $\bar{\Phi}(\lambda) = (1/4\pi) \int_{\omega=4\pi} \Phi(\lambda,\omega) \, d\omega$, and by use of (13-48) the $\bar{\Phi}(\lambda) = 1$. For this often-encountered situation of anisotropic scattering, (14-37) also reduces to (14-38).

14-5 EQUATIONS OF TRANSFER AND FLUX FOR PLANE LAYER

To evaluate the influence of some of the many variables in gas radiation problems, it is convenient to consider a simple geometry. A plane layer is often used, and there is

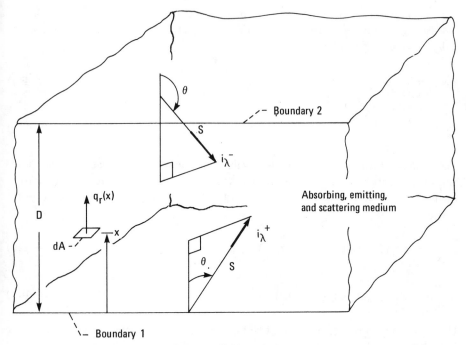

Figure 14-6 Plane layer between infinite parallel boundaries.

a considerable literature for this geometry in both engineering and astrophysical publications. The astrophysical interest [5-7] stems from the fact that the atmosphere of the earth and the outer radiating layers of the sun can be approximated as a plane layer. A plane layer of thickness D between two boundaries is shown in Fig. 14-6. The temperature and properties of the medium vary *only along the x direction.*

14-5.1 Equation of Transfer

The arbitrary paths S in Fig. 14-6 are at angles θ from the positive x direction. It will be convenient to adopt a new notation here. The prime denoting a directional quantity will be replaced by a $+$ or $-$, corresponding respectively to directions with positive or negative $\cos\theta$, that is, i^+ corresponds to $0 \leqslant \theta \leqslant 90°$, and i^- to $90° \leqslant \theta \leqslant 180°$.

The optical depth $\kappa_\lambda(x)$ is now defined *along the x coordinate* as

$$\kappa_\lambda(x) = \int_0^x K_\lambda(x^*)\, dx^* \tag{14-39}$$

For the $+$ directions the relation between optical positions along the S and x directions is given by

$$\kappa_\lambda(S) = \int_0^S K_\lambda(S^*)\, dS^* = \int_0^{x/\cos\theta} K_\lambda\left(\frac{x^*}{\cos\theta}\right) d\left(\frac{x^*}{\cos\theta}\right) = \frac{1}{\cos\theta}\int_0^x K_\lambda(x^*)\, dx^* = \frac{\kappa_\lambda(x)}{\cos\theta} \tag{14-40}$$

For the minus directions $dS = -dx/\cos(\pi - \theta) = dx/\cos\theta$, so the derivation in (14-40) again applies. Then, with $d\kappa_\lambda(S) = d\kappa_\lambda(x)/\cos\theta$, the equation of transfer (14-10) becomes, for i_λ^+ and i_λ^-,

$$\cos\theta \frac{\partial i_\lambda^+}{\partial \kappa_\lambda(x)} + i_\lambda^+(\kappa_\lambda(x),\theta) = I_\lambda'(\kappa_\lambda(x),\theta) \tag{14-41a}$$

$$\cos\theta \frac{\partial i_\lambda^-}{\partial \kappa_\lambda(x)} + i_\lambda^-(\kappa_\lambda(x),\theta) = I_\lambda'(\kappa_\lambda(x),\theta) \tag{14-41b}$$

The I_λ' depends on θ because of the angular dependency of the phase function Φ for anisotropic scattering. A partial derivative is used to emphasize that i_λ^+ and i_λ^- depend on both $\kappa_\lambda(x)$ and θ. A convenient substitution is to let $\mu = \cos\theta$; then (14-41) become

$$\mu \frac{\partial i_\lambda^+}{\partial \kappa_\lambda(x)} + i_\lambda^+(\kappa_\lambda(x),\mu) = I_\lambda'(\kappa_\lambda(x),\mu) \tag{14-42a}$$

$$\mu \frac{\partial i_\lambda^-}{\partial \kappa_\lambda(x)} + i_\lambda^-(\kappa_\lambda(x),\mu) = I_\lambda'(\kappa_\lambda(x),\mu) \tag{14-42b}$$

Using an integrating factor as in (14-11), Eqs. (14-42) are integrated subject to the boundary conditions

$$i_\lambda^+(\kappa_\lambda,\mu) = i_\lambda^+(0,\mu) \qquad \text{at } \kappa_\lambda = 0 \tag{14-43a}$$

$$i_\lambda^-(\kappa_\lambda,\mu) = i_\lambda^-(\kappa_{D\lambda},\mu) \qquad \text{at } \kappa_\lambda = \kappa_{D\lambda} \tag{14-43b}$$

where $\kappa_{D\lambda} = \int_0^D K_\lambda \, dx = \int_0^D (a_\lambda + \sigma_{s\lambda}) \, dx$. This results in the integrated forms

$$i_\lambda^+(\kappa_\lambda,\mu) = i_\lambda^+(0,\mu) \exp\frac{-\kappa_\lambda}{\mu} + \int_0^{\kappa_\lambda} I_\lambda'(\kappa_\lambda^*,\mu) \exp\frac{-(\kappa_\lambda - \kappa_\lambda^*)}{\mu} \frac{d\kappa_\lambda^*}{\mu} \tag{14-44a}$$

$$i_\lambda^-(\kappa_\lambda,\mu) = i_\lambda^-(\kappa_{D\lambda},\mu) \exp\frac{\kappa_{D\lambda} - \kappa_\lambda}{\mu} - \int_{\kappa_\lambda}^{\kappa_{D\lambda}} I_\lambda'(\kappa_\lambda^*,\mu) \exp\frac{\kappa_\lambda^* - \kappa_\lambda}{\mu} \frac{d\kappa_\lambda^*}{\mu} \tag{14-44b}$$

Note that in (14-44b) the range of θ is between $\pi/2$ and π so that $\mu = \cos\theta$ is a negative number.

14-5.2 Radiative Flux

The total radiative energy flux in the positive x direction is equal to the integral over all wavelengths of the spectral flux:

$$q_r(x) = \int_0^\lambda dq_{r\lambda}(x) = \int_0^\lambda dq_{r\lambda}(\kappa_\lambda) = \int_0^\lambda \frac{dq_{r\lambda}(\kappa_\lambda)}{d\lambda} \, d\lambda \tag{14-45}$$

Note that for a fixed x the κ_λ will vary with λ since K_λ is a function of λ.

The spectral flux in the positive x direction crossing the plane at x in Fig. 14-6 is found in two parts, one from $i_\lambda{}^+$ and one from $i_\lambda{}^-$. Since intensity represents energy per unit solid angle crossing an area normal to the direction of i', the projection of the area dA must be considered normal to either $i_\lambda{}^+$ or $i_\lambda{}^-$. The spectral energy flux in the positive x direction from $i_\lambda{}^+$ is then (using $d\omega = 2\pi \sin \theta \, d\theta$)

$$\frac{dq_{r\lambda}{}^+(\kappa_\lambda)}{d\lambda} = \int_{\theta=0}^{\pi/2} i_\lambda{}^+(\kappa_\lambda,\theta) \cos \theta \, 2\pi \sin \theta \, d\theta \qquad (14\text{-}46a)$$

Similarly the spectral flux in the negative x direction arising from $i_\lambda{}^-$ is

$$\frac{dq_{r\lambda}{}^-(\kappa_\lambda)}{d\lambda} = \int_{\pi-\theta=0}^{\pi-\theta=\pi/2} i_\lambda{}^-(\kappa_\lambda,\theta) \cos (\pi - \theta) \, 2\pi \sin (\pi - \theta) \, d(\pi - \theta)$$

$$= -2\pi \int_{\theta=\pi/2}^{\pi} i_\lambda{}^-(\kappa_\lambda,\theta) \cos \theta \sin \theta \, d\theta \qquad (14\text{-}46b)$$

The net flux in the positive x direction is

$$dq_{r\lambda}(\kappa_\lambda) = dq_{r\lambda}^+(\kappa_\lambda) - dq_{r\lambda}^-(\kappa_\lambda) \qquad (14\text{-}47)$$

Substituting Eqs. (14-46) into (14-47) gives

$$\frac{dq_{r\lambda}(\kappa_\lambda)}{d\lambda} = 2\pi \left[\int_{\theta=0}^{\pi/2} i_\lambda{}^+(\kappa_\lambda,\theta) \cos \theta \sin \theta \, d\theta + \int_{\theta=\pi/2}^{\pi} i_\lambda{}^-(\kappa_\lambda,\theta) \cos \theta \sin \theta \, d\theta \right]$$

$$(14\text{-}48)$$

Using the substitution $\mu = \cos \theta$ yields

$$\frac{dq_{r\lambda}(\kappa_\lambda)}{d\lambda} = 2\pi \left[\int_0^1 i_\lambda{}^+(\kappa_\lambda,\mu)\mu \, d\mu + \int_{-1}^0 i_\lambda{}^-(\kappa_\lambda,\mu)\mu \, d\mu \right]$$

$$= 2\pi \int_0^1 [i_\lambda{}^+(\kappa_\lambda,\mu) - i_\lambda{}^-(\kappa_\lambda,-\mu)]\mu \, d\mu \qquad (14\text{-}49)$$

The intensities are substituted from (14-44) to yield

$$\frac{dq_{r\lambda}(\kappa_\lambda)}{d\lambda} = 2\pi \int_0^1 i_\lambda{}^+(0,\mu) \exp\left(\frac{-\kappa_\lambda}{\mu}\right)\mu \, d\mu - 2\pi \int_0^1 i_\lambda{}^-(\kappa_{D\lambda}, -\mu) \exp\left(\frac{\kappa_{D\lambda} - \kappa_\lambda}{-\mu}\right)\mu \, d\mu$$

$$+ 2\pi \int_0^1 \int_0^{\kappa_\lambda} I_\lambda'(\kappa_\lambda^*, \mu) \exp\frac{\kappa_\lambda - \kappa_\lambda^*}{-\mu} \, d\kappa_\lambda^* \, d\mu$$

$$- 2\pi \int_0^1 \int_{\kappa_\lambda}^{\kappa_{D\lambda}} I_\lambda'(\kappa_\lambda^*, -\mu) \exp\frac{\kappa_\lambda^* - \kappa_\lambda}{-\mu} \, d\kappa_\lambda^* \, d\mu \qquad (14\text{-}50)$$

14-5.3 Divergence of Flux

For use in the energy equation (14-18), the term $\nabla \cdot \mathbf{q}_r$ is needed. For the plane layer $\mathbf{q}_r$ depends only on x so that

$$\nabla \cdot \mathbf{q}_r = \frac{dq_r}{dx} = \int_{\lambda=0}^{\infty} \frac{d^2 q_{r\lambda}}{dx\, d\lambda}\, d\lambda$$

With $d\kappa_\lambda = K_\lambda(x)\, dx$ this becomes

$$\nabla \cdot \mathbf{q}_r(x) = \int_0^{\infty} K_\lambda(\kappa_\lambda) \frac{d^2 q_{r\lambda}}{d\kappa_\lambda\, d\lambda}\, d\lambda \qquad (14\text{-}51)$$

where throughout the integration over λ, the κ_λ corresponds to the specified x. Differentiating (14-50) with respect to κ_λ yields

$$\frac{d^2 q_{r\lambda}}{d\kappa_\lambda\, d\lambda} = -2\pi \int_0^1 i_\lambda^{+}(0,\mu) \exp\frac{-\kappa_\lambda}{\mu}\, d\mu - 2\pi \int_0^1 i_\lambda^{-}(\kappa_{D\lambda},-\mu) \exp\frac{\kappa_{D\lambda}-\kappa_\lambda}{-\mu}\, d\mu$$

$$-2\pi \int_0^1 \int_0^{\kappa_\lambda} I_\lambda'(\kappa_\lambda^{*},\mu) \exp\frac{\kappa_\lambda-\kappa_\lambda^{*}}{-\mu}\, \frac{d\kappa_\lambda^{*}}{\mu}\, d\mu + 2\pi \int_0^1 I_\lambda'(\kappa_\lambda,\mu)\, d\mu$$

$$-2\pi \int_0^1 \int_{\kappa_\lambda}^{\kappa_{D\lambda}} I_\lambda'(\kappa_\lambda^{*},-\mu) \exp\frac{\kappa_\lambda^{*}-\kappa_\lambda}{-\mu}\, \frac{d\kappa_\lambda^{*}}{\mu}\, d\mu + 2\pi \int_0^1 I_\lambda'(\kappa_\lambda,-\mu)\, d\mu$$

$$(14\text{-}52)$$

14-5.4 Equation for Source Function

The remaining equation that is needed is that for the source function $I_\lambda'(\kappa_\lambda,\mu)$ which was defined in (14-9). As in (14-48) the integral over $d\omega_i$ can be written as

$$I_\lambda'(\kappa_\lambda,\theta) = (1-\Omega_{0\lambda}) i_{\lambda b}'(\kappa_\lambda) + \frac{\Omega_{0\lambda}}{2}\left[\int_{\theta=0}^{\pi/2} i_\lambda^{+}(\kappa_\lambda,\theta_i)\, \Phi(\lambda,\theta,\theta_i)\, \sin\theta_i\, d\theta_i\right.$$

$$\left. + \int_{\theta=\pi/2}^{\pi} i_\lambda^{-}(\kappa_\lambda,\theta_i)\, \Phi(\lambda,\theta,\theta_i)\, \sin\theta_i\, d\theta_i\right]$$

where $d\omega_i = 2\pi \sin\theta_i\, d\theta_i$ has been used. With $\mu = \cos\theta$ this becomes

$$I_\lambda'(\kappa_\lambda,\mu) = (1-\Omega_{0\lambda}) i_{\lambda b}'(\kappa_\lambda) + \frac{\Omega_{0\lambda}}{2}\left[\int_0^1 i_\lambda^{+}(\kappa_\lambda,\mu_i)\, \Phi(\lambda,\mu,\mu_i)\, d\mu_i\right.$$

$$\left. + \int_0^1 i_\lambda^{-}(\kappa_\lambda,-\mu_i)\, \Phi(\lambda,\mu,-\mu_i)\, d\mu_i\right]$$

The i_λ^+ and i_λ^- from (14-44) are inserted to yield

$$I_\lambda'(\kappa_\lambda,\mu) = (1 - \Omega_{0\lambda})i_{\lambda b}'(\kappa_\lambda) + \frac{\Omega_{0\lambda}}{2}\left[\int_0^1 i_\lambda^+(0,\mu_i) \exp\left(\frac{-\kappa_\lambda}{\mu_i}\right)\Phi(\lambda,\mu,\mu_i)\,d\mu_i\right.$$

$$+ \int_0^1 i_\lambda^-(\kappa_{D\lambda},-\mu_i) \exp\left(\frac{\kappa_{D\lambda}-\kappa_\lambda}{-\mu_i}\right)\Phi(\lambda,\mu,-\mu_i)\,d\mu_i$$

$$+ \int_0^1\int_0^{\kappa_\lambda} I_\lambda'(\kappa_\lambda^*,\mu_i) \exp\frac{\kappa_\lambda-\kappa_\lambda^*}{-\mu_i}\frac{d\kappa_\lambda^*}{\mu_i}\Phi(\lambda,\mu,\mu_i)\,d\mu_i$$

$$\left.+ \int_0^1\int_{\kappa_\lambda}^{\kappa_{D\lambda}} I_\lambda'(\kappa_\lambda^*,-\mu_i) \exp\frac{\kappa_\lambda^*-\kappa_\lambda}{-\mu_i}\frac{d\kappa_\lambda^*}{\mu_i}\Phi(\lambda,\mu,-\mu_i)\,d\mu_i\right] \tag{14-53}$$

This is a complicated integral equation for the source function. The equation involves $i_{\lambda b}'$, which depends on temperature and hence must be obtained by solving the energy equation. The energy equation is obtained by substituting (14-52) into (14-51) and then inserting this into the energy equation (14-18). It is evident that for a general case involving anisotropic scattering, the simultaneous solution of the energy equation with the source-function equation to obtain T (or $i_{\lambda b}'$) and I_λ' will be very difficult. For this reason a number of simplified cases will be considered.

14-5.5 Anisotropic Scattering Independent of Incidence Angle

For this type of scattering Φ becomes independent of μ_i but is still a function of μ. Equation (14-52) remains the same but (14-9) and (14-53) reduce to

$$I_\lambda'(\kappa_\lambda,\mu) = (1 - \Omega_{0\lambda})i_{\lambda b}'(\kappa_\lambda) + \Omega_{0\lambda}\Phi(\lambda,\mu)\bar{i}_\lambda(\kappa_\lambda) \tag{14-54}$$

$$I_\lambda'(\kappa_\lambda,\mu) = (1 - \Omega_{0\lambda})i_{\lambda b}'(\kappa_\lambda) + \frac{\Omega_{0\lambda}\Phi(\lambda,\mu)}{2}\left[\int_0^1 i_\lambda^+(0,\mu_i) \exp\frac{-\kappa_\lambda}{\mu_i}\,d\mu_i\right.$$

$$+ \int_0^1 i_\lambda^-(\kappa_{D\lambda},-\mu_i) \exp\frac{\kappa_{D\lambda}-\kappa_\lambda}{-\mu_i}\,d\mu_i$$

$$+ \int_0^1\int_0^{\kappa_\lambda} I_\lambda'(\kappa_\lambda^*,\mu_i) \exp\frac{\kappa_\lambda-\kappa_\lambda^*}{-\mu_i}\frac{d\kappa_\lambda^*}{\mu_i}\,d\mu_i$$

$$\left.+ \int_0^1\int_{\kappa_\lambda}^{\kappa_{D\lambda}} I_\lambda'(\kappa_\lambda^*,-\mu_i) \exp\frac{\kappa_\lambda^*-\kappa_\lambda}{-\mu_i}\frac{d\kappa_\lambda^*}{\mu_i}\,d\mu_i\right] \tag{14-55}$$

Another form for (14-52) can now be obtained by noting that the integrals in (14-55) are the same as in (14-52) and by comparison with (14-54) are equal to $2\bar{i}_\lambda(\kappa_\lambda)$. Then (14-52) becomes

$$\frac{d^2 q_{r\lambda}}{d\kappa_\lambda\,d\lambda} = 2\pi\left[\int_0^1 I_\lambda'(\kappa_\lambda,\mu)\,d\mu + \int_0^1 I_\lambda'(\kappa_\lambda,-\mu)\,d\mu - 2\bar{i}_\lambda(\kappa_\lambda)\right]$$

The $I'_\lambda(\kappa_\lambda, \mu)$ is now substituted from (14-54) to yield

$$\frac{d^2 q_{r\lambda}}{d\kappa_\lambda \, d\lambda} = 2\pi \left\{ 2(1 - \Omega_{0\lambda}) i'_{\lambda b}(\kappa_\lambda) + \Omega_{0\lambda} \bar{I}_\lambda(\kappa_\lambda) \left[\int_0^1 \Phi(\lambda, \mu) \, d\mu \right. \right.$$

$$+ \left. \left. \int_0^1 \Phi(\lambda, -\mu) \, d\mu \right] - 2 \bar{I}_\lambda(\kappa_\lambda) \right\}$$

(14-56)

From the definition of $\bar{\Phi}$ the two integrals on the right are equal to $2\bar{\Phi}$, and from Eq. (13-48) $\bar{\Phi} = 1$. Equation (14-56) then simplifies to

$$\frac{d^2 q_{r,\lambda}}{d\kappa_\lambda \, d\lambda} = 4\pi (1 - \Omega_{0\lambda}) [i'_{\lambda b}(\kappa_\lambda) - \bar{I}_\lambda(\kappa_\lambda)]$$

Using the definition of $\Omega_{0\lambda}$ gives

$$\frac{d^2 q_{r\lambda}}{d\kappa_\lambda \, d\lambda} = 4\pi \frac{a_\lambda}{a_\lambda + \sigma_{s\lambda}} [i'_{\lambda b}(\kappa_\lambda) - \bar{I}_\lambda(\kappa_\lambda)]$$

Since $d\kappa_\lambda = (a_\lambda + \sigma_{s\lambda}) \, dx$, this can be written

$$\frac{d^2 q_{r\lambda}}{dx \, d\lambda} = 4\pi a_\lambda [i'_{\lambda b}(x) - \bar{I}_\lambda(x)]$$

Integrate over all λ to obtain

$$\frac{dq_r}{dx} = 4\pi \int_0^\infty a_\lambda(\lambda) [i'_{\lambda b}(\lambda, x) - \bar{I}_\lambda(\lambda, x)] \, d\lambda$$

(14-57)

This is the form that was obtained in (14-38).

14-5.6 Isotropic Scattering

For isotropic scattering the phase function Φ becomes equal to unity as given by (13-48). The source function in (14-9) reduces to

$$I'_\lambda(\kappa_\lambda) = (1 - \Omega_{0\lambda}) i'_{\lambda b}(\kappa_\lambda) + \frac{\Omega_{0\lambda}}{4\pi} \int_{\omega=4\pi} i'_\lambda(\kappa_\lambda, \omega) \, d\omega$$

(14-58)

where it is no longer necessary to use an i subscript on ω. The I'_λ is *independent of angle* in this instance.

The spectral radiative flux (14-50) reduces to

$$\frac{dq_{r\lambda}(\kappa_\lambda)}{d\lambda} = 2\pi \int_0^1 i_\lambda^{\,+}(0, \mu) \exp\left(\frac{-\kappa_\lambda}{\mu}\right) \mu \, d\mu - 2\pi \int_0^1 i_\lambda^{\,-}(\kappa_{D\lambda}, -\mu) \exp\left(\frac{\kappa_{D\lambda} - \kappa_\lambda}{-\mu}\right) \mu \, d\mu$$

$$+ 2\pi \int_0^{\kappa_\lambda} I'_\lambda(\kappa_\lambda^*) \int_0^1 \exp\frac{\kappa_\lambda - \kappa_\lambda^*}{-\mu} \, d\mu \, d\kappa_\lambda^*$$

$$- 2\pi \int_{\kappa_\lambda}^{\kappa_{D\lambda}} I'_\lambda(\kappa_\lambda^*) \int_0^1 \exp\frac{\kappa_\lambda^* - \kappa_\lambda}{-\mu} \, d\mu \, d\kappa_\lambda^*$$

The exponential integral function can now be introduced. This function is defined as

$$E_n(\xi) = \int_0^1 \mu^{n-2} \exp\frac{-\xi}{\mu} d\mu \tag{14-59}$$

The $dq_{r\lambda}(\kappa_\lambda)$ can then be written as

$$\frac{dq_{r\lambda}(\kappa_\lambda)}{d\lambda} = 2\pi \int_0^1 i_\lambda^+(0,\mu) \exp\left(\frac{-\kappa_\lambda}{\mu}\right) \mu \, d\mu - 2\pi \int_0^1 i_\lambda^-(\kappa_{D\lambda}, -\mu) \exp\left(\frac{\kappa_{D\lambda} - \kappa_\lambda}{-\mu}\right) \mu \, d\mu$$

$$+ 2\pi \int_0^{\kappa_\lambda} I_\lambda'(\kappa_\lambda^*) E_2(\kappa_\lambda - \kappa_\lambda^*) d\kappa_\lambda^* - 2\pi \int_{\kappa_\lambda}^{\kappa_{D\lambda}} I_\lambda'(\kappa_\lambda^*) E_2(\kappa_\lambda^* - \kappa_\lambda) d\kappa_\lambda^* \tag{14-60}$$

The exponential integral functions are discussed in detail by Kourganoff [5] and Chandrasekhar [6]. The important relations are given in Appendix F.

Similarly, by use of the exponential integral function, Eq. (14-52) reduces to

$$\frac{d^2 q_{r\lambda}}{d\kappa_\lambda \, d\lambda} = -2\pi \int_0^1 i_\lambda^+(0,\mu) \exp\frac{-\kappa_\lambda}{\mu} d\mu - 2\pi \int_0^1 i_\lambda^-(\kappa_{D\lambda}, -\mu) \exp\frac{\kappa_{D\lambda} - \kappa_\lambda}{-\mu} d\mu$$

$$- 2\pi \int_0^{\kappa_{D\lambda}} I_\lambda'(\kappa_\lambda^*) E_1(|\kappa_\lambda - \kappa_\lambda^*|) d\kappa_\lambda^* + 4\pi I_\lambda'(\kappa_\lambda) \tag{14-61}$$

for isotropic scattering. The source-function equation (14-53) reduces to

$$I_\lambda'(\kappa_\lambda) = (1 - \Omega_{0\lambda}) i_{\lambda b}'(\kappa_\lambda) + \frac{\Omega_{0\lambda}}{2}\left[\int_0^1 i_\lambda^+(0,\mu) \exp\frac{-\kappa_\lambda}{\mu} d\mu \right.$$

$$+ \int_0^1 i_\lambda^-(\kappa_{D\lambda}, -\mu) \exp\frac{\kappa_{D\lambda} - \kappa_\lambda}{-\mu} d\mu$$

$$\left. + \int_0^{\kappa_{D\lambda}} I_\lambda'(\kappa_\lambda^*) E_1(|\kappa_\lambda^* - \kappa_\lambda|) d\kappa_\lambda^* \right] \tag{14-62}$$

Equation (14-57) is also valid for this case.

14-5.7 Diffuse Boundaries

For diffuse boundaries the $i_\lambda^+(0,\mu)$ and $i_\lambda^-(\kappa_{D\lambda}, -\mu)$ do not depend on angle (that is, they are independent of μ) and can be expressed in terms of the outgoing diffuse fluxes:

$$i_\lambda^+(0) = \frac{1}{\pi}\frac{dq_{\lambda o,1}}{d\lambda} \qquad i_\lambda^-(\kappa_{D\lambda}) = \frac{1}{\pi}\frac{dq_{\lambda o,2}}{d\lambda} \tag{14-63}$$

where boundaries 1 and 2 correspond to $\kappa_\lambda = 0$ and $\kappa_{D\lambda}$. Then, in Eqs. (14-60)-(14-62), the $i_\lambda^+(0)$ and $i_\lambda^-(\kappa_{D\lambda})$ can be taken out of the integrals over μ and the integrals expressed in terms of exponential integral functions:

$$\int_0^1 i_\lambda^+(0,\mu) \exp\left(\frac{-\kappa_\lambda}{\mu}\right) \mu \, d\mu = \frac{1}{\pi} \frac{dq_{\lambda o,1}}{d\lambda} E_3(\kappa_\lambda) \tag{14-64a}$$

$$\int_0^1 i_\lambda^-(\kappa_{D\lambda},-\mu) \exp\left(\frac{\kappa_{D\lambda}-\kappa_\lambda}{-\mu}\right) \mu \, d\mu = \frac{1}{\pi} \frac{dq_{\lambda o,2}}{d\lambda} E_3(\kappa_{D\lambda}-\kappa_\lambda) \tag{14-64b}$$

$$\int_0^1 i_\lambda^+(0,\mu) \exp\frac{-\kappa_\lambda}{\mu} \, d\mu = \frac{1}{\pi} \frac{dq_{\lambda o,1}}{d\lambda} E_2(\kappa_\lambda) \tag{14-64c}$$

$$\int_0^1 i_\lambda^-(\kappa_{D\lambda},-\mu) \exp\frac{\kappa_{D\lambda}-\kappa_\lambda}{-\mu} \, d\mu = \frac{1}{\pi} \frac{dq_{\lambda o,2}}{d\lambda} E_2(\kappa_{D\lambda}-\kappa_\lambda) \tag{14-64d}$$

To obtain equations for $dq_{\lambda o,1}$ and $dq_{\lambda o,2}$ as related to the boundary emissivities, Eq. (14-60) is evaluated at surface 1, $\kappa_\lambda = 0$, to yield

$$\frac{dq_{r\lambda,1}}{d\lambda} = \frac{dq_{\lambda o,1}}{d\lambda} - \left[2 \frac{dq_{\lambda o,2}}{d\lambda} E_3(\kappa_{D\lambda}) + 2\pi \int_0^{\kappa_{D\lambda}} I_\lambda'(\kappa_\lambda^*) E_2(\kappa_\lambda^*) \, d\kappa_\lambda^* \right]$$

By comparison with (10-5),

$$dq_{\lambda,1} = dq_{\lambda o,1} - dq_{\lambda i,1}$$

it is evident that the term in the square brackets is $dq_{\lambda i,1}/d\lambda$. Then, using (10-7),

$$\frac{dq_{\lambda o,1}}{d\lambda} = \epsilon_{\lambda,1} e_{\lambda b,1} + (1 - \epsilon_{\lambda,1}) \frac{dq_{\lambda i,1}}{d\lambda}$$

gives the equation for $dq_{\lambda o,1}$:

$$\frac{dq_{\lambda o,1}}{d\lambda} = \epsilon_{\lambda,1} e_{\lambda b,1} + 2(1 - \epsilon_{\lambda,1}) \left[\frac{dq_{\lambda o,2}}{d\lambda} E_3(\kappa_{D\lambda}) + \pi \int_0^{\kappa_{D\lambda}} I_\lambda'(\kappa_\lambda^*) E_2(\kappa_\lambda^*) \, d\kappa_\lambda^* \right] \tag{14-65a}$$

Similarly at surface 2 [by use of (14-60) evaluated at $\kappa_{D\lambda}$],

$$\frac{dq_{\lambda o,2}}{d\lambda} = \epsilon_{\lambda,2} e_{\lambda b,2} + 2(1 - \epsilon_{\lambda,2}) \left[\frac{dq_{\lambda o,1}}{d\lambda} E_3(\kappa_{D\lambda}) + \pi \int_0^{\kappa_{D\lambda}} I_\lambda'(\kappa_\lambda^*) E_2(\kappa_{\lambda D}-\kappa_\lambda^*) \, d\kappa_\lambda^* \right] \tag{14-65b}$$

Thus the boundary $dq_{\lambda o}$'s depend on $I_\lambda'(\kappa_\lambda)$, and hence they enter into the simultaneous solution of the system of equations.

The preceding general relations will now be further simplified for the special case of a gray medium.

14-6 THE GRAY MEDIUM WITH ISOTROPIC SCATTERING

A medium having absorption and scattering coefficients that are independent of wavelength is called a *gray medium*. From the discussion of gas-property spectral variations such as in connection with Fig. 13-2, it is evident that gases are usually far from gray. However, there are some instances when gases may be considered gray over all or a portion of the spectrum. When particles of soot or other material are present or are injected into a gas to enhance its absorption or emission of radiation, the absorption coefficient of the gas-particle mixture may act as if the mixture were nearly gray; in the case of soot there is little scattering as discussed in Chap. 17. Examination of the radiative behavior of a gray medium provides an understanding of some of the features of a real medium without the additional complicating features that real media introduce. These reasons, along with the mathematical simplifications introduced, account for the gray medium receiving a great deal of attention in the literature.

The equations for local flux and the source function will now be written for a *gray medium* with *isotropic scattering*. The equations are expressed in terms of the total quantities:

$$i' = \int_0^\infty i'_\lambda(\lambda)\, d\lambda \qquad \frac{\sigma T^4}{\pi} = \int_0^\infty i'_{\lambda b}(\lambda)\, d\lambda \qquad I' = \int_0^\infty I'_\lambda(\lambda)\, d\lambda$$

In Eq. (14-58) for the source function, the $\Omega_{0\lambda}$ becomes Ω_0 and κ_λ becomes κ. Then integrating over all wavelengths gives

$$I'(\kappa) = (1 - \Omega_0)\frac{\sigma T^4}{\pi} + \frac{\Omega_0}{4\pi}\int_{\omega=4\pi} i'(\kappa,\omega)\, d\omega \tag{14-66}$$

In the flux equation (14-60), note that the boundary values $i_\lambda^+(0,\mu)$ and $i_\lambda^-(\kappa_D,-\mu)$ can still have a spectral dependency (the boundaries have not been assumed to be gray). Integrating over all λ gives

$$q_r(\kappa) = 2\pi\int_0^1 i^+(0,\mu)\exp\left(\frac{-\kappa}{\mu}\right)\mu\, d\mu - 2\pi\int_0^1 i^-(\kappa_D,-\mu)\exp\left(\frac{\kappa_D - \kappa}{-\mu}\right)\mu\, d\mu$$

$$+ 2\pi\int_0^\kappa I'(\kappa^*)E_2(\kappa - \kappa^*)\, d\kappa^* - 2\pi\int_\kappa^{\kappa_D} I'(\kappa^*)E_2(\kappa^* - \kappa)\, d\kappa^* \tag{14-67}$$

Equation (14-61) for the divergence of the radiative flux becomes

$$\frac{dq_r}{d\kappa} = -2\pi\int_0^1 i^+(0,\mu)\exp\frac{-\kappa}{\mu}\, d\mu - 2\pi\int_0^1 i^-(\kappa_D,-\mu)\exp\frac{\kappa_D - \kappa}{-\mu}\, d\mu$$

$$- 2\pi\int_0^{\kappa_D} I'(\kappa^*)E_1(|\kappa - \kappa^*|)\, d\kappa^* + 4\pi I'(\kappa) \tag{14-68}$$

The integral equation (14-62) for the source function is

$$I'(\kappa) = (1 - \Omega_0) \frac{\sigma T^4(\kappa)}{\pi} + \frac{\Omega_0}{2} \left[\int_0^1 i^+(0,\mu) \exp \frac{-\kappa}{\mu} \, d\mu \right.$$

$$\left. + \int_0^1 i^-(\kappa_D, -\mu) \exp \frac{\kappa_D - \kappa}{-\mu} \, d\mu + \int_0^{\kappa_D} I'(\kappa^*) E_1(|\kappa^* - \kappa|) \, d\kappa^* \right] \tag{14-69}$$

EXAMPLE 14-2 For a gray isotropically scattering medium, derive equation (14-68) for $dq_r/d\kappa$ by use of (14-38).

For a gray one-dimensional layer, (14-38) becomes

$$\frac{1}{a} \nabla \cdot \mathbf{q}_r = \frac{a + \sigma_s}{a} \frac{dq_r}{(a + \sigma_s) \, dx} = \frac{1}{(1 - \Omega_0)} \frac{dq_r}{d\kappa} = 4 \int_0^\infty (e_{\lambda b} - \pi \bar{i}_\lambda) \, d\lambda = 4\pi (i_b' - \bar{i}) \tag{14-70}$$

From (14-24) and in a fashion similar to (14-49),

$$4\pi \bar{i} = \int_0^{4\pi} i' \, d\omega = \int_{\theta=0}^\pi i' 2\pi \sin \theta \, d\theta = 2\pi \left[\int_0^1 i^+(\mu) \, d\mu + \int_0^1 i^-(-\mu) \, d\mu \right]$$

For a gray isotropically scattering medium, (14-44) gives i^+ and i^- as

$$i^+(\mu) = i^+(0,\mu) \exp \frac{-\kappa}{\mu} + \int_0^\kappa I'(\kappa^*) \exp \frac{-(\kappa - \kappa^*)}{\mu} \frac{d\kappa^*}{\mu}$$

$$i^-(\mu) = i^-(\kappa_D,\mu) \exp \frac{\kappa_D - \kappa}{\mu} - \int_\kappa^{\kappa_D} I'(\kappa^*) \exp \frac{\kappa^* - \kappa}{\mu} \frac{d\kappa^*}{\mu}$$

Then

$$4\pi \bar{i} = 2\pi \left\{ \int_0^1 i^+(0,\mu) \exp \frac{-\kappa}{\mu} \, d\mu + \int_0^1 i^-(\kappa_D, -\mu) \exp \frac{\kappa_D - \kappa}{-\mu} \, d\mu \right.$$

$$+ \int_0^1 \int_0^\kappa I'(\kappa^*) \exp \left[\frac{-(\kappa - \kappa^*)}{\mu} \right] \frac{1}{\mu} \, d\mu \, d\kappa^*$$

$$\left. + \int_0^1 \int_\kappa^{\kappa_D} I'(\kappa^*) \exp \left(\frac{\kappa^* - \kappa}{-\mu} \right) \frac{1}{\mu} \, d\mu \, d\kappa^* \right\} \tag{14-71}$$

From (14-9) the $I'(\kappa)$ can be written as

$$I'(\kappa) = (1 - \Omega_0) i_b' + \Omega_0 \bar{i} = (1 - \Omega_0)(i_b' - \bar{i}) + \bar{i}$$

Combine this with (14-70) to eliminate i_b' and obtain

$$\frac{dq_r}{d\kappa} = 4\pi [I'(\kappa) - \bar{i}(\kappa)] \tag{14-72}$$

Equation (14-71) is inserted into (14-72), and the result is (14-68).

14-7 THE GRAY MEDIUM IN RADIATIVE EQUILIBRIUM

Heat transfer by radiation alone will now be considered; the effects of conduction and convection will be treated in Chap. 18. When energy transfer is *only by radiation*, the medium is said to be in *radiative equilibrium*.

For no convection, conduction, or internal heat sources, the energy equation (14-18) for steady state reduces to the following for a plane layer:

$$\nabla \cdot q_r = \frac{dq_r(x)}{dx} = K(x)\frac{dq_r(x)}{K(x)\,dx} = K(\kappa)\frac{dq_r(\kappa)}{d\kappa} = 0$$

where the extinction coefficient can be a function of x as a consequence of a and/or σ_s, depending on the local temperature and/or composition. Hence for radiative equilibrium with no internal heat sources,

$$\frac{dq_r}{d\kappa} = 0 \qquad (14\text{-}73)$$

The q_r is the total heat flux being transferred, since radiation is the only transfer mode present.

14-7.1 Absorbing Medium in Radiative Equilibrium with or without Scattering

In this instance Eqs. (14-67)–(14-69) apply with $dq_r/d\kappa = 0$ in (14-68). The $\kappa = \int_0^x (a + \sigma_s)\,dx$, where σ_s is zero for the case of absorption only. Equation (14-68) is then multiplied by $\Omega_0/4\pi$, and the result is substituted into (14-69) to give $I'(\kappa) = \Omega_0 I'(\kappa) + [(1 - \Omega_0)/\pi]\,\sigma T^4(\kappa)$. This reduces to

$$I'(\kappa) = \frac{\sigma T^4(\kappa)}{\pi} \qquad (14\text{-}74)$$

so that for radiative equilibrium the source function is equal to the local blackbody intensity. With this relation used to eliminate $I'(\kappa)$, and with $dq_r/d\kappa = 0$, the same equation for $T^4(\kappa)$ results from either (14-68) or (14-69):

$$4\sigma T^4(\kappa) = 2\pi \int_0^1 i^+(0,\mu) \exp\frac{-\kappa}{\mu}\,d\mu + 2\pi \int_0^1 i^-(\kappa_D,-\mu) \exp\frac{\kappa_D - \kappa}{-\mu}\,d\mu$$

$$+ 2\int_0^{\kappa_D} \sigma T^4(\kappa^*) E_1(|\kappa - \kappa^*|)\,d\kappa^* \qquad (14\text{-}75)$$

Equation (14-67), with $I' = \sigma T^4/\pi$, becomes

$$q_r = 2\pi \int_0^1 i^+(0,\mu) \exp\left(\frac{-\kappa}{\mu}\right)\mu\,d\mu - 2\pi \int_0^1 i^-(\kappa_D,-\mu) \exp\left(\frac{\kappa_D - \kappa}{-\mu}\right)\mu\,d\mu$$

$$+ 2\int_0^{\kappa} \sigma T^4(\kappa^*) E_2(\kappa - \kappa^*)\,d\kappa^* - 2\int_\kappa^{\kappa_D} \sigma T^4(\kappa^*) E_2(\kappa^* - \kappa)\,d\kappa^* \qquad (14\text{-}76)$$

Thus for a medium in which the absorption coefficient is nonzero, a single integral equation (14-75) governs the temperature distribution. After the temperature distribution has been obtained, it is inserted into (14-76) to obtain q_r. Since $dq_r/d\kappa = 0$, q_r is a constant and does not depend on κ. Hence (14-76) can be evaluated at any convenient κ, such as $\kappa = 0$. Now the case is considered in which there is scattering, but the absorption coefficient is zero.

14-7.2 Scattering Medium in Radiative Equilibrium with No Absorption

Consider a medium with zero absorption coefficient so that energy transfer is only by scattering, which is assumed isotropic. For pure scattering (no absorption), $\Omega_0 \to 1$ and the source function (14-66) becomes

$$I'(\kappa) = \frac{1}{4\pi} \int_{\omega = 4\pi} i'(\kappa, \omega) \, d\omega = \bar{i}(\kappa) \tag{14-77}$$

where $\bar{i}(\kappa)$ is the average scattered intensity at κ, and $\kappa = \int_0^x \sigma_s \, dx$. Equation (14-69) reduces to an integral equation for $\bar{i}(\kappa)$:

$$\bar{i}(\kappa) = \frac{1}{2} \left[\int_0^1 i^+(0, \mu) \exp \frac{-\kappa}{\mu} \, d\mu + \int_0^1 i^-(\kappa_D, -\mu) \exp \frac{\kappa_D - \kappa}{-\mu} \, d\mu \right.$$
$$\left. + \int_0^{\kappa_D} \bar{i}(\kappa^*) E_1(|\kappa - \kappa^*|) \, d\kappa^* \right] \tag{14-78}$$

The radiative flux equation (14-67) then reduces to

$$q_r(\kappa) = 2\pi \left[\int_0^1 i^+(0, \mu) \exp\left(\frac{-\kappa}{\mu}\right) \mu \, d\mu - \int_0^1 i^-(\kappa_D, -\mu) \exp\left(\frac{\kappa_D - \kappa}{-\mu}\right) \mu \, d\mu \right.$$
$$\left. + \int_0^\kappa \bar{i}(\kappa^*) E_2(\kappa - \kappa^*) \, d\kappa^* - \int_\kappa^{\kappa_D} \bar{i}(\kappa^*) E_2(\kappa^* - \kappa) \, d\kappa^* \right] \tag{14-79}$$

and (14-68) for the divergence of the radiative flux becomes

$$\frac{dq_r}{d\kappa} = -2\pi \left[\int_0^1 i^+(0, \mu) \exp \frac{-\kappa}{\mu} \, d\mu + \int_0^1 i^-(\kappa_D, -\mu) \exp \frac{\kappa_D - \kappa}{-\mu} \, d\mu \right.$$
$$\left. + \int_0^{\kappa_D} \bar{i}(\kappa^*) E_1(|\kappa - \kappa^*|) \, d\kappa^* \right] + 4\pi \bar{i}(\kappa) \tag{14-80}$$

For radiative equilibrium $dq_r/d\kappa = 0$ and (14-80) becomes the same as (14-78). The equation is also the same as (14-75), where $\bar{i}$ takes the place of $\sigma T^4/\pi$.

14-7.3 Gray Medium in Radiative Equilibrium ($dq_r/d\kappa = 0$) between Diffuse Gray Boundaries

For diffuse gray boundaries $i^+(0, \mu) = q_{o,1}/\pi$, $i^-(\kappa_D, -\mu) = q_{o,2}/\pi$ and (14-65) can be applied without the λ subscripts ($dq_{\lambda o}/d\lambda$ becomes simply q_o). For an absorbing medium with or without scattering, the relations (14-65) are used with (14-75) and (14-76). For a scattering medium without absorption, (14-78) and (14-79) are used. The result is the following equation for the net radiative flux in the x direction:

$$q_r = 2\left[q_{o,1}E_3(\kappa) - q_{o,2}E_3(\kappa_D - \kappa) + \int_0^\kappa G(\kappa^*)E_2(\kappa - \kappa^*)\,d\kappa^* \right.$$

$$\left. - \int_\kappa^{\kappa_D} G(\kappa^*)E_2(\kappa^* - \kappa)\,d\kappa^* \right] \tag{14-81}$$

where

$$G(\kappa) = \begin{cases} \sigma T^4(\kappa) \\ \\ \pi \bar{\imath}(\kappa) \end{cases} \qquad \kappa = \begin{cases} \int_0^x (a + \sigma_s)\,dx & \text{when } a > 0,\ \sigma_s \geq 0 \\ \\ \int_0^x \sigma_s\,dx & \text{when } a = 0,\ \sigma_s > 0 \end{cases}$$

Note that the heat flux is constant across the medium so that q_r can be evaluated at any convenient location such as $\kappa = 0$. The $G(\kappa)$ is found from the integral equation

$$G(\kappa) = \frac{1}{2}\left[q_{o,1}E_2(\kappa) + q_{o,2}E_2(\kappa_D - \kappa) + \int_0^{\kappa_D} G(\kappa^*)E_1(|\kappa - \kappa^*|)\,d\kappa^* \right] \tag{14-82}$$

The $q_{o,1}$ and $q_{o,2}$ are, from (14-65),

$$q_{o,1} = \epsilon_1 \sigma T_1^4 + 2(1 - \epsilon_1)\left[q_{o,2}E_3(\kappa_D) + \int_0^{\kappa_D} G(\kappa^*)E_2(\kappa^*)\,d\kappa^* \right] \tag{14-83a}$$

$$q_{o,2} = \epsilon_2 \sigma T_2^4 + 2(1 - \epsilon_2)\left[q_{o,1}E_3(\kappa_D) + \int_0^{\kappa_D} G(\kappa^*)E_2(\kappa_D - \kappa^*)\,d\kappa^* \right] \tag{14-83b}$$

By use of (14-81) evaluated at $\kappa = 0$ for (14-83a), and at $\kappa = \kappa_D$ for (14-83b), Eqs. (14-83) become

$$q_{o,1} = \epsilon_1 \sigma T_1^4 - (1 - \epsilon_1)(q_r - q_{o,1})$$

$$q_{o,2} = \epsilon_2 \sigma T_2^4 + (1 - \epsilon_2)(q_r + q_{o,2})$$

Solving for $q_{o,1}$ and $q_{o,2}$ gives

$$q_{o,1} = \sigma T_1^4 - \frac{1 - \epsilon_1}{\epsilon_1} q_r \tag{14-84a}$$

$$q_{o,2} = \sigma T_2^4 + \frac{1 - \epsilon_2}{\epsilon_2} q_r \tag{14-84b}$$

14-7.4 Solution for Gray Medium in Radiative Equilibrium between Black or Diffuse-Gray Walls at Specified Temperatures

Let the layer of gray medium with absorption coefficient $a(T)$ and scattering coefficient $\sigma_s(T)$ be between two infinite parallel plates, one at specified temperature T_1 and the other at T_2; the plates are a distance D apart (Fig. 14-7). It is desired to obtain the temperature distribution in the medium and the energy transfer q between the plates. Since the energy transfer is only by radiation, the total heat flow q is equal to q_r, and the r subscript can be omitted in what follows.

Medium between Black Walls. For black walls $\epsilon_1 = \epsilon_2 = 1$, and the $q_{o,1}$ and $q_{o,2}$ in (14-81) and (14-82) become σT_1^4 and σT_2^4. For energy transfer only by radiation in the geometry being considered here, $q(\kappa)$ is independent of κ as there are no energy sources or sinks in the medium. Evaluating q from (14-81) at the convenient location $\kappa = 0$ then gives the net heat flux from wall 1 to wall 2 as

$$q = \sigma T_1^4 - 2\sigma T_2^4 E_3(\kappa_D) - 2 \int_0^{\kappa_D} G(\kappa^*) E_2(\kappa^*) \, d\kappa^* \tag{14-85}$$

where from (14-82), $G(\kappa)$ is found from the integral equation

$$G(\kappa) = \frac{1}{2}\left[\sigma T_1^4 E_2(\kappa) + \sigma T_2^4 E_2(\kappa_D - \kappa) + \int_0^{\kappa_D} G(\kappa^*) E_1(|\kappa - \kappa^*|) \, d\kappa^*\right] \tag{14-86}$$

Equations (14-85) and (14-86) can be placed in useful dimensionless forms by

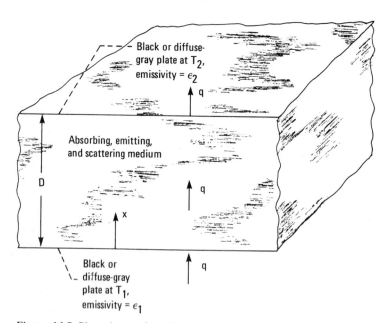

Figure 14-7 Plane layer of medium in radiative equilibrium between infinite parallel diffuse surfaces.

defining

$$\phi_b(\kappa) = \frac{G(\kappa)/\sigma - T_2^4}{T_1^4 - T_2^4} \qquad \psi_b = \frac{q}{\sigma(T_1^4 - T_2^4)} \tag{14-87}$$

where the b subscript emphasizes that this is for the black-wall case. This yields

$$\phi_b(\kappa) = \frac{1}{2}\left[E_2(\kappa) + \int_0^{\kappa_D} \phi_b(\kappa^*)E_1(|\kappa - \kappa^*|)\,d\kappa^*\right] \tag{14-88}$$

$$\psi_b = 1 - 2\int_0^{\kappa_D} \phi_b(\kappa^*)E_2(\kappa^*)\,d\kappa^* \tag{14-89}$$

The solution is obtained by solving (14-88) for $\phi_b(\kappa)$ by numerical or other means and then using $\phi_b(\kappa)$ in (14-89) to obtain ψ_b.

For the limiting case as both absorption and scattering in the medium become very small, $\kappa_D \to 0$ and $E_3(\kappa_D) \to 1/2$ so that (14-85) reduces to

$$q|_{\kappa_D \to 0} = \sigma(T_1^4 - T_2^4) \tag{14-90}$$

(or $\psi_b = 1$) which is the correct solution for black infinite parallel plates separated by a transparent medium. For this limit (14-86) yields

$$\left.\frac{G(\kappa)}{\sigma}\right|_{\kappa_D \to 0} = \frac{T_1^4 + T_2^4}{2} \tag{14-91}$$

(or $\phi_b = 1/2$) so that in a nearly transparent gray medium with $a > 0$ $[G(\kappa) = \sigma T^4(\kappa)]$ the medium temperature to the fourth power approaches the average of the fourth powers of the boundary temperatures. In a purely scattering medium, the $[G(\kappa)/\sigma]_{\kappa_D \to 0}$ in (14-91) is equal to $\pi \bar{i}(\kappa)/\sigma$ within the medium [Eq. (14-81)].

For convenience in the following discussion, consider the case in which $a > 0$ so that the results can be interpreted in terms of the medium temperature $[G(\kappa) = \sigma T^4(\kappa)]$. The same results will apply to $\pi \bar{i}(\kappa)$ for the case of scattering alone. Numerical results from (14-88) and (14-89) for the temperature distribution and heat flux in a gray gas with temperature-independent properties contained between infinite black parallel plates have been obtained by many researchers. Some of the integral-equation solution methods will be discussed in succeeding chapters. Heaslet and Warming [8] have presented results accurate to four significant figures for $[T^4(\kappa) - T_2^4]/(T_1^4 - T_2^4)$ and $q/[\sigma(T_1^4 - T_2^4)]$. These are shown in Fig. 14-8, and values for the dimensionless energy flux are given in Table 14-1.

The temperature distributions of Fig. 14-8a show that a discontinuity exists between the wall temperature and the gas temperature at the wall. This phenomenon is called the temperature "slip" or "jump." If the slip were not present, the curves would all go to 1 at $\kappa/\kappa_D = 0$, and 0 at $\kappa/\kappa_D = 1$. The slip disappears when heat conduction is included in the analysis. To determine the magnitude of the slip, the gas temperature is evaluated at $\kappa = 0$. This gives, using (14-86),

$$T^4(\kappa = 0) = \frac{1}{2}\left[T_1^4 + T_2^4 E_2(\kappa_D) + \int_0^{\kappa_D} T^4(\kappa^*)E_1(\kappa^*)\,d\kappa^*\right]$$

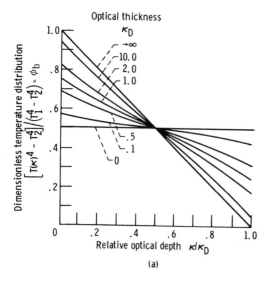

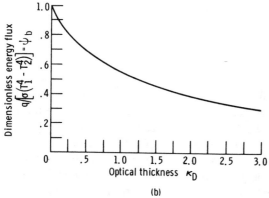

Figure 14-8 Temperature distribution and energy flux in gray gas contained between infinite black parallel plates [8]. (a) Temperature distribution; (b) energy flux.

Table 14-1 Dimensionless energy flux $\dfrac{q}{\sigma(T_1^4 - T_2^4)} = \psi_b$ [8][a]

Optical thickness κ_D	ψ_b	Optical thickness κ_D	ψ_b
0.1	0.9157	0.8	0.6046
0.2	0.8491	1.0	0.5532
0.3	0.7934	1.5	0.4572
0.4	0.7458	2.0	0.3900
0.5	0.7040	2.5	0.3401
0.6	0.6672	3.0	0.3016

[a] For $\kappa_D \gg 1$, $\psi_b = \dfrac{4/3}{1.42089 + \kappa_D}$

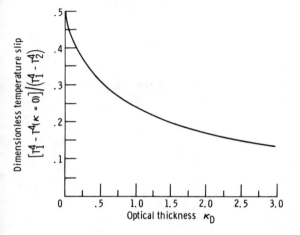

Figure 14-9 Discontinuity at wall between gray gas and black wall temperature [8].

which can be written as

$$\frac{T_1^4 - T^4(\kappa = 0)}{T_1^4 - T_2^4} = \frac{1}{2}\left[\frac{T_1^4}{T_1^4 - T_2^4} - \frac{T_2^4}{T_1^4 - T_2^4}E_2(\kappa_D) - \int_0^{\kappa_D}\frac{T^4(\kappa^*)}{T_1^4 - T_2^4}E_1(\kappa^*)\,d\kappa^*\right]$$

(14-92)

As κ_D approaches zero the integral vanishes and $E_2(\kappa_D) \to 1$ so that Eq. (14-92) reduces to

$$\left.\frac{T_1^4 - T^4(\kappa = 0)}{T_1^4 - T_2^4}\right|_{\kappa_D \to 0} = \frac{1}{2}$$

(14-93)

The magnitude of the slip for a gray gas with constant absorption coefficient is shown in Fig. 14-9 as a function of the optical thickness of the layer. [Note that from symmetry $T_1^4 - T^4(\kappa = 0) = T^4(\kappa = \kappa_D) - T_2^4$.]

Medium between Diffuse Gray Walls at Specified Temperatures. An extension can now be readily made to the situation in which the plates at T_1 and T_2 are gray rather than black. The integral equations have the same form as (14-85) and (14-86) except that the outgoing fluxes $q_{o,1}$ and $q_{o,2}$ replace σT_1^4 and σT_2^4. Hence, as in (14-87) for the case with gray walls let

$$\phi(\kappa) = \frac{G(\kappa) - q_{o,2}}{q_{o,1} - q_{o,2}} \qquad \psi = \frac{q}{q_{o,1} - q_{o,2}}$$

where

$$G(\kappa) = \begin{cases} \sigma T^4(\kappa) & \text{when } a > 0, \sigma_s \geqslant 0 \\ \pi\bar{\imath}(\kappa) & \text{when } a = 0, \sigma_s > 0 \end{cases}$$

The equations for $\phi(\kappa)$ and ψ are then the same as (14-88) and (14-89) so that $\phi = \phi_b$

and $\psi = \psi_b$, and for gray walls the $G(\kappa)$ and q are given by

$$G(\kappa) = \phi_b(\kappa)(q_{o,1} - q_{o,2}) + q_{o,2} \tag{14-94}$$

$$q = \psi_b(q_{o,1} - q_{o,2}) \tag{14-95}$$

Hence, assuming that ϕ_b and ψ_b have been obtained for the black case, it is only needed to find $q_{o,1}$ and $q_{o,2}$. From (14-84) with $q_r = q$,

$$q_{o,1} = \sigma T_1^4 - q \frac{1-\epsilon_1}{\epsilon_1} \tag{14-96a}$$

$$q_{o,2} = \sigma T_2^4 + q \frac{1-\epsilon_2}{\epsilon_2} \tag{14-96b}$$

Substitute these relations into (14-95) and solve for q to obtain

$$\frac{q}{\sigma(T_1^4 - T_2^4)} = \frac{\psi_b}{1 + \psi_b(1/\epsilon_1 + 1/\epsilon_2 - 2)} \tag{14-97}$$

Then substitute the $q_{o,1}$ and $q_{o,2}$ from (14-96) into (14-94) and eliminate the q by use of (14-97) to obtain

$$\frac{G(\kappa)/\sigma - T_2^4}{T_1^4 - T_2^4} = \frac{\phi_b(\kappa) + [(1-\epsilon_2)/\epsilon_2] \psi_b}{1 + \psi_b(1/\epsilon_1 + 1/\epsilon_2 - 2)} \tag{14-98}$$

These relations will also be obtained in Sec. 17-8.3 by use of the exchange-factor concept.

EXAMPLE 14-3 Let the medium between diffuse-gray walls have only scattering, $a = 0$, $\sigma_s > 0$, but in addition let it conduct heat with a constant thermal conductivity. Determine the heat transfer from plate 1 to plate 2 by combined heat conduction and pure scattering.

For pure scattering the temperature distribution within the layer does not enter into the radiative solution as given by (14-97) and (14-98). Hence the energy equation that determines the heat conduction is decoupled from the scattering process. The total heat flux being transferred is then found by adding to (14-97) the heat conduction as if scattering were not present. This gives

$$q = \frac{k(T_1 - T_2)}{D} + \frac{\sigma(T_1^4 - T_2^4) \psi_b}{1 + \psi_b(1/\epsilon_1 + 1/\epsilon_2 - 2)} \tag{14-99}$$

The ψ_b is given in Fig. 14-8b, where the abscissa would be interpreted as $\kappa_D = \int_0^D \sigma_s \, dx$ (also see Table 14-1).

EXAMPLE 14-4 Determine the heat transfer from plate 1 to plate 2 by heat conduction combined with absorption and scattering in the optically thin limit.

From Fig. 14-8b, as $\kappa_D \to 0$, $\psi_b \to 1$, so that (14-97) reduces to $q = \sigma(T_1^4 - T_2^4)/(1/\epsilon_1 + 1/\epsilon_2 - 1)$, which is the same as the result for no radiating medium between the plates. Equation (14-93) shows that in the optically thin limit, the radiative flux does not produce a temperature gradient in the medium and hence would not influence heat conduction. The ordinary heat conduction can thus be added to the radiation to give, in the optically thin limit,

$$q = \frac{k(T_1 - T_2)}{D} + \frac{\sigma(T_1^4 - T_2^4)}{1/\epsilon_1 + 1/\epsilon_2 - 1} \tag{14-100}$$

14-8 ENERGY RELATIONS BY USE OF PHOTON MODEL

The radiation field and transfer of radiation in a medium can also be expressed in terms of a photon model. This is sometimes helpful in providing a physical picture of the transport and is useful in Monte Carlo methods as will be considered in Chap. 20. Since photon energy is related to the frequency of the radiation, frequency will be used in this section rather than wavelength.

When the radiation is considered as a collection of photons, the conditions at any location in the medium are given by the photon distribution function f. Let

$$f(\nu, \mathbf{r}, S)\, d\nu\, dV\, d\omega \tag{14-101}$$

be the number of photons traveling in direction S in frequency interval $d\nu$ centered about ν, in volume dV at position $\mathbf{r}$, and within solid angle $d\omega$ about the direction S (see Fig. 14-10a). Each photon has energy $h\nu$. The energy per unit volume per unit frequency interval is then $h\nu f\, d\omega$ integrated over all solid angles. This is called the spectral radiant energy density:

$$U_\nu(\nu, \mathbf{r}) = h\nu \int_{\omega=0}^{4\pi} f(\nu, \mathbf{r}, S)\, d\omega \tag{14-102}$$

To obtain the intensity, the energy flux in the S direction is needed across the area dA in Fig. 14-10a, which is normal to the S direction. The photons have velocity c, and the particle density traveling in the normal direction across dA is $f\, d\nu\, d\omega$. The number of particles crossing dA in the S direction per unit time is then $cf\, d\nu\, d\omega\, dA$. The energy carried by these particles is $h\nu cf\, d\nu\, d\omega\, dA$. The spectral intensity is the energy in a single direction per unit time, unit frequency interval, and unit solid angle crossing a unit area normal to that direction. This gives the intensity at location $\mathbf{r}$ and in direction S as

$$i_\nu' = h\nu cf(\nu, \mathbf{r}, S) \tag{14-103}$$

The energy density and the intensity can then be related by using (14-103) to eliminate f from (14-102); that is,

$$U_\nu(\nu, \mathbf{r}) = \frac{1}{c} \int_{\omega=0}^{4\pi} i_\nu'\, d\omega = \frac{4\pi}{c} \bar{i}_\nu \tag{14-104}$$

Now consider the energy flux crossing an area within a medium. As shown in Fig. 14-10b, let dA be an arbitrary element whose unit normal vector is **n**. Energy

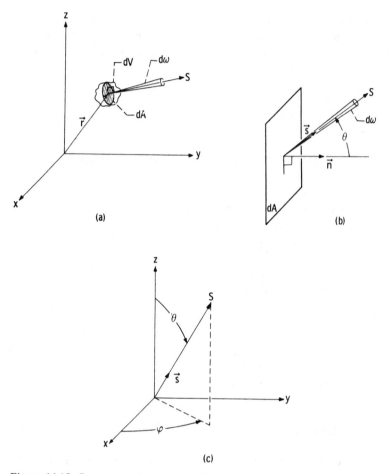

Figure 14-10 Geometries used in derivations of radiative energy quantities. (*a*) Quantities in intensity derivation; (*b*) quantities in flux derivation; (*c*) spherical coordinate system for radiative flux vector.

passes through dA from all directions; a typical direction is the S direction at angle θ to **n**. This energy is given by $h\nu c f \, d\nu \, d\omega \, dA \cos\theta$. To obtain the net energy flux crossing dA, integrate over all incident solid angles. The energy per unit of dA moving across dA in the direction of increasing **n** is then (note that $\cos\theta$ becomes negative for $\theta > \pi/2$ so that the sign of the portion of the energy flux traveling in the direction opposite to the positive **n** direction is automatically included)

$$dq_\nu = h\nu c \, d\nu \int_{\omega=0}^{4\pi} f \cos\theta \, d\omega = d\nu \int_{\omega=0}^{4\pi} i_\nu' \cos\theta \, d\omega \qquad (14\text{-}105)$$

The latter form of (14-105) was obtained by use of (14-103).

Let s be a unit vector in the direction S of the photons. Then $\cos\theta = \mathbf{s} \cdot \mathbf{n}$ and

(14-105) can be written as

$$dq_\nu = d\nu \int_{\omega=0}^{4\pi} i_\nu' \mathbf{s} \cdot \mathbf{n} \, d\omega \tag{14-106}$$

Thus dq_ν is the component in the **n** direction of a flux vector given by

$$d\mathbf{q}_\nu = d\nu \int_{\omega=0}^{4\pi} i_\nu' \mathbf{s} \, d\omega \tag{14-107}$$

that is, $dq_\nu = \mathbf{n} \cdot d\mathbf{q}_\nu$.

To further reveal the vector nature of $d\mathbf{q}_\nu$, consider a spherical coordinate system as shown in Fig. 14-10c. The unit vector **s** can then be written as

$$\mathbf{s} = \mathbf{i} \cos \varphi \sin \theta + \mathbf{j} \sin \varphi \sin \theta + \mathbf{k} \cos \theta \tag{14-108}$$

Substituting **s** and $d\omega = \sin \theta \, d\theta \, d\varphi$ into (14-107) gives the vector $d\mathbf{q}_\nu$ in terms of its three components:

$$
\begin{aligned}
d\mathbf{q}_\nu = d\nu \bigg[& \mathbf{i} \int_{\varphi=0}^{2\pi} \int_{\theta=0}^{\pi} i_\nu'(\theta,\varphi) \cos \varphi \sin^2 \theta \, d\theta \, d\varphi \\
& + \mathbf{j} \int_{\varphi=0}^{2\pi} \int_{\theta=0}^{\pi} i_\nu'(\theta,\varphi) \sin \varphi \sin^2 \theta \, d\theta \, d\varphi \\
& + \mathbf{k} \int_{\varphi=0}^{2\pi} \int_{\theta=0}^{\pi} i_\nu'(\theta,\varphi) \cos \theta \sin \theta \, d\theta \, d\varphi \bigg]
\end{aligned}
\tag{14-109}
$$

The divergence of the radiant heat-flux vector, $\nabla \cdot \mathbf{q}_r$, is used in the energy equation as discussed in Sec. 14-4.1.

REFERENCES

1. McCormick, N. J., and I. Kuščer: Singular Eigenfunction Expansions in Neutron Transport Theory, in "Advances in Nuclear Science and Technology," (E. J. Henley and J. Lewins, eds.), vol. 7, pp. 181–282, Academic Press, New York, 1973.
2. Gritton, E. C., and A. Leonard: Exact Solutions to the Radiation Heat Transport Equation in Gaseous Media Using Singular Integral Equation Theory, J. Quant. Spectrosc. Radiat. Transfer, vol. 10, pp. 1095–1118, 1970.
3. Yener, Y., M. N. Ozisik, and C. E. Siewert: Non-Gray Radiative Transfer in Plane-Parallel Media with Reflecting Boundaries, J. Quant. Spectrosc. Radiat. Transfer, vol. 16, pp. 165–175, 1976.
4. Williams, M. M. R.: The Wiener-Hopf Technique: An Alternative to the Singular Eigenfunction Method, in "Advances in Nuclear Science and Technology," (E. J. Henley and J. Lewins, eds.), vol. 7, pp. 283–327, Academic Press, New York, 1973.
5. Kourganoff, Vladimir: "Basic Methods in Transfer Problems," Dover Publications, Inc., New York, 1963.
6. Chandrasekhar, Subrahmanyan: "Radiative Transfer," Dover Publications, Inc., New York, 1960.

7. Goody, R. M.: "Atmospheric Radiation. Theoretical Basis," vol. I, Clarendon Press, Oxford, 1964.

8. Heaslet, Max A., and Robert F. Warming: Radiative Transport and Wall Temperature Slip in an Absorbing Planar Medium, *Int. J. Heat Mass Transfer,* vol. 8, pp. 979–994, 1965.

PROBLEMS

1 A slab of nonscattering solid material has a gray absorption coefficient of $a = 0.4$ cm^{-1} and a refractive index $n \approx 1$. It is 2 cm thick and has an approximately linear temperature distribution within it as established by heat conduction. What is the emitted intensity normal to the slab? What average slab temperature would give the same emitted normal intensity?

   Answer: 0.0195 W/cm^2, 370 K.

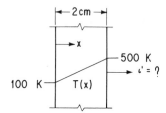

2 Estimate the Planck mean absorption coefficient for air at 12,000 K and 1 atm by use of Fig. 13-8.

3 A nonscattering stagnant gray medium with absorption coefficient $a = 0.2$ cm^{-1} is contained between black parallel plates 10 cm apart as shown (assume constant density and $n = 1$). Plot the temperature distribution $T(z)$ (neglect heat conduction). What is the net heat flux being transferred by radiation from the lower to the upper plate? If the plates are gray with $\epsilon_1 = 0.8$ and $\epsilon_2 = 0.4$, what is the heat flux being transferred?

   Answer: 0.15 W/cm^2, 0.089 W/cm^2.

$\epsilon_2 = 1$, $T_2 = 500$ K

10 cm

z

$\epsilon_1 = 1$, $T_1 = 600$ K

4 Consider the plane layer of absorbing and nonscattering gray gas between black parallel plates at temperatures T_1 and T_2 as in Sec. 14-7.4. A chemical reaction is producing a uniform heat generation per unit volume in the gas. Derive the equations of the temperature distribution in the gas and the local heat flux in the x direction. What is the equation to obtain the heat flux supplied to each plate? What does the temperature distribution become for the limiting case where $\kappa_D \to 0$?

5 An isothermal enclosure is filled with a nonscattering gas. Show that $\nabla \cdot \mathbf{q}_r$ must be zero for this condition.

6 A semi-infinite, scattering, absorbing-emitting gas at uniform temperature T_g is in contact with a diffuse-gray wall at T_w with emissivity ϵ_w. The medium is gray with constant scattering and absorption coefficients. The medium is not moving, and heat conduction is neglected. Show that the heat flux transferred to the wall is $\epsilon_w \sigma (T_g^4 - T_w^4)$.

7 A scattering, nonabsorbing, and nonconducting medium is contained between parallel plates 8 cm apart. The scattering is assumed isotropic and independent of wavelength, and the scattering coefficient is $\sigma_s = 0.2$ cm^{-1}. The plate temperatures are $T_1 = 700$ K and $T_2 = 600$ K. Compute the net heat flux transferred from plate 1 to plate 2 if the plates are black, or if the plates are gray with $\epsilon_1 = 0.7$ and $\epsilon_2 = 0.3$. How do these values compare with the transfers for the plates in vacuum?

   Answer: 0.280 W/cm^2, 0.126 W/cm^2; vacuum 0.630, 0.168.

8 In Prob. 7, for the situation of a scattering medium between gray plates, it is desired to double the amount of heat being transferred by having the scattering particles suspended in a heat conducting nonabsorbing medium. What thermal conductivity of the medium would be required to accomplish this?

   Answer: 0.01 W/(cm·K).

9 A gray absorbing and scattering medium is contained between gray parallel plates 5 cm apart and with $T_1 = 800$ K, $\epsilon_1 = 0.2$, $T_2 = 600$ K, $\epsilon_2 = 0.6$. The scattering is isotropic and independent of wavelength, and heat conduction is neglected. Compute the net heat flux being transferred from plate 1 to plate 2 if the scattering and absorption coefficients are respectively $\sigma_s = 0.1$ cm^{-1} and $a = 0.2$ cm^{-1}.

   Answer: 0.234 W/cm^2.

FIFTEEN

APPROXIMATE SOLUTIONS OF
THE EQUATIONS OF RADIATIVE TRANSFER

15-1 INTRODUCTION

Analytical or numerical solutions to the equations of radiative transfer and energy conservation to yield temperature distributions and heat flows in an absorbing, emitting, and scattering medium require considerable effort in most practical cases. Various approximate techniques can be usefully applied in some instances. One technique is to neglect one or more terms in the transfer equations when justified. For example, if the absorbing medium is cold, emission from the medium can be neglected.

Another technique is to examine limiting cases. For a medium with a small optical thickness the limiting forms of the equations show that the intensities leaving the wall can be considered unattenuated as the radiation travels through the medium. For an optically thick medium a diffusion approximation can be applied. In this approximation, since the medium is optically dense, the conditions at a local position in the medium are only influenced by closely surrounding conditions and the radiative transport becomes somewhat like heat conduction.

Other types of approximations involve using the complete transfer equations and then using various devices to obtain approximate solutions. In one method the exponential integral functions in the radiative equations are approximated by simple exponential functions. This enables the integral equations to be transformed into differential equations that can be readily solved. In other methods such as the Schuster-Schwarzschild approximation, the radiation is assumed to be diffuse in each coordinate direction.

15-2 SYMBOLS

A	area
A_l^m	coefficients in Eq. (15-121)
a	absorption coefficient
C_1, C_2	constants in Planck's spectral energy distribution
D	spacing between parallel planes; diameter
E	ratio $(1 - \epsilon)/\epsilon$
E_1, E_2, E_3	exponential integrals
e	emissive power
G	volumetric energy-generation rate
H	length over which temperature changes significantly
I	mean absorption value defined by Eq. (15-59); source function
i	radiation intensity
K	extinction coefficient
l_j, l_k	direction cosines
l_m	extinction mean free path, $1/(a_\lambda + \sigma_{s\lambda})$
$l_{m,s}$	scattering mean free path, $1/\sigma_{s\lambda}$
P	pressure
P_l^m	spherical harmonics, Eq. (15-123)
Q	energy per unit time
q	energy flux, energy per unit area and time
R	sphere radius
r	radial coordinate
$\mathbf{r}$	position vector
S	coordinate along path of radiation
$\mathbf{s}$	unit vector in S direction
T	absolute temperature
V	volume
$\left.\begin{array}{l} x,y,z \\ x_1,x_2,x_3 \end{array}\right\}$	distances measured along cartesian coordinates
Y_l^m	functions of angle in Eq. (15-122)
Γ	gamma function
δ_{kj}	Kronecker delta
ϵ	hemispherical emissivity
θ	cone angle, angle from normal of area
κ	optical depth
κ_D	optical thickness for path length D
λ	wavelength
μ	$\cos \theta$
σ	Stefan-Boltzmann constant
σ_s	scattering coefficient
φ	circumferential angle
ϕ	dimensionless temperature ratio, Tables 15-2 and 15-3
ψ	dimensionless heat flux, Tables 15-2 and 15-3

Ω	function defined by Eq. (15-43)
ω	solid angle

Subscripts

b	blackbody
D	mean absorption coefficient in Eq. (15-58)
e	emitted
$g\text{-}g$	evaluated at interface between gas regions 1 and 2
i	incident, incoming
o	outgoing
P	Planck mean value
R	Rosseland mean value in Eq. (15-48)
r	net value in r direction; radiative
s	sphere
w	evaluated on the wall
z	net value in the z direction
$+z, -z$	propagating in positive or negative z direction, respectively
λ	spectrally dependent
$\Delta\lambda$	value integrated over wavelength interval $\Delta\lambda$
0	evaluated at point of origin, initial value
$1, 2$	boundary 1 or 2, respectively, or region 1 or 2, respectively

Superscripts

*	dummy variable of integration
**	dummy variable of integration
$-$	(overbar) average over all incident solid angles
$'$	directional quantity
$(0), (1), (2)$	zeroth-, first-, or second-order term or moment
$+, -$	propagating in positive or negative direction

15-3 APPROXIMATE SOLUTIONS BY NEGLECTING TERMS IN THE EQUATION OF TRANSFER

In Chap. 14 the equation of transfer was integrated to give the intensity variation along the optical path κ [Eq. (14-13)]. In terms of the actual distance S along the path, this equation is

$$i'_\lambda(S) = i'_\lambda(0) \exp\left[-\int_0^S (a_\lambda + \sigma_{s\lambda})\, dS^*\right]$$

$$+ \int_0^S (a_\lambda + \sigma_{s\lambda}) I'_\lambda(S^*) \exp\left[-\int_{S^*}^S (a_\lambda + \sigma_{s\lambda})\, dS^{**}\right] dS^* \tag{15-1}$$

where a_λ and $\sigma_{s\lambda}$ may vary along the path S. The $I'_\lambda(S)$ as shown by (14-9) depends on the local blackbody intensity and on the locally scattered radiation, and is found from

Table 15-1 Approximations to equation of transfer

Approximation	Form of equation of transfer	Conditions
Strong transparent	$i'_\lambda(S) = i'_\lambda(0)$	The medium has such a low extinction coefficient that intensity does not change by absorption, emission, or scattering along a path within the medium.
Emission	$i'_\lambda(S) = \int_0^S a_\lambda(S^*) i'_{\lambda b}(S^*)\, dS^*$	No energy is incident from the boundaries, and the gas is relatively transparent so that emitted energy from the gas passes within the system without significant attenuation.
Cold medium with small scattering	$i'_\lambda(S) = i'_\lambda(0) \exp\left[-\int_0^S (a_\lambda + \sigma_{s\lambda})\, dS^*\right]$	Radiation emitted and scattered by the medium is negligible compared to that incident from boundaries or external sources.
Diffusion	$-\dfrac{4\pi}{3K_\lambda}\dfrac{\partial i'_{\lambda b}}{\partial S} = \dfrac{dq_\lambda(S)}{d\lambda}$	The optical depth of the gas is sufficiently large, and the temperature gradients sufficiently small so that the local intensity results only from local emission.

an integral equation such as (14-53). Since $I'_\lambda(S)$ depends on $i'_{\lambda b}(S)$, the temperature distribution must also be found by a simultaneous solution with the energy equation. As shown by (14-37) the energy equation involves an integration of incident intensities from all solid angles. The coupling of the equations for source function, energy conservation, and intensity can be quite complex. It is evident that a number of limiting cases can be examined, such as the medium being optically thick or thin, the scattering being strong or weak relative to absorption, and the internal emission being strong or weak relative to the radiation incident at the boundaries. A few simple approximations as summarized by the first three entries in Table 15-1 will be discussed in Secs. 15-3.1–15-3.3. In Sec. 15-4 a more formal discussion will be given for the optically thin limit, and in Sec. 15-5 the optically thick medium will be considered.

15-3.1 The Transparent-Gas Approximation

When the optical depth along a path in the gas is small, (15-1) can be simplified, as the two exponential attenuation terms each approach unity, thereby giving

$$i'_\lambda(S) = i'_\lambda(0) + \int_0^S (a_\lambda + \sigma_{s\lambda}) I'_\lambda(S^*)\, dS^* \qquad (15\text{-}2)$$

There is no attenuation along the path of either the intensity emitted or locally scattered within the medium. The intensity that enters at $S = 0$ is also not appreciably attenuated. The implications of this optically thin limit are given more fully in Sec. 5-4. If the stronger assumption is made that the optical depth is quite small, and if the $i_\lambda'(0)$ is finite, then the integral in (15-2) becomes small compared with the intensity at $S = 0$, and (15-2) further reduces to

$$i_\lambda'(S) = i_\lambda'(0) \tag{15-3}$$

This is the strong transparent approximation in Table 15-1. An incident intensity is thus essentially unchanged as it travels through the medium. The local energy balances based on this simple intensity relation are obviously much easier to carry out than those involving the complete equation of transfer. The use of the strong transparent approximation will now be demonstrated.

EXAMPLE 15-1 Two infinite parallel black plates at temperatures T_1 and T_2 as in Fig. 14-7 are a distance D apart, and the space between them is filled with non-scattering gas of absorption coefficient a_λ. Assuming that the strong transparent approximation holds, derive an expression for the gas temperature as a function of position between the plates. It is assumed the gas is in local thermodynamic equilibrium, although this assumption can sometimes break down in an optically thin gas (see Sec. 13-9).

Equation (14-32) is the general expression for energy conservation when there is radiative equilibrium in the gas:

$$a_P(T,P)\,\sigma T^4 = \pi \int_0^\infty a_\lambda(\lambda, T, P)\bar{\imath}_{\lambda,i}(\lambda)\,d\lambda \tag{15-4}$$

The $\bar{\imath}_{\lambda,i}$ is obtained from the contributions approaching a volume element from above and below:

$$4\pi\bar{\imath}_{\lambda,i}(\lambda) = \int_{\omega=0}^{4\pi} i_\lambda'(\lambda,\omega)\,d\omega = \int_\cap i_\lambda^+(\lambda,\omega)\,d\omega + \int_\cup i_\lambda^-(\lambda,\omega)\,d\omega \tag{15-5}$$

Since the walls are black, the strong transparent approximation gives, for any position between the plates,

$$i_\lambda^+(\lambda,\omega) = i_{\lambda b}'(\lambda,T_1) \qquad \text{and} \qquad i_\lambda^-(\lambda,\omega) = i_{\lambda b}'(\lambda,T_2)$$

Then, since the black intensity is independent of angle, (15-5) yields

$$4\pi\bar{\imath}_{\lambda,i}(\lambda) = 2\pi i_{\lambda b}'(\lambda,T_1) \int_0^{\pi/2} \sin\theta\,d\theta + 2\pi i_{\lambda b}'(\lambda,T_2) \int_0^{\pi/2} \sin\theta\,d\theta$$

$$= 2\pi[i_{\lambda b}'(\lambda,T_1) + i_{\lambda b}'(\lambda,T_2)] \tag{15-6}$$

Substituting (15-6) into (15-4) gives, at any x position between the plates,

$$\sigma T^4(x) = \frac{\pi}{2a_P(x)} \int_0^\infty a_\lambda(\lambda,x)[i_{\lambda b}'(\lambda,T_1) + i_{\lambda b}'(\lambda,T_2)]\,d\lambda \tag{15-7}$$

Equation (15-7) can be solved iteratively for $T(x)$. The necessity for an iterative solution arises because a_λ depends on local temperature.

If a_λ does not depend on gas temperature, using (14-31) results in

$$\int_0^\infty a_\lambda(\lambda) i'_{\lambda b}(\lambda, T_1)\, d\lambda = \frac{a_P(T_1) \sigma T_1^4}{\pi} \qquad \int_0^\infty a_\lambda(\lambda) i'_{\lambda b}(\lambda, T_2)\, d\lambda = \frac{a_P(T_2) \sigma T_2^4}{\pi}$$

and

$$a_P(T(x)) = \frac{\int_0^\infty a_\lambda(\lambda) e_{\lambda b}(T(x))\, d\lambda}{\sigma T^4(x)}$$

Then (15-7) reduces to

$$T^4(x) = \frac{1}{2 a_P(T(x))} [a_P(T_1) T_1^4 + a_P(T_2) T_2^4] \tag{15-8}$$

The local temperature solution, although still requiring an iterative solution on $T(x)$ and $a_P(T(x))$, is relatively easily found by use of tabulated values of $a_P(T)$. Note further that for a gray gas with temperature-independent properties, a_P is constant and (15-8) further reduces to

$$T^4(x) = \frac{T_1^4 + T_2^4}{2} \tag{15-9}$$

Hence, the entire gray gas approaches a fourth-power temperature that is the average of the fourth powers of the boundary temperatures. This limit was also found in Eq. (14-91).

15-3.2 The Emission Approximation

In the strong transparent approximation, the gas was optically thin and the local intensity was dominated by the intensities incident at the gas boundaries. In the emission approximation the gas is again optically thin but there is also negligible incoming energy at the gas boundaries. For these conditions there is only energy emission within the medium and no attenuation by either absorption or scattering [note that in (15-18b) scattering does not enter into the energy equation]. The equation of transfer becomes

$$i'_\lambda(S) = \int_0^S a_\lambda(S^*) i'_{\lambda b}(S^*)\, dS^* \tag{15-10}$$

The intensity $i'_\lambda(S)$ is the integrated contribution of all the emission along a path as the emitted energy travels through the gas without attenuation.

Equation (15-10) can be integrated over all wavelengths to give the total intensity

$$i'(S) = \int_0^\infty i'_\lambda(S)\, d\lambda = \int_0^S \left[\int_0^\infty a_\lambda(S^*) i'_{\lambda b}(S^*)\, d\lambda \right] dS^*$$

The definition of a_P [Eq. (14-31)] is now applied to give

$$i'(S) = \int_0^S a_P(S^*) \frac{\sigma T^4(S^*)}{\pi} dS^* \tag{15-11}$$

The fact that (15-11) contains the Planck mean absorption coefficient and was derived for optically thin conditions has sometimes led to the statement that the Planck mean absorption coefficient is applicable *only* in optically thin situations. However, the Planck mean was defined in general for the *emission* from a volume element in connection with (14-27), and hence can be applied for the emission term in a gas of *any* optical thickness.

EXAMPLE 15-2 Use the emission approximation to find the flux emerging from an isothermal slab of gas with Planck mean absorption coefficient 0.010 cm^{-1} and thickness $D = 1$ cm if the slab is bounded by transparent nonradiating walls [Fig. 15-1a].

If $i'(\theta)$ is the emerging total intensity in direction θ, the emerging flux is

$$q = \int_{\omega=0}^{2\pi} i'(\theta) \cos \theta \, d\omega = 2\pi \int_{\theta=0}^{\pi/2} i'(\theta) \cos \theta \sin \theta \, d\theta$$

Since the slab is isothermal with constant a_P, Eq. (15-11) can be integrated over any path $D/\cos \theta$ through the slab to yield

$$i'(\theta) = a_P \frac{\sigma T^4}{\pi} \frac{D}{\cos \theta}$$

Then

$$q = 2 \int_0^{\pi/2} a_P \sigma T^4 D \sin \theta \, d\theta = 2 a_P \sigma T^4 D = 0.02 \sigma T^4 \tag{15-12}$$

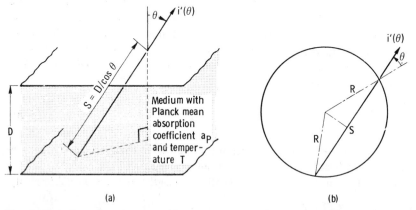

(a) (b)

Figure 15-1 Examples for emission approximation. (*a*) Slab geometry for Example 15-2; (*b*) emission from spherical gas-filled balloon with transparent skin.

It should be realized that (15-12) is really not a precise result even though the slab is optically thin in the sense that the optical thickness based on D is $a_p D = 0.01 \ll 1$. This is because the radiation reaching the slab boundary has passed through the thickness $D/\cos \theta$. Because for large θ directions this path length becomes *infinite*, the emission approximation cannot hold. A more accurate solution of the equation of transfer including the proper path lengths gives

$$q = 1.8 a_p \sigma T^4 D \qquad (15\text{-}13)$$

which is 10% less than Eq. (15-12) (see Sec. 17-5.4).

EXAMPLE 15-3 An inflated spherical balloon of radius R is in orbit and enters the earth's shadow. The balloon has a perfectly transparent wall and is filled with a gray gas of constant absorption coefficient a, such that $aR \ll 1$. Neglecting radiant exchange with the earth, derive a relation for the initial rate of energy loss from the balloon if the initial temperature of the gas is T_0.

From the emission approximation equation (15-11), Fig. 15-1b shows that the following can be written for the intensity at the surface:

$$i'(\theta) = \int_0^S \frac{a\sigma T_0^4}{\pi} \, dS = \frac{a\sigma T_0^4}{\pi} S$$

since a and T_0 are constants. From the geometry, $S = 2R \cos \theta$. Then q, the flux leaving the surface, is

$$q = 2\pi \int_0^{\pi/2} i'(\theta) \cos \theta \sin \theta \, d\theta$$

$$= 4 a\sigma T_0^4 R \int_0^{\pi/2} \cos^2 \theta \sin \theta \, d\theta = \tfrac{4}{3} a\sigma T_0^4 R$$

To obtain the energy loss Q per unit time from the entire sphere, multiply by the sphere surface area:

$$Q = \tfrac{4}{3} a\sigma T_0^4 R (4\pi R^2) = 4 a\sigma T_0^4 V_s \qquad (15\text{-}14)$$

where V_s is the sphere volume. This is what is expected—it was found that *any* isothermal gas volume radiates according to this formula (see Sec. 13-6) so long as there is no internal absorption; the emission approximation gives a compatible solution.

15-3.3 The Cold-Medium Approximation with Small Scattering

This approximation applies when the local blackbody emission and local scattering within the medium are small. Such a situation might arise in radiative transfer within a cold medium such as an absorbing cryogenic fluid. Although the intensity is attenuated by scattering, the scattering is small enough so that scattering into the S direction can be neglected. The equation of transfer (15-1) reduces to

$$i_\lambda'(S) = i_\lambda'(0) \exp\left[-\int_0^S (a_\lambda + \sigma_{s\lambda})\, dS*\right]$$ (15-15)

The local intensity thus consists only of the attenuated incident intensity.

EXAMPLE 15-4 100 W of radiant energy leave a spherical light bulb enclosed in a fixture having a flat glass plate as shown in Fig. 15-2. If the glass is 2 cm thick and has a gray extinction coefficient of 0.05 cm^{-1}, find the intensity directly from the bulb leaving the fixture at an angle of 60° to the bulb axis. (Assume the bulb diameter is 10 cm.) Neglect interface-interaction effects resulting from the difference in the index of refraction between the glass and surrounding air; these effects are discussed in Chap. 19.

Integrating (15-15) over λ and S results in the total intensity

$$i'(S,\theta) = i'(0,\theta) \exp\left[-(a + \sigma_s)S\right]$$

To obtain $i'(0,\theta)$, consider the light bulb a diffuse sphere. The energy flux (emissive power) at the surface of the sphere is 100 W divided by the sphere area. The intensity is this diffuse emissive power divided by π. Then

$$i'(0,\theta) = \frac{100 \text{ W}}{\pi 10^2 \text{ cm}^2 \times \pi \text{ sr}} = 0.101 \text{ W/(cm}^2\text{·sr)}$$

Then

$$i'(S,\theta) = 0.101 \exp\left[-0.05 \frac{2}{\cos 60°}\right] = 0.0827 \text{ W/(cm}^2\text{·sr)}$$

Note that this problem has involved only a simple attenuated-transmission solution. It was assumed that emission from the glass plate and scattering into the path direction are small, so that only the intensity originating at the source needed to be considered.

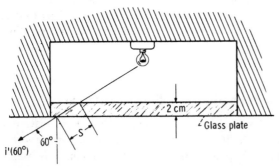

Figure 15-2 Intensity of beam from light fixture (Example 15-4).

15-4 OPTICALLY THIN LIMIT FOR TRANSFER BETWEEN PARALLEL PLATES

In Sec. 14-5 a general formulation including scattering was made for energy transfer between infinite parallel boundaries. These relations will now be considered for the optically thin limit; this is when

$$\kappa_{D\lambda} = \int_0^D (a_\lambda + \sigma_{s\lambda}) \, dx \ll 1$$

For simplicity the discussion here is limited to isotropic scattering with diffuse boundaries.

Consider first the heat-flux equation (14-60) used in conjunction with (14-65) for diffuse boundaries. From the series expansions in Appendix F the exponential integrals can be approximated for small arguments by

$$E_2(x) = 1 + O(x) \qquad E_3(x) = \tfrac{1}{2} - x + O(x^2)$$

Then (14-60) becomes [using (14-64a) and (14-64b) for diffuse boundaries in the first two integrals on the right]

$$\frac{dq_{r\lambda}(\kappa_\lambda)}{d\lambda} = \frac{dq_{\lambda o,1}}{d\lambda}(1 - 2\kappa_\lambda) - \frac{dq_{\lambda o,2}}{d\lambda}(1 - 2\kappa_{D\lambda} + 2\kappa_\lambda)$$

$$+ 2\pi \int_0^{\kappa_\lambda} I'_\lambda(\kappa_\lambda^*) \, d\kappa_\lambda^* - 2\pi \int_{\kappa_\lambda}^{\kappa_{D\lambda}} I'_\lambda(\kappa_\lambda^*) \, d\kappa_\lambda^*$$

If $\kappa_{D\lambda} \ll 1$, the terms of order κ_λ are neglected and this reduces to

$$\frac{dq_{r\lambda}(\kappa_\lambda)}{d\lambda} = \frac{dq_{\lambda o,1}}{d\lambda} - \frac{dq_{\lambda o,2}}{d\lambda} \tag{15-16}$$

Thus the local flux is the difference between the fluxes leaving the boundaries; the fluxes have been unattenuated by the medium. This corresponds to the strong transparent approximation discussed previously.

In a similar fashion the source-function equation (14-62) reduces to

$$I'_\lambda(\kappa_\lambda) = (1 - \Omega_{0\lambda}) i'_{\lambda b}(\kappa_\lambda) + \frac{\Omega_{0\lambda}}{2\pi} \left(\frac{dq_{\lambda o,1}}{d\lambda} + \frac{dq_{\lambda o,2}}{d\lambda} \right) \tag{15-17}$$

The flux derivative (14-61) simplifies to

$$\frac{d^2 q_{r\lambda}}{d\kappa_\lambda \, d\lambda} = -2 \left(\frac{dq_{\lambda o,1}}{d\lambda} + \frac{dq_{\lambda o,2}}{d\lambda} \right) + 4\pi I'_\lambda(\kappa_\lambda)$$

Substituting $I'_\lambda(\kappa_\lambda)$ from (15-17) and noting that $1 - \Omega_{0\lambda} = a_\lambda / K_\lambda$ yield

$$\frac{d^2 q_{r\lambda}}{d\kappa_\lambda \, d\lambda} = \frac{2a_\lambda}{K_\lambda} \left[2\pi i'_{\lambda b}(\kappa_\lambda) - \left(\frac{dq_{\lambda o,1}}{d\lambda} + \frac{dq_{\lambda o,2}}{d\lambda} \right) \right] \tag{15-18a}$$

Since $d\kappa_\lambda = K_\lambda\, dx$, this is also equal to

$$\frac{d^2 q_{r\lambda}}{dx\, d\lambda} = 2a_\lambda \left[2\pi i'_{\lambda b}(\kappa_\lambda) - \left(\frac{dq_{\lambda o,1}}{d\lambda} + \frac{dq_{\lambda o,2}}{d\lambda}\right)\right] \tag{15-18b}$$

which is *independent* of the scattering coefficient. Thus, in the optically thin approximation, scattering does not enter into the solution of the energy equation.

The diffuse fluxes at the boundaries (14-65) reduce, in the optically thin approximation, to

$$\frac{dq_{\lambda o,1}}{d\lambda} = \epsilon_{\lambda,1} e_{\lambda b,1} + (1 - \epsilon_{\lambda,1}) \frac{dq_{\lambda o,2}}{d\lambda}$$

$$\frac{dq_{\lambda o,2}}{d\lambda} = \epsilon_{\lambda,2} e_{\lambda b,2} + (1 - \epsilon_{\lambda,2}) \frac{dq_{\lambda o,1}}{d\lambda}$$

Solving simultaneously for the wall fluxes yields

$$\frac{dq_{\lambda o,1}}{d\lambda} = \frac{\epsilon_{\lambda,1} e_{\lambda b,1} + \epsilon_{\lambda,2} e_{\lambda b,2}(1 - \epsilon_{\lambda,1})}{1 - (1 - \epsilon_{\lambda,1})(1 - \epsilon_{\lambda,2})} \tag{15-19a}$$

$$\frac{dq_{\lambda o,2}}{d\lambda} = \frac{\epsilon_{\lambda,2} e_{\lambda b,2} + \epsilon_{\lambda,1} e_{\lambda b,1}(1 - \epsilon_{\lambda,2})}{1 - (1 - \epsilon_{\lambda,1})(1 - \epsilon_{\lambda,2})} \tag{15-19b}$$

EXAMPLE 15-5 An optically thin medium with absorption coefficient a_λ and isotropic scattering coefficient $\sigma_{s\lambda}$ is between two diffuse parallel plates a distance D apart. The plate tempratures are T_1 and T_2. What total heat flux is being transferred across the plates in the absence of conduction and convection?

For the optically thin approximation, Eq. (15-16) applies. Substituting $dq_{\lambda o,1}$ and $dq_{\lambda o,2}$ from (15-19) yields

$$\frac{dq_{r\lambda}}{d\lambda} = \frac{\epsilon_{\lambda,1} e_{\lambda b,1} + \epsilon_{\lambda,2} e_{\lambda b,2}(1 - \epsilon_{\lambda,1}) - \epsilon_{\lambda,2} e_{\lambda b,2} - \epsilon_{\lambda,1} e_{\lambda b,1}(1 - \epsilon_{\lambda,2})}{1 - (1 - \epsilon_{\lambda,1})(1 - \epsilon_{\lambda,2})}$$

$$= \frac{\epsilon_{\lambda,1} \epsilon_{\lambda,2}(e_{\lambda b,1} - e_{\lambda b,2})}{\epsilon_{\lambda,1} + \epsilon_{\lambda,2} - \epsilon_{\lambda,1} \epsilon_{\lambda,2}}$$

Then

$$q_r = \int_{\lambda=0}^\infty \frac{e_{\lambda b,1} - e_{\lambda b,2}}{1/\epsilon_{\lambda,1} + 1/\epsilon_{\lambda,2} - 1}\, d\lambda$$

This is the same as (10-11). The transferred flux is thus uninfluenced by the presence of the optically thin medium.

EXAMPLE 15-6 What is the medium temperature distribution in Example 15-5?

For radiative equilibrium $\nabla \cdot \mathbf{q}_r = 0$, and from (14-51)

$$\int_0^\infty K_\lambda(\kappa_\lambda) \frac{d^2 q_{r\lambda}}{d\kappa_\lambda\, d\lambda}\, d\lambda = 0$$

Substituting (15-18a) gives

$$\int_0^\infty a_\lambda \left[2\pi i'_{\lambda b}(\kappa_\lambda) - \left(\frac{dq_{\lambda o,1}}{d\lambda} + \frac{dq_{\lambda o,2}}{d\lambda} \right) \right] d\lambda = 0$$

Substituting (15-19) yields

$$2\pi \int_0^\infty a_\lambda i'_{\lambda b}(\kappa_\lambda)\, d\lambda = \int_0^\infty a_\lambda \frac{2(\epsilon_{\lambda,1} e_{\lambda b,1} + \epsilon_{\lambda,2} e_{\lambda b,2}) - \epsilon_{\lambda,1}\epsilon_{\lambda,2}(e_{\lambda b,1} + e_{\lambda b,2})}{\epsilon_{\lambda,1} + \epsilon_{\lambda,2} - \epsilon_{\lambda,1}\epsilon_{\lambda,2}} d\lambda$$

Using (14-31), the integral on the left can be expressed in terms of the Planck mean absorption coefficient to yield

$$\sigma T^4(x) = \frac{1}{2a_P(x)} \int_0^\infty a_\lambda(\lambda,x) \frac{2(\epsilon_{\lambda,1} e_{\lambda b,1} + \epsilon_{\lambda,2} e_{\lambda b,2}) - \epsilon_{\lambda,1}\epsilon_{\lambda,2}(e_{\lambda b,1} + e_{\lambda b,2})}{\epsilon_{\lambda,1} + \epsilon_{\lambda,2} - \epsilon_{\lambda,1}\epsilon_{\lambda,2}} d\lambda$$
$$(15\text{-}20)$$

This can be solved for $T(x)$ by iteration as discussed in connection with (15-7), and it reduces to (15-7) when $\epsilon_{\lambda,1} = \epsilon_{\lambda,2} = 1$. If all properties are independent of both wavelength and temperature, (15-20) reduces to

$$T^4 = \frac{1}{2} \frac{2(\epsilon_1 T_1^4 + \epsilon_2 T_2^4) - \epsilon_1\epsilon_2(T_1^4 + T_2^4)}{\epsilon_1 + \epsilon_2 - \epsilon_1\epsilon_2} \qquad (15\text{-}21)$$

This is a limiting form of (14-98) when $G = \sigma T^4$ and when $\phi_b \to \frac{1}{2}$ and $\psi_b \to 1$ for the optically thin approximation.

15-5 DIFFUSION METHODS IN RADIATIVE TRANSFER

In an optically dense medium, the radiation can travel only a short distance before being absorbed. Consider the situation in which this radiation penetration distance is small compared with the distance over which significant temperature changes occur. Then a local intensity will be the result of radiation coming only from nearby locations where the temperature is close to that of the location under consideration. Radiation emitted by locations where the temperature is appreciably different will be greatly attenuated before reaching the location being considered.

For these conditions it will be shown that it is possible to transform the integral-type equations that result from the radiative energy balance into a diffusion equation like that for heat conduction. The energy transfer depends only on the conditions in the immediate vicinity of the position being considered and can be described in terms of the gradient of the conditions at that position. The use of the diffusion approximation leads to a very great simplification in treating radiative transfer. Standard techniques, including well-developed finite-difference schemes, can be used for solving the resulting diffusion differential equations. Such methods for differential equations are developed to a much higher degree and are more familiar to most engineers, for example in the solution of heat conduction problems, than are the solution methods for the corresponding integral equations.

As will be shown in the derivations that follow, the diffusion approximation requires that the intensity within the medium be nearly isotropic [this will be revealed by (15-31)]. This can occur within the interior of an optically thick medium with small temperature gradients but cannot be valid near certain types of boundaries. For example, at a boundary adjacent to a vacuum at absolute zero temperature, radiation will leave the medium but there will be none incident from the vacuum. As a result of this large anisotropy, the diffusion approximation will not be valid near this type of boundary. To apply diffusion methods at the interfaces between such regions, "radiation slip" or "jump" boundary conditions are employed.

Real gases are usually transparent in some wavelength regions. The diffusion approach can only be applied in wavelength regions where the optical thickness of the medium is greater than about 2; the fact that some *mean* optical thickness meets this criterion is not sufficient. Wavelength-band applications of the diffusion method can be made in the optically thick regions.

15-5.1 Simplified Derivation of the Diffusion Equation

A simplified derivation of radiation diffusion in a one-dimensional layer will be carried out to illustrate the spirit of the diffusion approximation. The situation considered is for an absorbing-emitting medium in radiative equilibrium with isotropic scattering.

In the diffusion approximation, the medium is optically dense. Consequently, the radiation arriving at any location comes only from the immediate surroundings, as any other radiation is absorbed or scattered away before arriving at that location. Also, in the diffusion approximation the radiant energy density [Eq. (14-104)] changes slowly with position relative to distances for attenuation. This can be stated more rigorously by letting H be a path length over which the radiant energy density does change appreciably, and letting l_m be the extinction mean free path, $l_m = 1/(a_\lambda + \sigma_{s\lambda})$. Then, for the diffusion approximation to apply, $l_m/H \ll 1$.

Consider the layer of gas shown in Fig. 15-3a. The equation of transfer from (14-10) is

$$\frac{1}{\sigma_{s\lambda} + a_\lambda} \frac{di'_\lambda}{dS} + i'_\lambda(S) = I'_\lambda(S) \tag{15-22}$$

By using the relation $dS = dx/\cos \theta$, the equation of transfer giving the change of i'_λ with x for a fixed θ is written as

$$-\frac{\cos \theta}{\sigma_{s\lambda} + a_\lambda} \frac{\partial i'_\lambda(x,\theta)}{\partial x} = i'_\lambda(x,\theta) - I'_\lambda(x) \tag{15-23}$$

where the source function I'_λ does not depend on angle for isotropic scattering. Let H be a length over which radiant energy density changes by a significant amount. Then nondimensionalize (15-23) using H, and let $1/(\sigma_{s\lambda} + a_\lambda) = l_m$ to obtain

$$-\frac{l_m}{H} \cos \theta \frac{\partial i'_\lambda(x,\theta)}{\partial (x/H)} = i'_\lambda(x,\theta) - I'_\lambda(x) \tag{15-24}$$

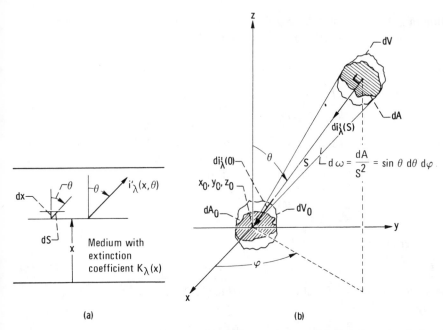

Figure 15-3 Geometry for derivation of diffusion equations. (*a*) One-dimensional plane layer of medium; (*b*) general three-dimensional region.

We now proceed to obtain a solution to (15-24) in the form of a series. Using the diffusion approximation that $l_m/H \ll 1$, the intensity can be written as a series of functions $i_\lambda'^{(n)}$ multiplied by powers of l_m/H:

$$i_\lambda' = i_\lambda'^{(0)} + \frac{l_m}{H} i_\lambda'^{(1)} + \left(\frac{l_m}{H}\right)^2 i_\lambda'^{(2)} + \cdots \tag{15-25}$$

Insert (15-25) into the equation of transfer (15-24) to obtain $[I_\lambda'$ is given by (14-9) with $\Phi = 1]$

$$-\mu \frac{l_m}{H} \left[\frac{\partial i_\lambda'^{(0)}}{\partial(x/H)} + \frac{l_m}{H} \frac{\partial i_\lambda'^{(1)}}{\partial(x/H)} + \cdots \right] = i_\lambda'^{(0)} + \frac{l_m}{H} i_\lambda'^{(1)} + \cdots - \left(1 - \frac{l_m}{l_{m,s}}\right) i_{\lambda b}'$$

$$- \frac{l_m}{l_{m,s}} \left[\frac{1}{4\pi} \int_{\omega_i=4\pi} \left(i_\lambda'^{(0)} + \frac{l_m}{H} i_\lambda'^{(1)} + \cdots \right) d\omega_i \right] \tag{15-26}$$

where $l_{m,s} = 1/\sigma_{s\lambda}$. In addition to the expansion parameter l_m/H, there has appeared an additional quantity $l_m/l_{m,s}$, which characterizes the extinction by scattering relative to the total extinction. For the diffusion approximation including absorption and scattering, this parameter should be of the order of 1/2. For small $l_m/l_{m,s}$ the problem would degenerate to one of diffusion by absorption alone; this could be the situation in a dense gas containing a few scattering particles. For $l_m/l_{m,s}$ approaching unity, there would be scattering alone such as in a thin carrier gas with many scattering particles.

In (15-26) collect the terms of zero order in l_m/H to obtain

$$i_\lambda'^{(0)} = \left(1 - \frac{l_m}{l_{m,s}}\right) i_{\lambda b}' + \frac{l_m}{l_{m,s}} \frac{1}{4\pi} \int_{\omega_i=4\pi} i_\lambda'^{(0)} \, d\omega_i \tag{15-27}$$

The terms $i_{\lambda b}'$ and $\int_{\omega_i=4\pi} i_\lambda'^{(0)} \, d\omega_i$ on the right in (15-27) do not depend on the incidence angle $d\omega_i$. Hence $i_\lambda'^{(0)}$ on the left cannot depend on incidence angle. By using this fact in the integral, (15-27) is reduced to

$$i_\lambda'^{(0)} = \left(1 - \frac{l_m}{l_{m,s}}\right) i_{\lambda b}' + \frac{l_m}{l_{m,s}} \frac{1}{4\pi} i_\lambda'^{(0)} \, 4\pi$$

which further reduces to

$$i_\lambda'^{(0)} = i_{\lambda b}' \tag{15-28}$$

Now collect the terms of first order in l_m/H in (15-26) to obtain

$$-\mu \frac{\partial i_\lambda'^{(0)}}{\partial(x/H)} = i_\lambda'^{(1)} - \frac{l_m}{l_{m,s}} \frac{1}{4\pi} \int_{\omega_i=4\pi} i_\lambda'^{(1)} \, d\omega_i$$

Substitute (15-28) for $i_\lambda'^{(0)}$ to obtain

$$-\mu \frac{d i_{\lambda b}'}{d(x/H)} = i_\lambda'^{(1)} - \frac{l_m}{l_{m,s}} \frac{1}{4\pi} \int_{\omega_i=4\pi} i_\lambda'^{(1)} \, d\omega_i \tag{15-29}$$

To find $i_\lambda'^{(1)}$ multiply by $d\omega_i = 2\pi \sin\theta \, d\theta = -2\pi \, d\mu$ and integrate over all solid angles:

$$\frac{d i_{\lambda b}'}{d(x/H)} \int_{\mu=-1}^{1} 2\pi\mu \, d\mu = \int_{\omega_i=4\pi} i_\lambda'^{(1)} \, d\omega_i - \frac{l_m}{l_{m,s}} \left(\frac{1}{4\pi} \int_{\omega_i=4\pi} i_\lambda'^{(1)} \, d\omega_i\right) \int_{\omega_i=4\pi} d\omega_i$$

The integral on the left is zero so

$$0 = \int_{\omega_i=4\pi} i_\lambda'^{(1)} \, d\omega_i - \frac{l_m}{l_{m,s}} \int_{\omega_i=4\pi} i_\lambda'^{(1)} \, d\omega_i$$

Hence, since $l_m/l_{m,s} \neq 1$, $\int_{\omega_i=4\pi} i_\lambda'^{(1)} \, d\omega_i = 0$ and (15-29) reduces to

$$i_\lambda'^{(1)} = -\mu \frac{d i_{\lambda b}'}{d(x/H)} \tag{15-30}$$

Substitute (15-28) and (15-30) into (15-25) to obtain

$$i_\lambda' = i_{\lambda b}' - \frac{\cos\theta}{a_\lambda + \sigma_{s\lambda}} \frac{d i_{\lambda b}'}{dx} \tag{15-31}$$

This result reveals the important feature that *in the diffusion solution the local intensity depends only on the magnitude and gradient of the local blackbody intensity.* Since temperature gradients are small and $a_\lambda + \sigma_{s\lambda}$ is large, the last term on the right is small and i'_λ is nearly isotropic.

The local spectral energy flux at x flowing in the x direction is found by multiplying i'_λ by $\cos\theta\, d\lambda$ and integrating over all solid angles as in (14-46a):

$$\frac{dq_\lambda(x)}{d\lambda} = 2\pi \int_{\theta=0}^{\pi} i'_\lambda(x,\theta) \cos\theta \sin\theta\, d\theta = 2\pi \int_{\mu=-1}^{1} i'_\lambda(x,\mu)\, \mu\, d\mu \qquad (15\text{-}32)$$

Using (15-31) in (15-32) gives, after noting that $i'_{\lambda b}$ does not depend on μ,

$$\frac{dq_\lambda(x)}{d\lambda} = 2\pi i'_{\lambda b}(x) \int_{\mu=-1}^{1} \mu\, d\mu - \frac{2\pi}{a_\lambda + \sigma_{s\lambda}} \frac{di'_{\lambda b}}{dx} \int_{\mu=-1}^{1} \mu^2\, d\mu = -\frac{4\pi}{3K_\lambda(x)} \frac{di'_{\lambda b}}{dx}$$

$$= \frac{-4}{3K_\lambda(x)} \frac{de_{\lambda b}}{dx} \qquad (15\text{-}33)$$

where $K_\lambda = a_\lambda + \sigma_{s\lambda}$. Equation (15-33) is an important result known as the *Rosseland diffusion equation.* This equation relates the local energy flux to local conditions only; it does not involve integrals of contributions from other regions and thus provides a considerable simplification over the exact formulation of the equation of transfer.

In a gray medium, (15-33) becomes

$$q(x) = -\frac{4\sigma}{3(a + \sigma_s)} \frac{d(T^4)}{dx} \qquad (15\text{-}34)$$

For $a = 0$ (no absorption), the local temperature is independent of the energy scattered from a volume element within the medium. In this instance, σT^4 in (15-34) is replaced by $\pi\bar{i}$ as discussed in Sec. 14-7.2.

15-5.2 The General Radiation-Diffusion Equation in an Absorbing-Emitting Medium

In the previous section the diffusion equation was derived for a simplified case. Only first-order terms were retained in the series of Eq. (15-25), and the region boundaries were not considered. The general equations will now be derived, including the second-order terms. Boundary conditions will be introduced into the theory so that the diffusion equations can be applied to finite regions. As demonstrated by the plane-layer solution on page 477, the boundary conditions must account for a jump in emissive power between the wall and the gas at the wall for a situation in which there is radiative transfer only. The derivation follows the general outline given by Deissler [1]. For simplicity, scattering is not included. The intermediate equations in the derivation become somewhat complex because of their general form. The final equations [such as Eq. (15-46)], however, are relatively simple and very useful.

The Rosseland equation for local radiative flux. Consider the geometry in Fig. 15-3b. There is a volume element dV_0 at x_0, y_0, z_0 having cross-sectional area dA_0 in the xy

plane. The energy flux crossing dA_0 originates from all surrounding volume elements such as dV. If the emission from dV produces an intensity $di'_\lambda(S)$, then the intensity reaching dV_0 is

$$di'_\lambda(0) = di'_\lambda(S) \exp\left[-a_\lambda(\lambda)S\right] \tag{15-35}$$

This accounts for attenuation along S but not emission along S, which will be accounted for later by integrating the contributions of (15-35) from all elements of the volume. A spatially constant a_λ has been used. This is not restrictive, as in the diffusion approximation the gas temperature does not change significantly over the region contributing significant radiation to a location. The solid angle subtended by dV when viewed from dA_0 is dA/S^2, where dA is the projected area of dV normal to S. The energy per unit time incident on dA_0 as a result of the intensity in (15-35) is then

$$d^4Q_{\lambda,i}(0) = di'_\lambda(S) \exp\left[-a_\lambda(\lambda)S\right] \frac{dA}{S^2} dA_0 \cos\theta \, d\lambda \tag{15-36}$$

From (13-41), the spectral energy emitted by dV per unit time is

$$di'_{\lambda,e}(S) = a_\lambda(\lambda,T,P) i'_{\lambda b}(\lambda,T) \, dS \tag{15-37}$$

Substituting (15-37) for $di'_\lambda(S)$ in (15-36) gives

$$d^4Q_{\lambda,i}(0) = a_\lambda(\lambda) i'_{\lambda b}(\lambda,T) \, dS \exp\left[-a_\lambda(\lambda)S\right] \frac{dA}{S^2} dA_0 \cos\theta \, d\lambda \tag{15-38}$$

If as a result of the gas being optically dense, the radiation field at dV_0 originates only from locations close to dV_0, then $i'_{\lambda b}(\lambda,T)$ in (15-38) can be obtained by expanding in a three-dimensional Taylor series about the origin $S = 0$ in the hope that truncation of the series after a few terms will give an adequate representation of the $i'_{\lambda b}$ distribution near dV_0. The general Taylor series in three dimensions can be written as

$$i'_{\lambda b}(\lambda,T) = \sum_{n=0}^{\infty} \left\{ \frac{1}{n!} \left[(z - z_0)\left(\frac{\partial}{\partial z}\right)_0 + (y - y_0)\left(\frac{\partial}{\partial y}\right)_0 + (x - x_0)\left(\frac{\partial}{\partial x}\right)_0 \right]^n i'_{\lambda b}(\lambda,T) \right\} \tag{15-39}$$

This series is carried for the next several steps and is then truncated to a few terms. By applying the binomial theorem twice to expand the factor in brackets, (15-39) becomes

$$i'_{\lambda b}(\lambda,T) = \sum_{n=0}^{\infty} \sum_{v=0}^{n} \sum_{s=0}^{v} \frac{(z - z_0)^{n-v}(y - y_0)^{v-s}(x - x_0)^s}{(n-v)!(v-s)!s!} \left(\frac{\partial^n i'_{\lambda b}}{\partial z^{n-v} \, \partial y^{v-s} \, \partial x^s} \right)_0 \tag{15-40}$$

This is substituted into (15-38), and the solid angle dA/S^2 set equal to $\sin\theta \, d\theta \, d\varphi$. The result is then integrated over the half space encompassing positive z values. This gives all energy traveling in the negative z direction that is incident on dA_0 as

$$\frac{d^2Q_{\lambda,-z}}{d\lambda} = a_\lambda(\lambda)\, dA_0 \sum_{n=0}^{\infty} \sum_{v=0}^{n} \sum_{s=0}^{v} \frac{1}{(n-v)!\,(v-s)!\,s!}$$

$$\times \left(\frac{\partial^n i'_{\lambda b}}{\partial z^{n-v}\,\partial y^{v-s}\,\partial x^s}\right)_0 \int_{\varphi=0}^{2\pi} \int_{\theta=0}^{\pi/2} \int_{S=0}^{\infty} (S\cos\theta)^{n-v}$$

$$\times (S\sin\varphi\sin\theta)^{v-s}(S\sin\theta\cos\varphi)^s \cos\theta\sin\theta$$

$$\times \exp\left[-a_\lambda(\lambda)S\right] dS\, d\theta\, d\varphi \qquad (15\text{-}41)$$

where the integral and summation signs have been interchanged, and spherical coordinates have been introduced: $x - x_0 = S\sin\theta\cos\varphi$; $y - y_0 = S\sin\theta\sin\varphi$; $z - z_0 = S\cos\theta$. The following assumptions have been used in integrating (15-41) over the entire half space: (1) a_λ is constant within the region that contributes significantly to the energy flux at dA_0, and (2) there are no bounding surfaces that contribute significant radiation energy at dA_0. Otherwise, a_λ would have to be retained as a variable in the integration, and the integration would have to be over a finite region with a specified intensity along the boundaries.

Carrying out the integration of (15-41) gives

$$\frac{d^2Q_{\lambda,-z}}{d\lambda} = \frac{dA_0}{4} \sum_{n=0}^{\infty} \sum_{v=0}^{n} \sum_{s=0}^{v} \Omega(n,v,s)\frac{1}{a_\lambda{}^n}\left(\frac{\partial^n i'_{\lambda b}}{\partial z^{n-v}\,\partial y^{v-s}\,\partial x^s}\right)_0 \qquad (15\text{-}42)$$

where

$$\Omega(n,v,s) = \frac{[1+(-1)^{v-s}]\,[1+(-1)^s]\,n!\,\Gamma[(n-v+2)/2]\,\Gamma[(v-s+1)/2]\,\Gamma[(s+1)/2]}{(n-v)!\,(v-s)!\,s!\,\Gamma[(n+4)/2]}$$

$$(15\text{-}43)$$

and Γ is the gamma function.

A similar derivation for the energy incident on dA_0 from below gives, for energy traveling in the positive z direction,

$$\frac{d^2Q_{\lambda,+z}}{d\lambda} = \frac{dA_0}{4} \sum_{n=0}^{\infty} \sum_{v=0}^{n} \sum_{s=0}^{v} (-1)^{n-v}\Omega(n,v,s)\frac{1}{a_\lambda{}^n}\left(\frac{\partial^n i'_{\lambda b}}{\partial z^{n-v}\,\partial y^{v-s}\,\partial x^s}\right)_0 \qquad (15\text{-}44)$$

The *net* energy flux passing through dA_0 in the positive z direction is then

$$\frac{dq_{\lambda,z}}{d\lambda} = \frac{d^2Q_{\lambda,+z} - d^2Q_{\lambda,-z}}{dA_0\, d\lambda}$$

$$= -\frac{1}{4} \sum_{n=0}^{\infty} \sum_{v=0}^{n} \sum_{s=0}^{v} [1-(-1)^{n-v}]\,\Omega(n,v,s)\frac{1}{a_\lambda{}^n}\left(\frac{\partial^n i'_{\lambda b}}{\partial z^{n-v}\,\partial y^{v-s}\,\partial x^s}\right)_0 \qquad (15\text{-}45)$$

Similar relations can be derived for the x and y directions.

In the diffusion approximation a region is considered where temperature changes slowly with optical depth. Hence derivatives such as $(1/a_\lambda{}^n)(\partial^n i'_{\lambda b}/\partial z^n)$ become small as n is increased and the series in (15-45) can be truncated. *Retaining only terms through the second derivative* causes the formidable-looking Eq. (15-45) to reduce to

$$\frac{dq_{\lambda,z}}{d\lambda} = -\frac{4\pi}{3a_\lambda}\left(\frac{\partial i'_{\lambda b}}{\partial z}\right)_0 = -\frac{4}{3a_\lambda}\left(\frac{\partial e_{\lambda b}}{\partial z}\right)_0 \tag{15-46}$$

This is the general relation for local energy flux in terms of the emissive power gradient and is in agreement with (15-33); it is called the *Rosseland diffusion equation* for radiative energy transfer. Retaining only first-order derivatives, as in the derivation of (15-33), also gives this same equation because the second-order terms have been found to cancel. Note that Eq. (15-46) has the same form as the Fourier law of heat conduction. This allows solution of some radiation problems by analogy with heat-conduction methods. As indicated by (15-33), if scattering were included in the present derivation, the only effect on the final result would be to replace a_λ by K_λ in (15-46).

To obtain the energy flux in a wavelength range, integrate (15-46) over the wavelength band $\Delta\lambda$ (the parentheses and 0 subscript will be dropped for simplicity):

$$q_{\Delta\lambda,z} = \int_{\Delta\lambda} -\frac{4}{3a_\lambda}\frac{\partial e_{\lambda b}}{\partial z}\,d\lambda \equiv -\frac{4}{3a_{R,\Delta\lambda}}\int_{\Delta\lambda}\frac{\partial e_{\lambda b}}{\partial z}\,d\lambda$$

$$= -\frac{4}{3a_{R,\Delta\lambda}}\frac{\partial}{\partial z}\int_{\Delta\lambda}e_{\lambda b}\,d\lambda = -\frac{4}{3a_{R,\Delta\lambda}}\frac{\partial e_{\Delta\lambda b}}{\partial z} \tag{15-47}$$

This defines the mean absorption coefficient $a_{R,\Delta\lambda}$ as

$$\frac{1}{a_{R,\Delta\lambda}} = \frac{\int_{\Delta\lambda}(1/a_\lambda)(\partial e_{\lambda b}/\partial z)\,d\lambda}{\int_{\Delta\lambda}(\partial e_{\lambda b}/\partial z)\,d\lambda}$$

By multiplying the numerator and denominator by $\partial z/\partial e_b$, this can be written as

$$\frac{1}{a_{R,\Delta\lambda}} = \frac{\int_{\Delta\lambda}(1/a_\lambda)(\partial e_{\lambda b}/\partial e_b)\,d\lambda}{\int_{\Delta\lambda}(\partial e_{\lambda b}/\partial e_b)\,d\lambda} \tag{15-48}$$

The a_R is called the *Rosseland mean absorption coefficient* after S. Rosseland, who first made use of diffusion theory in studying radiation effects in astrophysics [2]. The $\partial e_{\lambda b}/\partial e_b$ is found by differentiating Planck's law (2-11a) after letting $T = (e_b/\sigma)^{1/4}$:

$$\frac{\partial e_{\lambda b}}{\partial e_b} = \frac{\partial}{\partial e_b}\left(\frac{2\pi C_1}{\lambda^5\{\exp[(C_2/\lambda)(\sigma/e_b)^{1/4}] - 1\}}\right) = \frac{\pi}{2}\frac{C_1 C_2}{\lambda^6}\frac{\sigma^{1/4}}{e_b^{5/4}}\frac{\exp[(C_2/\lambda)(\sigma/e_b)^{1/4}]}{\{\exp[(C_2/\lambda)(\sigma/e_b)^{1/4}] - 1\}^2} \tag{15-49}$$

The emissive power jump as a boundary condition. Up to now the position considered in the gas was sufficiently far from any boundary so that the effect of the boundary did not enter the diffusion relations. Now the interaction of the radiating gas with a diffuse wall will be considered. Let the wall bounding the gas from above as shown in Fig. 15-4 have a hemispherical spectral emissivity $\epsilon_{\lambda w2}$. All quantities pertaining to the wall itself will have the subscript w to differentiate them from quantities *in the gas*

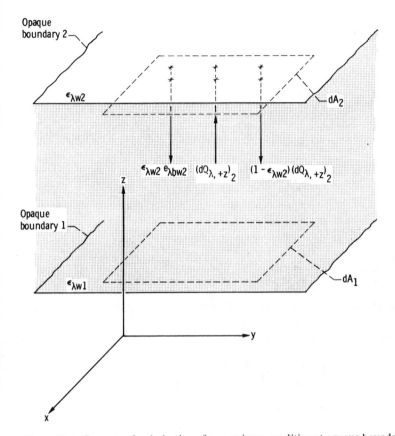

Figure 15-4 Geometry for derivation of energy-jump condition at opaque boundary.

at the wall, which will have no w subscript. Consider an area dA_2 *in the gas* parallel and immediately adjacent to the wall. The spectral energy passing through dA_2 in the negative z direction is

$$(d^2Q_{\lambda,-z})_2 = \epsilon_{\lambda w2}e_{\lambda bw2}\, d\lambda\, dA_2 + (1 - \epsilon_{\lambda w2})(d^2Q_{\lambda,+z})_2 \qquad (15\text{-}50)$$

where the terms on the right account for the emitted and reflected energy from wall 2, respectively. The net spectral flux across dA_2 in the positive z direction is then

$$(dq_{\lambda,z})_2 = \frac{(d^2Q_{\lambda,+z})_2 - (d^2Q_{\lambda,-z})_2}{dA_2} = \epsilon_{\lambda w2}\left[\frac{(d^2Q_{\lambda,+z})_2}{dA_2} - e_{\lambda bw2}\, d\lambda\right] \qquad (15\text{-}51)$$

This can be placed in the form

$$-e_{\lambda bw2} = \frac{(dq_{\lambda,z})_2}{\epsilon_{\lambda w2}\, d\lambda} - \frac{(d^2Q_{\lambda,+z})_2}{dA_2\, d\lambda} \qquad (15\text{-}52)$$

Now (15-44) is substituted for $(d^2Q_{\lambda,+z})_2$. The first term of (15-44) for $n = 0$ and

$dA_0 = dA_2$ is

$$\frac{dA_2}{4} 4\pi i'_{\lambda b2} = dA_2 e_{\lambda b2}$$

Then (15-52) becomes, at dA_2 in the gas adjacent to the wall,

$$e_{\lambda b2} - e_{\lambda bw2} = \frac{(dq_{\lambda,z})_2}{\epsilon_{\lambda w2} \, d\lambda} - \frac{1}{4\pi} \sum_{n=1}^{\infty} \sum_{v=0}^{n} \sum_{s=0}^{v} (-1)^{n-v} \Omega(n,v,s) \frac{1}{a_\lambda{}^n} \left(\frac{\partial^n e_{\lambda b}}{\partial z^{n-v} \, \partial y^{v-s} \, \partial x^s} \right)_2$$

(15-53)

Retaining only terms through the second order and using (15-46) to remove the first derivatives in terms of the spectral energy flux result in the following relation for the jump in emissive power at dA_2:

$$e_{\lambda b2} - e_{\lambda bw2} = \left(\frac{1}{\epsilon_{\lambda w2}} - \frac{1}{2} \right) \frac{(dq_{\lambda,z})_2}{d\lambda} - \frac{1}{2a_\lambda{}^2} \left(\frac{\partial^2 e_{\lambda b}}{\partial z^2} + \frac{1}{2} \frac{\partial^2 e_{\lambda b}}{\partial y^2} + \frac{1}{2} \frac{\partial^2 e_{\lambda b}}{\partial x^2} \right)_2$$

(15-54)

All the quantities that do not have a w subscript are evaluated at dA_2, which is *in the gas* adjacent to the wall. The quantities with a w subscript are evaluated on wall 2, and $dq_{\lambda,z}$ is the net flux in the positive z direction.

Similarly, the jump in emissive power at dA_1 in Fig. 15-4 is

$$e_{\lambda bw1} - e_{\lambda b1} = \left(\frac{1}{\epsilon_{\lambda w1}} - \frac{1}{2} \right) \frac{(dq_{\lambda,z})_1}{d\lambda} + \frac{1}{2a_\lambda{}^2} \left(\frac{\partial^2 e_{\lambda b}}{\partial z^2} + \frac{1}{2} \frac{\partial^2 e_{\lambda b}}{\partial y^2} + \frac{1}{2} \frac{\partial^2 e_{\lambda b}}{\partial x^2} \right)_1$$

(15-55)

where the quantities with a w subscript are on wall 1, and those without a w are *in the gas* adjacent to wall 1.

Equations (15-54) and (15-55) are boundary conditions that relate the emissive power $e_{\lambda b}$ in the gas immediately adjacent to the wall, to the wall emissive power $e_{\lambda bw}$. It is evident that there is a jump in emissive power in passing from the gas to the wall at each boundary. Some applications to clarify the use of these relations will be given in Sec. 15-5.3. The use of (15-42) and (15-44) in the derivation of these boundary relations implies that *the proportionality between local radiative flux and emissive power gradient in the gas holds even at points in the gas very near to a bounding surface.* Although this is not strictly true, the use of the jump boundary conditions corrects to a good approximation for the wall effects.

To apply (15-47) in a wavelength interval, the jump boundary conditions (15-54) and (15-55) must also be integrated over an increment of wavelength $\Delta\lambda$. The wall emissivities are assigned averaged values in this range, and the integration is carried out as in [1] to yield

$$e_{\Delta\lambda b2} - e_{\Delta\lambda bw2} = \left(\frac{1}{\epsilon_{\Delta\lambda w2}} - \frac{1}{2} \right) (q_{\Delta\lambda,z})_2 - \left\{ \frac{1}{2a_{D,\Delta\lambda}^2} \left(\frac{\partial^2 e_{\Delta\lambda b}}{\partial z^2} + \frac{1}{2} \frac{\partial^2 e_{\Delta\lambda b}}{\partial y^2} + \frac{1}{2} \frac{\partial^2 e_{\Delta\lambda b}}{\partial x^2} \right) \right.$$

$$\left. + \frac{I_{\Delta\lambda}}{2} \left[\left(\frac{\partial e_{\Delta\lambda b}}{\partial z} \right)^2 + \frac{1}{2} \left(\frac{\partial e_{\Delta\lambda b}}{\partial y} \right)^2 + \frac{1}{2} \left(\frac{\partial e_{\Delta\lambda b}}{\partial x} \right)^2 \right] \right\}_2$$

(15-56)

$$e_{\Delta\lambda bw1} - e_{\Delta\lambda b1} = \left(\frac{1}{\epsilon_{\Delta\lambda w1}} - \frac{1}{2}\right)(q_{\Delta\lambda,z})_1 + \left\{\frac{1}{2a_{D,\Delta\lambda}^2}\left(\frac{\partial^2 e_{\Delta\lambda b}}{\partial z^2} + \frac{1}{2}\frac{\partial^2 e_{\Delta\lambda b}}{\partial y^2} + \frac{1}{2}\frac{\partial^2 e_{\Delta\lambda b}}{\partial x^2}\right)\right.$$

$$\left. + \frac{I_{\Delta\lambda}}{2}\left[\left(\frac{\partial e_{\Delta\lambda b}}{\partial z}\right)^2 + \frac{1}{2}\left(\frac{\partial e_{\Delta\lambda b}}{\partial y}\right)^2 + \frac{1}{2}\left(\frac{\partial e_{\Delta\lambda b}}{\partial x}\right)^2\right]\right\}_1 \tag{15-57}$$

where $q_{\Delta\lambda} = \int_{\Delta\lambda} dq_\lambda$.

In these equations, two mean coefficients are given as originally derived in the discussion of [1]:

$$\frac{1}{a_{D,\Delta\lambda}^2} = \frac{\int_{\Delta\lambda}(1/a_\lambda^2)(\partial e_{\lambda b}/\partial e_b)\,d\lambda}{\int_{\Delta\lambda}(\partial e_{\lambda b}/\partial e_b)\,d\lambda} \tag{15-58}$$

$$I_{\Delta\lambda} = \frac{\int_{\Delta\lambda}(1/a_\lambda^2)(\partial^2 e_{\lambda b}/\partial e_b^2)\,d\lambda}{[\int_{\Delta\lambda}(\partial e_{\lambda b}/\partial e_b)\,d\lambda]^2} \tag{15-59}$$

The $I_{\Delta\lambda}$ has units of length squared times inverse energy flux.

The emissive power jump between two absorbing-emitting regions. When internal energy sources or sinks are present in absorbing-emitting media, it is possible in the absence of energy conduction to have a discontinuity in emissive power at the interface of two such adjacent media. This is obtained by considering a volume element at the interface between two regions. The lower region has absorption coefficient $a_{\lambda1}$, and the upper $a_{\lambda2}$. The net flux passing through the element per unit area normal to z is, by use of (15-42) and (15-44) in media 2 and 1, respectively,

$$\frac{(dq_{\lambda,z})_{g-g}}{d\lambda} = \frac{(d^2 Q_{\lambda,+z})_1 - (d^2 Q_{\lambda,-z})_2}{dA\,d\lambda}$$

$$= \frac{1}{4\pi}\sum_{n=0}^{\infty}\sum_{v=0}^{n}\sum_{s=0}^{v}\Omega(n,v,s)\left[\frac{(-1)^{n-v}}{a_{\lambda1}^n}\left(\frac{\partial^n e_{\lambda b}}{\partial z^{n-v}\,\partial y^{v-s}\,\partial x^s}\right)_1\right.$$

$$\left. - \frac{1}{a_{\lambda2}^n}\left(\frac{\partial^n e_{\lambda b}}{\partial z^{n-v}\,\partial y^{v-s}\,\partial x^s}\right)_2\right]_{g-g} \tag{15-60}$$

Neglecting terms of order higher than two gives the emissive power jump as

$$(e_{\lambda b2} - e_{\lambda b1})_{g-g} = -\frac{(dq_{\lambda,z})_{g-g}}{d\lambda} + \left\{-\frac{2}{3}\left[\frac{1}{a_{\lambda1}}\left(\frac{\partial e_{\lambda b}}{\partial z}\right)_1 + \frac{1}{a_{\lambda2}}\left(\frac{\partial e_{\lambda b}}{\partial z}\right)_2\right]\right.$$

$$+ \frac{1}{2a_{\lambda1}^2}\left(\frac{\partial^2 e_{\lambda b}}{\partial z^2} + \frac{1}{2}\frac{\partial^2 e_{\lambda b}}{\partial y^2} + \frac{1}{2}\frac{\partial^2 e_{\lambda b}}{\partial x^2}\right)_1$$

$$\left. - \frac{1}{2a_{\lambda2}^2}\left(\frac{\partial^2 e_{\lambda b}}{\partial z^2} + \frac{1}{2}\frac{\partial^2 e_{\lambda b}}{\partial y^2} + \frac{1}{2}\frac{\partial^2 e_{\lambda b}}{\partial x^2}\right)_2\right\}_{g-g} \tag{15-61}$$

The integrated form of (15-61) for a wavelength interval is given in [3]. As will be shown on page 512, the value of the jump $e_{b2} - e_{b1}$ is nonzero under certain conditions.

Summary. The general radiation-diffusion equation has been derived for a single wavelength as (15-45) or (15-46), and for a wavelength band as (15-47). The general boundary conditions at solid boundaries with normals into the gas in the negative and positive coordinate directions are given at a single wavelength by (15-54) and (15-55), Finally, a boundary condition for use at the interface between two absorbing-emitting media in the absence of heat conduction is given by (15-61).

15-5.3 Use of the Diffusion Solution

When the diffusion equation is utilized, it is assumed to apply throughout the entire medium including the region adjacent to a boundary. The effect of the boundary is imposed on the solution by utilizing a jump boundary condition. If a real gas is considered, three coefficients must be evaluated as given by (15-48), (15-58), and (15-59). However, each of these depends only on local conditions so they can be tabulated.

Gray stagnant gas between parallel plates. Most gases have strong variations of properties with wavelength, and it is necessary to solve the diffusion equation in a number of wavelength regions. For illustrative purposes it is not feasible to consider an involved spectral solution. There are some limited situations, such as soot-filled flames and high-temperature uranium gas, in which a gray-gas approximation can be made. The equations presented in Sec. 15-5.2 reduce considerably in this case. Let us examine then the case of a gray gas contained between infinite parallel gray plates at different temperatures (Fig. 15-5).

For a gray gas the absorption coefficient a_λ is independent of wavelength. Then the wavelength range for integration of Eqs. (15-48), (15-58), and (15-59) can be $0-\infty$. Letting $a_\lambda = a$, which can be taken out of the integrals, gives

$$\frac{1}{a_R} = \frac{1}{a_D} = \frac{1}{a}$$

and

$$I = \frac{(1/a^2)(\partial^2/\partial e_b^2) \int_0^\infty e_{\lambda b}\, d\lambda}{[(\partial/\partial e_b)(\int_0^\infty e_{\lambda b}\, d\lambda)]^2} = \frac{(1/a^2)(\partial^2 e_b/\partial e_b^2)}{(\partial e_b/\partial e_b)^2} = 0 \tag{15-62}$$

Equation (15-47) reduces to

$$q_z = -\frac{4}{3a}\frac{\partial}{\partial z}\left(\int_0^\infty e_{\lambda b}\, d\lambda\right) = -\frac{4}{3a}\frac{de_b}{dz}$$

This can be integrated directly because, with no sources or sinks in the gas, q_z is a constant for this geometry. Then, with the additional assumption that a does not

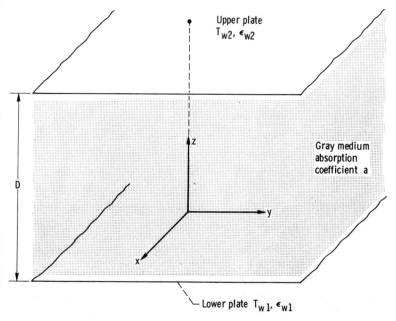

Figure 15-5   Radiant interchange between infinite parallel gray plates enclosing gray medium.

depend on temperature and is therefore independent of z, the result of integrating from 0 to z is

$$e_b(z) - e_{b1} = -\frac{3a}{4} q_z z \tag{15-63}$$

Evaluating (15-63) at $z = D$ yields

$$\frac{e_{b2} - e_{b1}}{q_z} = -\frac{3aD}{4} \tag{15-64}$$

The e_{b1} and e_{b2} are *in the gas* at the walls. To connect these unknown quantities with the specified wall conditions, the jump boundary conditions are applied. Differentiating (15-63) twice with respect to z shows that the second-derivative terms are zero in the boundary-condition equations (15-56) and (15-57). Equations (15-56) and (15-57) then become

$$\frac{e_{b2} - e_{bw2}}{q_z} = \frac{1}{\epsilon_{w2}} - \frac{1}{2} \tag{15-65}$$

and

$$\frac{e_{bw1} - e_{b1}}{q_z} = \frac{1}{\epsilon_{w1}} - \frac{1}{2} \tag{15-66}$$

To eliminate the unknown gas emissive powers e_{b1} and e_{b2}, add (15-65) and (15-66) to obtain

$$\frac{e_{bw1} - e_{bw2}}{q_z} + \frac{e_{b2} - e_{b1}}{q_z} = \frac{1}{\epsilon_{w1}} + \frac{1}{\epsilon_{w2}} - 1$$

Then substitute $e_{b2} - e_{b1}$ from (15-64) to obtain

$$\frac{e_{bw1} - e_{bw2}}{q_z} = \frac{1}{\epsilon_{w1}} + \frac{1}{\epsilon_{w2}} - 1 + \frac{3aD}{4} \tag{15-67}$$

or, taking the reciprocal,

$$\frac{q_z}{e_{bw1} - e_{bw2}} = \frac{1}{3aD/4 + 1/\epsilon_{w1} + 1/\epsilon_{w2} - 1} = \frac{1}{3\kappa_D/4 + 1/\epsilon_{w1} + 1/\epsilon_{w2} - 1} \tag{15-68}$$

Equation (15-68) gives the radiative energy transfer through a gray-gas layer as a function of the optical thickness aD and the plate emissivities. It is ratioed to the difference in the black emissive powers of the plates, which is the maximum possible energy transfer between plates. A comparison of this diffusion solution with the exact analytical solution of the same gray-gas problem by solution of the integral equations [4] is shown in Fig. 15-6 for equal wall emissivities. Agreement is found to be excellent for all optical thicknesses. The distribution of emissive power $e_b(z)$ is found from (15-63) by eliminating the unknown e_{b1} by use of (15-66), or in another form by eliminating e_{b1} and e_{b2} from (15-63)–(15-65). The result is in Table 15-2.

The discontinuity in emissive power between two gas regions. Consider two adjacent semi-infinite regions (Fig. 15-7). Let us determine the discontinuity in emissive power, if any, at the interface between the regions in the absence of heat conduction. First, consider the media in the two regions to have no internal heat sources or sinks. Both media are gray and stagnant; the lower region has a constant absorption coefficient a_1, and the upper a_2. The emissive power jump at the interface between the two media is

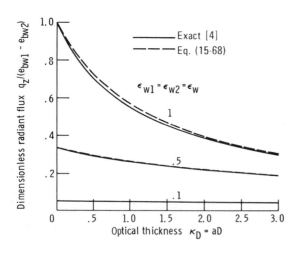

Figure 15-6 Validity of diffusion solution for energy transfer through gray gas between parallel gray plates.

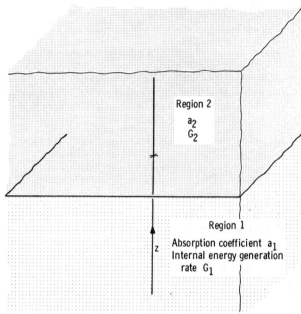

Figure 15-7 Geometry for derivation of interface emissive power discontinuity.

found by integrating (15-61) over all wavelengths. This gives, after noting that $a_{\lambda 1} = a_1$ and $a_{\lambda 2} = a_2$, and that derivatives with respect to x and y are zero for the one-dimensional layer,

$$(e_{b2} - e_{b1})_{g-g} = -(q_z)_{g-g} - \frac{2}{3}\left[\frac{1}{a_1}\left(\frac{de_b}{dz}\right)_1 + \frac{1}{a_2}\left(\frac{de_b}{dz}\right)_2\right]_{g-g} \tag{15-69}$$

Second derivatives with z have been taken to be zero by noting that in *either* region (15-47) gives

$$q_z = -\frac{4}{3a}\frac{de_b}{dz} \tag{15-70}$$

The q_z must be constant, since no heat sources or sinks are present. Therefore, in either region,

$$\frac{d^2 e_b}{dz^2} = 0 \tag{15-71}$$

Also, q_z must be the *same in either region* because the radiative flux is continuous across the interface. Therefore (15-70) can be substituted for the derivatives in (15-69) to give

$$(e_{b2} - e_{b1})_{g-g} = -(q_z)_{g-g} - \frac{2}{3}\left[\frac{1}{a_1}\left(-\frac{3a_1}{4}\right)q_z + \frac{1}{a_2}\left(-\frac{3a_2}{4}\right)q_z\right]_{g-g} \tag{15-72}$$

This reduces to

$$(e_{b2} - e_{b1})_{g-g} = -(q_z)_{g-g} + (q_z)_{g-g} = 0 \tag{15-73}$$

so that *no discontinuity in emissive power exists in this case.*

Consider now the presence of uniform volumetric energy sources of magnitude G_1 and G_2 in regions 1 and 2, respectively. Now the flux gradient in the z direction is given in either region by

$$\frac{dq_z}{dz} = G = -\frac{4}{3a}\frac{d^2 e_b}{dz^2} \tag{15-74}$$

In this case, the second derivatives of e_b with respect to z are obviously not zero, and (15-61) becomes

$$(e_{b2} - e_{b1})_{g-g} = -(q_z)_{g-g} + \left\{ -\frac{2}{3}\left[\frac{1}{a_1}\left(\frac{de_b}{dz}\right)_1 + \frac{1}{a_2}\left(\frac{de_b}{dz}\right)_2\right]\right.$$
$$\left. +\frac{1}{2}\left[\frac{1}{a_1{}^2}\left(\frac{d^2 e_b}{dz^2}\right)_1 - \frac{1}{a_2{}^2}\left(\frac{d^2 e_b}{dz^2}\right)_2\right]\right\}_{g-g} \tag{15-75}$$

Again, (15-70) must hold in either region. At the interface between the two media, since the flux is continuous,

$$q_{z,g-g} = -\frac{4}{3a_1}\left(\frac{de_b}{dz}\right)_{1,g-g} = -\frac{4}{3a_2}\left(\frac{de_b}{dz}\right)_{2,g-g} \tag{15-76}$$

Substituting (15-74) and (15-76) into (15-75) to eliminate the first and second derivatives of e_b gives

$$(e_{b2} - e_{b1})_{g-g} = -(q_z)_{g-g} + \left[-\frac{2}{3}\left(\frac{1}{a_1}\frac{-3a_1}{4}q_z + \frac{1}{a_2}\frac{-3a_2}{4}q_z\right)\right.$$
$$\left. +\frac{1}{2}\left(\frac{1}{a_1{}^2}\frac{-3a_1}{4}G_1 - \frac{1}{a_2{}^2}\frac{-3a_2}{4}G_2\right)\right]_{g-g}$$

which reduces to

$$(e_{b1} - e_{b2})_{g-g} = \frac{3}{8}\left(\frac{G_1}{a_1} - \frac{G_2}{a_2}\right) \tag{15-77}$$

The discontinuity in emissive power is seen to exist whenever the ratios G_1/a_1 and G_2/a_2 are unequal.

If the present problem is formulated in terms of integral equations [5] rather than by diffusion methods, there results

$$(e_{b1} - e_{b2})_{g-g} = \frac{1}{4}\left(\frac{G_1}{a_1} - \frac{G_2}{a_2}\right) \tag{15-78}$$

The diffusion solution, although giving the correct functional dependence of the emissive-power discontinuity, differs from the exact solution by a factor of $\frac{3}{2}$.

Other diffusion solutions for gray gases. In Table 15-2 solutions are gathered for the temperature distributions and energy transfer in simple geometries involving gray gases contained between gray walls (see Example 15-7 for further analytical details). These equations are derived from the diffusion equations; caution is advised in their application since real gases are usually not gray or optically thick in all wavelength regions. Agreement with exact solutions is sometimes not as good for cylindrical or spherical geometries as for the infinite parallel-plate case. Agreement has been found to be excellent in cylindrical and spherical geometries for all parametric variations so long as the optical thickness is greater than about 7, with better agreement as wall emissivity becomes lower and diameter ratios (D_{inner}/D_{outer}) become larger. A comparison for the cylindrical geometry will be discussed in connection with Fig. 15-11.

Table 15-2 Diffusion-theory predictions of energy transfer and temperature distributions for a gray gas between gray surfaces

Geometry	Relations[a]
Infinite parallel plates	$\psi = \dfrac{1}{(3aD/4) + E_1 + E_2 + 1}$ $\phi(z) = \psi\left[\dfrac{3a}{4}(D-z) + E_2 + \dfrac{1}{2}\right]$
Infinitely long concentric cylinders	$\psi = \dfrac{1}{\dfrac{3}{8}\left[aD_1 \ln\left(\dfrac{D_2}{D_1}\right) + \dfrac{1-(D_1/D_2)^2}{aD_1}\right] + (E_1 + \tfrac{1}{2}) + \dfrac{D_1}{D_2}(E_2 + \tfrac{1}{2})}$ $\phi(r) = \psi\left\{-\dfrac{3}{8}\left[aD_1 \ln\left(\dfrac{D}{D_2}\right) + \dfrac{D_1}{aD_2^2}\right] + (E_2 + \tfrac{1}{2})\dfrac{D_1}{D_2}\right\}$
Concentric spheres	$\psi = \dfrac{1}{\dfrac{3}{8}\left[aD_1\left(1-\dfrac{D_1}{D_2}\right) + 2\dfrac{1-(D_1/D_2)^3}{aD_1}\right] + (E_1 + \tfrac{1}{2}) + \dfrac{D_1^2}{D_2^2}(E_2 + \tfrac{1}{2})}$ $\phi(r) = \psi\left\{-\dfrac{3}{8}\left[aD_1\left(\dfrac{D_1}{D_2} - \dfrac{D_1}{D}\right) + \dfrac{2D_1^2}{aD_2^3}\right] + (E_2 + \tfrac{1}{2})\dfrac{D_1^2}{D_2^2}\right\}$

[a] Definitions: $E_N = (1 - \epsilon_{wN})/\epsilon_{wN}$, $\psi = Q_1/A_1\sigma(T_{w1}^4 - T_{w2}^4)$, $\phi(\xi) = [T^4(\xi) - T_{w2}^4]/(T_{w1}^4 - T_{w2}^4)$, $D = 2r$.

EXAMPLE 15-7 The space between two diffuse-gray spheres (Fig. 15-8) is filled with an optically dense stagnant medium having constant absorption coefficient a. Compute the heat flow Q_1 across the gap from sphere 1 to sphere 2 and the temperature distribution $T(r)$ in the gas, using the diffusion method with jump boundary conditions.

For a gray medium with constant a, Eq. (15-46) gives the net heat flux in the positive r direction as

$$q_r = -\frac{4}{3a}\frac{de_b}{dr} \tag{15-79}$$

From energy conservation, q_r varies with r as $q_r = Q_1/4\pi r^2$. Substitute into (15-79) and integrate from R_1 to R_2 to obtain

$$\frac{Q_1}{4\pi}\int_{R_1}^{R_2}\frac{dr}{r^2} = -\frac{4}{3a}\int_{e_{b1}}^{e_{b2}}de_b \tag{15-80}$$

$$\frac{Q_1}{4\pi}\left(\frac{1}{R_2}-\frac{1}{R_1}\right) = \frac{4}{3a}(e_{b2}-e_{b1}) \tag{15-81}$$

The e_{b1} and e_{b2} are *in the gas* at the boundaries, and the jump boundary conditions are needed to express these quantities in terms of wall values. The jump boundary conditions are given by (15-56) and (15-57) and involve second derivatives which will now be found. By integrating (15-80) from R_1 to r, one obtains

$$e_b(r) - e_{b1} = \frac{3aQ_1}{16\pi}\left(\frac{1}{r}-\frac{1}{R_1}\right) \tag{15-82}$$

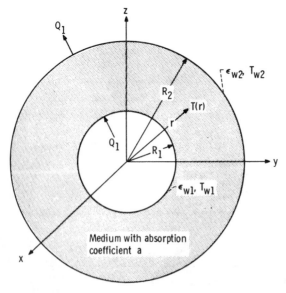

Figure 15-8 Radiation across gap between concentric spheres with intervening medium of constant absorption coefficient.

Substitute $r = (x^2 + y^2 + z^2)^{1/2}$ and differentiate twice with respect to x to obtain

$$\frac{\partial^2 e_b(r)}{\partial x^2} = -\frac{3aQ_1}{16\pi} \frac{(x^2 + y^2 + z^2)^{3/2} - 3x^2(x^2 + y^2 + z^2)^{1/2}}{(x^2 + y^2 + z^2)^3} \tag{15-83}$$

Similarly for the y and z directions.

In the boundary condition (15-56) the point 2 can be conveniently taken in Fig. 15-8 at $x = y = 0$ and $z = R_2$. This gives

$$\left[\frac{\partial^2 e_b(r)}{\partial x^2}\right]_2 = \left[\frac{\partial^2 e_b(r)}{\partial y^2}\right]_2 = -\frac{3aQ_1}{16\pi}\frac{1}{R_2{}^3}$$

$$\left[\frac{\partial^2 e_b(r)}{\partial z^2}\right]_2 = \frac{3aQ_1}{8\pi}\frac{1}{R_2{}^3}$$

Also $(q_z)_2 = Q_1/4\pi R_2{}^2$. Substituting into (15-56) gives

$$e_{b2} - e_{bw2} = \left(\frac{1}{\epsilon_{w2}} - \frac{1}{2}\right)\frac{Q_1}{4\pi R_2{}^2} - \frac{3Q_1}{32a\pi}\frac{1}{R_2{}^3} \tag{15-84}$$

Similarly, at the inner sphere boundary, from (15-57),

$$e_{bw1} - e_{b1} = \left(\frac{1}{\epsilon_{w1}} - \frac{1}{2}\right)\frac{Q_1}{4\pi R_1{}^2} + \frac{3Q_1}{32a\pi}\frac{1}{R_1{}^3} \tag{15-85}$$

Adding (15-84) and (15-85) gives

$$e_{b2} - e_{b1} = e_{bw2} - e_{bw1} + \frac{Q_1}{4\pi}\left[\frac{1}{R_2{}^2}\left(\frac{1}{\epsilon_{w2}} - \frac{1}{2}\right) + \frac{1}{R_1{}^2}\left(\frac{1}{\epsilon_{w1}} - \frac{1}{2}\right)\right.$$
$$\left. + \frac{3}{8a}\left(\frac{1}{R_1{}^3} - \frac{1}{R_2{}^3}\right)\right]$$

After substituting this into the right side of (15-81), the result is solved for Q_1 to give the ψ in the last entry in Table 15-2.

To obtain the temperature distribution, integrate (15-80) from R_2 to r to obtain

$$e_b(r) - e_{b2} = \frac{3aQ_1}{16\pi}\left(\frac{1}{r} - \frac{1}{R_2}\right)$$

Add (15-84) to eliminate e_{b2}:

$$e_b(r) - e_{bw2} = \frac{3aQ_1}{16\pi}\left(\frac{1}{r} - \frac{1}{R_2}\right) + \left(\frac{1}{\epsilon_{w2}} - \frac{1}{2}\right)\frac{Q_1}{4\pi R_2{}^2} - \frac{3Q_1}{32a\pi}\frac{1}{R_2{}^3} \tag{15-86}$$

This gives the last expression for ϕ in Table 15-2.

15-5.4 Final Remarks on Diffusion Method

The diffusion method is a powerful tool because of its usefulness in treating difficult problems by standard analytical techniques, and it is recommended for use whenever the assumptions used in its derivation are justified. The most stringent assumption is that of *optically thick conditions*, which usually is the assumption limiting application of the method. Because most gases have band spectra, the optically thick regions arise within the band limits. Here, if the radiation-absorption mean free path is quite small, the assumption that only local conditions affect the spectral radiant flux is quite good. At other wavelengths the gas can often be considered transparent, and diffusion methods are then not justified. Care must be taken in applying the diffusion equation only in geometrical and spectral regions where the assumption of an optically thick gas is valid.

The Rosseland mean absorption coefficient should not be used as the criterion for optical thickness. It may have a large value, but the spectral absorption coefficient may be very small in certain spectral regions. Use of the Rosseland mean in such cases may lead to large errors. The remedy is to use wavelength bands in which the spectral absorption coefficient is everywhere large and to evaluate a Rosseland mean for each of these regions. Howell and Perlmutter [6] applied the diffusion solution to a real gas and compared the results to those from an exact formulation by the Monte Carlo method. The agreement was generally not as good as for gray gases. Bobco [7] used a modified diffusion solution to find the directional emissivity for radiation from a semi-infinite slab of isothermal gray scattering-absorbing medium with isotropic scattering. The directional emissivities of the slab were found to differ considerably from a diffuse distribution.

15-6 APPROXIMATIONS BY USING MEAN ABSORPTION COEFFICIENTS

Before we continue to discuss solution methods in radiative transfer, some comments are warranted on the use of mean absorption coefficients formed by integrating over all wavelengths. Using a mean absorption coefficient eliminates carrying out a spectral analysis and integrating over all wavelengths to obtain total energy quantities. The question is whether it is possible to decide in advance what mean absorption coefficient will yield an accurate solution for a particular problem. Let us first examine in detail the mean coefficients that have been defined thus far and their relations to each other.

15-6.1 Some Mean Absorption Coefficients

To this point, three general types of mean absorption coefficient have been defined: the Planck mean (14-31)

$$a_P(T,P) = \frac{\int_0^\infty a_\lambda(\lambda,T,P)e_{\lambda b}(\lambda,T)\,d\lambda}{\sigma T^4} \tag{15-87}$$

the incident mean (14-33)

$$a_i(T,P) = \frac{\int_0^\infty a_\lambda(\lambda,T,P) i_{\lambda,i}(\lambda) \, d\lambda}{\int_0^\infty i_{\lambda,i}(\lambda) \, d\lambda} \tag{15-88}$$

and the Rosseland mean (15-48)

$$a_R(T,P) = \frac{\int_0^\infty [\partial e_{\lambda b}(\lambda,T)/\partial e_b(T)] \, d\lambda}{\int_0^\infty [1/a_\lambda(\lambda,T,P)][\partial e_{\lambda b}(\lambda,T)/\partial e_b(T)] \, d\lambda} \tag{15-89}$$

written here to include all wavelengths.

The incident mean can be conveniently utilized only under restrictive conditions when the incident intensity has a spectral form that remains fixed so that the a_i can be evaluated and tabulated. For example, the situation in which the incident energy is a solar spectrum occurs sufficiently often so that the a_i could be tabulated for this case. The a_i is useful for the transparent-gas approximation when the spectral intensity leaving the boundaries is known, as this spectrum will remain unchanged while traveling through the gas. If the mean intensity $i_{\lambda,i}$ is proportional to a blackbody spectrum *at the temperature of the position for which* $a_\lambda(\lambda,T,P)$ *is evaluated*, that is, $i_{\lambda,i} \propto i'_{\lambda b}(\lambda,T)$, then the incident mean becomes equal to a_P:

$$a_i(T,P) = \frac{\int_0^\infty a_\lambda(\lambda,T,P) i'_{\lambda b}(\lambda,T) \, d\lambda}{\int_0^\infty i'_{\lambda b}(\lambda,T) \, d\lambda} = a_P(T,P) \tag{15-90}$$

At first glance, the Rosseland mean appears to be entirely different in character from a_P and a_i, which are weighted by spectral distributions of energy or intensity. However, one may write (15-46) for a one-dimensional diffusion case,

$$\frac{dq_{\lambda,z}}{d\lambda} = -\frac{4}{3a_\lambda} \frac{de_{\lambda b}(\lambda,T)}{dz} = -\frac{4}{3a_\lambda} \left(\frac{\partial e_{\lambda b}}{\partial T} \frac{dT}{dz} + \frac{\partial e_{\lambda b}}{\partial \lambda} \frac{d\lambda}{dz} \right) \tag{15-91}$$

But $d\lambda/dz$ is zero since λ and z are independent variables, so that, for the *diffusion case only*,

$$\frac{\partial e_{\lambda b}}{\partial T} = \frac{\partial e_{\lambda b}}{\partial e_b} \frac{de_b}{dT} = \frac{(-3a_\lambda/4)(dq_{\lambda,z}/d\lambda)}{dT/dz} \tag{15-92}$$

Substituting (15-92) into (15-89) gives

$$a_R(T,P) = \frac{\int_0^\infty a_\lambda \, dq_{\lambda,z}}{\int_0^\infty dq_{\lambda,z}} \tag{15-93}$$

The Rosseland mean is thus seen to be an average value of a_λ weighted by the local spectral energy *flux* $dq_{\lambda,z}$ through the assumption that the local flux depends only on the local gradient of emissive power and the local a_λ.

For a gray gas the absorption coefficient is independent of wavelength, $a_\lambda(\lambda,T,P)$

$= a(T,P)$, and (15-87) to (15-89) reduce to

$$a_P(T,P) = a_i(T,P) = a_R(T,P) = a(T,P)$$

as would be expected.

Determination of any of the mean coefficients from spectral absorption coefficients usually requires tedious detailed numerical integrations. Even so, if appropriate mean values can be successfully applied to yield reasonably accurate solutions, the solution time can often be considerably decreased.

15-6.2 Approximate Solutions of the Transfer Equations Using Mean Absorption Coefficients

In this section some references are reviewed where mean absorption coefficients have been used in radiative transfer calculations. Solving the transfer equations is considerably simplified when a mean absorption coefficient is used because integrations over wavelength are not needed. Contrast this with exact solutions for real gases, which require that these integrations be performed during each solution. It is not possible to perform all the required integrations in advance so that integrations would not be needed during each specific calculation. For example, the incident energy absorbed at each location depends on the incident mean absorption coefficient a_i, which is weighted according to the incident spectrum. Since this spectrum can have an infinite variety of forms, the a_i cannot be conveniently tabulated in advance. Also the a_λ present in the exponential attenuation terms in Eq. (15-1), for example, cannot be conveniently averaged over wavelength [a possible averaging scheme is given in Eq. (15-94)].

To avoid having to carry out detailed spectral calculations certain approximations are often made. The most common is that the *gray* gas equations can be applied for a *real* gas by substituting an appropriate mean absorption coefficient in place of the a for the gray gas. In Sec. 14-4.4 it was shown that, although the Planck mean may indeed be used in the *part* of the energy balance equation dealing with local emission, the use of the same mean coefficient in the absorption and attenuation terms is invalid except in special cases. Patch [8, 9] has shown, by examination of 40 cases, that simple substitution of the Planck mean in the gray-gas equations leads to errors in total intensities that varied from -43 to 881% from the solutions using spectral properties in the transfer equations and then integrating the spectral results. Reductions in error were found by dividing the intensity into two or more spectral bands and using an individual Planck mean for each band.

In an effort to improve this situation, a number of other mean absorption coefficients have been introduced. Sampson [10] synthesized a coefficient that varies from the Planck mean to the Rosseland mean as the optical depth increases along a path. He found agreement, within a factor of two, with exact solutions for various problems. Abu-Romia and Tien [11] applied a weighted Rosseland mean over optically thick portions of the spectrum, and a Planck mean over optically thin regions and obtained relations for energy transfer between bounding surfaces. Planck and Rosseland mean absorption coefficients for carbon dioxide, carbon monoxide, and water vapor are also given as an aid to such computations.

Patch [8, 9] defined an *effective mean absorption* coefficient as

$$a_e(S,T,P) = \frac{\int_0^\infty a_\lambda(\lambda,T,P) i'_{\lambda b}(\lambda,T) \exp[-a_\lambda(\lambda,T,P)S]\, d\lambda}{\int_0^\infty i'_{\lambda b}(\lambda,T) \exp[-a_\lambda(\lambda,T,P)S]\, d\lambda} \tag{15-94}$$

The values of $a_e(S,T,P)$ can be tabulated as a function of temperature and pressure as for the other mean absorption coefficients. In addition, a_e depends on the path length and must be tabulated as a function of that variable. For S small, a_e approaches a_P. For large S, the exponential term in the integrals causes a_e to approach the minimum value of a_λ in the spectrum considered. In radiative transfer calculations the approximation is made that the real gas, with T and P known variables along S, is replaced along any path by an effective uniform gas with absorption coefficient a_e. The computations are then performed using a_e in the gray-gas equation of transfer. The a_e value used is found by equating $a_e S$ at the T and P of the point to which S is measured, to the optical depth of that point in the real gas. For 40 cases Patch [8, 9] shows agreement of total intensities within -25–28% of the integrated spectral solutions, as compared with the -43–881% agreement using a_P, as discussed above. This method has value in computer-oriented solutions, where the tabulated values of $a_e(S,T,P)$ can be effectively manipulated.

Other methods of using mean coefficients are given in [12–16].

15-7 APPROXIMATE SOLUTION USING EXPONENTIAL KERNEL

In section 14-7.4 the solution for a gray medium between diffuse gray plates was derived in terms of the functions ψ_b and ϕ_b that have been obtained numerically as given in Table 14-1 and Fig. 14-8. In this section an approximate solution will be found for ψ_b and ϕ_b by using an approximation for the exponential integrals in the integral equation for the radiative flux.

The ψ_b and ϕ_b are for radiative equilibrium between parallel black plates. Starting with (14-81) for these conditions, the radiative flux q_r is equal to the flux q being transferred between the plates, and $q_{o,1} = \sigma T_1^4$, $q_{o,2} = \sigma T_2^4$. Then with the dimensionless forms in (14-87), (14-81) becomes

$$\psi_b = 2E_3(\kappa) + 2\int_0^\kappa \phi_b(\kappa^*)E_2(\kappa - \kappa^*)\, d\kappa^* - 2\int_\kappa^{\kappa_D} \phi_b(\kappa^*)E_2(\kappa^* - \kappa)\, d\kappa^* \tag{15-95}$$

As given in Appendix F, the E_2 and E_3 can be approximated by the exponential functions

$$E_2(\kappa) \cong \frac{3}{4}\exp\left(-\frac{3}{2}\kappa\right) \qquad E_3(\kappa) \cong \frac{1}{2}\exp\left(-\frac{3}{2}\kappa\right) \tag{15-96}$$

These approximations are inserted into the integral equation (15-95) to yield

$$\psi_b = e^{-3\kappa/2} + \frac{3}{2}e^{-3\kappa/2}\int_0^\kappa \phi_b(\kappa^*)e^{3\kappa^*/2}\,d\kappa^* - \frac{3}{2}e^{3\kappa/2}\int_\kappa^{\kappa_D}\phi_b(\kappa^*)e^{-3\kappa^*/2}\,d\kappa^* \quad (15\text{-}97)$$

To solve for ψ_b and ϕ_b, Eq. (15-97) is differentiated twice with respect to κ, and to maintain the heat flux constant across the distance between the parallel plates the condition is used that $d\psi_b/d\kappa = 0$. This yields

$$0 = e^{-3\kappa/2} + \frac{4}{3}\frac{d\phi_b}{d\kappa} + \frac{3}{2}e^{-3\kappa/2}\int_0^\kappa \phi_b(\kappa^*)e^{3\kappa^*/2}\,d\kappa^* - \frac{3}{2}e^{3\kappa/2}\int_\kappa^{\kappa_D}\phi_b(\kappa^*)e^{-3\kappa^*/2}\,d\kappa^*$$

$$(15\text{-}98)$$

Equation (15-98) is subtracted from (15-97) and the integrals thereby eliminated to obtain

$$\psi_b = -\frac{4}{3}\frac{d\phi_b}{d\kappa}$$

Since ψ_b is a constant, this is integrated to yield

$$\phi_b = -\frac{3}{4}\psi_b\kappa + C \qquad\qquad (15\text{-}99)$$

where C is an integration constant. The ψ_b and C are found by substituting (15-99) back into the integral equation (15-97) to obtain

$$\psi_b = e^{-3\kappa/2} + \frac{3}{2}e^{-3\kappa/2}\int_0^\kappa \left(-\frac{3}{4}\psi_b\kappa^* + C\right)e^{3\kappa^*/2}\,d\kappa^*$$

$$-\frac{3}{2}e^{3\kappa/2}\int_\kappa^{\kappa_D}\left(-\frac{3}{4}\psi_b\kappa^* + C\right)e^{-3\kappa^*/2}\,d\kappa^*$$

The integrations are carried out, and after simplification this becomes

$$0 = e^{-3\kappa/2}\left(1 - \frac{1}{2}\psi_b - C\right) + e^{3(\kappa-\kappa_D)/2}\left(-\frac{3}{4}\psi_b\kappa_D - \frac{1}{2}\psi_b + C\right)$$

Thus there are two simultaneous equations for ψ_b and C:

$$1 - \frac{1}{2}\psi_b - C = 0 \qquad\qquad -\frac{3}{4}\psi_b\kappa_D - \frac{1}{2}\psi_b + C = 0$$

The solution yields

$$\psi_b = \frac{1}{\frac{3}{4}\kappa_D + 1} \qquad\qquad (15\text{-}100)$$

and $C = 1 - \psi_b/2$, which is substituted into (15-99) to give

$$\phi_b(\kappa) = 1 - \frac{\psi_b}{2}\left(1 + \frac{3}{2}\kappa\right) \tag{15-101}$$

The following table compares (15-100) with the exact results from Table 14-1 and the agreement is quite good for all κ_D.

	Dimensionless radiative flux, $\psi_b(\kappa_D)$	
κ_D	Eq. (15-100)	Table 14-1
0.2	0.8696	0.8491
0.4	0.7692	0.7458
0.6	0.6897	0.6672
1	0.5714	0.5532
1.5	0.4706	0.4572
2	0.4000	0.3900
3	0.3077	0.3016

Reference [17] discusses the use of the substitute kernel for nongray media and for cases in which radiative equilibrium is not present.

15-8 APPROXIMATE SOLUTION USING THE COMPLETE EQUATION OF TRANSFER

In Sec. 15-3, it was found that in certain situations it is possible to neglect one or more terms in the equation of transfer. Solutions using the resulting simplified equation are of course much easier than using the entire equation. In this section some analytical methods are presented that account for *all* terms in the equation of transfer. However, only approximate solutions will be sought so that, while some accuracy may be lost, the ability will be gained to obtain closed-form analytical solutions in many cases. This makes it possible to gain insight into the factors governing the radiative transfer, in addition to obtaining answers that are often of acceptable accuracy.

15-8.1 The Astrophysical Approximations

As mentioned in Chap. 13, much work has been done in the study of stellar structure by analysis of the observed radiation. Quite early in the twentieth century, astrophysicists considered the mathematical properties of the equation of transfer and applied some approximations that remain useful today. However, these approximations were developed for one-dimensional layers of an atmosphere, which is the case most useful in astrophysics, and the extensions to multidimensional problems are not always obvious or possible. In this section, two of these approximations will be examined briefly. For simplicity, scattering is not included. For more detailed treatments, see [18, 19].

The Schuster-Schwarzschild approximation. The simplest approximation is to assume that, for one-dimensional energy transfer, the intensity in the positive direction is isotropic and that in the negative direction has a different value but is also isotropic. This two-flux model is illustrated in Fig. 15-9.

Using (14-41) without scattering, the equation of transfer for the intensity in each hemisphere is written

$$-\frac{\cos\theta}{a_\lambda}\frac{di_\lambda^+(x)}{dx} = i_\lambda^+(x) - i'_{\lambda b}(x) \qquad 0 \leqslant \theta \leqslant \frac{\pi}{2} \tag{15-102a}$$

$$-\frac{\cos\theta}{a_\lambda}\frac{di_\lambda^-(x)}{dx} = i_\lambda^-(x) - i'_{\lambda b}(x) \qquad \frac{\pi}{2} \leqslant \theta \leqslant \pi \tag{15-102b}$$

From the isotropic assumption, the i_λ^+ and i_λ^- do not depend on θ. These equations are now integrated over their respective hemispheres to give

$$-\frac{1}{a_\lambda}\frac{di_\lambda^+(x)}{dx}\int_0^{\pi/2}\cos\theta\sin\theta\,d\theta = i_\lambda^+(x)\int_0^{\pi/2}\sin\theta\,d\theta - i'_{\lambda b}(x)\int_0^{\pi/2}\sin\theta\,d\theta$$

$$\tag{15-103a}$$

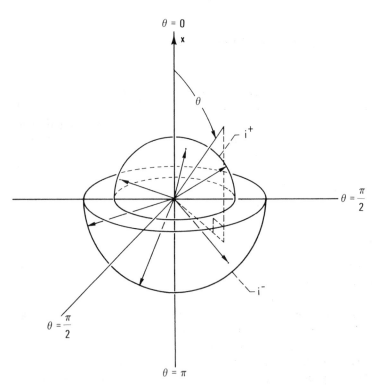

Figure 15-9 Approximation of intensities being isotropic in positive and negative directions.

$$-\frac{1}{a_\lambda}\frac{di_\lambda^-(x)}{dx}\int_{\pi/2}^\pi \cos\theta\sin\theta\,d\theta = i_\lambda^-(x)\int_{\pi/2}^\pi \sin\theta\,d\theta - i'_{\lambda b}(x)\int_{\pi/2}^\pi \sin\theta\,d\theta$$

$$(15\text{-}103b)$$

The integrations are carried out to yield

$$-\frac{1}{2a_\lambda}\frac{di_\lambda^+(x)}{dx} = i_\lambda^+(x) - i'_{\lambda b}(x) \qquad (15\text{-}104a)$$

$$\frac{1}{2a_\lambda}\frac{di_\lambda^-(x)}{dx} = i_\lambda^-(x) - i'_{\lambda b}(x) \qquad (15\text{-}104b)$$

Equations (15-104) with appropriate boundary conditions can be solved by use of an integrating factor as in Sec. 14-3.2. For the geometry of Fig. 15-5, if $i_\lambda^+(0)$ and $i_\lambda^-(D)$ are the spectral intensities at the walls for a gas layer between parallel planes, the $i_\lambda^+(x)$ and $i_\lambda^-(x)$ are (letting $\kappa_\lambda = \int_0^x a_\lambda\,dx$)

$$i_\lambda^+(\kappa_\lambda) = i_\lambda^+(0)\exp(-2\kappa_\lambda) + 2\int_0^{\kappa_\lambda} i'_{\lambda b}(\kappa_\lambda^*)\exp[2(\kappa_\lambda^* - \kappa_\lambda)]\,d\kappa_\lambda^* \qquad (15\text{-}105a)$$

$$i_\lambda^-(\kappa_\lambda) = i_\lambda^-(\kappa_{\lambda D})\exp[2(\kappa_\lambda - \kappa_{\lambda D})] + 2\int_{\kappa_\lambda}^{\kappa_{\lambda D}} i'_{\lambda b}(\kappa_\lambda^*)\exp[2(\kappa_\lambda - \kappa_\lambda^*)]\,d\kappa_\lambda^* \qquad (15\text{-}105b)$$

To illustrate the use of these relations, consider the simplified situation of a gray gas with no internal heat sources between parallel plates. Equations (15-105) retain the same form, but the λ subscripts are omitted. The temperature distribution and heat flux are found from (14-34) and (14-48). Since i^+ and i^- do not depend on θ in the present case, these equations become

$$\sigma T^4(\kappa) = \frac{1}{4}\left[i^+(\kappa)\int_0^{\pi/2} 2\pi\sin\theta\,d\theta + i^-(\kappa)\int_{\pi/2}^\pi 2\pi\sin\theta\,d\theta\right]$$

$$q = 2\pi\left[i^+(\kappa)\int_0^{\pi/2}\cos\theta\sin\theta\,d\theta + i^-(\kappa)\int_{\pi/2}^\pi \cos\theta\sin\theta\,d\theta\right]$$

The integrals are carried out to yield

$$\sigma T^4(\kappa) = \frac{\pi}{2}[i^+(\kappa) + i^-(\kappa)] \qquad (15\text{-}106a)$$

$$q = \pi[i^+(\kappa) - i^-(\kappa)] \qquad (15\text{-}106b)$$

The i^+ and i^- are substituted from (15-105), and the substitution is made that

$i_b' = \sigma T^4/\pi$, yielding

$$\sigma T^4(\kappa) = \frac{1}{2}\left[\pi i^+(0) \exp(-2\kappa) + 2\int_0^\kappa \sigma T^4(\kappa^*) \exp[2(\kappa^* - \kappa)]\, d\kappa^*\right.$$

$$\left. + \pi i^-(\kappa_D) \exp[2(\kappa - \kappa_D)] + 2\int_\kappa^{\kappa_D} \sigma T^4(\kappa^*) \exp[2(\kappa - \kappa^*)]\, d\kappa^*\right]$$

(15-107a)

$$q = \pi i^+(0) \exp(-2\kappa) + 2\int_0^\kappa \sigma T^4(\kappa^*) \exp[2(\kappa^* - \kappa)]\, d\kappa^*$$

$$- \pi i^-(\kappa_D) \exp[2(\kappa - \kappa_D)] - 2\int_\kappa^{\kappa_D} \sigma T^4(\kappa^*) \exp[2(\kappa - \kappa^*)]\, d\kappa^*$$

(15-107b)

Since in the absence of heat sources q does not vary with κ, (15-107b) can be evaluated at any convenient location. Choosing $\kappa = 0$ gives

$$q = \pi i^+(0) - \pi i^-(\kappa_D) \exp(-2\kappa_D) - 2\int_0^{\kappa_D} \sigma T^4(\kappa^*) \exp(-2\kappa^*)\, d\kappa^* \qquad (15\text{-}107c)$$

The integral equation (15-107a) for the gas temperature distribution, and the flux equation (15-107b) are analogous to the exact formulation given by (14-75) and (14-76).

If Eq. (15-106a) is differentiated with respect to κ, the result is

$$\frac{d(\sigma T^4)}{d\kappa} = \frac{\pi}{2}\left[\frac{di^+(\kappa)}{d\kappa} + \frac{di^-(\kappa)}{d\kappa}\right]$$

Now substitute (15-104) for the terms on the right:

$$\frac{d(\sigma T^4)}{d\kappa} = \pi[-i^+(\kappa) + i_b'(\kappa) + i^-(\kappa) - i_b'(\kappa)] = -\pi[i^+(\kappa) - i^-(\kappa)]$$

Comparing this with Eq. (15-106b) yields the diffusion-type relation for the Schuster-Schwarzschild approximation:

$$q = -\frac{d(\sigma T^4)}{d\kappa} = -\frac{1}{a}\frac{de_b}{dx}$$

Chandrasekhar [19] has extended this method as originally developed by Schuster [20] and Schwarzschild [21] by dividing the intensity into mean portions from discrete directions, terming this the *discrete-ordinate* method. The discrete-ordinate method has been shown to be equivalent to the differential-approximation or moment method, which will be described in Sec. 15-8.2 [22]. A comparison of the discrete-ordinate method with the two-flux and six-flux methods is given in [23]. The six-flux method was formed to provide a compromise between accuracy and convenience of use.

The Milne-Eddington approximation. With respect to the intensity, the approximation made independently by Eddington [24] and Milne [25] is the same as that of Schuster and Schwarzschild. For radiation crossing a unit area oriented normal to the x direction, all intensities with positive components in x have a constant value independent of angle, and all intensities with a negative x component have a different constant value; i.e., the local radiation in each direction can be considered isotropic (Fig. 15-9). However, in comparison with the Schuster-Schwarzschild method the approximimation is made one step later and with regard to computing heat fluxes rather than intensities.

Start with the one-dimensional equation of transfer (14-41) without scattering. Then multiply by $d\omega$ and by $\cos\theta\, d\omega$ to obtain the two equations

$$-\frac{\cos\theta}{a_\lambda}\frac{\partial i'_\lambda}{\partial x}\,d\omega = (i'_\lambda - i'_{\lambda b})\,d\omega \tag{15-108a}$$

$$-\frac{\cos^2\theta}{a_\lambda}\frac{\partial i'_\lambda}{\partial x}\,d\omega = \cos\theta(i'_\lambda - i'_{\lambda b})\,d\omega \tag{15-108b}$$

The reason for doing this is that $i'_\lambda \cos\theta$ is related to the heat flux, and Eqs. (15-108) will thus yield a pair of equations involving q_λ. If Eqs. (15-108) are integrated over all solid angles, the result is

$$-\frac{1}{a_\lambda}\int_{\omega=4\pi}\cos\theta\,\frac{\partial i'_\lambda(\theta,x)}{\partial x}\,d\omega = -\frac{1}{a_\lambda}\frac{d^2 q_\lambda}{d\lambda\,dx} = \int_{\omega=4\pi} i'_\lambda(\theta,x)\,d\omega - 4\pi i'_{\lambda b} \tag{15-109a}$$

$$\int_{\omega=4\pi} i'_\lambda(\theta,x)\cos\theta\,d\omega = \frac{dq_\lambda}{d\lambda} = -\frac{1}{a_\lambda}\int_{\omega=4\pi}\cos^2\theta\,\frac{\partial i'_\lambda(\theta,x)}{\partial x}\,d\omega \tag{15-109b}$$

The assumption is now introduced that the i'_λ is isotropic in each hemisphere. Then

$$-\frac{1}{a_\lambda}\frac{d^2 q_\lambda}{d\lambda\,dx} = i_\lambda^+\int_0^{\pi/2} 2\pi\sin\theta\,d\theta + i_\lambda^-\int_{\pi/2}^{\pi} 2\pi\sin\theta\,d\theta - 4\pi i'_{\lambda b} \tag{15-110a}$$

$$\frac{dq_\lambda}{d\lambda} = -\frac{1}{a_\lambda}\left(\frac{di_\lambda^+}{dx}\int_0^{\pi/2} 2\pi\cos^2\theta\sin\theta\,d\theta + \frac{di_\lambda^-}{dx}\int_{\pi/2}^{\pi} 2\pi\cos^2\theta\sin\theta\,d\theta\right) \tag{15-110b}$$

Integrating gives

$$-\frac{1}{a_\lambda}\frac{d^2 q_\lambda}{d\lambda\,dx} = 4\pi\left(\frac{i_\lambda^+ + i_\lambda^-}{2} - i'_{\lambda b}\right) \tag{15-111a}$$

$$\frac{dq_\lambda}{d\lambda} = -\frac{2\pi}{a_\lambda}\frac{d}{dx}\left(\frac{i_\lambda^+ + i_\lambda^-}{3}\right) \tag{15-111b}$$

Eliminating $i_\lambda{}^+ + i_\lambda{}^-$ between these two expressions gives

$$\frac{1}{a_\lambda{}^2}\frac{d^3q_\lambda(x)}{d\lambda\,dx^2} = 3\frac{dq_\lambda(x)}{d\lambda} + \frac{4\pi}{a_\lambda}\frac{di'_{\lambda b}(x)}{dx} \qquad (15\text{-}112)$$

or

$$\frac{d^3q_\lambda(\kappa_\lambda)}{d\lambda\,d\kappa_\lambda{}^2} = 3\frac{dq_\lambda(\kappa_\lambda)}{d\lambda} + 4\frac{de_{\lambda b}(\kappa_\lambda)}{d\kappa_\lambda} \qquad (15\text{-}113)$$

For the situation of a gray gas layer with no internal heat sources, (15-113) is integrated over all wavelengths and $d^2q/d\kappa^2 = 0$, giving

$$q(\kappa) = -\frac{4}{3}\frac{de_b(\kappa)}{d\kappa} \qquad (15\text{-}114)$$

This is the same relation as previously obtained by the diffusion approximation. The boundary conditions to be used in connection with this relation will not be discussed here. Rather they will be considered in the next section, which is a generalization of the Milne-Eddington approximation.

15-8.2 The Differential Approximation

The *differential* approximation reduces the integral equations of radiative transfer in absorbing-emitting media to differential equations by approximating the equation of transfer with a finite set of moment equations. The moments are generated by multiplying the equation of transfer by powers of the cosine between the coordinate direction and the direction of the intensity. This is a generalization of the method of Milne and Eddington, as Eqs. (15-108) are the equation of transfer multiplied by $(\cos\theta)^0$ and $(\cos\theta)^1$, respectively. As will be discussed, the first three moment equations have a definite physical significance so that developing a solution method by this technique has some physical basis. The development will be given in a three-dimensional coordinate system so that general geometries can be treated. The treatment due to Cheng [26–28] will be followed here. Other pertinent references are [22, 29–43].

A rectangular coordinate system with coordinates x_1, x_2, x_3 is shown in Fig. 15-10a. The variation in the intensity at position **r** along the S direction in the direction of the unit vector **s** is given by the equation of transfer (14-4) with $\sigma_{s\lambda} = 0$:

$$\frac{di'_\lambda}{dS} = a_\lambda(S)[i'_{\lambda b}(S) - i'_\lambda(S)]$$

Let a_λ be assumed constant, and integrate over all wavelengths to obtain

$$\frac{di'}{dS} = a[i'_b(S) - i'(S)] \qquad (15\text{-}115)$$

Although the simplified notation $i'(S)$ is used, the intensity is a function of position and angular direction vectors $i'(\mathbf{r},\mathbf{s})$, as shown in Fig. 15-10a. In terms of a three-

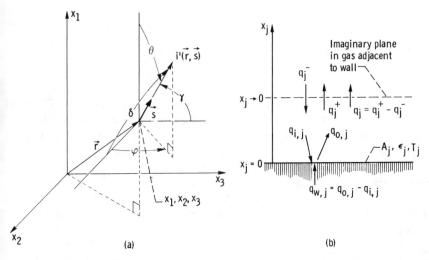

Figure 15-10 Differential approximation. (*a*) Coordinate system showing intensity as a function of position and angle for differential approximation; (*b*) heat fluxes in boundary condition.

dimensional coordinate system x_1, x_2, x_3, Eq. (15-115) can be written as

$$\sum_{j=1}^{3} l_j \frac{\partial i'(\mathbf{r}, \mathbf{s})}{\partial x_j} = a[i_b'(\mathbf{r}) - i'(\mathbf{r}, \mathbf{s})] \tag{15-116}$$

where the l_j's are the direction cosines (Fig. 15-10*a*), $l_1 = \cos\theta, l_2 = \cos\delta, l_3 = \cos\gamma$, and we have used the relation

$$\frac{di'}{dS} = \frac{\partial i'}{\partial x_1}\frac{\partial x_1}{\partial S} + \frac{\partial i'}{\partial x_2}\frac{\partial x_2}{\partial S} + \frac{\partial i'}{\partial x_3}\frac{\partial x_3}{\partial S}$$

The moments of i' are generated by multiplying i' by powers of the l_j and integrating over all solid angles. Some new notation is introduced to designate the moments

$$i'^{(0)}(\mathbf{r}) \equiv \int_{\omega=4\pi} i'(\mathbf{r}, \mathbf{s})\, d\omega$$

$$i_j'^{(1)}(\mathbf{r}) \equiv \int_{\omega=4\pi} l_j i'(\mathbf{r}, \mathbf{s})\, d\omega$$

$$i_{kj}'^{(2)}(\mathbf{r}) \equiv \int_{\omega=4\pi} l_k l_j i'(\mathbf{r}, \mathbf{s})\, d\omega \tag{15-117}$$

. . . .

$$i_{k^{n-1}j}'^{(n)}(\mathbf{r}) = \int_{\omega=4\pi} l_k^{n-1} l_j i'(\mathbf{r}, \mathbf{s})\, d\omega$$

Also

$$i_{kn}^{\prime(n)}(\mathbf{r}) = \int_{\omega=4\pi} l_k^n i'(\mathbf{r},s)\, d\omega$$

The zeroth-order moment $i'^{(0)}$ has the physical significance that dividing it by the speed of light gives the radiation energy density as shown by (14-104). The first moment $i_j^{\prime(1)}$ is the radiative energy flux in the j coordinate direction as shown in (14-106). The second moment $i_{jk}^{\prime(2)}$ divided by the speed of light can be shown to be the radiation stress and pressure tensor. The higher moments have no specific physical significance and are generated by analogy with the first three.

The moment equations are obtained by multiplying (15-116) by powers of the l_i and integrating over all solid angles ω. The zeroth-order moment equation is the integral of (15-116) itself, or noting that i_b' is independent of angle and applying the definitions of $i'^{(0)}$ and $i'^{(1)}$,

$$\sum_{j=1}^{3} \frac{\partial i_j^{\prime(1)}(\mathbf{r})}{\partial x_j} = a[4\pi i_b'(\mathbf{r}) - i'^{(0)}(\mathbf{r})] \tag{15-118}$$

Multiplying (15-116) by l_k ($k = 1, 2, 3$) and integrating give the first-order moment equation

$$\sum_{j=1}^{3} \int_{4\pi} l_k l_j \frac{\partial i'}{\partial x_j}\, d\omega = a\left[i_b' \int_{4\pi} l_k\, d\omega - \int_{4\pi} l_k i'\, d\omega \right]$$

which can be written as

$$\sum_{j=1}^{3} \frac{\partial i_{kj}^{\prime(2)}(\mathbf{r})}{\partial x_j} = -ai_k^{\prime(1)}(\mathbf{r}) \qquad k = 1, 2, 3 \tag{15-119}$$

This procedure is continued to generate, for example, the nth-order moment equation of the form

$$\sum_{j=1}^{3} \frac{\partial i_{kh_j}^{\prime(n+1)}(\mathbf{r})}{\partial x_j} = -ai_{kn}^{\prime(n)}(\mathbf{r}) \qquad k = 1, 2, 3 \tag{15-120}$$

By continuing the process used to obtain (15-118)–(15-120), an infinite set of moment equations can be generated as $n \to \infty$.

The next step is to approximate the infinite set of moment equations by a finite set. When such a truncation is carried out, there will in general be n equations in $n + 1$ unknowns. The additional equation needed to relate the moments and provide a determinate set is obtained by representing the unknown angular distribution of i' as a series of spherical harmonics and then truncating this series after a finite number of terms. This whole procedure becomes quite complicated and will only be briefly treated. It is the final differential approximation obtained in (15-128) that is of most importance here.

The series expansion used to represent i' is

$$i'(\mathbf{r},\mathbf{s}) = \sum_{l=0}^{\infty} \sum_{m=-l}^{+l} A_l^m(\mathbf{r}) Y_l^m(\omega) \tag{15-121}$$

where the $A_l^m(\mathbf{r})$ are coefficients to be determined, and the $Y_l^m(\omega)$ are the normalized spherical harmonics

$$Y_l^m(\omega) = \left[\frac{2l+1}{4\pi} \frac{(l-m)!}{(l+m)!}\right]^{1/2} e^{im\varphi} P_l^m(\cos\theta) \tag{15-122}$$

The $P_l^m(\cos\theta)$ are associated Legendre polynomials of the first kind [44], defined by

$$P_l^m(\cos\theta) = \frac{2^{m+1}}{\pi^{1/2}}(\sin\theta)^m \frac{\Gamma(l+m+1)}{\Gamma(l+\frac{3}{2})} \sum_{k=0}^{\infty} \frac{(m+\frac{1}{2})_k(l+m+1)_k}{k!\,(l+\frac{3}{2})_k}$$

$$\times \sin[(l+m+2k+1)\theta] \tag{15-123}$$

where $\Gamma(\xi)$ is the gamma function, and the notation $(\alpha)_k$ is Pochhammer's symbol

$$(\alpha)_0 = 1 \qquad \alpha \neq 0$$

$$(\alpha)_k = \alpha(\alpha+1)(\alpha+2)\cdots(\alpha+k-1)$$

Equations (15-122) and (15-123) are substituted into (15-121), and the resulting series of equations is truncated. If truncation is carried out by retaining only the $l = 0, 1$ terms, the result is known as the P_1 approximation; if terms $l = 0, 1, 2, 3$ are retained, it is known as the P_3 approximation; etc. (Note that the even-numbered P's are zero.) For the P_1 approximation the series is truncated by setting $A_l^m(\mathbf{r}) = 0$ for $l \geq 2$. This gives an equation for $i'(\mathbf{r},\mathbf{s})$ that is substituted into the first three moment equations to give

$$i'^{(0)}(\mathbf{r}) = 2\pi^{1/2} A_0^0(\mathbf{r}) \tag{15-124}$$

$$i'^{(2)}_{kj}(\mathbf{r}) = \tfrac{2}{3}\pi^{1/2} A_0^0(\mathbf{r})\delta_{kj} \tag{15-125}$$

where δ_{kj} is the Kronecker delta. The form of these equations has been considerably simplified by applying the orthogonality relations for spherical harmonics [45]. Further, note that the first moment of i' is shown by (14-106) to be the energy flux, or, for the j direction,

$$i'^{(1)}_j(\mathbf{r}) = \int_{\omega=4\pi} i'(\mathbf{r},\mathbf{s})l_j\,d\omega = q_j(\mathbf{r}) \tag{15-126}$$

Eliminating $A_0^0(\mathbf{r})$ by combining (15-124) and (15-125) gives

$$\delta_{kj}i'^{(0)}(\mathbf{r}) = 3i'^{(2)}_{kj}(\mathbf{r}) \tag{15-127}$$

Then take (15-127) and substitute (15-118), (15-119), and (15-126) to eliminate $i'^{(0)}$, $i'^{(2)}$, and $i'^{(1)}$, respectively. This results, with $\sigma T^4 = \pi i'_b$, in the *first*, or P_1

differential approximation to the equation of transfer

$$\frac{\partial}{\partial x_k}\left(\frac{1}{a}\sum_{j=1}^{3}\frac{\partial q_j}{\partial x_j}\right) - 4\sigma\frac{\partial T^4}{\partial x_k} - 3aq_k = 0 \qquad k = 1, 2, 3 \tag{15-128a}$$

or, more generally,

$$q_k = -\frac{4\sigma}{3a}\frac{\partial T^4}{\partial x_k} + \frac{1}{3a}\frac{\partial}{\partial x_k}\left(\frac{1}{a}\nabla\cdot\mathbf{q}\right) \qquad k = 1, 2, 3 \tag{15-128b}$$

Note that for radiative equilibrium $q = q_r$ and $\nabla\cdot\mathbf{q}_r = 0$, so that the final term vanishes. As mentioned by Cheng [26], Eq. (15-127) is equivalent to the assumption that the radiation pressure is isotropic, which in turn is equivalent to assuming radiative equilibrium in the gas.

The derivation briefly presented here by use of the moment equations can be developed in a more mathematically rigorous form by use of the spherical-harmonic method, as was done in [27]. The spherical-harmonic method requires considerably more algebraic manipulation and is equivalent to the moment method used here. Krook [22] has shown that the moment method, the spherical-harmonic method, and the discrete-ordinate method are all equivalent.

It is interesting that for $a \gg 1$, Eq. (15-128) reduces to the diffusion approximation as given by (15-47). For $a \ll 1$, Eq. (15-128) reduces to

$$\sum_{j=1}^{3}\frac{\partial q_j}{\partial x_j} = 4a\sigma T^4 + C \tag{15-129}$$

where C is a constant of integration. As pointed out by Cess [29], (15-129) is the correct optically thin limit only in certain cases.

Boundary conditions. Consider a gray boundary A_j that is perpendicular to the x_j direction as shown in Fig. 15-10b. The net radiative flux leaving A_j in the positive x_j direction is

$$q_{o,j} = \epsilon_j\sigma T_j^{\,4} + (1 - \epsilon_j)q_{i,j} \tag{15-130}$$

where q_o and q_i are the outgoing and incoming radiation fluxes. The $q_{i,j}$, however, is equal to the radiation flux in the gas traveling in the negative direction at the wall (Fig. 15-10b):

$$q_{i,j} = q_j^-(x_j \to 0)$$

The net flux in the gas in the positive x direction is $q_j = q_j^+ - q_j^-$ so that $q_{i,j} = -q_j(x_j \to 0) + q_j^+(x_j \to 0)$. Now note that $q_j^+(x_j \to 0)$ is equal to the outgoing flux from the wall $q_{o,j}$ so that

$$q_{i,j} = -q_j(x_j \to 0) + q_{o,j}$$

Substituting into (15-130) gives the boundary condition

$$q_{o,j} = \epsilon_j\sigma T_j^{\,4} + (1 - \epsilon_j)[-q_j(x_j \to 0) + q_{o,j}] \tag{15-131}$$

The outgoing flux can also be written in terms of the intensity leaving A_j as

$$q_{o,j} = \int_{\Omega} l_j i_j' \, d\omega \tag{15-132}$$

where l_j is the cosine of the angle between i_j' and the x_j direction.

A general form for the intensity is now found by substituting (15-122) and (15-123) into (15-121) and truncating as before. The moment equations are used to determine the $A_l^m(\mathbf{r})$, and after considerable manipulation the equation for $i'(\mathbf{r},\mathbf{s})$ is found to be

$$i'(\mathbf{r},\mathbf{s}) = \frac{1}{4\pi} [i'^{(0)}(\mathbf{r}) + 3q_3 \sin \varphi \sin \theta + 3q_1 \cos \theta + 3q_2 \cos \varphi \sin \theta] \tag{15-133}$$

As a specific case, assume the boundary surface is normal to the x_1 direction. Then (15-133) is substituted into (15-132) to give

$$q_{o,1} = \int_{\varphi=0}^{2\pi} \int_{\theta=0}^{\pi/2} \frac{1}{4\pi} [i'^{(0)}(\mathbf{r}) + 3q_3 \sin \varphi \sin \theta + 3q_1 \cos \theta$$

$$+ 3q_2 \cos \varphi \sin \theta]_{x_1 \to 0} \cos \theta \sin \theta \, d\theta \, d\varphi$$

which reduces to

$$q_{o,1} = \frac{i'^{(0)}(x_1 \to 0)}{4} + \frac{q_1(x_1 \to 0)}{2} \tag{15-134}$$

where $q_1(x_1 \to 0)$ is the net heat flux in the x_1 direction in the gas adjacent to the wall. The $i'^{(0)}$ can be eliminated by using (15-118). This gives

$$q_{o,1} = \left(\sigma T^4 - \frac{1}{4a} \sum_{j=1}^{3} \frac{\partial i_j'^{(1)}}{\partial x_j} + \frac{q_1}{2} \right)_{x_1 \to 0} \tag{15-135}$$

Combining (15-135) and (15-131) written for $j = 1$ to eliminate $q_{o,1}$, and using (15-126) to eliminate $i_j'^{(1)}$, gives the boundary condition

$$\left(\frac{1}{\epsilon_1} - \frac{1}{2} \right) q_1(x_1 \to 0) - \frac{1}{4a} \sum_{j=1}^{3} \frac{\partial q_j}{\partial x}\bigg|_{x_1 \to 0} = \sigma[T_1^4 - T^4(x_1 \to 0)] \tag{15-136a}$$

or, in general,

$$\left(\frac{1}{\epsilon_1} - \frac{1}{2} \right) q_S(S \to S_1) - \frac{1}{4a} \nabla \cdot \mathbf{q}|_{S \to S_1} = \sigma[T_1^4 - T^4(S \to S_1)] \tag{15-136b}$$

where S_1 is the location of the boundary. Note that, for the case of radiative equilibrium, the second term in (15-136) goes to zero, that is, $q = q_r$ and $\nabla \cdot \mathbf{q}_r = 0$.

Equations (15-128) and (15-136) comprise the governing equation and boundary condition for the P_1 differential approximation to radiative transfer. Stone and Gaustad [30] give a formulation for nongray gases for the astrophysical boundary condition of zero incident flux at one boundary.

Bayazitoglu and Higenyi [38] have derived the P_3 differential-approximation equations along with the boundary conditions necessary to apply the P_3 equation in radiative transfer. A significant increase in accuracy is reported for the P_3 over the P_1 approximation, in related work by Arpaci and Gozum [39] and Marshak [40]. Although there is an increase in the complexity of the solutions, the equations remain algebraic and of closed form for parallel flat plates. For concentric cylinders and spheres, numerical solution of the resulting fourth-order linear ordinary differential equations is necessary.

Applications of the differential approximation. Consider the case of infinite parallel gray plates (as in Fig. 15-5) with emissivities ϵ_{w1} and ϵ_{w2}, at temperatures T_{w1} and T_{w2}, separated by a distance D and having a gray gas between them. The heat flux through the gas is independent of x and y and by conservation of energy is constant with z, so all $\partial q_j / \partial x_j = 0$. Then the differential equation of transfer (15-128) reduces to

$$q_z = -\frac{4\sigma}{3a}\frac{\partial T^4}{\partial z} \tag{15-137}$$

The boundary condition at $z = 0$ becomes, from (15-136a),

$$\left(\frac{1}{\epsilon_{w1}} - \frac{1}{2}\right)q_z = \sigma[T_{w1}{}^4 - T_g{}^4(z \to 0)] \tag{15-138}$$

where the subscript g is used in T_g to emphasize that this temperature is in the gas. At $z = D$,

$$-\left(\frac{1}{\epsilon_{w2}} - \frac{1}{2}\right)q_z = \sigma[T_{w2}{}^4 - T_g{}^4(z \to D)] \tag{15-139}$$

the negative sign in (15-139) arising because the normal direction from the surface into the gas is in the negative z direction. These are precisely the equations found for the parallel-plate case in Sec. 15-5.3 by use of the diffusion approximation. Thus for infinite parallel plates the predicted temperature distributions and heat transfer by the P_1 differential approximation are the same as for the diffusion approximation. In Table 15-3 are also shown the P_1 differential predictions for concentric·cylinders and spheres. Comparison with Table 15-2 for the diffusion approximation with second-order slip shows that for these geometries the results of the two methods differ by the presence in the diffusion results of a term in the denominator with a factor of $1/a$. This term results from the second-order slip boundary condition, and causes the diffusion results for the cylindrical and spherical cases to approach $\psi = 0$ as the value of a approaches zero.

Figures 15-11 and 15-12 compare the exact solution and the diffusion, P_1 differential, and P_3 differential approximations for the cases of concentric cylinders and concentric spheres [38]. Physically, ψ cannot be larger than unity and should approach unity for small optical thicknesses when the bounding surfaces are black. This is the limit achieved by the Monte Carlo solution in Fig. 15-11 and the exact

Table 15-3 Differential P_1 approximations for energy transfer and temperature distribution for a gray gas between gray surfaces

Geometry	Relations†
Infinite parallel plates	$\psi = \dfrac{1}{(3aD/4) + E_1 + E_2 + 1}$
	$\phi(z) = \psi\left[\dfrac{3a}{4}(D-z) + E_2 + \dfrac{1}{2}\right]$
Infinitely long concentric cylinders	$\psi = \dfrac{1}{\dfrac{3}{8}aD_1 \ln\left(\dfrac{D_2}{D_1}\right) + (E_1 + \frac{1}{2}) + \dfrac{D_1}{D_2}(E_2 + \frac{1}{2})}$
	$\phi(r) = \psi\left[-\dfrac{3}{8}aD_1 \ln\left(\dfrac{D}{D_2}\right) + (E_2 + \frac{1}{2})\dfrac{D_1}{D_2}\right]$
Concentric spheres	$\psi = \dfrac{1}{\dfrac{3}{8}aD_1\left(1 - \dfrac{D_1}{D_2}\right) + (E_1 + \frac{1}{2}) + \dfrac{D_1{}^2}{D_2{}^2}(E_2 + \frac{1}{2})}$
	$\phi(r) = \psi\left[-\dfrac{3}{8}aD_1\left(\dfrac{D_1}{D_2} - \dfrac{D_1}{D}\right) + (E_2 + \frac{1}{2})\dfrac{D_1{}^2}{D_2{}^2}\right]$

† Definitions: $E_N = (1 - \epsilon_{wN})/\epsilon_{wN}$, $\psi = Q_1/A_1\sigma(T_{w_1}{}^4 - T_{w_2}{}^4)$, $\phi(\xi) = [T^4(\xi) - T_{w_2}{}^4]/(T_{w_1}{}^4 - T_{w_2}{}^4)$, $D = 2r$.

solution in Fig. 15-12. The diffusion approximation is based on the assumption of an optically thick medium; hence, it is in error for small optical thicknesses. It is found to give good results for optical thicknesses greater than unity for $D_{inner}/D_{outer} = 0.5$. For smaller D_{inner}/D_{outer} the diffusion results are not as good, especially for spheres, and larger optical thicknesses are required for good agreement with the exact solution. The P_3 differential approximation is better than the P_1 approximation and provides good results for $D_{inner}/D_{outer} = 0.5$. However, the results are poor for smaller diameter ratios, as shown in Fig. 15-12 (recall that physically $\psi \leqslant 1$). Additional information is given in [47–55].

Possibly a more accurate solution for ψ than provided by any of the approximate solutions alone can be found by using the optically thin solution, which is exact in the limit for small τ, and either the diffusion or differential approximations at large τ. A curve faired between these limiting solutions should provide acceptable accuracy at intermediate optical thicknesses.

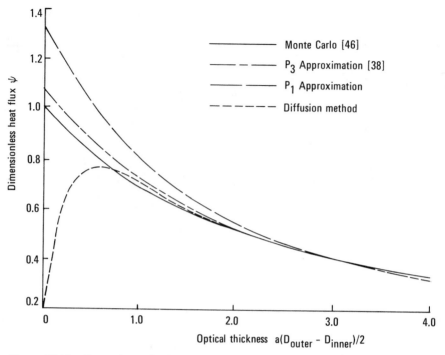

Figure 15-11 Comparison of solutions of energy transfer between infinitely long concentric black cylinders enclosing gray medium; $D_{inner}/D_{outer} = 0.5$.

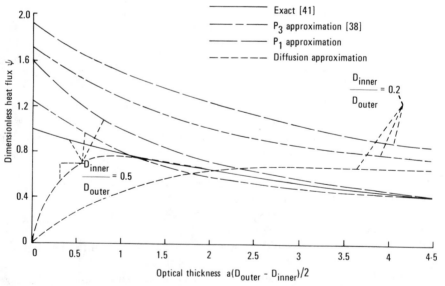

Figure 15-12 Comparison of solutions of energy transfer between black concentric spheres enclosing a gray medium.

REFERENCES

1. Deissler, R. G.: Diffusion Approximation for Thermal Radiation in Gases with Jump Boundary Condition, *J. Heat Transfer*, vol. 86, no. 2, pp. 240–246, 1964.
2. Rosseland, S.: "Theoretical Astrophysics; Atomic Theory and the Analysis of Stellar Atmospheres and Envelopes," Clarendon Press, Oxford, 1936.
3. Howell, John. R.: Radiative Interactions between Absorbing-Emitting and Flowing Media with Internal Energy Generation, *NASA* TN D-3614, 1966.
4. Heaslet, Max A., and Robert F. Warming: Radiative Transport and Wall Temperature Slip in an Absorbing Planar Medium, *Int J. Heat Mass Transfer*, vol. 8, no 7, pp. 979–994, 1965.
5. Howell, John R.: On the Radiation Slip between Absorbing-Emitting Regions with Heat Sources, *Int. J. Heat Mass Transfer*, vol. 10, no. 3, pp. 401–402, 1967.
6. Howell, John R., and Morris Perlmutter: Monte Carlo Solution of Radiant Heat Transfer in a Nongrey Nonisothermal Gas with Temperature Dependent Properties, *AIChE. J.*, vol. 10, no. 4 pp. 562–567, 1964.
7. Bobco, R. P.: "Directional Emissivities from a Two-Dimensional, Absorbing -Scattering Medium, The Semi-Infinite Slab, *J. Heat Transfer*, vol. 89, no. 4, pp. 313–320, 1967.
8. Patch, R. W.: Effective Absorption Coefficients for Radiant Energy Transport in Nongrey, Nonscattering Gases, *J. Quant. Spectrosc, Radiat. Transfer*, vol. 7, no. 4, pp. 611–637, 1967.
9. Patch, R. W.: Approximation for Radiant Energy Transport in Nongray, Non-scattering Gases, *NASA* TN D-4001, 1967.
10. Sampson, Douglas H.: Choice of an Appropriate Mean Absorption Coefficient for Use in the General Grey Gas Equations, *J. Quant. Spectrosc, Radiant. Transfer*, vol. 5, no. 1, pp. 211–225, 1965.
11. Abu–Romia, M. M., and C. L. Tien: Appropriate Mean Absorption Coefficients for Infrared Radiation in Gases, *J. Heat Transfer*, vol. 89, no. 4, pp. 321–327, 1967.
12. Grant, Ian P.: On the Representation of Frequency Dependence in Non-Grey Radiative Transfer, *J. Quant. Spectrosc. Radiat. Transfer*, vol. 5, no. 1, pp. 227–243, 1965.
13. Stewart, John C.: Non-Grey Radiative Transfer, *J. Quant. Spectrosc. Radiative Transfer*, vol. 4, no. 5, pp. 723–729, 1964.
14. Thomas, M., and W. S. Rigdon: A Simplified Formulation for Radiative Transfer, *AIAA J.*, vol. 2, no. 11, pp. 2052–2054, 1964.
15. Lick, Wilbert: Energy Transfer by Radiation and Conduction, *Proc. 1963 Heat Transfer Fluid Mech, Inst*, (Anatol Roshko, Bradford Sturtevant, and D. R. Bartz, ed.), 1963, pp. 14–26.
16. Howe, John T., and Yvonne S. Sheaffer: Spectral Radiative Transfer Approximations for Multicomponent Gas mixtures, *J. Quant. Spectrosc. Radiat, Transfer*, vol. 7, no. 4, pp. 695–701, 1967.
17. Gilles, Scott E., Allen C. Cogley, and Walter G. Vincenti: A Substitute-Kernel Approximation for Radiative Transfer in a Non-Grey Gas Near Equilibrium, with Application to Radiative Acoustics, *Int. J. Heat Mass Transfer*, vol. 12, pp. 445–458, 1969.
18. Kourganoff, Vladimir: "Basic Methods in Transfer Problems; Radiative Equilibrium and Neutron Diffusion," Dover Publications, Inc., New York, 1963.
19. Chandrasekhar, Subrahmanyan: "Radiative Transfer," Dover Publications, Inc., New York, 1960.
20. Schuster, A.: Radiation through a Foggy Atmosphere, *Astrophys. J.*, vol. 21, pp. 1–22, 1905.
21. Schwarzschild, K.: Equilibrium of the Sun's Atmosphere, *Ges. Wiss*, Gottingen, *Nachr.*, *Math-Phys. Klasse*, vol. 1, pp. 41–53, 1906.
22. Krook, Max: On the Solution of Equations of Transfer. I, *Astrophys. J.*, vol, 122, no. 3, pp. 488–497, 1955.
23. Daniel, K. J., N. M. Laurendeau and F. P. Incropera: Predictions of Radiation Absorption and Scattering in Turbid Water Bodies, *ASME J. Heat Transfer*, vol. 101, no. 1, pp. 63–67, 1979.
24. Eddington, A. S.: "The Internal Constitution of the Stars," Dover Publications, Inc., New York, 1959.

25. Milne, F. A.: Thermodynamics of the Stars, "Handbuch der Astrophysik," vol. 3, pp. 65–255, Springer-Verlag, OHG, Berlin, 1930.

26. Cheng, Ping: Two-dimensional Radiating Gas Flow by a Moment Method, *AIAA J.*, vol. 2, no. 9, pp. 1662–1664, 1964.

27. Cheng, Ping: Dynamics of a Radiating Gas with Application to Flow over a Wavy Wall, *AIAA J.*, vol. 4, no. 2, pp. 238–245, 1966.

28. Cheng, Ping: Exact Solutions and Differential Approximation for Multi-Dimensional Radiative Transfer in Cartesian Coordinate Configurations, *Prog. & Astronaut, & Aeronaut.*, vol. 31, p. 269, 1972.

29. Cess, Robert D.: On the Differential Approximation in Radiative Transfer, *Z. Angew. Math. Phys.*, vol 17, pp. 776–781, 1966.

30. Stone, Peter H., and John E. Gaustad: The Application of a Moment Method to the Solution of Non-Gray Radiative-Transfer Problems, *Astrophys. J.*, vol. 134, no. 2, pp. 456–468, 1961.

31. Traugott, S. C.: A Differential Approximation for Radiative Transfer with Application to Normal Shock Structure, *Proc, 1963 Heat Transfer Fluid Mech. Inst.* (Anatol Roshko, Bradford Sturtevant, and D. R. Bartz, eds.), 1963, pp. 1–13.

32. Adrianov, V. N., and G. I. Polyak: Differential Methods for Studying Radiant Heat Transfer, *Int. J. Heat Mass Transfer*, vol. 6, no. 5, pp. 355–362, 1963.

33. Traugott, S. C., and K. C. Wang: On Differential Methods for Radiant Heat Transfer, *Int. J. Heat Mass Transfer*, vol. 7, no. 2, pp. 269–273, 1964.

34. Dennar, E. A., and M. Sibulkin: An Evaluation of the Differential Approximation for Spherically Symmetric Radiative Transfer, *J. Heat Transfer*, vol. 91, no. 1, pp. 73–76, 1969.

35. Finkleman, David: Generalized Differential Approximations in One-dimensional Radiative Transfer, paper 69-WA/HT-45, *ASME*, November, 1969.

36. Selcuk, Nevin, and R. G. Siddall: Two-Flux Spherical Harmonic Modelling of Two-Dimensional Radiative Transfer in Furnaces, *Int. J. Heat Mass Transfer*, vol. 19, pp. 313–321, 1976.

37. Yuen, W. W., and C. L. Tien: A New Differential Formulation of Radiative Transfer and its Application to the One-Dimensional Problem, paper 76-HT-49, *ASME*, August, 1976.

38. Bayazitoglu, Y., and J. Higenyi: The Higher-Order Differential Equations of Radiative Transfer: P_3 Approximation, *AIAA J.*, vol. 17, no. 4, pp. 424–431, 1979.

39. Arpaci, V. S., and D. Gözüm: Thermal Stability of Fluids: The Benard Problem, *Phys. Fluids*, vol. 16, no. 5, pp. 581–588, 1973.

40. Marshek, R. E.: Note on the Spherical Harmonics Method as Applied to the Milne Problem for a Sphere, *Phy Rev.*, vol. 71, pp. 443–446, 1947.

41. Rhyming, I. L.: Radiative Transfer between Two Concentric Spheres Separated by an Absorbing-Emitting Gas, *Int. J. Heat Mass Transfer*, vol. 9, pp. 315–324, 1966.

42. Chou, Y. S., and C. L. Tien: A Modified Moment Method for Radiative Transfer in Non-Planar Systems, *J. Quant. Spectrosc. Radiat. Transfer*, vol. 8, pp. 919–933, 1968.

43. Shvartsburg, A. M., Error Estimation in Differential Approximation to Equation of Transfer, *Zh. Prikl. Mekh. Tekh. Fiz.* No. 5, pp. 9–13, 1976.

44. Abramowitz, Milton A., and Irene A. Stegun (eds.): "Handbook of Mathematical Functions with Formulas, Graphs, and Mathematical Tables," Applied Mathematics Series 55, National Bureau of Standards, 1965.

45. Wylie, Clarence R., Jr.: "Advanced Engineering Mathematics," 2 ed., McGraw-Hill Book Company, New York, 1960.

46. Perlmutter, M., and J. R. Howell: Radiant Transfer through a Gray Gas between Concentric Cylinders Using Monte Carlo, *J. Heat Transfer*, vol. 86, no. 2, pp. 169–179, 1964.

47. Heaslet, Max A., and R. F. Warming: Theoretical Predictions of Radiative Transfer in a Homogenous Cylindrical Medium, *J. Quant. Spectrosc. Radiat, Transfer*, vol. 6, pp. 751–774, 1966.

48. Schmid-Burgk, Johannes: Radiant Heat Flow through Cylindrically Symmetric Media, *J. Quant. Spectrosc. Radiat. Transfer*, vol. 14, pp. 979–987, 1974.

49. Modest, M. F., and D. S. Stevens: Two-Dimensional Radiative Equilibrium of a Gray Medium

between Concentric Cylinders, *J. Quant. Spectrosc. Radiat. Transfer,* vol. 19, pp. 353–365, 1978.

50. Loyalka, S. K.: Radiative Heat Transfer between Parallel Plates and Concentric Cylinders, *Int. J. Heat Mass Transfer,* vol. 12, pp. 1513–1517, 1969.

51. Kesten, Arthur A.: Radiant Heat Flux Distribution in a Cylindrically Symmetric Nonisothermal Gas with Temperature-Dependent Absorption Coefficient, *J. Quant. Spectrosc. Radiat. Transfer,* vol. 8, pp. 419–434, 1968.

52. Dua, Shyam S. and Ping, Cheng: Multi-dimensional Radiative Transfer in Non-Isothermal Cylindrical Media with Non-Isothermal Bounding Walls, *Int. J. Heat Mass Transfer,* vol, 18, pp. 254–259, 1975.

53. Canosa, Jose, and H. R. Penafiel: A Direct Solution of the Radiative Transfer Equation: Application to Rayleigh and Mie Atmospheres, *J. Quant, Spectrosc. Radiat. Transfer,* vol. 13, pp. 21–39, 1973.

54. Amlin, D. W., and S. A. Korpela: Influence of Thermal Radiation on the Temperature Distribution in a Semi-Transparent Solid, *ASME J. Heat Transfer,* vol. 101, no. 1, pp. 76–80, 1979.

55. Modest, M. F.: A Simple Differential Approximation for Radiative Transfer in Non-Gray Gases, *J. Heat Transfer,* vol. 101, no. 4, pp. 735–736, 1979.

PROBLEMS

1 Two infinite parallel gray plates at temperatures T_1 and T_2, having respective emissivities ϵ_1 and ϵ_2, are separated by a distance D. The space between them is filled with a gray gas having a constant absorption coefficient a. Obtain the heat flux being transferred and the temperature distribution in the gas using the strong transparent approximation.

$$Answer: \quad q = \frac{\sigma(T_1{}^4 - T_2{}^4)}{1/\epsilon_1 + 1/\epsilon_2 - 1} ; \quad \frac{T^4 - T_2{}^4}{T_1{}^4 - T_2{}^4} = \frac{1/\epsilon_2 - 1/2}{1/\epsilon_1 + 1/\epsilon_2 - 1}$$

2 A sphere of high-temperature optically thin gray gas of fixed volume is being cooled by radiation loss to a cold black boundary (neglect emission from the boundary). At any instant the entire gas can be considered isothermal and the emission approximation used to compute the radiative loss. Heat conduction is neglected. Write the transient energy equation and solve to obtain the gas temperature as a function of time, starting from an initial temperature T_i.

3 A spherical cavity 20 cm in diameter is filled with a gray medium having an absorption coefficient of 0.1 cm^{-1}. The cavity surface is black and is at a uniform temperature of 500 K. When the medium is first placed in the cavity, the medium is cold. For this condition use the cold medium approximation to estimate the heat flux radiated out of the small opening as shown.

Answer: 0.107 W/cm^2.

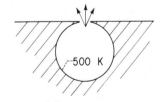

4 The space between two concentric diffuse-gray cylinders is filled with an optically dense stagnant medium having a constant absorption coefficient a. Compute the heat transfer across the gap from the inner to the outer cylinder and the radial temperature distribution in the medium by using the diffusion method with jump boundary conditions.

Answer: See Table 15-2.

5 Two parallel gray plates are 5 cm apart. Their temperatures and emissivities are $T_1 = 700$ K, $\epsilon_1 = 0.8$, $T_2 = 500$ K, $\epsilon_2 = 0.3$. Compute the heat flux being transferred across the gap between the plates by radiation when the gap is a vacuum and when the gap is filled with a gray medium having absorption coefficient $a = 0.6$ cm^{-1}. (Use diffusion theory as an approximation.)

   Answer: 0.284 W/cm^2, 0.174 W/cm^2.

6 Two gray parallel plates at temperatures T_1 and T_2 and with emissivities ϵ_1 and ϵ_2 are separated by an optically thick gray medium. The medium has a uniform volumetric source of G within it. Compute the temperature distribution within the gap by use of the diffusion method with jump boundary conditions.

7 Two infinite parallel plates are separated by a distance D. The plates are both locally at temperature T_0, and have emissivity ϵ. A gray gas with Rosseland mean absorption coefficient a_R is flowing in the gap between the plates with a uniform velocity U parallel to the plates. The gas has density ρ and heat capacity c_p. Heat is added at both plates at the rate q_w per unit area. Neglecting conduction in the gas, derive an equation for the temperature distribution in the gas by using the diffusion method. (*Hint:* See [1].)

8 A large plate of translucent glass is laid upon a sheet of polished aluminium. The aluminium is kept at a temperature of 500 K and has emissivity 0.03. The glass is 10 cm thick, and has a mean absorption coefficient $a_R = 4$ cm^{-1}. A transparent liquid flows over the exposed face of the glass and maintains that face at a temperature of 250 K.

   (*a*) What is the heat flux through the glass plate?
   (*b*) What is the temperature distribution in the glass plate?

   Neglect conduction in the glass in your calculations, and for simplicity assume that the index of refraction is unity for both the glass and the liquid.

9 Consider a gray absorbing medium between parallel diffuse gray plates at temperatures T_1 and T_2. In the optically thick limit, show that the exponential kernel approximation yields the same result for radiative heat transfer as obtained from the diffusion approximation with jump boundary conditions.

10 A long cylinder 10 cm in diameter is surrounded by another cylinder 20 cm in diameter. The surfaces are gray, the inner cylinder is at $T_1 = 850$ K, $\epsilon_1 = 0.4$, and the outer cylinder is at $T_2 = 1025$ K, $\epsilon_2 = 0.7$. What is the heat transfer to the inner cylinder per unit length for vacuum between the cylinders? The space between the cylinders is filled with a gray medium having absorption coefficient $a = 0.8$ cm^{-1}. Compute the heat flow using the P_1 differential approximation and the diffusion approximation.

   Answer: 3.858, 2.305, 2.287 kW/m.

SIXTEEN

AN INTRODUCTION TO THE MICROSCOPIC BASIS FOR GAS PROPERTIES AND SCATTERING BY PARTICLES

16-1 INTRODUCTION

In earlier chapters, thermal radiative transfer through absorbing, emitting, and scattering media has been treated chiefly from a macroscopic viewpoint. The atomic and molecular processes that govern the macroscopic effects have been only briefly described in Chap. 13. Since the physical phenomena are understood on atomic and molecular levels, much of the macroscopic material can be developed or at least interpreted on a fundamental basis. In the first portion of this chapter, the nomenclature of some of the atomic and molecular processes is introduced, and a qualitative discussion is given to relate these processes to the macroscopic approach. Only a little is done to give methods of quantitative analysis; rather, it is intended that for further information the reader will consult more specialized literature, having gained some knowledge of the background material from the present treatment.

The next portion of the chapter is on band absorption models and correlations for gases. The band behavior is developed by first describing the absorption of a single spectral line. Then models are presented of an absorption band constructed of many broadened lines. There are shown to be two effects of gas pressure: one is in broadening of the spectral lines, and the second is in increasing the number of gas molecules that interact with the radiation along a path. Limiting cases are considered for overlapping lines and for nonoverlapping lines that are either weak or strong absorbers. The trends indicated by the band models are shown to lead to practical correlations for band behavior.

The final section of the chapter is on scattering, which is the redirection of radiation by interaction with particles or molecules within the medium. The interaction can result from a combination of reflection, refraction, and diffraction. The amount of energy scattered, and its directional distribution, are given for various types of particles. A method for dealing with particles in a cloud is also developed.

The sections dealing with the Bohr model and the Schrödinger equation are presented in terms of *angular frequency* $\Omega = 2\pi\nu = 2\pi c/\lambda$ (rad/s), as this is the usual cyclical quantity related to these topics. The use of Ω gives some of the equations a shorter form resulting from the elimination of 2π factors. The development of band correlations, however, is given in terms of *wave number* $\eta = \nu/c = 1/\lambda$ as commonly used in that area of study. The information on scattering is given in terms of wavelength λ, as usual in this text.

16-2 SYMBOLS

A	area
$\bar{A}$	effective line- or bandwidth
A_{ij}	Einstein coefficient for spontaneous emission
A_0	width of band
a	absorption coefficient
B_{ij}	Einstein coefficient for absorption or induced emission
c	speed of light in a medium
c_0	speed of light in a vacuum
D	particle diameter
E	energy
e	electronic charge; emissive power
f	oscillator strength, Eq. (16-49)
$G(\bar{n})$	function of $\bar{n}$ in Rayleigh scattering relation, Eq. (16-93)
h	Planck's constant
$\hbar$	modified Planck's constant, $h/2\pi$
i	radiation intensity
J_1	Bessel function of first kind of order 1
k	Boltzmann's constant
l	separation constant in solution for ψ
M	mass of molecule or nucleus
m_e	mass of electron
m_l	separation constant in solution for ψ
m_p	mass of particle
N	number density, particles per unit volume
n	separation constant in solution for ψ; an integer; refractive index
$\bar{n}$	complex refractive index, $n - i\kappa$
P	total pressure
p	partial pressure, momentum of photon
Q	energy per unit time

R	radius
R_e	equilibrium distance between atoms
Ry	Rydberg constant
r	radial coordinate
r_e	radius of electron orbit
r_0	classical electron radius
S	coordinate along path of radiation
S_c	integrated absorption coefficient
S_{ij}	integrated line absorption
s	scattering cross section
T	absolute temperature
T_0	reference temperature of 100 K in Tables 16-3 and 16-4
t	time
V	potential energy; volume
v	velocity
X	mass path length, ρS
x,y,z	coordinates in cartesian system
α_p	polarizability
β	scattering angle; pressure broadening parameter
Δ	half-width of spectral line
δ	average spacing between lines of absorption band
ϵ	emittance
η	wave number
θ	cone angle
λ	wavelength
μ	reduced mass
ν	frequency
ρ	density of gas; reflectivity
σ_s	scattering coefficient
τ	time-dependent portion of Ψ
Φ	scattering phase function
φ	circumferential angle
Ψ	time-dependent wave function
ψ	time-independent wave function
Ω	angular frequency
ω	solid angle

Subscripts

a	absorbed
b	blackbody
c	for collisional broadening
D	for Doppler broadening
e	electron; equilibrium; emitted
i,j	energy state i or j
l	lth band
N_2	nitrogen

n	allowable particle orbits; for natural broadening
p	projected; photon; particle
s	scattered or scattering
ν	dependent upon frequency
Ω	dependent upon angular frequency
λ	dependent on wavelength

Superscripts

$'$	directional quantity
$+$	true value, not modified by addition of induced emission
$*$	complex conjugate

16-3 SOME ELEMENTS OF QUANTUM THEORY

16-3.1 Bohr Model of Hydrogen Atom

Classical physics is unable to account for the line-emission spectrum of gases. To account for line emission, Bohr in 1913 introduced his theory of the atom, and in so doing departed radically from the classical picture. Bohr's atom is constructed in its most simple form by considering the hydrogen atom and making three basic postulates:

1 An electron moves in a circular orbit without decay of energy, and the orbit is determined such that the electron is subject to a balance of dynamic and electrostatic forces.
2 Only stable orbits exist such that the angular momentum of the electron is *quantized*; that is, the angular momentum of the electron takes on only discrete values.
3 The difference in energy for electrons present in different stable orbits is equal to the energy of a photon required to produce a change in orbit.

To write these postulates in mathematical form, consider an electron of mass m_e and negative charge e in a circular orbit of radius r_e around a stationary hydrogen nucleus. The Coulomb force of attraction on the electron exerted by the nucleus is e^2/r_e^2, while the outward force from centripetal acceleration is $m_e r_e \Omega_e^2$, where Ω_e is the orbital angular frequency of the *electron*. This yields the force balance

$$\frac{e^2}{r_e^2} = m_e r_e \Omega_e^2 \tag{16-1}$$

The energy of the electron consists of potential [see Eq. (16-19)] and kinetic energy, giving

$$E = -\frac{e^2}{r_e} + \frac{m_e r_e^2 \Omega_e^2}{2} \tag{16-2}$$

By use of (16-1) this can be written as

$$E = -\frac{e^2}{2r_e} \qquad (16\text{-}3)$$

so that the electron energy has a reference level of zero as r_e becomes infinite.

Since the electron is accelerating, classical electrodynamic theory would dictate that it should radiate energy and consequently slow down and spiral into the nucleus. To provide radiation in the form of spectral lines, however, Bohr considered that the radiative energy loss must occur in finite steps so that the energy given by Eq. (16-3) would consist of a series of discrete levels. It was postulated that the allowable states would be those for which the electron angular momentum is a multiple of Planck's constant. Thus,

$$m_e r_{e,n}{}^2 \Omega_{e,n} = nh \qquad n = 1, 2, 3, \ldots, i, j, \ldots \qquad (16\text{-}4)$$

Equation (16-1) written for the nth orbit is combined with (16-4) to eliminate $\Omega_{e,n}$. This gives the allowable radii of the electron orbits as

$$r_{e,n} = \frac{n^2 \hbar^2}{m_e e^2} \qquad (16\text{-}5)$$

Equation (16-5) is used for the radius in (16-3) to yield the discrete energy states as

$$E_n = -\frac{e^4 m_e}{2n^2 \hbar^2} \qquad (16\text{-}6)$$

Now consider the transition between two energy states. The difference in energy between the jth and ith states is obtained from (16-6) as

$$E_j - E_i = \frac{e^4 m_e}{2\hbar^2} \left(\frac{1}{i^2} - \frac{1}{j^2} \right) \qquad (16\text{-}7)$$

The energy of the photon associated with a transition of the electron between the two stable orbits i and j is equal to $\hbar \Omega_{ij}$, where Ω_{ij} is the *photon* angular frequency. A spectral line at Ω_{ij} will be produced by an electron transition from E_j to E_i. Then (16-7) can be written as

$$E_j - E_i = \hbar \Omega_{ij} = \text{Ry} \left(\frac{1}{i^2} - \frac{1}{j^2} \right) \qquad (16\text{-}8)$$

where Ry is the Rydberg constant,*

$$\text{Ry} = \frac{e^4 m_e}{2\hbar^2} = 13.605 \text{ eV} \qquad (16\text{-}9)$$

and has units of energy. If the transition is considered between the lowest-energy orbit (ground state, $i = 1$) and the highest-energy orbit ($j = \infty$), it is seen that

$$E_\infty - E_1 = \text{Ry} \qquad (16\text{-}10)$$

*This should not be confused with another definition for the Rydberg constant that is also used: $\text{Ry} = e^4 m_e / 4\pi c_0 \hbar^3 = 1.0974 \times 10^7 \text{ m}^{-1}$.

This is the energy required to remove the electron from the atom, and Ry is considered to be the *ionization potential* for the hydrogen atom. Equation (16-8) is found to predict exactly the frequencies of the spectral line series of atomic hydrogen. The series corresponding to $i = 1, 2, 3$, and 4 have been named after their discoverers as follows: Lyman, $i = 1$, $j = 2, 3, 4, \ldots$; Balmer, $i = 2, j = 3, 4, 5, \ldots$; Paschen, $i = 3$, $j = 4, 5, 6, \ldots$; and Brackett, $i = 4, j = 5, 6, 7, \ldots$. For other atoms the prediction of line frequencies is not accurate, and in many cases fails completely. For atoms with a single electron in the outer shell, the theory can be patched up to yield adequate results.

16-3.2 Schrödinger Wave Equation

Because the Bohr theory is a rather curious mixture of classical and quantum ideas, and because the predictions of the theory are not adequate, a better formulation is required. This formulation is given by modern quantum theory. The price we pay for the more adequate predictions is a loss of the clear physical picture presented by the Bohr atom.

In 1924, Louis de Broglie suggested that matter could have wave properties associated with it in much the same way as a photon can be assigned a mass. The momentum of a *photon* is given by

$$p = \frac{\hbar\Omega}{c} = \frac{h}{\lambda} \tag{16-11}$$

Then by analogy, for a particle of mass m_p and velocity v, an associated wavelength can be found by letting $m_p v = h/\lambda$, giving the wavelength associated with the particle as

$$\lambda = \frac{h}{m_p v} \tag{16-12}$$

The idea that a particle of matter can have an associated wavelength seemed to be useless; however, experimental confirmation came in the form of diffraction patterns produced by the scattering of electrons from crystals. The patterns, which are a wave phenomenon were predictable if the electrons (assumed to be particles of matter) were given the wavelength predicted by (16-12).

If matter indeed has the properties of waves, then some form of equation should predict the behavior of the wave field. The intensity of the wave field is an indication of the particle density in the same manner as an electromagnetic field intensity is related to photon density. Where the waves interfere constructively, we expect to find a particle, and we expect this interference to occur over relatively small regions of space. The equation that is found to provide this behavior for the waves was derived by Schrödinger in 1926 and is known as the *Schrödinger wave equation*. In the *time-dependent form* it is, for a particle,

$$\frac{-h^2}{2m_p}\nabla^2\Psi + V\Psi = \frac{-h}{i}\frac{\partial\Psi}{\partial t} \tag{16-13}$$

where V is the time-dependent potential energy of the particle in the coordinates of ∇^2 and i is the imaginary root $i = \sqrt{-1}$. Equation (16-13) cannot be derived from a physical model as can the classical wave equation in Chap 4. Rather, the justification for this form is that it predicts observable effects. We are left to construct physical models to fit the mathematical equation if we desire them, rather than to fit an equation to the physical model, as is usual.

Schrödinger showed that the *wave function* Ψ has certain boundary conditions that are physically meaningful; it is a single-valued, finite, continuous, and vanishes at infinity. When these constraints on Ψ are observed, it is found that the solutions to (16-13) are eigenvalue-eigenfunction solutions. It is this fact that imposes quantization on a system through the mathematics: quantization is not *assumed*, but is a *result* of the boundary conditions on the Schrödinger equation.

Although the function Ψ has no direct physical interpretation, it corresponds in some ways to the amplitude in the classical wave equation. Since the intensity of the wave field is an indication of particle density, a more useful interpretation would be to consider Ψ as a probability density. However, since Ψ is in general a complex function, it is more convenient to treat the real quantity $\Psi\Psi^* = |\Psi|^2$ as the probability density, where Ψ^* is the complex conjugate of Ψ. The square of the magnitude of the wave function $|\Psi|^2$ gives the probability density of finding a particle of matter in a given location at any instant. This is analogous to the relation between radiative intensity [and hence photon density from (14-103)] and the square of the amplitude of the electric field intensity as given by (4-26).

To satisfy the boundary conditions, it is possible to obtain a solution to the time-dependent Schrödinger equation by separation of variables if the potential energy V does not depend on time. The separated product has the form

$$\Psi(x,y,z,t) = \psi(x,y,z)\tau(t) \tag{16-14}$$

Inserting this into Eq. (16-13) gives the two equations

$$\nabla^2\psi + \frac{2m_p}{\hbar^2}(\epsilon - V)\psi = 0 \tag{16-15}$$

and

$$\frac{d\tau}{dt} + \frac{\epsilon i}{\hbar}\tau = 0 \tag{16-16}$$

where ϵ is the separation constant.

Equation (16-16) has the solution (within an arbitrary multiplying constant)

$$\tau = \exp\left(-i\frac{\epsilon}{\hbar}t\right) = \cos\left(\frac{\epsilon}{\hbar}t\right) - i\sin\left(\frac{\epsilon}{\hbar}t\right) \tag{16-17}$$

and therefore substituting into (16-14) gives

$$\Psi = \psi(x,y,z)\exp\left(-i\frac{\epsilon}{\hbar}t\right) \tag{16-18}$$

[Note that the i in (16-17) and (16-18) is the imaginary number, not the ith energy state.] We now need to find ψ, the solution to the time-independent form of Schrödinger's wave equation (16-15), to determine the complete wave function Ψ.

The wave equation will be considered here specifically for determining the energy of an electron around the nucleus of the hydrogen atom. The potential energy of the electron is (based on zero potential energy at $r \to \infty$)

$$V = \int_{\infty}^{r_e} F \, dr = \int_{\infty}^{r_e} \frac{e^2}{r^2} \, dr = \frac{-e^2}{r_e} \tag{16-19}$$

where F is the Coulomb force between the electron and the nucleus. Using V from (16-19) results in the time-independent Schrödinger equation (16-15) becoming (the subscript of r_e is dropped for simplicity)

$$\nabla^2 \psi + \frac{2\mu}{\hbar^2} \left(\epsilon + \frac{e^2}{r} \right) \psi = 0 \tag{16-20}$$

The particle mass has been replaced by $\mu = M m_e / (M + m_e)$, the reduced mass of the nucleus-electron system, where M is the mass of the nucleus. The use of μ accounts, in the dynamics of the electron-nucleus system, for the slight motion of the nucleus around the center of mass, an effect that was neglected in (16-1).

In spherical coordinates, (16-20) is

$$\frac{1}{r^2} \frac{\partial}{\partial r} \left(r^2 \frac{\partial \psi}{\partial r} \right) + \frac{1}{r^2 \sin \theta} \frac{\partial}{\partial \theta} \left(\sin \theta \frac{\partial \psi}{\partial \theta} \right) + \frac{1}{r^2 \sin^2 \theta} \frac{\partial^2 \psi}{\partial \varphi^2} + \frac{2\mu}{\hbar^2} \left(\epsilon + \frac{e^2}{r} \right) \psi = 0 \tag{16-21}$$

A separation of variables can be applied to obtain ψ as a function of r, θ, and φ, where θ is the cone angle:

$$\psi = R(r)\Theta(\theta)\Phi(\varphi) \tag{16-22}$$

Substituting into (16-21) results in the three separated equations

$$\frac{-1}{\Phi} \frac{d^2\Phi}{d\varphi^2} = m_l^2 \tag{16-23}$$

$$\frac{1}{R} \frac{d}{dr} \left(r^2 \frac{dR}{dr} \right) + \frac{2\mu}{\hbar^2} \left(\epsilon + \frac{e^2}{r} \right) r^2 = l(l + 1) \tag{16-24}$$

$$\frac{-1}{\Theta \sin \theta} \frac{d}{d\theta} \left(\sin \theta \frac{d\Theta}{d\theta} \right) = l(l + 1) - \frac{m_l^2}{\sin^2 \theta} \tag{16-25}$$

where m_l and l are separation constants that are specified as $m_l = 0, \pm 1, \pm 2, \ldots, \pm l; l = 0, 1, 2, \ldots, n - 1$; and $n = 1, 2, \ldots, \infty$. The solution to (16-23) that is used is

$$\Phi = A \exp (i m_l \varphi) \tag{16-26}$$

Equation (16-24) has solutions for R in terms of Laguerre polynomials that involve the arbitrary constant n, and (16-25) has solutions for Θ in terms of Legendre polynomials. The solution for ψ thus depends on the three constants n, l, and m_l, each of which has discrete values. These constants are called the *quantum numbers*, and they define the possible discrete forms of ψ. The n is called the principle quantum number, l is the azimuthal or orbital angular-momentum quantum number, and m_l is the magnetic quantum number.

The radii for which the wave function has a large expectation value should correspond to the positions at which electrons are found with high probability. These radii are found by the usual spatial averaging techniques:

$$r_n = \frac{\int_{\text{all space}} \psi_n r \psi_n^* dV}{\int_{\text{all space}} \psi_n \psi_n^* dV} = \int_{\text{all space}} \psi_n r \psi_n^* dV \qquad (16\text{-}27)$$

where the denominator is unity by virtue of ψ being normalized as a probability density function. When this integration is carried out, the radii for various integer values of n are found to be exactly those predicted by Bohr [Eq. (16-5)] for the hydrogen atom. It is again emphasized that the discrete values of r are imposed by the mathematics of the Schrödinger equation, and not by assumption as in the Bohr theory.

Each of the linearly independent solutions for ψ specifies a *quantum state* of the electron. The energy of the electron in the hydrogen atom is found to be independent of the quantum numbers l and m_l. Thus a large number of quantum states that correspond to the various l and m_l have the same energy. Such states are called *degenerate*. By summing the number of such states that are present for a given energy, it is found that there are $2n^2$ degenerate states per energy level E_n. (Actually, the treatment given here predicts n^2 degenerate states; the inclusion of electron spin provides the factor of 2.)

In statistical mechanics, it is assumed that every quantum state in the atom is equally likely. Because there are $2n^2$ quantum states in a given energy level E_n, the factor $2n^2$ is called the *statistical weight* or *multiplicity* of energy level E_n in the hydrogen atom. Other atoms have other statistical weights. Knowing the statistical weight allows us to treat the total number of transitions per unit time occurring between two energy levels in terms of a transition rate averaged over all states in one level times the statistical weight of that level. Detailed examination of each degenerate state is unnecessary.

16-4 INDUCED EMISSION AND THE PLANCK DISTRIBUTION

The concept of induced emission was introduced in Sec. 13-5.5. It was noted that measuring the attenuation of a radiant beam traveling through a medium gives no distinct information about induced emission. This is because the induced emission physically combines with the true absorption to produce an effective absorption smaller than the true absorption. As far as radiation attenuation measurements are concerned, the true-absorption and induced-emission effects cannot be separated. Einstein [1, 2] showed, however, that induced emission must exist. Einstein's relatively simple arguments will now be given, as induced emission is employed in the course of a derivation of Planck's blackbody spectral distribution. Without induced emission, certain rules of statistical mechanics are also violated, but that will not be discussed here.

Consider bound-bound transitions in an absorbing medium exposed to incident radiation having spectral intensity i'_Ω. For simplicity let the system be a collection of noninteracting atoms. Since blackbody radiation is of interest, let the medium be in a black isothermal enclosure at uniform temperature — this is the condition for blackbody equilibrium (Sec. 2-5). An atom in the medium can absorb incident energy and thereby undergo a transition from energy state i to energy state j. State j will consequently have a larger energy than i, or in other words j is an "excited" state relative to i. The rate at which the transitions from i to j occur will depend on the intensity of the incident radiation field and the population of state i. Let N_i be the number of atoms per unit volume in state i. The Einstein coefficient B_{ij} is now introduced. This is defined as the probability per unit time and volume of a transition occurring from state i to state j per unit incident energy flux $i'_\Omega \, d\omega$, and is a function only of the particular atomic system being considered.* Then the number of transitions per unit time, considering the effect of incident energy from all directions, is

$$\left(\frac{dN_i}{dt}\right)_{i \to j} = B_{ij}N_i \int_{\omega = 4\pi} i'_\Omega \, d\omega \qquad (16\text{-}28)$$

Since the Einstein coefficient depends only on the states i and j for the particular atomic system, it is taken out of the integral over the solid angle.

The rate at which transitions will occur from the excited state j to the initial state i depends on two factors. These factors are *spontaneous emission*, which depends on the population N_j in the excited state, and *induced emission*, which depends on the population N_j *and* on the radiation field intensity. Thus let A_{ji} be the probability of a transition by spontaneous emission into a unit solid angle, and let B_{ji} be the transition probability for induced emission per unit of the incident quantity $i'_\Omega \, d\omega$. Then the rate of transitions from j to i is

$$\left(\frac{dn_j}{dt}\right)_{j \to i} = 4\pi N_j A_{ji} + N_j B_{ji} \int_{\omega = 4\pi} i'_\Omega \, d\omega \qquad (16\text{-}29)$$

Since on the average for a collection of randomly oriented emitting atoms in equilibrium the spontaneous emission is isotropic, $4\pi A_{ji}$ is the probability of transition from j to i by spontaneous emission into all directions.

For a system in equilibrium, the principle of *detailed balancing* must hold [3]. This principle states that at steady state the transition rates upward and downward between any two energy states must be equal when all transition processes are included. Using this principle, the dN/dt from (16-28) and (16-29) are equated:

$$B_{ij}N_i \int_{\omega = 4\pi} i'_{\Omega b} \, d\omega = 4\pi N_j A_{ji} + N_j B_{ji} \int_{\omega = 4\pi} i'_{\Omega b} \, d\omega \qquad (16\text{-}30)$$

---

*Other texts include or exclude various factors of 2 and π in the definitions. Sometimes the transition rate is written as proportional to the spectral energy density $(1/c) \int_{\omega = 4\pi} i'_\Omega \, d\omega$ rather than the intensity.

where at equilibrium in the assumed isothermal black enclosure the intensity becomes the blackbody intensity $i'_{\Omega b}$. For blackbody equilibrium conditions the incident intensity is also isotropic, so that $\int_{\omega=4\pi} i'_{\Omega b}\, d\omega = 4\pi i'_{\Omega b}$. Then, solving (16-30) for $i'_{\Omega b}$ gives

$$i'_{\Omega b} = \frac{A_{ji}}{(N_i/N_j)B_{ij} - B_{ji}} \tag{16-31}$$

At thermal equilibrium the populations of the energy states are related according to the Boltzmann distribution [3]. If E_i and E_j are the energies of the states, then the Boltzmann distribution gives

$$\frac{N_i}{N_j} = \exp\frac{-(E_i - E_j)}{kT} \tag{16-32}$$

where k is the Boltzmann constant. As discussed in Sec. 13-3.1 and in connection with Eq. (16-8), the energy difference $E_j - E_i$ is equal to the energy of the photon either absorbed to produce the transition from E_i to E_j or emitted when there is a transition from E_j to E_i. Then, in terms of angular frequency,

$$E_j - E_i = \hbar\Omega_{ij} \tag{16-33}$$

so that (16-32) becomes

$$\frac{N_i}{N_j} = e^{\hbar\Omega_{ij}/kT} \tag{16-34}$$

When (16-34) is applied to a system, the statistical weights discussed at the end of Sec. 16-3.2 must be included to account for all degenerate states in each energy level.

When (16-34) is substituted into (16-31), the result is

$$i'_{\Omega b} = \frac{A_{ji}}{B_{ij}(e^{\hbar\Omega_{ij}/kT} - B_{ji}/B_{ij})} \tag{16-35}$$

The Planck blackbody spectral intensity is given by (2-11b) as

$$i'_{\nu b} = \frac{e_{\nu b}}{\pi} = \frac{2C_1\nu^3}{c_0{}^4(e^{C_2\nu/c_0 T} - 1)}$$

which becomes, after substitution of $C_1 = hc_0{}^2$, $C_2 = hc_0/k$, $h = 2\pi\hbar$, and $\nu = \Omega_{ij}/2\pi$,

$$i'_{\Omega b} = \frac{\hbar\Omega_{ij}{}^3}{2\pi^2 c_0{}^2(e^{\hbar\Omega_{ij}/kT} - 1)} \tag{16-36}$$

Equation (16-35) has the same basic form as (16-36), and equating these two expressions for $i'_{\Omega b}$ gives the following relations between the Einstein coefficients:

$$B_{ij} = B_{ji} \tag{16-37}$$

and

$$\frac{A_{ji}}{B_{ij}} = \frac{\hbar\Omega_{ij}^{3}}{2\pi^2 c_0^2} \qquad (16\text{-}38)$$

(Note that degenerate states are not taken into account in these relations.)

Although, at the time of the derivation, induced emission had not been discerned by experiment, the analysis outlined in (16-28)–(16-38) gave strong evidence that it exists. If the induced-emission term of (16-29) is not included and the analysis is then carried through, the resulting equation by the Einstein approach is

$$i'_{\Omega b} = \frac{A_{ji}}{B_{ij}e^{\hbar\Omega_{ij}/kT}} \qquad (16\text{-}39)$$

To make (16-39) conform to Planck's distribution, the ratio of the Einstein coefficients must be, by comparison with (16-36) and by use of (16-34),

$$\frac{A_{ji}}{B_{ij}} = \frac{\hbar\Omega_{ij}^{3}}{2\pi^2 c_0^2} \frac{e^{\hbar\Omega_{ij}/kT}}{(e^{\hbar\Omega_{ij}/kT}-1)} = \frac{\hbar\Omega_{ij}^{3}}{2\pi^2 c_0^2} \frac{N_i}{N_i - N_j}$$

or

$$A_{ji} = \frac{\hbar\Omega_{ij}^{3}}{2\pi^2 c_0^2} \frac{N_i}{N_i - N_j} B_{ij} \qquad (16\text{-}40)$$

This would make A_{ji} dependent on the populations N_i and N_j of the i and j states. Because the transition probabilities A_{ji} and B_{ij} for a particular atomic system should depend only on the particular states i and j and not the populations of these states, Eq. (16-40) cannot be valid.

Suppose the properly specified Einstein coefficients given by (16-37) and (16-38) are substituted in (16-39), in which induced emission has been omitted. This will show what deviation should be expected from Planck's distribution as a result of not accounting for induced emission. Making this substitution gives

$$i'_{\Omega b} = \frac{\hbar\Omega_{ij}^{3}}{2\pi^2 c_0^2 e^{\hbar\Omega_{ij}/kT}} \qquad (16\text{-}41)$$

But [see Eq. (2-13)], this is Wien's distribution! A comparison of Planck's and Wien's spectral distributions in terms of wavelength, shown in Fig. 2-7, is thus a measure of the effect of induced emission on the spectral energy distribution. Wien's curve is slightly below the Planck curve as a result of omitting induced emission. It is evident that neglecting induced emission would introduce only a small error in most cases of engineering interest.

Note that this, and indeed any, derivation of the Planck blackbody distribution depends on the assumption of thermodynamic equilibrium. Also, Planck's distribution will not be consistent with the foregoing arguments unless the existence of induced emission is postulated.

16-5 THE EQUATION OF TRANSFER

The equation of transfer was derived in Sec. 14-3. It will now be considered from a microscopic view by using the concepts of the previous section. For simplicity, scattering will not be included. Consider a beam of radiation of intensity i'_Ω traveling through a gas along a path S. Let the gas atoms or molecules be in one of the two energy states i and j, with j being an excited state relative to state i; that is, $E_j > E_i$. Let the volume concentration of atoms in these states be N_i and N_j, respectively. Along a path distance dS the change in the intensity of the beam will be governed by the energy added or lost in the dS interval. Since scattering is neglected, the gains or losses are due to spontaneous emission, induced emission, and absorption. By use of the photon model discussed in Sec. 14-8, and considering only transitions between two energy states, the intensity added to the beam by spontaneous emission is

$$\left(\frac{di'_\Omega}{dS}\right)_{\substack{\text{spontaneous}\\ \text{emission}}} = \frac{\text{rate of transitions}}{\text{particle} - \text{solid angle}} \cdot \frac{\text{no. of particles}}{\text{volume}} \cdot \frac{\text{energy}}{\text{transition}}$$

$$= A_{ji}N_j\hbar\Omega_{ij} \tag{16-42}$$

Similar relations are derived for induced emission and absorption. The equation of transfer becomes

$$\frac{di'_\Omega}{dS} = A_{ji}N_j\hbar\Omega_{ij} + B_{ji}i'_\Omega N_j\hbar\Omega_{ij} - B_{ij}i'_\Omega N_i\hbar\Omega_{ij} \tag{16-43}$$

This can be rearranged into

$$\frac{di'_\Omega}{dS} = B_{ij}N_i\hbar\Omega_{ij}\left[\frac{A_{ji}N_j}{B_{ij}N_i} + \left(\frac{B_{ji}N_j}{B_{ij}N_i} - 1\right)i'_\Omega\right] \tag{16-44}$$

Although the system here is not in blackbody equilibrium, the Einstein coefficients can be used as previously obtained, as they depend only on the energy states and particular atomic system being considered. From (16-35) [noting that $B_{ij} = B_{ji}$ from (16-37)],

$$\frac{A_{ji}}{B_{ij}} = i'_{\Omega b}(e^{\hbar\Omega_{ij}/kT} - 1)$$

Substituting this, (16-34), and (16-37) into (16-44) gives, after simplification,

$$-\frac{di'_\Omega}{dS} = B_{ij}N_i\hbar\Omega_{ij}(1 - e^{-\hbar\Omega_{ij}/kT})(i'_\Omega - i'_{\Omega b}) \tag{16-45}$$

Note that $\hbar\Omega/kT = hc_0/k\lambda T$, and compare the quantity multiplying the i'_Ω on the right side of (16-45) with the absorption coefficient in (13-25). The true absorption coefficient is thus found to be

$$a_\Omega^+ = B_{ij}N_i\hbar\Omega_{ij} \tag{16-46}$$

and the absorption coefficient including induced emission is

$$a_\Omega = a_\Omega^+(1 - e^{-\hbar\Omega_{ij}/kT}) = B_{ij}N_i\hbar\Omega_{ij}(1 - e^{-\hbar\Omega_{ij}/kT}) \tag{16-47}$$

Equation (16-45) becomes

$$\frac{1}{a_\Omega}\frac{di_\Omega'}{dS} = i_{\Omega b}' - i_\Omega' \tag{16-48}$$

This is the form of the equation of transfer (14-10), obtained by the macroscopic derivation (for no scattering).

Rather than using the Einstein coefficient, it is customary to give the transition rate between bound electronic energy states in terms of a parameter called the *oscillator strength*, or *f number*. This is related to B_{ij} by

$$f_{ij} = \frac{\hbar m_e c_0 \Omega_{ij}}{\pi e^2} B_{ij} \tag{16-49}$$

where m_e is the mass of the electron, and e is the electronic charge. Substituting (16-49) into (16-46) gives a_Ω^+ in terms of the oscillator strength as

$$a_\Omega^+ = \frac{\pi e^2}{m_e c_0} f_{ij}N_i = \pi r_0 c_0 f_{ij}N_i \tag{16-50}$$

where r_0 is the classical electron radius (see Table A-1).

From (16-50), it is seen that the true absorption coefficient is directly proportional to two factors. These are N_i, the population of the initial state of the absorbing species, and f_{ij}, which through its connection with B_{ij} is related to the probability per unit time for transitions to occur from state i to j. Calculation of the population N_i, at least in the case of local thermodynamic equilibrium, is a problem in statistical mechanics. It is possible to derive f numbers for many electronic transitions by quantum mechanics and thus derive a_Ω^+ from first principles.

The determination of the spectral absorption coefficient by means of statistical and quantum mechanics requires a knowledge of the transition processes that can occur. In complicated atoms and molecules so many transitions are possible that calculations must either be restricted to only the important transitions, or else statistical or simplified models must be tried. Some discussion of the form of the spectral absorption coefficients for various types of transitions is given in the next section.

16-6 THE ABSORPTION PROPERTIES OF GASES

A gas can absorb energy by a variety of microscopic mechanisms. Each of these mechanisms involves adding the energy of the absorbed photon to the internal energy of a gas atom or molecule. A preliminary discussion of the types of absorption processes was given in Sec. 13-3.1.

16-6.1 Spectral Line Broadening

If the gas is not dissociated or ionized, its internal energy (not including translational energy) is contained in discrete vibrational, rotational, and electronic energy states of its atoms or molecules. The absorption of a photon can cause a transition of the atom or molecule to a higher energy state. Because only discrete energy states are involved in these transitions, photons of only certain energies can be absorbed. If the energies of the upper and lower discrete states are E_j and E_i, respectively, then only photons of energy $E_p = E_j - E_i$ can cause a transition. The energy of a photon is related to its frequency through the relation

$$E_p = E_j - E_i = h\nu_{ij} = hc\eta_{ij} \tag{16-51}$$

Consequently the discrete transitions result in the absorption of photons of only very definite frequencies, causing the appearance of dark lines in the transmission spectrum. Hence this process is termed *line absorption*. Because both the initial and final states of the atom or molecule are discrete bound states, these energy changes between states are called *bound-bound* transitions. The rates at which these transitions occur are available in tabular form for some molecules and atoms [4, 5]. The relations for the transition rates are often given by the semiclassical results describing radiating atoms multiplied by a modifying factor, called the *Gaunt factor*, that provides the correction for quantum-mechanical effects.

Equation (16-51) would predict that very little energy could be absorbed from the entire incident spectrum by any given absorption line, because only those photons having a single wave number could be absorbed. Other effects, however, cause the line to be broadened and consequently have a finite wave-number span around the transition wave number η_{ij} of (16-51). The wave-number span of the broadened spectral range, and the variation within it of the absorption ability, depends on the physical mechanism causing the broadening. Some of the important line-broadening mechanisms are natural broadening, Doppler broadening, collision broadening, and Stark broadening. For most engineering conditions involving infrared radiation, collision broadening is the most important.

The variation of the absorption coefficient with wave number within the broadened spectral line is called the *shape* of the spectral line. These shapes are important as they are related to the basic trends of gas absorption with temperature, pressure, and path length through the gas. The shape of a typical spectral line is illustrated by Fig. 16-1. The $a_{\eta, ij}(\eta)$ is the variation of absorption coefficient within the line broadened about the wave number η_{ij}, which is the transition wave number obtained from (16-51). The *integrated absorption coefficient* S_{ij} for a single line is found as the integral under the entire $a_{\eta, ij}(\eta)$ curve:

$$S_{ij} = \int_0^\infty a_{\eta, ij}(\eta)\, d\eta = \int_{-\infty}^\infty a_{\eta, ij}(\eta)\, d(\eta - \eta_{ij}) \tag{16-52}$$

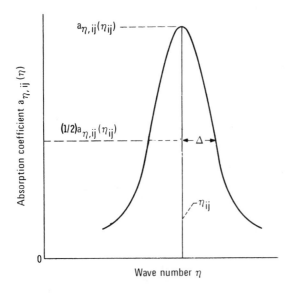

Figure 16-1 Absorption coefficient for broadened spectral line for transition between energy levels i and j.

The $a_{\eta,ij}(\eta)$ will be essentially zero except for η close to η_{ij}. The regions away from η_{ij}, where $a_{\eta,ij}$ becomes small, are called the "wings" of the line. The magnitudes of $a_{\eta,ij}$ and S_{ij} depend on the number of molecules in energy level i, and hence both depend on the gas density. Taking the ratio $a_{\eta,ij}(\eta)/S_{ij}$ tends to cancel the density effect somewhat, and the ratio shows the effect of density in changing the shape of the line.

The line shape depends on the governing line-broadening phenomenon. One characteristic of the line shape is given by Δ, the *half-width of the line*. This is the half-width (in units of wave number for the present discussion) evaluated at half the maximum line height, as shown in Fig. 16-1. It provides a definite width dimension to help describe the line. Since $a_{\eta,ij}$ goes to zero asymptotically as $|\eta - \eta_{ij}|$ increases, it is not possible to define a line width in terms of wave numbers at which $a_{\eta,ij}$ becomes zero.

Four phenomena that cause line broadening will now be discussed, along with the resulting line shapes.

Natural broadening. A perfectly stationary emitter unperturbed by all external effects is observed to emit energy over a finite spectral interval about a single transition wave number. This *natural line broadening* results from the uncertainty in the exact levels E_i and E_j of the transition energy states, which is related to the Heisenberg uncertainty principle. The natural line broadening produces a line shape

$$\frac{a_{\eta,ij}(\eta)}{S_{ij}} = \frac{\Delta_n/\pi}{\Delta_n{}^2 + (\eta - \eta_{ij})^2} \tag{16-53}$$

where Δ_n is the half-width for natural broadening. This form is called a *resonance* or *Lorentz profile*. In units of wave number it provides a profile that is symmetric about η_{ij} and that depends on Δ_n and the transition wave number η_{ij}.

For engineering applications the half-width produced by natural broadening is usually quite small compared with that caused by other line-broadening mechanisms. Natural line broadening is therefore usually neglected.

Doppler broadening. The atoms or molecules of an absorbing or emitting gas are not stationary, but have a distribution of velocities associated with their thermal energy. If an atom or molecule is emitting at wave number η_{ij} and at the same time is moving at velocity v toward an observer, the waves will arrive at the observer at an increased η given by

$$\eta = \eta_{ij}\left(1 + \frac{v}{c}\right) \tag{16-54}$$

If the emitter is moving away from the observer, v will be negative and the observed wave number will be less than η_{ij}. An example of such a wave-number decrease is the *red shift* of the radiation detected from galaxies in the universe. This provides evidence that the galaxies are moving away from earth, thereby indicating that the universe is expanding.

In thermal equilibrium the gas molecules will have a Maxwell-Boltzmann distribution of velocities. If an observer is detecting radiation along one coordinate direction, then the components of velocities of interest are those along the single direction either toward or away from the observer. The fraction of molecules moving in that direction within a velocity range between v and $v + dv$ is

$$\frac{dN}{N} = \sqrt{\frac{M}{2\pi kT}}\exp\left(-\frac{Mv^2}{2kT}\right)dv \tag{16-55}$$

where M is the mass of a molecule of the radiating gas, and k is the Boltzmann constant. Using (16-54) in (16-55) to eliminate v gives the fractional number of molecules providing radiation in each differential wave-number interval as a result of Doppler broadening. The result is a spectral line shape having a gaussian distribution:

$$\frac{a_{\eta,ij}(\eta)}{S_{ij}} = \frac{1}{\Delta_D}\sqrt{\frac{\ln 2}{\pi}}\exp\left[-(\eta - \eta_{ij})^2\frac{\ln 2}{\Delta_D^2}\right] \tag{16-56}$$

where Δ_D is the half-width of the line for Doppler broadening. The line shape depends only on Δ_D and η_{ij}. The Doppler half-width, however, is given by

$$\Delta_D = \frac{\eta_{ij}}{c}\left(\frac{2kT}{M}\ln 2\right)^{1/2} \tag{16-57}$$

thus depending on η_{ij}, T, and M. The dependency of Δ_D on $T^{1/2}$ shows that Doppler broadening is important at high temperatures.

Collision broadening. As the pressure of a gas is increased, the collision rate experienced by any given atom or molecule of the gas is also increased. The collisions can perturb the energy states of the atoms or molecules, resulting in collision broadening of the spectral lines. For noncharged particles, the line has a Lorentz profile [4]

$$\frac{a_{\eta,ij}(\eta)}{S_{ij}} = \frac{\Delta_c/\pi}{\Delta_c^2 + (\eta - \eta_{ij})^2} \tag{16-58}$$

which is the same shape as for natural broadening.

The collision half-width Δ_c is determined by the collision rate, and an approximate value can be found from kinetic theory. The Δ_c is given by

$$\Delta_c = \frac{1}{2\pi c} \frac{4\sqrt{\pi} D^2 P}{(MkT)^{1/2}} \tag{16-59}$$

where D is the diameter of the atoms or molecules, and P is the gas pressure for the single-component gas. Equation (16-59) shows that collision broadening becomes important at high pressures and low temperatures or, from the perfect gas law, high pressures and densities.

Collision broadening is often the chief contributor to line broadening for engineering infrared conditions, and the other line-broadening mechanisms can usually be neglected. The shapes of the Doppler and Lorentz broadened lines are compared in Fig. 16-2 for the same half-width and area under the curves. The Lorentz profile is lower at the line center, but remains of appreciable size further out in the wings of the line than the Doppler profile. Even when Doppler broadening is dominant near the line center, collision broadening is often the important mechanism far from the center.

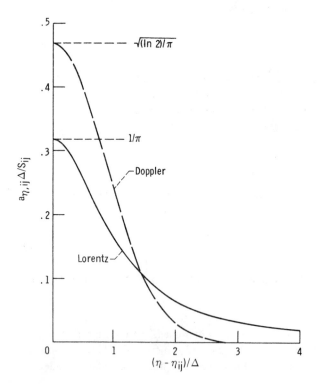

Figure 16-2 Line-shape parameter for Doppler and Lorentz broadened spectral lines (areas under two curves are equal).

Stark broadening. When strong electrical fields are present, the energy levels of the radiating gas particles can be greatly perturbed. This is the *Stark effect*, which can result in very large line broadening. It is often observed in ionized gases where radiating particle interactions with the electrons and protons give large Stark broadening effects. Calculation of the line shapes must be approached through quantum mechanics, and the resulting line shapes are quite unsymmetrical and complicated.

Stark and collision broadening are often lumped under the general heading of pressure broadening. Both effects depend on the pressure of the broadening component of the gas. When two or more broadening effects contribute simultaneously to the line broadening, calculation of the resulting line shape becomes more difficult. References [4, 6–8] can be consulted for additional information.

Broadening has been discussed here under the assumption that only one atomic or molecular species is present in the gas. If the gas consists of more than one component, then collision broadening in the radiation-absorbing gas is caused by collisions with like molecules (self-broadening) and by collisions with other species. Both collision processes must be included in calculating line shapes.

Although expressions for line shapes as given by $a_{\eta, ij}(\eta)/S_{ij}$ are now available, the absorption coefficient cannot be obtained without knowledge of S_{ij}. This will be briefly discussed at the end of the next section.

16-6.2 Absorption or Emission by a Spectral Line

By integrating Eq. (13-54) over the entire spectrum, the total energy absorbed along a path S per unit solid angle and projected area can be found, within a *uniform* gas:

$$\frac{d^2 Q_a'}{dA_p \, d\omega} = \int_0^\infty i_\eta'(0, \eta)[1 - \exp(-a_\eta S)] \, d\eta \tag{16-60a}$$

where $i_\eta'(0)$ is the incident spectral intensity at the origin of path S. Similarly, from the analogous forms of (13-58) and (13-60), the energy that is emitted to dA_p by a uniform gas in the region of solid angle $d\omega$ and path length S of Fig. 13-12 is

$$\frac{d^2 Q_e'}{dA_p \, d\omega} = \int_0^\infty i_{\eta b}'(\eta)[1 - \exp(-a_\eta S)] \, d\eta \tag{16-60b}$$

The integrals in (16-60) can be evaluated for a single broadened line, but first some simplifications can be made. Consider a spectral line centered about η_{ij}. The line absorption coefficient $a_{\eta, ij}(\eta)$ will be essentially zero except in a narrow wave-number range surrounding η_{ij}. Hence, unless S is large, the integrands in (16-60) will be of appreciable magnitude only within this narrow wave-number region, and the integration need be performed only over this narrow range. Within this range the $i_\eta'(0)$ or $i_{\eta b}'$ can be approximated as being constant, and since the largest absorption is at η_{ij}, the $i_\eta'(0)$ and $i_{\eta b}'$ are ordinarily taken at that wave number. Then Eqs. (16-60) become, for the spectral line,

$$\frac{d^2 Q_a'}{dA_p \, d\omega} = i_\eta'(0, \eta_{ij}) \int_{-\infty}^\infty \{1 - \exp[-a_{\eta, ij}(\eta)S]\} \, d(\eta - \eta_{ij}) \tag{16-61a}$$

$$\frac{d^2 Q'_e}{dA_p \, d\omega} = i'_{\eta b}(\eta_{ij}) \int_{-\infty}^{\infty} \left\{ 1 - \exp\left[-a_{\eta, ij}(\eta)S\right] \right\} d(\eta - \eta_{ij}) \tag{16-61b}$$

The absorbed and emitted energies for the line thus both involve the same integral. This integral will be called the *effective line width* $\bar{A}_{ij}$

$$\bar{A}_{ij}(S) \equiv \int_{-\infty}^{\infty} \left\{ 1 - \exp\left[-a_{\eta, ij}(\eta)S\right] \right\} d(\eta - \eta_{ij}) \tag{16-62}$$

which has the units of the spectral variable (η in this instance). By considering a spectral line within which the gas is perfectly absorbing ($a_{\eta, ij} \to \infty$), and having no absorption outside this line, it is found from (16-62) that $\bar{A}_{ij}$ can be interpreted as the width of a black line centered about η_{ij} that produces the same emission as the actual line.

The evaluation of $\bar{A}_{ij}$ will now be considered for some important limiting cases. First let the optical path length be small, $a_{\eta, ij}(\eta)S \ll 1$. The exponential term in Eq. (16-62) can be approximated in that case by

$$1 - \exp\left[-a_{\eta, ij}(\eta)S\right] \approx a_{\eta, ij}(\eta)S$$

so that

$$\bar{A}_{ij}(S) = S \int_{-\infty}^{\infty} a_{\eta, ij}(\eta) \, d(\eta - \eta_{ij})$$

With (16-52), this yields

$$\bar{A}_{ij}(S) = SS_{ij} \tag{16-63}$$

where the integrated absorption coefficient S_{ij} should not be confused with the path length S. The effective line width is thus linear with path length S *regardless of the line shape*. A line with this linear behavior is called a *weak line*.

Next consider asymptotic forms for the Lorentz line shape (16-58) for collision broadening, as this is the most important type of broadening for engineering applications in the infrared. Substituting (16-58) into (16-62) to obtain $\bar{A}_{ij}$ for the spectral line gives

$$\bar{A}_{ij}(S) = \int_{-\infty}^{\infty} \left\{ 1 - \exp\left[-\frac{S_{ij}}{\pi} \frac{\Delta_c S}{\Delta_c^2 + (\eta - \eta_{ij})^2}\right] \right\} d(\eta - \eta_{ij}) \tag{16-64}$$

This can be integrated to yield

$$\bar{A}_{ij}(S) = 2\pi\Delta_c \xi e^{-\xi}[J_0(i\xi) - iJ_1(i\xi)] \tag{16-65}$$

where $\xi = SS_{ij}/2\pi\Delta_c$ and J_n is the Bessel function of order n. For a strong line, $\xi \gg 1$, the asymptotic form of (16-65) is

$$\bar{A}_{ij}(S) = 2\sqrt{S_{ij}\Delta_c S} \tag{16-66}$$

For a weak line, $\xi \ll 1$, (16-65) reduces to (16-63). Equation (16-66) shows that, for a *strong Lorentz line,* the absorptance varies as the *square root* of the path length. This is in contrast to the results for *any weak line,* Eq. (16-63), where the absorptance varies *linearly* with path length. Experimental results bear out these functional dependencies.

Some formulas that approximate (16-65) quite well are given in [9]. One useful form is

$$\bar{A}_{ij}(S) = 2\pi\Delta_c \left(\frac{2\xi}{\pi}\right)^{1/2} \{1 - \exp\left[-(\pi\xi/2)^{1/2}\right]\} \tag{16-67}$$

Some expressions for $\bar{A}_{ij}$ have now been obtained; numerical values can be calculated if Δ_c and S_{ij} are known. An expression for Δ_c was given in (16-59), and it depends on gas pressure and temperature. The S_{ij} is also a function of these variables, as it depends on the number of gas molecules occupying an energy level and on the probability of a transition occurring. This gives S_{ij} in the form

$$S_{ij} \propto \rho \exp\left(\frac{-K_{ij}}{T}\right) \propto \frac{P}{T}\exp\left(\frac{-K_{ij}}{T}\right) \tag{16-68}$$

where K_{ij} is a coefficient depending on the particular quantum state involved in the transition. As a result, in (16-66), for a fixed gas temperature the $\bar{A}_{ij}$ will depend on ρS or PS (from $S_{ij}S$) and on ρ or P alone (from Δ_c). There are thus two effects of increasing gas density or pressure. One is the increase in absorption as a result of there being more molecules along the radiation path, and the second is the increased width of the spectral line. These trends will help guide the correlation of band behavior as many lines are superposed into an absorption band.

16-6.3 Continuum Absorption

Certain energy transition processes can result in the absorption of photons having a wide range of energies, as opposed to the relatively small range of energies that line absorption can encompass. The *continuum-absorption processes* are *bound-free* and *free-free*; these were discussed in Sec. 13-3.1 and are briefly reviewed here. Continuum absorption can also result from solid particles suspended in a gas (see Sec. 17-9.3).

Bound-free processes. Consider the situation in which a molecule absorbs a photon of energy sufficient to cause dissociation or ionization. A photon of *any* energy greater than the minimum necessary for these processes can be absorbed, giving rise to a continuous absorption spectrum. To produce ionization, ejection of an electron from a *bound* to a *free* state occurs upon absorption of a photon.

Free-free processes. A photon can be absorbed by a free electron as a result of an interaction of the electron with the electric field that exists in the vicinity of a positive ion. The energy of the absorbed photon is added to the kinetic energy of the absorbing electron, which remains in the free state. Since the initial and final states are not quantized, a continuous absorption spectrum results.

16-6.4 Band Absorption

Band structure. The gases that are commonly encountered in engineering calculations are diatomic or polyatomic, and therefore possess vibrational and rotational energy states that are absent in monatomic gases. The transitions between vibrational and rotational states usually provide the main contribution to the absorption coefficient in the significant thermal radiation spectral regions at moderate temperatures. As the temperature is raised, dissociation, electron transitions, and ionization become more probable, and the contributions of these additional processes to the absorption coefficient must be included.

When the absorption coefficient of a gas is determined experimentally, the contributions of all the line and continuum processes are superimposed. In computing such coefficients, each absorption process must be analyzed and then the complete coefficient obtained by combining the contributions from the various processes. In Fig. 16-3, the contributions to the spectral absorption coefficient as given by [10] are shown for air at a pressure of 1 atm and a range of temperatures. The ordinate is the ratio of the maximum value, regardless of wave number, of the contribution from a given

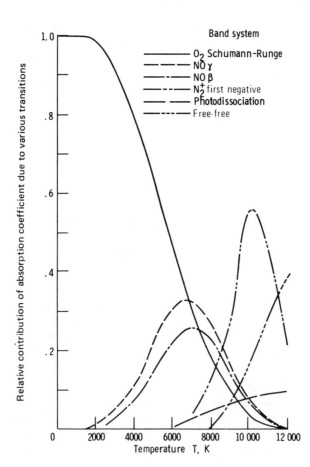

Figure 16-3 Relative contributions of energy transitions in bands of various species to absorption coefficient of air at 1 atm [10].

process to the sum of all such maxima at that same temperature. At low temperatures the entire absorption results from transitions of the oxygen between molecular states. As the temperature is increased, there is some formation of NO which provides additional bound-bound transitions. At high temperatures the continuous absorption processes are dominant; these are bound-free and free-free transitions.

The vibration-rotation bands are usually the most important absorbing and emitting spectral regions in engineering radiation calculations. The structure of such a band will now be examined in more detail, and this will indicate the complexity of computing band absorption coefficients from basic principles. Some simplified band models will then be discussed, by means of which some band-absorption features can be analyzed. The correlation of experimental band absorptance data will then be considered to illustrate how gas properties can be presented in a manner that is useful for engineering applications where band radiation quantities are required. Often in engineering heat transfer problems a reasonable approximation to the total radiation will suffice. It is then not necessary to go into the details of the radiation from individual bands. For total radiation calculations, charts of gas total emittance have been developed from total radiation measurements (see Chap. 17). Many of the functional dependencies of these charts had been developed empirically before the details of the radiation from the individual bands had been found. The information in the following sections will aid in understanding, from a microscopic viewpoint, how the physical variables influence gas radiation, but it is not intended to yield analytical predictions of the properties.

Let us now examine in more detail the vibration-rotation transitions governing the absorption coefficient of most polyatomic gases up to a temperature of about 3000 K. These transitions are strongly functions of wave number, and consequently the absorption coefficient is also strongly spectrally dependent. The spectral absorption in a vibration-rotation band consists of groups of very closely spaced spectral lines resulting from transitions between vibrational and rotational energy states. An example is shown in Fig. 16-4 for a portion of the carbon dioxide spectrum.* The absorption lines are so closely spaced in certain spectral regions that the individual lines are not resolved in most instances by experimental measurements. The lines appear to, or actually do, overlap as a consequence of broadening, and merge to form absorption bands. An example of absorption bands observed with low resolution are those for carbon dioxide in Fig. 13-2.

The large number of possible energy transitions that can produce an array of spectral lines, as in Fig. 16-4, is illustrated by the many energy levels and transition arrows in Fig. 16-5. This figure shows the potential energy for a diatomic molecule as a function of the separation distance between its two atoms. The two curves are each for a different electronic energy state where the electron may be shared by the two atoms. The distance R_e is the mean interatomic distance corresponding to each of the

---

*The notation $(01^1 0)$, etc., in Fig. 16-4 is a designation used to show the quantum state of a harmonic oscillator. In the general case $(v_1 v_2^l v_3)$, the v_i are the vibrational quantum numbers, and l is the quantum number for angular momentum. Transitions between two energy states, such as those denoted by $(00^0 0) \rightarrow (01^1 0)$, give rise to absorption lines. Certain selection rules govern the allowable transitions. A good introductory treatment is given in Chap. 3 of Goody [11].

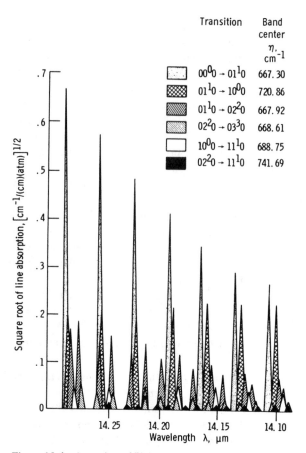

Figure 16-4 A portion of high-resolution spectrum of carbon dioxide [11].

electronic states. The long-dashed horizontal lines denote vibrational energy levels, while the short-dashed lines are rotational states superimposed on the vibrational states. Transitions between rotational levels of the *same* vibrational state involve small values of $E_j - E_i$. Hence from (16-33) these transitions give lines in band structures located at low wave numbers; that is, in the far infrared. Transitions between rotational levels in *different* vibrational states of the same electronic state give vibration-rotation bands at wave numbers in the near infrared. If transitions occur from a rotational level of an electronic and vibrational state to a rotational level in a different electronic and vibrational state, then large $E_j - E_i$ are involved and a band system can be formed in the high-wave-number visible and ultraviolet regions of the spectrum.

Band models. In modeling the absorption and emission behavior of bands of closely spaced lines, two types of models are employed. The first type, called *narrow-band* models, use characterizations of individual line shapes, widths, and spacings to derive the band characteristics within a defined wave-number interval. *Wide-band* models provide correlations of the band characteristics valid over the entire wave-number region of the band. These models thus account for the increasing importance of weakly absorbing lines in the wings of the band as the radiation path length becomes large. General discussions are given in [12-14] and in ref. [12] of Chapt. 17. In a detailed radiation-exchange calculation the absorbed and emitted energy will be needed in each band region, for example, in the four main CO_2 bands of Fig. 13-2. These spectral bands are separated by spectral regions that are nearly transparent. For the total absorbed or emitted energy in a uniform gas Eqs. (16-60a,b) are used; both involve the same type of integral. Since the absorption bands usually occupy a rather narrow spectral region, an average value of $i'_\eta(0)$ or $i'_{\eta b}$ can be taken out of the integral for each band. For example, for the total emitted energy, (16-60b) becomes

$$\frac{d^2 Q'_e}{dA_p \, d\omega} = \sum_l i'_{\eta b, l} \int_l [1 - \exp(-a_\eta S)] \, d\eta \qquad (16\text{-}69)$$

where the subscript l denotes a band, the integral is over each band, and the summation is over all the bands.

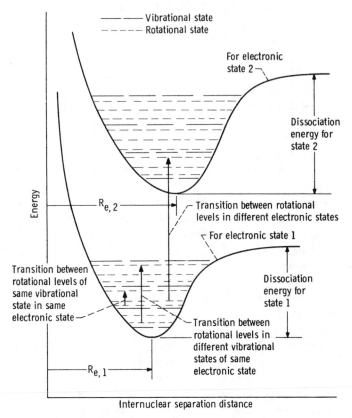

Figure 16-5 Potential energy diagram and transitions for a diatomic molecule.

In a fashion similar to the effective line width in (16-62), the integral in (16-69) is defined as the *effective bandwidth* $\bar{A}_l$

$$\bar{A}_l(S) \equiv \int_{\substack{\text{absorption} \\ \text{bandwidth}}} \{1 - \exp\left[-a_\eta(\eta)S\right]\} d\eta \tag{16-70}$$

and (16-69) becomes

$$\frac{d^2 Q_e'}{dA_p \, d\omega} = \sum_l i_{\eta b, l}' \bar{A}_l \tag{16-71}$$

The $\bar{A}_l$ has the units of the spectral variable, which is η in the case of (16-70), with units of per centimeter. The span of the absorption band that provides the upper and lower limits of the integral in Eq. (16-70) does not have a specific value that applies for all conditions. It can be defined as the spectral interval beyond which there is only a given small fractional contribution to $\bar{A}_l$. The width of this interval will increase slowly with path length as a result of proportionately more absorption taking place in the wings of the band.

From (16-71) and (13-60) the total emittance of a uniform gas is

$$\epsilon'(T,P,S) = \frac{\pi}{\sigma T^4} \sum_l i_{\eta b, l}' \bar{A}_l \tag{16-72}$$

The ϵ' can be used as described in Sec. 17-6.1 for engineering calculations of radiation from an isothermal gas to an enclosure boundary. The $\bar{A}_l$ can also be used as shown in (17-75) to obtain the band absorptance α_l for use in detailed spectral-exchange calculations in enclosures. It is evident, then, that for calculating such quantities as ϵ' and α_l for various conditions, correlations for $\bar{A}_l$ are needed as a function of path length, pressure, temperature, etc., for the important bands of radiating gases. The correlations of $\bar{A}_l$ will now be discussed.

By comparing (16-70) and (16-62) the $\bar{A}_l$ for the band is found to be the sum of the $\bar{A}_{ij}$ for all the spectral lines that occupy the band if all the $\bar{A}_{ij}$ act independently. Generally, the spectral lines overlap and consequently each line does not absorb as much energy as if it acted independently of adjacent lines. As was observed in Fig. 16-4, an absorption band is typically composed of many broadened lines. Hence the $a_\eta(\eta)$ in (16-70) is a complicated irregular function of wave number, and the integration for $\bar{A}_l$ is difficult. The integration would also require that the detailed shape of all the broadened lines be known. It is evident that a simplified model for the form of $a_\eta(\eta)$ must be devised if integration over the lines to obtain band radiation properties is to be a fruitful analytical approach. Two common narrow-band models represent extremes in specifying the individual line spacings and magnitudes (Fig. 16-6).

Elsasser [15] has modeled the lines as all having the same Lorentz shape and being of equal heights and equal spacings (and hence each having the same integrated absorption coefficient S_c). As shown in Fig. 16-6a this gives a_η as a periodic function that depends on the parameters governing the shape of the Lorentz lines and on the

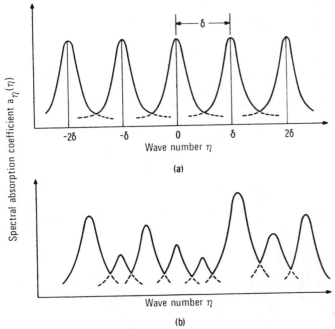

Figure 16-6 Models of absorption lines forming an absorption band. (a) Elsasser model, equally spaced Lorentz lines; (b) statistical model.

spacing δ between them. The absorption coefficient at a particular wave number is found by summing the contributions from all the adjacent lines. The distance of the line centers from a position η are $|\eta - 0|$, $|\eta - \delta|$, $|\eta - 2\delta|$, and so forth. Then, summing all the contributions by use of the Lorentz shape in (16-58) gives

$$a_\eta(\eta) = \frac{S_c}{\pi} \sum_{n=-\infty}^{\infty} \frac{\Delta_c}{\Delta_c^2 + (\eta - n\delta)^2}$$

Elsasser was able to sum this series to obtain

$$a_\eta(\eta) = \frac{S_c}{\delta} \frac{\sinh(\pi\beta/2)}{\cosh(\pi\beta/2) - \cos(\pi z/2)} \qquad \beta = 4\Delta_c/\delta \qquad z = 4(\eta - \eta_{ij})/\delta$$

where $\eta - \eta_{ij}$ varies between $-\delta/2$ and $\delta/2$ over each periodic line interval. The β is important as it specifies the effect of line structure. For large β the lines become broad compared to their spacing and the line structure is lost as the lines strongly overlap. This corresponds to large pressure [Eq. (16-59)]. The effective width for a band of m lines is

$$\bar{A}_l = m \int_{-\delta/2}^{\delta/2} \left\{ 1 - \exp\left[-\frac{SS_c}{\delta} \frac{\sinh(\pi\beta/2)}{\cosh(\pi\beta/2) - \cos(\pi z/2)} \right] \right\} d(\eta - \eta_{ij})$$

which can be written as

$$\frac{\bar{A}_l}{m\delta} = \frac{\bar{A}_l}{A_0} = 1 - \frac{1}{2} \int_0^2 \exp\left[-\frac{u \sinh(\pi\beta/2)}{\cosh(\pi\beta/2) - \cos(\pi z/2)}\right] dz \tag{16-73}$$

where $u = SS_c/\delta$. The $\bar{A}_l/A_0$ is the effective width relative to the actual width A_0 of the band.

Although the integral in (16-73) cannot be evaluated in closed form, some informative limiting results can be found. For weak nonoverlapping lines, $u \ll 1$, each line acts independently and $\bar{A}_l/A_0$ is then the same as for a single line. Using (16-63) with $S_{ij} = S_c$ in this band model gives

$$\frac{\bar{A}_l}{A_0} = \frac{\bar{A}_{ij}}{\delta} = \frac{SS_c}{\delta} = u \qquad u \ll 1 \tag{16-74}$$

For a long radiation path, $u \gg 1$, (16-73) yields

$$\frac{\bar{A}_l}{A_0} = 1 \qquad u \gg 1 \tag{16-75}$$

This represents total absorption within the *entire* width of the band. Three other limits, and the conditions giving them, are summarized in [13]. For strong nonoverlapping lines, (16-73) yields

$$\frac{\bar{A}_l}{A_0} = \text{erf}\left(\tfrac{1}{2}\sqrt{\pi\beta u}\right) \qquad \text{when} \qquad \beta \ll 1 \qquad \frac{u}{\beta} \gg 1 \tag{16-76}$$

If in addition βu is small so that lines are thin and spaced well apart,

$$\frac{\bar{A}_l}{A_0} = \sqrt{\beta u} \qquad \text{when} \qquad \beta \ll 1 \qquad \frac{u}{\beta} \gg 1 \qquad \beta u \ll 1 \tag{16-77}$$

This is also called the *square-root limit*. If β is very large, $\Delta_c \gg \delta$, the lines are very broad compared with the spacing between them, and the line structure becomes spread out over the band. This yields

$$\frac{\bar{A}_l}{A_0} = 1 - \exp(-u) \tag{16-78}$$

where, with the line structure spread out, the S_c/δ in u is the average absorption coefficient.

Another band model is a statistical array of lines as shown in Fig. 16-6b and presented by Goody [11]. There can be either a random spacing of identical lines or, more generally, the lines can differ from each other. A spacing that is essentially random is typical of bands of polyatomic molecules such as CO_2 and water vapor. To apply the model, probability distributions of line strengths and positions must be assumed. These statistical assumptions remove the necessity of calculating the exact properties of the individual lines in the band. The parameters u and β are based on average values of the line widths and strengths.

Many other band models have been proposed, some of them of more utility in certain cases than the Elsasser or Goody models. Several of these band models are discussed by Goody [11], Edwards and Menard [16], and Ludwig et al. [17]. Modifications to the Elsasser model have been made by Kyle [18] and Golden [19, 20], who treated evenly spaced lines with a Doppler profile, and Golden [21], who treated the same case with a Voigt profile. The Voigt profile is a combination of the Doppler and Lorentz profiles, thus accounting for both Doppler and collision broadening in the gas.

Band correlations for wide-band models. The band model of Edwards and Menard [16] consists of rotation lines in a band that are reordered in wave number so that they form an array with exponentially decreasing line intensities moving away from the band center. This is called the *exponential wide-band* model. The u and β are now based on average values over the band of the line integrated absorption coefficient, and the line width. For large absorption in the band, e.g., for large path length S, the center of the band becomes completely absorbing. As S is further increased, the $\bar{A}$ increases as a result of the wings of the band gradually becoming more absorbing. For this reason the limit in (16-75) is not reached for this model. The limit for large u is

$$\frac{\bar{A}_l}{A_0} = \ln u \qquad u \gg 1 \tag{16-79}$$

For this model the limit for the conditions in (16-77) becomes

$$\frac{\bar{A}_l}{A_0} = 2\sqrt{\beta u} \tag{16-80}$$

Relations (16-74)–(16-80) are limiting forms that show behavior for particular conditions of u and β. For the band correlations to be used, however, the $\bar{A}_l/A_0$ must be available for all values of the parameters u and β. Then these parameters must be related to the ordinary physical variables, such as path length, pressure, and temperature. Using the three limiting conditions in (16-75), (16-79), and (16-80), Edwards and Menard [16] constructed the following functions that smoothly join $\bar{A}_l/A_0$ for all regions of u and β and go to the proper limiting values:

$$\beta \leqslant 1: \quad \frac{\bar{A}_l}{A_0} = u \qquad\qquad \frac{\bar{A}_l}{A_0} \leqslant \beta$$

$$\frac{\bar{A}_l}{A_0} = 2\sqrt{\beta u} - \beta \qquad \beta \leqslant \frac{\bar{A}_l}{A_0} \leqslant (2 - \beta)$$

$$\frac{\bar{A}_l}{A_0} = \ln(\beta u) + 2 - \beta \qquad \frac{\bar{A}_l}{A_0} \geqslant (2 - \beta)$$

$$\beta > 1: \quad \frac{\bar{A}_l}{A_0} = u \qquad\qquad \frac{\bar{A}_l}{A_0} \leqslant 1$$

$$\frac{\bar{A}_l}{A_0} = \ln u + 1 \qquad\qquad \frac{\bar{A}_l}{A_0} \geqslant 1$$

Tien and Lowder [22] devised the continuous correlation

$$\frac{\bar{A}_l}{A_0} = \ln \left[uf(\beta) \frac{u+2}{u+2f(\beta)} + 1 \right] \tag{16-81}$$

where

$$f(\beta) = 2.94 \left[1 - \exp\left(-2.60\beta\right) \right]$$

This does not satisfy the square root limit (16-77), [23]. Other correlations have been constructed, such as those of Cess and Tiwari [23]

$$\frac{\bar{A}_l}{A_0} = 2 \ln \left[1 + \frac{u}{2 + u^{1/2}(1 + 1/\beta)^{1/2}} \right] \tag{16-82}$$

and Goody and Belton [24]

$$\frac{\bar{A}_l}{A_0} = 2 \ln \left[1 + \frac{\sqrt{\beta}\, u}{(u + 4\beta)^{1/2}} \right] \tag{16-83}$$

A more complicated expression that covers all ranges is given by Morizumi [14], who also summarizes some of the other correlations available in the literature. Tiwari [13] compares the results of using various narrow-band models in developing the exponential wide-band model. In addition, the correlations presented by Edwards and Bala-krishnan [25, 26], Edwards and Menard [16], Tien and Lowder [22], Tien and Ling [27], Cess and Tiwari [23], Felske and Tien [28], Goody and Belton [24], and Hsieh and Greif [29] are compared and analyzed for regions of accurate use.

Since the effective bandwidth $\bar{A}_l$ is calculated from models utilizing the line structure, it is evident that $\bar{A}_l$ will depend on the line spacing, the line half-width and the line integrated absorption, as well as other quantities, when a statistical model is used. To utilize these analytical results for radiative calculations involving a real gas mixture, it is necessary to know how all these factors are influenced by conditions such as the gas temperature, partial pressure of the absorbing gas, and the total pressure of the gas mixture. If relations between these quantities are specified, the correlation of experimental data based on the theoretically indicated dependencies of the band integration can be attempted. The background for these correlations is given by Goody [11]. Edwards and co-workers [30–38] assembled a large body of data on the common important radiating gases. By comparing their band-correlation relations with the data over a large range of pressure and temperature, they empirically determined how the required correlation quantities are related to the physical variables. This includes the effective bandwidth $A_0(T)$ and the pressure broadening parameter $\beta(T, P_e)$, where P_e is the effective broadening pressure. The u was expressed in terms of the mass path length $X = \rho S$ of the absorbing gas component. The data used are summarized in Table 16-1, which also contains some more recent data. Some of the correlation results will now be given.

The effective bandwidth $\bar{A}_l$ can be calculated from the relations in Table 16-2 as derived from the exponential wide-band model. These results are in units of wave

Table 16-1 Available band absorptance correlations for isothermal media

Gas	Bands	Reference	Comments	Type of correlation
CO_2	All important	31, 45	$300 \leqslant T \leqslant 1400$ K	Equivalent bandwidth
	2.7, 4.3, and 15 μm	32, 52[a]	$300 \leqslant T \leqslant 1400$ K $0.1 \leqslant X \leqslant 23,000$ g/m^2	Exponential wide band
	9.4 and 10.4 μm	33, 52[a]	$300 \leqslant T \leqslant 1400$ K $0.1 \leqslant X \leqslant 23,000$ g/m^2	Exponential wide band
	All important	53–56	$T \sim 300$ K	Equivalent bandwidth
	2.7 μm	39, 42	$T \sim 300$ K $0.1 \leqslant X \leqslant 23,000$ g/m^2	Basic spectroscopic theory
H_2O	All important	34, 52[a]	$300 \leqslant T \leqslant 1100$ K $1 \leqslant X \leqslant 38,000$ g/m^2	Exponential wide band
	2.7 and 6.3 μm	37	$300 \leqslant T \leqslant 1100$ K $1 \leqslant X \leqslant 21,000$ g/m^2	Equivalent line
	All important	53–56	$T \sim 300$ K	Equivalent bandwidth
	2.7 μm	39	$T \sim 300$ K $0.1 \leqslant X \leqslant 23,000$ g/m^2	Basic spectroscopic theory
CH_4	7.6 and 3.3 μm	32, 52[a]	$300 \leqslant T \leqslant 830$ K $0.1 \leqslant X \leqslant 1200$ g/m^2	Exponential wide band
CO	2.35 and 4.67 μm	35, 52[a]	$300 \leqslant T \leqslant 1800$ K $19 \leqslant X \leqslant 650$ g/m^2	Exponential wide band
	4.7 μm	29, 41, 42	$300 \leqslant T \leqslant 1800$ K $19 \leqslant X \leqslant 650$ g/m^2	Basic spectroscopic theory
HCl		57	Not correlated – presented in terms of spectral emittance	
H_2		58	Not correlated – presented in terms of spectral and total emittance	

Table 16-1 *(Continued)*

Gas	Bands	Reference	Comments	Type of correlation
Atmospheric gases – N_2, O_2, CO_2, O_3, H_2O, CH_3, and nitrogen oxides		11	Discussion of literature up to 1960	
Air	All important contributing bands	4 (Table 11-2)	References to literature up to 1965 for needed data to calculate band absorptance	
NH_3	3.0, 10.5, 2.9, and 6.15 μm	60	$T \sim 300$ K $2 \leqslant X \leqslant 312$ g/m²	Exponential wide band
NO	5.35 μm	41, 42	$T \sim 300$ K $0.845 \leqslant PS \leqslant 12.56$ atm-cm	Basic spectroscopic theory
N_2O	4.5 μm	42	$T = 303$ K $0.0186 \leqslant PS \leqslant 76.4$ atm-cm	Basic spectroscopic theory
CCl_4 (liquid)	All important	44		Two-parameter narrow-band models
H_2, N_2, O_2, CH_4, CO, A (liquids only)	All important far-infrared bands in 40–500-μm region; for H_2, 16.7–500 μm	59	T near normal boiling point at 1 atm, $S = 1.27$ and 2.54 cm, plus 3.25 cm for H_2	Data only of absorption coefficient versus wave number

[a]Correlations are given in Tables 16-2 and 16-3.

number, per centimeter. The correlations are in terms of mass path length $X = \rho S$ and the pressure broadening parameter β. These quantities account for the two effects of pressure; one is the number of molecules that interact with the radiation along a path in the gas, and the other is the broadening of the spectral lines. The quantities $b, n, C_1, C_2,$ and C_3* needed to evaluate these relations are given in Table 16-3 for CO_2, CH_4, H_2O, and CO, each in a mixture with nitrogen. The C_1 is the integrated band intensity (integral of S_{ij}/δ over the band), and C_3 is a bandwidth parameter. The method of using these band correlations will be shown by two examples.

*These C values should not be confused with the fundamental radiation constants in Table A-4.

Table 16-2 Effective bandwidth correlation equations for isothermal gas[a]

Pressure broadening parameter $\beta = \dfrac{C_2{}^2 P_e}{4C_1 C_3}$	Effective bandwidth $\bar{A}$, η, cm^{-1}	Limits of $\bar{A}$, η, cm^{-1}
$\beta \leqslant 1$	$\bar{A} = C_1 X$	$0 \leqslant \bar{A} \leqslant \beta C_3$
	$\bar{A} = C_2(XP_e)^{1/2} - \beta C_3$	$\beta C_3 \leqslant \bar{A} \leqslant C_3(2-\beta)$
	$\bar{A} = C_3\left(\ln\dfrac{C_2{}^2 XP_e}{4C_3{}^2} + 2 - \beta\right)$	$C_3(2-\beta) \leqslant \bar{A} \leqslant \infty$
$\beta > 1$	$\bar{A} = C_1 X$	$0 \leqslant \bar{A} \leqslant C_3$
	$\bar{A} = C_3\left(\ln\dfrac{C_1 X}{C_3} + 1\right)$	$C_3 \leqslant \bar{A} \leqslant \infty$

[a]C_1, C_2, C_3, b, and n are in Table 16-3. X is mass path length ρS, g/m^2. $P_e = [(bp + p_{N_2})/P_0]^n$, where $P_0 = 1$ atm, p is partial pressure of absorbing gas, and p_{N_2} is partial pressure of N$_2$ broadening gas in atmospheres. The l subscript on $\bar{A}$ has been dropped for convenience.

EXAMPLE 16-1. Find the effective bandwidth $\bar{A}$ of the 9.4-μm band of pure CO$_2$ at 1 atm and 500 K for a path length S of 0.364 m.

To obtain $\bar{A}$ from the relations in Table 16-2, the constant C_1 must be evaluated. From Table 16-3 at the 9.4-μm CO$_2$ band, $C_1 = 0.76\varphi_1(T)$, where

$$\varphi_1(T) = \left[1 - \exp\frac{-hc(\eta_3 - \eta_1)}{kT}\right]\left[\exp\left(-\frac{hc\eta_1}{kT}\right) - \tfrac{1}{2}\exp\left(-\frac{2hc\eta_1}{kT}\right)\right]$$

$$\times \left[1 - \exp\left(-\frac{hc\eta_1}{kT}\right)\right]^{-1}\left[1 - \exp\left(-\frac{hc\eta_3}{kT}\right)\right]^{-1}$$

Substitute the values $\eta_1 = 1351$ cm^{-1}, $\eta_2 = 667$ cm^{-1}, $\eta_3 = 2396$ cm^{-1}, $h = 6.626 \times 10^{-34}$ J $\cdot$ s, $k = 1.381 \times 10^{-23}$ J/K, $c = 2.998 \times 10^{10}$ cm/s, and $T = 500$ K. This gives $\varphi_1 = 0.0196$ so that $C_1 = 0.0149$ m^2/(cm $\cdot$ g). Table 16-2 gives the quantity β as $\beta = C_2{}^2 P_e/4C_1 C_3$. For the 9.4-$\mu$m CO$_2$ band, Table 16-3 gives C_2 and C_3 as

$$C_2 = 1.6\left(\frac{T}{T_0}\right)^{0.5} C_1{}^{0.5} \qquad \text{and} \qquad C_3 = 12.4\left(\frac{T}{T_0}\right)^{0.5}$$

Table 16-3 Exponential band-model correlation quantities[a]

Gas	Band, μm	Band center η, cm^{-1}	Pressure parameters b	n	C_1, cm$^{-1}/(g \cdot m^{-2})$	C_2,[b] cm$^{-1}/[(g \cdot m^{-2})]^{1/2}$	C_3,[b] cm^{-1}
CO_2[c]	15	667	1.3	0.7	19	$6.9(T/T_0)^{0.5}$	$12.9(T/T_0)^{0.5}$
	10.4	960	1.3	0.8	$0.76\varphi_1(T)$	$1.6(T/T_0)^{0.5}C_1^{0.5}$	$12.4(T/T_0)^{0.5}$
	9.4	1060	1.3	0.8	$0.76\varphi_1(T)$	$1.6(T/T_0)^{0.5}C_1^{0.5}$	$12.4(T/T_0)^{0.5}$
	4.3	2350	1.3	0.8	110	$31(T/T_0)^{0.5}$	$11.5(T/T_0)^{0.5}$
	2.7	3715	1.3	0.65	$4.0\varphi_2(T)$	$8.6\varphi_3(T)$	$24(T/T_0)^{0.5}$
CH_4	7.6	1310	1.3	0.8	28	$10(T/T_0)^{0.5}$	$23(T/T_0)^{0.5}$
	3.3	3020	1.3	0.8	46	$14.5(T/T_0)^{0.5}$	$55(T/T_0)^{0.5}$
H_2O[d]	6.3	1600	5.0	1.0	41.2	44	$52(T/T_0)^{0.5}$
	2.7	3750	5.0	1.0	23.3	39	$65(T/T_0)^{0.5}$
	1.87	5350	5.0	1.0	$3.0\varphi_{011}(T)$	$6.0C_1^{0.5}$	$46(T/T_0)^{0.5}$
	1.38	7250	5.0	1.0	$2.5\varphi_{101}(T)$	$8.0C_1^{0.5}$	$46(T/T_0)^{0.5}$
CO[e]	4.67	2143	1.1	0.8	20.9	$\varphi_5(T)$	$22(T/T_0)^{0.5}$
	2.35	4260	1.0	0.8	0.14	$0.08\varphi_5(T)$	$22(T/T_0)^{0.5}$

[a] For limits on T and X, see Table 16-1.
[b] T_0 is taken as 100 K for all cases.
[c] For CO_2,

$$\varphi_1 = \left\{1 - \exp\left[-\frac{hc}{kT}(\eta_3 - \eta_1)\right]\right\}\left[\exp\left(-\frac{hc\eta_1}{kT}\right) - \tfrac{1}{2}\exp\left(-\frac{2hc\eta_1}{kT}\right)\right]\left[1 - \exp\left(-\frac{hc\eta_1}{kT}\right)\right]^{-1}\left[1 - \exp\left(-\frac{hc\eta_3}{kT}\right)\right]^{-1}$$

$$\varphi_2 = \left\{1 - \exp\left[-\frac{hc}{kT}(\eta_1 + \eta_3)\right]\right\}\left[1 - \exp\left(-\frac{hc\eta_1}{kT}\right)\right]^{-1}\left[1 - \exp\left(-\frac{hc\eta_3}{kT}\right)\right]^{-1}$$

$$\varphi_3 = \left[1 + 0.053\left(\frac{T}{T_0}\right)^{3/2}\right]$$

where $\eta_1 = 1351$ cm^{-1}, $\eta_2 = 667$ cm^{-1}, and $\eta_3 = 2396$ cm^{-1}.
[d] For H_2O,

$$\varphi_{v_1v_2v_3} = \left[1 - \exp\left(-\frac{hc}{kT}\sum_{i=1}^{3} v_i\eta_i\right)\right]\prod_{i=1}^{3}\left[1 - \exp\left(-\frac{hc\eta_i}{kT}\right)\right]^{-1}$$

where $\eta_1 = 3652$ cm^{-1}, $\eta_2 = 1595$ cm^{-1}, and $\eta_3 = 3756$ cm^{-1}.
[e] For CO,

$$\varphi_5 = \left[15.15 + 0.22\left(\frac{T}{T_0}\right)^{3/2}\right]\left[1 - \exp\left(-\frac{hc\eta}{kT}\right)\right]$$

where $\eta = 2143$ cm^{-1}.

so that

$$\beta = \frac{(1.6)^2 P_e (T/T_0)^{0.5}}{4 \times 12.4} = 0.0516 P_e \left(\frac{T}{T_0}\right)^{0.5}$$

From Table 16-2, P_e for pure CO_2 at 1 atm is

$$P_e = \left(\frac{1.3 + 0}{1}\right)^{0.8} = 1.234$$

Then, since $T_0 = 100$ K, $\beta = 0.0516(1.234)(500/100)^{0.5} = 0.142$. Also $C_3 = 12.4(T/T_0)^{0.5} = 12.4(500/100)^{0.5} = 27.7$ cm^{-1}. Because $\beta \leqslant 1$, the correlation equations for the specified conditions are the first set in Table 16-2. The mass path length is given by $X = \rho S = 0.364\rho$ g/m^2. The gas density is

$$\rho = \frac{1}{22.42 l/g \cdot \text{mol}} \frac{44 \text{ g}}{\text{g} \cdot \text{mol}} \frac{1000 l}{m^3} \frac{273}{500} = 1.07 \times 10^3 \text{ g/m}^3$$

so that the mass path length is $X = 390$ g/m^2. The choice of correlation equation depends on the limits into which X causes $\bar{A}$ to fall. The first equation in Table 16.2 gives $\bar{A} = C_1 X = 0.0149 \times 390 = 5.8$ cm^{-1}, but this falls well outside the prescribed upper limit of the bandwidth given by $\beta C_3 = 0.142 \times 27.7 = 3.93$ cm^{-1} for the $\beta \leqslant 1$ part of the correlation. For intermediate X, the second line of Table 16-2 gives $\bar{A} = C_2(X P_e)^{1/2} - \beta C_3$ or $\bar{A} = 1.6(500/100)^{1/2}(0.0149)^{1/2} \times (481)^{1/2} - 3.93 = 5.6$ cm^{-1} and this lies within the range $\beta C_3 \leqslant \bar{A} \leqslant C_3(2 - \beta)$ or $3.93 \leqslant \bar{A} \leqslant 51.5$ cm^{-1} for this part of the correlation. The result for $\bar{A}$ compares reasonably well with an experimental value of 5.9 cm^{-1} from [31] for similar conditions.

EXAMPLE 16-2 Determine the energy per unit area and solid angle for the 9.4-μm band emitted from the end of a thin column of CO_2 gas at 1 atm pressure and 500 K if the column is 0.364 m long.

Using Eq. (16-60b) and integrating only over the 9.4-μm band give

$$\frac{d^2 Q_e'}{dA_p \, d\omega} = \int_{\Delta \eta} i_{\eta b}'(\eta)(1 - e^{-a_{\eta} S}) \, d\eta \approx \bar{A} i_{\eta b}'(\eta_{\text{band center}})$$

where (16-70) has been substituted and $\eta_{\text{band center}}$ is the wave number of the band center. For this band Table 16-3 gives $\eta_{\text{band center}} = 1060$ cm^{-1}. Using Eq. (2-11c) for $i_{\eta b}'$ and $\bar{A}$ from Example 16-1 give the result

$$\bar{A} i_{\eta b}'(\eta_{\text{band center}}) = 5.6 \left(\frac{2 C_1 \eta^3}{e^{C_2 \eta/T} - 1} \right)_{\text{band center}}$$

$$= 5.6 \frac{1}{\text{cm}} \frac{2 \times 0.59544 \times 10^{-12} \text{ W} \cdot \text{cm}^2 \, (1060)^3}{e^{1.4388 \times 1060/500} - 1} \frac{1}{\text{cm}^3}$$

$$\frac{d^2 Q_e'}{dA_p \, d\omega} = 3.95 \times 10^{-4} \text{ W/cm}^2$$

The band correlations in Tables 16-2 and 16-3 are convenient to use. They are, however, based somewhat on empirical results. To improve the theoretical foundation of the correlations, and for extrapolating beyond the range of experimental results, a new correlation was given in [26] based on the quantum-mechanical behavior of a vibration-rotation band. A brief description of the correlation will be given here for the four gases in Table 16-3. Information on NO and SO_2 is in [26] (also see [12]). (To be consistent with the notation used here, the $\eta, \beta,$ and ν in [26, 12] are changed to $\beta, \gamma,$ and η.)

The correlation is given in terms of three quantities: α, the integrated band intensity (called C_1 in Table 16-3); γ, the line-width parameter;* and ω, the bandwidth parameter (called C_3 in Table 16-3). The desired total band absorption $\bar{A}$ is found from the following correlations (the l subscript on $\bar{A}$ has been dropped for convenience):

For $\beta < 1$:

$$\bar{A} = \omega u \qquad\qquad\qquad 0 \leqslant u \leqslant \beta$$

$$\bar{A} = \omega(2\sqrt{\beta u} - \beta) \qquad\qquad \beta \leqslant u \leqslant \frac{1}{\beta}$$

$$\bar{A} = \omega(\ln \beta u + 2 - \beta) \qquad \frac{1}{\beta} \leqslant u \leqslant \infty$$

For $\beta \geqslant 1$:

$$\bar{A} = \omega u \qquad\qquad\qquad 0 \leqslant u \leqslant 1$$

$$\bar{A} = \omega(\ln u + 1) \qquad\qquad 1 \leqslant u \leqslant \infty$$

The β is π times the ratio of mean line width to spacing, including pressure broadening; $\beta = \gamma P_e$, where $P_e = [P/P_0 + (p/P_0)(b-1)]^n$, P is the total pressure of radiating and nonradiating gas, $P_0 = 1$ atm, and p is the partial pressure of the radiating gas ($P_e \to 1$ as $p \to 0$ and $P \to P_0$). The b and n are in Table 16-4 for each gas band. The $u = X\alpha/\omega$, where X is the mass path length of the radiating gas. The ω is found from $\omega = \omega_0(T/T_0)^{1/2}$, where ω_0 is in Table 16-4 and $T_0 = 100$ K. The table also gives the quantities necessary to obtain α and γ from the following:

$$\alpha(T) = \alpha_0 \frac{1 - \exp\left(-\sum\limits_{k=1}^{m} u_k \delta_k\right)}{1 - \exp\left(-\sum\limits_{k=1}^{m} u_{0,k}\delta_k\right)} \frac{\Psi(T)}{\Psi(T_0)}$$

$$\gamma(T) = \gamma_0 \left(\frac{T_0}{T}\right)^{1/2} \frac{\Phi(T)}{\Phi(T_0)}$$

where

$$\Psi(T) = \frac{\prod\limits_{k=1}^{m} \sum\limits_{v_k=v_{0,k}}^{\infty} [(v_k + g_k + |\delta_k| - 1)!/(g_k - 1)!v_k!] e^{-u_k v_k}}{\prod\limits_{k=1}^{m} \sum\limits_{v_k=0}^{\infty} [(v_k + g_k - 1)!/(g_k - 1)!v_k!] e^{-u_k v_k}}$$

$$\Phi(T) = \frac{\left(\prod\limits_{k=1}^{m} \sum\limits_{v_k=v_{0,k}}^{\infty} \left\{[(v_k + g_k + |\delta_k| - 1)!/(g_k - 1)!v_k!] e^{-u_k v_k}\right\}^{1/2}\right)^2}{\prod\limits_{k=1}^{m} \sum\limits_{v_k=v_{0,k}}^{\infty} [(v_k + g_k + |\delta_k| - 1)!/(g_k - 1)!v_k!] e^{-u_k v_k}}$$

*The γ is π times the ratio of mean line width to spacing for a dilute mixture of the radiating component at a total pressure of 1 atm so that there is negligible pressure broadening.

Table 16-4 Exponential wide-band parameters [26, 12]

Gas m, η (cm^{-1}), g	Band, μm	Band center η, cm^{-1}	$\delta_1 \cdots \delta_m$	Pressure parameters ($T_0 = 100$ K) b	n	α_0, cm^{-1}/g·m^{-2}	γ_0	ω_0, cm^{-1}
CO$_2$								
$m = 3$, $\eta_1 = 1351$, $g_1 = 1$	15	667	0, 1, 0	1.3	0.7	19.0	0.06157	12.7
$\eta_2 = 667$, $g_2 = 2$	10.4	960	-1, 0, 1	1.3	0.8	2.47×10^{-9}	0.04017	13.4
$\eta_3 = 2396$, $g_3 = 1$	9.4	1060	0, -2, 1[b]	1.3	0.8	2.48×10^{-9b}	0.11888[b]	10.1
	4.3	2410[a]	0, 0, 1	1.3	0.8	110.0	0.24723	11.2
	2.7	3660	1, 0, 1	1.3	0.65	4.0	0.13341	23.5
	2.0	5200	2, 0, 1	1.3	0.65	0.066	0.39305	34.5
CH$_4$								
$m = 4$, $\eta_1 = 2914$, $g_1 = 1$	7.66	1310	0, 0, 0, 1	1.3	0.8	28.0	0.08698	21.0
$\eta_2 = 1526$, $g_2 = 2$	3.31	3020	0, 0, 1, 0	1.3	0.8	46.0	0.06973	56.0
$\eta_3 = 3020$, $g_3 = 3$	2.37	4220	1, 0, 0, 1	1.3	0.8	2.9	0.35429	60.0
$\eta_4 = 1306$, $g_4 = 3$	1.71	5861	1, 1, 0, 1	1.3	0.8	0.42	0.68598	45.0
H$_2$O								
$m = 3$, $\eta_1 = 3652$, $g_1 = 1$	6.3	1600	0, 1, 0	$8.6(T_0/T)^{1/2} + 0.5$	1	41.2	0.09427	56.4
$\eta_2 = 1595$, $g_2 = 1$	2.7	3760[b]	0, 2, 0	$8.6(T_0/T)^{1/2} + 0.5$	1	0.19	0.13219	60.0
$\eta_3 = 3756$, $g_3 = 1$			1, 0, 0			2.30		
			0, 0, 1			22.40		
	1.87	5350	0, 1, 1	$8.6(T_0/T)^{1/2} + 0.5$	1	3.0	0.08169	43.1
	1.38	7250	1, 0, 1	$8.6(T_0/T)^{1/2} + 0.5$	1	2.5	0.11628	32.0
CO								
$m = 1$, $\eta_1 = 2143$, $g_1 = 1$	4.7	2143	1	1.1	0.8	20.9	0.07506	25.5
	2.35	4260	2	1.0	0.8	0.14	0.16758	20.0

[a]Upper band limit.
[b]See notes in [26].

575

in which

$$u_k = \frac{hc\eta_k}{kT} \qquad u_{0,k} = \frac{hc\eta_k}{kT_0}$$

and $v_{0,k} = 0$ if δ_k is positive, and is $|\delta_k|$ if δ_k is negative. Some illustrative numerical examples are given in [12].

Greif and coworkers [29, 39–42] developed wide-band correlations from basic spectroscopic theory, finding good agreement with experimental measurements in many cases. With this approach, no arbitrary constants are introduced, and no recourse is needed to experimental data to evaluate the constants.

To determine the effect of the band models on the final radiative transfer results, several band models have been applied to two problems [43] involving radiative transfer in gases with internal heat sources and heat transfer. In most instances good agreement was obtained by using the various models, but it is necessary to look at [43] to appreciate the detailed comparisons. A small amount of band-correlation information has been developed for liquids; Ref. [44] is an example that gives band absorption models for carbon tetrachloride. This information is important in evaluating the radiative contribution during the experimental determination of the thermal conductivity of an absorbing liquid.

The preceding discussion has concerned gases with a single radiating and absorbing component. If two gases are present that both absorb energy, then their band absorptances may overlap in some spectral regions. In this case, Hottel and Sarofim [45] show that, for two gases a and b in an overlapping band of width $\Delta\eta$, the following relation is valid:

$$\bar{A}_{a+b} = \Delta\eta \left[1 - \left(1 - \frac{\bar{A}_a}{\Delta\eta}\right)\left(1 - \frac{\bar{A}_b}{\Delta\eta}\right) \right] = \bar{A}_a + \bar{A}_b - \frac{\bar{A}_a \bar{A}_b}{\Delta\eta} \qquad (16\text{-}84)$$

Thus the simple sum of the two $\bar{A}$ is reduced by the quantity $\bar{A}_a\bar{A}_b/\Delta\eta$ (see also page 623). Restriction is to wave-number intervals over which both $\bar{A}_a$ and $\bar{A}_b$ are applicable average values, and in which there is no correlation between the positions of the individual lines of gases a and b.

Many additional complexities are introduced when a gas mixture is considered. For example, the partial pressure p of an absorbing gas in a multicomponent system varies with T and P, the populations of the energy states vary with T, and the overlapping of spectral lines changes with P. It is thus very complex to analytically formulate the dependence of $\bar{A}_l$ on T, p, and P for a real gas mixture. Useful results must depend heavily on experiment while theory is used as a guide. Some calculations for mixtures are given in [26]. Negrelli et al. [46] successfully applied the exponential wide-band model to a radiating flame layer of nonuniform composition and temperature. The predicted temperature distributions were in excellent agreement with those measured.

Hottel and Sarofim [45] discuss in detail total absorptance curves of the type shown in Fig. 13-13. Such curves are available for a number of gases, and their accuracy has been confirmed by many recent measurements. The use of total absorptances and effective bandwidths for various engineering problems will be discussed in Chap. 17.

16-7 SCATTERING FROM VARIOUS TYPES OF PARTICLES

The previous sections have been concerned with the absorption properties of gases. The analytical developments in Chaps. 13 and 14 also include the effects of scattering within a medium. Some information will be given here on particle scattering properties. As briefly discussed in Chap. 13, various phenomena may occur when incident radiation strikes a particle. Some of the radiation may be scattered by reflection from the surface, while the remaining energy penetrates into the particle where part of it can be absorbed. If the particle is not a strong internal absorber, some of this radiation will pass back out after traveling a single path through the particle or undergoing multiple internal reflections. When interacting with the particle boundary, the radiation will be refracted and will have its direction changed by subsequent internal reflections. The redirection, by these processes, of the energy penetrating into the particle and then escaping is scattering by refraction. Additional scattering is caused by diffraction, which is the result of slight bending of the propagation paths when radiation passes near the edges of an obstruction.

The reflection, refraction, and diffraction depend on the optical properties (that is, the complex refractive index $\bar{n} = n - i\kappa$) of the particle, and on the size of the particle relative to the wavelength of the incident radiation. An additional complication is the particle geometry. It is usually assumed that the medium surrounding the particles has a unity refractive index n and a zero extinction coefficient κ. In this instance the surrounding medium does not enter into the optical behavior of the medium-particle system.

In principle the scattering behavior can be obtained from the solution of the Maxwell electromagnetic equations that govern the radiation field for the medium-particle system. However, the solution provides very complicated relations for even simple particle geometries. Hence in many instances a number of simplifications are made, as will now be outlined.

One simplification is the geometric one of letting the scattering particles be spheres. This is not as restrictive an assumption as might appear, since, as discussed in [47], the results for spheres do have a wider geometric applicability. Consider an array of irregularly shaped particles, the surfaces of which are assumed composed of convex portions (no concave indentations). Because the particles are in a random orientation, an equal portion of surface elements will face each angular direction, which is the same angular distribution of surface elements as for a spherical particle. The net result is that the angular distribution of scattered radiation viewed at a distance from the actual particles will be the same as that scattered from spherical particles.

A second simplification is to consider the limiting solutions for scattering from large and small spheres. A parameter is $\pi D/\lambda$, where D is the sphere diameter. For large spheres ($\pi D/\lambda$ greater than about 5, where λ is the wavelength of the radiation in the particle material), the scattering is chiefly a reflection process and hence can be calculated from relatively simple geometric reflection relations. There is also diffraction of the radiation passing near the sphere, but this is accounted for separately, as will be discussed in Sec. 16-7.4. For small spheres ($\pi D/\lambda$ less than about $0.6/n$), the approximation of Rayleigh scattering can be used, as will be discussed in Sec. 16-7.5.

For the intermediate range of $\pi D/\lambda$, the general Mie scattering results are used, but the results of this general solution of Maxwell's equations are quite complicated.

A third type of simplification is to look at limiting cases of the optical constants of the particle. For metals the n and κ are often large, and simplifications can be introduced as in Sec. 4-6.2. For a dielectric ($\kappa = 0$) a limiting case is where $n \approx 1$. In this instance the reflectivity of the particle surface will be small.

Limiting types of surface conditions are also considered (that is, specular and diffuse surfaces). The particle surface can only act diffuse, however, if its dimensions are large compared with the wavelength of the incident radiation.

In what follows the theory and results will be discussed for several of the more useful scattering relations.

16-7.1 A Cloud of Large, Specularly Reflecting Spheres

One of the simplest scattering configurations is a cloud of large spherical particles ($\pi D/\lambda > {\sim}5$) that have specularly reflecting surfaces. Figure 13-9 shows a differential volume element of a particle cloud with cross section dA normal to the incident radiation and with thickness dS. The incident energy intercepted by the volume element is $i'_\lambda \, d\omega_i \, dA \, d\lambda$. It is assumed that the particle density is low enough so that each particle scatters independently and there is negligible shadowing of the particles by each other. Let the projected area of a particle normal to the direction of i'_λ be A_p so that the fraction of energy incident on dA that strikes the particle is A_p/dA. Part of this energy will be absorbed, and the remainder will be scattered by being reflected specularly.

The details of the reflection process are shown in Fig. 16-7. The energy intercepted by a band of width $R \, d\theta$ on the surface of the sphere is equal to the energy

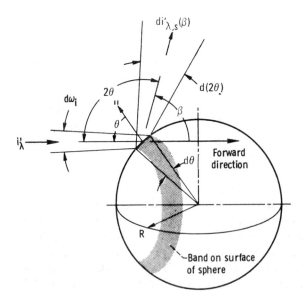

Figure 16-7 Reflection of incident radiation by surface of specular sphere.

intercepted by the particle multiplied by A_{band}/A_p, where A_{band} is the band area projected normal to i'_λ:

$$\text{Energy intercepted by band} = i'_\lambda \, d\omega_i \, d\lambda \, dA \, \frac{A_p}{dA} \frac{A_{band}}{A_p}$$

$$= i'_\lambda \, d\omega_i \, d\lambda \, 2\pi R^2 \sin\theta \cos\theta \, d\theta$$

The amount of reflected energy is $i'_\lambda \, d\omega_i \, d\lambda \, 2\pi R^2 \sin\theta \cos\theta \, d\theta \, \rho'_\lambda(\theta)$, where $\rho'_\lambda(\theta)$ is the directional specular reflectivity for incidence at angle θ. The amount of energy reflected from the entire sphere is found by integrating over the sphere area:

$$\text{Reflected energy} = i'_\lambda \, d\omega_i \, d\lambda \, \pi R^2 \int_0^{\pi/2} 2\rho'_\lambda(\theta) \sin\theta \, d(\sin\theta)$$

From Eq. (3-37) the integral is the hemispherical reflectivity ρ_λ. Hence the energy scattered by reflection from the entire sphere is $i'_\lambda \, d\omega_i \, d\lambda \, \pi R^2 \rho_\lambda$. By using the scattering cross section s_λ for the particle, the scattered energy is written $i'_\lambda \, d\omega_i \, d\lambda \, s_\lambda$. Hence $i'_\lambda \, d\omega_i \, d\lambda \, s_\lambda = i'_\lambda \, d\omega_i \, d\lambda \, \pi R^2 \rho_\lambda$, and the particle scattering cross section is

$$s_\lambda = \pi R^2 \rho_\lambda \tag{16-85}$$

Thus s_λ is equal to the projected area of the particle times the hemispherical reflectivity. Inserting (16-85) into (13-32) gives the scattering coefficient as

$$\sigma_{s\lambda} = \rho_\lambda \int_{R=0}^\infty \pi R^2 N_s(R) \, dR \tag{16-86}$$

If all the spheres are the same size with radius R, Eq. (16-86) gives

$$\sigma_{s\lambda} = \rho_\lambda \pi R^2 N_s \tag{16-87}$$

To obtain the phase function, Fig. 16-7 shows that the energy specularly reflected from the band of the sphere at angle θ will be reflected into direction 2θ and into a solid angle

$$d\omega_s = 2\pi \sin 2\theta \, d(2\theta) = 8\pi \sin\theta \cos\theta \, d\theta$$

The phase function is concerned only with the portion of the energy that is scattered. Since each particle is assumed to scatter independently, the *scattered portion* of the radiation emerging from a volume element dV of the particle cloud when observed at a distance large compared with the individual particle diameter will have the same phase function as for a single particle. Consider the energy incident on a single particle, $i'_\lambda \, d\omega_i \, d\lambda \, A_p$. Using the scattering cross section in (16-85), all the energy scattered away by a particle is written $i'_\lambda \, d\omega_i \, d\lambda \, \pi R^2 \rho_\lambda$. Then the scattered intensity is, from (13-43),

$$i'_{\lambda,s} = \frac{i'_\lambda \, d\omega_i \, d\lambda \, \pi R^2 \rho_\lambda}{d\omega_i \, A_p \, d\lambda} = i'_\lambda \rho_\lambda$$

The energy scattered away by a particle into $d\omega_s$ is $i'_\lambda \, d\omega_i \, d\lambda \, 2\pi R^2 \sin\theta \cos\theta \, d\theta \, \rho'_\lambda(\theta)$. The intensity scattered into direction 2θ [where this intensity is defined as in (13-44)] is then

$$i'_{\lambda,s}(2\theta) = \frac{i'_\lambda \, d\omega_i \, d\lambda \, 2\pi R^2 \sin\theta \cos\theta \, d\theta \, \rho'_\lambda(\theta)}{d\omega_i \, A_p \, d\omega_s \, d\lambda} = \frac{i'_\lambda \rho'_\lambda(\theta)}{4\pi} = \frac{i'_{\lambda,s}}{4\pi} \frac{\rho'_\lambda(\theta)}{\rho_\lambda}$$

Inserting this into (13-45) gives

$$\Phi(2\theta) = \frac{\rho'_\lambda(\theta)}{\rho_\lambda} \tag{16-88}$$

The angle 2θ is related to the angle β in Fig. 16-7 by $\beta = \pi - 2\theta$ so that, relative to the forward scattering direction,

$$\Phi(\beta) = \frac{\rho'_\lambda((\pi - \beta)/2)}{\rho_\lambda} \tag{16-89}$$

For unpolarized incident radiation the reflectivity $\rho'_\lambda(\theta)$ for a dielectric sphere can be found from Eq. (4-65). Also, the directional-hemispherical reflectivity is equal to unity minus the emissivity values in Fig. 4-6. As shown by this figure, the $\rho'_\lambda(\theta)$ for normal incidence is usually quite small compared with that at grazing angles ($\rho'_\lambda \to 1$ at

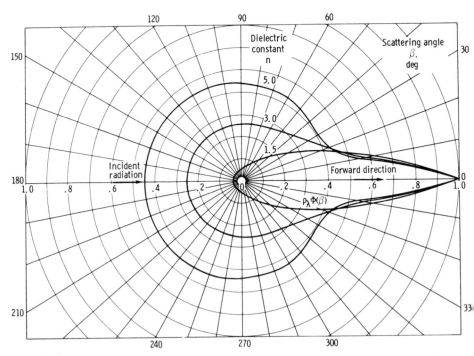

Figure 16-8 Scattering diagram for specular reflecting sphere that is large compared with wavelength of incident radiation.

$\theta = 90°$). Consequently, the forward scatter from the sphere (at $\beta = 0$) is unity, and the backward scatter (at $\beta = \pi$) is small. By use of Fig. 4-6 the quantity $\rho_\lambda \Phi(\beta)$ can be plotted for various indices of refraction n as shown in Fig. 16-8. The ρ_λ for a dielectric can be found by use of Fig. 4-7.

16-7.2 Reflection from a Diffuse Sphere

For a specularly reflecting sphere in Fig. 16-7, the energy scattered in each direction resulted from the reflection of energy at a single location on the sphere. If the sphere is diffuse, however, each surface element that intercepts incident radiation will reflect energy into the entire 2π solid angle above that element. Thus the radiation scattered into a specified direction will arise from the entire region of the sphere that receives radiation and is also visible from the specified direction. This is illustrated by Fig. 16-9a. The shaded portion of the sphere will not contribute radiation in the direction of the observer because it either does not receive radiation or is hidden from the direction of observation.

Consider the sphere of radius R in Fig. 16-9b. A typical surface-area element dA is located at angles ψ and φ. The observer is at angle β from the forward direction. The normal to dA is at θ and α relative to the directions of incidence and observation. The incident spectral energy flux within the incident solid angle $d\omega_i$ is $i'_\lambda\, d\omega_i\, d\lambda$. The projected area of dA normal to the incident direction is $dA \cos\theta$, so the energy received by dA is $i'_\lambda\, d\omega_i\, d\lambda\, dA \cos\theta$. The amount that is reflected is $\rho'_\lambda i'_\lambda\, d\omega_i\, d\lambda\, dA \cos\theta$, where ρ'_λ is the diffuse directional-hemispherical spectral reflectivity. The ρ'_λ is assumed independent of incidence angle and hence is equal to the hemispherical reflectivity ρ_λ. Using the cosine-law dependence for diffuse reflection gives the reflected energy per unit solid angle $d\omega_s$ in the direction of the observer as $\rho_\lambda i'_\lambda\, d\omega_i\, d\lambda\, dA \cos\theta \cos\alpha/\pi$. To integrate the reflected contributions that are received by the observer from all elements on the sphere surface, the dA, $\cos\theta$, and $\cos\alpha$ are expressed in terms of the spherical coordinates R, ψ, and φ, which gives $dA = R^2 \sin\varphi\, d\varphi\, d\psi$, $\cos\theta = \sin\varphi \cos\psi$, and $\cos\alpha = \sin\varphi \cos(\psi + \pi - \beta)$. Then the energy scattered by reflection into the β direction per unit solid angle $d\omega_s$ about that direction is

$$\frac{\rho_\lambda i'_\lambda\, d\omega_i\, d\lambda\, R^2}{\pi} \int_{\varphi=0}^{\pi} \int_{\psi=-\pi/2}^{\beta-(\pi/2)} \sin^3\varphi \cos\psi \cos(\psi + \pi - \beta)\, d\psi\, d\varphi$$

By integrating, this becomes

$$\frac{\rho_\lambda i'_\lambda\, d\omega_i\, d\lambda\, R^2}{\pi} \tfrac{2}{3}(\sin\beta - \beta \cos\beta)$$

The energy per unit $d\lambda$ scattered in direction β per unit $d\omega_s$ and per unit area and solid angle of the incident radiation is obtained by dividing the scattered energy by $\pi R^2\, d\omega_i\, d\lambda$, giving

$$i'_{\lambda,s}(\beta) = \frac{\rho_\lambda i'_\lambda}{\pi^2} \tfrac{2}{3}(\sin\beta - \beta \cos\beta)$$

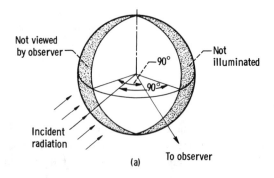

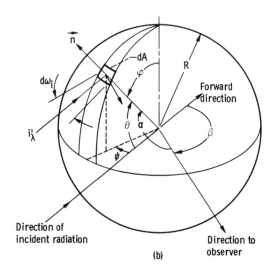

Figure 16-9 Scattering by reflection from diffuse sphere. (*a*) Illuminated region visible to observer; (*b*) geometry on sphere.

The entire amount of incident intensity that is scattered is $i'_{\lambda,s} = \rho_\lambda i'_\lambda$. Then from (13-45), the directional magnitude of the scattered intensity is equal to the entire scattered intensity times the phase function,

$$\frac{\rho_\lambda i'_\lambda}{\pi^2} \tfrac{2}{3}(\sin \beta - \beta \cos \beta) = \rho_\lambda i'_\lambda \frac{\Phi(\beta)}{4\pi}$$

so that the phase function for a diffuse sphere is

$$\Phi(\beta) = \frac{8}{3\pi} (\sin \beta - \beta \cos \beta) \tag{16-90}$$

The $\Phi(\beta)$ from (16-90) is plotted in Fig. 16-10. The largest scattering is for $\beta = 180°$, that is, toward an observer back in the same direction as the origin of the incident radiation. In this instance, the entire illuminated surface of the sphere is observed.

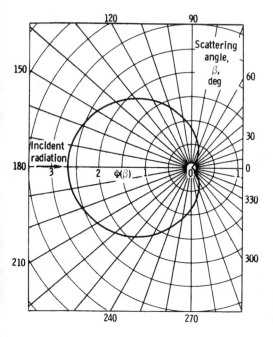

Figure 16-10 Scattering phase function for diffuse reflecting sphere, large compared with wavelength of incident radiation and with constant reflectivity.

16-7.3 Large Dielectric Sphere with Refractive Index Close to Unity

For a large dielectric ($\kappa = 0$) sphere with refractive index $n \approx 1$, the reflectivity of the particle surface approaches zero. The incident radiation can thus pass with unchanged amplitude into the sphere, and there is no scattering by reflection. With the extinction coefficient zero, the radiation will pass back out of the sphere with unchanged amplitude. However, the velocity $c = c_0/n$ inside the sphere medium is slightly less than that outside, so that radiation passing through different portions of the sphere and hence through different thicknesses will have different phase lags. The resulting interference of the waves passing out of the sphere yields a scattering cross section

$$s_\lambda = \frac{\pi D^2}{4}\left[2 - \frac{4}{W}\sin W + \frac{4}{W^2}(1 - \cos W)\right] \tag{16-91}$$

where $W = 2(\pi D/\lambda)(n - 1)$. Additional information for this situation is given in [47].

16-7.4 Diffraction from a Large Sphere

For large spheres there is diffraction of the radiation passing in the vicinity of the particle. The effects of the diffraction and reflection must be added to obtain the total scattering behavior. Fortunately the diffracton is predominantly in the forward scattering direction. This means that the diffraction can be included in the radiative transfer as if it were transmitted past the particle without interacting with the particle;

hence, diffraction can often be neglected in considering the energy exchange within a scattering medium.

The most familiar form of diffraction is when light passes through a small hole or slit. As shown in Fig. 16-11a, the result is a diffraction pattern of alternate illuminated and dark rings or strips. If a spherical particle is in the path of incident radiation,

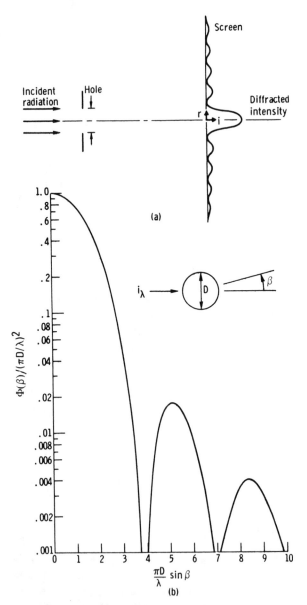

Figure 16-11   Diffraction by hole or large spherical particle. (a) diffraction of radiation by hole; (b) phase function for diffraction from large sphere.

Babinet's principle states that the diffracted intensities are the same as for a hole. This is a consequence of the fact that a hole and a particle produce complementary disturbances in the amplitude of an incident electromagnetic wave. The energy diffracted by a spherical particle is thus the same as that diffracted by a hole of the same diameter. As a consequence, the entire projected area of a sphere takes part in the diffraction process, and the scattering cross section for diffraction is equal to the projected area $\pi D^2/4$. Since diffraction and reflection occur simultaneously, the total scattering cross section can approach $2\pi D^2/4$ when the sphere is highly reflecting.

The phase function for diffraction by a large sphere is given in terms of a Bessel function of the first kind of order one (pages 107 and 108 of [47]):

$$\Phi(\beta) = \left(\frac{\pi D}{\lambda}\right)^2 \left\{\frac{2J_1[(\pi D/\lambda)\sin\beta]}{(\pi D/\lambda)\sin\beta}\right\}^2 \tag{16-92}$$

This function is plotted in Fig. 16-11b. Since the abscissa is $(\pi D/\lambda)\sin\beta$, for particles with large $\pi D/\lambda$ the diffracted radiation lies within a narrow angular region in the forward scattering direction. For small particles where $\pi D/\lambda$ is of order unity, the theory leading to (16-92) is invalid and the general Mie scattering theory must be applied. The integration to show that (16-92) satisfies (13-48) is discussed on page 398 of [48]; it is only necessary to integrate over small β, which simplifies the integration considerably.

16-7.5 Rayleigh Scattering

In many common situations, the scattering particles are considerably smaller in diameter than the wavelength of the incident radiation $(D \ll \lambda)$. Scattering from such particles is termed *Rayleigh* scattering after Lord Rayleigh, who examined this situation. Rayleigh scattering is important in the atmosphere, where the scattering is by gas molecules. The scattering cross section can be derived from quantum theory or electromagnetic theory. Originally, Rayleigh derived the functional dependence by dimensional analysis and arrived at the result

$$\frac{i'_{\lambda,s}}{i'_\lambda} \propto G^2(\bar{n})\frac{V^2}{\lambda^4} \tag{16-93}$$

where V is the volume of a particle, and $G(\bar{n})$ is a function of the complex refractive index of the scattering material. The important result is that for Rayleigh scattering, *the scattered energy in any direction is proportional to the inverse fourth power of the wavelength of the incident radiation*. This inverse dependence shows that when the incident radiation covers a wavelength spectrum, the shorter-wavelength radiation will be Rayleigh scattered with a strong preference.

Rayleigh scattering by the molecules of the atmosphere accounts for the background of the sky being blue, and for the sun becoming red in appearance at sunset. The blue portion of the incident sunlight is at the short-wavelength end of the visible spectrum. Hence it undergoes strong Rayleigh scattering into all directions, giving the sky its overall blue background. Without molecular scattering, the sky would appear black except for the direct view of the sun. As the sun is setting, the path length for direct radiation through the atmosphere becomes much longer than during the middle of the day. In traversing this longer path, more of the short-wavelength portion of the spectrum is scattered away from the direct path of the sun's rays. As a result, at sunset the sun takes on a red color as the longer-wavelength red rays are able to penetrate the atmosphere with less attenuation than the rest of the visible spectrum. If many dust particles are present, a deep red sunset may be seen.

If particles with a very limited range of sizes are present in the atmosphere, unusual scattering effects may be observed. Following the eruption of Krakatoa in 1883, the occurrence of blue and green suns and moons was noted over a period of many years. This effect was attributed to particles in the atmosphere of such a size range as to scatter only the red portion of the visible spectrum. On September 26, 1950, a blue sun and moon were observed in Europe, a phenomenon believed due to finely dispersed smoke particles of uniform size carried from forest fires burning in Canada.

Scattering cross sections for Rayleigh scattering. Equation (16-93) gives specifically only the dependence of the scattered radiation on wavelength and particle volume; additional information is needed for the particle scattering cross section and the angular distribution of the scattered intensity. Consider first small nonabsorbing ($\kappa = 0$) particles so that $\bar{n} = n$ and $\pi D/\lambda < \sim 0.6/n$, where λ is the wavelength in the particle material. The Rayleigh scattering cross section for unpolarized incident radiation is found from more advanced theory to be

$$s_\lambda = \frac{24\pi^3 V^2}{\lambda^4} \left(\frac{n^2 - 1}{n^2 + 2}\right)^2 = \frac{8}{3} \frac{\pi D^2}{4} \left(\frac{\pi D}{\lambda}\right)^4 \left(\frac{n^2 - 1}{n^2 + 2}\right)^2 \tag{16-94}$$

Often Rayleigh-scattering cross sections are given in terms of the polarizability α_p of the particles. This is a proportionality factor relating the forces induced in the molecules to the external electromagnetic field. Specifically, it relates the dipole moment per unit volume (defined as the polarization) produced in the material to the external field. For the case under discussion here, the polarizability is

$$\alpha_p = \frac{3}{4\pi} V \frac{n^2 - 1}{n^2 + 2} \tag{16-95}$$

so that (16-94) can be written

$$s_\lambda = \frac{2^7 \pi^5 \alpha_p^2}{3\lambda^4} \tag{16-96}$$

In this more general form, cross sections for various particles that follow the Rayleigh scattering relations can be introduced by substituting the requisite form for α_p into (16-96). Table 16-5 gives some quantities for individual particles and particles in a nonparticipating medium.

The actual scattering cross section for particles in a medium may vary with λ in a manner somewhat different from a $1/\lambda^4$ dependence. In air at standard temperature

Table 16-5 Polarizability for various scattering conditions

Scattering particles	Restrictions	Polarizability α_p, length^3
Electrons (Thomson scattering)	Energy of incident photon is small, $h\nu \ll m_e c_0^2$	$\dfrac{e^2}{m_e c_0^2}\left(\dfrac{\lambda}{2\pi}\right)^2$†
Small dielectric particle (Rayleigh)	Particle diameter is small compared with wavelength in medium and in particle	$\dfrac{n^2-1}{n^2+2}\left(\dfrac{D}{2}\right)^3$
Medium containing small particles (Lorentz-Lorenz)	Spacing between particles is small compared with wavelength ($<\lambda$). Particle diameter is very small ($D \ll \lambda$) compared with λ in both medium and particle. Spacing between particles $> D$	$\dfrac{3}{4\pi N}\left\|\dfrac{\bar{n}^2-1}{\bar{n}^2+2}\right\|$‡
Medium containing small particles	Spacing between particles is large ($\gg\lambda$). Particle diameter is very small ($D \ll \lambda$).	$\dfrac{\|\bar{n}^2-1\|}{4\pi N}$‡
Medium containing small particles	Spacing between particles is large ($\gg\lambda$). Particle diameter is very small ($D \ll \lambda$). The $\bar{n}$ is close to 1.	$\dfrac{\|\bar{n}-1\|}{2\pi N}$‡

† $e^2/m_e c_0^2$ = classical electron radius, 2.818×10^{-13} cm.

‡ In this instance the $\bar{n}$ is the refractive index of the particulate medium as a whole and depends on the particle number density and volume; see Ref. [47].

and pressure, for example, the restrictions are satisfied such that Rayleigh scattering from the gas molecules should govern. However, the variation of refractive index with wavelength causes the variation of the scattering cross section to depart somewhat from the $1/\lambda^4$ dependence. This is shown in Fig. 16-12, where the actual scattering dependence on wavelength is compared with $1/\lambda^4$.

When the particles are of a conducting material with a complex index of refraction $\bar{n} = n - i\kappa$, the scattering cross section has the following more general form than (16-94):

$$s_\lambda = \frac{24\pi^3 V^2}{\lambda^4}\left|\frac{\bar{n}^2-1}{\bar{n}^2+2}\right|^2 = \frac{8}{3}\frac{\pi D^2}{4}\left(\frac{\pi D}{\lambda}\right)^4\left|\frac{\bar{n}^2-1}{\bar{n}^2+2}\right|^2 \qquad (16\text{-}97)$$

$[(n^2$

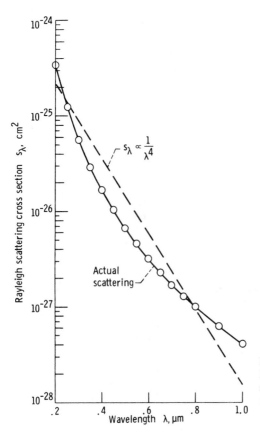

Figure **16-12** Comparison of actual Rayleigh scattering cross section for air at standard temperature and pressure with $1/\lambda^4$ variation (from [11]).

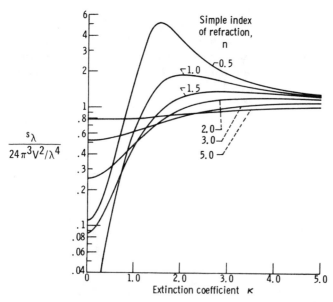

Figure **16-13** Rayleigh-scattering cross section as a function of simple index of refraction and extinction coefficient.

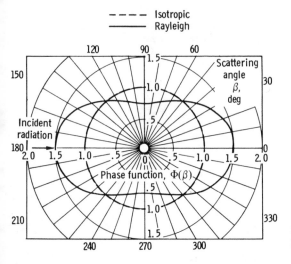

Figure 16-14 Phase functions for Rayleigh and isotropic scattering.

Phase function for Rayleigh scattering. For incident unpolarized radiation, electromagnetic theory gives, for Rayleigh scattering,

$$\Phi(\beta,\varphi) = \tfrac{3}{4}(1 + \cos^2 \beta) \qquad (16\text{-}99)$$

This is independent of the circumferential angle φ.

A plot of the phase function for Rayleigh and for isotropic scattering is given in Fig. 16-14. For Rayleigh scattering, the scattered energy is directed preferentially along the forward direction of the incident radiation, and strongly back toward the source of the radiation.

16-7.6 Mie Scattering Theory

When the scattering particles are not large, as in Secs. 16-7.1–16-7.4, and are not small enough to fall into the range that is adequately described by Rayleigh scattering, recourse must be taken to more complicated treatments. This is required for the approximate range $0.6/n < \pi D/\lambda < 5$, where λ is the wavelength inside the particle. Gustav Mie [49] originally applied electromagnetic theory to derive the properties of the electromagnetic field when a plane monochromatic wave is incident upon a spherical surface across which the optical properties n and κ change abruptly. The energy absorption by the medium, the absorption by the scattering particles, or both can be accounted for, and the results apply over the entire range of particle diameters. As might be expected, strong polarization effects can be present. In certain cases, the phase function becomes very complicated, as illustrated by Fig. 16-15.

Van der Hulst [47] gives an excellent detailed treatment of the Mie theory. The limiting cases for very small and very large particles are examined, and working formulas are presented for all size ranges. Cross sections and phase functions are discussed for dielectric and metallic particles of various shapes, including spheres and cylinders. Some further work on absorbing particles, using detailed Mie scattering theory, has been done by Plass [50].

———————————— Phase function
—————————— Portion of phase function due to
perpendicularly polarized component

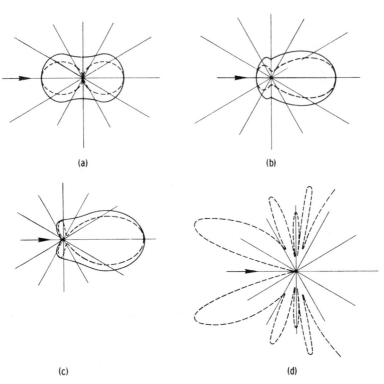

(a)   (b)

(c)   (d)

Figure 16-15 Phase functions for Mie scattering from metallic and dielectric spheres (on arbitrary scales) (from [47] and [61]). (a) $\pi D/\lambda \to 0$, metallic sphere, $\bar{n} = 0.57 - 4.29i$; (b) $\pi D/\lambda = 9.15$, metallic sphere, $\bar{n} = 0.57 - 4.29i$; (c) $\pi D/\lambda = 10.3$, metallic sphere, $\bar{n} = 0.57 - 4.29i$; (d) $\pi D/\lambda = 8$, dielectric sphere, $n = 1.25$.

One of the simpler results from the Mie theory is for small spheres. The general Mie equations can be expanded into a power series in terms of the parameter $\pi D/\lambda$, giving the scattering cross section as

$$s_\lambda = \frac{8}{3}\frac{\pi D^2}{4}\left(\frac{\pi D}{\lambda}\right)^4 \left|\frac{\bar{n}^2 - 1}{\bar{n}^2 + 2}\left[1 + \frac{3}{5}\frac{\bar{n}^2 - 2}{\bar{n}^2 + 2}\left(\frac{\pi D}{\lambda}\right)^2 + \cdots\right]\right|^2 \qquad (16\text{-}100)$$

The second term in brackets is the first correction to the Rayleigh scattering relation in (16-97), which was valid for very small particles.

For small spheres the limit where n becomes very large can also be considered. The scattering particles will then form an idealized cloud of very highly reflecting dielectric particles. The result for this case cannot be obtained by letting $\bar{n}$ in (16-100) approach $\infty + i0$. As n becomes large, the small part of the incident radiation that does penetrate the particle becomes almost totally internally reflected. This creates standing waves within the particle, which provide resonance peaks in the scattering. The expansion used to obtain Eq. (16-100) did not account for this behavior. In the limit for $n \to \infty$ the scattering cross section for small spheres is

$$s_\lambda = \frac{\pi D^2}{4} \left[\frac{10}{3} \left(\frac{\pi D}{\lambda} \right)^4 + \frac{4}{5} \left(\frac{\pi D}{\lambda} \right)^6 + \cdots \right] \tag{16-101}$$

If, in addition to $n \to \infty$, the particles are so small that only the first term within the brackets of (16-101) is significant, then for unpolarized incident radiation the phase function is given by

$$\Phi(\beta) = \tfrac{3}{5} [(1 - \tfrac{1}{2} \cos \beta)^2 + (\cos \beta - \tfrac{1}{2})^2] \tag{16-102}$$

A polar diagram of this function is given in Fig. 16-16. This shows that in contrast to Rayleigh scattering (Fig. 16-14) the highly reflecting particles produce a very strong scattering back toward the source.

The effect of phase function was investigated by Love et al. [51], who studied plane and cylindrical boundaries with given reflectivities adjacent to absorbing, emitting, and scattering gases. Some experimentally determined values of the scattering phase functions for glass beads and aluminum, carbon, iron, and silica particles were used for comparison of their effect in a variety of energy exchange cases. It is significant that little difference in energy transfer was found between the results using these experimental phase functions and the results using either the Rayleigh or isotropic phase functions. It appears therefore that *the assumption of isotropic scattering is often justified in energy exchange calculations in enclosures.*

In Fig. 16-17, results from [51] are shown for the fraction of emitted energy from a black disk that is scattered back to the base plane from a cylinder of gas adjacent to the disk. The results using various scattering phase functions in the gas are in very

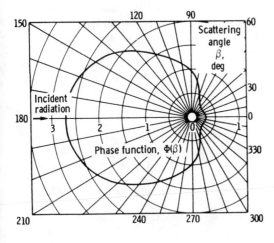

Figure 16-16 Phase function for scattering of unpolarized incident radiation from small nonabsorbing sphere with $n \to \infty$.

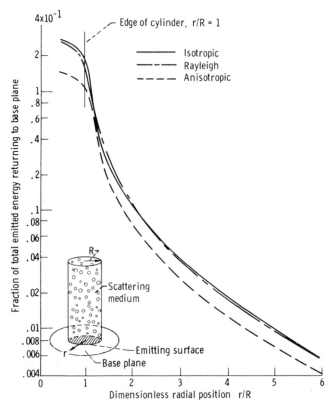

Figure 16-17 Effect of scattering phase function on energy scattered back to base plane by cylinder of scattering medium [51]. Optical diameter of cylinder, 2; height to diameter ratio, 5.

good agreement. In parallel-plane geometries the various phase functions gave energy transfer results that had less variation than in the cylindrical geometry. It should be emphasized that in some cases the insensitivity of results to the scattering phase function is not a reasonable assumption. Specifically, the phase function will be important for beam transmission or other situations in which strong sources transmit directionally into a scattering atmosphere.

REFERENCES

1. Einstein, Albert: On the Quanta Theory of Radiation, *Phys. Z.*, vol. 18, pp. 121–128, 1917.
2. Einstein, Albert: Emission and Absorption of Radiation according to the Quantum Theory, *Verh. Dtsch. Phys. Ges.*, vol. 18, pp. 318–323, 1916.
3. Heitler, Walter: "The Quantum Theory of Radiation," 3d ed., pp. 412–414, Clarendon Press, Oxford, 1954.
4. Bond, John W., Kenneth M. Watson, and Jasper A. Welch, Jr.: "Atomic Theory of Gas Dynamics," Addison-Wesley Publishing Company, Reading, Mass., 1965.

5. Wiese, W. L., M. W. Smith, and B. M. Glennon: Atomic Transition Probabilities, vol. I. Hydrogen through Neon – A Critical Data Compilation. *Rept.* NSRDS-NBS-4, *Natl. Bur. Std.*, vol. 1, May 20, 1966.

6. Griem, Hans R.: "Plasma Spectroscopy," McGraw-Hill Book Company, New York, 1964.

7. Breene, Robert G.: "The Shift and Shape of Spectral Lines," Pergamon Press, New York, 1961.

8. Penner, S. S.: "Quantitative Molecular Spectroscopy and Gas Emissivities," Addison-Wesley Publishing Company, Reading, Mass., 1959.

9. Tien, C. L.: Thermal Radiation Properties of Gases, in T. F. Irvine, Jr., and J. P. Hartnett (eds.), "Advances in Heat Transfer," vol. 5, pp. 253–324, Academic Press Inc., New York, 1968.

10. Armstrong, B. H., J. Sokoloff, R. W. Nicholls, D. H. Holland, and R. E. Meyerott: Radiative Properties of High Temperature Air. *J. Quant. Spectrosc. Radiat. Transfer*, vol. 1, no. 2, pp. 143–162, 1961.

11. Goody, R. M.: "Atmospheric Radiation, Theoretical Basis," vol. 1, Clarendon Press, Oxford, 1964.

12. Edwards, D. K.: Molecular Gas Band Radiation, in T. F. Irvine, Jr., and J. P. Hartnett (eds.), "Advances in Heat Transfer," vol. 12, pp. 115–193, Academic Press, Inc., New York, 1976.

13. Tiwari, S. N.: Band Models and Correlations for Infrared Radiation, in "Radiative Transfer and Thermal Control," vol. 49 of *Progress in Astronautics and Aeronautics Series*, pp. 155–182, *AIAA*, 1976.

14. Morizumi, S. J.: Comparison of Analytical Model with Approximate Models for Total Band Absorption and Its Derivative, *J. Quant. Spectrosc. Radiat. Transfer*, vol. 22, no. 5, pp. 467–474, 1979.

15. Elsasser, Walter M.: "Heat Transfer by Infrared Radiation in the Atmosphere," Harvard Meteorological Studies no. 6, Harvard University Press, Cambridge, Mass., 1942.

16. Edwards, D. K., and W. A. Menard: Comparison of Models for Correlation of Total Band Absorption, *Appl. Opt.*, vol. 3, no. 5, pp. 621–625, 1964.

17. Ludwig, C. B., W. Malkmus, J. E. Reardon, and J. A. L. Thomson: Handbook of Infrared Radiation from Combustion Gases, *NASA* SP-3080, 1973.

18. Kyle, T. G.: Absorption of Radiation by Uniformly Spaced Doppler Lines, *Astrophys. J.*, vol. 148, no. 3, pp. 845–848, 1967.

19. Golden, S. A.: The Doppler Analog of the Elsasser Band Model, *J. Quant. Spectrosc. Radiat. Transfer*, vol. 7, no. 3, pp. 483–494, 1967.

20. Golden, S. A.: The Doppler Analog of the Elsasser Band Model, II, *J. Quant. Spectrosc. Radiat. Transfer*, vol. 8, pp. 877–897, 1968.

21. Golden, S. A.: The Voigt Analog of an Elsasser Band, *J. Quant. Spectrosc. Radiat. Transfer*, vol. 9, no. 8, pp. 1067–1081, 1969.

22. Tien, C. L. and J. E. Lowder: A Correlation for Total Band Absorptance of Radiating Gases, *Int. J. Heat Mass Transfer*, vol. 9, no. 7, pp. 698–701, 1966.

23. Cess, R. D., and S. N. Tiwari: Infrared Radiative Energy Transfer in Gases, in T. F. Irvine, Jr., and J. P. Hartnett, (eds.), "Advances in Heat Transfer", vol. 8, pp. 229–283, Academic Press, Inc., New York, 1972.

24. Goody, R. M., and M. J. S. Belton: Radiative Relaxation Times for Mars (Discussion of Martian Atmospheric Dynamics), *Planet. Space Sci.*, vol. 15, no. 2, pp. 247–256, 1967.

25. Edwards, D. K., and A. Balakrishnan: Slab Band Absorptance for Molecular Gas Radiation, *J. Quant. Spectrosc. Radiat. Transfer*, vol. 12, pp. 1379–1387, 1972.

26. Edwards, D. K., and A. Balakrishnan: Thermal Radiation by Combustion Gases, *Int. J. Heat Mass Transfer*, vol. 16, no. 1, pp. 25–40, 1973.

27. Tien, C. L., and G. R. Ling: On a Simple Correlation for Total Band Absorptance of Radiative Gases, *Int. J. Heat Mass Transfer*, vol. 12, no. 9, pp. 1179–1181, 1969.

28. Felske, J. D., and C. L. Tien: A Theoretical Closed-Form Expression for the Total Band Absorptance of Infrared-Radiating Gases, *Int. J. Heat Mass Transfer*, vol. 17, pp. 155–158, 1974.

29. Hsieh, T. C., and R. Greif: Theoretical Determination of the Absorption Coefficient and the Total Band Absorptance Including a Specific Application to Carbon Monoxide, *Int. J. Heat Mass Transfer*, vol. 15, pp. 1477–1487, 1972.

30. Edwards, D. K.: Radiant Interchange in a Nongray Enclosure Containing an Isothermal Carbon-Dioxide–Nitrogen Gas Mixture, *J. Heat Transfer*, vol. 84, no. 1, pp. 1–11, 1962.

31. Edwards, D. K.: Absorption of Infrared Bands of Carbon Dioxide Gas at Elevated Pressures and Temperatures, *J. Opt. Soc. Am.*, vol. 50, no. 6, pp. 617–626, 1960.

32. Edwards, D. K., and W. A. Menard: Correlations for Absorption by Methane and Carbon Dioxide Gases, *Appl. Opt.*, vol. 3, no. 7, pp. 847–852, 1964.

33. Edwards, D. K., and W. Sun: Correlations for Absorption by the 9.4-μ and 10.4-μ CO_2 Bands, *Appl. Opt.*, vol. 3, no. 12, pp. 1501–1502, 1964.

34. Edwards, D. K., B. J. Flornes, L. K. Glassen, and W. Sun: Correlation of Absorption by Water Vapor at Temperatures from 300 K to 1100 K, *Appl. Opt.*, vol. 4, no. 6, pp. 715–721, 1965.

35. Edwards, D. K.: Absorption of Radiation by Carbon Monoxide Gas according to the Exponential Wide-band Model, *Appl. Opt.*, vol. 4, no. 10, pp. 1352–1353, 1965.

36. Edwards, D. K., and K. E. Nelson: Rapid Calculation of Radiant Energy Transfer between Nongray Walls and Isothermal H_2O or CO_2 Gas, *J. Heat Transfer*, vol. 84, no. 4, pp. 273–278, 1962.

37. Weiner, Michael M.: "Radiant Heat Transfer in Non-Isothermal Gases," Ph.D. thesis, University of California at Los Angeles, 1966.

38. Hines, W. S., and D. K. Edwards: Infrared Absorptivities of Mixtures of Carbon Dioxide and Water Vapor, *Chem. Eng. Prog. Symp. Ser.*, vol. 64, no. 82, pp. 173–180, 1968.

39. Lin, J. C., and R. Greif: Total Band Absorptance of Carbon Dioxide and Water Vapor Including Effects of Overlapping, *Int. J. Heat Mass Transfer*, vol. 17, pp. 793–795, 1974.

40. Lin, J. C., and R. Greif: Approximate Method for Absorption at High Temperatures, *Int. J. Heat Mass Transfer*, pp. 1805–1807, 1973.

41. Hashemi, A., T. C. Hsieh, and R. Greif: Theoretical Determination of Band Absorption with Specific Application to Carbon Monoxide and Nitric Oxide, *ASME J. Heat Transfer*, vol. 98, pp. 432–437, 1976.

42. Chu, K. H., and R. Greif: Theoretical Determination of Band Absorption for Nonrigid Rotation with Applications to CO, NO, N_2O and CO_2, *ASME J. Heat Transfer*, vol. 100, pp. 230–234, 1978.

43. Tiwari, S. N.: Applications of Infrared Band Model Correlations to Nongray Radiation, *Int. J. Heat Mass Transfer*, vol. 20, no. 7, pp. 741–751, 1977.

44. Novotny, J. L., D. E. Negrelli, and T. Van der Driessche: Total Band Absorption Models for Absorbing-Emitting Liquids: CCl_4, *ASME J. Heat Transfer*, vol. 96, pp. 27–31, 1974.

45. Hottel, Hoyt C., and Adel F. Sarofim: "Radiative Transfer," McGraw-Hill Book Company, New York, 1967.

46. Negrelli, D. E., J. R. Lloyd, and J. L. Novotny: A Theoretical and Experimental Study of Radiation-Convection Interaction in a Diffusion Flame, *J. Heat Transfer*, vol. 99, pp. 212–220, 1977.

47. Van der Hulst, Hendrick C.: "Light Scattering by Small Particles," John Wiley & Sons, Inc., New York, 1957.

48. Born, Max, and Emil Wolf: Principles of Optics, 2d ed., Pergamon Press, New York, 1964.

49. Mie, Gustav: Optics of Turbid Media, *Ann. Phys.*, vol. 25, no. 3, pp. 377–445, 1908.

50. Plass, Gilbert N.: Mie Scattering and Absorption Cross Sections for Absorbing Particles, *Appl. Opt.*, vol. 5, no. 2, pp. 279–285, 1966.

51. Love, Tom J., Leo W. Stockham, Fu C. Lee, William A. Munter, and Yih W. Tsai: Radiative Heat Transfer in Absorbing, Emitting and Scattering Media, ARL-67-0210, DDC no. AD-666427), Oklahoma University, December, 1967.

52. Edwards, D. K., L. K. Glassen, W. C. Hauser, and J. S. Tuchscher: Radiation Heat Transfer in Nonisothermal Nongray Gases, *J. Heat Transfer*, vol. 89, no. 3, pp. 219–229, 1967.

53. Howard, John N., Darrell E. Burch, and Dudley Williams: Near-Infrared Transmission through

Synthetic Atmospheres, *Geophys. Res.*, no. 40, AFCRL-TR-55-213, Air Force Cambridge Research Center, 1955.

54. Howard, J. N., D. E. Burch, and Dudley Williams: Infrared Transmission of Synthetic Atmospheres. I. Instrumentation, *J. Opt. Soc. Am.*, vol. 46, no. 3, pp. 186–190, 1956.

55. Howard, J. N., D. E. Burch, and Dudley Williams: Infrared Transmission of Synthetic Atmospheres. II. Absorption by Carbon Dioxide, *J. Opt. Soc. Am.*, vol. 46, no. 4, pp. 237–241, 1956.

56. Howard, J. N., D. E. Burch, and Dudley Williams: Infrared Transmission of Synthetic Atmospheres. IV. Application of Theoretical Band Models, *J. Opt. Soc. Am.*, vol. 46, no. 5, pp. 334–338, 1956.

57. Stull, V. Robert, and Gilbert N. Plass: Spectral Emissivity of Hydrogen Chloride from 1000–3400 cm^{-1}, *J. Opt. Soc. Am.*, vol. 50, no. 12, pp. 1279–1285, 1960.

58. Aroeste, Henry, and William C. Benton: Emissivity of Hydrogen Atoms at High Temperatures, *J. Appl. Phys.*, vol. 27, no. 2, pp. 117–121, 1956.

59. Jones, M. C.: Far Infrared Absorption in Liquefied Gases, NBS Technical Note 390, National Bureau of Standards, Boulder, Colorado, 1970.

60. Tien, C. L.: Band and Total Emissivity of Ammonia, *Int. J. Heat Mass Transfer*, vol. 16, pp. 856–857, 1973.

61. Kerker, M. (ed.): "Proceedings of the Interdisciplinary Conference on Electromagnetic Scattering," August, 1962, Pergamon Press, New York, 1963.

PROBLEMS

1 The Balmer series provides a visible series of line spectra for the hydrogen atom. This series corresponds to $i = 2, j = 3, 4, 5, \ldots$. Compute the emission angular frequencies and wavelengths for this series for $j = 3, 4, 5,$ and 6.

   Answer: 2.88 × 10^{15} rad/s   0.656 μm
   3.88 × 10^{15} rad/s   0.486 μm
   4.35 × 10^{15} rad/s   0.434 μm
   4.59 × 10^{15} rad/s   0.410 μm

2 Show that the spectral line frequencies for an atom are additive in such a manner that $\Omega_{ab} + \Omega_{bc} = \Omega_{ac}$.

3 For the transition Ω_{25} of the Balmer series, compute as a function of temperature the ratio of spontaneous to induced emission probabilities by using the relations between Einstein coefficients.

4 According to the "Handbuch der Physik," some of the quantities for the Lyman transition series of the hydrogen atom are

| Wavelength | Transition | | Transition probability | Oscillator strength |
λ, Å	i	j	$4\pi A_{ji}$, s^{-1}	$3f_{ij}$
1216	1	2	6.25 × 10^8	0.4162
1026	1	3	1.64 × 10^8	0.0791
973	1	4	0.68 × 10^8	0.0290
950	1	5	0.34 × 10^8	0.0139

Verify the values of λ and the relation between A_{ji} and f_{ij}. (*Note:* The factor 4π on the A_{ji} results from the A_{ji} being defined here on the basis of intensity rather than energy density; the factor 3 on the f_{ij} is to account for three degrees of freedom for the electron oscillation in the atom.)

5   Compute the half-width for Doppler broadening of neon at a wavelength of 0.6 μm and for $T = 300$ K.

   Answer: 0.023 cm^{-1}.

6   Two absorption lines have the same transition (centerline) wave number, $\eta_{ij} = 550$ cm^{-1}. Both have the same half-width, 0.15 cm^{-1}. One line has the Doppler profile; one has the Lorentz profile. Draw the two line shapes, $a_{\eta,ij}(\eta)/S_{ij}$, as a function of η, on the same plot.

7   A gas is composed of pure atomic hydrogen at a temperature of 500 K. Calculate the half-width of the hydrogen Lyman alpha line (the transition between $i = 1$ and $j = 2$) for the case of Doppler broadening. Then plot the line shape $a_{\eta,ij}(\eta)/S_{ij}$ for this line as a function of wave number. The mass of the hydrogen atom is 1.66×10^{-24} g.

   Answer: 0.659 cm^{-1}.

8   For the same gas and temperature, as in Prob. 7, compute the half-width of the line for collision broadening at a pressure of 1 atm. Assume the diameter of the hydrogen atom is about 1.06×10^{-8} cm. Plot $a_{\eta,ij}(\eta)/S_{ij}$ for collision broadening on the same wave number plot as Prob. 7.

   Answer: 0.0127 cm^{-1}.

9   For $\beta = 0.2$ prepare a plot comparing various band correlation functions of $\bar{A}_l/A_0$ for $0.01 \leqslant u \leqslant 100$. Compare the correlation functions of Edwards and Menard, Tien and Lowder, Cess and Tiwari, and Goody and Belton.

10 Find the effective bandwidth $\bar{A}$ of the 9.4-μm band of CO_2 at a partial pressure of 0.4 atm in a mixture with nitrogen at a total pressure of 1 atm. The temperature is 500 K, and the path length S is 0.364 m. Compare with the result in Example 16-1.

   Answer: 2.32 cm^{-1}.

11 For pure CO gas at 1 atm pressure, determine the effective bandwidth for the 4.67-μm band at $T = 600$ K for a path length S of 0.5 m.

   Answer: 221 cm^{-1}.

12 From Fig. 13-2 estimate the effective bandwidth for the 2.7-μm CO_2 band at 830 K, 10 atm, and a path length of 38.8 cm. Compare this with the result computed from the correlation in Tables 16-2 and 16-3.

   Answer: 420 cm^{-1}.

13 A collimated beam of red light ($\lambda \approx 0.65$ μm) is to be attenuated by scattering. A proposed scheme is to use very small spherical copper particles having a characteristic diameter of 200 Å (optical data for copper are in Table 4-2). The particles are to be suspended in a non-scattering, nonabsorbing medium. Assuming Rayleigh scattering is applicable, what is the particle scattering cross section s_λ? For the beam to be 10% attenuated by scattering in a path length of 1 m, approximately what number density of particles would be required? What are the volume fraction of the particles and the mass of the particles per cubic centimeter of the scattering medium?

   Answer: 14×10^{-8} μm^2; 75.5×10^{10} cm^{-3}; 3.16×10^{-6}; 28.4×10^{-6} gm.

14 The beam in Prob. 13 is changed to green light ($\lambda \approx 0.55$ μm) while the scattering particle size and number density are kept the same. Assuming the same optical constants apply at this wavelength, what is the percent attenuation of the beam for a 1-m path length?

   Answer: 18.6%.

15 Consider Rayleigh scattering for very small gold particles at the two wavelengths shown in Table 4-2. How do the scattering cross sections s_λ differ for these two λ (note Fig. 16-13)?

16 Verify that the phase function in Eq. (16-102) satisfies the normalization specified by (13-48).

SEVENTEEN

THE ENGINEERING TREATMENT
OF GAS RADIATION IN ENCLOSURES

17-1 INTRODUCTION

An extensive body of engineering literature deals with radiation exchange between solid surfaces when no absorbing and/or scattering medium is present between them. The methods for treating such problems are highly developed and have been examined in Chaps. 6–12. The additional complication of having an intervening absorbing-emitting and/or scattering gas present in problems of energy exchange between surfaces can be accounted for by building upon the foundation established for the simpler situation with no gas present. In this chapter, the relations in Chaps. 13–16 are used in the derivation of engineering methods for solving gas-radiation problems. In most engineering applications such as heat transfer in furnaces and combustion chambers, there is negligible scattering; hence only absorption and emission will be considered here. The methods developed are an extension of the surface-surface energy exchange methods for enclosures in Chaps. 7–10.

Most of the material in this chapter will be concerned with gas that is *isothermal*. This is often a realistic condition, as in many instances the products of combustion are well mixed. As compared with the development in Chaps. 14 and 15, this provides the simplification that the gas temperature distribution need not be computed to obtain the radiative behavior. The mean radiative beam length is introduced as an engineering approximation to account for the geometry of a gas radiating to its boundary. The beam length is found to be relatively insensitive to the wavelength dependency of the absorption coefficient. This insensitivity enables the effects of the geometry to be separated from the effects of spectral property variations. In Sec. 17-8 some of the methods developed for the isothermal gas will be carried over to nonisothermal-gas computations.

The chapter concludes with a section on flames, both nonluminous and those containing luminous particles, mainly soot. A nonluminous hydrocarbon flame contains carbon dioxide and water vapor as its chief radiating constituents. Radiation by these gases is fairly well understood. When soot is present and the flame thereby becomes luminous, the radiation is dependent on the radiative properties of the soot and the soot concentration within the flame. Some information is available on soot radiative properties, but the amount is insufficient. Determining the soot concentration is a serious difficulty in flame radiation computations. The concentration depends on the particular fuel, the flame geometry, and the mixing phenomena within the flame. At present there is no way of computing soot concentration from the basic parameters, such as the burner geometry, fuel-air ratio, and the particular fuel.

17-2 SYMBOLS

A	area
$AF\bar{\alpha}$	geometric absorption factor
$AF\bar{\tau}$	geometric transmission factor
$\bar{A}$	effective bandwidth
a	absorption coefficient
a, b, c	dimensions in system of two rectangles
C	ratio $L_e/L_{e,0}$; volume fraction of particles in medium
$\left.\begin{array}{l} C_{CO_2} \\ C_{H_2O} \end{array}\right\}$	pressure-correction coefficients
C_l	band coefficient in Eq. (17-91)
D	spacing between parallel plates; diameter
E_N	$(1 - \epsilon_N)/\epsilon_N$
E_n	exponential integral
E_λ	spectral absorption efficiency factor
e	emissive power
F	geometric configuration factor
$\bar{F}$	exchange factor
$\overline{gg}$	gas-gas direct exchange area
$\overline{gs}$	gas-surface direct exchange area
h	height of cylinder
i	radiation intensity
k, k_1, k_2	constants in equations for soot absorption
L_e	mean beam length of gas volume
$L_{e,0}$	mean beam length for limiting case of small absorption
N	total number of surfaces in enclosure
P	total pressure of gas or gas mixture
p	partial pressure
Q	energy per unit time

q	energy flux, energy per unit area and time
R	radius of hemisphere, semicylinder, cylinder, or sphere
S	coordinate along path of radiation
$\bar{S}$	geometric-mean beam length
$\overline{sg}$	surface-gas direct-exchange area
$\overline{ss}$	surface-surface direct-exchange area
T	absolute temperature
V	volume
W	width of plate
X	mass path length, ρS; shortest dimension of rectangular parallelepiped
$\alpha(S)$	absorptance
$\bar{\alpha}(S)$	geometric-mean absorptance
$\Delta\alpha, \Delta\epsilon$	correction for spectral overlap
δ_{kj}	Kronecker delta
ϵ	emissivity of surface
$\epsilon(S)$	emittance of medium
η	wave number
θ	cone angle, angle from normal of area
κ	optical depth
λ	wavelength
ρ	reflectivity; density
σ	Stefan-Boltzmann constant
$\tau(S)$	transmittance
$\bar{\tau}(S)$	geometric-mean transmittance
ϕ	dimensionless temperature ratio
ω	solid angle

Subscripts

b	blackbody
CO_2	carbon dioxide
g	gas
H_2O	water vapor
i	incident, incoming
j, k	surfaces j or k
$j-k$	from surface j to surface k
l	absorption band l
o	outgoing
u	uniform
w	wall
λ	spectrally (wavelength) dependent
η	wave-number–dependent

Superscripts

*, **	dummy variable of integration
+	quantities defined after Eq. (17-68)
'	directional quantity

17-3 NET-RADIATION METHOD FOR ENCLOSURE FILLED WITH ISOTHERMAL GAS–SPECTRAL RELATIONS

In Sec. 10-3 the radiation-exchange equations were developed for an enclosure that does not contain an absorbing-emitting medium, and that has surfaces with spectrally dependent properties. Since the absorption properties of gases and other absorbing media are almost always strongly wavelength dependent, the present development will be carried out at a single wavelength. Then in a later section integrations will be performed over all wavelengths to obtain the total radiative behavior. It will be assumed that surface directional property effects are sufficiently unimportant that the surfaces can be treated as diffuse emitters and reflectors.

Often in a gas-filled enclosure, such as in an engine combustion chamber or industrial furnace, there is sufficient mixing so that the entire gas is essentially iso-thermal. In this instance the analysis is simplified, as it is unnecessary to compute the gas-temperature distribution. Sometimes the gas temperature is specified, but if not, it is only necessary to compute a single gas temperature from the governing heat balances. Even with this simplification, however, a detailed radiation exchange computation between the gas and bounding surfaces is quite involved.

Consider an enclosure composed of N surfaces, each at a uniform temperature as shown in Fig. 17-1. Typical surfaces are designated by j and k. The enclosure is filled with an absorbing-emitting medium at uniform temperature T_g. The quantity Q_g is the amount of heat that it is necessary to *supply* by means other than radiation to the entire absorbing medium to maintain this temperature. A common source of Q_g would be combustion. If in the solution of a problem Q_g comes out to be negative, ' the medium is gaining a net amount of radiative energy from the enclosure walls, and

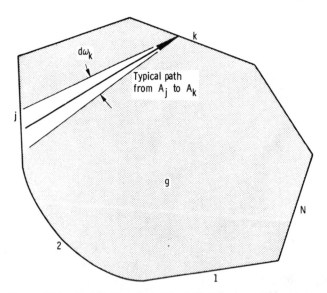

Figure 17-1 Enclosure composed of N discrete surface areas and filled with uniform gas g (enclosure shown in cross section for simplicity).

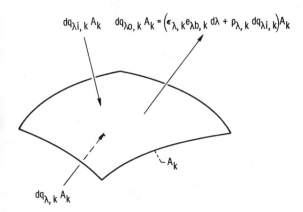

$dq_{\lambda i, k} A_k \qquad dq_{\lambda o, k} A_k = \left(\epsilon_{\lambda, k} e_{\lambda b, k} \, d\lambda + \rho_{\lambda, k} \, dq_{\lambda i, k} \right) A_k$

A_k

$dq_{\lambda, k} A_k$

Figure 17-2 Spectral energy quantities incident upon and leaving typical surface area of enclosure.

energy must be *removed* from the gas to maintain it at its steady temperature T_g. The Q_g is analogous to the Q_k at a surface, which is the energy supplied by nonradiative means to area A_k.

The enclosure theory will yield equations relating Q_k and T_k for each surface with Q_g and T_g for the gas or other absorbing isothermal medium filling the enclosure. Considering all the surfaces and the gas, if half the Q's and T's are specified, then the radiative heat balance equations can be solved for the remaining unknown Q or T values. If the heat input to the gas from external sources Q_g is given, the analysis will yield the steady gas temperature T_g. Conversely, if T_g is given, the analysis will yield the energy that must be supplied to maintain this gas temperature.

The net-radiation method developed in Chaps. 8 and 10 will now be extended to include gas-radiation terms. At the kth surface of an enclosure, as shown in Fig. 17-2 a heat balance gives

$$dQ_{\lambda,k} = dq_{\lambda,k} A_k = (dq_{\lambda o, k} - dq_{\lambda i, k}) A_k \qquad (17\text{-}1)$$

The $dq_{\lambda o, k}$ and $dq_{\lambda i, k}$ are respectively the outgoing and incoming radiative fluxes in wavelength interval $d\lambda$. The $dQ_{\lambda, k}$ is the energy supplied to the surface in wavelength region $d\lambda$. Note that, as discussed in connection with (10-4), the external energy supplied to A_k (by some means such as conduction or convection) equals $\int_{\lambda=0}^{\infty} dQ_{\lambda,k}$.

The outgoing spectral flux is composed of emitted and reflected energy:

$$dq_{\lambda o, k} = \epsilon_{\lambda, k}(\lambda, T_k) e_{\lambda b, k}(\lambda, T_k) \, d\lambda + \rho_{\lambda, k}(\lambda, T_k) \, dq_{\lambda i, k} \qquad (17\text{-}2)$$

The functional notation will usually be omitted to shorten the form of the equations that follow. The $e_{\lambda b, k}(\lambda, T_k) \, d\lambda$ is the blackbody spectral emission at T_k in the wavelength region $d\lambda$ about wavelength λ.

The $dq_{\lambda i, k}$ in (17-1) is the incoming spectral flux to A_k. It is equal to the sum of the contributions from all the surfaces that reach the kth surface after allowance for absorption in passing through the intervening gas, plus the contribution due to emission from the gas. The equation of transfer allows for both attenuation and emission as radiation passes along a path through the gas. A typical path from A_j to A_k within an incident solid angle $d\omega_k$ is shown in Fig. 17-1. If all such paths and solid angles by which radiation can pass from all the surfaces (including A_k if it is concave) to A_k are accounted for, the solid angles $d\omega_k$ will encompass all of the gas region that can radiate to A_k. Thus, if the equation of transfer, which includes the gas emission term, is used to compute the energy transported along all paths between surfaces, the gas emission will automatically be included. The radiation passing from one surface to another, including emission and absorption by the intervening gas, will now be considered.

A typical pair of surfaces is shown in Fig. 17-3. In the enclosure theory $dq_{\lambda o}$ is assumed uniform over each surface. Since the surfaces are assumed diffuse, the spectral intensity leaving dA_j is $i'_{\lambda o,j} = (1/\pi)dq_{\lambda o,j}/d\lambda$. By use of the equation of transfer [Eq. (14-13)], the intensity arriving at dA_k after traversing path S is

$$i'_{\lambda i,j-k} = i'_{\lambda o,j}\exp\left(-\kappa_\lambda\right) + \int_0^{\kappa_\lambda} i'_{\lambda b,g}(\kappa_\lambda^*)\exp\left[-(\kappa_\lambda - \kappa_\lambda^*)\right]d\kappa_\lambda^* \qquad (17\text{-}3)$$

where $\kappa_\lambda = \int_0^S a_\lambda(S^*)\,dS^*$ is the optical depth of the path S. The gas is assumed to be at uniform temperature and to have a constant spectral absorption coefficient. Then (17-3) reduces to

$$i'_{\lambda i,j-k} = i'_{\lambda o,j}\exp\left(-a_\lambda S\right) + a_\lambda i'_{\lambda b,g}\int_0^S \exp\left[-a_\lambda(S - S^*)\right]dS^*$$

which is integrated to give

$$i'_{\lambda i,j-k} = i'_{\lambda o,j}\exp\left(-a_\lambda S\right) + i'_{\lambda b,g}\left[1 - \exp\left(-a_\lambda S\right)\right] \qquad (17\text{-}4)$$

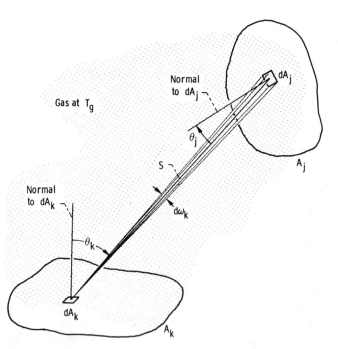

Figure 17-3 Radiation between two surfaces with isothermal gas between them.

For convenience, we introduce the following definitions: $\tau_\lambda(S) \equiv \exp(-a_\lambda S)$ is the spectral transmittance of the gas along path length S, and $\alpha_\lambda(S) \equiv 1 - \exp(-a_\lambda S)$ is the spectral absorptance along the path. Then (17-4) is written as

$$i'_{\lambda i,j-k} = i'_{\lambda o,j}\tau_\lambda(S) + i'_{\lambda b,g}(T_g)\alpha_\lambda(S) \tag{17-5}$$

This intensity arriving at dA_k in the solid angle $d\omega_k$ provides an arriving energy equal to $i'_{\lambda i,j-k}\, dA_k \cos\theta_k\, d\omega_k\, d\lambda$. But $d\omega_k = dA_j \cos\theta_j/S^2$, so the arriving spectral energy is

$$d^3Q_{\lambda i,j-k} = i'_{\lambda i,j-k}\, dA_k\, dA_j\, \frac{\cos\theta_k \cos\theta_j}{S^2}\, d\lambda$$

$$= [i'_{\lambda o,j}\tau_\lambda(S) + i'_{\lambda b,g}(T_g)\alpha_\lambda(S)] \frac{dA_k\, dA_j \cos\theta_k \cos\theta_j}{S^2}\, d\lambda \tag{17-6}$$

For a diffuse surface $dq_{\lambda o,j} = \pi i'_{\lambda o,j}\, d\lambda$ and $e_{\lambda b,g} = \pi i'_{\lambda b,g}$, so that (17-6) can be written as

$$d^3Q_{\lambda i,j-k} = [dq_{\lambda o,j}\tau_\lambda(S) + e_{\lambda b,g}(T_g)\, d\lambda\, \alpha_\lambda(S)] \frac{dA_k\, dA_j \cos\theta_k \cos\theta_j}{\pi S^2} \tag{17-7}$$

Equation (17-7) is now integrated over all of A_k and A_j to give the spectral energy along all paths from A_j that is incident upon A_k:

$$dQ_{\lambda i,j-k} \equiv \int_{A_k}\int_{A_j} [dq_{\lambda o,j}\tau_\lambda(S) + e_{\lambda b,g}(T_g)\, d\lambda\, \alpha_\lambda(S)] \frac{\cos\theta_k \cos\theta_j}{\pi S^2}\, dA_j\, dA_k \tag{17-8}$$

The first term of the double integral is the spectral energy leaving A_j that is transmitted to A_k. The second term is the spectral energy received at A_k as a result of emission by the constant-temperature gas in the envelope between A_j and A_k. This envelope is the volume occupied by all straight paths between any part of A_j and A_k.

17-3.1 Definitions of Spectral Transmission and Absorption Factors

The double integral in (17-8) has some similarity to the double integral in (7-22) for the configuration factor between two surfaces without an intervening gas. By analogy, define the factor $\bar{\tau}_{\lambda,j-k}$ such that

$$F_{j-k}\bar{\tau}_{\lambda,j-k} \equiv \frac{1}{A_j} \int_{A_k}\int_{A_j} \frac{\tau_\lambda(S)\cos\theta_k \cos\theta_j}{\pi S^2}\, dA_j\, dA_k \tag{17-9}$$

where F_{j-k} is the geometric configuration factor with no absorbing medium. With no absorbing medium; $\tau_\lambda(S) = 1$, and the right side of (17-9) becomes F_{j-k}. In this instance $\bar{\tau}_{j-k} = 1$. For complete absorption between A_j and A_k, $\bar{\tau}_{\lambda,j-k} = 0$. The $\bar{\tau}_{\lambda,j-k}$ is called the *geometric-mean transmittance* from A_j to A_k. Similarly, from the second quantity in brackets in Eq. (17-8), a *geometric-mean absorptance* $\bar{\alpha}_{\lambda,j-k}$ is defined so that

$$F_{j-k}\bar{\alpha}_{\lambda,j-k} \equiv \frac{1}{A_j} \int_{A_k} \int_{A_j} \frac{\alpha_\lambda(S)\cos\theta_k \cos\theta_j}{\pi S^2} \, dA_j \, dA_k \qquad (17\text{-}10)$$

For a nonabsorbing medium $\bar{\alpha}_{\lambda,j-k} = 0$, while for perfect absorption $\bar{\alpha}_{\lambda,j-k} = 1$. From the definitions of τ_λ and α_λ and Eqs. (17-9) and (17-10), the $\bar{\tau}_\lambda$ and $\bar{\alpha}_\lambda$ are related by

$$\bar{\alpha}_{\lambda,j-k} = 1 - \bar{\tau}_{\lambda,j-k} \qquad (17\text{-}11)$$

An alternative terminology is also used wherein the entire quantity $A_j F_{j-k}\bar{\tau}_{\lambda,j-k}$ is called the *geometrical transmission factor*, and $A_j F_{j-k}\bar{\alpha}_{\lambda,j-k}$ the *geometrical absorption factor*. Equation (17-8) can now be written as

$$dQ_{\lambda i,j-k} = (A_j F_{j-k}\bar{\tau}_{\lambda,j-k}) \, dq_{\lambda o,j} + (A_j F_{j-k}\bar{\alpha}_{\lambda,j-k}) e_{\lambda b,g}(T_g) \, d\lambda \qquad (17\text{-}12)$$

In computing the heat exchange in an enclosure, it is necessary to determine $\bar{\tau}_\lambda$ and $\bar{\alpha}_\lambda$. This usually involves some difficult double integrations. In the present discussion the enclosure-theory formulation will first be completed. Then the evaluation of $\bar{\tau}_\lambda$ and $\bar{\alpha}_\lambda$ will be considered. It is only necessary to perform one double integration to obtain both $\bar{\tau}_\lambda$ and $\bar{\alpha}_\lambda$ because of relation (17-11).

17-3.2 Matrix of Enclosure-Theory Equations

For an enclosure with N surfaces bounding an isothermal gas at T_g, the incident spectral energy on any surface A_k will equal that arriving from the directions of all surrounding surfaces:

$$dQ_{\lambda i,k} = A_k \, dq_{\lambda i,k} = \sum_{j=1}^{N} (dq_{\lambda o,j} A_j F_{j-k}\bar{\tau}_{\lambda,j-k} + e_{\lambda b,g} \, d\lambda \, A_j F_{j-k}\bar{\alpha}_{\lambda,j-k}) \qquad (17\text{-}13)$$

From reciprocity (7-25), $A_j F_{j-k} = A_k F_{k-j}$ so that the A_k can be eliminated to give

$$dq_{\lambda i,k} = \sum_{j=1}^{N} (dq_{\lambda o,j} F_{k-j}\bar{\tau}_{\lambda,j-k} + e_{\lambda b,g} \, d\lambda \, F_{k-j}\bar{\alpha}_{\lambda,j-k}) \qquad (17\text{-}14)$$

Equations (17-1), (17-2), and (17-14) form a set of three equations in three unknowns $dq_{\lambda o}$, $dq_{\lambda i}$, and dq_λ for each of the surfaces in the enclosure. The $dq_{\lambda i}$ is eliminated by combining (17-1) and (17-2) and by substituting (17-14) into (17-1). This yields the set of two equations for each surface

$$dq_{\lambda,k} = \frac{\epsilon_{\lambda,k}}{1 - \epsilon_{\lambda,k}} (e_{\lambda b,k} \, d\lambda - dq_{\lambda o,k}) \qquad (17\text{-}15)$$

$$dq_{\lambda,k} = dq_{\lambda o,k} - \sum_{j=1}^{N} (dq_{\lambda o,j} F_{k-j}\bar{\tau}_{\lambda,j-k} + e_{\lambda b,g} \, d\lambda \, F_{k-j}\bar{\alpha}_{\lambda,j-k}) \qquad (17\text{-}16)$$

Equation (17-15) is the same as for an enclosure without an absorbing gas [see (10-8), for example]. Equations (17-15) and (17-16) are analogous to (8-6) and (8-7) for the simpler case of a gray enclosure without an absorbing gas. From the symmetry of the integrals in (17-9) and (17-10) and the reciprocity relation $A_j F_{j-k} = A_k F_{k-j}$, it is found that

$$\bar{\tau}_{\lambda, j-k} = \bar{\tau}_{\lambda, k-j} \tag{17-17}$$

and

$$\bar{\alpha}_{\lambda, j-k} = \bar{\alpha}_{\lambda, k-j} \tag{17-18}$$

Then (17-16) can also be written as

$$dq_{\lambda, k} = dq_{\lambda o, k} - \sum_{j=1}^{N} (dq_{\lambda o, j} F_{k-j} \bar{\tau}_{\lambda, k-j} + e_{\lambda b, g} \, d\lambda \, F_{k-j} \bar{\alpha}_{\lambda, k-j}) \tag{17-19}$$

As in Sec. 8-3.1, the set of equations (17-15) and (17-19) can be further reduced by solving (17-15) for $dq_{\lambda o}$ and inserting it into (17-19). This results in the relation

$$\sum_{j=1}^{N} \left(\frac{\delta_{kj}}{\epsilon_{\lambda, j}} - F_{k-j} \frac{1 - \epsilon_{\lambda, j}}{\epsilon_{\lambda, j}} \bar{\tau}_{\lambda, k-j} \right) dq_{\lambda, j}$$

$$= \sum_{j=1}^{N} [(\delta_{kj} - F_{k-j} \bar{\tau}_{\lambda, k-j}) e_{\lambda b, j} \, d\lambda - F_{k-j} \bar{\alpha}_{\lambda, k-j} e_{\lambda b, g} \, d\lambda] \tag{17-20}$$

The Kronecker delta δ_{kj} has the values $\delta_{kj} = 1$ when $k = j$, and $\delta_{kj} = 0$ when $k \neq j$. This equation is analogous to (8-19). If (17-20) is written for each k from 1 to N, a set of N equations is obtained relating the $2N$ quantities dq_λ and $e_{\lambda b}$ for the surfaces, since the gas temperature (and hence $e_{\lambda b, g}$) is assumed known. One-half the dq_λ and $e_{\lambda b}$ values have to be specified, and the equations can then be solved for the remaining unknowns. To determine the total energy quantities, this set of equations must be solved in a number of wavelength intervals and the integration of each quantity performed over wavelength.

17-3.3 Heat Balance on Gas

Before discussing the solution of the enclosure equations in more detail, an additional heat balance is of interest. This is a heat balance on the gas, which will provide the energy required to maintain the gas at its specified temperature while it is exposed to the surrounding enclosure wall temperatures. From the energy balance on the entire enclosure, the energy that must be supplied to the gas by combustion, for example, is equal to the net quantity escaping from the boundaries. The total energy escaping from the enclosure at surface k is $-A_k \int_{\lambda=0}^{\infty} dq_{\lambda, k}$. Then the energy added to the gas is found by summing over all surfaces:

$$Q_g = - \sum_{k=1}^{N} A_k \int_{\lambda=0}^{\infty} dq_{\lambda, k} \tag{17-21}$$

This can be evaluated after the dq_λ are found for each surface in a sufficient number of wavelength intervals from the matrix of Eqs. (17-20).

EXAMPLE 17-1 As an example of the net-radiation method, consider the heat transfer in a system of two infinite parallel plates at temperatures T_1 and T_2 ($T_1 > T_2$) bounding a gas at uniform temperature T_g.

Equation (17-20) applied to a two-surface enclosure gives (note that $F_{1-1} = F_{2-2} = 0$) for $k = 1$:

$$\frac{1}{\epsilon_{\lambda,1}} dq_{\lambda,1} - F_{1-2} \frac{1 - \epsilon_{\lambda,2}}{\epsilon_{\lambda,2}} \bar{\tau}_{\lambda,1-2} dq_{\lambda,2}$$

$$= e_{\lambda b,1}\, d\lambda - F_{1-2}\bar{\tau}_{\lambda,1-2} e_{\lambda b,2}\, d\lambda - F_{1-2}\bar{\alpha}_{\lambda,1-2} e_{\lambda b,g}\, d\lambda \qquad (17\text{-}22a)$$

for $k = 2$:

$$-F_{2-1} \frac{1 - \epsilon_{\lambda,1}}{\epsilon_{\lambda,1}} \bar{\tau}_{\lambda,2-1} dq_{\lambda,1} + \frac{1}{\epsilon_{\lambda,2}} dq_{\lambda,2}$$

$$= -F_{2-1}\bar{\tau}_{\lambda,2-1} e_{\lambda b,1}\, d\lambda - F_{2-1}\bar{\alpha}_{\lambda,2-1} e_{\lambda b,g}\, d\lambda + e_{\lambda b,2}\, d\lambda \qquad (17\text{-}22b)$$

For the infinite parallel-plate geometry, $F_{1-2} = F_{2-1} = 1$, and from (17-17) and (17-18) $\bar{\tau}_{\lambda,2-1} = \bar{\tau}_{\lambda,1-2}$ and $\bar{\alpha}_{\lambda,2-1} = \bar{\alpha}_{\lambda,1-2}$. For simplicity the numerical subscripts on $\bar{\tau}$ and $\bar{\alpha}$ will be omitted. Then (17-22a) and (17-22b) become

$$\frac{1}{\epsilon_{\lambda,1}} dq_{\lambda,1} - \frac{1 - \epsilon_{\lambda,2}}{\epsilon_{\lambda,2}} \bar{\tau}_\lambda\, dq_{\lambda,2} = (e_{\lambda b,1} - \bar{\tau}_\lambda e_{\lambda b,2} - \bar{\alpha}_\lambda e_{\lambda b,g})\, d\lambda \qquad (17\text{-}23a)$$

$$-\frac{1 - \epsilon_{\lambda,1}}{\epsilon_{\lambda,1}} \bar{\tau}_\lambda\, dq_{\lambda,1} + \frac{1}{\epsilon_{\lambda,2}} dq_{\lambda,2} = (-\bar{\tau}_\lambda e_{\lambda b,1} + e_{\lambda b,2} - \bar{\alpha}_\lambda e_{\lambda b,g})\, d\lambda \qquad (17\text{-}23b)$$

Equations (17-23a) and (17-23b) are solved simultaneously for $dq_{\lambda,1}$ and $dq_{\lambda,2}$. After rearrangement and use of the relation $\bar{\alpha}_\lambda = 1 - \bar{\tau}_\lambda$, this yields

$$dq_{\lambda,1} = \frac{d\lambda}{1 - (1 - \epsilon_{\lambda,1})(1 - \epsilon_{\lambda,2})\bar{\tau}_\lambda{}^2}$$

$$\times \{\epsilon_{\lambda,1}\epsilon_{\lambda,2}\bar{\tau}_\lambda(e_{\lambda b,1} - e_{\lambda b,2}) + \epsilon_{\lambda,1}(1 - \bar{\tau}_\lambda)$$

$$\times [1 + (1 - \epsilon_{\lambda,2})\bar{\tau}_\lambda](e_{\lambda b,1} - e_{\lambda b,g})\} \qquad (17\text{-}24a)$$

$$dq_{\lambda,2} = \frac{d\lambda}{1 - (1 - \epsilon_{\lambda,1})(1 - \epsilon_{\lambda,2})\bar{\tau}_\lambda{}^2}$$

$$\times \{\epsilon_{\lambda,1}\epsilon_{\lambda,2}\bar{\tau}_\lambda(e_{\lambda b,2} - e_{\lambda b,1}) + \epsilon_{\lambda,2}(1 - \bar{\tau}_\lambda)$$

$$\times [1 + (1 - \epsilon_{\lambda,1})\bar{\tau}_\lambda](e_{\lambda b,2} - e_{\lambda b,g})\} \qquad (17\text{-}24b)$$

The total energy fluxes added to surfaces 1 and 2 are, respectively,

$$q_1 = \int_{\lambda=0}^{\infty} dq_{\lambda,1} \qquad \text{and} \qquad q_2 = \int_{\lambda=0}^{\infty} dq_{\lambda,2} \qquad (17\text{-}25)$$

The total energy added to the gas to maintain its specified temperature is equal to the net energy removed from the plates. Hence, per unit area of the plates,

$$q_g = -(q_1 + q_2) \tag{17-26}$$

When the medium between the plate does not absorb or emit radiation, then $\bar{\tau}_\lambda = 1$ and (17-24a) and (17-24b) reduce to (10-10). With an absorbing-radiating gas present the numerical integration of (17-24a) and (17-24b) over all wavelengths to obtain the q_1 and q_2 is difficult because of the very irregular variations of the gas absorption coefficient with wavelength. The integration will be performed in Sec. 17-7 by dividing the wavelength range into bands of finite width that are either absorbing or nonabsorbing.

17-4 EVALUATION OF SPECTRAL GEOMETRIC-MEAN TRANSMITTANCE AND ABSORPTANCE FACTORS

To compute values from the enclosure equations, the $\bar{\tau}$ and $\bar{\alpha}$ or $AF\bar{\tau}$ and $AF\bar{\alpha}$ must be evaluated. By use of the definitions in (17-9) and (17-10),

$$A_j F_{j-k} \bar{\tau}_{\lambda,j-k} = \int_{A_k} \int_{A_j} \frac{\exp(-a_\lambda S) \cos \theta_k \cos \theta_j}{\pi S^2} \, dA_j \, dA_k \tag{17-27}$$

$$A_j F_{j-k} \bar{\alpha}_{\lambda,j-k} = \int_{A_k} \int_{A_j} \frac{[1 - \exp(-a_\lambda S)] \cos \theta_k \cos \theta_j}{\pi S^2} \, dA_j \, dA_k$$

$$= A_j F_{j-k} (1 - \bar{\tau}_{\lambda,j-k}) \tag{17-28}$$

It is evident that it is the double integral in (17-27) that must be carried out for various geometric orientations of the surfaces A_j and A_k. The evaluation for some specific geometries will now be considered.

17-4.1 Hemisphere to Differential Area at Center of Its Base

As shown in Fig. 17-4, let A_j be the surface of a hemisphere of radius R, and dA_k be a differential area at the center of the hemisphere base. Then (17-27) becomes, since

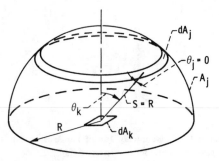

Figure 17-4 Hemisphere filled with isothermal gas.

$S = R$ and $\theta_j = 0$ (path R is normal to hemisphere surface),

$$A_j \, dF_{j-dk} \, \bar{\tau}_{\lambda,j-dk} = dA_k \int_{A_j} \frac{\exp(-a_\lambda R) \cos \theta_k \cos(0)}{\pi R^2} \, dA_j$$

The convenient dA_j is a ring element $dA_j = 2\pi R^2 \sin \theta_k \, d\theta_k$, and the factors involving R can be taken out of the integral as R is constant for the hemisphere geometry. This gives

$$A_j \, dF_{j-dk} \, \bar{\tau}_{\lambda,j-dk} = dA_k \frac{\exp(-a_\lambda R) \, 2\pi R^2}{\pi R^2} \int_{\theta_k=0}^{\pi/2} \cos \theta_k \sin \theta_k \, d\theta_k$$

$$= dA_k \exp(-a_\lambda R)$$

With $A_j \, dF_{j-dk} = dA_k F_{dk-j}$ and $F_{dk-j} = 1$, this reduces to

$$\bar{\tau}_{\lambda,j-dk} = \exp(-a_\lambda R) \tag{17-29}$$

This especially simple relation will be used later in conjunction with the concept of mean beam length. This is an approximate technique wherein the radiation from an actual gas volume is replaced by that from an effective hemispherical volume.

17-4.2 Top of Right Circular Cylinder to Center of Its Base

This geometry is shown in Fig. 17-5. Since $\theta_j = \theta_k = \theta$, the integral in Eq. (17-27) becomes, for the top of the cylinder A_j radiating to the element dA_k at the center of

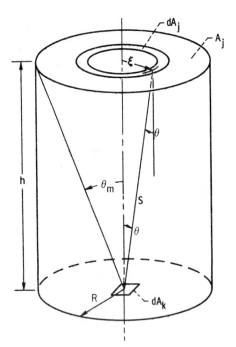

Figure 17-5 Geometry for exchange from top of gas-filled cylinder to center of its base.

its base,

$$A_j \, dF_{j-dk} \, \bar{\tau}_{\lambda,j-dk} = dA_k \int_{A_j} \frac{\exp(-a_\lambda S) \cos^2 \theta}{\pi S^2} \, dA_j \qquad (17\text{-}30)$$

A convenient change in the integral is made by noting that $dA_j \cos \theta / S^2$ is the solid angle by which the ring dA_j is viewed from dA_k. Consideration of the intersection of this solid angle with the surface of a unit hemisphere shows the solid angle is also equal to $2\pi \sin \theta \, d\theta$. By this substitution, the integral in (17-30) is transformed to

$$A_j \, dF_{j-dk} \, \bar{\tau}_{\lambda,j-dk} = dA_k \int_{A_j} \exp(-a_\lambda S) \, 2 \cos \theta \sin \theta \, d\theta \qquad (17\text{-}31)$$

Now let $a_\lambda S = \kappa_\lambda$. Then from Fig. 17-5 $\cos \theta = h/S = ha_\lambda/\kappa_\lambda$, and $\sin \theta \, d\theta = -d(\cos \theta) = (ha_\lambda/\kappa_\lambda^2) \, d\kappa_\lambda$. As θ goes from 0 to θ_m the limits on κ_λ are $\kappa_\lambda = a_\lambda h$ to $a_\lambda \sqrt{R^2 + h^2}$. Then (17-31) becomes

$$A_j \, dF_{j-dk} \, \bar{\tau}_{\lambda,j-dk} = dA_k \, 2h^2 a_\lambda^2 \int_{a_\lambda h}^{a_\lambda \sqrt{R^2+h^2}} \frac{\exp(-\kappa_\lambda)}{\kappa_\lambda^3} \, d\kappa_\lambda \qquad (17\text{-}32)$$

This integral can be found in terms of the exponential integral function defined in (14-59) by writing

$$\int_{a_\lambda h}^{a_\lambda \sqrt{R^2+h^2}} \frac{\exp(-\kappa_\lambda)}{\kappa_\lambda^3} \, d\kappa_\lambda = \int_{\infty}^{a_\lambda \sqrt{R^2+h^2}} \frac{\exp(-\kappa_\lambda)}{\kappa_\lambda^3} \, d\kappa_\lambda - \int_{\infty}^{a_\lambda h} \frac{\exp(-\kappa_\lambda)}{\kappa_\lambda^3} \, d\kappa_\lambda$$

$$(17\text{-}33)$$

Letting $\kappa_\lambda = (a_\lambda \sqrt{R^2 + h^2})/\mu$ and $a_\lambda h/\mu$, respectively, in the two integrals gives

$$-\frac{1}{(a_\lambda \sqrt{R^2 + h^2})^2} \int_0^1 \mu \exp\left(-\frac{a_\lambda \sqrt{R^2 + h^2}}{\mu}\right) d\mu + \frac{1}{(a_\lambda h)^2} \int_0^1 \mu \exp\left(-\frac{a_\lambda h}{\mu}\right) d\mu$$

The integral in (17-32) can then be written in terms of the exponential integral function as

$$\int_{a_\lambda h}^{a_\lambda \sqrt{R^2+h^2}} \frac{\exp(-\kappa_\lambda)}{\kappa_\lambda^3} \, d\kappa_\lambda = \frac{1}{(a_\lambda h)^2} E_3(a_\lambda h)$$

$$-\frac{1}{[a_\lambda h \sqrt{(R/h)^2 + 1}]^2} E_3 \left[a_\lambda h \sqrt{\left(\frac{R}{h}\right)^2 + 1} \right] \qquad (17\text{-}34)$$

so it can readily be evaluated for various values of the parameters R/h and $a_\lambda h$.

17-4.3 Side of Cylinder to Center of Its Base

Let dA_j be a ring around the wall of a cylinder as shown in Fig. 17-6, and note that $dA_j \cos \theta_j / S^2$ is the solid angle by which dA_j is viewed from dA_k. This solid angle is

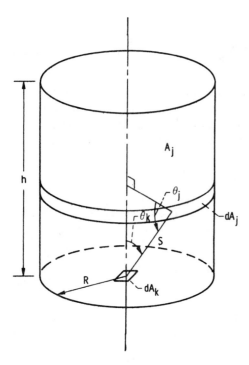

Figure 17-6 Geometry for exchange from side of gas-filled cylinder to center of its base.

also equal to $2\pi \sin \theta_k \, d\theta_k$. Then (17-27) can be written for the side of the cylinder to dA_k as

$$A_j \, dF_{j-dk} \, \overline{\tau}_{\lambda,j-dk} = 2dA_k \int_{A_j} \exp(-a_\lambda S) \cos \theta_k \sin \theta_k \, d\theta_k \tag{17-35}$$

This is of the same form as (17-31). Let $a_\lambda S = \kappa_\lambda$; then $\sin \theta_k = R/S = Ra_\lambda/\kappa_\lambda$ and $\cos \theta_k \, d\theta_k = d(\sin \theta_k) = -(Ra_\lambda/\kappa_\lambda^2) \, d\kappa_\lambda$. Making these substitutions and using (17-34) yield

$$A_j \, dF_{j-dk} \, \overline{\tau}_{\lambda,j-dk} = 2dA_k \, R^2 a_\lambda^2 \int_{a_\lambda R}^{a_\lambda \sqrt{R^2+h^2}} \frac{\exp(-\kappa_\lambda)}{\kappa_\lambda^3} \, d\kappa_\lambda$$

$$= 2dA_k \left(\frac{R}{h}\right)^2 (a_\lambda h)^2 \left\{ \frac{1}{[a_\lambda h(R/h)]^2} E_3\left[a_\lambda h\left(\frac{R}{h}\right)\right] \right.$$

$$\left. - \frac{1}{[a_\lambda h\sqrt{(R/h)^2+1}]^2} E_3\left[a_\lambda h\sqrt{\left(\frac{R}{h}\right)^2+1}\right] \right\} \tag{17-36}$$

As for (17-34), this can readily be evaluated for various values of the parameters R/h and $a_\lambda h$.

17-4.4 Entire Sphere to Any Element on Its Surface or to Its Entire Surface

From Fig. 17-7, since $\theta_k = \theta_j$ let them both be simply θ. Then $S = 2R \cos \theta$, and using the form in (17-31) gives

$$A_j \, dF_{j-dk} \, \bar{\tau}_{\lambda,j-dk} = \frac{2dA_k}{4R^2} \int_{S=0}^{2R} \exp\left(-a_\lambda S\right) S \, dS$$

Integrating gives

$$A_j \, dF_{j-dk} \, \bar{\tau}_{\lambda,j-dk} = \frac{2dA_k}{(2a_\lambda R)^2} \left[1 - (2a_\lambda R + 1)\exp\left(-2a_\lambda R\right)\right] \tag{17-37}$$

which is in terms of the single parameter $2a_\lambda R$, the optical diameter of the sphere.

Equation (17-37) can be integrated over any finite area A_k to give $\bar{\tau}_\lambda$ from the entire sphere to A_k as

$$A_j F_{j-k} \, \bar{\tau}_{\lambda,j-k} = \frac{2A_k}{(2a_\lambda R)^2} \left[1 - (2a_\lambda R + 1)\exp\left(-2a_\lambda R\right)\right]$$

Since $F_{j-k} = A_k/A_j$ [from (8-73)],

$$\bar{\tau}_{\lambda,j-k} = \frac{2}{(2a_\lambda R)^2} \left[1 - (2a_\lambda R + 1)\exp\left(-2a_\lambda R\right)\right] \tag{17-38}$$

which also holds for the entire sphere to its entire surface.

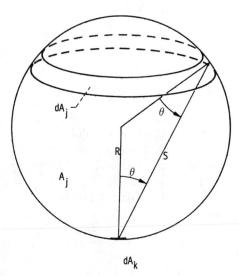

Figure 17-7 Geometry for exchange from surface of gas-filled sphere to itself.

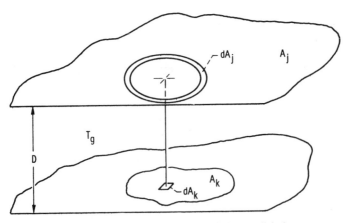

Figure 17-8 Isothermal gas layer between infinite parallel plates.

17-4.5 Infinite Plate to Area on Parallel Plate

Consider on one plate an element dA_k (Fig. 17-8), and on the other plate a concentric ring element dA_j centered about the normal to dA_k. Then the geometry is like that in Fig. 17-5 for a ring on the top of a cylinder to the center of its base. Then from (17-32)

$$A_j \, dF_{j-dk} \, \bar{\tau}_{\lambda,j-dk} = dA_k \, 2D^2 a_\lambda^2 \int_{a_\lambda D}^{\infty} \frac{\exp(-\kappa_\lambda)}{\kappa_\lambda^3} \, d\kappa_\lambda$$

where D is the spacing between the plates. By use of the procedure leading to (17-34), the integral is transformed to $E_3(a_\lambda D)/(a_\lambda D)^2$. Then integrating over any finite area A_k as shown in Fig. 17-8 gives $A_j F_{j-k} \bar{\tau}_{\lambda,j-k} = A_k 2E_3(a_\lambda D)$. With $A_j F_{j-k} = A_k F_{k-j}$ and $F_{k-j} = 1$, this reduces to

$$\bar{\tau}_{\lambda,j-k} = 2E_3(a_\lambda D) \tag{17-39}$$

17-4.6 Rectangle to a Directly Opposed Parallel Rectangle

Consider as in Fig. 17-9 the exchange from a rectangle to an area element on a directly opposed parallel rectangle. The upper rectangle has been divided into a circular region and a series of partial rings of small width. The contribution from the circle of radius R to $A_j \, dF_{j-dk} \, \bar{\tau}_{\lambda,j-dk}$ can be found from (17-32) and (17-34), for the top of a cylinder to the center of its base. For the nth partial ring, let f_n be the fraction it occupies of a full circular ring. Then by use of (17-31), the contribution of all the partial rings to $A_j \, dF_{j-dk} \, \bar{\tau}_{\lambda,j-dk}$ is approximated by

$$dA_k \sum_n f_n \exp(-a_\lambda S_n) \, 2 \cos \theta_n \sin \theta_n \, \Delta \theta_n$$

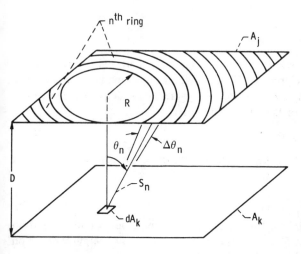

Figure 17-9 Geometry for exchange between two directly opposed parallel rectangles with intervening gas.

This evaluation of $A_j \, dF_{j-dk} \, \overline{\tau}_{\lambda,j-dk}$ is carried out for several area patches on A_k. This is usually sufficient so that the integration over A_k can be performed as indicated by (17-9) to yield

$$A_j F_{j-k} \overline{\tau}_{\lambda,j-k} = \int_{A_k} A_j \, dF_{j-dk} \, \overline{\tau}_{\lambda,j-dk}$$

17-5 MEAN BEAM LENGTH FOR RADIATION FROM AN ENTIRE GAS VOLUME TO ALL OR PART OF ITS BOUNDARY

In some practical situations it is desired to determine the energy radiated from a volume of isothermal gas to all or part of its boundaries, without considering emission and reflection from the boundaries. An example would be radiation from hot furnace gases to walls that are cool so that their emission is small, and that are rough and soot covered so that they are essentially nonreflecting. In Eq. (17-13) the $dq_{\lambda o,j}$, which is the spectral outgoing flux from a typical surface A_j, is then zero. The spectral incoming energy at A_k is then

$$A_k \, dq_{\lambda i,k} = \sum_{j=1}^{N} e_{\lambda b,g} \, d\lambda \, A_j F_{j-k} \overline{\alpha}_{\lambda,j-k} \qquad (17\text{-}40)$$

If the geometry is a hemisphere of gas radiating to an area element dA_k at the center of its base as shown in Fig. 17-4, Eq. (17-40) has an especially simple form. Since the hemispherical boundary is the only surface in view of dA_k, and dA_k is a differential element, (17-40) reduces to

$$dA_k \, dq_{\lambda i,k} = e_{\lambda b,g} \, d\lambda \, A_j \, dF_{j-dk} \, \overline{\alpha}_{\lambda,j-dk} \qquad (17\text{-}41)$$

From Eq. (17-29),

$$\bar{\alpha}_{\lambda, j-dk} = 1 - \bar{\tau}_{\lambda, j-dk} = 1 - \exp{(-a_\lambda R)}$$

Note also that for radiation between the surface of a hemisphere and the center of its base, $F_{dk-j} = 1$ so that from reciprocity $dF_{j-dk} = dA_k/A_j$. By combining these results, (17-41) reduces to the following simple expression giving the incident heat flux from a hemisphere of gas to the center of the hemisphere base:

$$dq_{\lambda i, k} = [1 - \exp{(-a_\lambda R)}]e_{\lambda b, g} \, d\lambda \tag{17-42}$$

From (13-59), $1 - \exp{(-a_\lambda R)}$ is the spectral emittance of the gas $\epsilon_\lambda(\lambda, T, P, R)$ for path length R.* Then (17-42) becomes

$$dq_{\lambda i, k} = \epsilon_\lambda(a_\lambda R)e_{\lambda b, g} \, d\lambda \tag{17-43}$$

Thus a very simple form is obtained for the spectral energy incident upon dA_k from the hemisphere of gas of radius R surrounding dA_k. The incident energy depends on the optical radius of the hemisphere $a_\lambda R$.

It would be most convenient if a relation having the simple form of (17-43) could be used to determine the value of $dq_{\lambda i, k}$ on A_k for *any* geometry of gas volume radiating to all or part of its boundary. Because the geometry of the gas enters (17-43) only through $\epsilon_\lambda(a_\lambda R)$, it is possible to define a fictitious value of R, say L_e, that would give a value of $\epsilon_\lambda(a_\lambda L_e)$ such that (17-43) would give the correct $dq_{\lambda i}$ for another geometry. This fictitious length L_e is called the *mean beam length*. Then for an arbitrary geometry of gas let

$$dq_{\lambda i, k} = \epsilon_\lambda(a_\lambda L_e)e_{\lambda b, g} \, d\lambda = [1 - \exp{(-a_\lambda L_e)}]e_{\lambda b, g} \, d\lambda \tag{17-44}$$

The mean beam length is thus the required radius of a gas hemisphere such that it radiates a flux to the center of its base equal to the average flux radiated to the area of interest by the actual volume of gas.

17-5.1 Mean Beam Length for Gas between Parallel Plates Radiating to Area on Plate

Consider two black infinite parallel plates at zero absolute temperature separated by a distance D. The plates enclose a uniform gas at temperature T_g with absorption coefficient a_λ. The rate at which spectral energy is incident upon A_k on one plate (Fig. 17-8) is, from (17-40) and (17-39),

$$dQ_{\lambda i, k} = A_k \, dq_{\lambda i, k} = e_{\lambda b, g} \, d\lambda \, A_j F_{j-k}\bar{\alpha}_{\lambda, j-k} = e_{\lambda b, g} \, d\lambda \, A_j F_{j-k}[1 - 2E_3(a_\lambda D)]$$
$$\tag{17-45}$$

Since the plates are infinite, $F_{k-j} = 1$. Then by reciprocity (7-25), $F_{j-k} = A_k/A_j$ and (17-45) reduces to

$$dq_{\lambda i, k} = [1 - 2E_3(a_\lambda D)]e_{\lambda b, g} \, d\lambda \tag{17-46}$$

---

*For simplicity the prime notation used for a directional quantity will be omitted; in this instance ϵ_λ is independent of direction.

Comparing (17-46) and (17-44) reveals the mean beam length to be

$$L_e = -\frac{1}{a_\lambda} \ln 2E_3(a_\lambda D)$$

or in terms of the optical thickness $a_\lambda D$,

$$\frac{L_e}{D} = -\frac{1}{a_\lambda D} \ln 2E_3(a_\lambda D) \tag{17-47}$$

17-5.2 Mean Beam Length for Sphere of Gas Radiating to Any Area on Boundary

Consider gas in a nonreflecting sphere of radius R where the sphere boundary A_j is at $T_j = 0$. By use of (17-40) and (17-37), the radiation flux incident on an element dA_k is found as

$$dq_{\lambda i, dk} = e_{\lambda b, g} \, d\lambda \frac{A_j}{dA_k} \, dF_{j-dk} \, \bar{\alpha}_{\lambda, j-dk}$$

$$= e_{\lambda b, g} \, d\lambda \frac{A_j}{dA_k} \, dF_{j-dk} \left\{ 1 - \frac{2dA_k}{(2a_\lambda R)^2 A_j \, dF_{j-dk}} [1 - (2a_\lambda R + 1) \exp(-2a_\lambda R)] \right\}$$

For a sphere, $dF_{j-dk} = dA_k / A_j$ [by use of (8-72)]. Then

$$dq_{\lambda i, dk} = e_{\lambda b, g} \, d\lambda \left\{ 1 - \frac{2}{(2a_\lambda R)^2} [1 - (2a_\lambda R + 1) \exp(-2a_\lambda R)] \right\}$$

Equate this to dq_λ from (17-44) and solve for L_e to obtain

$$\frac{L_e}{2R} = -\frac{1}{2a_\lambda R} \ln \left\{ \frac{2}{(2a_\lambda R)^2} [1 - (2a_\lambda R + 1) \exp(-2a_\lambda R)] \right\} \tag{17-48}$$

In view of the general applicability of (17-38), (17-48) gives the correct mean beam length for the entire sphere radiating to any portion of its boundary.

17-5.3 Radiation from Entire Gas Volume to Its Entire Boundary in Limit When Gas is Optically Thin

Because of the integrations involved, the mean beam length for an entire gas volume radiating to all or part of its boundary will usually be difficult to evaluate. It is fortunate that some practical approximations can be found by looking at the optically thin limit. By expanding the exponential term in a series for small $a_\lambda S$, the transmittance $\tau_\lambda = \exp(-a_\lambda S)$ becomes

$$\lim_{a_\lambda S \to 0} \tau_\lambda = \lim_{a_\lambda S \to 0} \left[1 - a_\lambda S + \frac{(a_\lambda S)^2}{2!} - \cdots \right] = 1$$

Any differential volume of the uniform-temperature gas emits spectral energy $4a_\lambda e_{\lambda b,g}\, d\lambda\, dV$. Since $\tau_\lambda = 1$, there is no attenuation of the emitted radiation, and all of it reaches the enclosure boundary. For the entire radiating volume the energy reaching the boundary is $4a_\lambda e_{\lambda b,g}\, d\lambda\, V$ so that the *average* spectral flux received at the boundary having entire area A is

$$dq_{\lambda i} = 4a_\lambda e_{\lambda b,g}\, d\lambda\, \frac{V}{A} \tag{17-49}$$

By use of the mean beam length the average flux reaching the boundary is, from (17-44),

$$dq_{\lambda i} = [1 - \exp(-a_\lambda L_e)] e_{\lambda b,g}\, d\lambda \tag{17-50}$$

For the special case of small absorption let L_e be designated by $L_{e,0}$. Then expand the exponential term in (17-50) in a series to obtain, for small $a_\lambda L_{e,0}$,

$$dq_{\lambda i} = \left\{ 1 - \left[1 - a_\lambda L_{e,0} + \frac{(a_\lambda L_{e,0})^2}{2!} - \cdots \right] \right\} e_{\lambda b,g}\, d\lambda = a_\lambda L_{e,0} e_{\lambda b,g}\, d\lambda$$

Equating this to the $dq_{\lambda i}$ in (17-49) gives the desired result for the mean beam length of an optically thin gas radiating to its entire boundary:

$$L_{e,0} = \frac{4V}{A} \tag{17-51}$$

To give a few examples, for a sphere of diameter D,

$$L_{e,0} = \frac{4\pi D^3/6}{\pi D^2} = \tfrac{2}{3}D \tag{17-52}$$

For an infinitely long cylinder of diameter D,

$$L_{e,0} = \frac{4\pi D^2/4}{\pi D} = D \tag{17-53}$$

For gas between infinite parallel plates spaced D apart,

$$L_{e,0} = \frac{4D}{2} = 2D \tag{17-54}$$

17-5.4 Correction for Mean Beam Length When Gas Is Not Optically Thin

For an optically thick gas it would be very convenient if L_e could be obtained by applying a simple correction factor to the $L_{e,0}$ computed from (17-51). It has been found that a useful technique is to introduce a correction coefficient C so that

$$L_e = CL_{e,0} \tag{17-55}$$

Then the incoming heat flux in (17-50) can be obtained as

$$dq_{\lambda i} = [1 - \exp(-a_\lambda C L_{e,0})]e_{\lambda b,g}\, d\lambda \qquad (17\text{-}56)$$

The coefficient C will now be examined by considering the example of a radiating gas between infinite parallel plates spaced D apart. Using (17-54) in (17-56) gives

$$dq_{\lambda i} = [1 - \exp(-a_\lambda C 2D)]e_{\lambda b,g}\, d\lambda$$

From (17-46) the actual flux received is

$$dq_{\lambda i} = [1 - 2E_3(a_\lambda D)]e_{\lambda b,g}\, d\lambda$$

To demonstrate how well these fluxes compare, the ratio

$$\frac{1 - 2E_3(a_\lambda D)}{1 - \exp(-2Ca_\lambda D)}$$

is plotted in Fig. 17-10 for a range of optical thicknesses $a_\lambda D$ using a value of $C = 0.9$. This value of C was found by trial to yield a ratio close to unity for all $a_\lambda D$, and hence it serves as a valid correction coefficient for this geometry.

Table 17-1 gives the mean beam length $L_{e,0}$ for a number of geometries, along with values of L_e that provide reasonably good radiative fluxes for nonzero optical thicknesses. The values of C are found to be in a range near 0.9 [1–3]. Hence, it is recommended that for a geometry for which L_e values have not been calculated, the approximation

$$L_e = 0.9 L_{e,0} = 0.9\, \frac{4V}{A} \qquad (17\text{-}57)$$

be used *for an entire gas volume radiating to its entire boundary.*

The different optical thicknesses of absorption bands may make it desirable to use a somewhat different mean beam length for each band. Mean beam lengths based on various band absorption models have been studied for slab geometries [4, 5] and in spheres and cylinders [6]. Variations in mean beam length from those predicted in Table 17-1 were greatest for the slab geometry, approaching $0.82 L_{e,0}$ at optical depths near 100.

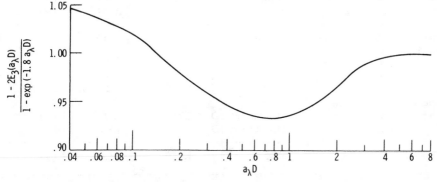

Figure 17-10 Ratio of emission by gas layer to that calculated using a mean beam length $L_e = 1.8D$.

Table 17-1 Mean beam lengths for radiation from entire gas volume

Geometry of gas volume	Characterizing dimension	Mean beam length for optical thickness $a_\lambda L_e \to 0$, $L_{e,0}$	Mean beam length corrected for finite optical thickness,[a] L_e	$C = L_e/L_{e,0}$
Hemisphere radiating to element at center of base	Radius R	R	R	1
Sphere radiating to its surface	Diameter D	$\tfrac{2}{3}D$	$0.65D$	0.97
Circular cylinder of infinite height radiating to concave bounding surface	Diameter D	D	$0.95D$	0.95
Circular cylinder of semi-infinite height radiating to:				
Element at center of base	Diameter D	D	$0.90D$	0.90
Entire base	Diameter D	$0.81D$	$0.65D$	0.80
Circular cylinder of height equal to diameter radiating to:				
Element at center of base	Diameter D	$0.77D$	$0.71D$	0.92
Entire surface	Diameter D	$\tfrac{2}{3}D$	$0.60D$	0.90
Circular cylinder of height equal to two diameters radiating to:				
Plane end	Diameter D	$0.73D$	$0.60D$	0.82
Concave surface	Diameter D	$0.82D$	$0.76D$	0.93
Entire surface	Diameter D	$0.80D$	$0.73D$	0.91
Circular cylinder of height equal to one-half the diameter radiating to:				
Plane end	Diameter D	$0.48D$	$0.43D$	0.90
Concave surface	Diameter D	$0.52D$	$0.46D$	0.88
Entire surface	Diameter D	$0.50D$	$0.45D$	0.90
Cylinder of infinite height and semicircular cross section radiating to element at center of plane rectangular face	Radius R		$1.26R$	
Infinite slab of gas radiating to:				
Element on one face	Slab thickness D	$2D$	$1.8D$	0.90
Both bounding planes	Slab thickness D	$2D$	$1.8D$	0.90
Cube radiating to a face	Edge X	$\tfrac{2}{3}X$	$0.6X$	0.90

Table 17-1 *(Continued)*

Geometry of gas volume	Characterizing dimension	Mean beam length for optical thickness $a_\lambda L_e \to 0$, $L_{e,0}$	Mean beam length corrected for finite optical thickness,[a] L_e	$C = L_e/L_{e,0}$
Rectangular parallelepipeds:				
$1 \times 1 \times 4$ {radiating to 1×4 face	Shortest edge X	$0.90X$	$0.82X$	0.91
radiating to 1×1 face		$0.86X$	$0.71X$	0.83
radiating to all faces		$0.89X$	$0.81X$	0.91
$1 \times 2 \times 6$ {radiating to 2×6 face		$1.18X$		
radiating to 1×6 face		$1.24X$		
radiating to 1×2 face		$1.18X$		
radiating to all faces		$1.20X$		
Gas volume surrounding an infinite tube bundle and radiating to a single tube:				
Equilateral triangular array:	Tube diameter			
$S = 2D$	D,	$3.4(S-D)$	$3.0(S-D)$	0.88
$S = 3D$	and	$4.45(S-D)$	$3.8(S-D)$	0.85
Square array:	spacing			
$S = 2D$	between tube centers, S	$4.1(S-D)$	$3.5(S-D)$	0.85

[a]Corrections are those suggested by Hottel et al. [1,2] or Eckert [3]. Corrections were chosen to provide maximum L_e where these references disagree.

17-6 TOTAL RADIATION EXCHANGE IN BLACK ENCLOSURE BETWEEN ENTIRE GAS VOLUME AND ENCLOSURE BOUNDARY BY USE OF MEAN BEAM LENGTH

In furnaces the walls are usually rough and soot-covered so they act practically as black surfaces. An important industrial problem is the radiant exchange between the furnace gas and the walls. In this section the simplified case of a black enclosure will be considered. The total radiation exchange will be considered between the entire gas volume and the enclosure boundary. This development will be carried out by application of the mean beam length. At industrial furnace and combustion chamber temperatures it is only heteropolar gases that absorb and emit significantly, such as CO_2, H_2O, CO, SO_2, NO, and CH_3. Gases with symmetric diatomic molecules, such as N_2, O_2, and H_2 are transparent to infrared radiation.

17-6.1 Radiation from Gas to All or Portion of Boundary

The mean beam length was found to be approximately independent of a_λ as evidenced by (17-57). This means that L_e can be used as a characteristic dimension of the gas volume and regarded as a constant for an integration over wavelength. The total heat flux from the gas incident on a surface is found by integrating (17-44) over λ:

$$q_i = \int_0^\infty [1 - \exp(-a_\lambda L_e)] e_{\lambda b,g} \, d\lambda \tag{17-58}$$

where L_e is independent of λ. Now define a gas total emittance ϵ_g such that

$$q_i = \epsilon_g \sigma T_g^4 \tag{17-59}$$

Equating the last two relations gives

$$\epsilon_g = \frac{\int_{\lambda=0}^\infty e_{\lambda b,g} [1 - \exp(-a_\lambda L_e)] \, d\lambda}{\sigma T_g^4} \tag{17-60}$$

The ϵ_g in (17-60) is a convenient quantity that can be presented in graphical form for each gas in terms of the variables L_e and T_g. Then for a particular geometry and gas condition, the ϵ_g is taken from the graphs and applied by use of (17-59).

The ϵ_g charts that will be presented here have been developed by Hottel [1] from many experimental measurements. The gas pressure will enter as a parameter because of the dependence of a_λ on gas density. If the gas is in a mixture, both the pressure of the mixture and the partial pressure of the radiating constituent under consideration will be parameters. A chart of ϵ_g was given for carbon dioxide (CO_2) in Fig. 13-13. Charts are presented here in more detail for CO_2 and water vapor (Figs. 17-11 and 17-13). Additional charts for sulfur dioxide, ammonia, carbon monoxide, methane, and a few other gases are in [2]. The discussion here will be limited to CO_2 and water vapor. Gases such as N_2, O_2 and H_2 do not emit or absorb significantly at industrial furnace temperatures.

In computing the radiation to area A by using (17-59),

$$Q_i = q_i A = A \epsilon_g \sigma T_g^4 \tag{17-61}$$

the mean beam length for the gas geometry is first obtained from Table 17-1 or Eq. (17-57). Then, with the partial pressure of the gas and its temperature known, the gas emittance is found by using Figs. 17-11–17-15. Figure 17-11 gives the total emittance of CO_2 obtained experimentally by using a mixture with nonabsorbing gases so that the total pressure of the mixture was at 1 atm while the partial pressure of the CO_2 was varied. The dashed lines indicate regions unverified by experimental data. For a mixture at total pressure other than 1 atm, there is a pressure-broadening correction that is to be applied [1]. This is given as a multiplying coefficient C_{CO_2} in Fig. 17-12. In the case of water vapor the emittance is influenced in a more complex manner by both the partial pressure of the water vapor and the total pressure of the gas mixture. For correlation purposes the values in Fig. 17-13 are emittances that were "reduced," by using a factor depending on p_{H_2O} and $p_{H_2O}L_e$, to limiting values as the partial pressure p_{H_2O} approached zero in a mixture having a total pressure $P = 1$ atm. A multiplying correction coefficient C_{H_2O} is given in Fig. 17-14 to account for the actual partial and total pressures involved.

If CO_2 and water vapor are both present in the gas mixture, an additional quantity

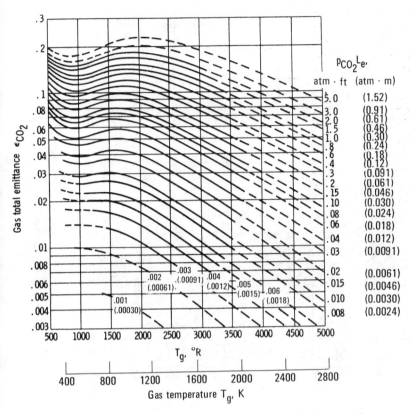

Figure 17-11   Total emittance of carbon dioxide in a mixture having a total pressure of 1 atm [1].

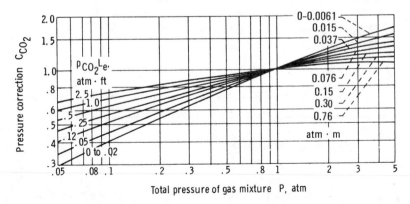

Figure 17-12   Pressure correction for CO_2 total emittance for values of P other than 1 atm [1].

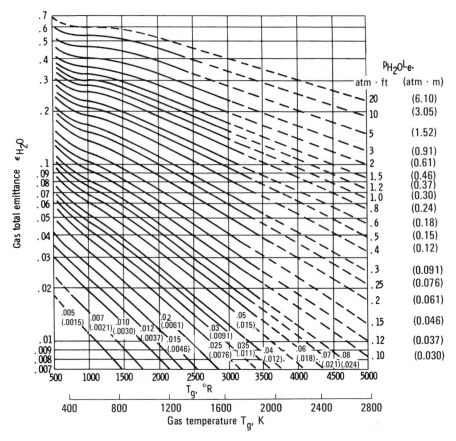

Figure 17-13 Total emittance of water vapor in limit of zero partial pressure in a mixture having a total pressure of 1 atm [1].

$\Delta\epsilon$ must be included to account for an emittance reduction resulting from spectral overlap of the CO_2 and H_2O absorption bands (for discussion of individual band behaviour, see [7, 8]. This correction is found from Fig. 17-15. For a mixture of CO_2 and water vapor in a nonabsorbing carrier gas, the emittance is then given by

$$\epsilon_g = C_{CO_2}\epsilon_{CO_2} + C_{H_2O}\epsilon_{H_2O} - \Delta\epsilon \tag{17-62}$$

To better understand (17-62), consider (17-60) when there are two radiating constituents having spectral absorption coefficients $a_{\lambda 1}(\lambda)$ and $a_{\lambda 2}(\lambda)$. Then

$$\epsilon_g = \frac{1}{\sigma T_g^4}\int_0^\infty e_{\lambda b,g}\left(1 - e^{-(a_{\lambda 1} + a_{\lambda 2})L_e}\right)d\lambda$$

$$= \frac{1}{\sigma T_g^4}\int_0^\infty e_{\lambda b,g}\left[1 - e^{-a_{\lambda 1}L_e} + 1 - e^{-a_{\lambda 2}L_e} - (1 - e^{-a_{\lambda 1}L_e})(1 - e^{-a_{\lambda 2}L_e})\right]d\lambda$$

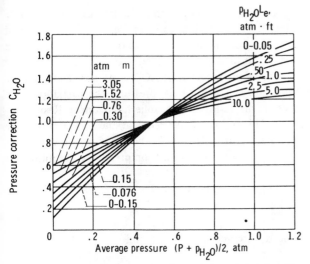

Figure 17-14   Pressure correction for water vapor total emittance for values of p_{H_2O} and P other than 0 and 1 atm, respectively [1].

The first four terms under the integral integrate to yield the total emittances of the individual radiating components so that

$$\epsilon_g = \epsilon_1 + \epsilon_2 - \frac{1}{\sigma T_g^4}\int_0^\infty e_{\lambda b,g}(1 - e^{-a_{\lambda 1}L e})(1 - e^{-a_{\lambda 2}L e})\, d\lambda$$

The last term on the right is the correction for spectral overlap. The integrand is non-zero only in the wavelength regions where both $a_{\lambda 1}$ and $a_{\lambda 2}$ are nonzero; these are the spectral overlap regions.

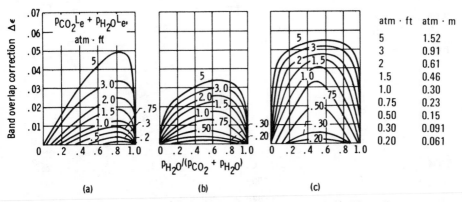

Figure 17-15   Correction on total emittance for band overlap when both CO_2 and water vapor are present [1]. (a) Gas temperature $T_g = 400$ K ($720°$R); (b) gas temperature $T_g = 810$ K ($1460°$R); (c) gas temperature, $T_g \geqslant 1200$ K ($2160°$R).

The charts in Figs. 17-11–17-15 have been subject to reevaluations to incorporate information from more recent detailed spectral data and to extend to larger pL_e and T_g [9–12]. Agreement was generally good in the central regions of Figs. 17-11 and 17-13, but some differences were found in the extrapolated (dashed) regions of those figures. Figure 17-16, from [12], shows the reevaluated CO_2 emittance calculated for line broadening at 1 bar total pressure and zero partial pressure (the partial-pressure effect is in reality not significant in the range of atmospheric pressure). Results for water vapor are given in [10] and [11], and the curves comparable to Fig. 17-13 are shown in Fig. 17-17. There is a significant difference in the shape of the curves for large $p_{H_2O}L_e$. The references also give recomputed results on pressure corrections and the effect of H_2O and CO_2 band overlap.

A useful analytical representation for ϵ_{H_2O} as given in Fig. 17-13 is given in [13] by

$$\epsilon_{H_2O} = a_0[1 - \exp(-a_1\sqrt{X})] \tag{17-63}$$

where a_0 and a_1 are given as functions of temperature by the following table:

T, K	a_0	a_1, $m^{-1/2} \cdot atm^{-1}$
300	0.683	1.17
600	0.674	1.32
900	0.700	1.27
1200	0.673	1.21
1500	0.624	1.15

For H_2O-air mixtures the parameter $X = p_{H_2O}L_e(300/T)(p_{air} + bp_{H_2O})$, where T is in kelvins, p in atmospheres and L_e in meters. The self-broadening coefficient b for water vapor is given by $b = 5.0(300/T)^{1/2} + 0.5$. A comparison with the Hottel chart is shown in Fig. 17-18.

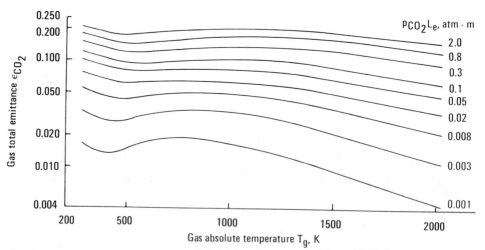

Figure 17-16 Total emittance or carbon dioxide in limit of zero partial pressure in a mixture at total pressure of 1 bar [12].

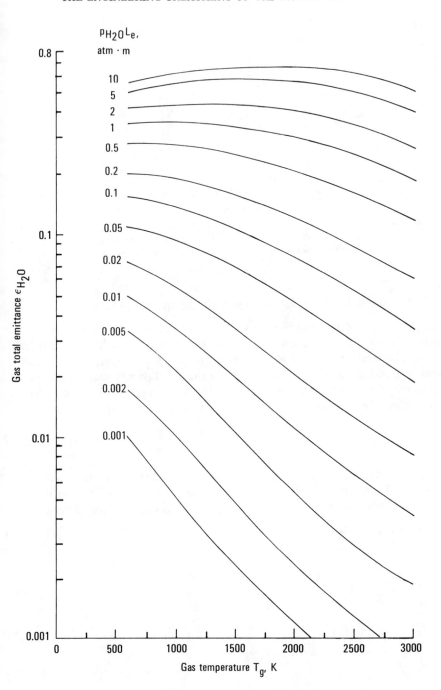

Figure 17-17 Total emittance of water vapor in limit of zero partial pressure in a mixture at total pressure of 1 atm [10, 11].

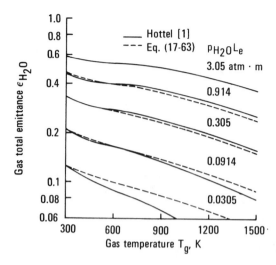

Figure 17-18 Comparison of Eq. (17-63) with Hottel chart [1] for $P = 1$ atm and $p_{H_2O} \to 0$.

17-6.2 Exchange between Entire Gas Volume and Boundary

Hottel [1] has provided a simple approximate procedure applicable when the cooled enclosure boundary is black and is at a temperature at which it will emit appreciable radiation. The total energy removed at the wall must equal the energy being supplied by some external means such as combustion to the gas. Taking a heat balance on the gas then shows that the net *average* heat flux being *removed* at the wall is the gas emission minus the emission from the wall that is absorbed by the gas; that is,

$$\frac{Q}{A} = \sigma[\epsilon_g T_g^{\,4} - \alpha_g(T_w)T_w^{\,4}]$$

(17-64)

The $\alpha_g(T_w)$ is the absorptance of the gas for radiation emitted from the wall at temperature T_w. The $\alpha_g(T_w)$ depends on T_w, as this determines the spectral distribution of the radiation received by the gas. According to [1], the α_g can be found from

$$\alpha_g = \alpha_{CO_2} + \alpha_{H_2O} - \Delta\alpha$$

(17-65)

where

$$\alpha_{CO_2} = C_{CO_2}\epsilon_{CO_2}^+ \left(\frac{T_g}{T_w}\right)^{0.65}$$

(17-66)

$$\alpha_{H_2O} = C_{H_2O}\epsilon_{H_2O}^+ \left(\frac{T_g}{T_w}\right)^{0.45}$$

(17-67)

$$\Delta\alpha = (\Delta\epsilon)_{\text{at } T_w}$$

(17-68)

The $\epsilon_{CO_2}^+$ and $\epsilon_{H_2O}^+$ are, respectively, ϵ_{CO_2} and ϵ_{H_2O} obtained from Figs. 17-11 and 17-13 evaluated at the abscissa T_w and at the respective parameters $p_{CO_2}L_eT_w/T_g$ and $p_{H_2O}L_eT_w/T_g$. For further information the reader is referred to [1, 2].

EXAMPLE 17-2 A cooled right cylindrical tank 4 ft in diameter and 4 ft long has a black interior surface and is filled with hot gas at a total pressure of 1 atm. The gas is composed of a transparent gas at partial pressure 0.75 atm, and carbon dioxide. The gas is uniformly mixed at a temperature of 2000°R. Compute how much energy must be removed from the surface of the tank to keep it cool if the tank walls are all sufficiently cooled so that only radiation from the gas is significant.

The geometry is a finite circular cylinder of gas, and the radiation to its walls will be computed. Using Table 17-1, the corrected mean beam length for this geometry is $L_e = 0.60D = 2.4$ ft. The partial pressure of the CO_2 is 0.25 atm, so that $p_{CO_2} L_e = 0.25 \times 2.4 = 0.6$ atm·ft. From Fig. 17-11, $\epsilon_{CO_2}(p_{CO_2} L_e, T_g) = 0.13$, and C_{CO_2} from Fig. 17-12 is 1.0, since the mixture total pressure is unity. Assuming the walls of the tank to be sufficiently cool so that their emitted energy is negligible, the energy to be removed is, from (17-61),

$$Q_i = \epsilon_{CO_2} \sigma T_g^4 A = 0.13 \times 0.173 \times 10^{-8}(2000)^4 24\pi = 271{,}000 \text{ Btu/h}$$

17-7 TOTAL RADIATION EXCHANGE IN ENCLOSURE BY INTEGRATION OF SPECTRAL EQUATIONS

The mean-beam-length approach in the previous section is concerned with either the radiation from a gas volume to all or a portion of a black enclosure boundary, or the average exchange between the gas and an entire black isothermal enclosure. For a more general analysis of radiation in an enclosure, the exchanges of total radiation must be considered between the various pairs of bounding surfaces having different temperatures. This involves integrating the exchange relations involving $\bar{\tau}_\lambda$ and $\bar{\alpha}_\lambda$ over all wavelengths. An illustration requiring such an integration was outlined in Example 17-1 for a parallel-plate geometry.

A form of the spectral equations was given by (17-20) that relates the gas black-body emissive power and the spectral fluxes dq_λ supplied in a differential wavelength interval to each surface. In the solution accounting for spectral effects as described in Sec. 10-3 for enclosures filled with nonabsorbing media, the set of enclosure equations is solved at each wavelength for the dq_λ (assume that the surface temperatures are specified), and the results then integrated over all wavelengths. For a gas radiation problem the gas properties vary so irregularly with wavelength that the detailed integration over λ would be a practical impossibility. This leads to a consideration of the use of finite wavelength bands.

17-7.1 Band Equations

An approach for integrating over wavelength is developed by dividing the spectrum into absorbing and nonabsorbing bands. For a typical band of width $\Delta\lambda$ the integration of Eq. (17-20) gives

$$\int_{\Delta\lambda} \sum_{j=1}^{N} \left(\frac{\delta_{kj}}{\epsilon_{\lambda,j}} - F_{k-j} \frac{1 - \epsilon_{\lambda,j}}{\epsilon_{\lambda,j}} \bar{\tau}_{\lambda,k-j} \right) dq_{\lambda,j}$$

$$= \int_{\Delta\lambda} \sum_{j=1}^{N} [(\delta_{kj} - F_{k-j}\bar{\tau}_{\lambda,k-j})e_{\lambda b,j} - F_{k-j}\bar{\alpha}_{\lambda,k-j}e_{\lambda b,g}] \, d\lambda \qquad (17\text{-}69)$$

It is assumed that the bands are sufficiently narrow so that $dq_{\lambda,j}$, $\epsilon_{\lambda,j}$, $\bar{\tau}_{\lambda,k-j}$, $\bar{\alpha}_{\lambda,k-j}$, $e_{\lambda b,j}$, and $e_{\lambda b,g}$ can be regarded as constants over the bandwidth, being characteristic of some mean wavelength within the band, or, in the case of $\bar{\tau}$ and $\bar{\alpha}$, being averaged over the band (see Sec. 17-7.2). (Then (17-69) can be written for band l as

$$\sum_{j=1}^{N} \left(\frac{\delta_{kj}}{\epsilon_{l,j}} - F_{k-j} \frac{1 - \epsilon_{l,j}}{\epsilon_{l,j}} \bar{\tau}_{l,k-j} \right) \Delta q_{l,j}$$

$$= \sum_{j=1}^{N} [(\delta_{kj} - F_{k-j}\bar{\tau}_{l,k-j})e_{lb,j} - F_{k-j}\bar{\alpha}_{l,k-j}e_{lb,g}] \, \Delta\lambda \qquad (17\text{-}70)$$

In a spectral region where the gas is essentially nonabsorbing, $\bar{\tau}_l = 1$ and $\bar{\alpha}_l = 0$ so that (17-70) reduces to

$$\sum_{j=1}^{N} \left(\frac{\delta_{kj}}{\epsilon_{l,j}} - F_{k-j} \frac{1 - \epsilon_{l,j}}{\epsilon_{l,j}} \right) \Delta q_{l,j} = \sum_{j=1}^{N} (\delta_{kj} - F_{k-j})e_{lb,j} \, \Delta\lambda \qquad (17\text{-}71)$$

which is of the form of (8-19).

17-7.2 Transmission and Absorption Factors

The $\bar{\tau}_{l,k-j}$ in (17-70) is found from (17-27) by taking an integrated average over the band:

$$\bar{\tau}_{l,k-j} = \frac{1}{A_k F_{k-j}} \int_{A_j} \int_{A_k} \frac{[(1/\Delta\lambda)\int_{\Delta\lambda} \tau_\lambda(S) \, d\lambda] \cos\theta_j \cos\theta_k}{\pi S^2} \, dA_k \, dA_j \qquad (17\text{-}72)$$

Similarly the $\bar{\alpha}_{l,k-j}$ is obtained as

$$\bar{\alpha}_{l,k-j} = \frac{1}{A_k F_{k-j}} \int_{A_j} \int_{A_k} \frac{[(1/\Delta\lambda)\int_{\Delta\lambda} \alpha_\lambda(S) \, d\lambda] \cos\theta_j \cos\theta_k}{\pi S^2} \, dA_k \, dA_j \qquad (17\text{-}73)$$

For each small bandwidth,

$$\bar{\alpha}_{l,k-j} = 1 - \bar{\tau}_{l,k-j}$$

and to evaluate $\bar{\alpha}_l$ and $\bar{\tau}_l$ only the single integral is needed:

$$A_k F_{k-j}\bar{\alpha}_{l,k-j} = \int_{A_l} \int_{A_k} \frac{\alpha_l(S) \cos\theta_j \cos\theta_k}{\pi S^2} \, dA_k \, dA_j \qquad (17\text{-}74)$$

where

$$\alpha_l(S) = \frac{1}{\Delta\lambda} \int_{\Delta\lambda} \alpha_\lambda(S)\, d\lambda = \frac{1}{\Delta\lambda} \int_{\Delta\lambda} [1 - \exp(-a_\lambda S)]\, d\lambda$$

From (16-62) the α_l can be expressed if desired in terms of the effective bandwidth as

$$\alpha_l(S) = \frac{\bar{A}_l(S)}{\Delta\lambda} \tag{17-75}$$

To obtain $\bar{\tau}$ and $\bar{\alpha}$ for use in (17-70), the integral in (17-74) must be evaluated between pairs of finite surfaces in the various wavelength bands involved. It is evident that, when there are more than a few bands that absorb appreciably, the solution involves considerable computational effort. A simplification has been developed by Dunkle [14] that saves considerable labor and yields good accuracy. Dunkle assumes that the integrated band absorption is a linear function of path length. This has some physical basis, as it holds exactly for a band of weak nonoverlapping lines [Eq. (16-74)]. Also, it is the form of some of the effective bandwidths in the correlations of Table 16-2 and those on page 574. As shown in [14] by means of a few examples, reasonable values of the energy exchange are obtained by use of this approximation. Hence, let α_l in (17-74) have the linear form from (16-74) [note that $\Delta\lambda = A_0$ in (16-74)]

$$\alpha_l(S) = \frac{\bar{A}_l}{\Delta\lambda} = \frac{\bar{A}_l}{A_0} = \frac{S_c}{\delta} S \tag{17-76}$$

where S_c and δ are the absorption coefficient and spacing of the individual weak lines as defined in Chap. 16.

Now define a mean path length S, called the *geometric-mean beam length* $\bar{S}_{k-j}$. This mean length is such that α_l evaluated from (17-76) by using $S = \bar{S}_{k-j}$ will yield $\bar{\alpha}_{l,k-j}$ as found from the integral in (17-74). After substitution of $\bar{\alpha}_{l,k-j} = (S_c/\delta)\bar{S}_{k-j}$ and $\alpha_l = (S_c/\delta)S$ into (17-74), the S_c/δ drops out and the relation to obtain $\bar{S}_{k-j}$ is

$$\bar{S}_{k-j} = \frac{1}{A_k F_{k-j}} \int_{A_j} \int_{A_k} \frac{\cos\theta_j \cos\theta_k}{\pi S}\, dA_k\, dA_j \tag{17-77}$$

which is only dependent on geometry. Dunkle [14] computed and tabulated $\bar{S}_{k-j}$ for parallel equal rectangles, rectangles at right angles, and a differential sphere and a rectangle. Results for equal opposed parallel rectangles are shown in Fig. 17-19. Values for parallel rectangles, and rectangles at right angles are given in Tables 17-2 and 17-3. Other $\bar{S}_{k-j}$ values are referenced by Hottel and Sarofim [2].

For a given gas at uniform conditions, the geometric-mean beam length can be used in the effective-bandwidth correlations discussed in Chap. 16 to obtain $\bar{A}_l$. Using the $\Delta\lambda$ as obtained in the next paragraph yields $\bar{\alpha}_l$ from Eq. (17-76) and $\bar{\tau}_l$ from $1 - \bar{\alpha}_l$. Then (17-70) and (17-71) can be solved for each wavelength band l. The total

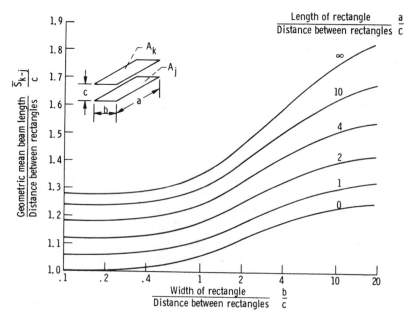

Figure 17-19 Geometric mean beam lengths for equal parallel rectangles [14].

energies at each surface k are found from

$$q_k = \sum_{\substack{\text{absorbing} \\ \text{bands}}} \Delta q_{l,k} + \sum_{\substack{\text{nonabsorbing} \\ \text{bands}}} \Delta q_{l,k} \qquad (17\text{-}78)$$

The wavelength span $\Delta\lambda$ of each band is needed to carry out the solution. As discussed after (16-70), this span can increase with path length. Edwards and Nelson [15] and Edwards [16] give recommended spans for CO_2 and H_2O vapor: these values are reproduced in Table 17-4 for the parallel-plate geometry. Note that these values are given in terms of wave number rather than wavelength. For other geometries, Edwards and Nelson give methods for choosing approximate spans for CO_2 and H_2O bands. Briefly, the method is to use approximate band spans based on the longest important mass path length in the geometry being studied. With this in mind, the limits of Table 17-4 are probably adequate for problems involving CO_2 and H_2O vapor.

If all surface temperatures are given in the problem at hand, the results found from (17-78) complete the solution. If q_k is given for n surfaces and T_k for the remaining $N - n$ surfaces, then the n unknown surface temperatures are guessed, the equations are solved for all the q's, and then the calculated q_k are compared to the given values. If they do not agree, new values of T_k for the n surfaces are assumed, and the calculation repeated. This procedure is continued until there is agreement between given and calculated q_k for all k. Equation (17-21) expressed as a sum over the wavelength bands gives the required energy input to the gas for the given T_g.

Two examples will be presented for an isothermal gas in an enclosure. Then we shall consider removing the restriction of uniform conditions in the gas.

EXAMPLE 17-3 Two black parallel plates are separated by a distance of $D = 1$ m. The plates are of width $W = 1$ m and of effectively infinite length (Fig. 17-20). The space between the plates is filled with carbon dioxide gas at $p_{CO_2} = 1$ atm and $T_g = 1000$ K. If plate 1 is maintained at 2000 K and plate 2 is maintained at 500 K, find the energy flux that must be supplied to plate 2 to maintain its temperature.

As shown by Fig. 17-20, the geometry is a four-boundary enclosure formed by the two plates and the two open bounding planes. The open bounding planes are perfectly absorbing (nonreflecting) and radiate no significant energy as the temperature of the surroundings is assumed low. The energy flux added to surface 2 will be found by using the enclosure equation (17-20) where $k = 2$ and $N = 4$. Since all surfaces are black, $\epsilon_{\lambda,j} = 1$ and (17-20) reduces to

$$\sum_{j=1}^{4} \delta_{2j}\, dq_{\lambda,j} = \sum_{j=1}^{4} [(\delta_{2j} - F_{2-j}\bar{T}_{\lambda,2-j})e_{\lambda b,j}\, d\lambda - F_{2-j}\bar{\alpha}_{\lambda,2-j}e_{\lambda b,g}\, d\lambda] \qquad (17\text{-}79)$$

The self-view factor $F_{2-2} = 0$ and $e_{\lambda b,3} = e_{\lambda b,4} = 0$, so the summations can be written as

$$dq_{\lambda,2} = [-F_{2-1}\bar{T}_{\lambda,2-1}e_{\lambda b,1} + e_{\lambda b,2} - (F_{2-1}\bar{\alpha}_{\lambda,2-1} + F_{2-3}\bar{\alpha}_{\lambda,2-3}$$
$$+ F_{2-4}\bar{\alpha}_{\lambda,2-4})e_{\lambda b,g}]\, d\lambda \qquad (17\text{-}80)$$

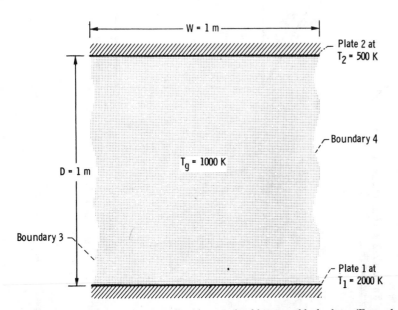

Figure 17-20 Isothermal carbon dioxide contained between black plates (Example 17-3).

Table 17-2 Geometric mean-beam-length ratios and configuration factors for parallel equal rectangles [14]

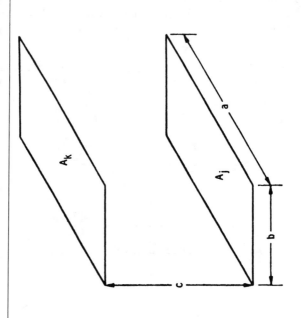

632

b/c

a/c		0	0.1	0.2	0.4	0.6	1.0	2.0	4.0	6.0	10.0	20.0
0	$\bar{S}_{k-j}/c$	1.000	1.001	1.003	1.012	1.025	1.055	1.116	1.178	1.205	1.230	1.251
	F_{k-j}											
0.1	$\bar{S}_{k-j}/c$	1.001	1.002	1.004	1.013	1.026	1.056	1.117	1.179	1.207	1.233	1.254
	F_{k-j}		0.00316	0.00626	0.01207	0.01715	0.02492	0.03514	0.04210	0.04463	0.04671	0.04829
0.2	$\bar{S}_{k-j}/c$	1.003	1.004	1.006	1.015	1.028	1.058	1.120	1.182	1.210	1.235	1.256
	F_{k-j}		0.00626	0.01240	0.02391	0.03398	0.04941	0.06971	0.08353	0.08859	0.09271	0.09586
0.4	$\bar{S}_{k-j}/c$	1.012	1.013	1.015	1.024	1.037	1.067	1.129	1.192	1.220	1.245	1.267
	F_{k-j}		0.01207	0.02391	0.04614	0.06560	0.09554	0.13513	0.16219	0.17209	0.18021	0.18638
0.6	$\bar{S}_{k-j}/c$	1.025	1.026	1.028	1.037	1.050	1.080	1.143	1.206	1.235	1.261	1.282
	F_{k-j}		0.01715	0.03398	0.06560	0.09336	0.13627	0.19341	0.23271	0.24712	0.25896	0.26795
1.0	$\bar{S}_{k-j}/c$	1.055	1.056	1.058	1.067	1.080	1.110	1.175	1.242	1.272	1.300	1.324
	F_{k-j}		0.02492	0.04941	0.09554	0.13627	0.19982	0.28588	0.34596	0.36813	0.38638	0.40026
2.0	$\bar{S}_{k-j}/c$	1.116	1.117	1.120	1.129	1.143	1.175	1.246	1.323	1.359	1.393	1.421
	F_{k-j}		0.03514	0.06971	0.13513	0.19341	0.28588	0.41525	0.50899	0.54421	0.57338	0.59563
4.0	$\bar{S}_{k-j}/c$	1.178	1.179	1.182	1.192	1.206	1.242	1.323	1.416	1.461	1.505	1.543
	F_{k-j}		0.04210	0.08353	0.16219	0.23271	0.34596	0.50899	0.63204	0.67954	0.71933	0.74990
6.0	$\bar{S}_{k-j}/c$	1.205	1.207	1.210	1.220	1.235	1.272	1.359	1.461	1.513	1.564	1.609
	F_{k-j}		0.04463	0.08859	0.17209	0.24712	0.36813	0.54421	0.67954	0.73258	0.77741	0.81204
10.0	$\bar{S}_{k-j}/c$	1.230	1.233	1.235	1.245	1.261	1.300	1.393	1.505	1.564	1.624	1.680
	F_{k-j}		0.04671	0.09271	0.18021	0.25896	0.38638	0.57338	0.71933	0.77741	0.82699	0.86563
20.0	$\bar{S}_{k-j}/c$	1.251	1.254	1.256	1.267	1.282	1.324	1.421	1.543	1.609	1.680	1.748
	F_{k-j}		0.04829	0.09586	0.18638	0.26795	0.40026	0.59563	0.74990	0.81204	0.86563	0.90785
∞	$\bar{S}_{k-j}/c$	1.272	1.274	1.277	1.289	1.306	1.349	1.452	1.584	1.660	1.745	1.832
	F_{k-j}		0.04988	0.09902	0.19258	0.27698	0.41421	0.61803	0.78078	0.84713	0.90499	0.95125

Table 17-3 Configuration factors and mean-beam-length functions for rectangles at right angles [14]

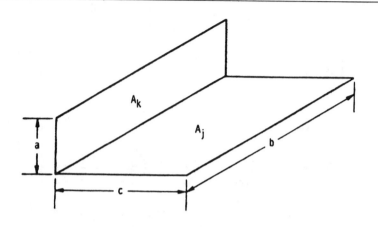

a/b		c/b					
		0.05	0.10	0.20	0.4	0.6	1.0
0.02	A_kF_{k-j}/b^2	0.007982	0.008875	0.009323	0.009545	0.009589	0.009628
	$A_kF_{k-j}\bar{S}_{k-j}/abc$	0.17840	0.12903	0.08298	0.04995	0.03587	0.02291
0.05	A_kF_{k-j}/b^2	0.014269	0.018601	0.02117	0.02243	0.02279	0.02304
	$A_kF_{k-j}\bar{S}_{k-j}/abc$	0.21146	0.18756	0.13834	0.08953	0.06627	0.04372
0.10	A_kF_{k-j}/b^2		0.02819	0.03622	0.04086	0.04229	0.04325
	$A_kF_{k-j}\bar{S}_{k-j}/abc$		0.20379	0.17742	0.12737	0.09795	0.06659
0.20	A_kF_{k-j}/b^2			0.05421	0.06859	0.07377	0.07744
	$A_kF_{k-j}\bar{S}_{k-j}/abc$			0.18854	0.15900	0.13028	0.09337
0.40	A_kF_{k-j}/b^2				0.10013	0.11524	0.12770
	$A_kF_{k-j}\bar{S}_{k-j}/abc$				0.16255	0.14686	0.11517
0.60	A_kF_{k-j}/b^2					0.13888	0.16138
	$A_kF_{k-j}\bar{S}_{k-j}/abc$					0.14164	0.11940
1.0	A_kF_{k-j}/b^2						0.20004
	$A_kF_{k-j}\bar{S}_{k-j}/abc$						0.11121
2.0	A_kF_{k-j}/b^2						
	$A_kF_{k-j}\bar{S}_{k-j}/abc$						
4.0	A_kF_{k-j}/b^2						
	$A_kF_{k-j}\bar{S}_{k-j}/abc$						
6.0	A_kF_{k-j}/b^2						
	$A_kF_{k-j}\bar{S}_{k-j}/abc$						
10.0	A_kF_{k-j}/b^2						
	$A_kF_{k-j}\bar{S}_{k-j}/abc$						
20.0	A_kF_{k-j}/b^2						
	$A_kF_{k-j}\bar{S}_{k-j}/abc$						

Table 17-3 *(Continued)*

a/b		c/b					
		2.0	4.0	6.0	10.0	20.0	∞
0.02	$A_k F_{k-j}/b^2$	0.009648	0.009653	0.009655	0.009655	0.009655	0.009655
	$A_k F_{k-j}\bar{S}_{k-j}/abc$	0.01263	0.006364	0.004288	0.002594	0.001305	
0.05	$A_k F_{k-j}/b^2$	0.02316	0.02320	0.02321	0.02321	0.02321	0.02321
	$A_k F_{k-j}\bar{S}_{k-j}/abc$	0.02364	0.01234	0.008342	0.005059	0.002549	
0.10	$A_k F_{k-j}/b^2$	0.04376	0.04390	0.04393	0.04394	0.04394	0.04395
	$A_k F_{k-j}\bar{S}_{k-j}/abc$	0.03676	0.01944	0.013184	0.008018	0.004049	
0.20	$A_k F_{k-j}/b^2$	0.07942	0.07999	0.08010	0.08015	0.08018	0.08018
	$A_k F_{k-j}\bar{S}_{k-j}/abc$	0.05356	0.02890	0.01972	0.012047	0.006103	
0.40	$A_k F_{k-j}/b^2$	0.13514	0.13736	0.13779	0.13801	0.13811	0.13814
	$A_k F_{k-j}\bar{S}_{k-j}/abc$	0.07088	0.03903	0.02666	0.01697	0.008642	
0.60	$A_k F_{k-j}/b^2$	0.17657	0.18143	0.18239	0.18289	0.18311	0.18318
	$A_k F_{k-j}\bar{S}_{k-j}/abc$	0.07830	0.04467	0.03109	0.02025	0.010366	
1.0	$A_k F_{k-j}/b^2$	0.23285	0.24522	0.24783	0.24921	0.24980	0.25000
	$A_k F_{k-j}\bar{S}_{k-j}/abc$	0.08137	0.04935	0.03502	0.02196	0.01175	
2.0	$A_k F_{k-j}/b^2$	0.29860	0.33462	0.34386	0.34916	0.35142	0.35222
	$A_k F_{k-j}\bar{S}_{k-j}/abc$	0.07086	0.04924	0.03670	0.02401	0.01325	
4.0	$A_k F_{k-j}/b^2$		0.40544	0.43104	0.44840	0.45708	0.46020
	$A_k F_{k-j}\bar{S}_{k-j}/abc$		0.04051	0.03284	0.02320	0.01300	
6.0	$A_k F_{k-j}/b^2$			0.46932	0.49986	0.51744	0.52368
	$A_k F_{k-j}\bar{S}_{k-j}/abc$			0.02832	0.02132	0.01272	
10.0	$A_k F_{k-j}/b^2$				0.5502	0.5876	0.6053
	$A_k F_{k-j}\bar{S}_{k-j}/abc$				0.01759	0.01146	
20.0	$A_k F_{k-j}/b^2$					0.6608	0.7156
	$A_k F_{k-j}\bar{S}_{k-j}/abc$					0.008975	

Table 17-4 Approximate band limits for parallel-plate geometry [15, 16, 20]

Gas	Band λ, μm	Band center η, cm^{-1}	Band limits η, cm^{-1} [a]	
			Lower	Upper
CO_2	15	667	$667 - (\bar{A}_{15}/1.78)$	$667 + (\bar{A}_{15}/1.78)$
	10.4	960	849	1013
	9.4	1060	1013	1141
	4.3	2350	$2350 - (\bar{A}_{4.3}/1.78)$	2430
	2.7	3715	$3715 - (\bar{A}_{2.7}/1.78)$	3750
H_2O	6.3	1600	$1600 - (\bar{A}_{6.3}/1.6)$	$1600 + (\bar{A}_{6.3}/1.6)$
	2.7	3750	$3750 - (\bar{A}_{2.7}/1.4)$	$3750 + (\bar{A}_{2.7}/1.4)$
	1.87	5350	4620	6200
	1.38	7250	6200	8100

[a] $\bar{A}$ are found for various bands from Tables 16-2 and 16-3. Terms such as $\bar{A}_{15}/1.78$ are $\bar{A}/2(1 - \tau_g)$ from Eq. (17) and Tables 1 and 2 of [15].

To simplify the example, it will be carried out by considering the entire wavelength region as a single band. To obtain the total energy supplied to plate 2, integrate over all wavelengths to obtain

$$q_2 = -F_{2-1}\int_0^\infty \bar{\tau}_{\lambda,2-1}e_{\lambda b,1}\,d\lambda + \sigma T_2^4$$

$$-\int_0^\infty (F_{2-1}\bar{\alpha}_{\lambda,2-1} + F_{2-3}\bar{\alpha}_{\lambda,2-3} + F_{2-4}\bar{\alpha}_{\lambda,2-4})e_{\lambda b,g}\,d\lambda$$

By use of the definitions of total transmission and absorption factors, which are

$$\bar{\tau}_{2-1}\sigma T_1^4 = \int_0^\infty \bar{\tau}_{\lambda,2-1}e_{\lambda b,1}\,d\lambda \qquad \bar{\alpha}_{2-1}\sigma T_g^4 = \int_0^\infty \bar{\alpha}_{\lambda,2-1}e_{\lambda b,g}\,d\lambda$$

and so forth, q_2 becomes

$$q_2 = \sigma T_2^4 - F_{2-1}\bar{\tau}_{2-1}\sigma T_1^4 - (F_{2-1}\bar{\alpha}_{2-1} + F_{2-3}\bar{\alpha}_{2-3} + F_{2-4}\bar{\alpha}_{2-4})\sigma T_g^4 \qquad (17\text{-}81)$$

To determine $\bar{\tau}$ and $\bar{\alpha}$, the geometric-mean beam length will be used. For opposing rectangles (Fig. 17-19) at an abscissa of 1.0 and on the curve for a length-to-spacing ratio of ∞, the $\bar{S}_{2-1}/D = 1.34$ or $\bar{S}_{2-1} = 1.34$ m. To determine $\bar{\alpha}_{2-1}$, which determines the emission of the gas, use the emittance chart in Fig. 17-11 at a pressure of 1 atm, a beam length of 1.34 m, and $T_g = 1000$ K. This gives $\bar{\alpha}_{2-1} = 0.22$. When obtaining $\bar{\tau}_{2-1}$, note from (17-80) that the radiation in the $\bar{\tau}_{2-1}$ term is $e_{\lambda b,1}$ and is coming from wall 1. Therefore it has a spectral distribution different from that of the gas radiation. To account for this nongray effect, (17-66) will be used with ϵ^+ evaluated at $p_{CO_2}\bar{S}_{2-1}(T_1/T_g) = 1.34(2000/1000) = 2.68$ atm·m and $T_1 = 2000$ K. Then, using Fig. 17-11 (extrapolated) and Eq. (17-66) results in $\bar{\tau}_{2-1} \approx 1 - 0.2(\tfrac{1}{2})^{0.65} = 0.87$.

From factor 10 in Appendix C the configuration factor F_{2-1} is given by

$$F_{2-1} = \frac{(D^2 + W^2)^{1/2} - D}{W} = \sqrt{2} - 1 = 0.414$$

Then $F_{2-3} = F_{2-4} = \tfrac{1}{2}(1 - 0.414) = 0.293$.

The $\bar{\alpha}_{2-3} = \bar{\alpha}_{2-4}$ remain to be found. For adjoint planes as in the geometry for Table 17-3, the following expression from Eq. (12) of [14] can be used, obtained for the present case where $b \to \infty, a = 1,$ and $c = 1$:

$$\bar{S}_{2-3} = \frac{1}{\pi F_{2-3}}(2\ln\sqrt{2}) = \frac{2 \times 0.347}{\pi \times 0.293} = 0.753 \text{ m}$$

Using Fig. 17-11 at $p\bar{S} = 0.753$ atm·m and $T_g = 1000$ K gives $\bar{\alpha}_{2-3} = \bar{\alpha}_{2-4} = 0.19$. Then

$$q_2 = \sigma T_2^4 - 0.414(0.87)\sigma T_1^4 - (0.414 \times 0.22 + 2 \times 0.293 \times 0.19)\sigma T_g^4$$

$$= 5.73 \times 10^{-12}(500^4 - 0.36 \times 2000^4 - 0.20 \times 1000^4) = -33.8 \text{ W/cm}^2$$

The solution is now complete. Note that the largest contribution to q_2 is by energy leaving surface 1 and being absorbed by surface 2. Emission from the gas to surface 2 and emission from surface 2 are negligible.

An alternative approach that is simpler for this particular example is to note that the term involving T_g in (17-81) is the flux received by surface 2 as a result of emission by the entire gas. This can be calculated from (17-61) using the mean beam length. Then $q_2 = \sigma T_2{}^4 - F_{2-1}\bar{\tau}_{2-1}\sigma T_1{}^4 - \epsilon_g \sigma T_g{}^4$. For this symmetric geometry the average flux from the gas to one side of the enclosure is the same as that to the entire enclosure boundary. Consequently the mean beam length can be obtained from (17-57), which gives $L_e = 0.9(4)V/A = 0.9(4)(1 \text{ m})^2/4 \text{ m} = 0.9$ m. Then from Fig. 17-11 at $T_g = 1000$ K and $p_{CO_2}L_e = 0.9 \text{ atm} \cdot \text{m}$, the $\epsilon_g = 0.20$. This gives the same q_2 as previously calculated.

EXAMPLE 17-4 Parallel nongray plates are 1 in apart and are at temperatures of $T_1 = 2000°\text{R}$ and $T_2 = 1000°\text{R}$. Pure CO_2 gas at 10 atm pressure and $T_g = 1000°\text{R}$ is between the plates. The plate hemispherical spectral emissivity as a function of wave number is approximated by the following table:

η, cm^{-1}	ϵ_η	η, cm^{-1}	ϵ_η
0–500	0.37	1150–2200	0.45
500–750	0.26	2200–2500	0.65
750–850	0.32	2500–3600	0.61
850–1000	0.37	3600–3750	0.69
1000–1150	0.46	3750–∞	0.73

Assume that only the 15-, 10.4-, 9.4-, 4.3-, and 2.7-μm CO_2 bands cause significant attenuation in the gas. Compute the total heat flux being added to plate 2.

In Example 17-1 the spectral exchange was found for radiation between infinite parallel plates with a gas between them. The total energy added to plate 2 is found by integrating (17-24b) over all wave numbers:

$$q_2 = \int_{\eta=0}^{\infty} \frac{\{\epsilon_{\eta,1}\epsilon_{\eta,2}\bar{\tau}_\eta(e_{\eta b,2} - e_{\eta b,1}) + \epsilon_{\eta,2}(1-\bar{\tau}_\eta)[1 + (1-\epsilon_{\eta,1})\bar{\tau}_\eta](e_{\eta b,2} - e_{\eta b,g})\} d\eta}{1 - (1-\epsilon_{\eta,1})(1-\epsilon_{\eta,2})\bar{\tau}_\eta{}^2}$$

In this example $\epsilon_{\eta,1} = \epsilon_{\eta,2}$ and $T_g = T_2$ so q_2 simplifies to

$$q_2 = -\int_0^\infty \frac{\epsilon_{\eta,1}{}^2\bar{\tau}_\eta(e_{\eta b,1} - e_{\eta b,2})}{1 - (1-\epsilon_{\eta,1})^2\bar{\tau}_\eta{}^2} d\eta$$

The integration can be expressed in finite-difference form as a sum over wavenumber bands. For the lth band let $\epsilon_{\eta,1} = \epsilon_l$ and $\bar{\tau}_\eta = \bar{\tau}_l$. Then

$$q_2 = -\sum_l \frac{\epsilon_l{}^2\bar{\tau}_l[e_b(T_1) - e_b(T_2)]_l \,\Delta\eta_l}{1 - (1-\epsilon_l)^2\bar{\tau}_l{}^2}$$

where $(e_b)_l \, \Delta\eta_l$ is the blackbody radiation in the lth band. From (17-75), $\bar{\tau}_l$ can be written as

$$\bar{\tau}_l = 1 - \bar{\alpha}_l = 1 - \frac{\bar{A}_l}{\Delta\eta_l}$$

where $\bar{A}_l$ is the integrated bandwidth that includes the integrated path-length variation for a parallel-plate geometry. The q_2 now becomes

$$q_2 = - \sum_l \frac{\epsilon_l^2 (1 - \bar{A}_l/\Delta\eta_l)[e_b(T_1) - e_b(T_2)]_l \, \Delta\eta_l}{1 - (1 - \epsilon_l)^2 (1 - \bar{A}_l/\Delta\eta_l)^2}$$

The needed quantities and results are shown in the tables that follow. Values of $\bar{A}_l$ were computed from the exponential wide-band correlations of Tables 16-2 and 16-3 using the mean beam length from Table 17-1 as the effective path length. The wave-number spans $\Delta\eta_l$ were computed from Table 17-4. For the nonabsorbing regions the $[e_b(T_1) - e_b(T_2)]_l \, \Delta\eta_l$ were computed using $F_{0-\lambda T}$ from Table A-5, that is, $[e_b(T_1)]_l \, \Delta\eta_l = F_{\lambda_1 T_1 - \lambda_2 T_1} e_{\lambda b}(T_1)$ where λ_1 and λ_2 correspond to the wave-number limits of the band. For the absorbing regions, $[e_b(T_1)]_l \, \Delta\eta_l = [e_b(T_1)]_{\text{band center}} \, \Delta\eta_l$. When the band correlations gave $\bar{A}_l > \Delta\eta_l$ then $\bar{A}_l/\Delta\eta_l = 1.0$ was used, as physically $\bar{A}_l$ cannot exceed $\Delta\eta_l$.

Band λ, μm	Band center η, cm^{-1}	C_1, cm^{-1}/(g·m^{-2})	C_2, cm^{-1}/[(g·m^{-2})]$^{1/2}$	C_3, cm^{-1}	β
15	667	19	16.3	30.4	0.691
10.4	960	0.0218	1.76	29.2	9.50
9.4	1060	0.0218	1.76	29.2	9.50
4.3	2350	110	73.0	27.1	3.49
2.7	3715	4.14	14.6	56.5	1.20

Band η, cm^{-1}	ϵ_l	$\bar{A}_l$, cm^{-1}	$\Delta\eta_l$, cm^{-1}	$[e_b(T_1) - e_b(T_2)]_l \, \Delta\eta_l$, Btu/(h·ft^2)	$-q_{l,2}$, Btu/(h·ft^2)
0–555	0.37	0	555	250	57
555–779	0.26	199	224	414	3
(15 μm)					
779–849	0.32	0	70	158	30
849–1013	0.37	9.6	164	590	117
(10.4 μm)					
1013–1141	0.46	9.6	128	480	125
(9.4 μm)					
1141–2221	0.45	0	1080	6758	1960
2221–2430	0.65	230	209	1550	0
(4.3 μm)					
2430–3573	0.61	0	1143	7389	3240
3573–3750	0.69	253	177	955	0
(2.7 μm)					
3750–∞	0.73	0	∞	7181	4110
					9642

The result for q_2 compares with a value of -9371 Btu/(h·ft^2) found for the same problem by Edwards and Nelson [15].* They use the network method of Oppenheim [17] in deriving the energy transfer equation, which gives the same result as the method used here. Partial emittances were used in place of the band correlations for computing gas properties, and these led to slightly different wavenumber spans for the bands used in [15]. Note that for this example most of the radiative transfer is in the transparent regions between the CO_2 absorption bands. Results for the exact and geometric-mean beam-length solutions are compared in [18] for hydrogen plasma between parallel plates.

17-8 RADIATION THROUGH NONISOTHERMAL GASES

Edwards and coworkers [19–21] have further extended the band and geometric-mean beam-length approaches to account for nonisothermal gases. Removing the isothermal gas restriction introduces considerable additional complication. In a nonisothermal case, the band absorption may vary strongly with position in the gas. Then, a linear absorption law may be valid in one portion of a gas, but a power law might be necessary in another portion. The Curtis-Godson technique in Sec. 17-8.1 is the basis for one engineering treatment of nonisothermal gases [19, 21–25]. Another engineering treatment, due chiefly in Hottel and coworkers [1, 2, 26], is the zoning method in Sec. 17-8.2. The methods of Chap. 15 can also be extended in an approximate way for use in simple geometries involving radiation in nonisothermal gases. In connection with this, the exchange-factor approximation is given in Sec. 17-8.3. The methods in this chapter and the Monte Carlo techniques of Chap. 20 are powerful enough to treat multidimensional problems.

17-8.1 The Curtis-Godson Approximation

An accurate and useful method for thermal radiation problems in nonuniform gases is the Curtis-Godson approximation [19, 22–25]. In this method, the transmittance of a given path through a nonisothermal gas is related to the transmittance through an equivalent isothermal gas. Then the solution can be obtained by using isothermal-gas methods. The relation between the nonisothermal and the isothermal gas is carried out by assigning an equivalent amount of isothermal absorbing material to act in place of the nonisothermal gas. The amount is based on a scaling temperature and a mean density or pressure that is obtained in the analysis. These mean quantities are found by specifying that the transmittance of the uniform gas be equal to the transmittance of the nonuniform gas in the weak and strong absorption limits.

Goody [24], Krakow et al. [22], and Simmons [25] have discussed the Curtis-Godson method for the case of attenuation in a narrow vibration-rotation band. Excellent comparisons with exact numerical results were obtained. Weiner and

---

*The results of [15] have an error in $q_{l,2}$ for the $\eta = 2430$–3573-cm^{-1} range. The comparison described here is after correction of that error.

Edwards [21] have applied the method for environments with steep temperature gradients in gases with overlapping band structures. Comparison of the analysis with experimental data was again excellent. In the following development, spectral variations will be expressed in terms of wave number, as is usual for band correlations. The Curtis-Godson technique is most useful when the gas temperature distribution is specified. If the gas temperature distribution is not known, an iterative procedure would have to be developed for its determination. This is not considered here, as the method is not too practical for that type of calculation.

For a nonuniform gas the absorption coefficient a_η is variable along the path. An effective bandwidth $\bar{A}_l(S)$ is defined, analogous to Eq. (16-70) but using an integrated absorption coefficient:

$$\bar{A}_l(S) = \int_{\substack{\text{absorption} \\ \text{bandwidth}}} \left\{ 1 - \exp\left[-\int_0^S a_\eta(\eta,S^*)\, dS^* \right] \right\} d\eta$$

$$= \Delta\eta_l - \int_l \left\{ \exp\left[-\int_0^S a_\eta(\eta,S^*)\, dS^* \right] \right\} d\eta \qquad (17\text{-}82)$$

Similarly, for a path length extending from S^* to S, the effective bandwidth is

$$\bar{A}_l(S - S^*) = \int_{\substack{\text{absorption} \\ \text{bandwidth}}} \left\{ 1 - \exp\left[-\int_{S^*}^S a_\eta(\eta,S^{**})\, dS^{**} \right] \right\} d\eta \qquad (17\text{-}83)$$

The equation of transfer will now be placed in a form utilizing $\bar{A}_l(S)$ and $\bar{A}_l(S - S^*)$.

The integrated form of the equation of transfer for intensity at S as a result of radiation traveling along a path from 0 to S is given, from Eq. (15-1), by

$$i'_\eta(\eta,S) = i'_\eta(\eta,0) \exp\left[-\int_0^S a_\eta(\eta,S^*)\, dS^* \right]$$

$$+ \int_0^S a_\eta(\eta,S^*) i'_{\eta b}(\eta,S^*) \exp\left[-\int_{S^*}^S a_\eta(\eta,S^{**})\, dS^{**} \right] dS^* \qquad (17\text{-}84)$$

Now note that

$$-\frac{\partial}{\partial S^*} \left\{ 1 - \exp\left[-\int_{S^*}^S a_\eta(\eta,S^{**})\, dS^{**} \right] \right\} = a_\eta(\eta,S^*) \exp\left[-\int_{S^*}^S a_\eta(\eta,S^{**})\, dS^{**} \right]$$

$$(17\text{-}85)$$

Insert (17-85) into (17-84) to obtain

$$i'_\eta(\eta,S) = i'_\eta(\eta,0) \exp\left[-\int_0^S a_\eta(\eta,S^*)\, dS^* \right]$$

$$- \int_0^S i'_{\eta b}(\eta,S^*) \frac{\partial}{\partial S^*} \left\{ 1 - \exp\left[-\int_{S^*}^S a_\eta(\eta,S^{**})\, dS^{**} \right] \right\} dS^* \qquad (17\text{-}86)$$

Equation (17-86) is integrated over the bandwidth $\Delta\eta_l$ of the lth band, and the order of integration changed on the last term. The $i'_\eta(\eta,S)$, $i'_\eta(\eta,0)$, and $i'_{\eta b}(\eta,S)$ are approximated by average values within the band to yield

$$i'_l(S)\,\Delta\eta_l = i'_l(0)\int_l \left\{ \exp\left[-\int_0^S a_\eta(\eta,S^*)\,dS^*\right]\right\}d\eta$$

$$-\int_0^S i'_{l,b}(S^*)\frac{\partial}{\partial S^*}\int_l\left\{1-\exp\left[-\int_{S^*}^S a_\eta(\eta,S^{**})\,dS^{**}\right]\right\}d\eta\,dS^* \qquad (17\text{-}87)$$

Equations (17-82) and (17-83) are substituted into (17-87) to obtain the equation of transfer in terms of the $\bar{A}_l$:

$$i'_l(S)\,\Delta\eta_l = i'_l(0)[\Delta\eta_l - \bar{A}_l(S)] - \int_0^S i'_{l,b}(S^*)\frac{\partial\bar{A}_l(S-S^*)}{\partial S^*}\,dS^* \qquad (17\text{-}88)$$

An alternative form can be found by integrating (17-88) by parts to obtain

$$i'_l(S)\,\Delta\eta_l = i'_l(0)[\Delta\eta_l - \bar{A}_l(S)] + i'_{l,b}(0)\bar{A}_l(S) + \int_0^S \bar{A}_l(S-S^*)\frac{di'_{l,b}(S^*)}{dS^*}\,dS^* \qquad (17\text{-}89)$$

Equations (17-88) and (17-89) are nearly exact forms of the integrated equation of transfer in terms of the band properties. The only approximation is that the intensity in each term does not vary significantly across the wave-number span of the band.

Note that, for a *uniform* gas, (17-89) gives (since $di'_{l,b}/dS = 0$)

$$i'_{l,u}(S)\,\Delta\eta_l = i'_l(0)[\Delta\eta_l - \bar{A}_{l,u}(S)] + i'_{l,b,u}\bar{A}_{l,u}(S) \qquad (17\text{-}90)$$

where the u subscript denotes a uniform gas.

To compute $i'_l(S)$ or $i'_{l,u}(S)$ from (17-88), (17-89), or (17-90), expressions are needed for the effective bandwidth $\bar{A}_l$ for nonuniform and uniform gases. From (16-74) and (16-77), the limiting cases of $\bar{A}_l$ for bands of independent weak or strong absorption lines in a uniform gas have the form

$$\bar{A}_{l,u}(S) = C_{1,l}\rho_u S_u \qquad \text{weak} \qquad (17\text{-}91a)$$

$$\bar{A}_{l,u}(S) = C_{2,l}\rho_u S_u^{1/2} \qquad \text{strong} \qquad (17\text{-}91b)$$

where $C_{1,l}$ and $C_{2,l}$ are coefficients of proportionality for the lth band, and S_c and Δ_c for the lines have been taken as proportional to gas density.

For the nonuniform gas the effective bandwidth will depend on the variation of properties along the path. The effective bandwidths are then obtained by applying (17-91a) and (17-91b) locally along the path. This gives, for a band of weak lines,

$$\bar{A}_l(S) = C_{1,l}\int_0^S \rho(S^*)\,dS^* \qquad \text{weak} \qquad (17\text{-}92a)$$

Similarly, for a band of strong lines, after first squaring (17-91b),

$$\bar{A}_l{}^2(S) = C_{2,l}{}^2 \int_0^S \rho^2(S^*)\, dS^*$$

so that

$$\bar{A}_l(S) = C_{2,l} \left[\int_0^S \rho^2(S^*)\, dS^* \right]^{1/2} \qquad \text{strong} \tag{17-92b}$$

It has been assumed that the $C_{1,l}$ and $C_{2,l}$ do not vary along the path.

In the Curtis-Godson method the *nonuniform gas is replaced by an effective amount of uniform gas such that the correct intensity is obtained at the weak and strong absorption limits.* To have the uniform intensity equal the nonuniform intensity, equate the results from (17-90) and (17-89) to obtain

$$i_l'(0)[\Delta\eta_l - \bar{A}_{l,u}(S)] + i_{l,b,u}'\bar{A}_{l,u}(S)$$

$$= i_l'(0)[\Delta\eta_l - \bar{A}_l(S)] + i_{l,b}'(0)\bar{A}_l(S) + \int_0^S \bar{A}_l(S-S^*)\frac{di_{l,b}'(S^*)}{dS^*}\, dS^*$$

which simplifies to

$$[i_{l,b,u}'(T_u) - i_l'(0)]\bar{A}_{l,u}(S)$$

$$= [i_{l,b}'(0) - i_l'(0)]\bar{A}_l(S) + \int_0^S \bar{A}_l(S-S^*)\frac{di_{l,b}'(S^*)}{dS^*}\, dS^* \tag{17-93}$$

To have (17-93) valid at the weak absorption limit, substitute $\bar{A}_{l,u}$ from (17-91a) and $\bar{A}_l$ from (17-92a) to obtain the following after canceling the $C_{1,l}$:

$$[i_{l,b,u}'(T_u) - i_l'(0)]\rho_u S_u$$

$$= [i_{l,b}'(0) - i_l'(0)] \int_0^S \rho(S^*)\, dS^* + \int_0^S \left[\int_{S^*}^S \rho(S^{**})\, dS^{**} \right] \frac{di_{l,b}'(S^*)}{dS^*}\, dS^* \tag{17-94a}$$

Similarly, at the strong absorption limit, insert (17-91b) and (17-92b) into (17-93) to obtain

$$[i_{l,b,u}'(T_u) - i_l'(0)]\rho_u S_u{}^{1/2}$$

$$= [i_{l,b}'(0) - i_l'(0)] \left[\int_0^S \rho^2(S^*)\, dS^* \right]^{1/2} + \int_0^S \left[\int_{S^*}^S \rho^2(S^{**})\, dS^{**} \right]^{1/2} \frac{di_{l,b}'(S^*)}{dS^*}\, dS^*$$

$$\tag{17-94b}$$

For a *known* distribution of temperature and density in a nonuniform gas, (17-94a) and (17-94b) can be solved simultaneously for ρ_u and S_u, which are the equivalent uniform gas density and path length for that particular band. The $i'_{\lambda,b,u}(T_u)$ is not an additional unknown since the temperature T_u corresponds to ρ_u through the perfect gas law. Then (17-90) can be used for any effective bandwidth dependency on ρ_u and S_u (that is, not only at the weak and strong limits) to solve for $i'_{\lambda,u}(S)$. This will exactly equal the intensity $i'_\lambda(S)$ in the nonuniform gas in the weak and strong limits and will usually be a good approximation for intermediate absorption values. Once the intensities are found, the heat transfer can be obtained by using the relations for a uniform gas. The evaluation of (17-94a) and (17-94b) will usually require numerical integration. Because the Curtis-Godson method requires evaluation of at least two integrals for each band along each path, it may in many cases be equally feasible to evaluate the exact equation (17-88) or (17-89) by computer.

As originally formulated (see, for example, the discussion in Goody [24]), the Curtis-Godson approximation was limited to application over a small frequency span in an absorption band. The limitation was due to line overlapping and the change in the spectral position of important lines with temperature. It has been shown, however (see, for example, Weiner and Edwards [21] and Plass [27]), that the method gives good results even when applied to situations with large temperature gradients with the use of fairly wide frequency spans. These references also account for overlapping absorption bands.

The Curtis-Godson technique appears to have application even in multi-dimensional problems, although it was originally applied to one-dimensional atmospheric problems. It should be possible to proceed as follows. For a known field of temperature and density, the boundaries are subdivided into convenient, nearly isothermal zones. Between each two zones, an equivalent uniform path length and density are found for each important band by the use of (17-94a) and (17-94b). Based on these parameters, the $\overline{A}_l$ can be obtained from one of the correlations of gas properties. The uniform gas analysis of Sec. 17-7 can then be carried through to obtain intensities and heat flows.

A band absorptance formulation analogous to the Curtis-Godson approximation but involving three parameters has been developed by Cess and Wang [28]. The additional parameter enables the equivalent isothermal gas to give the correct behavior not only in the linear and square-root limits, but also in the logarithmic limit [see (16-79)] for very strong absorption.

17-8.2 The Zoning Method

The zoning method consists of subdividing nonisothermal enclosures filled with nonisothermal gas into areas and volumes that can be considered essentially isothermal. An energy balance is then written for each division of area and volume. This leads to a set of simultaneous equations for the unknown heat fluxes or temperatures in the same manner as the procedure in Sec. 17-3 for an isothermal gas. The method is practical and powerful; Hottel and Sarofim [2] discuss it at some length. Multi-dimensional applications have been carried out by Hottel and Cohen [26] and Einstein [29, 30]. The discussion in this section is limited to radiation exchange only; extensions to include conduction and convection are found in Chap. 18 and in [2]. The zoning method has an advantage over the Curtis-Godson method because unknown temperature distributions in the gas can be treated. The Curtis-Godson technique is most useful where the temperature distribution is known; if the distribution is not known, an iteration on the gas temperature is required.

The basic concepts of the zoning method are now developed for a gas with constant absorption coefficient. Consider volume V_γ in Fig. 17-21 and surface A_k. From (13-39) the emissive power (not including induced emission) from a volume element dV_γ is $4\pi a_\lambda i'_{\lambda b}\, dV_\gamma\, d\lambda$, or per unit solid angle around dV_γ it is $a_\lambda i'_{\lambda b}\, dV_\gamma\, d\lambda$. The surface element dA_k subtends the solid angle $dA_k \cos\theta_k/S^2_{\gamma-k}$ when viewed from dV_γ. The fraction of radiation transmitted through the path length $S_{\gamma-k}$ is $\exp\left[-\int_{S_\gamma}^{S_k} a_\lambda(S^*)\, dS^*\right]$. Multiplying these factors together and integrating over V_γ and A_k gives the spectral energy arriving at A_k from gas volume V_γ as

$$dq_{\lambda i,\, \gamma-k}A_k = d\lambda \int_{V_\gamma}\int_{A_k} \frac{a_\lambda(\gamma)i'_{\lambda b}(\gamma)\cos\theta_k}{S^2_{\gamma-k}} \exp\left[-\int_{S_\gamma}^{S_k} a_\lambda(S^*)\, dS^*\right] dA_k\, dV_\gamma$$

$$(17\text{-}95)$$

If $a_\lambda(\gamma)$ is uniform, the exponential factor becomes $\exp\left[-a_\lambda(S_k - S_\gamma)\right] = \tau_\lambda(S_{\gamma-k})$. The entire gas volume is divided into finite subvolumes V_γ and the assumption is made

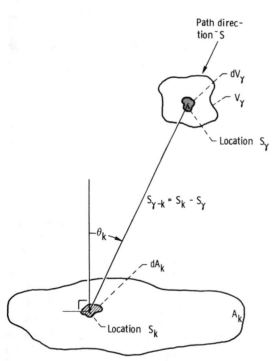

Path direction $\tilde{}S$

dV_γ

V_γ

Location S_γ

$S_{\gamma-k} = S_k - S_\gamma$

θ_k

dA_k

A_k

Location S_k

Figure 17-21 Radiation from gas volume V_γ to area A_k.

that conditions are uniform over each V_γ. Then (17-95) simplifies to

$$dq_{\lambda i, \gamma-k} A_k = d\lambda \, a_\lambda i'_{\lambda b}(\gamma) \int_{V_\gamma} \int_{A_k} \frac{\cos \theta_k}{S^2_{\gamma-k}} \tau_\lambda(S_{\gamma-k}) \, dA_k \, dV_\gamma \tag{17-96}$$

If the gas is gray, (17-96) integrated over all wavelengths gives the total energy incident on A_k as

$$q_{i, \gamma-k} A_k = a \, \frac{\sigma T_\gamma^4}{\pi} \int_{V_\gamma} \int_{A_k} \frac{\cos \theta_k}{S^2_{\gamma-k}} \tau(S_{\gamma-k}) \, dA_k \, dV_\gamma \tag{17-97}$$

Now define the *gas-surface direct exchange area* $\overline{g_\gamma s_k}$ as

$$\overline{g_\gamma s_k} \equiv \frac{a}{\pi} \int_{V_\gamma} \int_{A_k} \frac{\cos \theta_k}{S^2_{\gamma-k}} \tau(S_{\gamma-k}) \, dA_k \, dV_\gamma \tag{17-98}$$

Equation (17-97) can then be written as

$$q_{i, \gamma-k} A_k = \overline{g_\gamma s_k} \, \sigma T_\gamma^4 \tag{17-99}$$

Thus the energy $q_{i, \gamma-k} A_k$ arriving at A_k can be regarded as the blackbody emissive power σT_γ^4 of the gas in V_γ that is radiated from an *effective* area $\overline{g_\gamma s_k}$. Let the entire gas volume be divided into Γ finite regions. The energy flux incident upon A_k from all the gas-volume regions is then

$$(q_{i,k})_{\text{from gas}} = \frac{1}{A_k} \sum_{\gamma=1}^{\Gamma} \overline{g_\gamma s_k} \, \sigma T_\gamma^4 \tag{17-100}$$

Now consider the interchange between the bounding areas of the enclosure. The energy leaving surface area A_j and reaching A_k is, for a nonisothermal gas with uniform gray properties,

$$q_{i,j-k} A_k = \frac{q_{o,j}}{\pi} \int_{A_k} \int_{A_j} \tau(S_{j-k}) \frac{\cos \theta_j \cos \theta_k \, dA_j \, dA_k}{S^2_{j-k}} \tag{17-101}$$

where, as in the usual enclosure theory, $q_{o,j}$ is assumed to be uniform over A_j. Define the *surface-surface direct exchange area* as

$$\overline{s_j s_k} \equiv \int_{A_k} \int_{A_j} \tau(S_{j-k}) \frac{\cos \theta_j \cos \theta_k \, dA_j \, dA_k}{\pi S^2_{j-k}} \tag{17-102}$$

Equation (17-101) can then be written as

$$q_{i,j-k} A_k = \overline{s_j s_k} q_{o,j} \tag{17-103}$$

Thus the energy $q_{i,j-k} A_k$ from A_j arriving at A_k is the energy flux $q_{o,j}$ leaving A_j times an *effective* area $\overline{s_j s_k}$. The energy flux incident upon A_k as a result of fluxes leaving all N surfaces of the enclosure is then

$$(q_{i,k})_{\text{from surfaces}} = \frac{1}{A_k} \sum_{j=1}^{N} \overline{s_j s_k} q_{o,j} \tag{17-104}$$

Now the total energy flux incident upon surface A_k can be obtained as

$$q_{i,k} = (q_{i,k})_{\text{from surfaces}} + (q_{i,k})_{\text{from gas}} = \frac{1}{A_k} \left(\sum_{j=1}^{N} \overline{s_j s_k} q_{o,j} + \sum_{\gamma=1}^{\Gamma} \overline{g_\gamma s_k} \sigma T_\gamma^{\,4} \right) \quad \text{17-105)}$$

The usual net-radiation equations [(8-1) and (8-2)] also apply at surface A_k:

$$q_k = q_{o,k} - q_{i,k} \tag{17-106}$$

$$q_{o,k} = \epsilon_k \sigma T_k^{\,4} + (1 - \epsilon_k) q_{i,k} \tag{17-107}$$

For problems in which T_γ is given for all gas volume elements V_γ, (17-105)–(17-107) are sufficient to solve for N unknown values of T_k and q_k; the other N values of T_k and q_k must be provided as known boundary conditions. The methods of Sec. 8-3 can be directly applied.

When the T_γ of the Γ gas elements are unknowns, then Γ additional equations must be found. These are obtained by taking an energy balance on each gas zone. In radiative equilibrium, for each gas element V_γ the emission and absorption of energy are equal (no heat sources or sinks in gas). Then for a gray gas with constant properties a heat balance on the volume region V_γ gives

$$4a\sigma T_\gamma^{\,4} V_\gamma = \sum_{\text{all } V_{\gamma^*}} \int_{V_\gamma} \int_{V_{\gamma^*}} \frac{4a\sigma T_{\gamma^*}^{\,4}\, dV_{\gamma^*}}{4\pi}\, \tau(S_{\gamma^*-\gamma}) \frac{a\, dV_\gamma}{S_{\gamma^*-\gamma}^2}$$

$$+ \sum_{\text{all } A_k} \int_{V_\gamma} \int_{A_k} \frac{q_{o,k} \cos \theta_k}{\pi}\, dA_k\, \tau(S_{k-\gamma}) \frac{a\, dV_\gamma}{S_{k-\gamma}^2}$$

$$= a^2 \sum_{\gamma^*=1}^{\Gamma} \sigma T_{\gamma^*}^{\,4} \int_{V_\gamma} \int_{V_{\gamma^*}} \frac{\tau(S_{\gamma^*-\gamma})\, dV_{\gamma^*}\, dV_\gamma}{\pi S_{\gamma^*-\gamma}^2}$$

$$+ a \sum_{k=1}^{N} q_{o,k} \int_{V_\gamma} \int_{A_k} \frac{\cos \theta_k}{\pi S_{k-\gamma}^2}\, \tau(S_{k-\gamma})\, dA_k\, dV_\gamma \tag{17-108}$$

It is assumed that a is uniform throughout the enclosure, and that each V_γ is isothermal. As is usual in enclosure calculation methods, $q_{o,k}$ is taken as constant across A_k. Define the *surface-gas direct exchange area* as

$$\overline{s_k g_\gamma} \equiv \frac{a}{\pi} \int_{V_\gamma} \int_{A_k} \frac{\cos \theta_k}{S_{k-\gamma}^2}\, \tau(S_{k-\gamma})\, dA_k\, dV_\gamma \tag{17-109}$$

Comparing (17-109) with (17-98) shows that there is reciprocity between the surface-gas and gas-surface direct exchange areas:

$$\overline{s_k g_\gamma} = \overline{g_\gamma s_k} \tag{17-110}$$

Define the *gas-gas direct exchange area* as

$$\overline{g_{\gamma^*} g_\gamma} \equiv \frac{a^2}{\pi} \int_{V_\gamma} \int_{V_{\gamma^*}} \frac{\tau(S_{\gamma^*-\gamma})\, dV_{\gamma^*}\, dV_\gamma}{S_{\gamma^*-\gamma}^2} \tag{17-111}$$

Substituting (17-109) to (17-111) into (17-108) gives

$$4a\sigma T_\gamma^4 V_\gamma = \sum_{\gamma^*=1}^{\Gamma} \sigma T_{\gamma^*}^4 \overline{g_{\gamma^*}g_\gamma} + \sum_{k=1}^{N} q_{o,k} \overline{g_\gamma s_k} \qquad (17\text{-}112)$$

The $\overline{g_{\gamma^*}g_\gamma}$ have also been tabulated [26] so that (17-112), written for each V_γ, provides the additional set of Γ equations required to obtain the gas temperature distribution.

The notation developed by Hottel and coworkers has been used in this section with only slight modification. A comparison with Sec. 17-3 shows that, in terms of the notation used there [Eq. (17-9)], the following identity exists:

$$F_{j-k}\overline{\tau}_{j-k}A_j = \overline{s_j s_k} \qquad (17\text{-}113)$$

The gas absorptance factor in (17-10), $F_{j-k}\overline{\alpha}_{j-k}A_j$, is generally not related in a useful way to $\overline{g_\gamma s_k}$. The latter quantity is derived for an element of gas volume, while $F_{j-k}\overline{\alpha}_{j-k}A_j$ is concerned with the entire gas volume.

Hottel and coworkers [1, 2, 26] have further developed the approach outlined in this section. It is possible to allow for spectral variations in gas properties, approximately but easily. Variations in properties with position in the enclosure are handled by defining a suitable mean absorption coefficient between each set of zones. Einstein [29, 30] modified the $\overline{gs}$ and $\overline{gg}$ factors to give better accuracy when strong gradients are present. All these approximations become difficult to carry through if the absorption coefficient is a strong function of temperature. Noble [31] shows how matrix theory can be used as a computational aid in the zone method using exchange areas.

Values of $\overline{s_j s_k}$ and $\overline{g_\gamma s_k}$ have been tabulated for cubical isothermal volumes and square isothermal boundary elements by Hottel and Cohen [26]. Exchange areas are available for elements in rectangular [5], cylindrical [5, 30], and conical [32] enclosures. Hottel and Sarofim [2] present a table of references to factors for several other geometries, and an extensive tabulation of factors for the cylindrical geometry. If the gas in the enclosure is nonhomogeneous in temperature and/or composition, then the exchange areas must account for variations in absorption coefficient in the gas between any two elements. In [33], exchange areas are derived for cubes in nonhomogeneous media. Values are presented for narrow bands with overlapped lines and with nonoverlapped lines for the cases of adjacent and remote pairs of cubes, assuming a linear temperature profile between the cubes.

17-8.3 The Exchange-Factor Approximation

In this section we consider regions of *nonisothermal* gray gas between parallel plates, concentric cylinders, and concentric spheres. The purpose here is to show how the engineering concepts of exchange factors can be applied to extend the results for black bounding surfaces to surfaces that are diffuse-gray. The approach is that of Perlmutter and Howell [34], who have shown that once results are available for an analysis with black boundaries, the diffuse-gray-wall results can be obtained from simple algebraic relations. This was also demonstrated in Sec. 14-7.4 for a gas layer between parallel plates.

The theory follows the same general development as the net-radiation method (Sec. 8-3.1). A heat balance at surface A_k gives

$$Q_k = q_k A_k = (q_{o,k} - q_{i,k})A_k \qquad (17\text{-}114)$$

The energy flux leaving A_k is composed of emitted and reflected energy:

$$q_{o,k} = \epsilon_k \sigma T_k^4 + (1 - \epsilon_k)q_{i,k} \qquad (17\text{-}115)$$

If $q_{i,k}$ is eliminated from (17-114) and (17-115), the result is

$$Q_k = A_k \frac{\epsilon_k}{1 - \epsilon_k} (\sigma T_k^4 - q_{o,k}) \qquad (17\text{-}116)$$

The $q_{i,k}$ in (17-115) can be found in terms of Hottel's exchange areas as in (17-105). However, here we choose to define a different quantity, called the *exchange factor* $\overline{F}_{j-k}$, as *the fraction of the energy leaving surface j that is incident on surface k when all boundaries are black and the intervening medium is in radiative equlibrium* (transfer in the gas is only by radiation without any heat sources or sinks); When the gas is transparent, $\overline{F}_{j-k}$ becomes the configuration factor F_{j-k} (Sec. 7-4.3). Because the gas is in radiative equlibrium, energy conservation requires that energy leaving surface 1 must finally reach other enclosure surfaces or return to surface 1. Any energy absorbed in the gas must be reemitted by the gas to maintain equilibrium, and the $\overline{F}_{j-k}$ includes all interactions with the gas by means of which energy leaving A_j arrives at A_k.

For a general enclosure of N surfaces surrounding a gas in radiative equilibrium, the incident energy on surface k can be written in terms of exchange factors as

$$Q_{i,k} = \sum_{j=1}^{N} Q_{o,j}\overline{F}_{j-k} \qquad (17\text{-}117)$$

Note that using exchange areas as in Sec. 17-8.2 would require an additional term in the $Q_{i,k}$ to account for energy emitted by the gas and reaching the wall. This energy is included by definition within the $\overline{F}_{j-k}$.

Substituting (17-117) into (17-115) to eliminate $q_{i,k}$ gives

$$Q_{o,k} = q_{o,k} A_k = \epsilon_k \sigma T_k^4 A_k + (1 - \epsilon_k) \sum_{j=1}^{N} Q_{o,j}\overline{F}_{j-k} \qquad (17\text{-}118)$$

Because the medium is in radiative equilibrium, all energy leaving a given surface must finally reach an enclosure surface. Thus,

$$\sum_{k=1}^{N} \overline{F}_{j-k} = 1 \qquad (17\text{-}119)$$

Note that no energy is being supplied to the gas by any external means such as combustion.

Because $\overline{F}_{j-k}$ is defined as the fraction of energy leaving A_j that arrives at A_k for *black* boundaries enclosing a gas, it can be obtained from the black-walled solution, which we assume has already been found. Thus,

$$\bar{F}_{j-k} = \left(\frac{Q_{i,k(\text{from } j)}}{A_j \sigma T_j^4}\right)_{\text{black surfaces}} \equiv \psi_{j-k,b} \tag{17-120}$$

where the notation $\psi_{j-k,b}$ is used to emphasize that this is a quantity obtained from the black solution.

For an enclosure of two surfaces, it is possible to obtain closed-form solutions. The procedure will now be outlined for infinite parallel plates. With the gas absorption coefficient assumed constant, the fraction $\bar{F}_{1-2}$ of energy leaving surface 1 that reaches surface 2 must equal the fraction $\bar{F}_{2-1}$ from surface 2 to 1. This arises from the symmetry in radiation paths for energy leaving either surface. In radiative equilibrium with no heat sources or sinks in the gas, the radiation absorbed at a position must be reemitted at that position. The radiation leaving either plate will undergo the same absorption-emission history while traveling to the other plate.

The $\bar{F}_{1-2}$ is found from the black solution (17-120) as

$$\bar{F}_{1-2} = \frac{Q_{i,2}}{A_1 \sigma T_1^4} = \left(\frac{\sigma T_2^4 - q_2}{\sigma T_1^4}\right)_{\text{black surfaces}} = \psi_{1-2,b}$$

where $A_1 = A_2$ has been used. Then $\bar{F}_{2-1} = \bar{F}_{1-2}$, and for simplicity call them ψ_b. From Eq. (17-119), $\bar{F}_{1-1} = \bar{F}_{2-2} = 1 - \psi_b$, and (17-118) becomes

$$q_{o,1} = \epsilon_1 \sigma T_1^4 + (1 - \epsilon_1)(q_{o,1} - q_{o,1}\psi_b + q_{o,2}\psi_b)$$

$$q_{o,2} = \epsilon_2 \sigma T_2^4 + (1 - \epsilon_2)(q_{o,1}\psi_b + q_{o,2} - q_{o,2}\psi_b)$$

Solving simultaneously for $q_{o,1}$ and $q_{o,2}$ yields the symmetric relations

$$q_{o,1} = \frac{\epsilon_1 \epsilon_2 \sigma T_1^4 + \epsilon_1(1 - \epsilon_2)\psi_b \sigma T_1^4 + \epsilon_2(1 - \epsilon_1)\psi_b \sigma T_2^4}{\psi_b(\epsilon_1 + \epsilon_2 - 2\epsilon_1\epsilon_2) + \epsilon_1\epsilon_2} \tag{17-121a}$$

$$q_{o,2} = \frac{\epsilon_1 \epsilon_2 \sigma T_2^4 + \epsilon_2(1 - \epsilon_1)\psi_b \sigma T_2^4 + \epsilon_1(1 - \epsilon_2)\psi_b \sigma T_1^4}{\psi_b(\epsilon_1 + \epsilon_2 - 2\epsilon_1\epsilon_2) + \epsilon_1\epsilon_2} \tag{17-121b}$$

The $q_{o,1}$ is substituted into (17-116) to yield, after rearrangement,

$$\frac{q_1}{\sigma(T_1^4 - T_2^4)} = \frac{\psi_b}{\psi_b(1/\epsilon_1 + 1/\epsilon_2 - 2) + 1} \tag{17-122a}$$

which can be written in the alternative form

$$\frac{q_1}{\sigma(T_1^4 - T_2^4)} = \frac{\psi_b}{(E_2 + E_1)\psi_b + 1} \tag{17-122b}$$

where

$$E_1 = \frac{1 - \epsilon_1}{\epsilon_1} \qquad E_2 = \frac{1 - \epsilon_2}{\epsilon_2}$$

Evaluating (17-122b) for the black case, $E_1 = E_2 = 0$, shows that

$$\psi_b = \frac{q_{1b}}{\sigma(T_1^4 - T_2^4)} \tag{17-122c}$$

Equation (17-122b) gives the energy supplied to surface 1 and removed from surface 2. Since the gas absorption coefficient is independent of temperature, energy leaving a location on the boundary and reaching a given point in the gas will be attenuated by the same amount *regardless* of the gas temperature distribution. Further, any portion of the energy absorbed along the path is balanced by isotropic emission at each location. With these facts, a synthesis of black-wall enclosure solutions and gas-element exchange factors can be used to find the temperature distribution in the gas for diffuse-gray walls.

The energy emitted by a local volume element of gas of area A and thickness dx between the parallel plates (Fig. 17-22) is given by $Q_e = 4a\sigma T^4(x)A\,dx$. For radiative equilibrium, this must equal the heat absorbed by the volume element, which can be written as (per unit area)

$$4a\sigma T^4(x)\,dx = \sum_{j=1}^{N} q_{o,j}\,d\bar{F}_{j-dx} = q_{o,1}\,d\bar{F}_{1-dx} + q_{o,2}\,d\bar{F}_{2-dx} \qquad (17\text{-}123a)$$

or

$$\phi = \frac{T^4(x) - T_2^4}{T_1^4 - T_2^4} = \frac{1}{4a\sigma(T_1^4 - T_2^4)}\left(\frac{q_{o,1}d\bar{F}_{1-dx}}{dx} + \frac{q_{o,2}d\bar{F}_{2-dx}}{dx}\right) - \frac{T_2^4}{T_1^4 - T_2^4}$$

$$(17\text{-}123b)$$

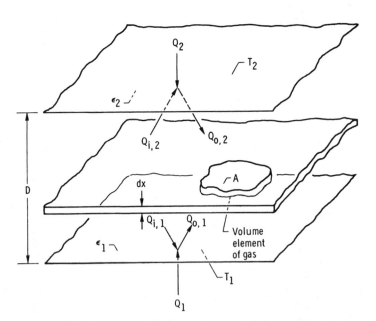

Figure 17-22 Energy quantities for gas between infinite parallel gray plates.

Here $d\bar{F}_{j-dx}$ is the fraction of energy flux leaving boundary surface A_j that is *absorbed* in volume element dx when the boundaries are black. Again, the $\bar{F}$ factors include the effect of energy absorption and reemission by the gas as the energy travels from the surface to the volume element. Because radiative equilibrium is the condition being studied here, no energy is lost during these processes since all absorbed energy at a location must be reemitted.

To determine the $d\bar{F}_{j-dx}$, consider the case in which the entire system is isothermal. Equation (17-123a) then reduces to

$$4a\,dx = d\bar{F}_{1-dx} + d\bar{F}_{2-dx} \tag{17-124}$$

This relation is used to eliminate $d\bar{F}_{2-dx}$ from (17-123b) written for black surfaces. The resulting equation is solved for $d\bar{F}_{1-dx}$, giving

$$d\bar{F}_{1-dx} = 4a\,dx\,\phi_b \tag{17-125}$$

where, from (17-123b),

$$\phi_b = \frac{T_1{}^4 d\bar{F}_{1-dx} + T_2{}^4 d\bar{F}_{2-dx}}{4a\,dx(T_1{}^4 - T_2{}^4)} - \frac{T_2{}^4}{T_1{}^4 - T_2{}^4}$$

Then substituting (17-125) into (17-124) gives

$$d\bar{F}_{2-dx} = 4a\,dx(1 - \phi_b) \tag{17-126}$$

Substituting (17-125), (17-126), (17-121a), and (17-121b) to eliminate $d\bar{F}_{1-dx}$, $d\bar{F}_{2-dx}$, $q_{o,1}$, and $q_{o,2}$ from (17-123b) results, after much manipulation, in the gas temperature distribution

$$\phi = \frac{T^4(x) - T_2{}^4}{T_1{}^4 - T_2{}^4} = \frac{\phi_b + E_2\psi_b}{1 + \psi_b(E_1 + E_2)} \tag{17-127}$$

Equations (17-122) and (17-127) relate the energy transfer and fourth-power temperature distribution for the case of a gray gas between gray walls to the case of a gray gas between black walls for the geometry of infinite parallel plates. Similar relations for the geometry of infinitely long concentric cylinders are given in [34, 35], and these plus the relations for concentric spheres (that can be used with the results of [36–38]) are given in Table 17-5. Reference [34] also includes exchange-factor relations when there are heat sources in the gas.

EXAMPLE 17-5 A gray gas with absorption coefficient 0.5 cm^{-1} is contained between gray parallel plates 2 cm apart. The plate temperatures and emissivities are $T_1 = 1000$ K, $T_2 = 840$ K, $\epsilon_1 = 0.1$, and $\epsilon_2 = 0.2$. What are the energy transfer

Table 17-5 Relations between gray and black-wall solutions for radiation between surfaces enclosing a gray gas in radiative equilibrium

Geometry	Relation[a]
Infinite parallel plates (ψ_b and ϕ_b are given in [8] of Chap. 14 and Fig. 14-8a and b)	$$\psi = \frac{\psi_b}{(E_2 + E_1)\psi_b + 1}$$ $$\phi(x) = \frac{\phi_b(x) + E_2\psi_b}{(E_2 + E_1)\psi_b + 1}$$
Infinitely long concentric cylinders (ψ_b and ϕ_b are given in [34, 35])	$$\psi = \frac{\psi_b}{[(D_1/D_2)E_2 + E_1]\psi_b + 1}$$ $$\phi(r) = \frac{\phi_b(r) + E_2(D_1/D_2)\psi_b}{[(D_1/D_2)E_2 + E_1]\psi_b + 1}$$
Concentric spheres (ψ_b and ϕ_b are given in [36–38])	$$\psi = \frac{\psi_b}{[(D_1/D_2)^2 E_2 + E_1]\psi_b + 1}$$ $$\phi(r) = \frac{\phi_b(r) + E_2(D_1/D_2)^2\psi_b}{[(D_1/D_2)^2 E_2 + E_1]\psi_b + 1}$$

[a] Definitions: $E_N = (1 - \epsilon_{wN})/\epsilon_{wN}$, $\quad \psi = Q_1/A_1\sigma(T_{w1}^4 - T_{w2}^4)$, $\psi_b = Q_{1b}/A_1\sigma(T_{w1}^4 - T_{w2}^4)$, $\quad \phi(\xi) = [T^4(\xi) - T_{w2}^4]/(T_{w1}^4 - T_{w2}^4)$.

between the plates and the temperature of the gas at a point 0.5 cm from surface 1? For black walls, Fig. 14-8b gives for $aD = 0.5 \times 2 = 1.0$, $q_{1b}/\sigma(T_1^4 - T_2^4) = 0.56$ so that, from (17-122c), $\psi_b = 0.56$. From Table 17-5,

$$\psi = \frac{q_1}{\sigma(T_1^4 - T_2^4)} = \frac{\psi_b}{(E_2 + E_1)\psi_b + 1}$$

and, for this example, $E_1 = (1 - 0.1)/0.1 = 9$ and $E_2 = (1 - 0.2)/0.2 = 4$ so that

$$q_1 = \sigma(T_1^4 - T_2^4)\frac{0.56}{13 \times 0.56 + 1}$$

$$= 5.73 \times 10^{-12}(1 - 0.5) \times 10^{12} \times \frac{0.56}{8.28} = 0.194\,\text{W/cm}^2$$

The temperature at $x = 0.5$ cm can be calculated from the result in Table 17-5:

$$\phi(x) = \frac{T^4(x) - T_2^4}{T_1^4 - T_2^4} = \frac{\phi_b(x) + E_2\psi_b}{(E_2 + E_1)\psi_b + 1}$$

From Fig. 14-8a, at the abscissa $\kappa/\kappa_D = ax/aD = 0.25$ and on the curve for $\kappa_D = aD = 1$, $\phi_b = 0.62$, so that

$$\phi|_{x=0.5} = \frac{0.62 + 4 \times 0.56}{13 \times 0.56 + 1} = 0.345$$

Then

$$T^4(0.5) = T_2^4 + 0.345(T_1^4 - T_2^4)$$

$$= [0.5 + 0.345(1 - 0.5)] \times 10^{12}$$

$$= 0.673 \times 10^{12}$$

which gives

$$T(0.5) = 906\ \text{K}$$

Note that for gray walls a curve of ϕ against ax/aD will only have an antisymmetrical shape about $ax/aD = 0.5$ as in Fig. 14-8a when $\epsilon_1 = \epsilon_2$.

17-9 FLAMES, LUMINOUS FLAMES, AND PARTICLE RADIATION

Under certain conditions, gases emit much more radiation in the visible region of the spectrum than would be expected from the absorption coefficients discussed to this point. For example, the typical almost transparent blue flame of a bunsen burner can be made into a smokey yellow-orange flame by changing only the fuel-air ratio. Such luminous emission is usually ascribed to hot carbon (soot) particles that are formed because of incomplete combustion in hydrocarbon flames. There is room for argument even here. Echigo, Nishiwaki, and Hirata [39] and others have advanced the hypothesis, supported by some experimental facts, that the luminous emission from some flames is due to the emission from vibration-rotation bands of chemical species that appear during the combustion process *prior* to the formation of soot particles. However, since soot formation is the most widely accepted view, soot radiation will be emphasized in this discussion of luminous flames.

Combustion is very complicated, often consisting of chemical reactions in series and parallel involving a variety of intermediate species. The composition and concentration of these species cannot be predicted very well unless complete knowledge is available of the reaction kinetics of the flame; this knowledge will not usually be at hand. Because the radiation properties of the flame depend on the distributions of temperature and species within the flame, a detailed prediction of radiation from flames is not often possible from knowledge only of the original combustible constituents and the flame geometry. Because of these difficulties, it is usually necessary to resort to empirical methods for predicting radiation from systems involving combustion.

To facilitate the present discussion, let us separately examine two facets: (1) The calculation of a theoretical flame temperature by considering the chemical energy release and without accounting for heat loss by radiation, and (2) the more complex problem of radiation from a gas containing solid particles that will alter the theoretical flame temperature.

17-9.1 Theoretical Flame Temperature

To present empirical correlations of radiation from flames, a characteristic parameter is the average temperature of a well-mixed flame as a result of the chemical energy addition. Well-developed methods exist [40–42] for computing the theoretical flame temperature of a given combustion system from available thermodynamic data. The effect of preheating either the fuel or oxidizer or both can be included. Complete combustion and no heat losses are assumed. The computation of flame temperature is conveniently shown by an example.

> **EXAMPLE 17-6** Using the mean heat-capacity data of Fig. 17-23 (adapted from [40]) and the heat of combustion from Table 17-6, calculate the theoretical temperature of an ethane flame burning with 100% excess air (by volume). The ethane is supplied at room temperature (25°C), and the air feed is preheated to 500°C. The flame is burning in an environment at a pressure of 1 atm.
>
> The theoretical flame temperature T is computed using energy conservation with no heat losses. The energy in the combustion constituents plus the energy of combustion is equated to the energy of the combustion products. This gives
>
> $$T - T_{\text{ref}} = \frac{\left(\begin{array}{c}\text{energy in feed air and} \\ \text{fuel above } T_{\text{ref}}\end{array}\right) + \left(\begin{array}{c}\text{energy released} \\ \text{by combustion}\end{array}\right)}{\left(\begin{array}{c}\text{total mass of} \\ \text{products}\end{array}\right) \times \left(\begin{array}{c}\text{mean heat capacity} \\ \text{of products}\end{array}\right)} \qquad (17\text{-}128)$$
>
> For ethane, assuming complete combustion, the reaction is
>
> $$2C_2H_6 + 7O_2 \rightarrow 4CO_2 + 6H_2O$$
>
> Assume that 2 kg-mol of ethane is burned; then 7 mol of oxygen is consumed. Since there is 100% excess air, only one-half the feed air contributes oxygen to the combustion process, so 14 mol of oxygen is introduced in the feed air.

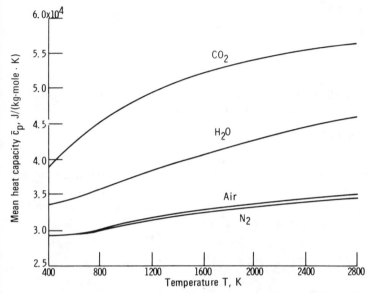

Figure 17-23 Mean heat capacity of various gases averaged between T and 298 K.

Oxygen makes up 21% by volume of the feed air, and since mole fraction is equal to part by volume, the moles of air are $m_{air} = 14/0.21 = 66.67$ mol. The sensible heat in the feed components above a reference temperature is $H_i = \Sigma_k \ [m\bar{c}_p(T_i - T_{ref})]_k$ where $\bar{c}_p$ is the mean heat capacity between the reference temperature and the input temperature T_i. Using Fig. 17-23, which has a reference temperature

Table 17-6 Heat of combustion and flame temperature for hydrocarbon fuels [42, 43]

Fuel	Heat of combustion J/kg	Maximum flame temperature, K (combustion with dry air at 298 K)		
		Theoretical (complete combustion)	Theoretical (with dissociation and ionization)	Experimental
Carbon monoxide (CO)	4.83×10^7	2615		
Hydrogen (H_2)	12.0×10^7	2490		
Methane (CH_4)	5.0×10^7	2285	2191	2158
Ethane (C_2H_6)	4.74×10^7	2338	2222	2173
Propane (C_3H_8)	4.64×10^7	2629	2240	2203
n-Butane (C_4H_{10})	4.56×10^7	2357	2246	2178
n-Pentane (C_5H_{12})	4.53×10^7	2360		
Ethylene (C_2H_4)	4.72×10^7	2523	2345	2253
Propylene (C_3H_6)	4.57×10^7	2453	2323	2213
Butylene (C_4H_8)	4.53×10^7	2431	2306	2208
Amylene (C_5H_{10})	4.50×10^7	2477		
Acetylene (C_2H_2)	4.82×10^7	2859		
Benzene (C_6H_6)	4.06×10^7	2484		
Toluene ($C_6H_5CH_3$)	4.09×10^7	2460		

of 298 K (25°C), gives

$$H_i = [m\bar{c}_p(T_i - T_{ref})]_{ethane} + [m\bar{c}_p(T_i - T_{ref})]_{air}$$

$$= 0 + 66.67 \times 2.93 \times 10^4(773 - 298) = 9.28 \times 10^8 \, J$$

where the ethane contributes nothing, as it is supplied at T_{ref}. The heat ΔH released by combustion is found by using the heat of combustion from Table 17-6 and the fact that the molecular weight of ethane is 30:

$$\Delta H = 2 \text{ mol} \times 30 \text{ kg/mol} \times 4.74 \times 10^7 \, J/kg = 28.4 \times 10^8 \, J$$

The numerator of (17-128) is then $H_i + \Delta H = 37.7 \times 10^8 \, J$.

Nitrogen, which makes up about 79% by volume of air, remains from that portion of the feed air that supplied the oxygen for combustion. This amount of nitrogen in the combustion products is then $33.3 \times 0.79 = 26.3$ mol. The total quantity of products after combustion is (in moles):

Product	Quantity, mol
Carbon dioxide (CO_2)	4
Water vapor (H_2O)	6
Air	33.3 (half of feed air)
Nitrogen (N_2)	26.3

To find the denominator of (17-128), the individual quantities are summed: $H_{prod} = \Sigma_j \, m_j \bar{c}_{p,j}$. However, $\bar{c}_p$ depends on the product temperature T, which is the flame temperature and is not yet known. We must estimate T to determine the $\bar{c}_{p,j}$ values and then substitute the quantities into (17-128). If the calculated flame temperature agrees with the assumed value, the solution is finished. Otherwise, a new temperature is estimated, and H_{prod} is recalculated and substituted into (17-128); this procedure is continued until the assumed and calculated temperatures agree. A table of calculations is as follows:

Assumed flame temperature, K	Mean heat capacity $\bar{c}_p$, J/(kg-mol·K)				Product enthalpy H_{prod}, J	Calculated temperature, K
	H_2O	CO_2	Air	N_2		
2400	4.44×10^4	5.51×10^4	3.35×10^4	3.30×10^4	2.47×10^6	1825
2000	4.25×10^4	5.36×10^4	3.33×10^4	3.28×10^4	2.44×10^6	1845

Because a 400 K change in the assumed flame temperature produced only about a 20 K change in the calculated flame temperature, a value of 1853 K is estimated as within a few degrees of the converged result.

In this example, it was assumed there was complete combustion and no dissociation of combustion products occurs. No consideration was given to energy loss by radiation, which would lower the flame temperature. Methods for including these effects are discussed in [42]. A list of theoretical flame temperatures (no radiation included) is shown in Table 17-6 for various hydrocarbon flames. Results for complete combustion with dry air are shown, followed by calculated results modified to allow for product dissociation and ionization. The latter are compared with experimental results. In addition, the heats of combustion of the substances are shown. All data are from [42, 43]. Extensive tabulations of similar data for over 200 hydrocarbons are given in [40, 43].

Let us proceed to a consideration of the radiation emitted by a nonluminous flame now that its average temperature is known.

17-9.2 Radiation from Nonluminous Flames

The phenomena involved in radiation from the nonluminous portion of the combustion products are fairly well understood. The complexities of the chemical reaction are not too important here, since it is the gaseous end products situated above the active burning region that are being considered. In most instances hydrocarbon combustion is being considered, and the radiation is from the CO_2 and H_2O bands in the infrared. For flames a few or more feet thick, as in commercial furnaces, the emission leaving the flame within the CO_2 and H_2O vibration-rotation bands can be close to blackbody emission. The gaseous radiation properties and methods in this Chapter can be used to compute the radiative heat transfer. The analysis is greatly simplified if the gas is well mixed so that it can be assumed isothermal. For a nonisothermal condition the gas can be divided into approximately isothermal zones, and convection can be included if the circulation pattern in the combustion chamber is known. A nonisothermal analysis with convection was carried out in [44] for cylindrical flames. In [45–49] radiation from nonluminous flames is treated for various types of flames (laminar or turbulent, mixed or diffusion). The flame shape for an open diffusion flame is considered in [50]. An example of a nonluminous flame radiation computation will now be given.

EXAMPLE 17-7 In Example 17-6 the combustion products were 4 mol of CO_2, 6 mol of H_2O vapor, 33.3 mol of air, and 26.3 mol of N_2. Assume these products are in a cylindrical region 4 ft high and 2 ft in diameter and are uniformly mixed at the theoretical flame temperature 1853 K. The pressure is 1 atm. Compute the radiation from the gaseous region.

The partial pressure of each constituent is equal to its mole fraction:

$$p_{CO_2} = \left(\frac{4}{69.6}\right)(1 \text{ atm}) = 0.0575 \text{ atm}$$

$$p_{H_2O} = \left(\frac{6}{69.6}\right)(1 \text{ atm}) = 0.0862 \text{ atm}$$

The gas mean beam length for negligible self-absorption is, from (17-51),

$$L_{e,0} = \frac{4V}{A} = \frac{4[\pi(2^2/4)]4}{(2\pi \times 4) + 2\pi(2^2/4)} = \frac{16\pi}{10\pi} = 1.6 \text{ ft}$$

To include self-absorption, a correction factor of 0.9 is applied to give $L_e = 0.9(1.6) = 1.44$ ft. Then

$$p_{CO_2}L_e = 0.0575 \times 1.44 = 0.0828 \text{ atm ft}$$

$$p_{H_2O}L_e = 0.0862 \times 1.44 = 0.124 \text{ atm ft}$$

Using Figs. 17-11–17-15 at the flame temperature ($3340°R$) results in $\epsilon_{CO_2} = 0.039$ and $\epsilon_{H_2O} = 0.029 \times 1.08 = 0.031$. The 1.08 factor in ϵ_{H_2O} is a correction for the partial pressure of the water vapor not being zero. In addition, there is a negative correction resulting from spectral overlap of the CO_2 and H_2O radiation bands. This is obtained from Fig. 17-15 at the values of the parameters

$$\frac{p_{H_2O}}{p_{CO_2} + p_{H_2O}} = \frac{0.0862}{0.0575 + 0.0862} = 0.60$$

$$p_{CO_2}L_e + p_{H_2O}L_e = 0.0828 + 0.124 = 0.207 \text{ atm ft}$$

The correction is $\Delta\epsilon = 0.002$. Then the gas emittance is

$$\epsilon_g = \epsilon_{CO_2} + \epsilon_{H_2O} - \Delta\epsilon = 0.039 + 0.031 - 0.002 = 0.068$$

The radiation from the gas region at the theoretical flame temperature is then

$$Q = \epsilon_g A\sigma T_g^4 = 0.068(10\pi)0.173 \times 10^{-8}(3340)^4 = 0.46 \times 10^6 \text{ Btu/h}$$

17-9.3 Radiation from and through Luminous Flames and Gases with Particles

Several factors complicate the radiative transfer in the region of the flame that is actively burning. The simultaneous production and loss of energy produces a temperature variation and thus a variation of properties and emission within the flame. The intermediate combustion products resulting from the complex reaction chemistry can significantly alter the radiation characteristics from those of the final products. Soot is the most important radiating product formed in burning hydrocarbons. Soot emits in a continuous spectrum in the visible and infrared regions and can often double or triple the heat radiated by the gaseous products alone. A method for increasing the flame emission, if desired, is to promote slow initial mixing of the oxygen with the fuel so that large amounts of soot will form at the base of the flame. Ash particles in the combustion gases can also contribute to the radiation [51].

Determining the effect of soot on flame radiation resolves into two requirements. One of these is to somehow obtain the soot distribution in the flame; this is the biggest obstacle to the calculations. The distribution depends on the type of fuel, the mixing of fuel and oxidant, and the flame temperature. The soot distribution is too complicated to calculate from basic principles, so some experimental knowledge of a given combustion system is needed. The second requirement is to know the radiative properties of the soot. Then if the soot concentration and distribution are known, a radiation computation can be attempted. At present the radiant properties of soot are only approximately known.

If flames found in both the laboratory and industry are included, the individual soot particles produced in hydrocarbon flames generally range in diameter from 50 Å to more than 3000 Å. The soot can be in the form of spherical particles, agglomerated masses, or sometimes long filaments. The experimental determination of the physical form of the soot is difficult, as any type of probe used to gather soot for photomicrographic analysis may cause agglomeration of particles or otherwise alter the soot characteristics. The nucleation and growth of the soot particles are not well understood. Some of the soot can be nucleated in less than a millisecond after the fuel enters the flame, and the rate at which soot continues to form does not seem influenced much by the residence time of the fuel in the flame. An unknown precipitation mechanism must govern soot production. In typical gaseous diffusion flames, the volume of soot per total volume of combustion products has been found experimentally to be about 10^{-8}-10^{-6} [51–53].

For a beam of radiation passing through a transparent carrier gas containing suspended soot, it has been found experimentally that the attenuation obeys Bouguer's law,

$$i'_\lambda(S) = i'_\lambda(0) \exp(-a_\lambda S) \tag{17-129}$$

From the Mie theory the radiative properties of soot depend on the size parameter $\pi D/\lambda$ (where D is the particle diameter) and the optical constants n and κ of the particles, which depend on the chemical composition of the soot. The n and κ depend somewhat on λ, as will be shown later, but do not depend strongly on temperature [54, 55]. At the temperatures in combustion systems, the radiation is mostly in a wavelength range of unity and larger; hence, for the small soot particles $\pi D/\lambda$ is generally much less than 0.25. In this range of size parameter the Mie theory implies in Eq. (16-100) that the scattering cross section depends on $(\pi D/\lambda)^4$, and that the absorption cross section [see Eq. (17-135)] depends on $\pi D/\lambda$ to the first power. Thus scattering is negligible compared with absorption, and a_λ in (17-129) is actually the absorption coefficient of the soot rather than the extinction coefficient of (13-13). Then, as a consequence of (17-44), the spectral emittance of an isothermal luminous gas volume, composed of soot in a nonradiating carrier gas, is written as

$$\epsilon_\lambda = 1 - \exp(-a_\lambda L_e) \tag{17-130}$$

where L_e is the mean beam length for the volume. Radiation by the carrier gas will be included in a later section.

Experimental correlation of soot spectral absorption A relatively simple empirical relation for a_λ, found experimentally in some instances, is

$$a_\lambda = Ck\lambda^{-\alpha} \tag{17-131}$$

where C is the soot volume concentration (average volume of particles per unit volume of cloud), and k is a constant. The λ will always be in micrometers for the numerical quantities given here. Hottel, on page 100 of [1], recommends the use of

$$a_\lambda = \frac{Ck_1}{\lambda^{0.95}} \tag{17-132a}$$

in the infrared region down to $\lambda = 0.8\ \mu m$. In some more recent experiments, Siddall and McGrath [56] also found the functional relation of (17-132a) to hold approximately. They give, in the range $\lambda = 1$-$7\ \mu m$, the following mean values of α:

Source of soot	Mean α for $\lambda = 1$-$7\ \mu m$
Amyl acetate	0.89, 1.04
Avtur kerosene	0.77
Benzene	0.94, 0.95
Candle	0.93
Furnace samples	0.96, 1.14, 1.25
Petrotherm	1.06
Propane	1.00

Thus the 0.95 exponent recommended by Hottel appears reasonable.

In [56] the data were inspected in more detail to see if α had a functional variation with λ that would provide a more accurate correlation than a simple constant α. In some instances α took the form

$$\alpha = c_1 + c_2 \ln \lambda$$

where c_1 and c_2 are positive constants. Examples are shown in Fig. 17-24. In other cases, as in Fig. 17-25, a more general polynomial was required to express α as a function of λ. Thus, as a generalization of (17-132a) for the infrared region,

$$a_\lambda = \frac{Ck_1}{\lambda^{\alpha(\lambda)}} \qquad (17\text{-}132b)$$

and letting α be a constant is only an approximation.

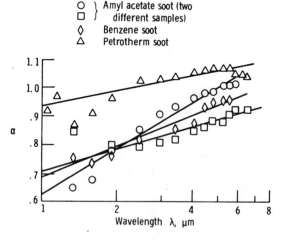

○ } Amyl acetate soot (two
□ }   different samples)
◇   Benzene soot
△   Petrotherm soot

Figure 17-24 Experimental values of the exponent α plotted against λ for cases where α varies approximately linearly with $\ln \lambda$ [56].

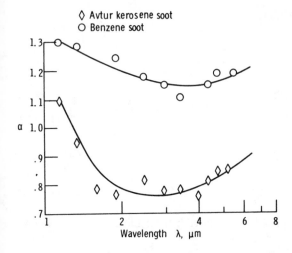

Figure 17-25 Experimental values of the exponent α as a function of λ where α does not vary linearly with $\ln \lambda$ [56].

In the visible range an inspection by Hottel [1] of experimental data led to the recommended form

$$a_\lambda = \frac{Ck_2}{\lambda^{1.39}} \qquad (17\text{-}133)$$

for the wavelength region around $\lambda = 0.6 \ \mu m$ (say $\lambda \approx 0.3\text{-}0.8 \ \mu m$).

Electromagnetic-theory prediction of soot spectral absorption To try to understand, from a more fundamental basis, the absorption coefficient of a soot cloud in a non-absorbing gas, electromagnetic theory can be employed [55–61]. The absorption coefficient is written as

$$a_\lambda = E_\lambda AN \qquad (17\text{-}134)$$

The product $E_\lambda A$ is the spectral absorption cross section, defined in the same manner as the scattering cross section s_λ used in (13-30). The N is the number of particles per unit volume, and A is the projected area of a particle ($A = \pi D^2/4$, as particles are assumed spherical). The E_λ by itself is the spectral absorption efficiency factor, which is the ratio of the spectral absorption cross section to the actual physical cross section of the particle (ratio of energy absorbed to that incident on the particle). For the limit of small $\pi D/\lambda$, the Mie equations give E_λ for a small absorbing sphere as

$$E_\lambda = \frac{24\pi D}{\lambda} \frac{n\kappa}{(n^2 - \kappa^2 + 2)^2 + 4n^2\kappa^2} \qquad (17\text{-}135)$$

where n and κ are the simple index of refraction and the extinction coefficient of the sphere material when the complex index of refraction is $\bar{n} = n - i\kappa$. Since n and κ are functions of λ, (17-135) can be written as

$$E_\lambda = \frac{24\pi D}{\lambda} F(\lambda) \tag{17-136}$$

Then from (17-134)

$$a_\lambda = \frac{24\pi D}{\lambda} F(\lambda)AN = \frac{36\pi C}{\lambda} F(\lambda) \tag{17-137}$$

Where $C = N\pi D^3/6$ is the volume of particles per unit volume of cloud. The ratio

$$\frac{a_\lambda}{C} = \frac{36\pi}{\lambda} F(\lambda) = \frac{36\pi}{\lambda} \frac{n\kappa}{(n^2 - \kappa^2 + 2)^2 + 4n^2\kappa^2} \tag{17-138}$$

is then a function of wavelength and can be evaluated if the n and κ of soot are known as functions of λ.

In [55] the n and κ of acetylene and propane soot were measured by collecting soot and then compressing it on a brass plate. The values obtained are given in Table 17-7. By using these values a_λ/C was evaluated from (17-138), yielding the points in Fig. 17-26. Although the a_λ/C decreases with λ as expected from the form of (17-131), it is evident than an approximate curve fit by straight lines (on the logarithmic plot) would yield exponents on λ somewhat different than those of (17-132a) and (17-133).

Table 17-7 Optical constants of acetylene and propane soots [55]

	Acetylene soot			Propane soot		
Wave-length λ, μm	Index of re-fraction n	Extinc-tion co-efficient κ	Absorption coefficient per particle volume frac-tion a_λ/C μm^{-1}	Index of re-fraction n	Extinc-tion co-efficient κ	Absorption coefficient per particle volume frac-tion a_λ/C μm^{-1}
0.4358	1.56	0.46	9.37	1.57	0.46	9.29
0.4500	1.56	0.48	9.45	1.56	0.50	9.83
0.5500	1.56	0.46	7.42	1.57	0.53	8.44
0.6500	1.57	0.44	5.96	1.56	0.52	7.07
0.8065	1.57	0.46	5.02	1.57	0.49	5.34
2.5	2.31	1.26	1.97	2.04	1.15	2.34
3.0	2.62	1.62	1.44	2.21	1.23	1.75
4.0	2.74	1.64	0.998	2.38	1.44	1.24
5.0	2.88	1.82	0.747	2.07	1.72	1.30
6.0	3.22	1.84	0.505	2.62	1.67	0.727
7.0	3.49	2.17	0.270	3.05	1.91	0.484
8.5	4.22	3.46	0.213	3.26	2.10	0.357
10.0	4.80	3.82	0.143	3.48	2.46	0.271

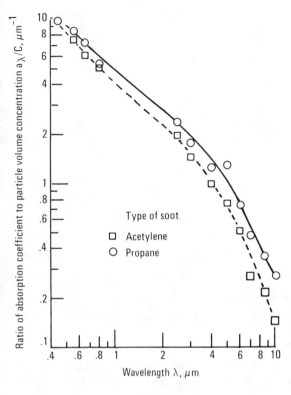

Figure 17-26 Ratio of spectral absorption coefficient to particle volume concentration for soot particles.

The form of $\alpha(\lambda)$ predicted by Mie theory can now be examined in the infrared region by equating the expressions in (17-132b) and (17-137). This gives

$$\frac{k_1}{\lambda^{\alpha(\lambda)}} = \frac{36\pi}{\lambda} F(\lambda) \tag{17-139}$$

By evaluating this relation at $\lambda = 1$, the constant k_1 for the infrared region is found as

$$k_1 = 36\pi F(1) \tag{17-140}$$

Then $\lambda^{\alpha(\lambda)} = \lambda F(1)/F(\lambda)$. Taking the logarithm of both sides and solving for $\alpha(\lambda)$ give

$$\alpha(\lambda) = 1 + \frac{\ln[F(1)/F(\lambda)]}{\ln \lambda} \tag{17-141}$$

The optical properties of the soot can then be used in $F(1)$ and $F(\lambda)$ as defined in (17-138), and $\alpha(\lambda)$ can be found. This was done in [56] using the properties of a baked electrode carbon at 2250 K. The results are shown in Fig. 17-27. The trend is the same as in the experimental curves of Fig. 17-24, but the α values are larger than the experimental values. They are also larger than the average value of 0.95 recommended in (17-132a). The discrepancy is probably partly due to the optical properties

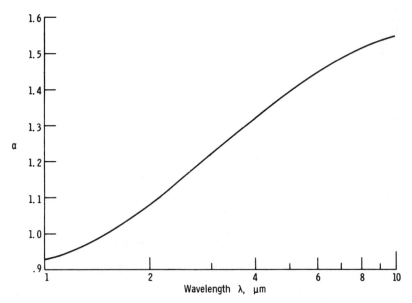

Figure 17-27 Calculated variation of the exponent α with wavelength using properties of baked electrode carbon at 2250 K [56].

of the baked electrode carbon being different from those of soot. This is further discussed in [62], in which optical properties of carbonaceous materials more like real soot are given, and good comparisons of predictions with experiment are obtained in some instances.

Total emittance of soot cloud A total emittance can be found for a path S through an isothermal cloud of suspended soot having a uniform concentration. The emittance here accounts for soot absorptance alone; the suspending gas is assumed to be non-emitting. The effect of an emitting gas will be included later. The total emittance is found from (13-60) as

$$\epsilon(T,S) = \frac{\int_0^\infty e_{\lambda b}[1 - \exp(-a_\lambda S)]\,d\lambda}{\sigma T^4}$$

which can also be written as

$$\epsilon(T,CS) = \frac{\int_0^\infty e_{\lambda b}\{1 - \exp[-(a_\lambda/C)\,CS]\}\,d\lambda}{\sigma T^4} \tag{17-142}$$

By using a_λ/C from (17-138) or Fig. 17-26, the ϵ can be evaluated numerically and will be a function of the cloud temperature and the product of concentration and path length CS. This is shown in Fig. 17-28 for propane soot. By integrating over a distribution of particle sizes, it was found in [56] that the individual particle sizes were unimportant, and thus at a fixed T and S the ϵ depends only on the soot-volume concentration in the cloud.

With certain assumptions a convenient expression can be derived for the total emittance of a soot cloud in a nonemitting gas. As indicated by (17-138), if n and κ are weak functions of λ, then $a_\lambda/C = k/\lambda$, where k is a constant depending on the type of soot. As shown by the previous discussion, this is only a rough approximation, as the exponent α can deviate appreciably from unity. However, in some instances this was found to be a good approximation, for example, for the soot data in [63], and in Fig. 17-26 for λ up to about 5. The coefficient k is found as the value of a_λ/C at $\lambda = 1\ \mu m$, and Fig. 17-26 yields k (propane) ≈ 4.9 and k (acetylene) ≈ 4. For soot from coal flames, k has been found to range from 3.7 to 7.5; for oil flames $k \approx 6.3$ [64].

The above approximation for a_λ is inserted into (17-142). Also, for the wavelength regions of interest here, the $e_{\lambda b}$ can be approximated quite well by Wien's formula, $e_{\lambda b} \approx 2\pi C_1/\lambda^5 \exp(C_2/\lambda T)$. Equation (17-142) then becomes

$$\epsilon(T, CS) = \frac{\int_0^\infty (2\pi C_1/\lambda^5)\, e^{-C_2/\lambda T}\, (1 - e^{-kCS/\lambda})\, d\lambda}{\int_0^\infty (2\pi C_1/\lambda^5)\, e^{-C_2/\lambda T}\, d\lambda}$$

where, to be consistent in the degree of approximation, Wien's formula has also been used in the denominator. For an integral of the form $\int_0^\infty \lambda^{-5}\, e^{-K/\lambda}\, d\lambda$, where K is independent of λ, the substitution of $\eta = K/\lambda$ yields $K^{-4} \int_0^\infty \eta^3\, e^{-\eta}\, d\eta$. Then ϵ can be written as

$$\epsilon(T, CS) = \frac{[(C_2/T)^{-4} - (C_2/T + kCS)^{-4}] \int_0^\infty \eta^3\, e^{-\eta}\, d\eta}{(C_2/T)^{-4} \int_0^\infty \eta^3\, e^{-\eta}\, d\eta}$$

The integrals cancel, giving ϵ as

$$\epsilon(T, CS) = 1 - \frac{1}{[1 + (kCS/C_2)\,T]^4} \tag{17-143}$$

With $k = 4.9$ for propane soot, a comparison with values from the numerical integration in [55] is shown in Table 17-8; Eq. (17-143) is shown to be a useful approxi-

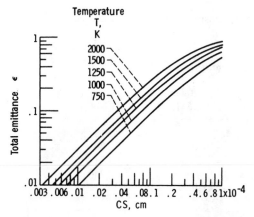

Figure 17-28 Total emittance of soot suspensions as a function of temperature and volume fraction path-length product for propane soot [55].

mation. The $k/C_2 = 4.9/0.01439 \, \text{m}^{-1} \cdot \text{K}^{-1} = 341 \, \text{m}^{-1} \cdot \text{K}^{-1}$ for this soot. Reference [51] recommends $k/C_2 = 350 \, \text{m}^{-1} \cdot \text{K}^{-1}$ as a mean value for all types of soot, and this coefficient does not vary significantly with temperature over the temperature range of interest in combustion chambers.

A somewhat more complicated expression results from not using the Wien formula in (17-142). Then, again letting $a_\lambda = Ck/\lambda$ and using (2-11a) give

$$\epsilon(T, CS) = 1 - \frac{\int_0^\infty e_{\lambda b} \, e^{-kCS/\lambda} \, d\lambda}{\sigma T^4} = 1 - \frac{2\pi C_1}{\sigma T^4} \int_0^\infty \frac{e^{-kCS/\lambda} \, d\lambda}{\lambda^5 (e^{C_2/\lambda T} - 1)}$$

If we let $z = 1 + kCST/C_2$ and $t = C_2/\lambda T$, this becomes

$$\epsilon(T, CS) = 1 - \frac{2\pi C_1}{\sigma C_2^4} \int_0^\infty \frac{t^3 e^{-tz}}{1 - e^{-t}} \, dt$$

From pages 260 and 271 of [65], the integral is the pentagamma function $\psi^{(3)}(z)$ for which tabulated values are available. Then, with (2-22), ϵ becomes

$$\epsilon(T, CS) = 1 - \frac{15}{\pi^4} \, \psi^{(3)}\left(1 + \frac{kCST}{C_2}\right) \tag{17-144}$$

Results from this expression are also included in Table 17-8; they are not better than those from the simpler expression (17-143). Yuen and Tien [66] show that (17-144) can be approximated by the convenient form $\epsilon(T, CS) = 1 - \exp(-3.6kCST/C_2)$.

Stull and Plass [57] give results for the scattering coefficient of soot, which show (in agreement with the Mie theory) that in most instances scattering has little effect on emittance in the wavelength range that contains significant energy at hydrocarbon combustion temperatures. Erickson et al. [67] have experimentally studied scattering from a luminous benzene-air flame. Their results agreed with the predictions of Stull and Plass if the soot particles are taken to be of two predominant diameters. This indicates that small particles on the order of 250 Å in diameter are formed along with agglomerated particles with an equivalent diameter of 1850 Å. These sizes were observed by gathering soot with a probe and using electron microscopy. A comparison of some of the experimental results of [67] with the analysis of [57] is shown in Fig. 17-29.

Table 17-8 Comparison of approximate suspension emittances for propane soot with values from reference [55]

Concentration–path-length product CS, μm	Suspension temperature T, K	Suspension emittance ϵ		
		Equation (17-143)	Reference [55]	Equation (17-144)
0.01	1000	0.0135	0.013	0.0127
0.10	1000	0.125	0.125	0.120
1.00	1000	0.690	0.64	0.671
0.01	2000	0.0268	0.030	0.0254
0.10	2000	0.232	0.250	0.223
1.00	2000	0.875	0.90	0.857

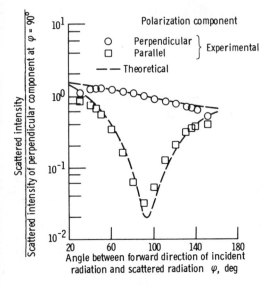

Figure 17-29 Comparison of experiment with Mie scattering theory for radiation scattered from benzene-air flame at wavelength $\lambda = 5461$ Å. Theoretical curves based on spheres of diameter 250 Å with 0.002% spheres of diameter 1850 Å, all with complex refractive index $n - i\kappa = 1.79 - 0.79i$. (From [67].)

Thring, Beér, and Foster [68] have put some of the results of Stull and Plass [57], along with their own extensive experimental results, into useful graphs of emittance, extinction coefficient, and soot concentration for flames applicable in industrial practice. They note, however, that soot concentration can only be predicted for flames geometrically similar and with the same control variables as those that have already been studied. Their paper contains a useful review of the worldwide effort to gather information on and give methods for the prediction of radiation from luminous industrial flames. Other such information is found in [51, 69–73].

In addition to the uncertainties in optical properties and hence in a_λ and ϵ for soot, it is noted that a_λ and ϵ are given in terms of the soot concentration. To use Eq. (17-131), the Ck is needed. To use Fig. 17-28 to determine ϵ for a given flame size, the C in the abscissa must be known. At present there is no way to compute C from first principles, knowing the fuel and burner geometry. Hence some indication of C or Ck must be obtained by examining flames experimentally. It may be possible to extrapolate performance for a particular application by examining a similar flame. A recent mathematical model of soot formation is given in [74].

One technique to obtain information on the soot-concentration quantity Ck_2 [as in Eq. (17-133) for the visible region] is to sight through the flame onto a cold black background with a pyrometer and match the brightness of the pyrometer filament to that of the flame while using first a red filter and then a green filter. With each of the filters, the pyrometer is also sighted on a blackbody source, and the source temperatures are obtained that produce the same brightness as when the flame was viewed. As a convenient simplification at the small λT for red and green wavelengths when considering typical flame temperatures, the blackbody intensity can be approximated very well by Wien's formula, Eq. (2-13):

$$i'_{\lambda b} = \frac{2C_1}{\lambda^5 \exp(C_2/\lambda T)} \tag{17-145}$$

Then if T_r is the blackbody temperature producing the same brightness as the flame did when the red filter was used, this intensity is

$$i'_{\lambda b,r} = \frac{2C_1}{\lambda_r^5 \exp\left(C_2/\lambda_r T_r\right)} \tag{17-146}$$

where λ_r is the red wavelength, $0.665\,\mu m$. This intensity can also be written as a spectral emittance of the flame times the blackbody intensity at the flame temperature, which gives

$$\frac{2C_1}{\lambda_r^5 \exp\left(C_2/\lambda_r T_r\right)} = \epsilon_{\lambda_r} \frac{2C_1}{\lambda_r^5 \exp\left(C_2/\lambda_r T_f\right)}$$

By grouping the exponential terms and taking the logarithm, the result can be rearranged into

$$\frac{1}{T_f} - \frac{1}{T_r} = \frac{\lambda_r}{C_2} \ln \epsilon_{\lambda_r} \tag{17-147}$$

Similarly, with a green filter,

$$\frac{1}{T_f} - \frac{1}{T_g} = \frac{\lambda_g}{C_2} \ln \epsilon_{\lambda_g} \tag{17-148}$$

where the green wavelength λ_g is $0.555\,\mu m$.

Now as a simple approximation, (17-133) is used for a_λ in the visible region. Then ϵ_λ from (17-130) is

$$\epsilon_\lambda = 1 - \exp\left(-\frac{Ck_2 S}{\lambda^{1.39}}\right) \tag{17-149}$$

where S is the path length sighted through the flame. Substitute (17-149) into (17-147) and (17-148) to obtain

$$\frac{1}{T_f} - \frac{1}{T_r} = \frac{\lambda_r}{C_2} \ln \left[1 - \exp\left(-\frac{Ck_2 S}{\lambda_r^{1.39}}\right)\right] \tag{17-150}$$

$$\frac{1}{T_f} - \frac{1}{T_g} = \frac{\lambda_g}{C_2} \ln \left[1 - \exp\left(-\frac{Ck_2 S}{\lambda_g^{1.39}}\right)\right] \tag{17-151}$$

These two equations are solved for T_f and Ck_2 to yield the needed measure of the soot concentration as well as the flame temperature.

As an approximation, Ck_2 is assumed independent of wavelength and is used in (17-130), (17-132a) and (17-133) to yield, for a path length S,

$$\epsilon_\lambda = 1 - \exp\left(-\frac{Ck_2 S}{\lambda^{1.39}}\right) \qquad \text{visible, } \lambda < 0.8\,\mu m \tag{17-152}$$

$$\epsilon_\lambda = 1 - \exp\left(-\frac{Ck_2 S}{\lambda^{0.95}}\right) \qquad \text{infrared, } \lambda > 0.8\,\mu m \tag{17-153}$$

Then with these spectral emittances the definition in Eq. (17-142) can be used to evaluate the total emittance of the flame as

$$\epsilon(T_f,S) = \frac{\int_0^\infty e_{\lambda b}(T_f)\epsilon_\lambda(\lambda, T_f, S)\, d\lambda}{\sigma T_f^4} \tag{17-154}$$

Some convenient graphs for use in this procedure are given in [1]. The hope is that the Ck_2 obtained in this way can be applied to "similar" flames. This is a very rough approximation; there are so many variables affecting the flow and mixing in the flame that it is difficult to know when the flames will have a similar character. The detailed nature of flames is a continuing area of active research [49, 74–78].

Soot cloud in an isothermal radiating gas Generally in a flame or combustion products containing soot, there will be gaseous constituents that radiate energy (the previous treatment was for soot in a nonradiating carrier gas). To simplify the present discussion, it will be assumed that only three radiating constituents are present — carbon dioxide, water vapor and soot — but the method can be extended to more constituents. As a monochromatic beam of radiation passes through the gas-soot suspension, the local attenuation depends on the sum of the absorption coefficients so that from (13-56) and (13-59) the spectral emittance is

$$\epsilon_\lambda = \alpha_\lambda = 1 - e^{-(a_{\lambda C} + a_{\lambda H} + a_{\lambda S})S} \tag{17-155}$$

where the subscripts C, H, and S refer to CO_2, H_2O, and soot. Then from (17-142) the total emittance is

$$\epsilon = \frac{1}{\sigma T^4} \int_0^\infty (1 - e^{-(a_{\lambda C} + a_{\lambda H} + a_{\lambda S})S})e_{\lambda b}\, d\lambda \tag{17-156}$$

This can be rewritten in the equivalent forms

$$\epsilon = \frac{1}{\sigma T^4} \int_0^\infty [1 - (1 - \epsilon_{\lambda C})(1 - \epsilon_{\lambda H})(1 - \epsilon_{\lambda S})]e_{\lambda b}\, d\lambda$$

$$= \frac{1}{\sigma T^4} \int_0^\infty (\epsilon_{\lambda C} + \epsilon_{\lambda H} + \epsilon_{\lambda S} - \epsilon_{\lambda C}\epsilon_{\lambda H} - \epsilon_{\lambda C}\epsilon_{\lambda S} - \epsilon_{\lambda S}\epsilon_{\lambda H} + \epsilon_{\lambda C}\epsilon_{\lambda H}\epsilon_{\lambda S})e_{\lambda b}\, d\lambda$$

where $\epsilon_{\lambda J} = 1 - e^{-a_{\lambda J}S}$ with $J = C, H, S$. The first three terms in the last integral yield the total emittances of the three constituents, so that

$$\epsilon = \epsilon_C + \epsilon_H + \epsilon_S - \frac{1}{\sigma T^4} \int_0^\infty (\epsilon_{\lambda C}\epsilon_{\lambda H} + \epsilon_{\lambda C}\epsilon_{\lambda S} + \epsilon_{\lambda S}\epsilon_{\lambda H} - \epsilon_{\lambda C}\epsilon_{\lambda H}\epsilon_{\lambda S})e_{\lambda b}\, d\lambda \tag{17-157}$$

A term such as $\epsilon_{\lambda C}\epsilon_{\lambda H}$ is nonzero only in spectral regions where both $a_{\lambda C}$ and $a_{\lambda H}$ are nonzero, that is, where both constituents are radiating. Thus the terms in the integral represent overlap regions in the spectrum where two or three components are radiating. The total emittance ϵ is then the sum of the three individual emittances, computed as if the other constituents were absent (see Figs. 17-11 and 17-13) minus a correction term for spectral overlap. This is the same type of correction that was discussed in connection with Eq. (17-62). For the simplified case of gray constituents,

$$\epsilon = 1 - (1 - \epsilon_C)(1 - \epsilon_H)(1 - \epsilon_S) \tag{17-158}$$

Results for the spectral-overlap terms were calculated in [59] by using existing information on the form of the CO_2 and H_2O vapor absorption bands and the soot absorption coefficient $a_\lambda = Ck/\lambda$ discussed earlier. Typical emittances are shown in Fig. 17-30. For low values of CS in the figure, the soot concentration is low especially when S is high. Hence the left sides of the curves are dominated by the gas emittance, and the vertical displacement of the curves shows the increase in gas emittance with path length. As CS is increased from 0.001, the curves are somewhat horizontal (especially at high S which corresponds to low C) as the soot concentration is not sufficient to increase ϵ for the mixture significantly. For larger CS the soot begins to dominate, and for all the path lengths shown, ϵ approaches unity when CS is about 3×10^{-4} cm. This is consistent with the results in Fig. 17-28. Some calculations are given in [77] for a combustion chamber filled with H_2O, CO_2 and soot. Kunitomo [78] gives results for the ratio, in a flame, of the soot-cloud emittance to the emittance of the nonluminous suspending gas; these results are for a liquid fuel. The ratio increases as the fuel carbon-hydrogen ratio is increased, and as the excess air is decreased.

Gases containing particles other than soot In addition to hydrocarbon flames containing soot, there are applications involving radiation from gases containing other types of particles. A common example is ash particles in pulverized coal flames [51]. Another example is the luminosity in the exhaust plume of solid-fueled and some liquid-fueled rockets. For a solid fuel the luminosity may be caused by metal particles added to promote combustion stability. Williams and Dudley [79] present calculations for a rocket exhaust with entrained liquid aluminium oxide particles.

The presence of particles in an otherwise weakly absorbing medium can cause the mixture to be strongly absorbing. *Seeding* of a gas with particles, such as finely divided carbon, has been proposed to increase gas absorption and heating by incident radiation [80], or as a means of shielding a surface from incident radiation [81, 82]. These

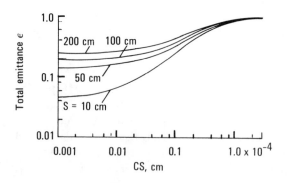

Figure 17-30 Total emittance of gas-soot suspension as a function of volume fraction of soot times path length. Gas temperature, 1600 K; total pressure, 1 atm; partial pressures: $p_{H_2O} = 0.19$ atm, $p_{CO_2} = 0.09$ atm [59].

techniques have possible application in connection with advanced energy systems, and for the collection of solar energy [83].

Another use for seeding is in the direct determination of flame temperatures for a nonluminous flame by the *line-reversal technique.* In this method, a seeding material such as a sodium or cadmium salt is introduced into an otherwise transparent flame. These materials produce a strong line in the visible spectrum because of an electronic transition; cadmium gives a red line, and sodium a bright yellow line. A continuous source such as a tungsten lamp is placed so that it may be viewed *through* the seeded flame with a spectroscope. The intensity seen in the spectroscope at the line wavelength is, from (14-13),

$$i'_{\lambda,\text{scope}} = i'_{\lambda,\text{cont source}} \exp\left(-\kappa_\lambda\right) + \int_0^{\kappa_\lambda} i'_{\lambda b}(\kappa_\lambda^*) \exp\left[-(\kappa_\lambda - \kappa_\lambda^*)\right] d\kappa_\lambda^* \qquad (17\text{-}159)$$

If the flame is assumed isothermal and of diameter D and no attenuation occurs along the remainder of the path between the continuous source and the spectroscope, (17-159) becomes

$$i'_{\lambda,\text{scope}} = i'_{\lambda,\text{cont source}} \exp\left(-\kappa_{\lambda,D}\right) + i'_{\lambda b,\text{flame}}\left[1 - \exp\left(-\kappa_{\lambda,D}\right)\right] \qquad (17\text{-}160)$$

where

$$\kappa_{\lambda,D} = \int_0^D a_\lambda(S^*)\, dS^*$$

In the wavelength region adjacent to the absorbing and emitting spectral line, the flame is essentially transparent so the background radiation observed adjacent to the line is $i'_{\lambda,\text{cont source}}$. Hence, by subtracting $i'_{\lambda,\text{cont source}}$ from (17-160), the line intensity relative to the adjacent background is found as

$$i'_{\lambda,\text{scope}} - i'_{\lambda,\text{cont source}} = \left(i'_{\lambda b,\text{flame}} - i'_{\lambda,\text{cont source}}\right)\left[1 - \exp\left(-\kappa_{\lambda,D}\right)\right] \qquad (17\text{-}161)$$

If the flame is at a higher temperature than the continuous source, (17-161) shows that the line intensity in the spectroscope will be greater than the continuous background intensity. The line will appear as a bright line imposed upon a less bright continuous spectrum. Increasing the temperature of the continuous source causes the source term to override. The line then appears as a dark line on a brighter continuous spectrum. If the continuous source is a blackbody and its temperature is made *equal* to the flame temperature, then $i'_{\lambda,\text{cont source}} = i'_{\lambda b,\text{flame}}$, and (17-161) reduces to $i'_{\lambda,\text{scope}} = i'_{\lambda,\text{cont source}}$. The line will then disappear into the continuum in the spectroscope. This is because the absorption by the flame and the flame emission exactly compensate. If the continuous source is a tungsten lamp, the source temperature measurement is usually made with an optical pyrometer.

In the derivation of (17-161) it was assumed that the flame is transparent except within the spectral line produced by the cadmium or sodium seeding. If soot is in the flame, the soot particles, emit, and scatter radiation in a continuous spectrum along the path of the incident beam. The line-reversal technique is of less practical utility in this instance, as it then depends on the soot behavior. The effect of soot is analyzed in [84].

Another instance of radiation attenuation by means of particles is found in the effect of dust or "grains" that are believed to exist in interstellar space and cause reductions in the observed intensity of radiation from stars [85–87].

REFERENCES

1. Hottel, H. C.: Radiant-heat Transmission, in William H. McAdams (ed.), "Heat Transmission," 3d ed., chap. 4, McGraw-Hill Book Company, New York, 1954.
2. Hottel, Hoyt C., and Adel F. Sarofim: "Radiative Transfer," McGraw-Hill Book Company, New York, 1967.
3. Eckert, E. R. G., and Robert M. Drake, Jr.: "Heat and Mass Transfer," 2d ed., McGraw-Hill Book Company, New York, 1959.
4. Edwards, D. K., and Balakrishnan, A.: Slab Band Absorptance for Molecular Gas Radiation, *J. Quant. Spectrosc. Radiat. Transfer*, vol. 12, pp. 1379–1397, 1972.
5. Nelson, D. A.: A Study of Band Absorption Equations for Infrared Radiative Transfer in Gases. I, Transmission and Absorption Functions for Planar Media, *J. Quant. Spectrosc. Radiat. Transfer*, vol. 14, pp. 69–80, 1974.
6. Wassel, A. T., and D. K. Edwards: Mean Beam Lengths for Spheres and Cylinders, *ASME J. Heat Transfer*, vol. 98, pp. 308–309, 1976.
7. Felske, J. D., and C. L. Tien: Wide Band Characterization of the Total Band Absorptance of Overlapping Infrared Gas Bands, *Combust. Sci. Technol.*, vol. 11, p. 111, 1975.
8. Saido, K., and W. H. Giedt: "Spectral Absorption of Water Vapor and Carbon Dioxide Mixtures in the 2.7 Micron Band, *J. Heat Transfer*, vol. 99, no. 1, pp. 53–59, 1977.
9. Leckner, Bo: Spectral and Total Emissivity of Water Vapor and Carbon Dioxide, *Combust. Flame*, vol. 19, pp. 33–48, 1972.
10. Boynton, Frederick P., and Claus B. Ludwig: Total Emissivity of Hot Water Vapor – II, Semi-Empirical Charts Deduced from Long-Path Spectral Data, *Int. J. Heat Mass Transfer*, vol. 14, pp. 963–973, 1971.
11. Ludwig, C. B., W. Malkmus, J. E. Reardon, and J. A. L. Thomson, Handbook of Infrared Radiation from Combustion Gases, *NASA* SP-3080, 1973.
12. Sarofim, A. F., I. H. Farag, and H. C. Hottel: Radiative Heat Transmission from Non-Luminous Gases. Computational Study of the Emissivities of Carbon Dioxide, *ASME* paper 78-HT-16, May, 1978.
13. Cess, R. D., and M. S. Lian: A Simple Parameterization for the Water Vapor Emissivity, *ASME J. Heat Transfer*, vol. 98, no. 4, pp. 676–678, 1976.
14. Dunkle, R. V.: Geometric Mean Beam Lengths for Radiant Heat-Transfer Calculations, *J. Heat Transfer*, vol. 86, no. 1, pp. 75–80, 1964.
15. Edwards, D. K., and K. E. Nelson: Rapid Calculation of Radiant Energy Transfer between Nongray Walls and Isothermal H_2O and CO_2 Gas, *J. Heat Transfer*, vol. 84, no. 4, pp. 273–278, 1962.
16. Edwards, D. K.: Radiation Interchange in a Nongray Enclosure Containing an Isothermal Carbon-Dioxide–Nitrogen Gas Mixture, *J. Heat Transfer*, vol. 84, no. 1, pp. 1–11, 1962.
17. Oppenheim, A. K.: Radiation Analysis by the Network Method, *Trans. ASME*, vol. 78, no. 4, pp. 725–735, 1956.
18. Mandell, David A., and Forrest A. Miller: Comparison of Exact and Mean Beam Length Results for a Radiating Hydrogen Plasma, *J. Quant. Spectrosc. Radiat. Transfer*, vol. 13, pp. 49–56, 1973.
19. Edwards, D. K., and M. M. Weiner: Comment on Radiative Transfer in Non-isothermal Gases, *Combust. Flame*, vol. 10, no. 2, pp. 202–203, 1966.
20. Edwards, D. K., L. K. Glassen, W. C. Hauser, and J. S. Tuchscher: Radiation Heat Transfer in Nonisothermal Nongray Gases, *J. Heat Transfer*, vol. 89, no. 3, pp. 219–229, 1967.

21. Weiner, M. M., and D. K. Edwards: Non-isothermal Gas Radiation in Superposed Vibration-Rotation Bands, *J. Quant. Spectrosc. Radiat. Transfer*, vol. 8, no. 5, pp. 1171–1183, 1968.
22. Krakow, Burton, Harold J. Babrov, G. Jordan Maclay, and Abraham L. Shabott: Use of the Curtis-Godson Approximation in Calculations of Radiant Heating by Inhomogeneous Hot Gases, *Appl. Opt.*, vol. 5, no. 11, 1791–1800, 1966.
23. Simmons, F. S.: Band Models for Non-isothermal Radiating Gases, *Appl. Opt.*, vol. 5, no. 11, pp. 1801–1811, 1966.
24. Goody, R. M.: "Atmospheric Radiation, Theoretical Basis," vol. I, Clarendon Press, Oxford, 1964.
25. Simmons, F. S.: Application of Band Models to Inhomogeneous Gases. Molecular Radiation and its Application to Diagnostic Techniques (R. Goulard, ed.), *NASA* TM X-53711, pp. 113–133, 1968.
26. Hottel, H. C., and E. S. Cohen: Radiant Heat Exchange in a Gas-filled Enclosure: Allowance for Nonuniformity of Gas Temperature, *AIChEJ.*, vol. 4, no. 1, pp. 3–14, 1958.
27. Plass, Gilbert N.: Radiation from Nonisothermal Gases, *Appl. Opt.*, vol. 6, no. 11, pp. 1995–1999, 1967.
28. Cess, R. D., and L. S. Wang: A Band Absorptance Formulation for Nonisothermal Gaseous Radiation, *Int. J. Heat Mass Transfer*, vol. 13, no. 3, pp. 547–555, 1970.
29. Einstein, Thomas H.: Radiant Heat Transfer to Absorbing Gases Enclosed between Parallel Flat Plates with Flow and Conduction, *NASA* TR R-154, 1963.
30. Einstein, Thomas H.: Radiant Heat Transfer to Absorbing Gases Enclosed in a Circular Pipe with Conduction, Gas Flow, and Internal Heat Generation, *NASA* TR R-156, 1963.
31. Noble, James J.: The Zone Method: Explicit Matrix Relations for Total Exchange Areas, *Int. J. Heat Mass Transfer*, vol. 18, no. 2, pp. 261–269, 1975.
32. Bannerot, R. B., and F. Wierum: Approximate Configuration Factors for a Gray Nonisothermal Gas-Filled Conical Enclosure, pp. 41–63, in "Thermophysics and Spacecraft Thermal Control," vol. 35 of *Progress in Astronautics and Aeronautics*, MIT Press, Cambridge, Mass., 1974.
33. Edwards, D. K., and A. Balakrishnan: Volume Interchange Factors for Nonhomogeneous Gases, *ASME J. Heat Transfer*, vol. 94, no. 2, pp. 181–188, 1972.
34. Perlmutter, M., and J. R. Howell: Radiant Transfer through a Gray Gas between Concentric Cylinders Using Monte Carlo, *J. Heat Transfer*, vol. 86, no. 2, pp. 169–179, 1964.
35. Greif, Ralph, and Gean P. Clapper: Radiant Heat Transfer between Concentric Cylinders, *Appl. Sci. Res.*, sec. A, vol. 15, pp. 469–474, 1966.
36. Rhyming, I. L.: Radiative Transfer between Two Concentric Spheres Separated by an Absorbing and Emitting Gas, *Int. J. Heat Mass Transfer*, vol. 9, no. 4, pp. 315–324, 1966.
37. Sparrow, E. M., C. M. Usiskin, and H. A. Hubbard: Radiation Heat Transfer in a Spherical Enclosure Containing a Participating, Heat-generating Gas, *J. Heat Transfer*, vol. 83, no. 2, pp. 199–206, 1961.
38. Viskanta, R., and R. L. Merriam: Heat Transfer by Combined Conduction and Radiation between Concentric Spheres Separated by Radiating Medium, *J. Heat Transfer*, vol. 90, no. 2, pp. 248–256, 1968.
39. Echigo, R., N. Nishiwaki, and M. Hirata: A Study on the Radiation of Luminous Flames, *Eleventh Symp. (Int.) Combustion,* The Combustion Institute, 1967, pp. 381–389.
40. Perry, Robert H., Cecil H. Chilton, and Sidney D. Kirkpatrick (eds.): "Chemical Engineers' Handbook," 4th ed., McGraw-Hill Book Company, New York, 1963.
41. Hougen, Olaf A., Kenneth M. Watson, and Roland A. Ragatz: Material and Energy Balance, in "Chemical Process Principles," 2d ed., vol. 1, John Wiley & Sons, Inc., New York, 1954.
42. Gaydon, A. G., and H. G. Wolfhard: "Flames, Their Structure, Radiation, and Temperature," 2d ed., Macmillan and Company, New York, 1960.
43. Barnett, Henry C., and Robert R. Hibbard (eds.): Basic Considerations in the Combustion of Hydrocarbon Fuels with Air, *NACA Rept.* 1300, 1957.
44. Hottel, H. C., and A. F. Sarofim: The Effect of Gas Flow Patterns on Radiative Transfer in Cylindrical Furnaces, *Int. J. Heat Mass Transfer*, vol. 8, no. 8, pp. 1153–1169, 1965.

45. Dayan, A., and C. L. Tien: Radiant Heating from a Cylindrical Fire Column, *Combust. Sci. Technol.*, vol. 9, pp. 41–47, 1974.

46. Edwards, D. K., and A. Balakrishnan: Thermal Radiation by Combustion Gases, *Int. J. Heat Mass Transfer*, vol. 16, pp. 25–40, 1973.

47. Modak, Ashok: Thermal Radiation from Pool Fires, *Combust. Flame*, vol. 29, no. 2, pp. 177–192, 1977.

48. Modak, Ashok: Nonluminous Radiation from Hydrocarbon-Air Diffusion Flames, *Combust. Sci. Technol.*, vol. 10, pp. 245–259, 1975.

49. Taylor, P. B., and P. J. Foster: The Total Emissivities of Luminous and Non-luminous Flames, *Int. J. Heat Mass Transfer*, vol. 17, pp. 1591–1605, 1974.

50. Annamali, K., and P. Durbetaki: Characteristics of an Open Diffusion Flame, *Combust. Flame*, vol. 25, pp. 137–139, 1975.

51. Sarofim, A. F., and H. C. Hottel: Radiative Transfer in Combustion Chambers: Influence of Alternative Fuels, *Sixth Int. Heat Transfer Conf.*, Toronto, vol. 6, pp. 199–217, August, 1978.

52. Sato, T., T. Kunitomo, S. Yoshi, and T. Hashimoto: On the Monochromatic Distribution of the Radiation from the Luminous Flame, *Bull. Jpn. Soc. Mech. Eng.*, vol. 12, pp. 1135–1143, 1969.

53. Kunugi, M., and H. Jinno: Determination of Size and Concentration of Soot Particles in Diffusion Flames by a Light-Scattering Technique, *Eleventh Symp. (Int.) Combust.*, pp. 257–266, 1966.

54. Howarth, C. R., P. J. Foster, and M. W. Thring: The Effect of Temperature on the Extinction of Radiation by Soot Particles, *Third Int. Heat Transfer Conf. AIChE*, vol. 5, pp. 122–128, 1966.

55. Dalzell, W. H., and A. F. Sarofim: Optical Constants of Soot and Their Application to Heat-flux Calculations, *J. Heat Transfer*, vol. 91, no. 1, pp. 100–104, 1969.

56. Siddall, R. G., and I. A. Mc Grath: The Emissivity of Luminous Flames, *Ninth Symp. (Int.) Combust.* (W. G. Berl, ed.), 1963, pp. 102–110.

57. Stull, V. Robert, and Gilbert N. Plass: Emissivity of Dispersed Carbon Particles, *J. Opt. Soc. Am.*, vol. 50, no. 2, pp. 121–129, 1960.

58. Hawksley, P. G. W.: The Methods of Particle Size Measurements, pt. 2, Optical Methods and Light Scattering, *Brit. Coal Utilization Res. Assoc. Monthly Bull.*, vol. 16, nos. 4 and 5, pp. 134–209, 1952.

59. Felske, J. D., and C. L. Tien: Calculation of the Emissivity of Luminous Flames, *Combust. Sci. Technol.*, vol. 7, no. 1, pp. 25–31, 1973.

60. Felske, J. D., and C. L. Tien: The Use of the Milne-Eddington Absorption Coefficient for Radiative Heat Transfer in Combustion Systems, *Trans. ASME, J. Heat Transfer*, vol. 99, no. 3, pp. 458–465, 1977.

61. Buckius, R. O., and C. L. Tien: Infrared Flame Radiation, *Int. J. Heat Mass Transfer*, vol. 20, no. 2, pp. 93–106, 1977.

62. Kunitomo, Takeshi, and Takashi Sato, Experimental and Theoretical Study on the Infrared Emission of Soot Particles in Luminous Flame, *4th Int. Heat Transfer Conf.*, Paris-Versailles, September, 1970.

63. Liebert, Curt H., and Robert R. Hibbard: Spectral Emittance of Soot, *NASA* TN D-5647, 1970.

64. Gray, William A., and R. Muller: "Engineering Calculations in Radiative Heat Transfer," Pergamon Press, New York, 1974.

65. Abramowitz, Milton, and Irene A. Stegun: "Handbook of Mathematical Functions," NBS Applied Mathematics Series, no. 55, 1967.

66. Yuen, W. W., and C. L. Tien: A Simple Calculation Scheme for the Luminous-Flame Emissivity, *Sixteenth Symp. (Int.) Combustion,* The Combustion Institute, 1977, pp. 1481–1487.

67. Erickson, W. D., G. C. Williams, and H. C. Hottel: Light Scattering Measurements on Soot in a Benzene-Air Flame, *Combust. Flame*, vol. 8, no. 2, pp. 127–132, 1964.

68. Thring, M. W., J. M. Beer, and P. J. Foster: The Radiative Properties of Luminous Flames, *Third Int. Heat Transfer Conf. AIChE*, vol. 5, pp. 101–111, 1966.

69. Thring, M. W., P. J. Foster, I. A. Mc Grath, and J. S. Ashton: Prediction of the Emissivity of Hydrocarbon Flames, *Int. Develop. Heat Transfer, ASME*, 1963, pp. 796–803.
70. Sato, Takashi, and Ryuichi Matsumoto: Radiant Heat Transfer from Luminous Flame, *Int. Develop. Heat Transfer, ASME*, 1963, pp. 804–811.
71. Yagi, S., and H. Inoue: Radiation from Soot Particles in Luminous Flames, *Eighth Symp. (Int.) Combust.*, pp. 288–293, 1962.
72. Bone, William A., and Donald T. A. Townsend: "Flame and Combustion in Gases," Long-Mans, Green and Co., Ltd., London, 1927.
73. Leckner, B.: Radiation from Flames and Gases in a Cold Wall Combustion Chamber, *Int. J. Heat Mass Transfer*, vol. 13, no. 1, pp. 185–197, 1970.
74. Magnussen, B. F., and B. H. Hjertager: On Mathematical Modeling of Turbulent Combustion with Special Emphasis on Soot Formation and Combustion, *Sixteenth Symp. (Int.) Combustion,* The Combustion Institute, 1977, pp. 719–729.
75. Gibson, M. M., and J. A. Monahan: A Simple Model of Radiation Heat Transfer from a Cloud of Burning Particles in a Confined Gas Stream, *Int. J. Heat Mass Transfer*, vol. 14, pp. 141–147, 1971.
76. Selcuk, Nevin, and R. G. Siddall: Two-Flux Spherical Harmonic Modelling of Two-Dimensional Radiative Transfer in Furnaces, *Int. J. Heat Mass Transfer*, vol. 19, pp. 313–321, 1976.
77. Leckner, Bo: Radiative Heat Transfer Calculations in Real Gases, *Warme Stoffubertrag.*, vol. 7, pp. 236–247, 1974.
78. Kunitomo, Takeshi: Luminous Flame Emission under Pressure up to 20 Atm, pp. 271–281, in N. H. Afgan and J. M. Beer (eds.), "Heat Transfer in Flames," Scripta Book Co., Washington, D.C., 1974.
79. Williams, J. J., and D. P. Dudley: The Radiative Contribution to Heat Transfer in Metalized Propellant Exhausts, *Fifth Symp. Thermophys. Properties, ASME*, Boston, Sept. 30–Oct. 2, 1970.
80. Lanzo, Chester D., and Robert G. Ragsdale: Heat Transfer to a Seeded Flowing Gas from an Arc Enclosed by a Quartz Tube, *Proc. 1964 Heat Transfer Fluid Mech. Inst.* Warren H. Giedt and Salomon Levy, eds.), pp. 226–244, 1964.
81. Howell, John R., and Harold E. Renkel: Analysis of the Effect of a Seeded Propellant Layer on Thermal Radiation in the Nozzle of a Gaseous-core Nuclear Propulsion System, *NASA* TN D-3119, 1965.
82. Siegel, Robert: Radiative Behavior of a Gas Layer Seeded with Soot, *NASA* TN D-8278, July, 1976.
83. Abdelrahman, M., P. Fumeaux, and P. Suter: Study of Solid-Gas Suspensions Used for Direct Absorption of Concentrated Solar Radiation, *Sol. Energy*, vol. 22, no. 1, pp. 45–48, 1979.
84. Thomas, D. Lyddon: Problems in Applying the Line Reversal Method of Temperature Measurement to Flames, *Combust. Flame*, vol. 12, no. 6, pp. 541–549, 1968.
85. Greenberg, J. M., and T. P. Roark (eds.): Interstellar Grains, *NASA* SP-140, 1967.
86. Donn, B., and K. S. Krishna Swamy: Extinction by Interstellar Grains, Mie Particles and Polycyclic Aromatic Molecules, *Physica*, vol. 41, no. 1, pp. 144–150, 1969.
87. Huffman, Donald R.: Optical Properties of Particulates, *Astrophys. Space Sci.*, vol. 34, pp. 175–184, 1975.

PROBLEMS

1 Evaluate the geometric mean transmittance $\bar\tau_{d1-2}$ from the element dA_1 to the area A_2 in Prob. 2 of Chap. 7. The region between the areas is filled with a gray medium at uniform temperature having an absorption coefficient $a = 0.025$ in^{-1}.

   Answer: 0.53.

2 A spherical cavity is filled with an isothermal gray medium having an absorption coefficient a. Set up the relations needed to obtain the geometric mean transmittance $\bar\tau_{1-d2}$ from the

cavity surface A_1 to an area element at the center of the opening A_2. (*Note:* A similar situation is treated by Koh, *Int. J. Heat Mass Transfer*, vol. 8, no. 2, 1965.)

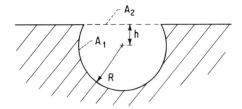

3   A sphere of gray gas at uniform temperature is situated above a surface with the region between the sphere and surface being nonabsorbing. Derive a relation for the radiative energy incident on the circular area as shown. (*Hint:* Consider the circular area as being a cut through a concentric sphere surrounding the gas sphere, and make use of the symmetry of the geometry.)

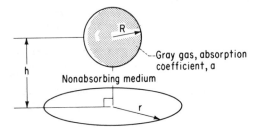

4   For the radiating sphere of gas in Prob. 3, derive an expression for the local energy flux received along the plane surface as a function of distance r.

5   Pure carbon dioxide at 1 atm and 2220 K is contained between parallel plates 0.15 m apart. What is the heat flux received at the plates as a result of radiation by the gas? (Use the CO_2 total emittance chart.)

   Answer: 100,000 W/m^2.

6   A rectangular furnace of dimensions $1 \times 1 \times 4$ ft has soot-covered interior walls that can be considered black. The furnace is filled with combustion products at a uniform temperature of 3500°R composed of 40% by volume CO_2, 30% by volume water vapor, and the remainder N_2. The total pressure is 2 atm. Compute the energy radiated by the gas to the walls using the CO_2 and H_2O total emittance charts.

   Answer: 812,000 Btu/h.

7   A pipe 10 cm in diameter is carrying superheated steam at 1.2 atm pressure and at a uniform temperature of 1110 K. What is the radiative flux from the steam received at the pipe wall?

   Answer: 18,100 W/m^2.

8   A furnace at atmospheric pressure with interior in the form of a cube having a 2-ft edge dimension is filled with a 50:50 mixture by volume of CO_2 and N_2. The gas temperature is uniform at 3000°R, and the walls are cooled to 2000°R. The interior wall surfaces are black. At what rate is energy being supplied to the gas (and removed from the walls) to maintain these temperatures? Use the method in Sec. 17-6.2.

   Answer: 216,000 Btu/h.

9   Consider the same conditions and furnace volume as in Prob. 8. The furnace is now a cylinder with height equal to two times its diameter. What is the energy rate?

   Answer: 216,000 Btu/h.

10   Carbon dioxide at 5 atm pressure and 3000°R is contained between parallel nongray plates 5 cm apart. The plates are both at 2000°R and have hemispherical spectral emissivity as a function of wave number as given in Example 17-4. Considering the five CO_2 absorption bands as in Example 17-4, compute the heat loss by radiation from the gas to the plates.

11   A gray gas having absorption coefficient 0.2 cm^{-1} is contained between concentric gray spheres. The inner sphere has a diameter of 5 cm, a temperature of 1100 K, and emissivity 0.2. The outer sphere has a diameter of 10 cm, temperature of 830 K, and emissivity 0.4. Compute the heat transfer from the inner to the outer sphere by using the result of the exchange-factor approximation. (*Hint:* ψ_b can be found from Table 1 of [36].)

12   A cubical black-walled furnace with 1-m edges contains a mixture of gases at $T = 1800$ K and pressure of 2 atm. The gas has volume fractions of 0.4/0.4/0.2 for $CO_2/H_2O/N_2$. Cooling water is passed over one face of the cube. If the water is initially at 30°C and has a maximum allowable temperature rise of 10°C, what flow rate of water (kilogram/hour) is required? (Neglect reradiation from the wall, and assume all other walls are cool.)

13   A very large furnace has a large number of tubes arranged in an equilateral triangular array. The tubes are of 1 in outside diameter, and the tube centers are spaced 2.5 in apart. The furnace gas is composed of 75% CO_2 and 25% H_2O by volume, and the combustion process in the furnace keeps the gas at a temperature of 2000°F and a total pressure of 0.9 atm. What is the radiative heat flux incident on a tube in the interior of the bundle (i.e., completely surrounded by other tubes) per foot of tube length?

   Answer: 2100 Btu/h

14   Two opposed parallel rectangles are separated by a distance of 0.3 m. The rectangles are of size 0.6 × 0.9 m. The space between the rectangles is filled with H_2O vapor at $P = 1$ atm and $T = 1100$ K. (Assume for this problem that only the 2.7-μm band of the H_2O participates in the radiative absorption and emission of the gas, and use the data of Tables 16-2 and 16-3 to compute α_g.) Rectangle 1 has $T_1 = 1360$ K, $\epsilon_1 = 1.0$. Rectangle 2 has $T_2 = 530$ K, $\epsilon_2 = 0.5$. Assume that the surroundings are at low temperature. Compute the total energy being added to each plate. (Use the method of Sect. 17-7.)

15   A furnace is in the form of a cube with an edge length of 50 cm and with black walls that are cooled so that they emit negligible energy. The furnace is filled with nitrogen maintained at 1500 K. It is desired that the average energy flux received at the walls be 20% of the blackbody flux emitted at the gas temperature. To accomplish this the gas is seeded with propane soot (Fig. 17-28). What volume fraction of soot is required? If the edge dimension of the furnace is doubled, what soot volume fraction is required? For the edge length of 50 cm, what soot volume fraction is required to double the flux received at the walls?

   Answer: 0.0033, 0.0017, 0.0077.

16   A wall is to be shielded from normally incident radiation by flowing along it a layer of cool nitrogen seeded with soot particles. The seeded layer is $\frac{1}{2}$ cm thick. For design purposes it is desired to examine the attenuation of the incident radiation as a function of soot volume concentration in the nitrogen. For incident green light and for infrared radiation at $\lambda = 5$ μm, prepare a plot of percent transmission as a function of soot volume concentration. [Use the Mie theory, Eq. (17-138), and the optical properties of propane soot. Neglect scattering.]

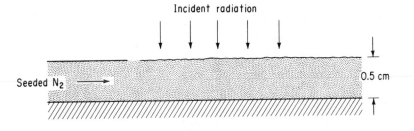

17 Heat is being transferred by radiation from a gray plate with $T_1 = 1000$ K, $\epsilon_1 = 0.8$, to a second plate parallel to it with $T_2 = 900$ K, $\epsilon_2 = 0.5$. The plates are 10 cm apart, and the space between them is filled with stagnant nitrogen gas. Heat conduction by the gas is neglected. It is desired to reduce the net heat transfer from plate 1 to plate 2 by 35%. One idea is to suspend propane soot uniformly in the gas. Estimate the volume concentration of soot required. (*Hint:* See Sec. 14-7.4.)

EIGHTEEN

ENERGY TRANSFER BY RADIATION COMBINED WITH CONDUCTION AND/OR CONVECTION

18-1 INTRODUCTION

When appreciable heat conduction and/or convection occur simultaneously with radiation in an absorbing-emitting medium, there are mathematical complications in addition to those discussed for radiation alone. Unless it can be shown that both the conduction and convection have a negligible effect compared with the radiation or vice versa, a nonlinear integrodifferential equation will result for the energy equation of the general problem having combined modes of heat transfer. Fortunately, the formulation can be simplified for some circumstances when all modes must be included. For example, when the gas is optically thick the diffusion approximation can be applied. The radiation integrals are replaced by differential terms and a nonlinear differential equation will then result. Other approximations, such as the transparent-gas or optically thin approximation (Sec. 15-3.1), can be applied under suitable conditions to simplify the radiative terms.

Because the combined-mode problems treated in this chapter are generally mathematically complex, it is not usually possible to obtain an analytical solution even for seemingly simple physical cases. Consequently, for each physical situation discussed here the analysis will be formulated and some of the intermediate steps in the solution outlined; then the results of a numerical solution will be given and discussed. The situations considered will be a stationary conducting and radiating gas layer between two parallel planes, a boundary-layer flow, a channel flow of radiating and heat-conducting gas, and transient heating of a slab of material by radiation absorption.

18-2 SYMBOLS

A	area
a	absorption coefficient
c	propagation speed in medium
c_p, c_v	heat capacity
D	distance between parallel planes; tube diameter
E_n	exponential integral function
e	emissive power
$\bar{F}$	exchange factor
f	function of η in Blasius boundary-layer solution; gas-to-gas exchange factor
g	surface-to-gas exchange factor; gravity
i	radiation intensity
K_v	function defined in Eq. (18-24)
k	thermal conductivity
l	length
N_j	conduction-radiation parameter based on temperature T_j
P	pressure
Pr	Prandtl number
Q	energy per unit time
q	energy flux, energy per unit area and time
q'''	heat generation per unit volume and time
$\mathbf{q}_r$	radiative heat flux vector
r	radial coordinate
$\mathbf{r}$	position vector
S	surface area
T	absolute temperature
U	radiant energy density
u, v	velocities in x, y directions
$\bar{u}$	mean velocity
V	volume
x, y	rectangular coordinates
α	thermal diffusivity
β	coefficient of volume expansion
δ	boundary-layer thickness
ϵ	emissivity
ϵ_h	eddy diffusivity for turbulent heat transfer
η	Blasius similarity variable
Θ	dimensionless temperature, T/T_1
θ	cone angle, angle from normal of area
κ	optical depth
λ	wavelength
μ	$\cos\theta$; fluid viscosity
ν	kinematic viscosity
ρ	density of fluid

σ	Stefan-Boltzmann constant
τ	time
τ_0	surface transmissivity
ϕ	dimensionless temperature group, $(T^4 - T_2^4)/(T_1^4 - T_2^4)$
ψ	slip coefficient defined by Eq. (18-22); boundary-layer stream function
ω	solid angle

Subscripts

b	blackbody
c	conduction
D	evaluated at $x = D$
e	emitted
i	incident; inlet
j	jth surface
m	mean value
o	outlet; outgoing
P	Planck mean value
R	Rosseland mean value
r	radiative; in radial direction
S	surface
V	volume
w	evaluated at wall
λ	spectrally dependent
0	free stream value; initial value
$1, 2$	surface 1 or 2

Superscripts

$'$	directional quantity
$*$	dummy variable of integration

18-3 RADIATION WITH CONDUCTION

There are some important situations in which heat is transported within a medium by only radiation and conduction. These usually involve solid or highly viscous media so that convection in the medium is not important. In a liquid or gas, forced and/or free convection are usually of sufficient importance that they must be included. To develop the theory, only radiation and conduction are considered in this section; convection will be added later. The following are three practical situations in which combined radiation-conduction transport is important.

One application is in the glass industry. Glass can absorb significant amounts of radiation in certain wavelength regions (see Figs. 5-29–5-31). The absorbed radiation is then reemitted within the glass, thereby providing a radiative transport traveling layer by layer through the medium. The ordinary thermal conduction is thus augmented by a *radiative conduction*. Radiative effects are quite important in influencing the temperature distribution within molten glass in a furnace and during heat treatment of glass plates. These effects have been analyzed in [1-7].

A second application is in glassy materials sometimes used as an ablating coating to protect the interior of a body from high external temperatures by sacrifice of the ablating coating. The radiation-conduction process is important in regulating the temperature distribution within the ablating layer [8-10]. The temperature distribution is influential in determining how the ablating material will soften, melt, or vaporize. These processes ultimately govern how efficiently the material will protect the surface.

A third application is radiation within cryodeposits of solidified gas formed on a very cold surface. The surface may be on a space vehicle orbiting at the upper fringe of the atmosphere, or may be part of a cryopump used to produce a high vacuum by condensing the gas within a chamber. The cryodeposit coating changes the radiative properties of the cold surface and can thus significantly influence the radiative exchange with this surface. Radiative transfer in cryodeposits is considered in [11-13].

In this section, some methods are examined for treating energy transfer by combined radiation and conduction. The conduction-radiation parameter is introduced, and the energy equation formulated. Then some approximations are considered, the most simple being the addition of separately computed radiation and conduction transfers to obtain the combined transfer. Where applicable, the approximation methods to the equation of transfer presented in Chap. 15 can be applied to simplify the radiation terms in these multimode problems, and the diffusion method is applied here as an example. Also, application of the Monte Carlo technique to combined-mode problems is mentioned in Sec. 20-6.

18-3.1 The Conduction-Radiation Parameter

When conduction is present, a dimensionless conduction-radiation parameter N is introduced. Its definition can be developed by analyzing the one-dimensional layer in Fig. 18-1. The material has thermal conductivity k and absorption coefficient a, and is one radiation-mean-free-path thick. If it is assumed that the temperature profile within the slab is nearly linear, the conduction through the layer for area A is

$$Q_c = -kA \frac{T_2 - T_1}{1/a} \tag{18-1}$$

The total radiation emitted by the layer of area A can be written by using Eq. (13-40) as (not including induced emission)

$$Q_e = 4a\sigma T_m{}^4 A \frac{1}{a} \tag{18-2}$$

by suitably defining a mean temperature T_m and neglecting attenuation in the volume.

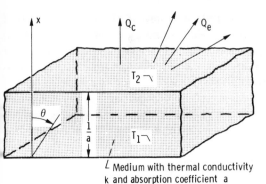

L Medium with thermal conductivity k and absorption coefficient a

Figure 18-1 Conduction through and radiation from plane volume element.

The ratio of conducted to emitted flux is then

$$\frac{Q_c}{Q_e} = \frac{ka(T_1 - T_2)}{4\sigma T_m^4} \tag{18-3}$$

Dividing by T_1 and letting $\Theta = T/T_1$ give

$$\frac{Q_c}{Q_e} = \frac{ka}{4\sigma T_1^3} \frac{1 - \Theta_2}{\Theta_m^4} = N_1 \frac{1 - \Theta_2}{\Theta_m^4} \tag{18-4}$$

The $N_j \equiv ka/4\sigma T_j^3$ is the *conduction-radiation parameter* (or *Stark number*) based on the jth temperature.

For the special case when $\Theta_m^4 = 1 - \Theta_2$, $Q_c/Q_e = N_1$, and the parameter gives a measure of the relative energy amounts carried by conduction and emitted radiation for this layer of thickness $1/a$. Generally, however, N_j does *not* directly give the relative values of conduction to emission because the ratio of these values depends also upon both the temperature difference and temperature level as shown by (18-4).

18-3.2 Energy Balance

To obtain a solution for combined conduction-radiation energy transfer in an absorbing-emitting medium, the energy equation (14-18) can be used. After neglecting terms that do not apply, this is solved subject to the boundary conditons, to obtain the temperature distribution in the medium; the heat flow can then be found. Without fluid motion, the terms involving convection, viscous dissipation, and volume expansion are omitted from (14-18) to yield

$$\rho c_p \frac{\partial T}{\partial \tau} = \nabla \cdot (k \nabla T - \mathbf{q}_r) + q''' \tag{18-5}$$

The $\nabla \cdot \mathbf{q}_r$ is then substituted from (14-28) to obtain

$$\rho c_p \frac{\partial T}{\partial \tau} = \nabla \cdot (k \nabla T) + q''' - 4 \int_{\lambda=0}^{\infty} a_\lambda e_{\lambda b}(\lambda, T)\, d\lambda + \int_{\lambda=0}^{\infty} \int_{\omega=0}^{4\pi} a_\lambda i_\lambda'(\lambda, \omega)\, d\omega\, d\lambda \tag{18-6}$$

Since the radiation terms in (18-6) depend not only on the local temperature but also on the entire surrounding radiation field, the energy equation is an integro-differential equation for the temperature distribution in the medium. The conduction and heat-storage terms depend on a different power of the temperature than the radiation terms, and the energy equation is thus nonlinear.

Numerical solutions of the energy equation have been carried out by Gardon (see Sec. 19-6.3), Viskanta and Grosh [14, 15], and others. Most solutions have been for media in a plane-slab geometry, although some solutions have been carried out for other configurations [16-18]. The following discussion will show how the energy equation is treated more specifically for a plane layer.

18-3.3 Plane-Layer Geometry

Consider a layer of conducting-radiating material between parallel black plates as illustrated by Fig. 18-1. Plate 1 is at temperature T_1, plate 2 is at T_2, and the plates are a distance D apart. (The specific spacing of $1/a$ shown in the figure was used in Sec. 18-3.1.) The medium between the plates is gray and has a constant thermal conductivity k and absorption coefficient a. The integrodifferential equation governing steady energy transfer will now be developed for this geometry.

For one-dimensional heat conduction and constant k, the $\nabla \cdot (k \nabla T)$ in (18-5) reduces to $k d^2 T/dx^2$. The $\nabla \cdot \mathbf{q}_r$ becomes $dq_{r,x}/dx$. The temperature distribution is at steady state, $\partial T/\partial \tau = 0$, and $q''' = 0$. Then with $\kappa = ax$, (18-5) reduces to

$$ka \frac{d^2 T}{d\kappa^2} = \frac{dq_{r,\kappa}}{d\kappa} \tag{18-7}$$

The $dq_{r,\kappa}/d\kappa$ is given by (14-68). For black walls, $i^+(0, \mu) = \sigma T_1^4/\pi$ and $i^-(\kappa_D, -\mu) = \sigma T_2^4/\pi$. For zero scattering $I'(\kappa) = \sigma T^4(\kappa)/\pi$ from (14-69). Then (18-7) becomes

$$ka \frac{d^2 T}{d\kappa^2} = -2\sigma T_1^4 \int_0^1 \exp \frac{-\kappa}{\mu} d\mu - 2\sigma T_2^4 \int_0^1 \exp \frac{\kappa_D - \kappa}{-\mu} d\mu$$

$$-2 \int_0^{\kappa_D} \sigma T^4(\kappa^*) E_1(|\kappa - \kappa^*|) d\kappa^* + 4\sigma T^4(\kappa) \tag{18-8}$$

where $\kappa_D = aD$ and $\mu = \cos \theta$. Two boundary conditions are needed to solve (18-8), as this equation contains a second derivative:

$$T(x) = T_1 \qquad \text{at } x = 0 \qquad\qquad T(x) = T_2 \qquad \text{at } x = D \tag{18-9}$$

Some further simplifications in form are possible. Define the nondimensional quantities $\Theta = T/T_1$, $\Theta_2 = T_2/T_1$, and $N_1 = ka/4\sigma T_1^3$. Then by using the exponential integral function defined in (14-59) and in Appendix F, (18-8) can be written as

$$N_1 \frac{d^2 \Theta(\kappa)}{d\kappa^2} = \Theta^4(\kappa) - \frac{1}{2}\left[E_2(\kappa) + \Theta_2^4 E_2(\kappa_D - \kappa) + \int_0^{\kappa_D} \Theta^4(\kappa^*) E_1(|\kappa - \kappa^*|) d\kappa^* \right]$$

$$\tag{18-10}$$

Equation (18-10) is the desired integrodifferential equation for the temperature distribution $\Theta(\kappa)$. It is a nonlinear equation since Θ is raised to the first power in the conduction term while it is raised to the fourth power in the radiation terms. The boundary conditions in dimensionless form are

$$\Theta = 1 \qquad \text{at } \kappa = 0 \qquad\qquad \Theta = \Theta_2 \qquad \text{at } \kappa = \kappa_D \qquad\qquad (18\text{-}11)$$

Examining (18-10) and (18-11) shows that the solution depends on the parameters N_1, κ_D, and Θ_2.

In addition to the temperature distribution, the heat transfer across the layer from plate 1 to plate 2 is usually of interest. Equation (14-85) with $G = \sigma T^4$ gives the net-heat-flux expression for radiation alone across a gray gas between black plates. This radiative flux equation was obtained for convenience at $\kappa = 0$. In addition, at the same location there is now a conduction flux $-k(dT/dx)|_{x=0} = -ka(dT/d\kappa)|_{\kappa=0}$, so that the heat-flux equation becomes (note that from energy conservation q will not depend on κ for the situation being considered here)

$$q = -ka\frac{dT}{d\kappa}\bigg|_{\kappa=0} + \sigma T_1^{\,4} - 2\sigma T_2^{\,4} E_3(\kappa_D) - 2\int_0^{\kappa_D} \sigma T^4(\kappa^*) E_2(\kappa^*)\, d\kappa^* \qquad (18\text{-}12)$$

On the right, the first term is the conduction away from wall 1, the second is the radiation leaving wall 1, the third is the radiation leaving wall 2 that is then attenuated by the medium and reaches wall 1, and the last term is the radiation from the medium to wall 1. By using the previously defined dimensionless variables, the heat flux can be written as

$$\frac{q}{\sigma T_1^{\,4}} = -4N_1\frac{d\Theta}{d\kappa}\bigg|_{\kappa=0} + 1 - 2\left[\Theta_2^{\,4} E_3(\kappa_D) + \int_0^{\kappa_D} \Theta^4(\kappa^*) E_2(\kappa^*)\, d\kappa^*\right] \qquad (18\text{-}13)$$

Since there are no heat sources in the medium, this q computed at the lower wall is the same at all κ within the medium.

Viskanta and Grosh [14, 15] obtained solutions of (18-10) by numerical integration and iteration. Some of their temperature distributions are shown in Fig. 18-2. For $N_1 \to \infty$, conduction dominates, and the solution reduces to the linear profile for conduction through a plane later. When $N_1 = 0$, the conduction term drops out, and the temperature profile has a discontinuity (temperature slip) at each wall as discussed for the case of radiation alone in Sec. 14-7.4. When conduction is present, there is no temperature slip. Some heat-flux results from [15] are shown in Table 18-1 as obtained from Eq. (18-13). Timmons and Mingle [19] have carried out solutions for the same problem with specular, rather than diffuse, boundaries. Results are within a few percent of those for diffuse boundaries.

Some experimental results are given by Schimmel et al. [20] for layers of CO_2, N_2O, and mixtures CO_2–CH_4 and CO_2–N_2O. The conduction heat fluxes at the walls were compared with analytical results using gray-gas, box, and wide-band gas models. The wide-band model of Edwards and Menard (see Sec. 16-6.4) was found to give the best results.

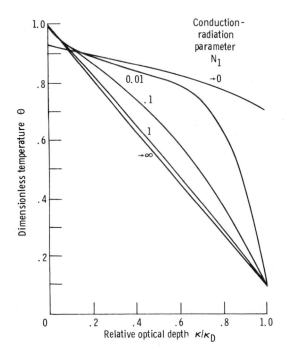

Figure 18-2 Dimensionless temperature distribution in gray gas between infinite parallel black plates with conduction and radiation. Plate temperature ratio $\Theta_2 = 0.1$; optical spacing $\kappa_D = 1.0$. *(Data from [14].)*

Table 18-1 Heat flux between parallel black plates by combined radiation and conduction through a gray medium [15]

Optical thickness κ_D	Plate-temperature ratio Θ_2	Conduction-radiation parameter N_1	Dimensionless energy flux $q/\sigma T_1{}^4$
0.1	0.5	0	0.859
		0.01	1.074
		0.1	2.880
		1	20.88
		10	200.88
1.0	0.5	0	0.518
		0.01	0.596
		0.1	0.798
		1	2.600
		10	20.60
1.0	0.1	0	0.556
		0.01	0.658
		0.1	0.991
		1	4.218
		10	36.60
10	0.5	0	0.102
		0.01	0.114
		0.1	0.131
		1	0.315
		10	2.114

18-3.4 Simple Addition of Radiation and Conduction Energy Transfers

A relatively simple idea to obtain the combined energy transfer by radiation and conduction is to assume that the interaction between the two transfer processes is so weak that each process can be considered to act independently. Then the conduction and radiation transfers are each formulated as if the other mechanism were not present. Einstein [21] and Cess [22] investigated this approximation for an absorbing-emitting gray medium between infinite parallel plates. When the plates were black, the energy transfer was within 10% of the exact solution. Exact results are approached in the optically thin and thick limits. Larger errors are possible if highly reflecting surfaces are present. Howell [23] shows that the additive solution is also fairly accurate for a gray gas between black concentric cylinders.

An additive solution cannot be used to predict temperature profiles. It is a simple method for predicting energy transfer by combined modes, although the accuracy of the solutions so obtained becomes doubtful in some situations. The use of the additive method is not advisable for problems where the accuracy has not been established by comparison with more exact solutions.

> **EXAMPLE 18-1** By using the addition approximation, obtain a relation for the energy transfer from a gray infinite plate at temperature T_1 and emissivity ϵ_1 to a parallel infinite gray plate at T_2 with emissivity ϵ_2. The spacing between the plates is D, and the region between the plates is filled with gray material having constant absorption coefficient a and thermal conductivity k. Use the diffusion approximation for the radiative transfer.
>
> The energy flux by pure conduction from surface 1 to 2 to be used in the additive solution is
>
> $$q_c = \frac{k(T_1 - T_2)}{D} \tag{18-14}$$
>
> The diffusion solution for pure radiation from 1 to 2 is found from Table 15-2 as
>
> $$q_r = \frac{\sigma(T_1^4 - T_2^4)}{3aD/4 + 1/\epsilon_1 + 1/\epsilon_2 - 1} \tag{18-15}$$
>
> Since the two modes are assumed independent, the additive solution gives
>
> $$q = q_c + q_r \tag{18-16}$$
>
> Using the dimensionless variables defined in connection with Eq. (18-10) gives q as
>
> $$\frac{q}{\sigma T_1^4} = \frac{4N_1(1 - \Theta_2)}{\kappa_D} + \frac{1 - \Theta_2^4}{3\kappa_D/4 + 1/\epsilon_1 + 1/\epsilon_2 - 1} \tag{18-17}$$
>
> Equation (18-17) must give correct results at $N_1 = 0$ (pure radiation) within the accuracy of the diffusion solution and at $N_1 \to \infty$ (pure conduction), because it simply adds these two limiting cases. As shown by (15-33), the effect of scattering

is included in (18-17) if $\kappa_D = (a + \sigma_s)D$. Examples 14-3 and 14-4 show that superposition provides the exact solution for the limits of a purely scattering medium or an optically thin medium.

A comparison of $q/\sigma T_1^4$ from (18-17) with exact numerical solutions for $\epsilon_1 = \epsilon_2 = 1$ and $\Theta_2 = 0.5$ from the work of Viskanta and Grosh [14, 15] is shown in Fig. 18-3. For this geometry and black surfaces, the results of the additive solution are very accurate. The additive method appears even better here because of the fortuitous benefit that the diffusion solution gives a pure-radiation heat transfer that is slightly above the exact pure-radiation solution (Fig. 15-6) while the pure-conduction result is too low. This is because the conduction solution is based on the linear gradient of T while the actual gradients at the boundaries are larger when radiation is present (see Fig. 18-7a, for example). The errors in the two solutions tend to cancel, thus giving an accurate combined solution for this geometry. Nelson [24] shows that superposition provides good results for a non-gray gas with a single absorption band.

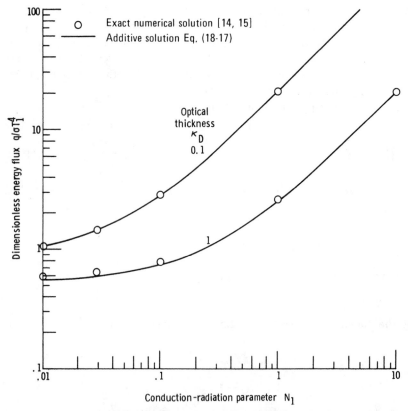

Figure 18-3 Comparison of simple additive and exact numerical solutions of combined conduction-radiation energy transfer between black parallel plates. Plate temperature ratio $\Theta_2 = 0.5$.

18-3.5 The Diffusion Method

This method has an advantage over the additive method in that a solution is obtained to the coupled energy equation and this yields the temperature distribution in the medium. In Sec; 15-5.2, it was shown that the diffusion heat-flux relation for radiative transfer has the same form as the Fourier conduction law. By using the Rosseland mean absorption coefficient defined in (15-48), the radiative flux vector can be written as

$$q_r = -\frac{4}{3a_R} \nabla e_b = -\frac{16\sigma T^3}{3a_R} \nabla T = -k_r \nabla T \tag{18-18}$$

where k_r is the *radiative conductivity* defined by

$$k_r \equiv \frac{16\sigma T^3}{3a_R} \tag{18-19}$$

Then the energy flux vector by combined radiation and conduction at any position in the medium can be expressed as

$$q = q_r + q_c = -(k_r + k) \nabla T = -\left(\frac{16\sigma T^3}{3a_R} + k\right) \nabla T \tag{18-20}$$

This can be used, as in the heat conduction equation, to obtain an energy balance on a differential volume element within the absorbing-emitting medium. For example, in two-dimensional rectangular coordinates, with no internal heat sources, the energy equation is

$$-\nabla \cdot q = \frac{\partial}{\partial x}\left[\left(\frac{16\sigma T^3}{3a_R} + k\right)\frac{\partial T}{\partial x}\right] + \frac{\partial}{\partial y}\left[\left(\frac{16\sigma T^3}{3a_R} + k\right)\frac{\partial T}{\partial y}\right] = 0 \tag{18-21}$$

The medium behaves like a conductor with thermal conductivity dependent on temperature.

To obtain the temperature distribution in the medium, an equation such as (18-21) must be integrated subject to the imposed boundary conditions. These conditions would often be specified temperatures on the enclosure surfaces. However, near a boundary the diffusion approximation is not valid; consequently, the solution is incorrect near the wall and it cannot be matched directly to the boundary conditions. To overcome this difficulty, the boundary condition at the edge of the absorbing-emitting medium will be modified so that the resulting solution to the diffusion equation with this effective boundary condition will be correct in the region away from the boundaries where the diffusion approximation is valid.

In the pure radiation case, a temperature slip was introduced to overcome the difficulty of matching the diffusion solution in the medium to the wall temperature. For combined conduction-radiation, a similar concept was introduced by Goldstein and Howell [25, 26]. By using asymptotic expansions to match the linearized solutions for intensity, flux, and temperature near the wall with the diffusion solution for these quantities far from the wall, an effective slip condition was derived. As shown by Fig.

18-4, this slip gives the boundary condition $T(x \to 0)$ that the diffusion solution must have if it is to extend all the way to the wall. The slip is given in terms of the *slip coefficient* ψ, which is a function only of the conduction-radiation parameter N. In terms of quantities at wall 1, ψ_1 is given by

$$\psi_1 = \frac{\sigma[T_1{}^4 - T^4(x \to 0)]}{q_{r,1}} \tag{18-22}$$

where $q_{r,1}$ is the radiative flux at the boundary as evaluated by the diffusion approximation; T_1 is the wall temperature; $T(x \to 0)$ is the extrapolated temperature in the medium at the wall, which is the effective slip temperature to be used in the diffusion solution. The ψ_1 is computed from the relations of [25] as

$$\psi_1 = \frac{3}{4\pi} \int_0^1 \tan^{-1} \frac{1}{K_v} \, dv \tag{18-23}$$

where K_v is the function

$$K_v = \frac{1}{\pi} \left(\frac{N_1}{2v^3} - \frac{2}{v} - \ln \frac{1-v}{1+v} \right) \tag{18-24}$$

A graph of ψ as a function of N that can be used for any geometry is shown in Fig. 18-5.

With the diffusion approximation, results can be obtained for both the energy transfer and temperature profiles, as will be shown by an example. Other solutions of this general type have been presented in [27, 28].

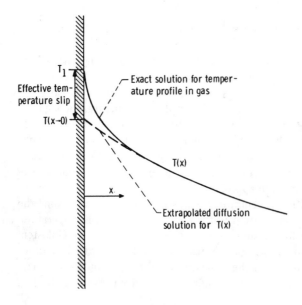

Figure 18-4   Use of effective temperature slip as boundary condition for diffusion solution in combined conduction and radiation.

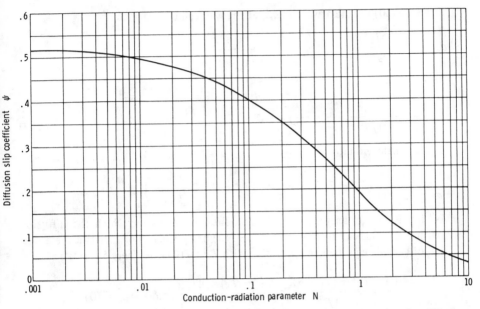

Figure 18-5   Slip coefficient for combined conduction-radiation solutions by the diffusion approximation.

EXAMPLE 18-2 Using the diffusion method, find an equation for the temperature profile in a medium of constant absorption coefficient a and thermal conductivity k, contained between infinite parallel black plates at T_1 and T_2 spaced D apart with the lower plate 1 at $x = 0$. What is the heat transfer across the layer?

For this geometry (18-20) becomes in dimensionless form (note that $a_R = a$ in this case)

$$\frac{q}{\sigma T_1{}^4} = -\left(\frac{4}{3a}\frac{d\Theta^4}{dx} + \frac{k}{\sigma T_1{}^3}\frac{d\Theta}{dx}\right) = -\left(\frac{4}{3}\frac{d\Theta^4}{d\kappa} + 4N_1\frac{d\Theta}{d\kappa}\right) \tag{18-25}$$

From energy conservation, with no internal heat sources, the q is constant across the space between plates. Equation (18-25) can then be integrated from 0 to κ_D to yield

$$\frac{q}{\sigma T_1{}^4}\kappa_D = -\{\frac{4}{3}[\Theta^4(\kappa_D) - \Theta^4(0)] + 4N_1[\Theta(\kappa_D) - \Theta(0)]\} \tag{18-26}$$

where $\Theta(0)$ and $\Theta(\kappa_D)$ are *in the medium* at the lower and upper boundaries. These two temperatures are eliminated by using the slip boundary conditions to relate them to the specified wall temperatures T_1 and T_2.

Consider first the boundary condition at wall 1. For the particular $N = N_1$ of the problem, the ψ_1 is found from Fig. 18-5 and set equal to

$$\psi_1 = \frac{\sigma[T_1{}^4 - T^4(0)]}{q_{r,1}}$$

From (18-20) the radiative flux $q_{r,1}$ at the wall can be written as

$$q_{r,1} = -\frac{16\sigma T_1^3}{3a} \frac{dT}{dx}\bigg|_1 = \frac{16\sigma T_1^3}{3a} \frac{q}{16\sigma T_1^3/3a + k}$$

Then ψ_1 becomes

$$\psi_1 = \frac{\sigma[T_1^4 - T^4(0)]}{(4\sigma/3a)q/(4\sigma/3a + k/4T_1^3)}$$

This is rearranged into

$$\frac{4}{3a}q = \frac{1}{\psi_1}\left\{\frac{4\sigma}{3a}[T_1^4 - T^4(0)] + \frac{k}{4T_1^3}[T_1^4 - T^4(0)]\right\} \tag{18-27}$$

As shown in the derivation of ψ [25], the conditions for which the diffusion solution is valid lead to the jump $T_1 - T(0)$ being small. For convenience a portion of (18-27) can then be linearized. With $T_1 - T(0) = \delta$ where δ is small,

$$\frac{T_1^4 - T^4(0)}{4T_1^3} = \frac{T_1^4 - (T_1 - \delta)^4}{4T_1^3} \approx \frac{T_1^4 - T_1^4 + 4T_1^3\delta}{4T_1^3} = \delta = T_1 - T(0)$$

Then (18-27) becomes $(4/3a)q \approx (1/\psi_1)\{(4\sigma/3a)[T_1^4 - T^4(0)] + k[T_1 - T(0)]\}$ or, in dimensionless form,

$$\frac{4}{3}\psi_1 \frac{q}{\sigma T_1^4} = \frac{4}{3}[1 - \Theta^4(0)] + 4N_1[1 - \Theta(0)] \tag{18-28}$$

Similarly, at wall 2 (note that ψ_2 corresponds to $N = N_2$ in Fig. 18-5)

$$\frac{4}{3}\psi_2 \frac{q}{\sigma T_1^4} = \frac{4}{3}[\Theta^4(\kappa_D) - \Theta_2^4] + 4N_1[\Theta(\kappa_D) - \Theta_2] \tag{18-29}$$

Now add (18-26), (18-28), and (18-29) to eliminate the unknown temperatures in the medium $\Theta(0)$ and $\Theta(\kappa_D)$. This yields the energy flux transferred across the layer as

$$\frac{q}{\sigma T_1^4} = \frac{1 - \Theta_2^4 + 3N_1(1 - \Theta_2)}{\frac{3}{4}\kappa_D + \psi_1 + \psi_2} \tag{18-30}$$

The results of (18-30) are plotted in Fig. 18-6 and compared with the exact and additive solutions (the exchange-factor approximation shown in the figure will be discussed in the next section). At $\kappa_D = 1$, the results compare very well with the exact solution. For a small optical thickness $\kappa_D = 0.1$, however, the diffusion-slip procedure breaks down for intermediate values of N_1, and the simple additive solution provides much better energy transfer values.

An advantage of the diffusion solution is that it yields the temperature distribution in the medium. Temperature profiles can be predicted by integrating (18-25) from 0 to κ (note that q is constant) and then using (18-28) and (18-30)

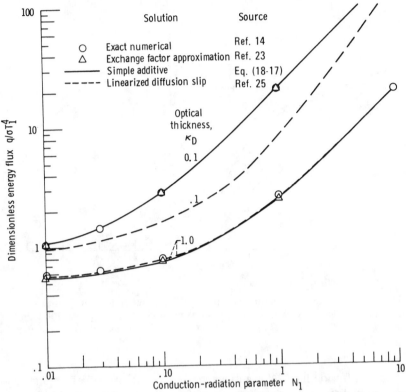

Figure 18-6 Comparison of various methods for predicting energy transfer by conduction and radiation across a layer between parallel black plates. Plate temperature ratio $T_2/T_1 = \Theta_2 = 0.5$.

to eliminate $\Theta(0)$ and q. This yields

$$\frac{1 - \Theta^4(\kappa) + 3N_1[1 - \Theta(\kappa)]}{1 - \Theta_2{}^4 + 3N_1(1 - \Theta_2)} = \frac{3\kappa/4 + \psi_1}{3\kappa_D/4 + \psi_1 + \psi_2} \tag{18-31}$$

Profiles are shown in Fig. 18-7. For $\kappa_D = 1$ (Fig. 18-7a), the profiles are poor except for the largest N_1 shown. Better results are obtained for all N_1 in Fig. 18-7b for $\kappa_D = 10$ because the assumptions in the diffusion solution become more valid at larger κ_D. On the basis of the assumptions used in the diffusion-slip analysis and the comparisons with exact analytical solutions, good temperature-distribution results are expected for $\kappa_D \gtrsim 2$. For $N_1 \to 0$ and $N_1 \to \infty$, the diffusion-slip method goes to the correct limiting solutions.

Within their limits of applicability, diffusion methods provide a different interpretation of the conduction-radiation parameter from that in Sec. 18-3.1. The ratio of molecular conductivity to radiative conductivity given by (18-19) is

$$\frac{k}{k_r} = \frac{k}{16\sigma T^3/3a_R} = \frac{3}{4}\frac{ka_R}{4\sigma T^3} = \frac{3}{4}N \tag{18-32}$$

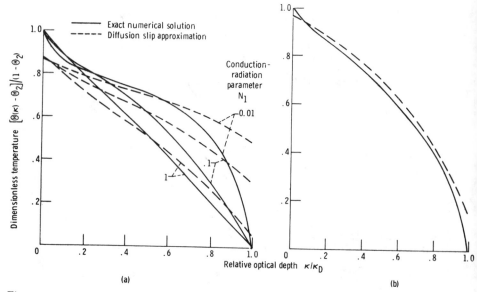

Figure 18-7 Comparison of temperature profile by exact solution [14] with diffusion-slip approximation. Plate temperature ratio $\Theta_2 = 0.5$; plate emissivities $\epsilon_1 = \epsilon_2 = 1.0$. (a) Optical thickness $\kappa_D = 1$; (b) optical thickness $\kappa_D = 10$, conduction-radiation parameter $N_1 = 0.02916$.

Therefore, in the diffusion limit, N is a direct measure of the ratio of k to k_r and consequently is also a direct measure of the ratio of the energy transferred by the two modes.

18-3.6 The Exchange-Factor Approximation

In Sec. 17-8.3 the exchange-factor approximation was introduced for pure radiation problems. Exchange factors are applied here to combined-mode problems, and it is shown that their use provides a convenient technique. In Sec. 17-8.3, the exchange factor $\bar{F}_{j-k}$ is defined as the fraction of the energy leaving surface j that is incident on surface k when all boundaries are black and the intervening medium is in radiative equilibrium (that is, when radiation is the only means of energy exchange). The $\bar{F}_{j-k}$ includes the effect of absorption and reemission of energy by the medium while the energy is in transit from A_j to A_k. A similar exchange factor $d\bar{F}_{j-dx}$ between a surface area and volume element was also defined. The reader may find it helpful to review Sec. 17-8.3 before proceeding.

The total energy emitted by a volume element in the *presence* of heat conduction can be equated to three terms: (1) the energy that would be emitted in the absence of conduction, (2) the *net* energy supplied to the element by conduction (and thus that must be radiated away), and (3) any additional energy d^2Q_{extra} added to the element by radiation over that given by the radiative-equilibrium (zero-conduction) case because of the change in temperature profile by conduction being included. This can be written as

$$4a_p \sigma T^4 \, dV = 4a_p \sigma T_R{}^4 \, dV + k \nabla^2 T \, dV + d^2 Q_{extra} \tag{18-33}$$

The temperature T_R for radiative equilibrium is written, as in (17-123a), in terms of exchange factors:

$$4a_p \sigma T_R{}^4 \, dV = \sum_j Q_{o,j} \, d^2 \bar{F}_{j-dV}$$

The $Q_{o,j}$ is the energy leaving the jth surface of the enclosure surrounding the gas. The $d^2 \bar{F}_{j-dV}$ is the fraction of energy leaving A_j that is *absorbed* in dV for radiative equilibrium in a black enclosure, and it includes the effects of gas absorption and emission while the energy is in transit from A_j to dV. Substituting this relation and assuming that $d^2 Q_{extra}$ is small reduce (18-33) to

$$4a_p \sigma T^4 \approx \frac{1}{dV} \sum_j Q_{o,j} \, d^2 \bar{F}_{j-dV} + k \nabla^2 T \tag{18-34}$$

Note that the exchange factors include the effect of gas-to-gas volume-element radiant interchange based on the temperature profile for *no conduction*. The approximation in (18-34) is that the gas-to-gas radiant exchange is not significantly affected by the new temperature profile that results because of the presence of gas conduction (that is, $d^2 Q_{extra} \approx 0$). The similar approximation is also made that the $Q_{o,j}$ from the solution without conduction can be used. If radiation predominates, then this is a good assumption; if radiation is small, then it will not matter that the radiative terms are somewhat inaccurate.

Equation (18-34) is a nonlinear differential equation for T, the local gas temperature. Howell [23] applied this approach to gray gases in annular enclosures and between infinite parallel plates. Accuracy is found to be comparable to the simple additive solution when computing the heat flux through the gas as shown in Fig. 18-6. The chief advantage of the method is that accurate temperature distributions can be obtained for combined-mode energy transfer problems with little effort. The procedure will be demonstrated with an example.

EXAMPLE 18-3 Find an expression that will yield the temperature profile in a gray gas contained between infinite parallel plates D apart if the gas has absorption coefficient a, thermal conductivity k, and the plates are black at T_1 and T_2. Use the exchange-factor approximation.

Using a layer dx thick as the volume element results in (18-34) becoming, in this geometry,

$$k \frac{d^2 T(x)}{dx^2} \, dx + d\bar{F}_{1-dx} \, \sigma T_1{}^4 + d\bar{F}_{2-dx} \, \sigma T_2{}^4 = 4a\sigma T^4(x) \, dx \tag{18-35}$$

The exchange factors are given by (17-125) and (17-126) as

$$d\bar{F}_{1-dx} = 4a \, dx \, \phi_b(x) \qquad \text{and} \qquad d\bar{F}_{2-dx} = 4a \, dx \, (1 - \phi_b)$$

The ϕ_b can be obtained to good accuracy for this geometry from the diffusion relations, Table 15-2, as

$$\phi_b(\kappa) = \frac{\frac{3}{4}(\kappa_D - \kappa) + \frac{1}{2}}{\frac{3}{4}\kappa_D + 1}$$

Substituting into (18-35) and using the usual nondimensional quantities give

$$\Theta^4(\kappa) - N_1 \frac{d^2\Theta(\kappa)}{d\kappa^2} - \frac{1}{3\kappa_D/4 + 1} \left[\tfrac{3}{4}(\kappa_D - \kappa) + \tfrac{1}{2} + \Theta_2{}^4(\tfrac{3}{4}\kappa + \tfrac{1}{2})\right] = 0 \qquad (18\text{-}36)$$

which is to be solved subject to the boundary conditions $\Theta = 1$ at $\kappa = 0$ and $\Theta = \Theta_2$ at $\kappa = \kappa_D$. This relation gives the correct limiting diffusion solution for $N_1 \to 0$ and the correct conduction solution for $N_1 \to \infty$. Howell [23] has solved for $[\Theta(\kappa) - \Theta_2]/(1 - \Theta_2)$ numerically, using exchange factors from the numerical solutions of the pure-radiation problem, which are more accurate than the diffusion exchange factors used in (18-36). Agreement with the numerical solution of the coupled conduction-radiation problem is shown in Fig. 18-8.

The energy transfer by conduction was found by numerically evaluating $d\Theta/d\kappa|_{\kappa=0}$ and using this to evaluate the conduction flux at the boundary. The radiation flux was assumed to be unaffected by the conduction process in the spirit of the exchange-factor approximation. The results obtained in this approximate way agree quite well with the numerical solution, as shown in Fig. 18-6.

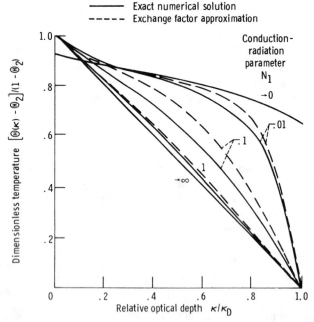

Figure 18-8 Comparison of temperature profiles by exact solution with exchange-factor approximation. Optical thickness $\kappa_D = 1.0$; plate temperature ratio $\Theta_2 = 0.1$; plate emissivities $\epsilon_1 = \epsilon_2 = 1.0$.

Since $d\Theta/d\kappa$ varies with κ while the radiative flux without conduction is constant with κ, evaluating $d\Theta/d\kappa$ at a location other than $\kappa = 0$ would give different results. For the most accurate calculation of heat flux, it was found in [23] that the temperature gradient should be evaluated at the boundary with highest temperature. Using the gradient at the coldest wall leads to a flux prediction that is always too large.

The exchange-factor approximation is not limited to any range of geometry, surface emissivity, or optical thickness; it can be applied to any situation where the exchange factors have been previously obtained or where they can be obtained by some simplified pure-radiation solution. The resulting nonlinear energy equation can usually be cast in the form of a matrix of nonlinear difference equations that can be solved by the numerical technique outlined by Ness [29] (see Sec. 12-3.2).

In situations where the boundaries are not the chief contributors to the radiant energy flux in the gas, the gas-temperature profile becomes important to the radiative flux distribution. In such a case the exchange-factor approximation may become inaccurate; however, [23] gives results for parallel plates and concentric cylinders that compare well with exact numerical solutions for both energy transfer and temperature profiles. Fulkerson and Bannerot [30] have extended the approximation to cases involving both convection and conduction. Limiting cases for heat transfer through optically thick and thin media are presented along with an asymptotic solution uniformly valid at all optical thicknesses.

Lick [31] has also presented various approximations for solving conduction-radiation problems. He develops methods for treating gases with spectral and temperature-dependent properties. Goldstein and Howell [25] outline methods of treating temperature-dependent properties using the apparent-slip technique.

Although numerical methods are generally the only way to obtain exact solutions to combined-mode problems, the approximate methods outlined in this chapter should provide acceptable accuracy for most engineering conduction-radiation problems. All the methods presented here can be applied to problems in two and three dimensions.

18-4 CONVECTION, CONDUCTION, AND RADIATION

Examples of the interaction of convection, conduction, and radiation in absorbing-emitting media are found in combustion chambers, atmospheric phenomena, shock wave problems, rocket nozzles, and industrial furnaces. As a consequence, a large amount of literature is available. Review articles and comprehensive books are given in [22, 32–36]. In spite of this material, these problems remain difficult to solve. In this section, some of the methods in use are outlined.

18-4.1 Boundary-Layer Problems

Radiant emission and absorption can affect the heat transfer in a convection boundary layer [37–43]. Such boundary-layer problems are approached by writing the usual continuity and momentum equations; these do not contain any radiation terms and are not influenced by heat transfer if constant fluid properties are assumed, as is the case here. The energy equation includes an energy source term to account for the net thermal radiation gained by a volume element; from (14-16) this can be written as $-\nabla \cdot \mathbf{q}_r$.

For a two-dimensional laminar boundary layer flowing over a flat plate (Fig. 18-9), assuming a gray fluid with constant properties, negligible viscous dissipation, and no internal heat sources, the energy equation is

$$\rho c_p \left(u \frac{\partial T}{\partial x} + v \frac{\partial T}{\partial y} \right) = k \frac{\partial^2 T}{\partial y^2} - \frac{\partial q_{r,y}}{\partial y} \tag{18-37}$$

where $q_{r,y}$ is the radiation flux in the positive y direction. It is assumed that the convection term in the x direction is dominant over the radiative contribution in that direction, that is, $\rho c_p u\, \partial T/\partial x \gg \partial q_{r,x}/\partial x$, where $q_{r,x}$ is the radiative flux in the x direction; hence the $-\partial q_{r,x}/\partial x$ term has been neglected on the right side of (18-37). In the diffusion limit this states that

$$\rho c_p u \frac{\partial T}{\partial x} \gg \frac{\partial}{\partial x} \left(-\frac{16\sigma T^3}{3 a_R} \frac{\partial T}{\partial x} \right) \tag{18-38}$$

must be true. The problem becomes one of introducing into (18-37) one of the formulations for $q_{r,y}$ that have been developed, and then solving the resulting energy equation together with the momentum and continuity equations. In [38, 39] the diffusion solution is used for $q_{r,y}$, and the last two terms in (18-37) can then be combined as was done in (18-20).

Various techniques can be used to solve the resulting energy equation. References [30, 37, 44] use matched asymptotic expansions of the energy equation assuming a known flow field. A linearized energy equation near the surface is matched to an asymptotic solution far from the surface. Viskanta and Grosh [38] had earlier applied the diffusion approach; they assumed that the diffusion solution was valid all the way to the boundary. Cess [39] and others have assumed the boundary layer to be optically thin, so that it emits but does not absorb radiation. By introducing some other assumptions, Cess was able to treat nongray gas effects. Neglecting absorbed radiation can be a useful approximation when the boundary layer is heated by frictional dissipation while the surface and surrounding gas are cool.

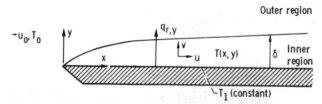

Figure 18-9 Boundary-layer flow over flat plate.

Fritsch et al. [40] studied the shielding of a body by a radiant absorbing layer. The boundary layer was assumed to absorb, but not to emit radiation. The effects of transpiration and an external radiation field were included. Howe [42] had treated a similar problem under somewhat different conditions for the external radiation field.

References [43, 45–48] examine the effect of a radiation on the boundary-layer development in free convection of an absorbing-emitting gas. Reference [45] treats the development of the boundary layer on a horizontal cylinder, while [43, 46–48] examine layer growth on a vertical plate. References [43, 49] experimentally determine the onset of free convection in a gas exposed to thermal radiation.

Optically thin thermal layer. To analyze laminar forced-flow heat transfer on a flat plate, an expression is needed for the radiative source term $\partial q_{r,y}/\partial y$ in (18-37). Within the boundary layer it is assumed that the thermal conditions are changing slowly enough in the x direction, as compared with the y direction, so that the conditions contributing to $q_{r,y}$ at a specific x, say x^+, are all at that x^+ and hence at the temperature distribution $T(x^+, y)$. Then $\partial q_{r,y}/\partial y$ can be evaluated from the one-dimensional relations derived previously. From (18-7) and (18-8), which are written for a region between two black walls,

$$\frac{\partial q_{r,y}}{\partial y} = 4a\sigma T^4 - 2a\sigma \left[T_1{}^4 E_2(\kappa) + T_2{}^4 E_2(\kappa_D - \kappa) + \int_0^{\kappa_D} T^4(\kappa^*) E_1(|\kappa - \kappa^*|)\, d\kappa^* \right]$$

$$(18\text{-}39)$$

For only one bounding wall, the T_2 term is not present and the upper limit of the integral is extended to infinity. Also the $T^4(\kappa^*)$ is replaced by $T^4(x, \kappa^*)$ to emphasize the approximation in which, for the radiation term, the surroundings of position x are taken to be at $T(x, y)$. Then the laminar boundary-layer equation (18-37) becomes, for the temperature $T(x, y)$,

$$\rho c_p \left(u\frac{\partial T}{\partial x} + v\frac{\partial T}{\partial y} \right) = k\frac{\partial^2 T}{\partial y^2} - 4a\sigma T^4$$

$$+ 2a\sigma \left[T_1{}^4 E_2(\kappa) + \int_0^\infty T^4(x, \kappa^*) E_1(|\kappa - \kappa^*|)\, d\kappa^* \right] \qquad (18\text{-}40)$$

where $\kappa = ay$.

The temperature field can be considered as composed of two regions. Near the wall in the usual thermal boundary-layer region of thickness δ that would be present in the absence of radiation, there are large temperature gradients, and heat conduction is important. This region is usually of small thickness; hence, it can be assumed optically thin so that radiation will pass through it without attenuation. For larger y than this region, the temperature gradients are small and heat conduction is neglected compared with radiation transfer. The approximate analysis can now proceed, for example, along the path developed in [22].

In the outer region the velocity in the x direction has the free-stream value u_0, and with the neglect of heat conduction the boundary-layer equation reduces to

$$\rho c_p u_0 \frac{\partial T}{\partial x} = -4a\sigma T^4 + 2a\sigma \left[T_1^4 E_2(\kappa) + \int_0^\infty T^4(x, \kappa^*) E_1(|\kappa - \kappa^*|) \, d\kappa^* \right] \qquad (18\text{-}41)$$

To obtain an approximate solution by iteration, substitute the incoming free-stream temperature T_0 for the temperature on the right side as a first approximation, and then carry out the integral to obtain a second approximation. This yields, for the outer region to first-order terms,

$$T(x, \kappa) = T_0 + \sigma(T_1^4 - T_0^4) E_2(\kappa) \frac{2ax}{\rho c_p u_0} + \cdots \qquad (18\text{-}42)$$

where at $x = 0$, $T = T_0$.

At the edge of the thermal layer $\kappa = a\delta$, which is small, so that $E_2(a\delta) \approx E_2(0) = 1$. Hence, at $y = \delta$, (18-42) becomes

$$T(x, \delta) = T_0 + \sigma(T_1^4 - T_0^4) \frac{2ax}{\rho c_p u_0} + \cdots \qquad (18\text{-}43)$$

Equation (18-43) is the edge boundary condition that the outer radiation layer imposes on the inner thermal layer. The outer temperature is, to this approximation, increasing linearly with x. This is the result of the flowing gas absorbing a net radiation from the plate in proportion to the difference $T_1^4 - T_0^4$ and the absorption coefficient a.

To solve the boundary-layer equation in the inner thermal-layer region, the last integral in (18-40) is divided into two parts, one from $\kappa = 0$ to $a\delta$ and the second from $a\delta$ to ∞. The first portion is neglected as the thermal layer is optically thin, and the second is evaluated by using the outer solution (18-42). By retaining only first-order terms, the boundary-layer energy equation is reduced to

$$u \frac{\partial T}{\partial x} + v \frac{\partial T}{\partial y} = \alpha \frac{\partial^2 T}{\partial y^2} + \frac{2a\sigma}{\rho c_p} (T_1^4 + T_0^4 - 2T^4) \qquad (18\text{-}44)$$

The boundary conditions are given by (18-43) at $y = \delta$, and the specified wall temperature $T = T_1$ at $y = 0$.

The solution becomes quite complex and will not be developed further here. The reader interested in this specialized topic is referred to [22, 50] for additional information.

Optically thick thermal layer. At the opposite extreme from the situation in the previous section, if the thermal layer has become very thick or the medium is highly absorbing, then the boundary layer can be optically thick. The analysis is then considerably simplified, as the diffusion approximation can be employed. With reference to (18-20), the radiative diffusion adds a radiative conductivity to the ordinary thermal conductivity. Then the laminar boundary-layer energy equation (18-37) can be written as

$$\rho c_p \left(u \frac{\partial T}{\partial x} + v \frac{\partial T}{\partial y} \right) = \frac{\partial}{\partial y} \left[\left(\frac{16\sigma T^3}{3a_R} + k \right) \frac{\partial T}{\partial y} \right] \tag{18-45}$$

With the assumption of constant fluid properties, the boundary-layer momentum equation and the continuity equation do not depend on temperature. Consequently, the flow is unchanged by the heat transfer, and the velocity distribution is given by the Blasius solution [51]. The Blasius solution is in terms of a similarity variable $\eta = y\sqrt{u_0/\nu x}$; the stream-function and velocity components are given by

$$\psi = \sqrt{\nu x u_0}\, f(\eta) \qquad u = \frac{\partial \psi}{\partial y} = u_0 \frac{df}{d\eta} \qquad v = -\frac{\partial \psi}{\partial x} = \frac{1}{2}\sqrt{\frac{\nu u_0}{x}} \left(\eta \frac{df}{d\eta} - f \right) \tag{18-46}$$

These quantities are substituted into the energy equation, which can then be placed in the form

$$-\frac{\Pr}{2} f \frac{dT}{d\eta} = \frac{d}{d\eta} \left[\left(\frac{16\sigma T^3}{3a_R k} + 1 \right) \frac{dT}{d\eta} \right] \tag{18-47}$$

The boundary conditions that have been used in the numerical solution are

$$T = T_1 \quad \text{at } \eta = 0 \qquad T = T_0 \quad \text{at } \eta = \infty$$

To be more precise, a diffusion-slip condition should be used at the wall, but this has not been formulated for the combined radiation, convection, and conduction situation.

Let $\Theta = T/T_0$ and $N_0 = k a_R/4\sigma T_0^3$; then (18-47) becomes

$$-\frac{\Pr}{2} f \frac{d\Theta}{d\eta} = \frac{d}{d\eta} \left[\left(\frac{4\Theta^3}{3N_0} + 1 \right) \frac{d\Theta}{d\eta} \right] \tag{18-48}$$

Numerical solutions were carried out in [38], and some typical temperature profiles are shown in Fig. 18-10. For $N_0 = 10$ the profile was found to be within 2%

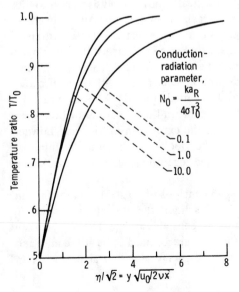

Conduction-radiation parameter, $N_0 = \dfrac{k a_R}{4\sigma T_0^3}$

0.1
1.0
10.0

Temperature ratio T/T_0

$\eta/\sqrt{2} = y\sqrt{u_0/2\nu x}$

Figure 18-10 Boundary-layer temperature profiles for laminar flow on flat plate [38]. Prandtl number $\Pr = 1.0$; temperature ratio $T_1/T_0 = 0.5$.

of that for conduction and convection alone (that is, for $N_0 \to \infty$). The effect of radiation is found to thicken the thermal boundary layer similarly to the effect of decreasing the Prandtl number. This would be expected, since the Prandtl number is the ratio of viscous to thermal diffusion ν/α. The radiation has supplied an additional means for thermal diffusion, thereby effectively increasing the α.

18-4.2 Channel Flows

Of engineering interest in some high-temperature heat exchange devices, is combined radiative and convective energy transfer for the flow of an absorbing-emitting gas in a channel. For laminar flow, the energy equation (18-37) (with $\nu = 0$ for fully developed flow) applies within the flow. References [21, 52-61] treat problems with varying degrees of approximation. Viskanta [52] gives approximate numerical solutions for the flow of a gray absorbing-emitting medium in a parallel-plate channel. Temperature-independent properties are assumed. In addition to κ_D, N, and temperature ratios, a new parameter enters; it is a Nusselt number defined to include a radiative contribution and thus differing from the usual parameter.

Einstein [21, 53] applied the gas-to-surface and gas-to-gas view-factor methods of Hottel (Sec. 17-8.2) to a solution of the energy equation in a finite-length parallel-plate channel and circular tube. Internal heat generation in the gas was included. Comparison was made with the work of Adrianov and Shorin [54], who had used the cold-material approximation (Sec. 15-3.3) so that absorption but not emission from the gas was included. Chen [55] included scattering in his analysis of flow between parallel plates, but assumed a slug-flow velocity profile.

The analyses mentioned so far in this section have all been for gray gases flowing in channels with gray or black walls, and for temperature-independent properties. DeSoto and Edwards [62] presented a tube-flow heat transfer analysis that accounts for nongray gases with temperature-dependent radiative properties. An exponential band model (Sec. 16-6.4) was used to account for the spectral effects. Entrance-region flows were included. A similar study made by Pearce and Emery [63] employed a *box model* for the nongray absorption properties. In this model an absorption band is approximated by having a constant absorption coefficient within an effective bandwidth, and zero absorption elsewhere. Jeng et al. [64, 65] also analyzed the laminar flow of a radiating gas in a tube at constant wall temperature for both gray and nongray media. In the latter case they used the expression of Tien and Lowder ([22] in Chap. 16) for band radiation. In [65] they looked at the optically thin and thick limits for real gases. In the thick limit the logarithmic limit was used for a vibration-rotation band as given by (16-79).

To examine a specific situation, consider the analysis of Einstein [53] for flow in a tube of diameter D as shown in Fig. 18-11a. Gas enters the tube at temperature T_i and leaves at T_o. The tube wall temperature is constant at T_w. The surrounding environments at the inlet and exit ends of the tube are assumed to be at the inlet and exit gas temperatures, respectively. The governing energy equation at position $\mathbf{r}$ in

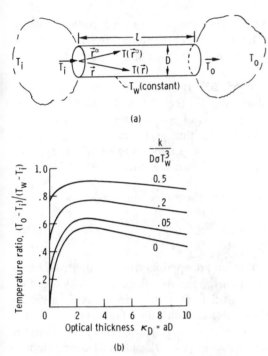

Figure 18-11 Combined radiation and convection for absorbing gas flowing in tube with constant wall temperature [53]. (a) Tube geometry and boundary conditions; (b) exit temperature for $T_i/T_w = 0.4$, $l/D = 5$, and $\rho u c_p/\sigma T_w^3 = 33$.

the tube for laminar flow can be written as

$$\rho c_p u \left.\frac{\partial T}{\partial x}\right|_r = \frac{k}{r}\frac{\partial}{\partial r}\left(r\frac{\partial T}{\partial r}\right)\bigg|_r - 4a\sigma T^4(r)$$

$$+ a\left[\iiint_V \sigma T^4(r^*)f(r^* - r)\,dV + \iint_S \sigma T_s^4(r^*)g(r^* - r)\,dS\right] \tag{18-49}$$

The triple integral is the radiation absorbed at r as a result of emission from all the other gas in the tube. The $f(r^* - r)$ is a gas-to-gas exchange factor from position r^* to position r. The double integral is the radiation absorbed at r as a result of emission from the boundaries, which include the tube wall and the end planes of the tube. The $g(r^* - r)$ is a surface-to-gas exchange factor; the f and g are given in [53].

Some typical results of the numerical solution are given in Fig. 18-11b for a Poiseuille flow velocity profile. These results show how well the gas obtains energy from the wall, since the ordinate is a measure of how close the exit gas approaches the wall temperature. The results are given in terms of gas optical thickness based on tube diameter and a conduction-radiation parameter based on wall temperature. As the optical thickness is increased from zero, the amount of radiated energy from the wall that is absorbed by the gas increases to a maximum value. Then for large κ_D the heat absorbed by the gas decreases. The decrease is caused by the self-shielding of the gas, which means that for high κ_D most of the direct radiation from the tube wall is absorbed in a thin gas layer near the wall. Since the gas emission is isotropic, about one-half the energy reemitted by this thin layer goes back toward the wall. Thus the gas in the center of the tube is shielded from direct radiation, and the heat transfer efficiency decreases.

In some of the tube-flow analyses [55, 62], the radiative terms in the energy equation (18-49) have been simplified by assuming that the chief radiative contribution to the gas at an axial location is the result of temperatures in the immediate surroundings. The axial temperature variation is thus neglected in determining the radiative terms. The radiative fluxes are calculated as those from an infinite cylinder of gas having a radial temperature distribution the same as that at the axial location for which the flux is being evaluated.

For fully developed turbulent flow in a tube, the energy equation is

$$\rho c_p u \frac{\partial T}{\partial x} = \frac{1}{r} \frac{\partial}{\partial r} \left[(k + \rho c_p \epsilon_h) r \frac{\partial T}{\partial r} \right] - \nabla \cdot \mathbf{q}_r \qquad (18\text{-}50)$$

Landram et al. [66] used the optically thin limit to approximate $-\nabla \cdot \mathbf{q}_r$. It was evaluated for a volume element, as equal to the energy absorbed from the wall by use of an incident mean absorption coefficient, minus the energy emitted by use of a Planck mean absorption coefficient (see Sec. 14-4.4). If axial diffusion of radiation is neglected as discussed in connection with (18-37), then $\nabla \cdot \mathbf{q}_r = (1/r) \partial(r q_{r,r})/\partial r$. Expressions for $q_{r,r}$ (radiative flux in radial direction) are given in [68, 69], and detailed temperature-distribution and Nusselt-number results are given in [68] for heat transfer to turbulent flow in a tube. Detailed calculations using an exponential band absorption model are given in [70] for flow in the thermal entrance region of a flat-plate duct downstream of a step in wall temperature. Both laminar and turbulent flow are considered

For laminar flow in a vertical tube, free convection can be significant, and a buoyancy force enters the momentum equation. For fully developed tube flow the momentum equation is

$$\frac{dP}{dx} + \rho g = \mu \frac{1}{r} \frac{d}{dr} \left(r \frac{du}{dr} \right) \qquad (18\text{-}51)$$

This has to be solved in conjunction with (18-50), as the buoyancy term ρg is temperature dependent; the solution is given in [69]. As is usual in free convection, the density was linearized by letting $\rho = \rho_w[1 + \beta(T_w - T)]$. In the same spirit the spectral blackbody function was linearized by letting $e_{\lambda b}(T) = e_{\lambda b}(T_w) - (\partial e_{\lambda b}/\partial T)_w(T_w - T)$.

18-4.3 Stability Problems with Free Convection

Arpaci and Gozum [47] used the P_1 differential approximation (Sec. 15-8.2) to include radiation in an analysis of the thermal convective stability of a nongray fluid between horizontal parallel confining boundaries (the Benard problem with radiation.) Following [47], the radiative flux equation (15-128a) for a nongray fluid is

$$\int_0^\infty \left[\frac{\partial}{\partial x_k} \left(\frac{1}{a_\lambda} \sum_{j=1}^3 \frac{\partial^2 q_{r\lambda,j}}{\partial x_j \, d\lambda} \right) - 4 \frac{\partial e_{\lambda b}}{\partial x_k} - 3a_\lambda \frac{\partial q_{r\lambda,k}}{\partial \lambda} \right] d\lambda = 0 \qquad (18\text{-}52)$$

In the limit $a_\lambda x \to 0$, (18-52) reduces to the thin-gas approximation with the term $-3a_\lambda \, \partial q_{r\lambda, k}/\partial\lambda \to 0$ so that

$$\int_0^\infty \left[\frac{\partial}{\partial x_k} \left(\frac{1}{a_\lambda} \sum_{j=1}^3 \frac{\partial^2 q_{r\lambda, j}}{\partial x_j \, \partial\lambda} \right) - 4 \frac{\partial e_{\lambda b}}{\partial x_k} \right] d\lambda = 0$$

By setting the integrand equal to zero, integrating with respect to x_k, and then integrating with respect to λ, this is transformed to

$$\sum_{j=1}^3 \frac{\partial q_{r,j}}{\partial x_j} = \int_0^\infty 4 a_\lambda e_{\lambda b} \, d\lambda + C = 4 a_p \sigma T^4 + C \qquad (18\text{-}53)$$

as in (15-129). The $a_P = \int_0^\infty a_\lambda e_{\lambda b} \, d\lambda / \sigma T^4$ from (14-31). For the optically thick limit, Eq. (18-52) reduces to [as in (15-34) for no scattering]

$$q_{r, k} = -\frac{4\sigma}{3 a_R} \frac{\partial T^4}{\partial x_k} \qquad (18\text{-}54)$$

where (15-93) has been used.

Because (18-52) should apply at both the optically thin and thick limits, (18-53) and (18-54) can be substituted into (18-52) to yield

$$\frac{\partial}{\partial x_k} \left(\sum_{j=1}^3 \frac{\partial q_{r,j}}{\partial x_j} \right) - 4 a_P \sigma \frac{\partial T^4}{\partial x_k} - 3 a_P a_R q_{r, k} = 0 \qquad (18\text{-}55)$$

where from (18-53) the Planck mean was used to approximate a_λ in the first term of (18-52). For the one-dimensional case

$$\frac{d^2 q_{r, x}}{dx^2} - 3 a_P a_R q_{r, x} = 4 a_P \sigma \frac{dT^4}{dx} \qquad (18\text{-}56)$$

For the problem of the initial steady temperature profile in the fluid (no free convection, heat generation or dissipation), the general energy equation (14-18) becomes

$$\frac{d}{dx} \left(k \frac{dT}{dx} \right) - \frac{dq_{r, x}}{dx} = 0 \qquad (18\text{-}57)$$

Equations (18-57) and (18-56) and the boundary conditions on $q_{r, x}$ specified by (15-136), along with the prescribed wall temperatures, constitute the formulation for the initial fluid-temperature profile. This is used in the stability problem in [47], which requires only the temperature gradient in the initial state to determine whether the fluid will become unstable. In [47], the P_3 and P_5 approximations were also used, with increasing accuracy in comparison with an exact solution.

18-4.4 Other Multimode Problems

Radiation effects in rocket exhaust plumes have been examined by de Soto [71]. References [8-10] treat radiation effects in ablating bodies. A very large body of literature exists that deals with reentry of bodies into the atmosphere [72], and radiation within and from hypersonic shocks. A rigorous treatment of these problems is difficult because of the nonequilibrium chemical reactions that are coupled with the radiation effects. References [33, 35, 36, 72] give a good introduction and discussion of shock problems. Radiation interaction with a layer of gas including transpiration is treated in [73].

18-5 TRANSIENT PROBLEMS

In a transient situation there can be two types of internal energy storage per unit volume and time; one is the local variation with time of the radiant energy density, and the second is the ordinary heat capacity of the material as encountered in conventional heat conduction and convection problems. From (14-104) the first of these is

$$\frac{\partial}{\partial \tau} \int_0^\infty U_\lambda \, d\lambda = \frac{\partial}{\partial \tau} \left(\frac{1}{c} \int_0^\infty \int_0^{4\pi} i_\lambda' \, d\omega \, d\lambda \right)$$

The second is $\rho c_v \partial T / \partial \tau$. Because of the large value of the propagation speed c in the medium, the storage of radiant energy is usually neglected. There are, however, some problems dealing with the effects of nuclear weapons ([18, 19] of Chap. 20) and some situations in astrophysics that require consideration of transient variations in the radiant energy.

Consider first only the effect of transient radiation. The equation of transfer derived in Chap. 14 neglected changes in radiation intensity with time. The equation of transfer is written for a beam of radiation of intensity i_λ' traveling in the S direction. As the radiation travels through the differential length from S to $S + dS$, its intensity is increased by emission and decreased by absorption. Also during the residence of the radiation within dS, the intensity can change with time. The residence time is $d\tau = dS/c$. Hence, the change in i_λ' can be written as

$$di_\lambda' = \frac{\partial i_\lambda'}{\partial \tau} \frac{dS}{c} + \frac{\partial i_\lambda'}{\partial S} dS$$

By substitution for di_λ', the equation of transfer (14-4) for the case of no scattering becomes

$$\frac{1}{c} \frac{\partial i_\lambda'(S, \tau)}{\partial \tau} + \frac{\partial i_\lambda'(S, \tau)}{\partial S} = a_\lambda(S, \tau) [i_{\lambda b}'(S, \tau) - i_\lambda'(S, \tau)] \tag{18-58}$$

Since the conditions such as temperature within the medium are changing with time, the absorption coefficient is a function of time as well as position.

To better understand the transient term, consider as a simple illustration what the radiative behavior would be if a thick medium at temperature T_1 instantaneously had its temperature increased to a higher uniform value T_2. The medium would then be at T_2 but the intensity within the medium would have to change from $i_{\lambda b}'(T_1)$ to $i_{\lambda b}'(T_2)$.

During this process, the radiation would not be in equilibrium. The equation of transfer reduces to (assuming as an approximation that a_λ can be used in the emission term, which is an equilibrium assumption)

$$\frac{1}{c}\frac{\partial i_\lambda'(\tau)}{\partial \tau} = a_\lambda(T_2)[i_{\lambda b}'(T_2) - i_\lambda'(\tau)] \tag{18-59}$$

After integration with the condition $i_\lambda' = i_{\lambda b}'(T_1)$ at $\tau = 0$, the result is

$$\frac{i_{\lambda b}'(T_2) - i_\lambda'(\tau)}{i_{\lambda b}'(T_2) - i_{\lambda b}'(T_1)} = \exp[-ca_\lambda(T_2)\tau] \tag{18-60}$$

The radiation relaxation time (time to change by a factor of $e = 2.718$) for equilibrium to be reestablished is thus $1/ca_\lambda(T_2)$, which is usually very short for reasonable values of a_λ in view of the large value of the propagation velocity c in the medium.

In the preceding illustration, it was assumed that the medium temperature could be instantaneously raised so that at the beginning of the transient the radiation intensity was not in equilibrium at the black radiation value corresponding to T_2. Generally the temperature change of a medium would be governed by the heat capacity of the medium, and consequently transient temperature changes would be much slower than the radiation relaxation time. Hence, when used with the transient energy-conservation equation containing the heat-capacity term, the unsteady-radiation term in the equation of transfer would be negligible. This is why the steady form of the equation of transfer, as derived in Chap. 14, can be instantaneously applied (as in the following example) during almost all transient heat-transfer processes involving radiation.

EXAMPLE 18-4 A gray medium is in a slab configuration originally at a uniform temperature T_0. The absorption coefficient is a, and the slab half-thickness is D. The heat capacity of the medium is c_v and its density is ρ. At $\tau = 0$, the slab is placed in surroundings at zero temperature. Neglecting conduction and convection, discuss the solutions for the temperature profiles for radiative cooling when a is very large and when a is very small.

At the slab center, which is located at $x = 0$, the condition of symmetry provides the relation for any time:

$$\frac{\partial T}{\partial x} = 0 \qquad x = 0, \tau$$

At time $\tau = 0$ there is the condition for any x:

$$T = T_0 \qquad x, \tau = 0$$

As discussed in Sec. 14-7.4 for radiation only being included, there will be a temperature slip at the boundaries $x = \pm D$, so that the temperature at the boundaries will be finite rather than being equal to the zero outside temperature. If heat conduction were present at the boundary, the temperature slip would not exist.

For large a the diffusion approximation can be employed, and from (15-34) the heat flux in the x direction is

$$q(x, \tau) = -\frac{4}{3a}\frac{\partial e_b(x, \tau)}{\partial x} = -\frac{4\sigma}{3a}\frac{\partial T^4(x, \tau)}{\partial x}$$

By conservation of energy

$$-\frac{\partial q(x, \tau)}{\partial x} = \rho c_v \frac{\partial T}{\partial \tau}$$

Combining these two equations to eliminate q gives the transient energy diffusion equation for the temperature distribution in the slab with constant absorption coefficient:

$$\rho c_v \frac{\partial T}{\partial \tau} = \frac{4\sigma}{3a}\frac{\partial^2 T^4(x, \tau)}{\partial x^2}$$

Defining dimensionless variables as $\tau^* = a\,\sigma T_0^3 \tau/\rho c_v$, $\kappa = ax$, and $\Theta = T/T_0$, gives

$$\frac{\partial \Theta}{\partial \tau^*} = \frac{4}{3}\frac{\partial^2 \Theta^4(\kappa, \tau^*)}{\partial \kappa^2}$$

The initial condition and the boundary condition at $x = 0$ become, respectively, $\Theta(\kappa, 0) = 1$ and $(\partial \Theta/\partial \kappa)(0, \tau^*) = 0$. At the boundary $\kappa = aD$, a slip condition must be used. From (15-54), when the surroundings are empty space at zero temperature, $e_{bw} = 0$ and $\epsilon_w = 1$, so that at the exposed boundary of the medium for any time

$$\sigma T^4 \big|_{x=D} = \frac{1}{2}\left(-\frac{4}{3}\frac{\sigma}{a}\frac{\partial T^4}{\partial x}\right)_{x=D} - \frac{\sigma}{2a^2}\frac{\partial^2 T^4}{\partial x^2}\bigg|_{x=D}$$

or

$$0 = \left(2\Theta^4 + \frac{4}{3}\frac{\partial \Theta^4}{\partial \kappa} + \frac{\partial^2 \Theta^4}{\partial \kappa^2}\right)_{\kappa=aD}$$

Similar relations apply at $x = -D$. For these conditions, solution by numerical techniques is probably necessary.

For a small abosrption coefficient, and since there are no enclosing radiating boundaries present, the emission approximation (Sec. 15-3.2) can be applied. For very small a the medium is optically so thin that it is at uniform temperature throughout its thickness at any instant. From the results of Example 15-2, the heat flux emerging from each boundary of the layer is $q = 4a\sigma T^4 D$. The energy equation then becomes $\rho c_v\, dT/d\tau = -4a\sigma T^4$, or, in dimensionless form, $d\Theta/d\tau^* = -4\Theta^4$. Integrating with the condition that $\Theta = 1$ at $\tau^* = 0$ gives the transient temperature throughout the slab as

$$\Theta = \frac{1}{(1 + 12\tau^*)^{1/3}}$$

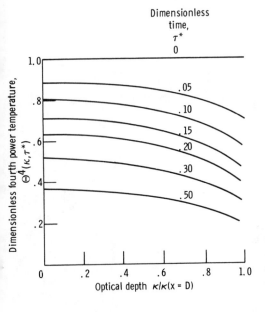

Figure 18-12 Dimensionless temperature profiles as a function of time for radiative cooling of a gray slab; optical thickness $\kappa\,(x = D) = 1.0$. *(From [74].)*

Viskanta and Bathla [74] have obtained numerical solutions to the transient form of the complete equation of transfer, along with the limiting solutions derived here. Some of their results for intermediate optical thickness are shown in Fig. 18-12. Numerical solutions for spherical geometries are found in [75, 76], and results for the region between coaxial cylinders are given in [77]. Wendlandt [7] derives expressions for the transient temperature distribution in a semi-infinite absorbing medium exposed to a laser pulse.

The transient heating of a number of semitransparent solid geometries is reviewed in [78]. Consider, for example, a unidirectional flux F that is incident at angle θ on a semi-infinite gray solid as shown in Fig. 18-13a. A fraction $\tau_0(\theta)$ is transmitted through the interface. The absorption coefficient of the solid is a. The fraction of entering radiation that reaches depth x is $\exp(-ax)$, and the amount $a\exp(-ax)$ is absorbed at x per unit volume. The solid is assumed to be in its initial state of heating so that its temperature is low, and hence emission within the solid and heat loss from its surface can be neglected. The energy equation, initial condition, and boundary conditions are then

$$\rho c_v \frac{\partial T}{\partial \tau} = k \frac{\partial^2 T}{\partial x^2} + F \cos\theta \; \tau_0(\theta) a e^{-ax} \tag{18-61}$$

$$T(x, 0) = T_0 \qquad \text{(initial condition)}$$

$$\frac{\partial T}{\partial x}(0, \tau) = 0 \qquad \lim_{x \to \infty} T(x, \tau) = T_0$$

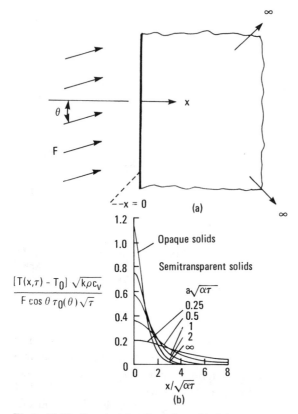

Figure 18-13 Transient heating of semi-infinite solid by incident unidirectional radiative flux. (*a*) Flux incident on semi-infinite slab; (*b*) dimensionless temperature distribution.

The solution is in [79] and can be placed in the form

$$\frac{[T(x,\tau) - T_0]\sqrt{k\rho c_v}}{F\cos\theta\ \tau_0(\theta)\sqrt{\tau}} = \frac{2}{\sqrt{\pi}}e^{-X^2} - 2X\operatorname{erfc}X - \frac{1}{A}e^{-2AX}$$

$$+ \frac{1}{2A}\left[e^{A(A-2X)}\operatorname{erfc}(A - X) + e^{A(A+2X)}\operatorname{erfc}(A + X)\right]\ (18\text{-}62)$$

where $X = x/2\sqrt{\alpha\tau}$, $A = a\sqrt{\alpha\tau}$, and $\alpha = k/\rho c_v$. (Note that τ = time, τ_0 = surface transmissivity.)

Some results are plotted in Fig. 18-13*b*. Since the parameter on the curves -depends on the absorption coefficient a, an increase of a will increase the surface temperature that is reached at any time. A small a increases the penetration of the temperature distribution into the solid. As $a\sqrt{\alpha\tau} \to \infty$, the temperature distribution approaches that for an opaque solid. Thus no matter what a is, given sufficient time, the semi-infinitely thick material will act as an opaque solid.

REFERENCES

1. Kellett, B. S.: Transmission of Radiation through Glass in Tank Furnaces, *J. Soc. Glass Tech.*, vol. 36, pp. 115–123, 1952.
2. Gardon, Robert: The Emissivity of Transparent Materials, *J. Am. Ceram. Soc.*, vol. 39, no. 8, pp. 278–287, 1956.
3. Condon, Edward U.: Radiative Transport in Hot Glass, *J. Quant. Spectrosc. Radiat. Transfer*, vol. 8, no. 1, pp. 369–385, 1968.
4. Eryou, N. D., and L. R. Glicksman: An Experimental and Analytical Study of Radiative and Conductive Heat Transfer in Molten Glass, *J. Heat Transfer*, vol. 94, no. 2, pp. 224–230, 1972.
5. Anderson, E. E., R. Viskanta, and W. H. Stevenson: Heat Transfer through Semitransparent Solids, *J. Heat Transfer*, vol. 95, no. 2, pp. 179–186, 1973.
6. Kuriyama, Masaaki, et al.: The Effect of Radiation Heat Transfer in the Measurement of Thermal Conductivity for the Semitransparent Medium, *Bull. JSME*, vol. 19, no. 134, pp. 973–979, 1976.
7. Wendlandt, B. C. H.: Temperature in an Irradiated Thermally Conducting Medium, *J. Phys. D*, vol. 6, pp. 657–660, 1973.
8. Kadanoff, Leo P.: Radiative Transport within an Ablating Body, *J. Heat Transfer*, vol. 83, no. 2, pp. 215-225, 1961.
9. Nelson, H. F.: Radiative Transfer through Carbon Ablation Layers, *J. Quant. Spectrosc. Radiat. Transfer*, vol. 13, pp. 427–445, 1973.
10. Boles, M. A., and M. N. Özişik: Simultaneous Ablation and Radiation in an Absorbing, Emitting and Isotropically Scattering Medium, *J. Quant. Spectrosc. Radiat. Transfer*, vol. 12, pp. 838–847, 1972.
11. Merriam, R. L., and R. Viskanta: Radiative Characteristics of Cryodeposits for Room Temperature Black Body Radiation, *Cryogenic Eng. Conf.*, Case-Western Reserve University, Cleveland, August, 1968.
12. McConnell, Dudley G.: Radiant Energy Transport within Cryogenic Condensates, *Inst. Environmental Sci. Ann. Meeting Equipment Exposition*, San Diego, Calif., Apr. 11–13, 1966.
13. Gilpin, R. R., R. B. Roberton, and B. Singh: Radiative Heating in Ice, *ASME J. Heat Transfer*, vol. 99, pp. 227–232, 1977.
14. Viskanta, R., and R. J. Grosh: Heat Transfer by Simultaneous Conduction and Radiation in an Absorbing Medium, *J. Heat Transfer*, vol. 84, no. 1, pp. 63–72, 1962.
15. Viskanta, R., and R. J. Grosh: Effect of Surface Emissivity on Heat Transfer by Simultaneous Conduction and Radiation, *Int. J. Heat Mass Transfer*, vol. 5, pp. 729–734, 1962.
16. Viskanta, R., and R. L. Merriam: Heat Transfer by Combined Conduction and Radiation between Concentric Spheres Separated by Radiating Medium, *J. Heat Transfer*, vol. 90, no. 2, pp. 248–256, 1968.
17. Greif, Ralph, and Gean P. Clapper: Radiant Heat Transfer between Concentric Cylinders, *Appl. Sci. Res.*, sec. A, vol. 15, pp. 469–474, 1966.
18. Men, A. A.: Radiative-Conductive Heat Transfer in a Medium with a Cylindrical Geometry. I, *Inz. Fiz. Zh.*, vol. 24, no. 6, pp. 984–991, 1973.
19. Timmons, D. H., and J. O. Mingle: Simultaneous Radiation and Conduction with Specular Reflection, paper 68-28, *AIAA*, January, 1968.
20. Schimmel, W. P., J. L. Novotny, and F. A. Olsofka: Interferometric Study of Radiation-Conduction Interaction, *Fourth Int. Heat Transfer Conf.*, Paris, September, 1970.
21. Einstein, Thomas H.: Radiant Heat Transfer to Absorbing Gases Enclosed between Parallel Flat Plates with Flow and Conduction, *NASA* TR R-154, 1963.
22. Cess, R. D.: The Interaction of Thermal Radiation with Conduction and Convection Heat Transfer, in Thomas F. Irvine, Jr., and James P. Hartnett (eds.), "Advances in Heat Transfer," vol. 1, pp. 1–50, Academic Press, Inc., New York, 1964.
23. Howell, John R.: Determination of Combined Conduction and Radiation of Heat through

Absorbing Media by the Exchange Factor Approximation, *Chem. Eng. Progr. Symp. Ser.,* vol. 61, no. 59, pp. 162-171, 1965.

24. Nelson, D. A.: On the Uncoupled Superposition Approximation for Combined Conduction-Radiation through Infrared Radiating Gases, *Int. J. Heat Mass Transfer,* vol. 18, no. 5, pp. 711-713, 1975.

25. Goldstein, Marvin E., and John R. Howell: Boundary Conditions for the Diffusion Solution of Coupled Conduction-Radiation Problems, *NASA* TN D-4618, 1968.

26. Howell, J. R., and M. E. Goldstein: Effective Slip Coefficients for Coupled Conduction-Radiation Problems, *J. Heat Transfer,* vol. 91, no. 1, pp. 165-166, 1969.

27. Taitel, Yehuda, and J. P. Hartnett: Application of Rosseland Approximation and Solution Based on Series Expansion of the Emission Power to Radiation Problems, *AIAA J.,* vol. 6, no. 1, pp. 80-89, 1968.

28. Wang, L. S., and C. L. Tien: A Study of Various Limits in Radiation Heat-transfer Problems, *Int. J. Heat Mass Transfer,* vol. 10, no. 10, pp. 1327-1338, 1967.

29. Ness, A. J.: Solution of Equations of a Thermal Network on a Digital Computer, *Sol. Energy,* vol. 3, no. 2, p. 37, 1959.

30. Fulkerson, G. D., and R. B. Bannerot: An Approximation for Combined Heat Transfer in a Radiatively Absorbing and Emitting Gas, paper 73-750, *AIAA,* July 16, 1973.

31. Lick, Wilbert: Energy Transfer by Radiation and Conduction, *Proc. 1963 Heat Transfer Fluid Mech. Inst.* (Anatol Roshko, Bradford Sturtevant, and D. R. Bartz, eds.), 1963, pp. 14-26.

32. Viskanta, R.: Radiation Transfer and Interaction of Convection with Radiation Heat Transfer, in (Thomas F. Irvine, Jr., and James P. Hartnett (eds.), "Advances in Heat Transfer" vol. 3, pp. 175-251, Academic Press, Inc., New York, 1966.

33. Pai, Shih-I: "Radiation Gas Dynamics," Springer-Verlag OHG, Berlin, 1966.

34. Bond, John W., Jr., Kenneth M. Watson, and Jasper, A. Welch, Jr.: "Atomic Theory of Gas Dynamics," chaps. 10-13, Addison-Wesley Publishing Company, Reading, Mass., 1965.

35. Zel'dovich, Ya. B., and Yu. P. Raizek: "Physics of Shock Waves and High-temperature Hydrodynamic Phenomena," vol. I, pt. II, Academic Press, Inc., New York, 1966.

36. Vincenti, Walter G., and Charles H. Kruger, Jr.: "Introduction to Physical Gas Dynamics," chaps. 11-12, John Wiley & Sons, Inc., New York, 1965.

37. Novotny, J. L., and Kwang-Tzu Yang: The Interaction of Thermal Radiation in Optically Thick Boundary Layers, *J. Heat Transfer,* vol. 89, no. 4, pp. 309-312, 1967.

38. Viskanta, R., and R. J. Grosh: Boundary Layer in Thermal Radiation Absorbing and Emitting Media, *Int. J. Heat Mass Transfer,* vol. 5, pp. 795-806, 1962.

39. Cess, R. D.: Radiation Effects Upon Boundary-layer Flow of an Absorbing Gas, *J. Heat Transfer,* vol. 86, no. 4, pp. 469-475, 1964.

40. Fritsch, C. A., R. J. Grosh, and M. W. Wildin: Radiative Heat Transfer through an Absorbing Boundary Layer, *J. Heat Transfer,* vol. 88, no. 3, pp. 296-304, 1966.

41. Kubo, Syozo: Stagnation Point Flow of a Radiating Gas of a Large Optical Thickness II, *Trans. Jpn. Soc. Aeronaut. Space Sci.,* vol. 18, no. 41, pp. 141-151, 1975.

42. Howe, John T.: Radiation Shielding of the Stagnation Region by Transpiration of an Opaque Gas, *NASA* TN D-329, 1960.

43. Hasegawa, Shu, Ryozo Echigo, and Kenji Fukuda: Analytical and Experimental Studies on Simultaneous Radiative and Free Convective Heat Transfer along a Vertical Plate, *Proc. Jpn. Soc. Mech. Eng.,* vol. 38, no. 315, pp. 2873-2882, 1972; vol. 39, no. 317, pp. 250-257, 1973.

44. Doura, S., and J. R. Howell: An Approximate Solution for the Energy Equation with Radiant Participating Media, paper 77-HT-70, *ASME,* August, 1977.

45. Novotny, J. L., and Kelleher, M. D.: Free-convection Stagnation Flow of an Absorbing-Emitting Gas, *Int. J. Heat Mass Transfer,* vol. 10, no. 9, pp. 1171-1178, 1967.

46. Cess, R. D.: The Interaction of Thermal Radiation with Free Convection Heat Transfer, *Int. J. Heat Mass Transfer,* vol. 9, no. 11, pp. 1269-1277, 1966.

47. Arpaci, Vedat S., and Doğan Gőzűm: Thermal Stability of Radiating Fluids: The Benard Problem, *Phys. Fluids,* vol. 16, no. 5, pp. 581-588, 1973.

48. Arpaci, V. S., and Y. Bayazitoglu: Thermal Stability of Radiating Fluids: Asymmetric Slot Problem, *Phys. Fluids,* vol. 16, no. 5, pp. 589-593, 1973.

49. Gille, John, and Richard Goody: Convection in a Radiating Gas, *J. Fluid Mech.*, vol. 20, pt. 1, pp. 47–79, 1964.

50. Cess, Robert D.: The Interaction of Thermal Radiation in Boundary Layer Heat Transfer, *Third Int. Heat Transfer Conf. AIChE*, vol. 5, pp. 154–163, 1966.

51. Schlichting, Hermann (J. Kestin, trans.): "Boundary Layer Theory," 4th ed., p. 116, McGraw-Hill Book Company, New York, 1960.

52. Viskanta, R.: Interaction of Heat Transfer by Conduction, Convection, and Radiation in a Radiating Fluid, *J. Heat Transfer,* vol. 85, no. 4, pp. 318–328, 1963.

53. Einstein, Thomas H.: Radiant Heat Transfer to Absorbing Gases Enclosed in a·Circular Pipe with Conduction, Gas Flow, and Internal Heat Generation, *NASA* TR R-156, 1963.

54. Adrianov, V. N., and S. N. Shorin: Radiative Transfer in the Flow of a Radiating Medium, *Trans.* TT-1, Purdue University, February, 1961.

55. Chen, John C.: Simultaneous Radiative and Convective Heat Transfer in an Absorbing, Emitting, and Scattering Medium in Slug Flow between Parallel Plates, *Rept.* BNL-6876-R, Brookhaven National Laboratory, March 18, 1963.

56. Edwards, D. K., and A. Balakrishnan: Self-Absorption of Radiation in Turbulent Molecular Gases, *Combust. Flame,* vol. 20, pp. 401–417, 1973.

57. Chiba, Z., and R. Greif: Heat Transfer to Steam Flowing Turbulently in a Pipe, *Int. J. Heat Mass Transfer,* vol. 16, pp. 1645–1648, 1973.

58. Nakra, N. K., and T. F. Smith: Combined Radiation-Convection for a Real Gas, paper 76-HT-58, *ASME,* August, 1976.

59. Bergquam, J. B., and N. S. Wang: Heat Transfer by Convection and Radiation in an Absorbing, Scattering Medium Flowing between Parallel Plates, paper 76-HT-50, *ASME,* August, 1976.

60. Greif, Ralph, and D. R. Willis: Heat Transfer between Parallel Plates including Radiation and Rarefaction Effects, *Int. J. Heat Mass Transfer,* vol. 10, pp. 1041–1048, 1967.

61. Martin, J. K., and C. C. Hwang: Combined Radiant and Convective Heat Transfer to Laminar Steam Flow between Gray Parallel Plates with Uniform Heat Flux, *J. Quant. Spectrosc. Radiat. Transfer,* vol. 15, pp. 1071–1081, 1975.

62. De Soto, Simon, and D. K. Edwards: Radiative Emission and Absorption in Non-isothermal Nongray Gases in Tubes, *Proc. 1965 Heat Transfer Fluid Mech. Inst.* (A. F. Charwat, ed.), 1965, pp. 358–372.

63. Pearce, B. L., and A. F. Emery: Heat Transfer by Thermal Radiation and Laminar Forced Convection to an Absorbing Fluid in the Entry Region of a Pipe, *J. Heat Transfer,* vol. 92, no. 2, pp. 221–230, 1970.

64. Jeng, D. R., E. J. Lee, and K. J. DeWitt: Simultaneous Conductive and Radiative Heat Transfer for Laminar Flow in Circular Tubes with Constant Wall Temperature, *Proc. 5th Int. Heat Transfer Conf.,* Tokyo, vol. I, pp. 118–122, 1974.

65. Jeng, D. R., E. J. Lee, and K. J. DeWitt: A Study of Two limiting Cases in Convective and Radiative Heat Transfer with Nongray Gases, *Int. J. Heat Mass Transfer,* vol. 19, no. 6, pp. 589–596, 1976.

66. Landram, C. S., R. Greif, and I. S. Habib: Heat Transfer in Turbulent Pipe Flow with Optically Thin Radiation, *J. Heat Transfer,* vol. 91, no. 3, pp. 330–336, 1969.

67. Wassel, A. T., and D. K. Edwards: Molecular Gas Band Radiation in Cylinders, *J. Heat Transfer,* vol. 96, no. 1, pp. 21–26, 1974.

68. Wassel, A. T., and D. K. Edwards: Molecular Gas Radiation in a Laminar or Turbulent Pipe Flow, *J. Heat Transfer,* vol. 98, no. 1, pp. 101–107, 1976.

69. Greif, R.: Laminar Convection with Radiation: Experimental and Theoretical Results, *Int. J. Heat Mass Transfer,* vol. 21, no. 4, pp. 447–480, 1978.

70. Balakrishnan, A., and D. K. Edwards: Molecular Gas Radiation in the Thermal Entrance Region of a Duct, *J. Heat Transfer,* vol. 101, no. 3, pp. 489–495, 1979.

71. De Soto, Simon: The Radiation from an Axisymmetric, Real Gas System with a Non-isothermal Temperature Distribution, *Chem. Eng. Progr. Symp. Ser.,* vol. 61, no. 59, pp. 138–154, 1965.

72. Penner, S. S., and D. B. Olfe: "Radiation and Reentry," Academic Press, Inc., New York, 1968.

73. Viskanta, R., and R. L. Merriam: Shielding of Surfaces in Couette Flow against Radiation by Transpiration of an Absorbing-Emitting Gas, *Int. J. Heat Mass Transfer,* vol. 10, no. 5, pp. 641–653, 1967.

74. Viskanta, Raymond, and Pritam S. Bathla: Unsteady Energy Transfer in a Layer of Gray Gas by Thermal Radiation, *Z. Angew. Math. Phys.,* vol. 18, no. 3, pp. 353–367, 1967.

75. Viskanta, R., and P. S. Lall: Transient Cooling of a Spherical Mass of High-temperature Gas by Thermal Radiation, *J. Appl. Mech.,* vol. 32, no. 4, pp. 740–746, 1965.

76. Viskanta, R., and P. S. Lall: Transient Heating and Cooling of a Sperical Mass of Gray Gas by Thermal Radiation, *Proc. Heat Transfer Fluid Mech. Inst.* (M. A. Saad and J. A. Miller, eds.), 1966, p. 181–197.

77. Chang, Yan-Po, and R. Scott Smith, Jr.: Steady and Transient Heat Transfer by Radiation and Conduction in an Medium Bounded by Two Coaxial Cylindrical Surfaces, *Int. J. Heat Mass Transfer,* vol. 13, no. 1, pp. 69–80, 1970.

78. Viskanta, R., and E. E. Anderson: Heat Transfer in Semitransparent Solids, in Thomas F. Irvine, Jr., and James P. Hartnett (eds.), "Advances in Heat Transfer," vol. 11, pp. 317–441, Academic Press, Inc., New York, 1975.

79. Carslaw, H. S., and J. C. Jaeger: "Conduction of Heat in Solids," 2nd ed., p. 80, Clarendon Press, Oxford, 1959.

PROBLEMS

1 The region between two infinite black parallel plates is filled with a stationary gray medium. The plate temperatures are 1000 and 500 K, the spacing between the plates is 5 cm, and the absorption coefficient of the medium is 0.2 cm^{-1}. If the medium thermal conductivity is 0.011 W/(cm K), what is the net energy flux being transferred between the plates?

   Answer: 4.5 W/cm^2

2 A long chamber has two sides at T_1 and T_2, and the other two sides insulated. The chamber is filled with a stationary medium that is scattering and is optically thin. The medium conducts heat with thermal conductivity k. The surfaces are diffuse-gray with emissivities as shown; the insulated walls both have the same emissivity. Develop an approximate expression for the heat transfer from A_1 to A_2. (*Hint:* See Example 14-4.)

   Answer: $$\frac{Q}{al} = \frac{k}{b}(T_1 - T_2) + \frac{\sigma(T_1^4 - T_2^4)}{1/\epsilon_1 + 1/\epsilon_2 - 2F_{1-2}/(1 + F_{1-2})}$$

   where $F_{1-2} = \sqrt{1 + B^2} - B; \; B = b/a$

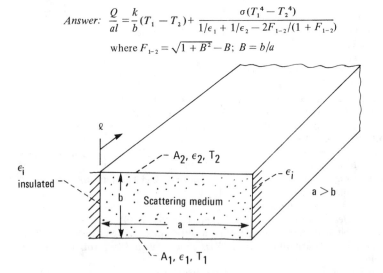

3   Two infinite plates are separated by an optically thick absorbing-emitting and conducting gas in which a chemical reaction is occurring that produces a uniform heat generation q''' per unit volume and time. The plates are gray and have temperatures and emissivities T_1, ϵ_1 and T_2, ϵ_2. Determine the heat fluxes q_1 and q_2 that must be supplied to each of the plates as a result of combined radiation exchange and conduction between them, to maintain the specified temperatures T_1 and T_2. Use the radiative diffusion approximation. Assume the gas is not moving and that the thermal conductivity and absorption coefficient are both constant.

4   A chamber wall consists of three parallel plates with properties as shown. The surfaces are all gray, and plate 2 is sufficiently thin so that it can be considered of uniform temperature throughout its thickness. What is the net heat flux from plate 1 to plate 3 if the regions between the plates are both a vacuum? An absorbing-emitting and conducting medium with properties shown is now introduced between plates 2 and 3. Use the additive approximation and diffusion theory to estimate the net flux from 1 to 3 for this situation.

Answer: 0.876 W/cm², 0.964 W/cm²

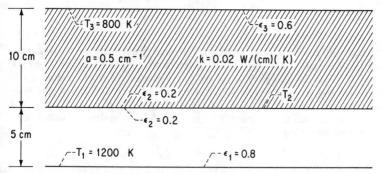

5   Repeat Prob. 7 of Chap. 15, but include conduction in the gas if the thermal conductivity is k. For this problem, also assume that $a_R(T) = a_{R,0}(T/T_0)^3$.

6   A stagnant gray gas with absorption coefficient $a = 2$ ft⁻¹ is contained between two parallel plates spaced 6 in apart. The plates are nongray and have hemispherical spectral emissivities and temperatures as shown. The gas thermal conductivity is $k = 0.1$ Btu/(h ft° R). Compute the energy flux transferred from plate 1 to plate 2. Use the additive approximation and the diffusion solution for the radiative transfer.

Answer: 610 Btu/(h ft²)

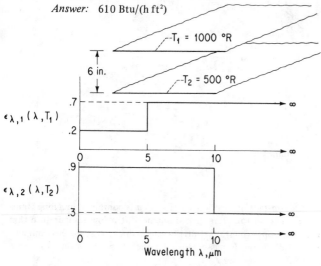

7   Two gray parallel plates with emissivities $\epsilon_1 = 0.6$ and $\epsilon_2 = 0.9$ are spaced 6 in apart. The plate temperatures are $T_1 = 1000°R$ and $T_2 = 600°R$. A stagnant nongray gas with absorption coefficient as shown is in the space between the plates. The thermal conductivity of the gas is $k = 0.1$ Btu/(h ft°R). Using the additive and diffusion approximations, find the heat transfer from plate 1 to plate 2. How significant is the heat conduction?

   Answer: 823 Btu/(h ft²)

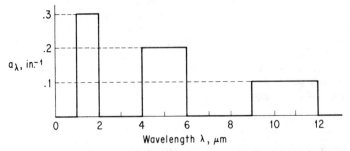

8   An absorbing liquid with absorption coefficient a and thermal conductivity k is flowing down a vertical flat plate. The plate is diffuse-gray with absorptivity α and is insulated on the back as shown. A solar flux is incident on the plate at angle η. Formulate the relations necessary to determine the mean temperature of the liquid at the bottom of the plate. This should be in terms of liquid flow rate, plate length, plate absorptivity, and the insulation thickness and conductivity. The temperatures are low enough so that as a first approximation radiation by the liquid and plate can be neglected.

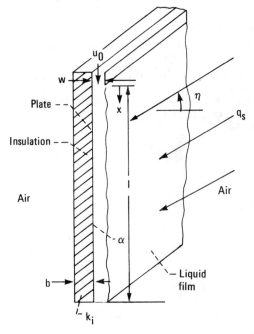

9   An optically thin gray gas with constant absorption coefficient a is contained in a long transparent cylinder of diameter D. The surrounding environment is at a low temperature that can be considered zero for a first approximation. Initially the cylinder is at the environment temperature. Then an electrical discharge is passed through the cylinder continuously pro-

ducing in the gas a uniform heat source q''' per unit volume and time. Derive a relation for the transient gas-temperature variation if radiation is assumed to be the only significant mode of heat transfer. What is the maximum temperature T_m that the gas will achieve?

$$\textit{Answer:} \quad -\frac{t}{\rho c_v / 8 a \sigma T_m{}^3} = \frac{1}{2} \ln \frac{1 + \Theta}{1 - \Theta} + \tan^{-1} \Theta, \; T_m = \left(\frac{q'''}{4 a \sigma}\right)^{1/4}, \; \Theta = \frac{T}{T_m}$$

10 A partially transparent solid absorbing sphere is placed in a black cavity at T_b. The sphere temperature is initially low. The sphere has surface transmissivity τ_0, thermal conductivity k, and absorption coefficient a. Transparent gas at T_g is circulating in the cavity, providing a heat transfer coefficient h at the sphere surface. Give the equations and boundary conditions to compute the transient temperature distribution in the sphere for the initial period during which radiation by the sphere is small. Neglect any refraction effects and any dependence of τ_0 on angle.

NINETEEN

RADIATIVE BEHAVIOR OF WINDOWS, COATINGS, AND SEMITRANSPARENT SOLIDS

19-1 INTRODUCTION

In previous chapters most of the discussion of radiative transfer within a medium has been concerned with dielectric materials that have a refractive index n of unity. This has considerable application, as the absorbing-emitting medium is often a gas, and almost all gases have a refractive index that is very close to unity (see Table 19-1). However, many important effects result from the nonunity refractive indexes possessed by many common materials, such as those listed in the table. The most obvious of these effects are reflection and refraction at an interface; these phenomena determine the behavior of coatings, thin films, and multiple windows. Another interesting effect of nonunity n, as discussed in Sec. 2-4.12, is to increase the blackbody emission *within* the *medium* by a factor of n^2.

The previous discussions did not include enclosures in which radiation passed through windows. The analytical development was for enclosures composed of opaque walls or, in cavity geometries, with an opening through which the enclosure could directly interact with the environment. In some instances, single or multi-layered windows form a major portion of the enclosure (the flat-plate solar collector is a prime example). Multiple reflections from the window surfaces can be appreciable, especially if more than one window-layer is present, as in a two-cover-plate solar collector. Transmission losses through the windows can also be significant. The windows are usually thin, and hence it is often unnecessary to account for a temperature variation through the window thickness.

If one or more reflecting partially transparent layers are directly attached to an opaque surface or to other reflecting partially transparent layers, a coating system can be formed that has desirable radiative properties. The coating can be either thick or thin compared to the wavelength of the radiation. Thin films produce wave interference effects between incident and reflected waves in the film, thus influencing their reflection and transmission characteristics. Thin-film coating systems with high or low reflectivity can be produced. A composite window can be formed that is selectively

718

Table 19-1 Refractive indices of some common substances (from [1])

Material	Refractive index n
Gases (at $\lambda = 0.589$ μm)	
Air	1.00029
Argon	1.00028
Carbon dioxide	1.00045
Chlorine	1.00077
Hydrogen	1.00013
Methane	1.00044
Nitrogen	1.00030
Oxygen	1.00027
Water vapor	1.00026
Liquids (at $\lambda = 0.589$ μm)	
Chlorine	1.385
Ethyl alcohol	1.36–1.34 (16–76°C)
Oxygen	1.221
Water	1.33–1.32 (15–100°C)
Solids (λ given)	
Glass:	
Crown	1.50–1.55 ⎫ (2–0.36 μm)
Flint	1.55–1.95 ⎭
Ice	1.31 (0.589 μm)
Quartz	1.52–1.69 (2.3–0.19 μm)
Rock salt	1.5–1.9 (8.8–0.19 μm)

transmitting, or a surface that is selectively absorbing can be produced. Examples are the transparent heat mirror and solar selective surfaces as discussed in Sec. 5-5. Films that are thick relative to the radiation wavelength can also be used to modify surface characteristics.

The development in this chapter will provide analytical tools for treating enclosures with windows and obtaining the radiative behavior of coatings. Since the windows and coatings are usually thin, temperature variations within them are not considered in most of the development presented here. There are applications, however, in which the temperature distribution must be obtained within a partially transmitting medium, including the effect of interface reflections. This requires solving the transfer equations within the medium as in Chap. 14, but with the inclusion of reflection boundary conditions at the interfaces. Some important applications are heat treating of glass plates, temperature distributions in glass melting tanks, high-temperature solar components, laser heating of windows and lenses, and heating of spacecraft and aircraft windows. The theory for these applications is introduced in the last section of this chapter; a detailed treatment is given in [2].

19-2 SYMBOLS

A	ratio of energy absorbed in plate system to energy incident; area
a	absorption coefficient
c	speed of light
E	amplitude of electric intensity wave; overall emittance
F	configuration factor
i	intensity
L	plate thickness
N	number of surfaces in enclosure
n	simple index of refraction
q	energy flux, rate of energy per unit area
R	ratio of energy reflected from plate system to energy incident
r	ratio of reflected to incident electric intensity
S	path length
T	ratio of energy transmitted through plate system to energy incident; temperature
t	ratio of transmitted to incident electric intensity
x	coordinate in medium
α	absorptivity
γ	the quantity $4\pi n L/\lambda_0$
κ	extinction coefficient
θ	angle measured away from normal of surface
λ	wavelength
ν	frequency
ρ	reflectivity
τ	transmissivity of layer; interface transmissivity; time
φ	circumferential angle
χ	angle of refraction
ω	solid angle; angular frequency

Subscripts

c	at collector plate
e	emitted
i	incoming; incident direction at boundary in Fig. 19-16
j	at surface j
k	at surface k
l	lost from outer surface
M	maximum value
m,n	number of identical plates in system
max	maximum refraction angle
o	outgoing
r	reflected
s	surroundings
sub	substrate
w	window

λ, ν	spectrally dependent
0	in vacuum
1,2,3,4	interface or medium 1,2,3, or 4
$\perp$	perpendicular component
$\parallel$	parallel component

Superscripts

$'$	for reciprocal path; directional quantity
$''$	bidirectional quantity
$*$	complex conjugate
$+$	in positive x direction
$-$	in negative x direction

19-3 TRANSMISSION, ABSORPTION, AND REFLECTION OF WINDOWS

Enclosures can have windows that are partially transparent to radiation. A window can be of a single material, or it can have one or more transmitting coatings on it. The transparency is a function of the window and coating thicknesses and can be quite wavelength dependent as illustrated by the transmittance of glass in Fig. 5-30. Some applications are glass or plastic cover plates for flat-plate solar collectors, and thick and thin film coatings to provide modified reflection and transmission properties for camera lenses and solar cells. In these and many other instances the temperatures within the windows and coatings are often unimportant, and it is not necessary to consider the energy equation within the transmitting media.

As discussed in Sec. 4-5 and illustrated in Fig. 19-1, radiation incident on the surface of a layer will be reflected and refracted. For a layer of thickness L, the refracted portion will travel a distance $L/\cos \chi$ and then be partially reflected from the inside of the second surface. To analyze the radiative behavior of a plate such as that shown in Fig. 19-1, the reflectivities are needed for radiation striking the outside and the inside of an interface. For a smooth interface, reflectivity relations are given by Eqs. (4-62), since windows are generally dielectrics with a small extinction coefficient. Since these relations depend only on the square of the terms containing $\theta - \chi$, the θ and χ can be interchanged and hence the reflectivity is the same for radiation incident on an interface from either the outside or from within the material. For a constant absorption coefficient within the material, the transmittance along the path length within the window is $\tau = \exp(-aL/\cos \chi)$ where a is a function of wavelength as shown by the transmission behavior of glass in Figs. 5-29 and 5-30.*

The Fresnel reflection relations in (4-62) make it evident that the reflectivity at a smooth interface is not only a function of incidence angle, but is different for the two components of polarization. In what follows the path of radiation will be traced through partially transparent plates where the interface reflections will be specular. If precise results are required, the resulting formulas should be applied at each incidence

---

*For simplicity in notation, the functional dependence on λ is omitted in this section.

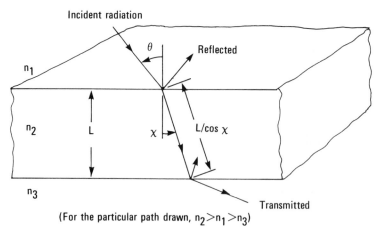

Figure 19-1 Reflection and transmission of radiation by a window.

angle, and for each component of polarization. If the incident radiation is nonpolarized and diffuse, then half the energy is in each component of polarization. The fraction of diffuse energy incident in each θ direction within the increment $d\theta$ is $2 \sin\theta \cos\theta \, d\theta$. Then, in integrating to find the total reflectivity, the results for each direction are weighted in the integration according to the amount of incident energy in each $d\theta$ at θ, and in each component of polarization.

19-3.1 Single-Layer Window or Transmitting Layer with Thickness $L > \lambda$ (No Interference Effects)

When a window or transmitting layer is thin such that the thickness is comparable to the radiation wavelength, there can be interference of incident and reflected waves. This will be discussed later. For the present discussion consider a window where L is at least several wavelengths thick so that interference effects are usually lost. The actual thickness required for this to occur depends on the coherence length of the radiation in the material, which is beyond the scope of the present discussion.

Ray-Tracing Method. Referring to Fig. 19-2, consider a unit intensity incident on the upper boundary, and apply a ray-tracing technique. At contact with the first interface, an amount ρ is reflected so that $1 - \rho$ enters the material. Of this, $(1 - \rho)\tau$ is transmitted [hence $(1 - \rho)(1 - \tau)$ is absorbed along the path] to interface 3 where $\rho(1 - \rho)\tau$ is reflected, and consequently $(1 - \rho)^2\tau$ passes out of the plate through the lower boundary. As the process continues as shown in Fig. 19-2, the fraction of incident energy reflected by the plate is the sum of the terms leaving surface 1:

$$R = \rho\left[1 + (1 - \rho)^2\tau^2(1 + \rho^2\tau^2 + \rho^4\tau^4 + \cdots)\right] = \rho\left[1 + \frac{(1 - \rho)^2\tau^2}{1 - \rho^2\tau^2}\right] \tag{19-1}$$

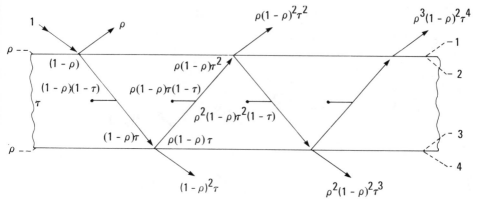

Figure 19-2 Multiple internal reflections for radiation incident on an isothermal window.

The fraction transmitted is the sum of terms leaving surface 4:

$$T = \tau(1-\rho)^2 [1 + \rho^2\tau^2 + \rho^4\tau^4 + \cdots] = \frac{\tau(1-\rho)^2}{1-\rho^2\tau^2} = \tau \frac{1-\rho}{1+\rho} \frac{1-\rho^2}{1-\rho^2\tau^2} \qquad (19\text{-}2)$$

The last term on the right is often close to unity, so in this instance

$$T \approx \tau \frac{1-\rho}{1+\rho} \qquad (19\text{-}3)$$

The fraction of energy absorbed is

$$A = (1-\rho)(1-\tau)[1 + \rho\tau + \rho^2\tau^2 + \rho^3\tau^3 + \cdots] = \frac{(1-\rho)(1-\tau)}{1-\rho\tau} \qquad (19\text{-}4)$$

In the limit when absorption in the plate can be neglected, $\tau = 1$ and $R = 2\rho/(1+\rho)$, $T = (1-\rho)/(1+\rho)$, and $A = 0$.

Net-Radiation Method. From the enclosure theory in Chap. 8, it is evident that the net radiation method is a powerful analytical tool that can, in many situations, be much less difficult to apply than the ray-tracing method. The net-radiation method [3] will now be applied to derive the radiation characteristics of a partially absorbing layer. Referring to Fig. 19-3, the outgoing flux at each interface can be written in terms of the incoming fluxes to yield the following equations for the conditions of a unit incoming flux at surface 1 and a zero incoming flux at surface 4:

$$q_{o,1} = \rho q_{i,1} + (1-\rho)q_{i,2} = \rho + (1-\rho)q_{i,2} \qquad (19\text{-}5a)$$

$$q_{o,2} = (1-\rho)q_{i,1} + \rho q_{i,2} = (1-\rho) + \rho q_{i,2} \qquad (19\text{-}5b)$$

$$q_{o,3} = \rho q_{i,3} + (1-\rho)q_{i,4} = \rho q_{i,3} \qquad (19\text{-}5c)$$

$$q_{o,4} = (1-\rho)q_{i,3} + \rho q_{i,4} = (1-\rho)q_{i,3} \qquad (19\text{-}5d)$$

The transmittance of the layer is used to relate the internal q_i's and q_o's to give

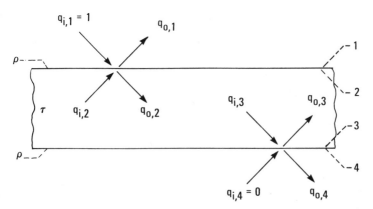

Figure 19-3 Net-radiation method applied to partially transmitting layer.

$q_{i,2} = q_{o,3}\tau$; $q_{i,3} = q_{o,2}\tau$. These are used to eliminate the q_i's from Eqs. (19-5), and the resulting equations are solved for the q_o's to yield

$$q_{o,1} = \rho \left[1 + \frac{(1-\rho)^2 \tau^2}{1-\rho^2 \tau^2} \right] \qquad q_{o,2} = \frac{1-\rho}{1-\rho^2 \tau^2}$$

$$q_{o,3} = \frac{\rho\tau(1-\rho)}{1-\rho^2 \tau^2} \qquad\qquad q_{o,4} = \frac{\tau(1-\rho)^2}{1-\rho^2 \tau^2}$$

For $q_{i,1} = 1$, the fractions reflected and transmitted by the plate are $q_{o,1}$ and $q_{o,4}$ so that

$$R = q_{o,1} = \rho \left[1 + \frac{(1-\rho)^2 \tau^2}{1-\rho^2 \tau^2} \right] \qquad\qquad (19\text{-}6a)$$

$$T = q_{o,4} = \frac{\tau(1-\rho)^2}{1-\rho^2 \tau^2} \qquad\qquad (19\text{-}6b)$$

The fraction absorbed is

$$A = (q_{o,2} + q_{o,3})(1-\tau) = \frac{(1-\rho)(1-\tau)}{1-\rho\tau} \qquad\qquad (19\text{-}6c)$$

These results agree, as they should, with those obtained by the ray-tracing method.

EXAMPLE 19-1 What is the fraction of externally incident unpolarized radiation that is transmitted through a glass plate in air? The plate is 0.75 cm thick, and the radiation is incident at $\theta = 50°$. $n_{\text{glass}} = 1.53$, and $a_{\text{glass}} = 0.1$ cm^{-1}.

To find the path length through the glass write $\chi = \sin^{-1}(\sin\theta/n) = \sin^{-1}(\sin 50°/1.53) = \sin^{-1} 0.500 = 30°$. The path length is $S = 0.75/\cos\chi = 0.866$ cm. The transmittance is $\tau = \exp(-aS) = \exp(-0.1 \times 0.866) = 0.917$. The surface reflectivities for the two components of polarization are

$$\rho_\| = \frac{\tan^2(\theta - \chi)}{\tan^2(\theta + \chi)} = 0.00412 \qquad \rho_\perp = \frac{\sin^2(\theta - \chi)}{\sin^2(\theta + \chi)} = 0.1206$$

Then the overall transmittance for each component is

$$T_\| = \tau \frac{1 - \rho_\|}{1 + \rho_\|} \frac{1 - \rho_\|^2}{1 - \rho_\|^2 \tau^2} = 0.917 \frac{0.9959}{1.0041} \frac{1 - 0.00002}{1 - 0.00001} = 0.9095$$

$$T_\perp = \tau \frac{1 - \rho_\perp}{1 + \rho_\perp} \frac{1 - \rho_\perp^2}{1 - \rho_\perp^2 \tau^2} = 0.917 \frac{0.8794}{1.1206} \frac{1 - 0.0145}{1 - 0.0122} = 0.7179$$

For unpolarized incident radiation, one-half the energy is in each component. Hence $T = (T_\| + T_\perp)/2 = 0.814$.

19-3.2 Multiple Parallel Windows

Ray-Tracing Method. As a more general case, consider a system of windows composed of a group of m identical plates and a group of n identical plates; the m and n plates can be different from each other. Since there are so many possible reflection paths, it might seem that ray tracing would be very complicated. Figure 19-4 shows, however, how the analysis by ray tracing can be organized so that it can be readily carried out. An amount R_m of the incident unit energy is reflected from the group of m plates, and an amount T_m is transmitted. The R_m and T_m can be found by building up results for a system of m plates with the formulas that will now be obtained. Continuing the reflection and transmission process yields the terms shown in Fig. 19-4.

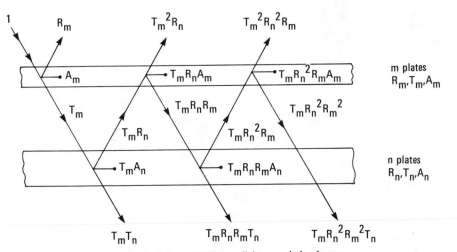

Figure 19-4 Ray-tracing method for multiple parallel transmitting layers.

Summing the reflected transmitted terms gives, for the system of $m + n$ plates,

$$R_{m+n} = R_m + R_n T_m^2 (1 + R_m R_n + R_m^2 R_n^2 + \cdots) = R_m + \frac{R_n T_m^2}{1 - R_m R_n} \qquad (19\text{-}7a)$$

$$T_{m+n} = T_m T_n (1 + R_m R_n + R_m^2 R_n^2 + \cdots) = \frac{T_m T_n}{1 - R_m R_n} \qquad (19\text{-}7b)$$

The absorbed energy can be found by summing the terms in Fig. 19-4, and the total absorbed energy will equal $1 - R_{m+n} - T_{m+n}$.

To illustrate how R_m (or R_n) is obtained, R_1 is first found from (19-1), which is written for a single plate. Then, from (19-7a), the R_m for two plates is

$$R_2 = R_{1+1} = R_1 + \frac{R_1 T_1^2}{1 - R_1 R_1}$$

where T_1 is from (19-2) for a single plate. Then, for three plates,

$$R_3 = R_{1+2} = R_1 + \frac{R_2 T_1^2}{1 - R_1 R_2}$$

The process can be continued to any value m.

Net-Radiation Method. The system of m and n plates will now be analyzed by the net-radiation method. Using the notation in Fig. 19-5, the outgoing radiation terms are written in terms of the incoming fluxes as

$$q_{o,m1} = R_m + q_{i,m2} T_m \qquad (19\text{-}8a)$$

$$q_{o,m2} = q_{i,m2} R_m + T_m \qquad (19\text{-}8b)$$

$$q_{o,n1} = q_{i,n1} R_n \qquad (19\text{-}8c)$$

$$q_{o,n2} = q_{i,n1} T_n \qquad (19\text{-}8d)$$

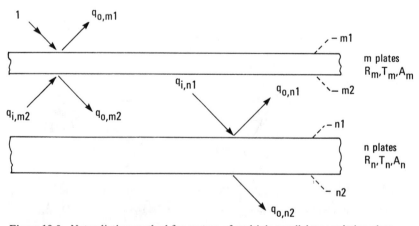

Figure 19-5 Net-radiation method for system of multiple parallel transmitting plates.

The q_i are further related to the q_o by using the relations $q_{i,m2} = q_{o,n1}$ and $q_{i,n1} = q_{o,m2}$. The q_i are then eliminated, and solving for the q_o yields

$$T_{m+n} = q_{o,n2} = \frac{T_m T_n}{1 - R_m R_n} \tag{19-9}$$

$$R_{m+n} = q_{o,m1} = R_m + \frac{R_n T_m{}^2}{1 - R_m R_n} \tag{19-10}$$

The fractions of energy absorbed in the group of m plates and in the group of n plates within the system of $m + n$ plates are

$$A_m{}^{(m+n)} = A_m + q_{i,m2} A_m = A_m \left(1 + \frac{T_m R_n}{1 - R_m R_n}\right) \tag{19-11a}$$

$$A_n{}^{(m+n)} = q_{i,n1} A_n = \frac{T_m A_n}{1 - R_m R_n} \tag{19-11b}$$

Note that T_{m+n} is symmetric, that is, the m and n subscripts can be reversed and the expression remains the same; hence $T_{m+n} = T_{n+m}$. The T_{m+n} is the transmission for the system of $m + n$ plates for radiation incident first on the m plates, while T_{n+m} is for incidence first on the n plates. From (19-10), however, $R_{m+n} \neq R_{n+m}$; that is, the system reflectance depends on whether the radiation is incident first on the group of m plates or on the group of n plates.

19-3.3 Results for Transmission through Multiple Parallel Glass Plates

Transmission through multiple glass plates is of interest in the design of flat-plate solar collectors. The glass surface reflectivity as given by (4-62a) and (4-62b) depends on the angle of incidence and the component of polarization. Since the reflections within the glass are all assumed to be specular, the same angles of reflection and refraction are maintained throughout the multiple reflection process. Equation (19-9) was used to calculate the overall transmittance through one and three parallel glass plates with an index of refraction $n = 1.5$. Neglecting absorption within the glass, results are shown in Fig. 19-6 for the two components of polarization, and as a function of the incidence angle of the radiation. As $\theta \to 90°$, the T goes to zero; this is because dielectrics have perfect reflectivity at grazing incidence (Fig. 4-6). For incidence at Brewster's angle, the transmittance for the parallel component becomes unity, as there is zero reflection at this angle, and the transmission losses are being neglected.

Incident solar radiation is unpolarized and hence has equal amounts of energy in the parallel and perpendicular components. The transmittance is then the average of the two T values computed by individually using $\rho_\parallel$ and $\rho_\perp$. This is shown in Fig. 19-7 for the limiting case of nonabsorbing plates and for absorbing plates having a product of absorption coefficient and thickness of 0.0524 per plate. For angles near normal incidence, the effect of absorption reduces the transmission by about 5% for each plate.

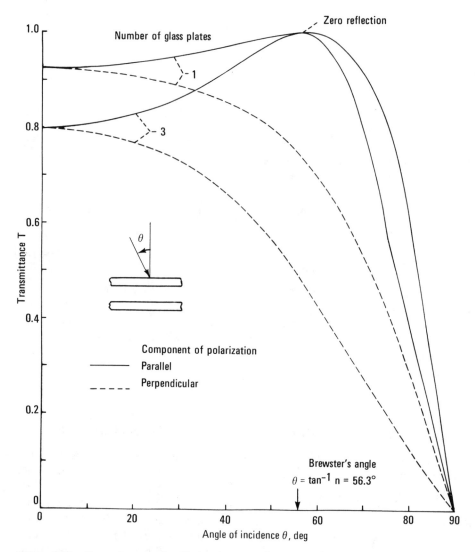

Figure 19-6 Transmittance of radiation in two components of polarization for nonabsorbing parallel glass plates; $n = 1.5$ (from [4]).

19-3.4 Interaction of Transmitting Plates with Absorbing Plate

A flat-plate solar collector usually consists of a few parallel transmitting windows covering an opaque absorber plate as shown in Fig. 19-8. It is desired to obtain the fraction A_c of incident energy that is absorbed by the opaque plate. At the collector plate,

$$A_c = q_{i,c} - q_{o,c} \tag{19-12a}$$

$$q_{o,c} = (1 - \alpha_c)q_{i,c} \tag{19-12b}$$

Across the space between the transmitting plates and the opaque collector plate,

$$q_{o,c} = q_{i,n} \qquad (19\text{-}12c)$$

$$q_{o,n} = q_{i,c} \qquad (19\text{-}12d)$$

For the system of n plates,

$$q_{o,n} = T_n + q_{i,n} R_n \qquad (19\text{-}12e)$$

The system of equations (19-12) is solved to yield the fraction of incident energy

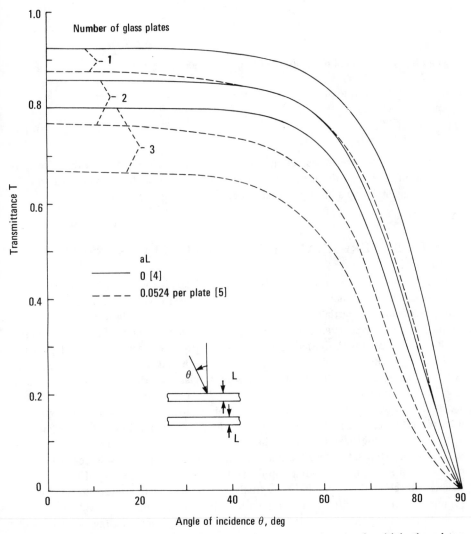

Figure 19-7   Effect of incidence angle and absorption on transmittance of multiple glass plates; $n = 1.5$.

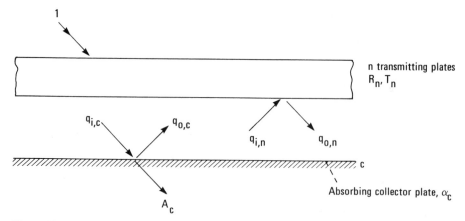

Figure 19-8 Interaction of transmitting windows and an absorbing plate.

absorbed by the opaque plate:

$$A_c = \frac{\alpha_c T_n}{1 - (1 - \alpha_c) R_n} \tag{19-13}$$

In this section, the fundamentals have been developed for analyzing the reflection transmission, and absorption behavior of window systems, and some numerical results have been given. Many additional aspects are treated in the literature [3-9].

19-4 ENCLOSURE ANALYSIS WITH PARTIALLY TRANSPARENT WINDOWS

The enclosures considered in Chap. 7-10 have had opaque walls, or in some instances have had an opening such as at the end of a cylindrical cavity. A more general enclosure could contain partially transmitting windows as shown in Fig. 19-9. Only a simplified case for such enclosure will be considered here, and more information is given in [8]. For this simplified case, the window properties are assumed independent of wavelength, and the radiation transmitted through and reflected from the window is assumed diffuse. A window such as a smooth glass plate will reflect in a specular fashion so that the reflected portion of an individual beam of radiation leaving the surface will not be diffuse. However, within an enclosure there usually are extensive multiple reflections, and the directionality of each reflection loses its importance in contributing to the heat fluxes on the boundaries. Hence the assumption of diffuse reflection is often satisfactory when the enclosure has multiple surfaces. When the window is hot enough to radiate appreciably, the analysis is restricted to a window that is thin enough so that it is essentially at a uniform temperature throughout. A symmetric window will be considered so that the radiative properties will be the same on both sides; for example, if the window is coated, the same coating exists on both sides. A two-band semigray analysis is given in [8] to account for the large change in transmission properties on either side of the infrared cutoff wavelength as shown for glass in Fig. 5-30.

For the quantities in Fig. 19-9, an overall heat balance at window k yields

$$q_k = q_{o,k} - q_{i,k} + q_{l,k} - q_{e,k} \qquad (19\text{-}14a)$$

The q_k is energy supplied to the window by a means other than radiation, such as by convective heating or by heating wires within the semitransparent material. For convective cooling of the window, q_k is negative. The radiative flux leaving the inside surface of the window consists of emitted energy, reflected incoming energy, and transmitted externally incident flux:

$$q_{o,k} = E_{w,k}\, \sigma T_k^4 + R_{w,k} q_{i,k} + T_{w,k} q_{e,k} \qquad (19\text{-}14b)$$

where E_w, R_w, and T_w are the overall emittance, reflectance and transmittance of the window (T without a w subscript is the window temperature). Similarly the flux lost from the outside surface of the window is given by

$$q_{l,k} = E_{w,k}\, \sigma T_k^4 + R_{w,k} q_{e,k} + T_{w,k} q_{i,k} \qquad (19\text{-}14c)$$

For a gray window $E_{w,k} = A_{w,k} = 1 - T_{w,k} - R_{w,k}$. The use of an emittance means that the window has a temperature that can be considered uniform throughout its thickness. The $q_{l,k}$ is eliminated from Eqs. (19-14a) and (19-14c) to give

$$q_{o,k} = q_k - E_{w,k}\, \sigma T_k^4 + (1 - T_{w,k}) q_{i,k} + (1 - R_{w,k}) q_{e,k} \qquad (19\text{-}14d)$$

where q_k and $q_{e,k}$ are specified. The incoming flux is obtained from the usual enclosure relation

$$q_{i,k} = \sum_{j=1}^{N} q_{o,j} F_{k-j} \qquad (19\text{-}14e)$$

Equations (19-14b), (19-14d) and (19-14e) provide three equations relating q_o, q_i, and T for each partially transparent surface. These can be combined with the equations for opaque surfaces as given in Sec. 8-3 to obtain a complete set of equations that are sufficient to solve for all the unknown surface temperatures and heat fluxes.

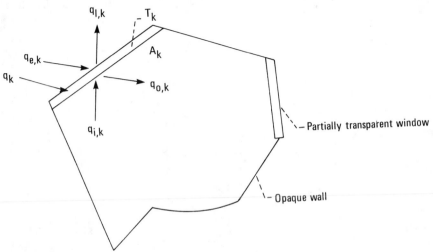

Figure 19-9 Enclosure with partially transparent windows.

19-5 EFFECT OF COATINGS OR THIN FILMS ON SURFACES

The radiative behavior of a surface can be modified by depositing on it one or more thin layers of other materials. These layers can be dielectric or metallic and can be thick or thin relative to the wavelength of the radiation. As will be shown, it is possible to obtain high or low absorption by the coated surface, and these characteristics will depend on the wavelength of the radiation. Thus it is possible to use thin film coatings to tailor surfaces with a desired wavelength selective behavior.

19-5.1 Coating without Wave Interference Effects

The geometry is shown in Fig. 19-10, and it consists of a coating of thickness L on a thick substrate. The coating has a transmittance τ and reflectivities ρ_1 and ρ_2 at the first and second interfaces. The film is thick enough so that it is not necessary to consider interference between waves reflected from the two interfaces. It will be determined to what extent the film alters the reflection characteristics from that of the substrate alone. By the net-radiation method, as derived for multiple layers, the fraction of incident radiation that is reflected is

$$R = \frac{\rho_1 + \rho_2(1 - 2\rho_1)\tau^2}{1 - \rho_1\rho_2\tau^2} \tag{19-15}$$

This result will now be used to calculate the behavior of a few different types of coatings.

Nonabsorbing Dielectric Coating. To illustrate the performance of a coated surface, consider the relatively simple case of a dielectric film on a dielectric substrate. There is normally incident radiation through a surrounding dielectric medium having an index of refraction n_s. The film and substrate have refractive indexes n_1 and n_2. Although the film is thick relative to the radiation wavelength, it is still physically quite thin, and since it is also a dielectric, the effect of absorption within it is generally quite

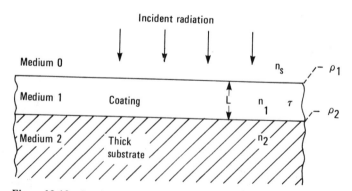

Figure 19-10 Coating of thickness L on a thick substrate.

small. Then $\tau \cong 1$, and R in (19-15) reduces to

$$R = \frac{\rho_1 + \rho_2(1 - 2\rho_1)}{1 - \rho_1\rho_2} \tag{19-16}$$

For normal incidence the interface reflectivities are given by

$$\rho_1 = \left(\frac{n_1 - n_s}{n_1 + n_s}\right)^2 \tag{19-17a}$$

$$\rho_2 = \left(\frac{n_2 - n_1}{n_2 + n_1}\right)^2 \tag{19-17b}$$

Substituting (19-17a) and (19-17b) into (19-16) gives, after simplification,

$$R = 1 - \frac{4n_s n_1 n_2}{(n_1^2 + n_s n_2)(n_2 + n_s)} \tag{19-18}$$

Dielectric coatings can be used to provide a reduced reflection at the surface and hence maximize the amount of radiation passing into the substrate. The proper n_1 to minimize reflection is obtained by letting $dR/dn_1 = 0$. This yields

$$n_1 = \sqrt{n_s n_2} \tag{19-19}$$

Hence, the minimum reflection is obtained if the n_1 of the coating is the geometric mean of the n values on either side of the coating. Using this n_1 in (19-18) gives the R for minimum reflection:

$$R = 1 - \frac{2\sqrt{n_s n_2}}{n_2 + n_s} \tag{19-20}$$

If there is no coating, the reflectivity of the substrate by itself is $\rho_{sub} = (n_2 - n_s)^2/(n_2 + n_s)^2$. The ratio of ρ_{sub} to the minimum R can be simplified to

$$\frac{\rho_{sub}}{R} = 1 + \frac{2\sqrt{n_s n_2}}{n_2 + n_s} \tag{19-21}$$

If an antireflection coating on glass in air is considered, then $n_s \approx 1$ and $n_2 \approx 1.53$, which yields $\rho_{sub}/R = 1.98$. Thus for these materials, at best the dielectric coating can reduce surface reflection to about half the uncoated value. As will be seen later, better results can be obtained by using thin films. The square root of $n_2 = 1.53$ is 1.24, which from (19-19) is the optimum value of n_1, and it is difficult to find a suitable coating material with a refractive index this low. Some commonly used materials are magnesium flouride, with $n = 1.38$, or cryolite (sodium aluminium fluoride), with $n = 1.36$. Also used are lithium fluoride with $n = 1.36$, and aluminium fluoride with $n = 1.39$.

Theory of Attenuating Film on Metal Substrate. Now the same geometry is considered as in Fig. 19-10, but the coating is attenuating ($\tau < 1$) and is on a metallic substrate. An example is a coated metal absorber plate for a flat-plate solar collector.

The external radiation is normally incident in air, with $n_s \approx 1$. The complex index of refraction for the coating is $n_1 - i\kappa_1$, and for the substrate $n_2 - i\kappa_2$. The film transmittance is $\tau = \exp(-a_1 L)$ where $a_1 = 4\pi\kappa_1/\lambda$. The a, κ, and n are wavelength dependent so the following can be regarded as a spectral calculation. For normal incidence [see Eqs. (4-55) and (4-54)]

$$\rho_1 = \frac{(n_1 - 1)^2 + \kappa_1{}^2}{(n_1 + 1)^2 + \kappa_1{}^2} \tag{19-22a}$$

$$\rho_2 = \frac{(n_2 - n_1)^2 + (\kappa_2 - \kappa_1)^2}{(n_2 + n_1)^2 + (\kappa_2 + \kappa_1)^2} \tag{19-22b}$$

Equation (19-15) applies, and this becomes

$$R = \frac{\rho_1 + \rho_2(1 - 2\rho_1)\exp(-2a_1 L)}{1 - \rho_1\rho_2 \exp(-2a_1 L)} \tag{19-23}$$

As an example, if we let $a_1 = 10^3 \, \text{cm}^{-1}$ and $\lambda = 0.7 \, \mu\text{m}$, the results shown in Fig. 19-11 are obtained for various coating thicknesses, for two substrate reflectivities and two indices of refraction of the coating. For high reflectance R of the coated metal, it is evident that the substrate reflectivity should be high and the absorptivity of the

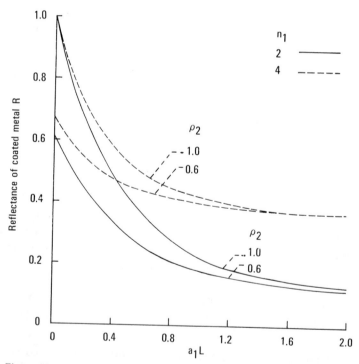

Figure 19-11 Reflectivity of thick attenuating film on metal substrate; no interference effects. $a_1 = 10^3 \, \text{cm}^{-1}$; $\lambda = 0.7 \, \mu\text{m}$.

coating low; the index of refraction of the coating is not important. To obtain a low reflectance, $a_1 L > 1$ and $n_1 < 2$ are desired; the substrate reflectivity ρ_2 is not important when $a_1 L$ becomes greater than about 1.

19-5.2 Thin Nonabsorbing Dielectric Film with Wave Interference Effects

When a film coating is thin, of order λ thick, there are interference effects between waves reflected from the first and second surfaces of the film. As given by Eqs. (4-43) and (4-44), the amplitude reflection coefficients for the two components of polarization are

$$\frac{E_{\|,r}}{E_{\|,i}} = r_\| = \frac{\tan(\theta - \chi)}{\tan(\theta + \chi)} \tag{19-24a}$$

$$\frac{E_{\perp,r}}{E_{\perp,i}} = r_\perp = -\frac{\sin(\theta - \chi)}{\sin(\theta + \chi)} \tag{19-24b}$$

When $n_2 > n_1$ for a wave incident from medium 1, then $\theta > \chi$ and $r_\perp$ is negative; that is, there is a phase change of π upon reflection. The $\tan(\theta - \chi)$ is positive, but $\tan(\theta + \chi)$ becomes negative for $\theta + \chi > \pi/2$, and $r_\|$ then yields a phase change of π. In similar fashion, for transmitted radiation,

$$\frac{E_{\|,t}}{E_{\|,i}} = t_\| = \frac{2 \sin \chi \cos \theta}{\sin(\theta + \chi) \cos(\theta - \chi)} \tag{19-25a}$$

$$\frac{E_{\perp,t}}{E_{\perp,i}} = t_\perp = \frac{2 \sin \chi \cos \theta}{\sin(\theta + \chi)} \tag{19-25b}$$

In going from medium 2 to medium 1, the χ and θ are interchanged in these relations (the χ and θ remain the angles in 2 and 1, respectively), and the reflection coefficients are equal to $-r_\|$ and $-r_\perp$. In this instance, the transmission coefficients are called t' and are equal to

$$t'_\| = \frac{2 \sin \theta \cos \chi}{\sin(\chi + \theta) \cos(\chi - \theta)} \tag{19-26a}$$

$$t'_\perp = \frac{2 \sin \theta \cos \chi}{\sin(\chi + \theta)} \tag{19-26b}$$

For a simplified case of normal incidence, and for radiation going from medium 1 to medium 2, these expressions reduce to

$$\frac{E_r}{E_i} = r_\| = r_\perp = r = \frac{n_1 - n_2}{n_1 + n_2} \tag{19-27a}$$

$$\frac{E_t}{E_i} = t_\parallel = t_\perp = t = \frac{2n_1}{n_1 + n_2} \tag{19-27b}$$

Formally the $r_\parallel$ will have a negative sign as $E_{\parallel,r}$ and $E_{\parallel,i}$ point in opposite directions as shown in Fig. 4-4. This sign is not significant in the present discussion. For normally incident radiation going from medium 2 into medium 1,

$$r = \frac{n_2 - n_1}{n_2 + n_1} \tag{19-28a}$$

$$t' = \frac{2n_2}{n_2 + n_1} \tag{19-28b}$$

For simplicity the following discussion is limited to normal incidence on the thin film. Figure 19-12 shows the radiation reflected from the first and second interfaces. The beams **a** and **b** can interfere with each other. For normal incidence beam **b** reflected from the second interface travels $2L$ farther than beam **a**, which is reflected from the first interface. Hence reflected beam **b** originated at time $2L/c_1$ earlier than reflected beam **a**, where c_1 is the propagation speed in the film. If beam **a** originated at time 0, then beam **b** originated at time $-2L/c_1$. If the two waves originated from the same vibrating source, the phase of **b** relative to **a** is $e^{i\omega\tau} = e^{-i\omega 2L/c_1}$. The circular frequency can be written as $\omega = 2\pi c_0/\lambda_0$ where λ_0 is the wavelength in vacuum. Then $e^{i\omega\tau} = e^{-i4\pi n_1 L/\lambda_0}$, where $n_1 = c_0/c_1$ is the refractive index of the film.

These results will now be applied to analyze the performance of a thin film. Consider a thin nonabsorbing film of refractive index n_1 on a substrate with index n_2. For a normally incident wave of unit amplitude, the reflected radiation is shown in Fig. 19-13. Taking into account the phase relationships, the reflected amplitude is

$$R_M = r_1 + t_1 t_1' r_2 \exp\left(-i\frac{4\pi n_1 L}{\lambda_0}\right) - t_1 t_1' r_1 r_2^2 \exp\left(-i\frac{8\pi n_1 L}{\lambda_0}\right)$$

$$+ t_1 t_1' r_1^2 r_2^3 \exp\left(-i\frac{12\pi n_1 L}{\lambda_0}\right) - \cdots = r_1 + \frac{t_1 t_1' r_2 \exp\left(-i\,4\pi n_1 L/\lambda_0\right)}{1 + r_1 r_2 \exp\left(-i\,4\pi n_1 L/\lambda_0\right)} \tag{19-29}$$

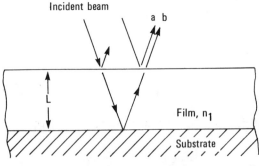

Figure 19-12 Reflection from the first and second interfaces of a thin film. This is for normal incidence; paths are drawn at an angle for clarity.

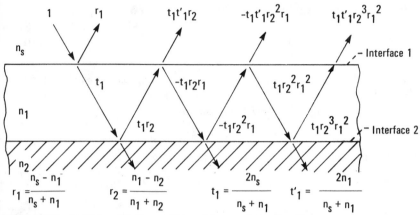

Figure 19-13 Multiple reflections within a thin nonattenuating film.

Note that $t_1 t_1' = 1 - r_1^2$ so this can be further reduced to

$$R_M = \frac{r_1 + r_2 \exp(-i4\pi n_1 L/\lambda_0)}{1 + r_1 r_2 \exp(-i4\pi n_1 L/\lambda_0)} \qquad (19\text{-}30)$$

One application for this type of thin coating is to obtain low reflection from a surface for use in reducing reflection losses during transmission through a series of lenses in optical equipment. To have zero reflected amplitude, $R_M = 0$, requires $r_1 = -r_2 \exp(-i4\pi n_1 L/\lambda_0)$. This can be obtained if $r_1 = r_2$ and $\exp(-i4\pi n_1 L/\lambda_0) = -1$. Since $\exp(-i\pi) = -1$, this yields $L = \lambda_0/4n_1$. The quantity λ_0/n_1 is the wavelength of the radiation within the film. Hence the film thickness for zero reflection at normal incidence is one-quarter of the wavelength within the film. The required condition $r_1 = r_2$ gives $(n_s - n_1)/(n_s + n_1) = (n_1 - n_2)/(n_1 + n_2)$ which reduces to $n_1 = \sqrt{n_s n_2}$. Thus for normal incidence onto a quarter-wave film from a dielectric medium with index of refraction n_s, the index of refraction of the film should be $n_1 = \sqrt{n_s n_2}$, that is, the geometric mean of the n values on either side of the film. The thin film provides better performance than the thick film, as it is possible to achieve zero reflectivity. However, this result is only for normal incidence and at one wavelength. To obtain more than one condition of zero reflectivity, it is necessary to use multilayer films. The optimization of the design of a low-reflectivity multilayer coating is discussed in [10]. For a system of two nonabsorbing quarter wave films zero reflection is obtained for the condition $n_1^2 n_3 = n_2^2 n_s$, where n_1 and n_2 are for the coatings (coating with n_2 is next to the substrate), and n_3 is for the substrate.

The previous expressions have been derived for the reflected amplitude from a thin film; now the reflected energy for normal incidence will be considered. From Sec. 4-4.3 the reflected energy depends on $|\vec{E}|^2$. Since $R_M = E_r/E_i$, the reflectivity for energy is $R = |R_M|^2 = R_M R_M^*$, where R_M^* is the complex conjugate of R_M. From (19-30), the reflectivity of the film is

$$R = \frac{r_1 + r_2 e^{-i\gamma_1}}{1 + r_1 r_2 e^{-i\gamma_1}} \frac{r_1 + r_2 e^{i\gamma_1}}{1 + r_1 r_2 e^{i\gamma_1}} \qquad \gamma_1 = \frac{4\pi n_1 L}{\lambda_0}$$

After multiplication and simplification this becomes

$$R = \frac{r_1^2 + r_2^2 + 2r_1r_2 \cos \gamma_1}{1 + r_1^2 r_2^2 + 2r_1r_2 \cos \gamma_1} \tag{19-31}$$

Another form of this equation that is found in the literature is obtained by inserting $r_1 = (n_s - n_1)/(n_s + n_1)$ and $r_2 = (n_1 - n_2)/(n_1 + n_2)$ and using the identity $\cos \gamma_1 = 1 - 2 \sin^2(\gamma_1/2)$. After simplification R takes the form

$$R = \frac{n_1^2 (n_s - n_2)^2 - (n_s^2 - n_1^2)(n_1^2 - n_2^2) \sin^2(2\pi n_1 L/\lambda_0)}{n_1^2 (n_s + n_2)^2 - (n_s^2 - n_1^2)(n_1^2 - n_2^2) \sin^2(2\pi n_1 L/\lambda_0)} \tag{19-32}$$

For a one-quarter-wave film, $L = \lambda_0/4n_1$, and this reduces to

$$R = \left(\frac{n_s n_2 - n_1^2}{n_s n_2 + n_1^2}\right)^2 \tag{19-33}$$

As deduced before by looking at R_M, the reflectivity becomes zero when $n_1 = \sqrt{n_s n_2}$. If n_1 is high, R is increased, and this behavior can be used to obtain dielectric mirrors. For various film materials of refractive index n_1 on glass ($n_2 = 1.5$), the reflectivity becomes, for incidence in air ($n_s \approx 1$),

Film	n_1	n_2	R
None	1	1.5	0.04
ZnS	2.3	1.5	0.31
Ge	4.0	1.5	0.69
Te	5.0	1.5	0.79

Multilayer films can be used to obtain reflectivities very close to 1.

This development has shown how ray tracing and the net-radiation method can be used to analyze the radiative behaviour of partially transparent layers, including windows and coatings. The material presented is quite brief, in view of the extensive information available. In addition to the references already discussed, the reader is referred to [11–17] for further information.

19-6 EFFECTS OF NONUNITY REFRACTIVE INDEX ON RADIATIVE BEHAVIOR WITHIN THE MEDIUM

The partially transmitting layers considered in Sec. 19.3–19.5 have a uniform temperature throughout their thickness. Layers with internal temperature variation are analyzed using the equation of transfer as discussed in Sec. 14-5–14-7, along with the appropriate conditions at the boundaries that include refractive index effects. Some introductory material is now given for this type of more detailed heat transfer analysis.

19-6.1 Effect of Refractive Index on Intensity

Consider radiation with intensity $i'_{\lambda,1}$ in a dielectric medium of refractive index n_1. Let the radiation in solid angle $d\omega_1$ pass into a dielectric medium of refractive index n_2 as

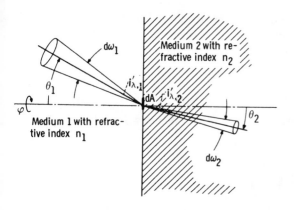

Figure 19-14 Beam with initial intensity $i'_{\lambda,1}$ crossing interface between two dielectric media with unequal refractive indices.

in Fig. 19-14. As a result of the differing indices of refraction, the rays will change direction as they pass into medium 2. The radiation in solid angle $d\omega_1$ at incidence angle θ_1 will pass into solid angle $d\omega_2$ at angle of refraction θ_2. If reflection or scattering at the interface is neglected for the moment, the energy of the radiation is conserved in crossing the interface. From the definition of intensity this conservation of energy is given by

$$i'_{\lambda,1} \cos \theta_1 \, dA \, d\omega_1 d\lambda_1 = i'_{\lambda,2} \cos \theta_2 \, dA \, d\omega_2 d\lambda_2 \tag{19-34}$$

where dA is an area element in the plane of the interface. Using the relation for solid angle

$$d\omega = \sin \theta \, d\theta \, d\varphi \tag{19-35}$$

results in (19-34) becoming (noting that the increment of circumferential angle $d\varphi$ is not changed in crossing the interface)

$$i'_{\lambda,1} \sin \theta_1 \cos \theta_1 \, d\theta_1 d\lambda_1 = i'_{\lambda,2} \sin \theta_2 \cos \theta_2 \, d\theta_2 d\lambda_2 \tag{19-36}$$

From (4-36) Snell's law relates the indices of refraction to the angles of incidence and refraction by

$$\frac{n_1}{n_2} = \frac{\sin \theta_2}{\sin \theta_1} \tag{19-37}$$

Then by differentiation

$$n_1 \cos \theta_1 \, d\theta_1 = n_2 \cos \theta_2 \, d\theta_2 \tag{19-38}$$

Substituting (19-37) and (19-38) into (19-36) gives

$$\frac{i'_{\lambda,1} d\lambda_1}{n_1^2} = \frac{i'_{\lambda,2} d\lambda_2}{n_2^2} \tag{19-39}$$

Although (19-39) was derived for radiation crossing the interface of two media, the equation also holds for intensity at any point in a transparent medium with variable refractive index so long as the local properties of the medium are independent of direction, that is, are isotropic. This isotropy will be the case except in certain plasma-physics applications. Thus, in general in a transparent isotropic medium, for either the spectral intensity or by integrating over all wavelengths, the total intensity in a medium with spectrally independent refractive index, there is the relation

$$\frac{i'_\lambda \, d\lambda}{n^2} = \text{const} \qquad \frac{i'}{n^2} = \text{const} \tag{19-40}$$

In terms of frequency, which does not change with n, we can write either

$$\frac{i'_\nu d\nu}{n^2} = \text{const} \quad \text{or} \quad \frac{i'_\nu}{n^2} = \text{const} \tag{19-41}$$

Hence for spectral calculations with variable n, it is more convenient to work with frequency than wavelength.

19-6.2 The Effect of Angle for Total Reflection

Consider a volume element dV inside a semi-infinite region of refractive index n_2 as shown in Fig. 19-15. Suppose that diffuse radiation of intensity i'_1 is incident upon the boundary of this region from a region having refractive index n_1, where $n_1 < n_2$. Radiation incident at grazing angles to the interface ($\theta_1 \approx 90°$) will be refracted into medium 2 at a maximum value of θ_2 given by

$$\sin \theta_{2,\max} = \frac{n_1}{n_2} \sin 90° = \frac{n_1}{n_2} \tag{19-42}$$

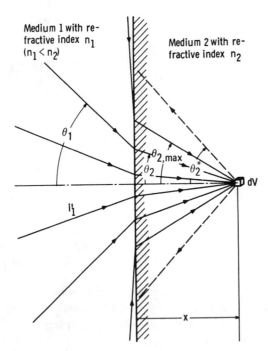

Medium 1 with refractive index n_1
$(n_1 < n_2)$

Medium 2 with refractive index n_2

θ_1

$\theta_{2,\max}$

θ_2

θ_2^*

dV

i'_1

x

Figure 19-15 Effect of refraction on radiation transport in media with non-unity refractive index.

Hence the volume element in medium 2 will receive direct radiation from medium 1 only at angular directions within the range

$$0 \leqslant \theta_2 \leqslant \theta_{2,\max} \left(= \sin^{-1} \frac{n_1}{n_2} \right) \tag{19-43}$$

Now consider emission from dV. The portion of this emission that enters region 1 will be along paths found by reversing the arrows on the solid lines in Fig. 19-15. However, there is also radiation from dV along paths such as those shown by the dashed lines in Fig. 19-15 that are incident on the interface at angles θ_2^*, where

$$\sin \theta_2^* > \frac{n_1}{n_2} \tag{19-44}$$

From (19-37), this means that such a ray would enter medium 1 at an angle given by

$$\sin \theta_1 = \frac{n_2}{n_1} \sin \theta_2^* > \frac{n_2 n_1}{n_1 n_2} = 1 \tag{19-45}$$

But $\sin \theta_1$ *cannot* be greater than unity for real values of θ_1. This result is interpreted to mean that any ray incident upon the interface from medium 2 at any angle greater than

$$\theta_{2,\max} = \sin^{-1} \frac{n_1}{n_2} \tag{19-46}$$

cannot enter medium 1, and must be totally reflected at the interface. The angle defined by (19-46) is called the *angle for total reflection*.

From Sec. 2-4.12 the blackbody emission inside a medium with a refractive index that is constant but not unity has an intensity given by

$$i'_{b,m} = n^2 i'_b \tag{19-47}$$

Consequently, for an absorbing-emitting gray medium with absorption coefficient a, the total energy emitted by a volume element is

$$dQ_e = 4n^2 a \sigma T^4 \, dV \tag{19-48}$$

If spectral variations of n are known to be important, then an integration over wave length must be included, provided of course that the data for n as a function of wave-length are available. The integration is best accomplished by using $i'_{\nu b}$, that is, frequency as the spectral variable.

From (19-47) it might appear that because $n > 1$, the intensity radiated from a dielectric medium into air could be larger than the usual blackbody radiation i'_b. This is not the case, as some of the energy emitted within the medium is reflected back into the emitting body at the medium-air interface. Consider a thick dielectric medium ($\kappa = 0$) at uniform temperature and with refractive index n. The maximum intensity received at an element dA on the interface from all directions within the medium is $n^2 i'_b$. Only the energy received within a cone having a vertex angle $\theta_{\max}$ relative to the normal of dA will penetrate through the interface; for incidence angles larger than

θ_{max} the energy will be reflected back into the medium. Hence, the energy received at dA that leaves the medium is

$$\int_{\theta=0}^{\theta_{max}} n^2 i_b' \, dA \cos\theta \, 2\pi \sin\theta \, d\theta = 2\pi n^2 i_b' \, dA \, \frac{\sin^2\theta_{max}}{2}$$

From (19-42), with $n_1 = 1$ and $n_2 = n$ in this case, $\sin\theta_{max} = 1/n$ so that the total hemispherical emissive power leaving the interface is $2\pi n^2 i_b' \, dA/2n^2 = \pi i_b' \, dA$. Dividing by πdA gives i_b' as the maximum diffuse intensity leaving the interface, which is the expected blackbody radiation.

EXAMPLE 19-2 An elemental volume dV is located in a glass plate at $x = 1$ cm from the plate interface with air (as in Fig. 19-15). Diffuse-gray radiation of intensity 10 W/(cm^2 sr) in the air is entering the glass (note that this intensity enters the glass, so the reflectivity of the interface has already been accounted for). If the absorption coefficient of the glass is 0.005 cm^{-1} and its refractive index is $n = 1.75$, determine the temperature at dV as a result of only this incident intensity and the emitted energy. Assume the radiation absorbed from the surrounding glass is small, and neglect heat conduction.

An energy balance on dV states that the emitted energy will equal the incident energy that is transmitted through the glass and absorbed by dV; that is,

$$4n^2 a\sigma T^4(x) \, dV = a \, dV \int_{\omega=4\pi} i'(x,\theta) \, d\omega$$

From (19-40) the intensity $i'_{glass}(0,\theta)$ in the glass at the glass surface is related to the entering diffuse intensity in air $i'_{air}(0)$, by $i'_{glass}(0,\theta) = n^2 i'_{air}(0)$. Since the path length from the glass surface to dV is $x/\cos\theta$, the intensity $i'_{glass}(x,\theta)$ at dV is given by Bouguer's law as

$$i'_{glass}(x,\theta) = i'_{glass}(0,\theta) \exp\left(-\frac{ax}{\cos\theta}\right) = n^2 i'_{air}(0) \exp\left(-\frac{ax}{\cos\theta}\right)$$

Substituting into the energy balance and solving for T^4 give

$$T^4 = \frac{i'_{air}(0)}{4\sigma} \int_{\omega=4\pi} \exp\left(-\frac{ax}{\cos\theta}\right) d\omega$$

Over part of the 4π solid angle surrounding dV there will be no energy incident on dV, so that the intergration limits on θ must be derived by considering the restriction of (19-42). This gives (with $\mu = \cos\theta$)

$$T^4 = \frac{2\pi i'_{air}(0)}{4\sigma} \int_0^{\theta_{max}=\sin^{-1}(1/1.75)} \exp\left(-\frac{ax}{\cos\theta}\right) \sin\theta \, d\theta$$

$$T^4 = \frac{2\pi i'_{air}(0)}{4\sigma} \int_{\cos\theta_{max}}^1 \exp\left(-\frac{ax}{\mu}\right) d\mu$$

From Eq. (F-1) in Appendix F this becomes

$$T^4 = \frac{2\pi i'_{\text{air}}(0)}{4\sigma} \left[E_2(ax) - \int_0^{\cos\theta_{\max}} \exp\left(-\frac{ax}{\mu}\right) d\mu \right]$$

Now let $\gamma = \mu/\cos\theta_{\max}$ to obtain

$$T^4 = \frac{2\pi i'_{\text{air}}(0)}{4\sigma} \left[E_2(ax) - \cos\theta_{\max} E_2\left(\frac{ax}{\cos\theta_{\max}}\right) \right]$$

Substituting numerical values gives

$$T^4 = \frac{2\pi 10 \text{ W/cm}^2}{4 \times 5.729 \times 10^{-12} \text{W/(cm}^2\text{K}^4)} \left[E_2(0.005) - 0.821 E_2\left(\frac{0.005}{0.821}\right) \right]$$

The E_2 values can be interpolated from Table F-1 to give $T^4 = (5\pi \times 10^{12}/5.729)$ (0.18) K^4 or $T = 840$ K.

In this example, the effect of the surface reflectivity in determining the intensity of the radiation that is able to cross the interface and enter the material is not explicitly treated. It can be introduced for optically smooth surfaces by using the reflectivity relations from Chap. 4 and in the next section.

19-6.3 Interface Conditions for Analysis of Radiation within a Plane Layer

In Sec. 14-5 a plane layer of partially transmitting medium was analyzed, and the solution depended on boundary conditions (14-43) giving the intensities within the medium at the boundaries. Later results of Chap. 14 were for the layer bounded by diffuse gray or black plates, so that the boundary intensities were $1/\pi$ times the outgoing diffuse fluxes from the plates. Here we consider a plane layer that is within a medium having a different refractive index, such as a glass plate in air. Radiation is incident from the surrounding environment, and some of it will cross the boundaries and enter the medium. We shall obtain here expressions for the intensities *inside* the medium at the boundaries. The analysis of Chap. 14 can then be applied using these internal boundary conditions.

The geometry is shown in Fig. 19-16, which is a more general situation than in Fig. 14-6. As mentioned in connection with (19-41), it is convenient to use frequency as the spectral variable in situations where radiation crosses an interface between media with different n. Both the plane layer and surrounding medium are assumed to be dielectrics. A superscript s is used to designate conditions outside the layer. The angles θ and φ give the direction within the medium except at the boundaries. The θ_i and φ_i fix incident directions at the boundaries. The $dq_{v,s}$ are the spectral fluxes *incident* on the layer from within the exterior medium; these fluxes are assumed uniform over the layer boundary.

The intensity $i_v^+(0, \theta, \varphi)$ leaving boundary 1 inside the medium is composed of the transmitted portion of $i'_{v,s}(0, \theta_i, \varphi_i)$ and the reflected portion of $i_v^-(0, \theta_i, \varphi_i)$. The

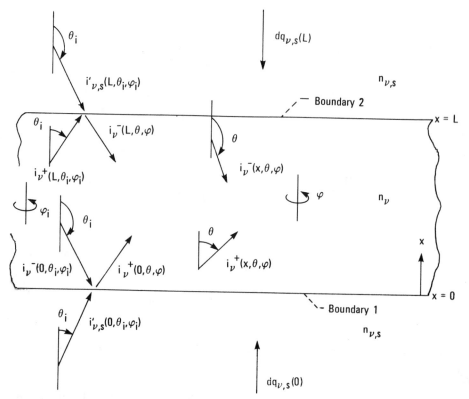

Figure 19-16 Intensities in a plane layer surrounded by a medium with a different refractive index.

bidirectional spectral reflectivity (3-20) relates the reflected and incident intensities as

$$\rho_\nu''(\theta, \varphi, \theta_i, \varphi_i) = \frac{i_{\nu,r}''(\theta, \varphi, \theta_i, \varphi_i)}{i_{\nu,i}'(\theta_i, \varphi_i) \cos \theta_i \, d\omega} \tag{19-49}$$

Similarly, a bidirectional transmissivity of the interface can be defined as

$$\tau_\nu''(\theta, \varphi, \theta_i, \varphi_i) = \frac{i_{\nu,\tau}''(\theta, \varphi, \theta_i, \varphi_i)}{i_{\nu,i}'(\theta_i, \varphi_i) \cos \theta_i \, d\omega} \tag{19-50}$$

where $i_{\nu,\tau}''$ is the transmitted intensity in the direction θ, φ. Then by integrating over all incident solid angles at boundary 1, the intensity inside the layer leaving the boundary is found:

$$i_\nu^+(0, \theta, \varphi) = \int_{\varphi_i=0}^{2\pi} \int_{\theta_i=0}^{\pi/2} \tau_{\nu,1}''(\theta, \varphi, \theta_i, \varphi_i) i_{\nu,s}'(0, \theta_i, \varphi_i) \cos \theta_i \sin \theta_i \, d\theta_i \, d\varphi_i$$

$$+ \int_{\varphi_i=0}^{2\pi} \int_{\theta_i=\pi/2}^{\pi} \rho_{\nu,1}''(\theta, \varphi, \theta_i, \varphi_i) i_\nu^-(0, \theta_i, \varphi_i) \cos \theta_i \sin \theta_i \, d\theta_i \, d\varphi_i \tag{19-51}$$

Similarly, inside the medium at boundary 2,

$$i_\nu^-(L,\theta,\varphi) = \int_{\varphi_i=0}^{2\pi} \int_{\theta_i=\pi/2}^{\pi} \tau_{\nu,2}''(\theta,\varphi,\theta_i,\varphi_i) i_{\nu,s}'(L,\theta_i,\varphi_i) \cos\theta_i \, \sin\theta_i \, d\theta_i \, d\varphi_i$$

$$+ \int_{\varphi_i=0}^{2\pi} \int_{\theta_i=0}^{\pi/2} \rho_{\nu,2}''(\theta,\varphi,\theta_i,\varphi_i) i_\nu^+(L,\theta_i,\varphi_i) \cos\theta_i \, \sin\theta_i \, d\theta_i \, d\varphi_i \qquad (19\text{-}52)$$

As in Chap. 14, it is convenient to use the variable $\mu = \cos\theta$ to obtain

$$i_\nu^+(0,\mu,\varphi) = \int_{\varphi_i=0}^{2\pi} \left\{ \int_{\mu_i=0}^{1} [\tau_{\nu,1}''(\mu,\varphi,\mu_i,\varphi_i) i_{\nu,s}'(0,\mu_i,\varphi_i) \right.$$

$$\left. - \rho_{\nu,1}''(\mu,\varphi,-\mu_i,\varphi_i) i_\nu^-(0,-\mu_i,\varphi_i)] \, \mu_i \, d\mu_i \right\} d\varphi_i \qquad (19\text{-}53)$$

$$i_\nu^-(L,\mu,\varphi) = \int_{\varphi_i=0}^{2\pi} \left\{ \int_{\mu_i=0}^{1} [-\tau_{\nu,2}''(\mu,\varphi,-\mu_i,\varphi_i) i_{\nu,s}'(L,-\mu_i,\varphi_i) \right.$$

$$\left. + \rho_{\nu,2}''(\mu,\varphi,\mu_i,\varphi_i) i_\nu^+(L,\mu_i,\varphi_i)] \, \mu_i \, d\mu_i \right\} d\varphi_i \qquad (19\text{-}54)$$

As a special case, let the layer have optically smooth surfaces and the incident fluxes $dq_{\nu,s}$ at 0 and L be diffuse. As discussed in Chap. 4, the reflection at an interface depends on the component of polarization. The analysis should consider the portion of radiation in each polarization component and then add the two energies to obtain the total quantity. For simplicity this effect will be neglected here, and average values of the surface properties will be used.

The internal reflections will be specular from the smooth interfaces; hence $\theta = \pi - \theta_i$ and $\varphi = \varphi_i + \pi$, and from (4-65a) at $x = L$,

$$\frac{i_\nu^-(L,\theta,\varphi)}{i_\nu^+(L,\theta_i,\varphi_i)} = \frac{1}{2} \frac{\sin^2(\theta_i-\chi)}{\sin^2(\theta_i+\chi)} \left[1 + \frac{\cos^2(\theta_i+\chi)}{\cos^2(\theta_i-\chi)} \right] \qquad (19\text{-}55)$$

and similarly for $i_\nu^+(0,\theta,\varphi)/i_\nu^-(0,\theta_i,\varphi_i)$ where $\pi - \theta_i$ is used on the right in place of θ_i. The χ is determined from Snell's law:

$$\frac{\sin\chi}{\sin(\pi-\theta_i)} = \frac{\sin\chi}{\sin\theta_i} = \frac{n_\nu}{n_{\nu,s}} \qquad \text{at 1 } (x=0)$$

$$\frac{\sin\chi}{\sin\theta_i} = \frac{n_\nu}{n_{\nu,s}} \qquad \text{at 2 } (x=L)$$

For the transmitted intensity a factor of $(n_\nu/n_{\nu,s})^2$ is used to account for the effect discussed in Sec. 19-6.1, and incidence is from the surrounding medium onto the layer. Then, using (4-65a) for either boundary 1 or 2 gives

$$\frac{i_\nu^+(0,\theta,\varphi)}{i_{\nu,s}'(0,\theta_i,\varphi_i)} = \frac{i_\nu^-(L,\theta,\varphi)}{i_{\nu,s}'(L,\theta_i,\varphi_i)} = \left\{ 1 - \frac{1}{2} \frac{\sin^2(\theta_i-\theta)}{\sin^2(\theta_i+\theta)} \left[1 + \frac{\cos^2(\theta_i+\theta)}{\cos^2(\theta_i-\theta)} \right] \right\} \left(\frac{n_\nu}{n_{\nu,s}} \right)^2 \qquad (19\text{-}56)$$

where θ and θ_i are related by:

$\sin \theta / \sin \theta_i = n_{\nu,s}/n_\nu$

The externally incident intensities are given by

$$i'_{\nu,s}(0, \theta_i, \varphi_i) = dq_{\nu,s}(0)/\pi \, d\nu \tag{19-57a}$$

$$i'_{\nu,s}(L, \theta_i, \varphi_i) = dq_{\nu,s}(L)/\pi \, d\nu \tag{19-57b}$$

The analysis will not be carried further, as it goes beyond the scope of the intended treatment. A detailed development is given in [2], where the literature is also reviewed. A few noteworthy references are those by Gardon [18–21], who treated problems of thermal radiation in glass, where effects of the refractive index are significant. Reference [18] includes some analysis of perpendicular and parallel polarization contributions; in [19, 20] a comprehensive analysis of the heat treatment of glass is given. This analysis of heat treatment includes the effects of conduction within the glass and convection at the surface. In [21], a review of radiant heat transfer as studied by researchers in the glass industry is given, and a digest of much of the literature on the subject up to 1961 is presented. Condon [22] gives a more recent review of radiation problems in the glass industry. In [23–26], the emittance is given from layers of material with nonunity refractive index, both semi-infinite in extent [25] and bounded by a substrate [23, 24, 26]. The radiation and conduction heat transfer through a high temperature semi-transparent layer of slag is analyzed in [27]. Inhomogeneous films may provide useful absorption properties for solar energy collection [28, 29].

REFERENCES

1. Hodgman, Charles D. (ed.): "Handbook of Chemistry and Physics," 38th ed., Chemical Rubber Publishing Company, Cleveland, Ohio, 1957.
2. Viskanta, R. and E. E. Anderson: Heat Transfer in Semi-Transparent Solids, in J. P. Hartnett and T. F. Irvine, Jr. (eds.), "Advances in Heat Transfer," vol. 11, pp. 317–441, Academic Press, New York, 1975.
3. Siegel, Robert: Net Radiation Method for Transmission through Partially Transparent Plates, *Sol. Energy*, vol. 15, pp. 273–276, 1973.
4. Shurcliff, William A: Transmittance and Reflection Loss of Multi-Plate Planar Window of a Solar-Radiation Collector: Formulas and Tabulations of Results for the Case of $n = 1.5$, *Sol. Energy*, vol. 16, pp. 149–154, 1974.
5. Duffie, J. A., and William A. Beckman: "Solar Energy Thermal Processes," John Wiley & Sons, Inc., New York, 1974.
6. Wijeysundera, N. E.: A Net Radiation Method for the Transmittance and Absorptivity of a Series of Parallel Regions, *Sol. Energy*, vol. 17, pp. 75–77, 1975.
7. Edwards, D. K.: Solar Absorption by Each Element in an Absorber-Coverglass Array, *Sol. Energy*, vol. 19, pp. 401–402, 1977.
8. Siegel, Robert: Net Radiation Method for Enclosure Systems Involving Partially Transparent Walls, *NASA* TN D-7384, August, 1973.
9. Viskanta, R., D. L. Siebers, and R. P. Taylor: Radiation Characteristics of Multiple-Plate Glass Systems, *Int. J. Heat Mass Transfer*, vol. 21, pp. 815–818, 1978.
10. Thornton, B. S. and Q. M. Tran: Optimum Design of Wideband Selective Absorbers with Provisions for Specified Included Layers, *Sol. Energy*, vol. 20, no. 5, pp. 371–378, 1978.

11. Hsieh, C. K., and R. Q. Coldewey: Study of Thermal Radiative Properties of Antireflection Glass for Flat-Plate Solar Collector Covers, *Sol. Energy*, vol. 16, pp. 63–72, 1974.
12. Musset, A., and A. Thelen: Multilayer Antireflection Coatings, in "Progress in Optics," vol. 8, pp. 203–237, American Elsevier Publishing Company, Inc., New York, 1970.
13. Forsberg, C. H., and G. A. Domoto: Thermal-Radiation Properties of Thin Metallic Films on Dielectrics, *J. Heat Transfer*, vol. 94, no. 4, pp. 467–472, 1972.
14. Taylor, R. P., and R. Viskanta: Spectral and Directional Radiation Characteristics of Thin-Film Coated Isothermal Semitransparent Plates, *Wärme Stoffübertrag.*, vol. 8, pp. 219–227, 1975.
15. Heavens, O. S.: "Optical Properties of Thin Solid Films," Dover Publications, Inc., New York, 1965.
16. Roux, J. A., A. M. Smith, and F. Shahrokhi: Effect of Boundary Conditions on the Radiative Reflectance of Dielectric Coatings, pp. 131–144, in "Thermophysics and Spacecraft Thermal Control," vol. 35 of *Progress in Astronautics and Aeronautics*, MIT Press, Cambridge, Mass., 1974.
17. Palik, E. D., N. Ginsburg, H. B. Rosenstock, and R. T. Holm: Transmittance and Reflectance of a Thin Absorbing Film on a Thick Substrate, *Appl. Opt.*, vol. 17, no. 21, pp. 3345–3347, 1978.
18. Gardon, Robert: The Emissivity of Transparent Materials, *J. Am. Ceram. Soc.*, vol. 39, no. 8, pp. 278–287, 1956.
19. Gardon, Robert: Calculation of Temperature Distributions in Glass Plates Undergoing Heat-Treatment, *J. Am. Ceram. Soc.*, vol. 41, no. 6, pp. 200–209, 1958.
20. Gardon, Robert: Appendix to Calculation of Temperature Distributions in Glass Plates Undergoing Heat Treatment, Ref. [19], Mellon Institute, Pittsburgh, 1958.
21. Gardon, Robert: A Review of Radiant Heat Transfer in Glass, *J. Am. Ceram. Soc.*, vol. 44, no. 7, pp. 305–312, 1961.
22. Condon, Edward U.: Radiative Transport in Hot Glass, *J. Quant. Spectrosc. Radiat. Transfer*, vol. 8, no. 1, pp. 369–385, 1968.
23. Caren, R. P., and C. K. Liu: Effect of Inhomogeneous Thin Films on the Emittance of a Metal Substrate, *ASME J. Heat Transfer*, vol. 93, no. 4, pp. 466–468, 1971.
24. Baba, H., and A. Kanayama: Directional Monospectral Emittance of Dielectric Coating on a Flat Metal Substrate, paper 75-664, *AIAA*, May, 1975.
25. Armaly, B. F., T. T. Lam, and A. L. Crosbie: Emittance of Semi-Infinite Absorbing and Isotropically Scattering Medium with Refractive Index Greater than Unity, *AIAA J.*, vol. 11, no. 11, pp. 1498–1502, 1973.
26. Anderson, E. E.: Estimating the Effective Emissivity of Nonisothermal Diatherminous Coatings, *ASME J. Heat Transfer*, vol. 97, pp. 480–482, 1975.
27. Viskanta, R., and D. M. Kim: Heat Transfer Through Irradiated, Semi-transparent Layers at High Temperature, *J. Heat Transfer*, vol. 102, no. 1, pp. 182–184, 1980.
28. Heavens, O. S.: Optical Properties of Thin Films—Where To?, *Thin Solid Films*, vol. 50, pp. 157–161, 1978.
29. Fan, J. C. C.: Selective-Black Absorbers Using Sputtered Cermet Films, *Thin Solid Films*, vol. 54, pp. 139–148, 1978.

PROBLEMS

1 A horizontal glass plate 0.3 cm thick is covered by a plane layer of water 0.6 cm thick. A beam of radiation is incident from air onto the upper surface of the water at an incidence angle of 45°. What is the path length of the radiation through the glass? $n_{H_2O} = 1.33$, and $n_{glass} = 1.53$.)

   Answer: 0.338 cm

2 Redo Example 15-4 including surface reflection and refraction effects.

   Answer: 0.0750 W/(cm² sr)

3 Radiation is normally incident on a series of two glass plates, each $\frac{1}{8}$ in thick, in air. What is the fraction T that is transmitted? ($n_{glass} = 1.53$, and $a_{glass} = 0.3$ in^{-1}.) What is the fraction transmitted for a single plate $\frac{1}{4}$ in thick in air?

  Answer: 2 plates: $T = 0.783$; 1 plate: $T = 0.849$

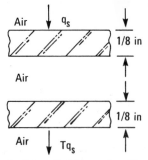

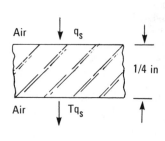

4 Prove that for an absorbing transmitting layer with $L > \lambda$, $T + A + R = 1$.

5 Derive an analytical expression that shows whether or not the total absorption in a system of n plates and m plates $A_{(m+n)}$ is independent of whether radiation is first incident on the n or the m plates.

6 As a result of surface treatment, a partially transparent plate has a different reflectivity at each surface. For radiation incident on surface 1, obtain an expression for the overall reflectance and transmittance in terms of ρ_1, ρ_2, and τ.

  Answer: $R = \dfrac{\rho_1 + \rho_2(1 - 2\rho_1)\tau^2}{1 - \rho_1\rho_2\tau^2}$ $T = \dfrac{(1 - \rho_1)(1 - \rho_2)\tau}{1 - \rho_1\rho_2\tau^2}$

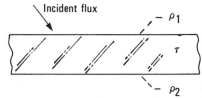

7 Two parallel, partially transparent plates have different τ values and a different ρ at each surface. Obtain an expression for the overall transmittance of the two-plate system.

  Answer: $T = \dfrac{T_1 T_2}{1 - R_{12}R_{21}}$, where $T_n = \dfrac{(1 - \rho_{n1})(1 - \rho_{n2})\tau_n}{1 - \rho_{n1}\rho_{n2}\tau_n^2}$

  and $R_{nm} = \dfrac{\rho_{nm} + \rho_{nn}(1 - 2\rho_{nm})\tau_n^2}{1 - \rho_{nm}\rho_{nn}\tau_n^2}$

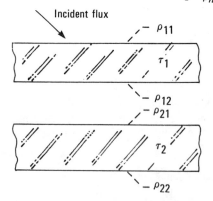

8 In a solar still a thin layer of condensed water is flowing down a glass plate. Consider the glass plate to have transmittance τ, and the water layer to be nonabsorbing. Obtain an expression for the overall transmittance T of the system for radiation incident on the glass.

$$\text{Answer:} \quad T = \frac{(1 - \rho_1)(1 - \rho_2)(1 - \rho_3)\tau}{(1 - \rho_2\rho_3)(1 - \rho_1\rho_2\tau^2) - \rho_1\rho_3(1 - \rho_2)^2\tau^2}$$

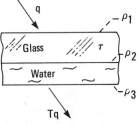

9 The two glass plates in Prob. 3 are followed by a collector surface with absorptivity $\alpha_c = 0.95$. What is the fraction A_c absorbed by the collector surface for normally incident radiation?

Answer: $A_c = 0.749$

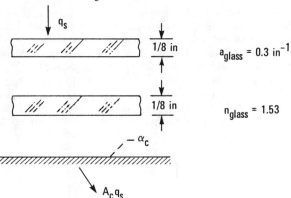

$a_{glass} = 0.3 \text{ in}^{-1}$

$n_{glass} = 1.53$

10 Using the net-radiation method, derive Eq. (19-15) for a thick dielectric film on a dielectric substrate.

11 Extend Prob. 10 to a system of two differing dielectric layers coated onto a dielectric substrate.

12 Two opaque gray plates have a nonabsorbing transparent plate between them. The transparent plate has surface reflectivities ρ. Derive a relation for the heat transfer from plate 1 to plate 2. Neglect heat conduction in the transparent plate.

$$\text{Answer:} \quad q = \frac{\sigma(T_1^4 - T_2^4)}{1/\epsilon_1 + 1/\epsilon_2 - 1 + 2\rho/(1 - \rho)}$$

Opaque

$- T_2, \epsilon_2$

$- \rho$

$\tau = 1$

$- \rho$

$- T_1, \epsilon_1$

Opaque

13 Water is flowing between two identical glass plates adjacent to an opaque plate. Derive an expression for the fraction of incident energy that is absorbed by the water in terms of the τ and ρ values for the layers and interfaces. (Include only radiative heat transfer.)

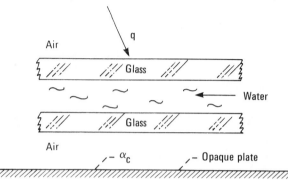

14 Do Prob. 8 of Chap. 15 with the index of refraction for glass equal to 1.53, and for the liquid equal to 1.35.

15 The sun shines on a flat-plate solar collector that has a single $\frac{1}{2}$-cm-thick glass cover (use the data in Fig. 5-31 and assume properties constant below 1 μm) over a silicon oxide coated aluminum absorber plate (use data in Fig. 5-26 and assume properties constant below 0.4 μm). Determine the effective solar absorptivity of the collector for normal incidence (i.e. the fraction of incident solar energy that is actually absorbed by the absorber plate).

16 Radiation in air is incident from all directions with uniform intensity i'_{air} on a dielectric medium having an optically smooth surface and absorption coefficient a. The simple index of refraction of the medium is n. The reflectivity as a function of incidence angle is given by the electromagnetic-theory relations in Chap. 4. Give the relations and method to determine the amount of the incident energy directly absorbed per unit volume in the medium as a function of depth x from the surface.

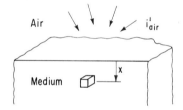

TWENTY

THE MONTE CARLO TECHNIQUE
FOR ABSORBING, EMITTING,
AND SCATTERING MEDIA

20-1 INTRODUCTION

The Monte Carlo technique is a method of statistical sampling of events to determine the average behavior of a system. In Chap. 11 the method was applied to radiative transfer between surfaces without a participating intervening medium. The information given in that chapter is necessary to that presented here. Based on the model of radiative transfer in Chap. 11, extensions are made here to absorbing, emitting, and scattering media. The model consists in following a finite number of energy bundles through their transport histories. The radiative behavior of the system is then determined from the average behavior of these bundles.

Monte Carlo has more obvious utility in solving problems of radiative transfer through absorbing, emitting, and scattering media than for surface-radiation interchange problems. This is because a complete definition of the local radiation balance in a gas or other participating media requires an integration of the incoming radiation, not only from the surrounding surfaces, but from all volume elements of the surrounding medium. Such problems are difficult to solve analytically. As described in previous chapters, much effort has been expended in attempting to develop analytical solution methods. This is often done by making as many assumptions as are necessary to obtain an answer and philosophically accepting the resulting loss of accuracy, if not validity. Surfaces that are black, gray, diffuse, or specular, and media that are optically dense, almost transparent, gray, or isothermal are typical assumptions that fall into this category. Few problems in radiative transfer are solved analytically without explicitly or implicitly making one or more of these assumptions.

By extending the Monte Carlo model of radiative energy exchange outlined for surface-interaction problems in Chap. 11, it is possible to account for a large variety of effects in problems with radiating media. This can be done without resorting to the simplifying assumptions that are often necessary in the analytical approaches.

20-2 SYMBOLS

a	absorption coefficient
D	separation distance between parallel plates
e	emissive power
$F_{0-\lambda}$	fraction of total blackbody emission in spectral region $0-\lambda$
j	volume increment index
K	extinction coefficient
k	number of volume increments
L	dimensionless path length, l/D
l	path length to absorption
N	total number of Monte Carlo bundles per unit time
n	bundle index
P	probability density function
p	increment index
Q	energy per unit time
q'''	internal energy source rate per unit volume
q	energy flux, energy per unit area per unit time
R	randomly chosen number in range $0-1$
r	radial coordinate
S	coordinate along path of radiation (will not have a subscript)
S	number of events at some position per unit time (will always have a subscript to avoid confusion with path-length coordinate).
T	absolute temperature
V	volume
w	energy carried by sample Monte Carlo bundle
X	dimensionless distance, x/D
x	distance normal to surface
ϵ	emissivity
Θ	dimensionless temperature, T/T_1
θ	cone angle (measured from normal of area)
κ_D	optical thickness, aD
λ	wavelength
σ	Stefan-Boltzmann constant
Φ	scattering phase function
φ	circumferential angle
ω	solid angle

Subscripts

b	blackbody
e	emitted
i	inner

j	volume increment j
l	path length
max	maximum
min	minimum
n	normal direction
o	original, or outer
P	Planck mean value
r	radiant
dV	elemental volume dV
w	wall
1, 2	surface or region 1 or 2
θ	for cone angle
φ	for circumferential angle
λ	spectrally dependent

Superscript

*	dummy variable of integration

20-3 DISCUSSION OF THE METHOD

The additional factor introduced to the model discussed in Chap. 11 is the path length traveled in the medium by an individual energy bundle before it is absorbed or leaves the system. The required relations are given in Table 20-1 as related to a random number (see also Example 20-1). It is possible to allow for variations in medium properties along the bundle path; indeed, it is in principle possible to account for variations in the refractive index of the medium by causing the bundles to travel curved paths.

If a problem is solved in which radiative equilibrium can be assumed, then whenever a bundle is absorbed in the medium, a new bundle must be emitted from the same point in the medium to ensure no accumulation of energy. The functions required for determination of the angles and wavelengths of emission are shown in Table 20-1. The new emitted bundle in the medium may be considered as merely the continuation of the history of the absorbed bundle, and the history is continued until the energy reaches a bounding surface.

Under conditions of radiative equilibrium, the total energy d^2Q_e emitted by a volume element dV is given by Eq. (13-40) integrated over all λ:

$$d^2Q_e = 4dV \int_0^\infty a_\lambda e_{\lambda b}\, d\lambda \tag{20-1}$$

For radiative equilibrium the energy contained in the bundles emitted by the volume must be equal to the energy contained in the bundles absorbed, or

$$d^2Q_e = wS_{dV} \tag{20-2}$$

where w is the energy per bundle and S_{dV} is the number of bundles absorbed per unit

Table 20-1 Useful relations for Monte Carlo solution of radiation problems in a medium

Phenomenon	Variable	Relation
Emission from a volume element with absorption coefficient a_λ	Cone angle θ	$\cos\theta = 1 - 2R_\theta$
	Circumferential angle φ	$\varphi = 2\pi R_\varphi$
	Wavelength λ 　Gray medium	$F_{0-\lambda} = R_\lambda$
	Nongray medium	$\dfrac{\int_0^\lambda a_\lambda(\lambda^*)i'_{\lambda b}(\lambda^*)\,d\lambda^*}{\int_0^\infty a_\lambda(\lambda^*)i'_{\lambda b}(\lambda^*)\,d\lambda^*} = R_\lambda$
Attenuation by medium with extinction coefficient K_λ	Path length l	
	Uniform medium properties	$l = -\dfrac{1}{K_\lambda}\ln R_l$
	Nonuniform medium properties	$-\int_0^l K_\lambda(S)\,dS = \ln R_l$
Isotropic scattering from a volume element	Cone angle θ	$\cos\theta = 1 - 2R_\theta$
	Circumferential angle φ	$\varphi = 2\pi R_\varphi$
Anisotropic scattering in a gray medium with phase function Φ independent of incidence angle and circumferential scattering angle	Cone angle θ	$R_\theta = \dfrac{1}{2}\int_0^\theta \Phi(\theta^*)\sin\theta^*\,d\theta^*$
	Circumferential angle φ	$\varphi = 2\pi R_\varphi$

time in dV. Then, if we note from Eq. (14-31) that

$$a_P \equiv \frac{\int_0^\infty a_\lambda e_{\lambda b}\,d\lambda}{\sigma T_{dV}{}^4} \tag{20-3}$$

where a_P is the Planck mean absorption coefficient, (20-3) can be substituted into (20-1) to eliminate the integral. Then equating (20-1) and (20-2) gives

$$T_{dV} = \left(\frac{wS_{dV}}{4a_{P}\sigma\,dV}\right)^{1/4} \tag{20-4}$$

This allows determination of the local temperature in the gas from the gas properties and the Monte Carlo quantities found in the solution. If a_P depends on local temperature T_{dV}, an iteration is required. A temperature distribution is assumed for a first iteration to obtain the bundle histories. The Monte Carlo quantities are used in (20-4) to obtain a new temperature distribution, which is then used for the second iteration. The process is repeated until the temperatures converge.

If the medium is not in radiative equilibrium, then the temperature distribution must be found by solution of the complete energy equation (14-18). The radiant flux vector q_r for use in (14-18) is found by Monte Carlo, in terms of its components from (14-14),

$$q_{r,n} = \int_{\omega = 4\pi} i' \cos \theta \, d\omega = w(S^+ - S^-) \tag{20-5}$$

The value of $q_{r,n}$ must be found for each point in the medium at which the energy equation is to be applied. The S^+ and S^- are the number of sample bundles per unit time crossing an area normal to $q_{r,n}$ in directions with positive and negative $\cos \theta$, respectively, where θ is measured from the normal of the area. When $q_{r,n}$ is thus determined, the energy equation can be solved numerically for the temperature distribution in the medium. For many problems the temperature distribution must be known *a priori* to find S^+ and S^- and thus $q_{r,n}$. In that case, a temperature distribution is assumed, S^+ and S^- and thus $q_{r,n}$ are determined, the energy equation (14-18) is solved for a new temperature distribution, and the process is repeated to convergence.

There are so many variations possible on the Monte Carlo model, many of which might lead to increased efficiency, that they cannot all be mentioned here. One of the most frequently suggested is the fractional absorption of energy when a bundle reaches a surface of known absorptivity. In such a scheme, the bundle energy is reduced after each reflection. The bundle history is then followed until a sufficient number of reflections have occurred to reduce the bundle energy below some predetermined level. This level is chosen so that the effect of the bundle in succeeding reflections would be negligible. The history is then terminated. Such a procedure leads to better accuracy for many problems because a bundle history extends on the average through many more events, and a given number of bundles provides a larger number of events for compiling averages. Haji-Shiekh and Sparrow [1] have suggested some other shortcuts for reducing the programming difficulties of problems involving spectral and directional properties. The obvious rule of thumb is to use whatever shortcuts can be applied to the case in question and not be bound by cookbook rules.

EXAMPLE 20-1 A gray gas with constant absorption coefficient a is contained between infinite parallel black plates. Plate 1 is at T_1, and plate 2 is at $T_2 = 0$. The plates are separated by a distance D. Construct a Monte Carlo flow chart for determining the energy transfer and the gas temperature distribution.

The emission per unit time and area from surface 1 is $q_{e,1} = \sigma T_1^4$. If N energy bundles are to be emitted per unit time, then each one must carry an amount of energy w given by $w = q_{e,1}/N = \sigma T_1^4/N$. The bundles are emitted at cone angles θ given by the first line of Table 11-1, $\sin \theta = \sqrt{R_\theta}$, where R_θ is a random number in the range 0-1. A typical bundle will travel a path length l after emission. The probability of traveling a given distance S before absorption in a medium of constant absorption coefficient a is

$$P(S) = \frac{e^{-aS}}{\int_0^\infty e^{-aS} \, dS} = ae^{-aS}$$

because of Bouguer's law, Eq. (13-12). Using Eq. (11-4), this is put in the form of a cumulative distribution to obtain

$$R_l = \frac{\int_0^l e^{-aS}\,dS}{\int_0^\infty e^{-aS}\,dS} = 1 - e^{-al} \qquad \text{or} \qquad l = -\frac{1}{a}\ln\,(1 - R_l)$$

However, because R_l is uniformly distributed between 0 and 1, this relation may as well be written as $l = -(1/a)\ln R_l$ or $L = -(1/\kappa_D)\ln R_l$, where $L = l/D$ and $\kappa_D = aD$.

The dimensionless distance normal to the plate $X = x/D$ that a bundle will travel when moving through a path length L is then $X = L\cos\theta = -(\cos\theta/\kappa_D)\ln R_l$. Divide the distance D between the plates into k equal increments of dimensionless width $\Delta X = \Delta x/D$, and number the increments with an increment number j, where $j = 1, 2, 3, \ldots, k$. Then the increment number at which absorption occurs is $j = \text{TRUNC}\,(X/\Delta X) + 1$, where TRUNC denotes the operation of truncating the value of $X/\Delta X$ to its integer. At each absorption, a tally is kept of the increment in which absorption occurs by increasing a counter S_j in the memory of the computer by one unit. This operation is denoted by $S_j = S_j + 1$.

If the bundle is absorbed in a gas element, it is immediately emitted from the same element to conserve energy in this steady-state problem. This is done by choosing an angle of emission θ from the probability for emission into all cone angles of a unit sphere surrounding dV:

$$P(\theta) = \frac{\sin\theta}{\int_0^\pi \sin\theta\,d\theta}$$

Using the cumulative distribution function

$$R_\theta = \int_0^\theta P(\theta^*)\,d\theta^* = \frac{1 - \cos\theta}{2}$$

gives the emission angle in terms of a random number as $\theta = \cos^{-1}(1 - 2R_\theta)$. The distance from the wall to the next absorption point is then given by $X = X_0 - (\cos\theta/\kappa_D)\ln R_l$, where X_0 is the position of the previous absorption.

The process of absorptions and emissions is continued until the energy bundle reaches a black boundary. This occurs when $X \geqslant 1$ or $X \leqslant 0$, and a counter S_{w1} or S_{w2} is then increased by one unit to record the absorption at the black surface.

A new bundle is emitted, and the process is repeated until all N bundles have been emitted. The dimensionless net energy flux leaving surface 1 is then found from the total bundles emitted minus those reabsorbed at surface 1; that is,

$$\frac{q_1}{\sigma T_1^4} = \frac{q_{e,1} - wS_{w1}}{\sigma T_1^4} = \frac{w(N - S_{w1})}{wN} = 1 - \frac{S_{w1}}{N}$$

The net energy flux *arriving* at surface 2, $-q_2$, is given by

$$-\frac{q_2}{\sigma T_1^4} = \frac{wS_{w2}}{wN} = \frac{S_{w2}}{N}$$

The temperature at each gas increment is found from (20-4) as

$$\Theta_j = \frac{T_j}{T_1} = \left(\frac{wS_j}{4\kappa_D \sigma \Delta X T_1^4}\right)^{1/4} = \left(\frac{S_j}{4\kappa_D N \Delta X}\right)^{1/4}$$

and the formulation is complete.

A flow chart is shown in Fig. 20-1. Note that, since $S_{w1} + S_{w2} = N$,

$$\frac{q_1}{\sigma T_1^4} = 1 - \frac{S_{w1}}{N} = \frac{S_{w2}}{N} = -\frac{q_2}{\sigma T_1^4}$$

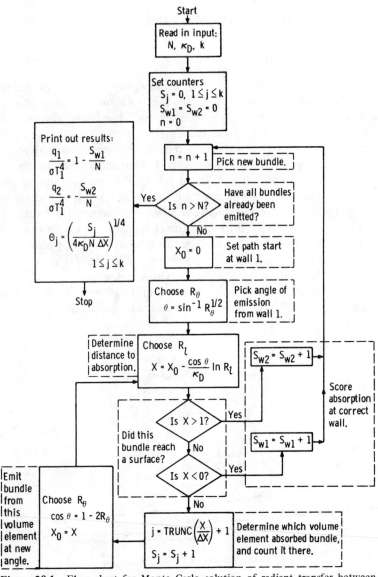

Figure 20-1 Flow chart for Monte Carlo solution of radiant transfer between infinite parallel black plates.

and, as expected, $q_1 = -q_2$; the only reason for printing out both quantities is to check on the results.

By noting the linearity with T^4 of this problem, it is possible to gain solutions for any combination of surface temperatures by use of this flow chart [2]. Also, by use of the exchange-factor relations of Sec. 17-8.3, solutions can be obtained for any combination of gray surface emissivities.

Some results obtained by the Monte Carlo method will now be examined.

20-4 RADIATION THROUGH GRAY GASES

20-4.1 Infinite Parallel Planes

Because of the wealth of solutions available in the literature for a gray gas between infinite parallel plates, almost every new method of solution is tried in this configuration and then compared with the results of one or more of the analytical approaches typified by [3, 4]. The Monte Carlo method is no exception, and in [2] the local gas emissive power and the net energy transfer between diffuse-gray plates is calculated in a manner quite similar to Example 20-1. Parameters are the plate emissivity ϵ (taken equal for both plates) and various values of the gas-layer optical thickness $\kappa_D = aD$, where a is constant. Two cases are examined, the first being a gas with no internal energy generation contained between plates at different temperatures. The second case is a gas with uniformly distributed energy sources between plates at equal temperatures. Figure 20-2 indicates the accuracy that can be obtained by Monte Carlo solutions in such idealized situations. The calculated energy transfer values have a 99.99% probability of lying within ±5% of the midpoints shown.

In Fig. 20-3, the emissive power distribution within the gas is shown. Comparison with the exact solutions of [3, 4] is quite good; however, some trends common to all

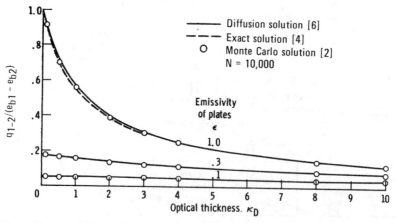

Figure 20-2 Net heat transfer between infinite parallel gray plates separated by gray gas.

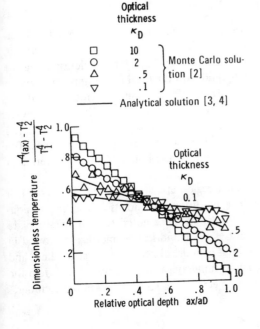

Figure 20-3 Emissive power distribution in gray gas between infinite parallel black plates.

straightforward Monte Carlo solutions in gas radiation problems are as follows:

First, the calculated individual points in Fig. 20-3 reveal increasing error with decreasing optical thickness. This reflects the smaller fraction of energy bundles being absorbed in a given volume element as the optical thickness of the gas decreases. As the number of absorbed bundles decreases, the expected accuracy of the local emissive power becomes lower, and more error naturally appears in the results. Conversely, as the optical thickness increases, error becomes smaller; and in Fig. 20-3, the curve of results for an optical thickness of 10 is quite smooth.

A second effect mentioned in [2] is not evident from Fig. 20-3, and that is that the computing-machine time required for solution of problems involving large optical thickness, say larger than 10, becomes quite large. This is simply because the free path of an energy bundle, $L = -(1/\kappa_D) \ln R_l$, becomes very short for large optical thickness; therefore, many absorptions occur during a typical bundle history.

Two limits are now obvious. For small optical thickness, accuracy becomes poor; for large optical thickness, computer running time becomes excessive. From a practical viewpoint, these are not serious limitations, as the transparent and diffusion approximations to the exact analytical formulation become valid in just those regions where the straightforward Monte Carlo approach begins to fail. In addition, the range of optical thickness over which a Monte Carlo solution can be effectively utilized can be extended by a variety of techniques, including those with the graphic names of *biasing,* *splitting,* and *Russian roulette,* and a large number of specialized schemes for specific solutions. Many of these involve biasing the path length to increase the number of bundles absorbed in otherwise weakly absorbing regions.

20-4.2 Infinitely Long Concentric Cylinders

A more difficult problem to treat analytically than infinite parallel plates is the determination of the emissive power distribution and local energy flux in a gray gas within the annulus between concentric cylinders. The energy equation for determining the emissive power at any radius involves an integral with the local radius appearing in one of the limits. When the energy equation is written for each incremental layer of gas, the resulting set of equations have integrals with limits that are different for every equation in the set. This set of integral equations must be solved simultaneously. The Monte Carlo approach, however, differs only slightly from that for parallel planes. The only additional complication is the determination of the bundle position in terms of cylindrical coordinates.

Some Monte Carlo results for an annular region are shown in Fig. 20-4 as taken from [5]. The results are compared with a modified diffusion solution [6] as considered in Sec. 15-5. Trends in accuracy similar to those noted for the infinite-plate case are evident. Other Monte Carlo analyses of cylindrical geometries are found in [7], where the absorptance of infinitely long cylindrical geometries was studied, and in [8], where finite cylinders of homogeneous nongray gas were analyzed to obtain the local source function and the profiles of the spectral lines emerging from the cylinder.

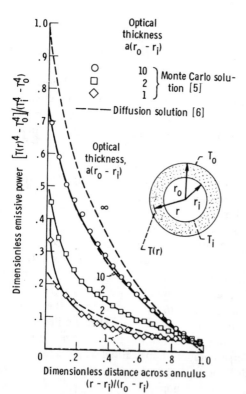

Figure 20-4 Dimensionless emissive power distribution in gray gas in annulus between black concentric cylinders of radius ratio $r_i/r_o = 0.1$.

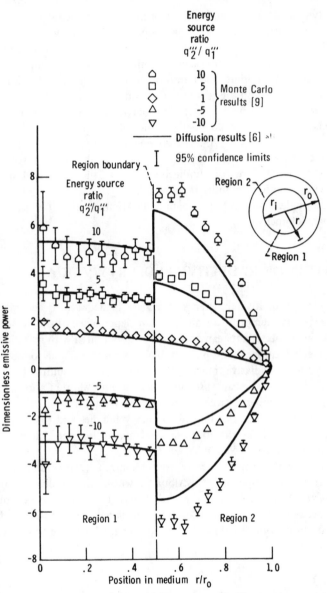

Figure 20-5 Dimensionless emissive power distribution in coaxial gas regions with differing internal energy sources. Radius ratio $r_i/r_o = 0.5$; optical thickness $= 2$ (in each region).

20-4.3 Radiation between Adjacent Gray Regions

A Monte Carlo formulation has been applied to study the interaction of radiative energy between two regions, each of which has individual gray radiative properties and internal energy generation rates [9]. This is of interest here because it gives another insight into sources of error due to geometrical effects. Figure 20-5 shows the emissive power distribution in two concentric cylindrical regions of the same optical thickness but different rates of energy generation. The dimensionless emissive power shown in Fig. 20-5 has been adjusted for the wall slip (Sec. 14-7.4) so that the curves all go to zero at the outer wall. The emissive power was made dimensionless by dividing by the local internal energy generation rate times the radius of the inner cylinder. The vertical bars give the 95% confidence limits on the Monte Carlo results.

The emissive power in the gas bears a direct proportionality to the number of energy bundles absorbed in a given volume element. The volume elements used in the calculation of the results shown in Fig. 20-5 are of equal radius and, therefore, of differing volumes. The elements near the center ($r/r_o \rightarrow 0$) have the smallest volume and, consequently, the smallest number of bundle absorptions. This is reflected in the increasing width of the 95% confidence limits at these points.

Taniguchi [10] has applied Monte Carlo to radiative transfer in a gray gas contained in rectangular parallelepipeds.

20-5 CONSIDERATION OF RADIATIVE PROPERTY VARIATIONS

Many methods for treating radiative transfer in gases are inadequate to accurately account for the strong spectral, temperature, and pressure dependence of the radiative absorption coefficient. Such coefficients can sometimes be computed with reasonable accuracy by quantum-mechanical methods, but few analyses have been able to include the effect of all variables in the radiative transfer. Many treatments are limited to gray gases or the use of various types of mean absorption coefficients. Monte Carlo is well suited to the consideration of property variations with many variables. It involves very little extra effort to assign wavelengths to individual energy bundles and to allow the paths of the bundles to depend on the local spectral absorption coefficient. The relations necessary to achieve this are given in Table 20-1.

If property variations with temperature are considered, an iterative solution is usually necessary because the temperature distribution within the medium is not generally known a priori. Determination of the path length to absorption becomes more difficult also, because the absorption coefficient varies with position. By applying the formalism outlined in Sec. 11-3.2, the path length l is found to be given by

$$\ln R_l = -\int_0^l a_\lambda(S)\, dS \tag{20-6}$$

To evaluate this integral to determine l along a fixed line after choosing a random number R_l is time-consuming but at least feasible. Howell and Perlmutter [11] used this approach, reducing the complexity somewhat by considering temperature and wavelength-dependent absorption coefficients for hydrogen in the simple geometry of the gas contained between infinite parallel plates. They considered energy transfer through the gas between plates at different temperatures, and the case of internal energy generation with a parabolic distribution of source strength in the gas. To evaluate the path length, Eq. (20-6) was approximated by dividing the gas into plane increments of thickness Δx. The path length through a given increment was then $\Delta l = \Delta x/\cos\theta$, where θ is the angle between the bundle path and the perpendicular to the plates. Equation (20-6) was then replaced by the computational form

$$\ln R_l + \Delta l \sum_{j=1}^{p} a_{\lambda,j} > 0 \qquad (20\text{-}7)$$

and the summation was carried out until a value of the integer p was reached that satisfied the inequality. The value of p would be related to the increment number in which absorption occurs. The values of $a_{\lambda,j}$ were assumed for the first iteration and then recalculated in successive interations on the basis of the newly computed local temperatures. This procedure was continued until convergence was obtained.

Figure 20-6 shows the property variations used, and Fig. 20-7 shows a set of emissive power distributions calculated as outlined. The accuracy becomes poorer, as evidenced by increased scatter, in the regions of low temperature because of the decrease with temperature of the absorption coefficient and, therefore, number of absorptions in the low-temperature regions.

Taniguchi [12] used an incident mean absorption coefficient to account for property variations. An improved zone method is given in [26].

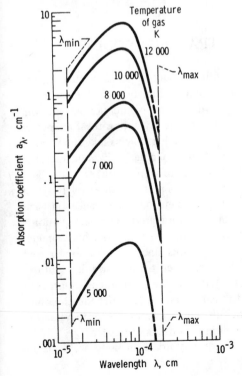

Figure 20-6   Spectral absorption coefficient of hydrogen at 1000 atm [11].

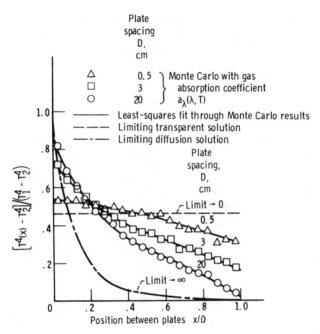

Figure 20-7 Emissive power distribution in hydrogen between infinite parallel plates at temperatures $T_1 = 9500\,\mathrm{K}$ and $T_2 = 4500\,\mathrm{K}$ [11].

20-6 COUPLING WITH OTHER HEAT TRANSFER MODES

When radiative transfer in a gas is coupled with conductive or convective energy transfer, solution becomes even more difficult. Combined energy transfer occurs by radiation, where fourth-power temperatures govern, and by conduction and/or convection, where derivatives or differences of temperature to about the first power govern. The radiative terms in the energy equation for such problems take the form of multiple integrals, while the conduction terms contain second derivatives. Further, the radiative surface properties that appear may be functions of wavelength, direction, and temperature. When gases are involved, these variables plus pressure can strongly affect the local gas radiation physical properties. The complete energy balance on each element of the system then takes the form of a nonlinear integrodifferential equation.

In the solution of coupled-mode problems [13–16], the convective and conductive terms were treated as lumped energy sources or sinks, and Monte Carlo was used to evaluate the radiative terms on the basis of an assumed temperature distribution. With the radiative terms evaluated and substituted into the original equations, conventional numerical techniques were applied to the resulting differential equations and a local temperature distribution was generated. This was used as a basis for reevaluating the radiative integrals, and the procedure was continued until convergence.

Two examples of the power of the Monte Carlo approach in these problems are given by [16] and [17]. In [16] the local temperature distribution as a function of length and radius, and the axial heat-flux distribution in a conical rocket nozzle, were determined under conditions expected in a gas-core nuclear propulsion system. Variations in physical properties with local temperature, pressure, and wavelength were examined, albeit not simultaneously, and coupled radiation and convection were considered. In addition, the ability was demonstrated of a layer of optically thick gas injected along the nozzle wall to attenuate the extreme predicted radiative fluxes to the wall. In [17] the temperature profiles and heat-flux distributions were calculated in an end-fired cylindrical furnace for various flow patterns. The results compared very well with previous calculations using zone interchange methods. The Monte Carlo method was found to be more flexible than the zone method in dealing with changes in concentration of the radiating gases or changes in the flow pattern.

20-7 TRANSIENT RADIATION PROBLEMS

Monte Carlo techniques for radiative transfer under transient conditions, such that the change in the radiation field with time becomes important, have been developed by Fleck [18, 19] and Campbell [20]. The model is essentially that outlined in previous sections, with the additional proviso that the flight times between events in the history of each bundle are computed, which adds some complexity to the problem. Energy bundles are followed along their paths, and their position at some time t is used to determine the distribution of energy at that time. The bundles of course travel at the speed of light in the medium.

The review article by Fleck [18] gives a comprehensive discussion of applications, including the effects of scattering under transient conditions.

20-8 INCORPORATION OF SCATTERING PHENOMENA

Scattering of radiation is easily treated by Monte Carlo for any given distribution of scattering angles. It is analyzed exactly as absorption and nonisotropic reemission in a gas volume.

Collins and Wells [21] have used a modified Monte Carlo neutron diffusion code and other more specialized Monte Carlo codes to study the transmission of thermal radiation from a nuclear explosion. They examined the effects of Rayleigh scattering and of Mie scattering (see Chap. 16) from particles of a given size distribution. Multiple scattering within an atmosphere of arbitrarily described density distribution and effects due to ground and cloud reflections were included. Love et al. [22], Stockham and Love [23] and Scofield and Love [24] have used Monte Carlo to study problems with combined absorption and scattering including the transfer in fog. House and Avery [25] give a general discussion of Monte Carlo applications in nonequilibrium radiative transport.

20-9 CONCLUDING REMARKS

The Monte Carlo approach to radiation in attenuating media has been outlined. Perhaps a sufficient comment as to the power of the method is made by referring to Example 20-1, or more specifically to Fig. 20-1. This figure gives a rather complete diagram of the logic required for programming the problem of energy transfer through a nonisothermal gray gas between infinite parallel black plates at different temperatures. A comparison of this diagram with the analyses of, say, [3, 4] or Chap. 14 will show the simplifications in both concept and formulation that may be inherent in the Monte Carlo method. This becomes even more evident in a two- or three-dimensional geometry.

REFERENCES

1. Haji-Sheikh, A., and E. M. Sparrow: Probability Distributions and Error Estimates for Monte Carlo Solutions of Radiation Problems, *Prog. Heat Mass Transfer*, vol. 2, pp. 1–22, 1969.
2. Howell, J. R., and M. Perlmutter: Monte Carlo Solution of Thermal Transfer through Radiant Media between Gray Walls, *J. Heat Transfer*, vol. 86, no. 1, pp. 116–122, 1964.
3. Usiskin, C. M., and E. M. Sparrow: Thermal Radiation between Parallel Plates Separated by an Absorbing-Emitting Nonisothermal Gas, *Int. J. Heat Mass Transfer*, vol. 1, no. 1, pp. 28–36, 1960.
4. Heaslet, Max A., and Robert F. Warming: Radiative Transport and Wall Temperature Slip in an Absorbing Planar Medium, *Int. J. Heat Mass Transfer*, vol. 8, no. 7, pp. 979–994, 1965.
5. Perlmutter, M., and J. R. Howell: Radiant Transfer through a Gray Gas between Concentric Cylinders Using Monte Carlo, *J. Heat Transfer*, vol. 86, no. 2, pp. 169–179, 1964.
6. Deissler, R. G.: Diffusion Approximation for Thermal Radiation in Gases with Jump Boundary Condition, *J. Heat Transfer*, vol. 86, no. 2, pp. 240–246, 1964.
7. Progelhof, R. C., and J. L. Thorne: Determination of the Radiation Properties of a Semi-Transparent Cylindrical Body Using the Monte Carlo Method, paper 70-WA/HT-13, *ASME*, November, 1970.
8. Avery, L. W., L. L. House, and A. Skumanich: Radiative Transport in Finite Homogeneous Cylinders by the Monte Carlo Technique, *J. Quant. Spectrosc. Radiat. Transfer*, vol. 9, pp. 519–531, 1969.
9. Howell, John R.: Radiative Interactions between Absorbing, Emitting, and Flowing Media with Internal Energy Generation, paper 66-434, *AIAA*, June, 1966. (See also *NASA TN* D-3614, 1966.)
10. Taniguchi, Hiroshi: The Radiative Heat Transfer of Gas in a Three Dimensional System Calculated by Monte Carlo Method, *Bull. JSME*, vol. 12, no. 49, pp. 67–78, 1969.
11. Howell, John R., and Morris Perlmutter: Monte Carlo Solution of Radiant Heat Transfer in a Nongrey Nonisothermal Gas with Temperature Dependent Properties, *AIChE J.*, vol. 10, no. 4, pp. 562–567, 1964.
12. Taniguchi, Hiroshi: Temperature Distributions of Radiant Gas Calculated by Monte Carlo Method, *Bull. JSME*, vol. 10, no. 42, pp. 975–988, 1967.
13. Howell, John R., Mary K. Strite, and Harold Renkel: Heat-transfer Analysis of Rocket Nozzles Using Very High Temperature Propellants, *AIAA J.*, vol. 3, no. 4, pp. 669–673, 1965.
14. Howell, John R., and Mary Kern Strite: Heat Transfer in Rocket Nozzles Using High-temperature Hydrogen Propellant with Real Property Variations, *J. Spacecr. Rockets*, vol. 3, no. 7, pp. 1063–1068, 1966.
15. Howell, John R., and Harold E. Renkel: Analysis of the Effect of a Seeded Propellant Layer on Thermal Radiation in the Nozzle of a Gaseous-core Nuclear Propulsion System, *NASA TN* D-3119, 1965.
16. Howell, John R., Mary K. Strite, and Harold Renkel: Analysis of Heat-transfer Effects in Rocket Nozzles Operating with Very High-temperature Hydrogen, *NASA TR* R-220, 1965.

17. Steward, F. R., and P. Cannon: The Calculation of Radiative Heat Flux in a Cylindrical Furnace Using the Monte Carlo Method, *Int. J. Heat Mass Transfer,* vol. 14, no. 2, pp. 245–262, 1971.
18. Fleck, Joseph A., Jr.: The Calculation of Nonlinear Radiation Transport by a Monte Carlo Method. Statistical Physics, *Meth. Computational Phys.* (Berni Alder, Sidney Fernbach, and Manuel Rotenberg, eds.), vol. 1, pp. 43–65, 1963.
19. Fleck, Joseph A., Jr.: The Calculation of Nonlinear Radiation Transport by a Monte Carlo Method, *Rept. UCRL-*6698 (Del.), Lawrence Radiation Laboratory, Nov. 13, 1961.
20. Campbell, Philip M., and Robert G. Nelson: Numerical Methods for Nonlinear Radiation Transport Calculations, *Rept. UCRL-*7838, Lawrence Radiation Laboratory, Sept. 29, 1964.
21. Collins, David G., and Michael B. Wells: Monte Carlo Codes for Study of Light Transport in the Atmosphere. Description of Codes, *Rept. RRA-*T54-1, Radiation Research Associates, Inc. (ECOM-00240-F, vol. 1, DDC no. AD-625115), vol. 1, August, 1965.
22. Love, Tom J., Leo W. Stockham, Fu. C. Lee, William A. Munter, and Yih W. Tsai: Radiative Heat Transfer in Absorbing, Emitting and Scattering Media, Oklahoma University (ARL-67-0210, DDC No. AD-666427), December, 1967.
23. Stockham, Leo W., and Tom J. Love: Radiative Heat Transfer from a Cylindrical Cloud of Particles, *AIAA J.,* vol. 6, no. 10, pp. 1935–1940, 1968.
24. Scofield, Gordon L., and Tom J. Love: Radiative Transfer Analysis from a Heated Airport Runway to Fog, *Int. J. Heat Mass Transfer,* vol. 13, no. 2, pp. 345–358, 1970.
25. House, L. L., and L. L. Avery: The Monte Carlo Technique Applied to Radiative Transfer, *J. Quant. Spectrosc. Radiat. Transfer,* vol. 9, pp. 1579–1591, 1969.
26. Vercammen, H. A. J., and G. F. Froment: An Improved Zone Method using Monte Carlo Techniques for the Simulation of Radiation in Industrial Furnaces, *Int. J. Heat Mass Transfer,* vol. 23, no. 3., pp. 329–337, 1980.

PROBLEMS

1 For a nongray nonscattering gas with nonuniform properties, derive the following relations for the wavelength of emission λ and the path length to absorption l:

$$R_\lambda = \frac{\int_0^\lambda a_\lambda i'_{\lambda b}\, d\lambda}{\int_0^\infty a_\lambda i'_{\lambda b}\, d\lambda} \qquad \ln R_l = -\int_0^l a_\lambda(S)\, dS$$

2 In Example 20-1 let both plate temperatures T_1 and T_2 be nonzero. Plate 1 is black but plate 2 has a spectrally varying hemispherical emissivity $\epsilon_{\lambda,2}(\lambda, T_2)$. Modify the flow chart in Fig. 20-1 to account for this in determining the energy transfer and gas temperature distribution.

3 A gray nonscattering gas with constant absorption coefficient a is between parallel black plates of finite width L and infinite length. The plate temperatures are T_1 and $T_2 = 0$, and the gas is at constant temperature T_g. The plates are a distance D apart, and the side boundaries are black and are at zero temperature as shown. Construct a Monte Carlo flow chart to obtain the energy transfer to plate 2.

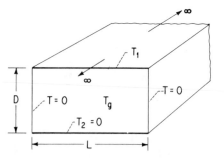

4 Derive the cumulative distribution function for the angle of scatter for the cases of isotropic and Rayleigh scattering.

5 Two infinite black parallel plates are spaced a distance D apart. The plates are at temperatures T_1 and T_2, and the space between them is filled with a gray, isotropically scattering medium. The medium does not absorb or emit radiation. Draw a Monte Carlo flow chart to obtain the energy transfer from plate 1 to 2.

6 Generalize Prob. 5 to account for anisotropic scattering with a phase function that is independent of incidence angle and circumferential scattering angle.

CONVERSION FACTORS, RADIATION CONSTANTS, AND BLACKBODY FUNCTIONS

Table A-1 lists some fundamental constants of radiation physics. Tables of conversion factors between the International System of Units (SI) and other systems of units are given in Tables A-2 and A-3. In Table A-4, values of the various radiation constants are given in both SI and English engineering units. Table A-5 lists blackbody emission properties as functions of the variable λT, again in both SI and English engineering units.

With regard to Table A-5, Pivovonsky and Nagel [2] and Wiebelt [3] have presented polynomial curves fitted to the function $F_{0-\lambda T}$. These curve fits can be quite useful for computer solutions of various types of radiation problems. Wiebelt recommends use of the following polynomials:

$$F_{0-\lambda T} = \frac{15}{\pi^4} \sum_{m=1,2,\ldots} \frac{e^{-mv}}{m^4} \{[(mv + 3)mv + 6]mv + 6\} \qquad v \geq 2$$

$$F_{0-\lambda T} = 1 - \frac{15}{\pi^4} v^3 \left(\frac{1}{3} - \frac{v}{8} + \frac{v^2}{60} - \frac{v^4}{5040} + \frac{v^6}{272,160} - \frac{v^8}{13,305,600} \right) \qquad v < 2$$

where

$$v = \frac{C_2}{\lambda T}$$

and C_2 is given in Table A-4 of this appendix. The series is carried out to a sufficient number of terms to gain the desired accuracy.

Table A-1 Fundamental numerical values [1]

First Bohr electron radius	$a_0 = \hbar^2/m_e e^2 = 0.5292 \times 10^{-10}$ m
Speed of light in vacuum	$c_0 = 2.9979 \times 10^8$ m/s
Electronic charge	$e = 1.6022 \times 10^{-19}$ C
	$e^2 = 2.3071 \times 10^{-28}$ J $\cdot$ m
Planck's constant	$h = 6.6262 \times 10^{-34}$ J $\cdot$ s
	$\hbar = h/2\pi = 1.0546 \times 10^{-34}$ J $\cdot$ s
Boltzmann constant	$k = 1.3806 \times 10^{-23}$ J/K
Electron rest mass	$m_e = 9.1096 \times 10^{-31}$ kg
Classical electron radius	$r_0 = e^2/m_e c_0^2 = 2.8179 \times 10^{-15}$ m
Atomic unit cross section	$\pi a_0^2 = 0.8797 \times 10^{-20}$ m^2
Thomson cross section	$\sigma_T = 8\pi r_0^2/3 = 6.6525 \times 10^{-29}$ m^2
Electron volt	1 eV $= 1.6022 \times 10^{-19}$ J
Temperature associated with 1 eV	1 eV$/k = 11{,}605$ K
Rest energy of electron	$m_e c_0^2 = 8.1873 \times 10^{-14}$ J $= 5.1100 \times 10^5$ eV
Ionization potential of hydrogen atom	$e^2/2a_0 = e^4 m_e/2\hbar^2 = 13.606$ eV

Table A-2 Conversion factors for lengths

	Mile, mi	Kilometer, km	Meter, m	Foot, ft	Inch, in
1 mile =	1	1.609	1609	5280	6.336×10^4
1 kilometer =	0.6214	1	10^3	3.281×10^3	3.937×10^4
1 meter =	6.214×10^{-4}	10^{-3}	1	3.281	39.37
1 foot =	1.894×10^{-4}	3.048×10^{-4}	0.3048	1	12
1 inch =	1.578×10^{-5}	2.540×10^{-5}	2.540×10^{-2}	8.333×10^{-2}	1
1 centimeter =	6.214×10^{-6}	10^{-5}	10^{-2}	3.281×10^{-2}	0.3937
1 millimeter =	6.214×10^{-7}	10^{-6}	10^{-3}	3.281×10^{-3}	0.03937
1 micrometer =	6.214×10^{-10}	10^{-9}	10^{-6}	3.281×10^{-6}	3.937×10^{-5}
1 nanometer =	6.214×10^{-13}	10^{-12}	10^{-9}	3.281×10^{-9}	3.937×10^{-8}
1 angstrom =	6.214×10^{-14}	10^{-13}	10^{-10}	3.281×10^{-10}	3.937×10^{-9}

	Centimeter, cm	Millimeter, mm	Micrometer, μm	Nanometer, nm	Angstrom, Å
1 mile =	1.609×10^5	1.609×10^6	1.609×10^9	1.609×10^{12}	1.609×10^{13}
1 kilometer =	10^5	10^6	10^9	10^{12}	10^{13}
1 meter =	10^2	10^3	10^6	10^9	10^{10}
1 foot =	30.48	3.048×10^2	3.048×10^5	3.048×10^8	3.048×10^9
1 inch =	2.540	25.40	2.540×10^4	2.540×10^7	2.540×10^8
1 centimeter =	1	10	10^4	10^7	10^8
1 millimeter =	10^{-1}	1	10^3	10^6	10^7
1 micrometer =	10^{-4}	10^{-3}	1	10^3	10^4
1 nanometer =	10^{-7}	10^{-6}	10^{-3}	1	10
1 angstrom =	10^{-8}	10^{-7}	10^{-4}	10^{-1}	1

Table A-3 Useful conversion factors

Area	Mass
$1 \text{ ft}^2 = 0.0929030 \text{ m}^2$	$1 \text{ lbm} = 0.453592 \text{ kg}$
$1 \text{ in}^2 = 6.4516 \times 10^{-4} \text{ m}^2$	$1 \text{ kg} = 2.20462 \text{ lbm}$
$1 \text{ m}^2 = 10.7639 \text{ ft}^2$	

Volume	Density
$1 \text{ ft}^3 = 0.028317 \text{ m}^3$	$1 \text{ lbm/ft}^3 = 16.0185 \text{ kg/m}^3$
$1 \text{ m}^3 = 35.315 \text{ ft}^3$	$1 \text{ kg/m}^3 = 0.062428 \text{ lbm/ft}^3$

Heat	Heat rate per unit area
$(1 \text{ kJ} = 1 \text{ kW·s})$	$1 \text{ W/m}^2 = 0.31700 \text{ Btu}^a/(\text{h·ft}^2)$
$1 \text{ kJ} = 0.94782 \text{ Btu}^a = 0.23885 \text{ kcal}^a$	$= 0.85985 \text{ kcal}^a/(\text{h·m}^2)$
$1 \text{ Btu} = 1.0551 \text{ kJ} = 0.25200 \text{ kcal}$	$1 \text{ Btu/(h·ft}^2) = 3.1546 \text{ W/m}^2$
$1 \text{ kcal} = 4.1868 \text{ kJ} = 3.9683 \text{ Btu}$	$= 2.7125 \text{ kcal/(h·m}^2)$
$1 \text{ kW·h} = 3.60 \times 10^6 \text{ J}$	$1 \text{ kcal/(h·m}^2) = 1.1630 \text{ W/m}^2$
	$= 0.36867 \text{ Btu/(h·ft}^2)$

Heat rate	Heat transfer coefficient
$1 \text{ W} = 3.4121 \text{ Btu}^a/\text{h} = 0.85985 \text{ kcal}^a/\text{h}$	$1 \text{ W/(m}^2 \cdot \text{K}) = 0.17611 \text{ Btu}^a/(\text{h·ft}^2 \cdot {}^\circ\text{R})$
$1 \text{ Btu/h} = 0.29307 \text{ W} = 0.25200 \text{ kcal/h}$	$= 0.85985 \text{ kcal}^a/(\text{h·m}^2 \cdot \text{K})$
$1 \text{ kcal/h} = 1.1630 \text{ W} = 3.9683 \text{ Btu/h}$	$1 \text{ Btu/(h·ft}^2 \cdot {}^\circ\text{R}) = 5.6783 \text{ W/(m}^2 \cdot \text{K})$
	$= 4.8824 \text{ kcal/(h·m}^2 \cdot \text{K})$
	$1 \text{ kcal/(h·m}^2 \cdot \text{K}) = 1.1630 \text{ W/(m}^2 \cdot \text{K})$
	$= 0.20482 \text{ Btu/(h·ft}^2 \cdot {}^\circ\text{R})$

Thermal conductivity	Specific heat
$1 \text{ W/(m·K)} = 0.57779 \text{ Btu/(h·ft·}^\circ\text{R})$	$1 \text{ kJ/(kg·K)} = 0.23885 \text{ Btu/(lb·}^\circ\text{R})$
$= 0.85985 \text{ kcal/(h·m·K)}$	$= 0.23885 \text{ kcal/(kg·K)}$
$1 \text{ Btu/(h·ft·}^\circ\text{R}) = 1.7307 \text{ W/(m·K)}$	$1 \text{ Btu/(lb·}^\circ\text{R}) = 4.1868 \text{ kJ/(kg·K)}$
$= 1.4882 \text{ kcal/(h·m·K)}$	$= 1.0000 \text{ kcal/(kg·K)}$
$1 \text{ kcal/(h·m·K)} = 1.1630 \text{ W/(m·K)}$	$1 \text{ kcal/(kg·K)} = 4.1868 \text{ kJ/(kg·K)}$
$= 0.67197 \text{ Btu/(h·ft·}^\circ\text{R})$	$= 1.0000 \text{ Btu/(lb·}^\circ\text{R})$

Energy per unit mass	Temperature
$1 \text{ kJ/kg} = 0.42992 \text{ Btu/lb} = 0.23885 \text{ kcal/kg}$	$\text{K} = \frac{5}{9}{}^\circ\text{R} = \frac{5}{9}({}^\circ\text{F} + 459.67) = {}^\circ\text{C} + 273.15$
$1 \text{ Btu/lb} = 2.3260 \text{ kJ/kg} = 0.55556 \text{ kcal/kg}$	${}^\circ\text{R} = \frac{9}{5}\text{K} = \frac{9}{5}({}^\circ\text{C} + 273.15) = {}^\circ\text{F} + 459.67$
$1 \text{ kcal/kg} = 4.1868 \text{ kJ/kg} = 1.8000 \text{ Btu/lb}$	${}^\circ\text{F} = \frac{9}{5}{}^\circ\text{C} + 32$
	${}^\circ\text{C} = \frac{5}{9}({}^\circ\text{F} - 32)$

^{a}International Steam Table (for all Btu and kcal).

Table A-4 Radiation constants

Symbol	Definition	Value
C_1	Constant in Planck's spectral energy distribution	0.18892×10^8 Btu$\cdot\mu$m^4/(h$\cdot$ft^2) 0.59544×10^8 W$\cdot\mu$m^4/m^2 0.59544×10^{-16} W$\cdot$m^2
C_2	Constant in Planck's spectral energy distribution	$25,898$ μm $\cdot^\circ$R $14,388$ μm$\cdot$K
C_3	Constant in Wien's displacement law	5216.0 μm$\cdot^\circ$R 2897.8 μm$\cdot$K
C_4	Constant in equation for maximum blackbody intensity	6.879×10^{-14} Btu/(h$\cdot$ft$^2 \cdot\mu$m$\cdot^\circ$R^5) 4.095×10^{-12} W/(m$^2 \cdot\mu$m$\cdot$K^5)
$\sigma_{\text{calculated}}$	Calculated Stefan-Boltzmann constant	0.1712×10^{-8} Btu/(h$\cdot$ft$^2 \cdot^\circ$R^4) 5.6696×10^{-8} W/(m$^2 \cdot$K^4)
$\sigma_{\text{experimental}}$	Experimental Stefan-Boltzmann constant	0.173×10^{-8} Btu/(h$\cdot$ft$^2 \cdot^\circ$R^4) 5.729×10^{-8} W/(m$^2 \cdot$K^4)
q_{solar}	Solar constant	429 ± 7 Btu/(h$\cdot$ft^2) 1353 ± 21 W/m^2 [older measurements: 442 Btu/(h$\cdot$ft^2), 1395 W/m^2]
T_{solar}	Effective surface radiating temperature of the sun	5780 K, $10,400^\circ$R

Table A-5 Blackbody functions

Wavelength-temperature product λT		Blackbody hemispherical spectral emissive power divided by fifth power of temperature $e_{\lambda b}/T^5$		Blackbody fraction $F_{0-\lambda T}$	Difference between successive $F_{0-\lambda T}$ values ΔF
$\mu\text{m·}^\circ\text{R}$	$\mu\text{m·K}$	$\text{Btu/(h·ft}^2\cdot\mu\text{m·}^\circ\text{R}^5)$	$\text{W/(m}^2\cdot\mu\text{m·K}^5)$		
1000	555.6	0.000671×10^{-15}	0.400×10^{-16}	0.170×10^{-7}	0.
1100	611.1	0.00439	0.261×10^{-15}	0.136×10^{-6}	0.119×10^{-6}
1200	666.7	0.0202	0.120×10^{-14}	0.756×10^{-6}	0.620×10^{-6}
1300	722.2	0.0713	0.424×10^{-14}	0.317×10^{-5}	0.241×10^{-5}
1400	777.8	0.204	0.00122×10^{-11}	0.106×10^{-4}	0.748×10^{-5}
1500	833.3	0.496×10^{-15}	0.00296×10^{-11}	0.301×10^{-4}	0.194×10^{-4}
1600	888.9	1.057	0.00630	0.738×10^{-4}	0.437×10^{-4}
1700	944.4	2.023	0.01205	0.161×10^{-3}	0.876×10^{-4}
1800	1000.0	3.544	0.02111	0.321×10^{-3}	0.00016
1900	1055.6	5.767	0.03434	0.589×10^{-3}	0.00027
2000	1111.1	8.822×10^{-15}	0.05254×10^{-11}	0.00101	0.00042
2100	1166.7	12.805	0.07626	0.00164	0.00063
2200	1222.2	17.776	0.10587	0.00252	0.00089
2300	1277.8	23.746	0.14142	0.00373	0.00121
2400	1333.3	30.686	0.18275	0.00531	0.00158
2500	1388.9	38.526×10^{-15}	0.22945×10^{-11}	0.00733	0.00202
2600	1444.4	47.167	0.28091	0.00983	0.00250
2700	1500.0	56.483	0.33639	0.01285	0.00302
2800	1555.6	66.334	0.39505	0.01643	0.00358
2900	1611.1	76.571	0.45602	0.02060	0.00417
3000	1666.7	87.047×10^{-15}	0.51841×10^{-11}	0.02537	0.00477
3100	1722.2	97.615	0.58135	0.03076	0.00539
3200	1777.8	108.14	0.64404	0.03677	0.00600
3300	1833.3	118.50	0.70573	0.04338	0.00661
3400	1888.9	128.58	0.76578	0.05059	0.00721
3500	1944.4	138.29×10^{-15}	0.82362×10^{-11}	0.05838	0.00779
3600	2000.0	147.56	0.87878	0.06672	0.00834
3700	2055.6	156.30	0.93088	0.07559	0.00887
3800	2111.1	164.49	0.97963	0.08496	0.00936
3900	2166.7	172.08	1.0248	0.09478	0.00982
4000	2222.2	179.04×10^{-15}	1.0663×10^{-11}	0.10503	0.01025
4100	2277.8	185.36	1.1039	0.11567	0.01064
4200	2333.3	191.05	1.1378	0.12665	0.01099
4300	2388.9	196.09	1.1678	0.13795	0.01130
4400	2444.4	200.51	1.1942	0.14953	0.01158
4500	2500.0	204.32×10^{-15}	1.2169×10^{-11}	0.16135	0.01182
4600	2555.6	207.55	1.2361	0.17337	0.01202
4700	2611.1	210.20	1.2519	0.18556	0.01219
4800	2666.7	212.32	1.2645	0.19789	0.01233
4900	2722.2	213.93	1.2741	0.21033	0.01244
5000	2777.8	215.06×10^{-15}	1.2808×10^{-11}	0.22285	0.01252
5100	2833.3	215.74	1.2848	0.23543	0.01257
5200	2888.9	216.00	1.2864	0.24803	0.01260
5300	2944.4	215.87	1.2856	0.26063	0.01260
5400	3000.0	215.39	1.2827	0.27322	0.01259

Table A-5 Blackbody functions *(Continued)*

Wavelength-temperature product λT		Blackbody hemispherical spectral emissive power divided by fifth power of temperature $e_{\lambda b}/T^5$		Blackbody fraction $F_{0-\lambda T}$	Difference between successive values ΔF
$\mu m \cdot {}^\circ R$	$\mu m \cdot K$	Btu/(h·ft²·μm·°R⁵)	W/(m²·μm·K⁵)	$F_{0-\lambda T}$	$F_{0-\lambda T}$
5500	3055.6	214.57×10^{-15}	1.2779×10^{-11}	0.28576	0.01255
5600	3111.1	213.46	1.2713	0.29825	0.01249
5700	3166.7	212.07	1.2630	0.31067	0.01242
5800	3222.2	210.43	1.2532	0.32300	0.01233
5900	3277.8	208.57	1.2422	0.33523	0.01223
6000	3333.3	206.51×10^{-15}	1.2299×10^{-11}	0.34734	0.01211
6100	3388.9	204.28	1.2166	0.35933	0.01199
6200	3444.4	201.88	1.2023	0.37118	0.01185
6300	3500.0	199.35	1.1872	0.38289	0.01171
6400	3555.6	196.69	1.1714	0.39445	0.01156
6500	3611.1	193.94×10^{-15}	1.1550×10^{-11}	0.40585	0.01140
6600	3666.7	191.09	1.1380	0.41708	0.01124
6700	3722.2	188.17	1.1206	0.42815	0.01107
6800	3777.8	185.18	1.1029	0.43905	0.01089
6900	3833.3	182.15	1.0848	0.44977	0.01072
7000	3888.9	179.08×10^{-15}	1.0665×10^{-11}	0.46031	0.01054
7100	3944.4	175.98	1.0481	0.47067	0.01036
7200	4000.0	172.86	1.0295	0.48085	0.01018
7300	4055.6	169.74	1.0109	0.49084	0.01000
7400	4111.1	166.60	0.99221	0.50066	0.00981
7500	4166.7	163.47×10^{-15}	0.97357×10^{-11}	0.51029	0.00963
7600	4222.2	160.35	0.95499	0.51974	0.00945
7700	4277.8	157.25	0.93650	0.52901	0.00927
7800	4333.3	154.16	0.91813	0.53809	0.00909
7900	4388.9	151.10	0.89990	0.54700	0.00891
8000	4444.4	148.07×10^{-15}	0.88184×10^{-11}	0.55573	0.00873
8100	4500.0	145.07	0.86396	0.56429	0.00855
8200	4555.6	142.10	0.84629	0.57267	0.00838
8300	4611.1	139.17	0.82884	0.58087	0.00821
8400	4666.7	136.28	0.81163	0.58891	0.00804
8500	4722.2	133.43×10^{-15}	0.79467×10^{-11}	0.59678	0.00787
8600	4777.8	130.63	0.77796	0.60449	0.00771
8700	4833.3	127.87	0.76151	0.61203	0.00754
8800	4888.9	125.15	0.74534	0.61941	0.00738
8900	4944.4	122.48	0.72944	0.62664	0.00723
9000	5000.0	119.86×10^{-15}	0.71383×10^{-11}	0.63371	0.00707
9100	5055.6	117.29	0.69850	0.64063	0.00692
9200	5111.1	114.76	0.68346	0.64740	0.00677
9300	5166.7	112.28	0.66870	0.65402	0.00662
9400	5222.2	109.85	0.65423	0.66051	0.00648

Table A-5 Blackbody functions *(Continued)*

Wavelength-temperature product λT		Blackbody hemispherical spectral emissive power divided by fifth power of temperature $e_{\lambda b}/T^5$		Blackbody fraction $F_{0-\lambda T}$	Difference between successive $F_{0-\lambda T}$ values ΔF
$\mu\mathrm{m}\cdot{}^\circ\mathrm{R}$	$\mu\mathrm{m}\cdot\mathrm{K}$	$\mathrm{Btu}/(\mathrm{h}\cdot\mathrm{ft}^2\cdot\mu\mathrm{m}\cdot{}^\circ\mathrm{R}^5)$	$\mathrm{W}/(\mathrm{m}^2\cdot\mu\mathrm{m}\cdot\mathrm{K}^5)$		
9500	5277.8	107.47×10^{-15}	0.64006×10^{-11}	0.66685	0.00634
9600	5333.3	105.14	0.62617	0.67305	0.00620
9700	5388.9	102.86	0.61257	0.67912	0.00607
9800	5444.4	100.62	0.59925	0.68506	0.00594
9900	5500.0	98.431	0.58621	0.69087	0.00581
10,000	5555.6	96.289×10^{-15}	0.57346×10^{-11}	0.69655	0.00568
10,100	5611.1	94.194	0.56098	0.70211	0.00556
10,200	5666.7	92.145	0.54877	0.70754	0.00544
10,300	5722.2	90.141	0.53684	0.71286	0.00532
10,400	5777.8	88.181	0.52517	0.71806	0.00520
10,500	5833.3	86.266×10^{-15}	0.51376×10^{-11}	0.72315	0.00509
10,600	5888.9	84.394	0.50261	0.72813	0.00498
10,700	5944.4	82.565	0.49172	0.73301	0.00487
10,800	6000.0	80.777	0.48107	0.73777	0.00477
10,900	6055.6	79.031	0.47067	0.74244	0.00466
11,000	6111.1	77.325×10^{-15}	0.46051×10^{-11}	0.74700	0.00456
11,100	6166.7	75.658	0.45059	0.75146	0.00446
11,200	6222.2	74.031	0.44089	0.75583	0.00437
11,300	6277.8	72.441	0.43143	0.76010	0.00427
11,400	6333.3	70.889	0.42218	0.76429	0.00418
11,500	6388.9	69.373×10^{-15}	0.41315×10^{-11}	0.76838	0.00409
11,600	6444.4	67.892	0.40434	0.77238	0.00401
11,700	6500.0	66.447	0.39573	0.77630	0.00392
11,800	6555.6	65.036	0.38732	0.78014	0.00384
11,900	6611.1	63.658	0.37912	0.78390	0.00376
12,000	6666.7	62.313×10^{-15}	0.37111×10^{-11}	0.78757	0.00368
12,100	6722.2	60.999	0.36328	0.79117	0.00360
12,200	6777.8	59.717	0.35565	0.79469	0.00352
12,300	6833.3	58.465	0.34819	0.79814	0.00345
12,400	6888.9	57.242	0.34091	0.80152	0.00338
12,500	6944.4	56.049×10^{-15}	0.33380×10^{-11}	0.80482	0.00331
12,600	7000.0	54.884	0.32687	0.80806	0.00324
12,700	7055.6	53.747	0.32009	0.81123	0.00317
12,800	7111.1	52.636	0.31348	0.81433	0.00310
12,900	7166.7	51.552	0.30702	0.81737	0.00304
13,000	7222.2	50.493×10^{-15}	0.30071×10^{-11}	0.82035	0.00298
13,100	7277.8	49.459	0.29456	0.82327	0.00292
13,200	7333.3	48.450	0.28855	0.82612	0.00286
13,300	7388.9	47.465	0.28268	0.82892	0.00280
13,400	7444.4	46.502	0.27695	0.83166	0.00274

Table A-5 Blackbody functions *(Continued)*

Wavelength-temperature product λT		Blackbody hemispherical spectral emissive power divided by fifth power of temperature $e_{\lambda b}/T^5$		Blackbody fraction $F_{0-\lambda T}$	Difference between successive values ΔF
$\mu\text{m}\cdot{}^\circ\text{R}$	$\mu\text{m}\cdot\text{K}$	Btu/(h·ft²·μm·${}^\circ$R⁵)	W/(m²·μm·K⁵)		$F_{0-\lambda T}$
13,500	7500.0	45.563×10^{-15}	0.27135×10^{-11}	0.83435	0.00269
13,600	7555.6	44.645	0.26589	0.83698	0.00263
13,700	7611.1	43.749	0.26055	0.83956	0.00258
13,800	7666.7	42.874	0.25534	0.84209	0.00253
13,900	7722.2	42.019	0.25024	0.84457	0.00248
14,000	7777.8	41.184×10^{-15}	0.24527×10^{-11}	0.84699	0.00243
14,100	7833.3	40.368	0.24042	0.84937	0.00238
14,200	7888.9	39.572	0.23567	0.85171	0.00233
14,300	7944.4	38.794	0.23104	0.85399	0.00229
14,400	8000.0	38.033	0.22651	0.85624	0.00224
14,500	8055.6	37.291×10^{-15}	0.22209×10^{-11}	0.85843	0.00220
14,600	8111.1	36.565	0.21777	0.86059	0.00216
14,700	8166.7	35.856	0.21354	0.86270	0.00211
14,800	8222.2	35.163	0.20942	0.86477	0.00207
14,900	8277.8	34.487	0.20539	0.86681	0.00203
15,000	8333.3	33.825×10^{-15}	0.20145×10^{-11}	0.86880	0.00199
15,100	8388.9	33.179	0.19760	0.87075	0.00196
15,200	8444.4	32.547	0.19383	0.87267	0.00192
15,300	8500.0	31.929	0.19016	0.87455	0.00188
15,400	8555.6	31.326	0.18656	0.87640	0.00185
15,500	8611.1	30.736×10^{-15}	0.18305×10^{-11}	0.87821	0.00181
15,600	8666.7	30.159	0.17961	0.87999	0.00178
15,700	8722.2	29.595	0.17625	0.88173	0.00174
15,800	8777.8	29.043	0.17297	0.88344	0.00171
15,900	8833.3	28.504	0.16976	0.88512	0.00168
16,000	8888.9	27.977×10^{-15}	0.16662×10^{-11}	0.88677	0.00165
16,100	8944.4	27.462	0.16355	0.88839	0.00162
16,200	9000.0	26.957	0.16055	0.88997	0.00159
16,300	9055.6	26.464	0.15761	0.89153	0.00156
16,400	9111.1	25.982	0.15474	0.89306	0.00153
16,500	9166.7	25.510×10^{-15}	0.15193×10^{-11}	0.89457	0.00150
16,600	9222.2	25.049	0.14918	0.89604	0.00148
16,700	9277.8	24.597	0.14649	0.89749	0.00145
16,800	9333.3	24.156	0.14386	0.89891	0.00142
16,900	9388.9	23.723	0.14129	0.90031	0.00140
17,000	9444.4	23.301×10^{-15}	0.13877×10^{-11}	0.90168	0.00137
17,100	9500.0	22.887	0.13630	0.90303	0.00135
17,200	9555.6	22.482	0.13389	0.90435	0.00132
17,300	9611.1	22.085	0.13153	0.90565	0.00130
17,400	9666.7	21.697	0.12922	0.90693	0.00128

Table A-5 Blackbody functions *(Continued)*

Wavelength-temperature product λT		Blackbody hemispherical spectral emissive power divided by fifth power of temperature $e_{\lambda b}/T^5$		Blackbody fraction $F_{0-\lambda T}$	Difference between successive $F_{0-\lambda T}$ values ΔF
$\mu\text{m}\cdot{}^\circ\text{R}$	$\mu\text{m}\cdot\text{K}$	$\text{Btu}/(\text{h}\cdot\text{ft}^2\cdot\mu\text{m}\cdot{}^\circ\text{R}^5)$	$\text{W}/(\text{m}^2\cdot\mu\text{m}\cdot\text{K}^5)$		
17,500	9722.2	21.318×10^{-15}	0.12696×10^{-11}	0.90819	0.00126
17,600	9777.8	20.946	0.12475	0.90942	0.00123
17,700	9833.3	20.582	0.12258	0.91063	0.00121
17,800	9888.9	20.226	0.12046	0.91182	0.00119
17,900	9944.4	19.877	0.11838	0.91299	0.00117
18,000	10,000.0	19.536×10^{-15}	0.11635×10^{-11}	0.91414	0.00115
18,100	10,055.6	19.201	0.11435	0.91527	0.00113
18,200	10,111.1	18.874	0.11240	0.91638	0.00111
18,300	10,166.7	18.553	0.11049	0.91748	0.00109
18,400	10,222.2	18.239	0.10862	0.91855	0.00107
18,500	10,277.8	17.931×10^{-15}	0.10679×10^{-11}	0.91961	0.00106
18,600	10,333.3	17.630	0.10500	0.92064	0.00104
18,700	10,388.9	17.335	0.10324	0.92166	0.00102
18,800	10,444.4	17.045	0.10151	0.92267	0.00100
18,900	10,500.0	16.762	0.09983	0.92365	0.00099
19,000	10,555.6	16.484×10^{-15}	0.09817×10^{-11}	0.92462	0.00097
19,100	10,611.1	16.212	0.09655	0.92558	0.00095
19,200	10,666.7	15.945	0.09496	0.92652	0.00094
19,300	10,722.2	15.684	0.09341	0.92744	0.00092
19,400	10,777.8	15.428	0.09188	0.92835	0.00091
19,500	10,833.3	15.177×10^{-15}	0.09039×10^{-11}	0.92924	0.00089
19,600	10,888.9	14.931	0.08892	0.93012	0.00088
19,700	10,944.4	14.690	0.08749	0.93098	0.00086
19,800	11,000.0	14.453	0.08608	0.93183	0.00085
19,900	11,055.6	14.221	0.08470	0.93267	0.00084
20,000	11,111.1	13.994×10^{-15}	0.08334×10^{-11}	0.93349	0.00082
20,200	11,222.2	13.553	0.08071	0.93510	0.00161
20,400	11,333.3	13.128	0.07819	0.93666	0.00156
20,600	11,444.4	12.720	0.07575	0.93816	0.00151
20,800	11,555.6	12.327	0.07341	0.93963	0.00146
21,000	11,666.7	11.949×10^{-15}	0.07116×10^{-11}	0.94104	0.00142
21,200	11,777.8	11.585	0.06899	0.94242	0.00137
21,400	11,888.9	11.234	0.06691	0.94375	0.00133
21,600	12,000.0	10.897	0.06490	0.94504	0.00129
21,800	12,111.1	10.572	0.06296	0.94629	0.00125
22,000	12,222.2	10.258×10^{-15}	0.06109×10^{-11}	0.94751	0.00122
22,200	12,333.3	9.956	0.05930	0.94869	0.00118
22,400	12,444.4	9.665	0.05756	0.94983	0.00115
22,600	12,555.6	9.384	0.05589	0.95094	0.00111
22,800	12,666.7	9.114	0.05428	0.95202	0.00108

Table A-5 Blackbody functions *(Continued)*

Wavelength-temperature product λT		Blackbody hemispherical spectral emissive power divided by fifth power of temperature $e_{\lambda b}/T^5$		Blackbody fraction $F_{0-\lambda T}$	Difference between successive $F_{0-\lambda T}$ values ΔF
$\mu m \cdot °R$	$\mu m \cdot K$	Btu/(h·ft²·$\mu m \cdot °R^5$)	W/(m²·$\mu m \cdot K^5$)		
23,000	12,777.8	8.852×10^{-15}	0.05272×10^{-11}	0.95307	0.00105
23,200	12,888.9	8.600	0.05122	0.95409	0.00102
23,400	13,000.0	8.357	0.04977	0.95508	0.00099
23,600	13,111.1	8.122	0.04837	0.95604	0.00096
23,800	13,222.2	7.895	0.04702	0.95698	0.00093
24,000	13,333.3	7.676×10^{-15}	0.04572×10^{-11}	0.95788	0.00091
24,200	13,444.4	7.465	0.04446	0.95877	0.00088
24,400	13,555.6	7.260	0.04324	0.95963	0.00086
24,600	13,666.7	7.063	0.04206	0.96046	0.00084
24,800	13,777.8	6.872	0.04092	0.96128	0.00081
25,000	13,888.9	6.687×10^{-15}	0.03982×10^{-11}	0.96207	0.00079
25,200	14,000.0	6.508	0.03876	0.96284	0.00077
25,400	14,111.1	6.336	0.03773	0.96359	0.00075
25,600	14,222.2	6.169	0.03674	0.96432	0.00073
25,800	14,333.3	6.007	0.03577	0.96503	0.00071
26,000	14,444.4	5.850×10^{-15}	0.03484×10^{-11}	0.96572	0.00069
26,200	14,555.6	5.699	0.03394	0.96639	0.00067
26,400	14,666.7	5.552	0.03307	0.96705	0.00066
26,600	14,777.8	5.410	0.03222	0.96769	0.00064
26,800	14,888.9	5.273	0.03140	0.96831	0.00062
27,000	15,000.0	5.139×10^{-15}	0.03061×10^{-11}	0.96892	0.00061
27,200	15,111.1	5.010	0.02984	0.96951	0.00059
27,400	15,222.2	4.885	0.02909	0.97009	0.00058
27,600	15,333.3	4.764	0.02837	0.97065	0.00056
27,800	15,444.4	4.646	0.02767	0.97120	0.00055
28,000	15,555.6	4.532×10^{-15}	0.02699×10^{-11}	0.97174	0.00054
28,200	15,666.7	4.422	0.02633	0.97226	0.00052
28,400	15,777.8	4.315	0.02570	0.97277	0.00051
28,600	15,888.9	4.211	0.02508	0.97327	0.00050
28,800	16,000.0	4.110	0.02448	0.97375	0.00049
29,000	16,111.1	4.012×10^{-15}	0.02389×10^{-11}	0.97423	0.00047
29,200	16,222.2	3.917	0.02333	0.97469	0.00046
29,400	16,333.3	3.824	0.02278	0.97514	0.00045
29,600	16,444.4	3.735	0.02224	0.97558	0.00044
29,800	16,555.6	3.648	0.02172	0.97601	0.00043
30,000	16,666.7	3.563×10^{-15}	0.02122×10^{-11}	0.97644	0.00042
30,200	16,777.8	3.481	0.02073	0.97685	0.00041
30,400	16,888.9	3.401	0.02026	0.97725	0.00040
30,600	17,000.0	3.324	0.01979	0.97764	0.00039
30,800	17,111.1	3.248	0.01935	0.97802	0.00038

Table A-5 Blackbody functions (Continued)

Wavelength-temperature product λT		Blackbody hemispherical spectral emissive power divided by fifth power of temperature $e_{\lambda b}/T^5$		Blackbody fraction $F_{0-\lambda T}$	Difference between successive $F_{0-\lambda T}$ values ΔF
$\mu m \cdot {}^\circ R$	$\mu m \cdot K$	Btu/(h·ft² ·μm·$^\circ R^5$)	W/(m² ·μm·K^5)		
31,000	17,222.2	3.175×10^{-15}	0.01891×10^{-11}	0.97840	0.00037
31,200	17,333.3	3.104	0.01849	0.97877	0.00037
31,400	17,444.4	3.035	0.01807	0.97912	0.00036
31,600	17,555.6	2.967	0.01767	0.97947	0.00035
31,800	17,666.7	2.902	0.01728	0.97982	0.00034
32,000	17,777.8	2.838×10^{-15}	0.01690×10^{-11}	0.98015	0.00033
32,200	17,888.9	2.776	0.01653	0.98048	0.00033
32,400	18,000.0	2.716	0.01618	0.98080	0.00032
32,600	18,111.1	2.657	0.01583	0.98111	0.00031
32,800	18,222.2	2.600	0.01549	0.98142	0.00031
33,000	18,333.3	2.545×10^{-15}	0.01515×10^{-11}	0.98172	0.00030
33,200	18,444.4	2.490	0.01483	0.98201	0.00029
33,400	18,555.6	2.438	0.01452	0.98230	0.00029
33,600	18,666.7	2.386	0.01421	0.98258	0.00028
33,800	18,777.8	2.336	0.01392	0.98286	0.00028
34,000	18,888.9	2.288×10^{-15}	0.01363×10^{-11}	0.98313	0.00027
34,200	19,000.0	2.240	0.01334	0.98339	0.00026
34,400	19,111.1	2.194	0.01307	0.98365	0.00026
34,600	19,222.2	2.149	0.01280	0.98390	0.00025
34,800	19,333.3	2.105	0.01254	0.98415	0.00025
35,000	19,444.4	2.062×10^{-15}	0.01228×10^{-11}	0.98440	0.00024
35,200	19,555.6	2.021	0.01203	0.98463	0.00024
35,400	19,666.7	1.980	0.01179	0.98487	0.00023
35,600	19,777.8	1.940	0.01156	0.98510	0.00023
35,800	19,888.9	1.902	0.01133	0.98532	0.00022
36,000	20,000.0	1.864×10^{-15}	0.01110×10^{-11}	0.98554	0.00022
36,200	20,111.1	1.827	0.01088	0.98576	0.00022
36,400	20,222.2	1.791	0.01067	0.98597	0.00021
36,600	20,333.3	1.756	0.01046	0.98617	0.00021
36,800	20,444.4	1.722	0.01026	0.98638	0.00020
37,000	20,555.6	1.689×10^{-15}	0.01006×10^{-11}	0.98658	0.00020
37,200	20,666.7	1.656	0.00986	0.98677	0.00020
37,400	20,777.8	1.624	0.00967	0.98696	0.00019
37,600	20,888.9	1.593	0.00949	0.98715	0.00019
37,800	21,000.0	1.563	0.00931	0.98734	0.00018
38,000	21,111.1	1.533×10^{-15}	0.00913×10^{-11}	0.98752	0.00018
38,200	21,222.2	1.505	0.00896	0.98769	0.00018
38,400	21,333.3	1.476	0.00879	0.98787	0.00017
38,600	21,444.4	1.449	0.00863	0.98804	0.00017
38,800	21,555.6	1.422	0.00847	0.98821	0.00017

Table A-5 Blackbody functions *(Continued)*

Wavelength-temperature product λT		Blackbody hemispherical spectral emissive power divided by fifth power of temperature $e_{\lambda b}/T^5$		Blackbody fraction $F_{0-\lambda T}$	Difference between successive values ΔF
$\mu m \cdot {}^\circ R$	$\mu m \cdot K$	$Btu/(h \cdot ft^2 \cdot \mu m \cdot {}^\circ R^5)$	$W/(m^2 \cdot \mu m \cdot K^5)$		
39,000	21,666.7	1.396×10^{-15}	0.00831×10^{-11}	0.98837	0.00016
39,200	21,777.8	1.370	0.00816	0.98853	0.00016
39,400	21,888.9	1.345	0.00801	0.98869	0.00016
39,600	22,000.0	1.320	0.00786	0.98885	0.00016
39,800	22,111.1	1.296	0.00772	0.98900	0.00015
40,000	22,222.2	1.273×10^{-15}	0.00758×10^{-11}	0.98915	0.00015
42,000	23,333.3	1.065	0.00634	0.99051	0.00136
44,000	24,444.4	0.898	0.00535	0.99165	0.00114
46,000	25,555.6	0.762	0.00454	0.99262	0.00097
48,000	26,666.7	0.651	0.00388	0.99344	0.00082
50,000	27,777.8	0.560×10^{-15}	0.00333×10^{-11}	0.99414	0.00071
52,000	28,888.9	0.484	0.00288	0.99475	0.00061
54,000	30,000.0	0.420	0.00250	0.99528	0.00053
56,000	31,111.1	0.367	0.00218	0.99574	0.00046
58,000	32,222.2	0.321	0.00191	0.99614	0.00040
60,000	33,333.3	0.283×10^{-15}	0.00168×10^{-11}	0.99649	0.00035
62,000	34,444.4	0.250	0.00149	0.99680	0.00031
64,000	35,555.6	0.222	0.00132	0.99707	0.00027
66,000	36,666.7	0.197	0.00117	0.99732	0.00024
68,000	37,777.8	0.176	0.00105	0.99754	0.00022
70,000	38,888.9	0.158×10^{-15}	0.940×10^{-14}	0.99773	0.00019
72,000	40,000.0	0.142	0.844	0.99791	0.00017
74,000	41,111.1	0.128	0.760	0.99806	0.00016
76,000	42,222.2	0.115	0.687	0.99820	0.00014
78,000	43,333.3	0.104	0.622	0.99833	0.00013
80,000	44,444.4	0.0948×10^{-15}	0.564×10^{-14}	0.99845	0.00012
82,000	45,555.6	0.0862	0.513	0.99855	0.00010
84,000	46,666.7	0.0786	0.468	0.99865	0.00010
86,000	47,777.8	0.0718	0.428	0.99874	0.00009
88,000	48,888.9	0.0657	0.391	0.99882	0.00008
90,000	50,000.0	0.0603×10^{-15}	0.359×10^{-14}	0.99889	0.00007
92,000	51,111.1	0.0554	0.330	0.99896	0.00007
94,000	52,222.2	0.0510	0.304	0.99902	0.00006
96,000	53,333.3	0.0470	0.280	0.99908	0.00006
98,000	54,444.4	0.0434	0.259	0.99913	0.00005
100,000	55,555.6	0.0402×10^{-15}	0.239×10^{-14}	0.99918	0.00005

REFERENCES

1. Mechtly, E. A.: The International System of Units. Physical Constants and Conversion Factors, *NASA* SP-7012, 2d rev., 1973.
2. Pivovonsky, Mark, and Max R. Nagel: "Tables of Blackbody Radiation Functions," The Macmillan Company, New York, 1961.
3. Wiebelt, John A.: "Engineering Radiation Heat Transfer," Holt, Rinehart and Winston, Inc., New York, 1966.

SOURCES OF DIFFUSE
CONFIGURATION FACTORS

This appendix contains tables of references to about 200 configuration factors that are available in the literature. The table is composed of three parts. Part *A* is for configuration factors between two elemental surfaces; part *B* gives references for factors between an elemental and a finite surface; part *C* is for factors between two finite areas. More than one reference is given for some factors, and in certain cases the reference in which a factor was originally derived is not given because of the difficulty in obtaining such earlier works.

The factors are arranged in the following manner: Factors involving only plane surfaces are given first, followed by those involving cylindrical bodies, conical bodies, spherical bodies, and more complex bodies. Within each such category, progression is made from simpler to more complex geometries.

Table B-1 Table of references for configuration factors

(A) Factors for two differential elements

Configuration number	Geometry	Configuration	Source
A-1	Two elemental areas in arbitrary configuration		Eq. (7-8)
A-2	Two elemental areas lying on parallel generating lines		Example 7-3
A-3	Elemental area to infinitely long strip of differential width lying on parallel generating line		Appendix C and Ref. [1]
A-4	Infinitely long strip of differential width to similar strip on parallel generating line		Appendix C and Ref. [1]
A-5	Strip of finite length and differential width to strip of same length on parallel generating line		Appendix C and Ref. [2]
A-6	Corner element of end of square channel to sectional wall element on channel		Example 7-20
A-7	Exterior element on tube surface to exterior element on adjacent parallel tube of same diameter		Ref. [3]
A-8	Exterior element on partitioned tube to similar element on adjacent parallel tube of same diameter		Ref. [3]

Table B-1 Table of references for configuration factors

(A) Factors for two differential elements *(continued)*

Configuration number	Geometry	Configuration	Source
A-9	Two ring elements on interior of right circular cylinder		Appendix C and Refs. [4,5]
A-10	Band of differential length on inside of cylinder to differential ring on cylinder base		Ref. [6]
A-11	Differential area on ring element to ring element on adjacent fin with view partially blocked by solid coaxial cylinder		Ref. [7]
A-12	Ring element on fin to ring element on adjacent fin		Refs. [7,8]

Table B-1 Table of references for configuration factors

(A) Factors for two differential elements *(continued)*

Configuration number	Geometry	Configuration	Source
A-13	Two elements on interior of right circular cone		Refs. [9,10]
A-14	Two differential elements on interior of spherical cavity		Appendix C and Refs. [1,5,10–12]
A-15	Band on outside of sphere to band on another sphere		Ref. [13] (equal radii); [14] (unequal radii)
A-16	Two differential elements on exterior of toroid		Ref. [15]
A-17	Element on exterior of toroid to ring element on exterior of toroid		Ref. [15]

Table B-1 Table of references for configuration factors

(A) Factors for two differential elements *(continued)*

Configuration number	Geometry	Configuration	Source
A-18	Element on exterior of toroid to hoop element on exterior of toroid		Ref. [15]

Table B-1 Table of references for configuration factors

(B) Factors for exchange between differential element and finite area

Configuration number	Geometry	Configuration	Source
B-1	Plane element to plane extending to infinity and intersecting plane of element at angle Φ		Refs. [16–18]
B-2	Plane strip element of any length to plane of finite width and infinite length		Example 7-7
B-3	Plane element to infinitely long surface of arbitrary shape generated by line moving parallel to itself and plane of element		Appendix C and Refs. [16–20]

Table B-1 Table of references for configuration factors

(B) Factors for exchange between differential element and finite area *(continued)*

Configuration number	Geometry	Configuration	Source
B-4	Strip element of finite length to rectangle in plane parallel to strip; strip is opposite to one edge of rectangle		Appendix C and Refs. [5, 16–18]
B-5	Strip element of finite length to plane rectangle that intercepts plane of strip at angle Φ and with one edge parallel to strip		Appendix C and Ref. [5] for Φ = 90° only; [16–18]
B-6	Plane element to plane rectangle; normal to element passes through corner of rectangle; surfaces are on parallel planes		Appendix C and Refs. [1, 5, 16–19, 21, 22]
B-7	Area element to any parallel rectangle		Ref. [1]
B-8	Plane element to plane rectangle; planes containing two surfaces intersect at angle Φ		Appendix C and Refs. [1, 5], for Φ = 90° only; [16–18]
B-9	Plane element to right triangle in plane parallel to plane of element; normal to element passes through vertex of triangle		Example 7-18

Table B-1 Table of references for configuration factors

(B) Factors for exchange between differential element and finite area *(continued)*

Configuration number	Geometry	Configuration	Source
B-10	Plane element to a parallel isosceles triangle; normal to element passes through vertex		Appendix C and Ref. [23]
B-11	Plane element to parallel regular coaxial polygon; normal to element passes through polygon center		Appendix C and Ref. [23]
B-12	Plane element to plane area with added triangular area; element is on corner of rectangle with one side in common with plane at angle Φ		Refs. [16–18, 24]
B-13	Same geometry as preceding with triangle reversed relative to plane element		Refs. [16–18, 24]
B-14	Plane element to circular disk on plane parallel to that of element		Appendix C and Refs. [1, 5, 7, 16-18, 25]
B-15	Plane element to segment of parallel coaxial disk		Appendix C and Ref. [23]

Table B-1 Table of references for configuration factors

(B) Factors for exchange between differential element and finite area *(continued)*

Configuration number	Geometry	Configuration	Source
B-16	Plane element to segment of disk in plane parallel to element		Ref. [5]
B-17	Plane element to circular disk; planes containing element and disk intersect at 90°, and centers of element and disk lie in plane perpendicular to those containing areas		Appendix C, Refs. [5, 16–18, 20, 26], and Example 7-6
B-18	Strip element of finite length to perpendicular circular disk located at one end of strip		Refs. [20, 26]
B-19	Plane element to a thin coaxial ring parallel to the element		Ref. [23]
B-20	Plane element to ring area in plane perpendicular to element		Example 7-9
B-21	Radial and wedge elements on circular disk to disk in parallel plane		Refs. [21, 26]

Table B-1 Table of references for configuration factors

(B) Factors for exchange between differential element and finite area *(continued)*

Configuration number	Geometry	Configuration	Source
B-22	Plane element to ring sector on circular disk parallel to element		Ref. [25]
B-23	Plane element to annular ring on circular disk parallel to element with cylindrical blockage of view		Ref. [7]
B-24	Plane element to sector of circular disk parallel to element		Ref. [25]
B-25	Area element to parallel elliptical plate		Appendix C and Refs. [19, 24]
B-26	Plane element to annular disk with conical blockage of view		Ref. [27]
B-27	Infinitely long cylinder to parallel infinitely long strip element, $a \geqslant r$		Appendix C and Refs. [16–18, 28][a]

[a]Factor incorrect in Refs. [16-18], see [28].

Table B-1 Table of references for configuration factors

(B) Factors for exchange between differential element and finite area *(continued)*

Configuration number	Geometry	Configuration	Source
B-28	Plane of infinite width and infinite length to infinitely long strip on the surface of a parallel cylinder		Appendix C and Refs. [16–18]
B-29	Plane element to right circular cylinder of finite length; normal to element passes through center of one end of cylinder and is perpendicular to cylinder axis		Appendix C and Refs. [5, 16–18]
B-30	Plane vertical element to circular cylinder tilted toward the element		Ref. [29]
B-31	Plane element to interior of coaxial right circular cylinder		Ref. [23]
B-32	Element is at end of wall on inside of finite-length cylinder enclosing concentric cylinder of same length; factor is from element to inside surface of outer cylinder		Refs. [16–18, 20, 26]

Table B-1 Table of references for configuration factors

(B) Factors for exchange between differential and finite area *(continued)*

Configuration number	Geometry	Configuration	Source
B-33	Elemental strip of finite length to parallel cylinder of same length, normals at ends of strip pass through cylinder axis		Refs. [16–18, 20, 26]
B-34	Strip or element on plane parallel to cylinder axis to cylinder of finite length		Refs. [20, 26]
B-35	Infinitely long strip of differential width to parallel semi-cylinder		Ref. [30]
B-36	Infinite strip on any side of any of three fins to tube or environment, and infinite strip on tube to fin or environment		Ref. [31]
B-37	Element and strip element on interior of finite cylinder to interior of cylindrical surface		Refs. [20, 26]

Table B-1 Table of references for configuration factors

(B) Factors for exchange between differential element and finite area *(continued)*

Configuration number	Geometry	Configuration	Source
B-38	Elemental strip on inner surface of outer concentric cylinder to interior surface of outer concentric cylinder		Refs. [16–18, 20, 26]
B-39	Elemental strip on inner surface of outer concentric cylinder to either annular end		Refs. [16–18, 20, 26]
B-40	Element on inside of outer finite concentric cylinder to inside cylinder or annular end		Refs. [20, 26]
B-41	Strip element on exterior of inner finite-length concentric cylinder to inside of outer cylinder or to annular end		Refs. [20, 26]
B-42	Strip on plane inside cylinder of finite length to inside of cylinder		Refs. [20, 26]

Table B-1 Table of references for configuration factors
(B) Factors for exchange between differential element and finite area *(continued)*

Configuration number	Geometry	Configuration	Source
B-43	Area element on interior of cylinder to base of second concentric cylinder; cylinders are one atop other		Refs. [20, 26]
B-44	Ring element on fin to tube		Ref. [8]
B-45	Ring element on interior of right circular cylinder to end of cylinder		Appendix C and Ref. [4]
B-46	Exterior element on tube surface to finite area on adjacent parallel tube of same diameter		Ref. [3]
B-47	Exterior element on tube surface of partitioned tube to finite area on adjacent parallel tube of same diameter		Ref. [3]

Table B-1 Table of references for configuration factors

(B) Factors for exchange between differential element and finite area *(continued)*

Configuration number	Geometry	Configuration	Source
B-48	Element on wall of right circular cone to base of cone		Ref. [32]
B-49	Plane element on a ring to an inverted cone; ring and cone have the same axis and plane of ring does not intersect cone		Ref. [33]
B-50	Plane element to trunicated cone		Ref. [27]
B-51	Plane element on a ring to an inverted cone; ring and cone have the same axis and plane of ring intersects cone		Refs. [33–35]
B-52	Any infinitesimal element on interior of sphere to any finite element on interior of same sphere		Appendix C, Ref. [1], and Sec. 8-4.2

Table B-1 Table of references for configuration factors

(B) Factors for exchange between differential element and finite area *(continued)*

Configu-ration number	Geometry	Configuration	Source
B-53	Spherical point source to rectangle; point source is on one corner of rectangle that intersects with receiving rectangle at angle Φ		Refs. [1, 16–18, 24]
B-54	Plane element to sphere; normal to element passes through center of sphere		Appendix C and Ref. [23]
B-55	Plane element to sphere; tangent to element passes through center of sphere		Appendix C and Ref. [23]
B-56	Area element to sphere; in [73] the element can be on the surface of a sphere or cylinder		Refs. [17, 36–38, 73]
B-57	Sphere to ring element oriented normal to sphere axis		Ref. [28]
B-58	Spherical element to sphere		Appendix C and Ref. [23]

Table B-1 Table of references for configuration factors

(B) Factors for exchange between differential element and finite area *(continued)*

Configu-ration number	Geometry	Configuration	Source
B-59	Elemental area on sphere to finite area on second sphere		Ref. [14]
B-60	Area element to axisymmetric surface —paraboloid, cone, cylinder (formulation given, factors are not evaluated)		Ref. [39]
B-61	Element on interior (or exterior) of any axisymmetric body of revolution to band of finite length on interior (or exterior)		Refs. [40[b], 41[b], 42]
B-62	Element on exterior of toroid to toroidal segment of finite width		Ref. [15]
B-63	Element on exterior of toroid to toroidal band of finite width		Ref. [15]
B-64	Element and ring element on exterior of toroid to entire exterior of toroid		Refs. [15, 43]

[b]Kernels of integrals and limits are formulated in terms of appropriate variables, but integrations are not carried out explicitly.

Table B-1 Table of references for configuration factors

(B) Factors for exchange between differential element and finite area *(continued)*

Configuration number	Geometry	Configuration	Source
B-65	Slender torus to element on perpendicular axis		Ref. [19]

Table B-1 Table of references for configuration factors

(C) Factors for two finite areas

Configuration number	Geometry	Configuration	Source
C-1	Two infinitely long plates of equal finite width W and one common edge of included angle Φ		Appendix C
C-2	Two infinitely long plates of unequal width with one common edge and included angle $\Phi = 90°$		Appendix C and Ref. [17]
C-3	Finite rectangle to infinitely long rectangle of same width and with one common edge		Ref. [44]

Table B-1 Table of references for configuration factors

(C) Factors for two finite areas *(continued)*

Configu-ration number	Geometry	Configuration	Source
C-4	Two finite rectangles of same width with common edge and included angle Φ		Appendix C and Refs. [1, 21, 22] for $\Phi = 90°$ only; Refs. [5, 16-18, 20, 44[c]]
C-5	Two rectangles with common edge and included angle Φ		Ref. [17]
C-6	Two rectangles with one edge of each parallel, and with one corner touching; planes containing rectangles intersect at angle Φ		Ref. [17]
C-7	Two rectangles of same width with two parallel edges; planes containing rectangles intersect at angle Φ		Ref. [5] for $\Phi = 90°$ only; Ref. [17]
C-8	Two rectangles with two parallel edges; planes containing rectangles intersect at angle Φ		Refs. [16,17, 20]; [71] for $\Phi = 90°$
C-9	Two infinitely long directly opposed parallel strips of same finite width		Appendix C and Refs. [17, 21]

[c]Ref. [44] indicates that tabulated values for this case are incorrect in all other references. Corrected values are listed in Ref. [44].

Table B-1 Table of references for configuration factors

(C) Factors for two finite areas *(continued)*

Configuration number	Geometry	Configuration	Source
C-10	Parallel, directly opposed rectangles of same width and length		Appendix C and Refs. [1, 5, 16–18, 21, 22, 24, 45]
C-11	Two rectangles in parallel planes with one rectangle directly opposite portion of other		Refs. [17, 19, 45]
C-12	Two unequal parallel squares that are coaxial and have parallel sides		Ref. [46]
C-13	Two rectangles of arbitrary size in parallel planes; all edges lie parallel to either x or y axis		Refs. [16, 17, 20, 45, 71]
C-14	Rectangle to arbitrarily oriented rectangle of arbitrary size		Refs. [47, 48][d]
C-15	Two flat plates of arbitrary shape and arbitrary orientation		Ref. [49][e]

[d] Available as general computer program only.

[e] Kernels of integrals and limits are formulated in terms of appropriate variables, but integrations are not carried out explicitly.

Table B-1 Table of references for configuration factors
(C) Factors for two finite areas *(continued)*

Configuration number	Geometry	Configuration	Source
C-16	Finite areas on interior of square channel		Ref. [20]
C-17	Factor between bases of right convex prism of regular triangular, square, pentagonal, hexagonal, or octagonal cross section		Ref. [44]
C-18	Factors between various sides, and sides and bases of regular hexagonal prism		Ref. [44]
C-19	Circular disk to arbitrarily placed rectangle in parallel plane (using configuration-factor algebra with configuration C-22)		Ref. [50]
C-20	Circle to arbitrarily placed rectangle in plane parallel to normal to circle (using configuration-factor algebra with configuration C-22)		Ref. [50]

Table B-1 Table of references for configuration factors

(C) Factors for two finite areas *(continued)*

Configuration number	Geometry	Configuration	Source
C-21	Disk to arbitrarily oriented rectangle or disk of arbitrary size		Ref. [47[f]]
C-22	Circular disk to parallel right triangle; normal from center of circle passes through one acute vertex		Ref. [50]
C-23	Parallel, directly opposed plane circular disks[g]		Appendix C and Refs. [1, 5, 10, 16–18, 20, 21, 24, 26, 51[g]]
C-24	Directly opposed ring and disk of arbitrary radi		Refs. [17, 26]
C-25	Parallel, directly opposed plane ring areas		Refs. [16, 26] and Example 7–10
C-26	Sector of circular disk to sector of parallel circular disk		Ref. [25]

[f]Available as general computer program only
[g]Reference [51] also has nonparallel disks that can both be inscribed in a sphere.

Table B-1 Table of references for configuration factors

(C) Factors for two finite areas *(continued)*

Configuration number	Geometry	Configuration	Source
C-27	Entire inner wall of finite cylinder to ends		Refs. [52, 53]
C-28	Internal surface of cylindrical cavity to cavity opening		Refs. [17 (Fig. 6-14), 54]
C-29	Inner surface of cylinder to annulus on one end		Refs. [24, 26, 53]
C-30	Inner surface of cylinder to disk at one end of cylinder		Refs. [24, 26, 53]
C-31	Portion of inner surface of cylinder to remainder of inner surface		Refs. [21, 26, 53]
C-32	Finite ring areas on interior of right circular cylinders to separate similar areas and to ends		Refs. [20, 21, 53]
C-33	Finite areas on interior of right circular cylinder		Ref. [20]

Table B-1 Table of references for configuration factors

(C) Factors for two finite areas *(continued)*

Configu-ration number	Geometry	Configuration	Source
C-34	Infinite cylinder to parallel infinitely long plane of finite width		Appendix C and Refs. [5, 16, 17, 28][h]
C-35	Infinitely long plane of finite width to infinitely long cylinder		Refs. [26, 28]
C-36	Rows of infinitely long parallel cylinders in square or equilateral-triangular arrays		Ref. [55]
C-37	Infinite plane to first, second, and first plus second rows of infinitely long parallel tubes of equal diameter		Refs. [1, 21, 22, 24, 72]
C-38	Finite-length cylinder to rectangle with two edges parallel to cylinder axis and of length equal to cylinder		Refs. [5, 24]
C-39	Finite cylinder to finite rectangle of same length		Ref. [56]

[h]This factor is given incorrectly in all references except [5, 28] and Appendix C.

Table B-1 Table of references for configuration factors

(C) Factors for two finite areas *(continued)*

Configuration number	Geometry	Configuration	Source
C-40	Cylinder to any rectangle in plane perpendicular to cylinder axis (using configuration-factor algebra with configuration C-44)		Ref. [50]
C-41	Cylinder to any rectangle in plane parallel to cylinder axis (using configuration-factor algebra with configuration C-44)		Ref. [50]
C-42	Finite area on exterior of cylinder to finite area on plane parallel to cylinder axis		Ref. [20]
C-43	Finite area on exterior of cylinder to finite area on skewed plane		Ref. [20]
C-44	Outside surface of cylinder to perpendicular right triangle; triangle is in plane of cylinder base with one vertex of triangle at center of base		Ref. [50]

Table B-1 Table of references for configuration factors

(C) Factors for two finite areas *(continued)*

Configuration number	Geometry	Configuration	Source
C-45	Cylinder and plane of equal length parallel to cylinder axis; plane inside cylinder; all factors between plane and inner surface of cylinder		Refs. [20, 26]
C-46	Inner surface of cylinder to disk of same radius		Refs. [20, 26]
C-47	Interior surface of circular cylinder of radius R to disk of radius r where $r < R$; disk is perpendicular to axis of cylinder, and axis passes through center of disk (using configuration-factor algebra with configuration C-23)		Example 7-11
C-48	Outer surface of cylinder to annular disk at end of cylinder		Refs. [57, 58]
C-49	Annular ring to similar annular ring each at end of cylinder		Refs. [8, 20, 26]

Table B-1 Table of references for configuration factors

(C) Factors for two finite areas *(continued)*

Configuration number	Geometry	Configuration	Source
C-50	Annular ring to annular ring of different outer diameter, each at end of cylinder		Ref. [7]
C-51	Factors for interchange between fins and tube (given in algebraic form, untabulated)		Refs. [8, 58]
C-52	Finite area on exterior of cylinder to finite area on exterior of parallel cylinder		Ref. [20]
C-53	Cylinder of arbitrary length and radius to rectangle, disk, or cylinder of arbitrary size and orientation		Refs. [47, 48][i]
C-54	Cylinder and plate with arbitrary orientation		Ref. [49[j]]
C-55	Concentric cylinders of infinite length; inner to outer cylinder; outer to inner cylinder; outer cylinder to itself		Appendix C and Ref. [16]

[i] Available as general computer program only.

[j] Kernels of integrals and limits are formulated in terms of appropriate variables, but integrations are not carried out explicitly.

Table B-1 Table of references for configuration factors

(C) Factors for two finite areas *(continued)*

Configuration number	Geometry	Configuration	Source
C-56	Nonconcentric infinitely long circular cylinders, inner cylinder to area on outer cylinder		Ref. [24]
C-57	Inside surface of outer concentric cylinder of finite length to inner cylinder of same length		Appendix C and Refs. [5, 16, 21, 24, 26, 57, 59]
C-58	Inside surface of outer concentric cylinder to itself		Appendix C and Refs. [5, 8, 16, 24, 26, 59]
C-59	Inside surface of outer concentric cylinder to either end of annulus		Refs. [5, 16, 24, 26, 58, 59]
C-60	Concentric cylinders of different finite lengths—portion of inner cylinder to entire outer cylinder		Refs. [26, 57]
C-61	Concentric cylinders of different finite lengths—portion of inside of outer to outside of entire inner cylinder		Refs. [20, 26, 49[k]]

[k]Kernels of integrals and limits are formulated in terms of appropriate variables, but integrations are not carried out explicitly.

Table B-1 Table of references for configuration factors

(C) Factors for two finite areas *(continued)*

Configuration number	Geometry	Configuration	Source
C-62	Parallel cylinders of different radii and length—any portions of outer curved surfaces		Ref. [49[l]]
C-63	Concentric cylinders of different radii, one atop the other; factors between inside of upper cylinder and inside or base of lower cylinder		Refs. [20, 26]
C-64	Inside of right circular cylinder to outside of right circular cylinder of the same height and on the same axis		Ref. [60]
C-65	Infinitely long parallel semicylinders of same diameter		Example 7–16 and Ref. [5]
C-66	Finite area on exterior of inner cylinder to finite area on interior of concentric outer cylinder		Ref. [20]

[l]Kernels of integrals and limits are formulated in terms of appropriate variables, but integrations are not carried out explicitly.

Table B-1 Table of references for configuration factors

(C) Factors for two finite areas *(continued)*

Configuration number	Geometry	Configuration	Source
C-67	Two tubes connected with fin of finite thickness; length can be finite or infinite; all factors between finite surfaces formulated in terms of integrations between differential strips		Ref. [2]
C-68	Two tubes connected with tapered fins of finite thickness; tube length can be finite or infinite; all factors between finite surfaces formulated in terms of integrations between differential strips		Ref. [2]
C-69	Sandwich tube and fin structure of infinite or finite length; all factors between finite surfaces formulated in terms of integrations between differential strips		Ref. [2]
C-70	Concentric cylinders connected by fin of finite thickness; length finite or infinite; all factors between finite surfaces formulated in terms of integrations between differential strips		Ref. [2]

Table B-1 Table of references for configuration factors

(C) Factors for two finite areas *(continued)*

Configuration number	Geometry	Configuration	Source
C-71	Exterior of infinitely long cylinder to interior of concentric semicylinder		Example 7–23
C-72	Interior of infinitely long semicylinder 1 to interior of semicylinder 2 when concentric parallel cylinder 3 is present		Example 7–23
C-73	Between axisymmetrical sections of right circular cone		Refs. [20, 53]
C-74	Between axisymmetrical sections of right circular cone and base or ring or disk on base		Refs. [21, 53]
C-75	Internal surface of conical cavity to cavity opening		Example 7–14 and Ref. [53]; Refs. [17 (Fig. 6–14), 54][m]
C-76	Entire inner surface of frustum of cone to ends		Refs. [52, 53]
C-77	Circular disk and right circular cone with axis normal to the disk and with vertex at center of disk		Ref. [61]

[m]Results are not the configuration factor, and abscissa of Fig. 6–14 of [17] is ten times too large.

Table B-1 Table of references for configuration factors

(C) Factors for two finite areas *(continued)*

Configuration number	Geometry	Configuration	Source
C-78	Two nested right circular cones having common axis and vertex, and closed by a ring in the base		Ref. [61]
C-79	Right circular cone of arbitrary size to rectangle, disk, cylinder, or cone of arbitrary size and orientation		Refs. [47, 48][n]
C-80	Cone to arbitrarily skewed plate		Ref. [49][o]
C-81	Internal surface of spherical cavity to cavity opening		Example 7-14 and Refs. [17 (Fig. 6-14), 54][p]
C-82	Any finite area on interior of sphere to any other finite area on interior		Appendix C and Ref. [1]

[n] Available as general computer program only.

[o] Kernels of integrals and limits are formulated in terms of appropriate variables, but integrations are not carried out explicitly.

[p] Results are not the configuration factor, and abscissa of Fig. 6-14 of [17] is 10 times too large.

Table B-1 Table of references for configuration factors

(C) Factors for two finite areas *(continued)*

Configuration number	Geometry	Configuration	Source
C-83	Sphere to coaxial disk		Appendix C and Refs. [24, 28]
C-84	Sphere to segment of coaxial disk		Appendix C and Refs. [24, 28]
C-85	Sphere to sector of coaxial disk		Appendix C and Refs. [24, 28]
C-86	Sphere to rectangle		Ref. [50]
C-87	Sphere to rectangle normal to sphere axis		Ref. [28]
C-88	Sphere to arbitrary rectangle (using configuration-factor algebra and configuration C-86)		Refs. [47,[q] 50]

[q] Available as general computer program only.

Table B-1 Table of references for configuration factors

(C) Factors for two finite areas *(continued)*

Configuration number	Geometry	Configuration	Source
C-89	Sphere to regular polygon normal to sphere axis		Ref. [28]
C-90	Sphere to disk not on axis		Ref. [28]
C-91	Sphere of arbitrary diameter to disk or cone of arbitrary size and orientation		Refs. [47 48][r]
C-92	Sphere to arbitrarily skewed plate		Ref. [49][s]
C-93	Sphere to cylinder		Refs. [47,[r] 48,[r] 62]
C-94	Cone to sphere having same diameter as base of cone; axis of cone passes through center of sphere		Refs. [13, 47,[r] 48[r]]

[r]Available as general computer program only.
[s]Kernels of integrals and limits are formulated in terms of appropriate variables, but integrations are not carried out explicitly.

Table B-1 Table of references for configuration factors

(C) Factors for two finite areas *(continued)*

Configuration number	Geometry	Configuration	Source
C-95	Concentric spheres; inner to outer sphere; outer to inner sphere; outer sphere to itself		Appendix C, Refs. [16, 17], and Example 7–13
C-96	Area on surface of sphere to rectangle in plane perpendicular to axis of sphere		Ref. [20]
C-97	Sphere to sphere		Refs. [13 (equal radii), 47,[t] 48,[t] 62–66]
C-98	Area on sphere to cap on another sphere		Ref. [14]
C-99	Cap on sphere to cap on another sphere		Ref. [14]
C-100	Area on sphere to area on another sphere		Ref. [14]

[t]Available as general computer program only.

Table B-1 Table of references for configuration factors

(C) Factors for two finite areas *(continued)*

Configuration number	Geometry	Configuration	Source
C-101	Cap on sphere to band on another sphere		Ref. [14]
C-102	Band on one sphere to band on another sphere		Ref. [14]
C-103	Internal surface of hemispherical cavity to cavity opening		Example 7-14 and Ref [53]; Refs. [17 (Fig. 6–14), 54][u]
C-104	Between axisymmetrical section of hemisphere and base or ring or disk on base		Refs. [24, 53]
C-105	Between axisymmetrical sections of hemisphere		Refs. [20, 53]
C-106	Annular ring around base of hemisphere to hemisphere		Ref. [67]
C-107	Annular ring around base of hemisphere to section of hemisphere		Ref. [67]

[u]Results are not the configuration factor, and abscissa of Fig. 6-14 of [17] is 10 times too large.

Table B-1 Table of references for configuration factors

(C) Factors for two finite areas *(continued)*

Configuration number	Geometry	Configuration	Source
C-108	Hemisphere to coaxial hemisphere in contact		Ref. [68]
C-109	Sphere to hemisphere		Ref. [62]
C-110	Sphere to ellipsoid		Refs. [47,[v] 62]
C-111	Ellipsoid of arbitrary major and minor axes to rectangle, disk, cylinder, cone, or ellipsoid of arbitrary size and orientation		Ref. [47][v]
C-112	From Moebius strip to itself		Ref. [69]
C-113	Exterior of toroid to itself		Refs. [15, 43]
C-114	Segment of finite width on toroid to exterior of toroid		Ref. [15]
C-115	Toroidal band of finite width to exterior of toroid		Ref. [15]

[v] Available as general computer program only.

Table B-1 Table of references for configuration factors

(C) Factors for two finite areas *(continued)*

Configu-ration number	Geometry	Configuration	Source
C-116	Toroid of arbitrary size to rectangle, disk, cylinder, sphere, cone, ellipsoid, or toroid of arbitrary size and orientation		Ref. [47]^{W}
C-117	Arbitrary polynomial of revolution to rectangle, disk, cylinder, sphere, cone, ellipsoid, toroid, or other arbitrary polynomial of revolution of arbitrary size and orientation (poly-nomials of fifth order or less)		Ref [47]^{W}
C-118	General plane polygon to any general plane polygon or two or more intersecting or adjoining polygons		Ref. [70]^{W}

^{W}Available as general computer program only.

REFERENCES

1. Jakob, Max: "Heat Transfer," vol. 2, John Wiley & Sons, Inc., New York, 1957.
2. Sotos, Carol J., and Norbert O. Stockman: Radiant-Interchange View Factors and Limits of Visibility for Differential Cylindrical Surfaces with Parallel Generating Lines, *NASA* TN D-2556, 1964.

3. Sparrow, E. M., and V. K. Jonsson: Angle-Factors for Radiant Interchange between Parallel-oriented Tubes, *J. Heat Transfer,* vol. 85, no. 4, pp. 382–384, 1963.

4. Usiskin, C. M., and R. Siegel: Thermal Radiation from a Cylindrical Enclosure with Specified Wall Heat Flux, *J. Heat Transfer,* vol. 82, no. 4, pp. 369–374, 1960.

5. Sparrow, E.M., and R. C. Cess: "Radiation Heat Transfer," Augmented edition, Hemisphere Publishing Corp., Washington, D.C., 1978.

6. Sparrow, E. M., L. U. Albers, and E. R. G. Eckert: Thermal Radiation Characteristics of Cylindrical Enclosures, *J.Heat Transfer*, vol. 84, no. 1, pp. 73–81, 1962.

7. Minning, C. P.: Shape factors Between Coaxial Annular Discs Separated by a Solid Cylinder, *AIAA J.*, vol. 17, no. 3, pp. 318-320, March 1979.

8. Sparrow, E. M., G. B. Miller, and V. K. Jonsson: Radiative Effectiveness of Annular-finned Space Radiators, Including Mutual Irradiation between Radiator Elements, *J. Aerospace Sci.*, vol. 29, no. 11, pp. 1291–1299, 1962.

9. Sparrow, E. M., and V. K. Jonsson: Radiant Emission Characteristics of Diffuse Conical Cavities, *J. Opt. Soc. Am.,* vol. 53, no. 7, pp. 816–821, 1963.

10. Kezios, Stothe P., and Wolfgang Wulff: Radiative Heat Transfer through Openings of Variable Cross Sections, *Proc. Third Int. Heat Transfer Conf. AIChE,* vol. 5, 1966, pp. 207–218.

11. Sparrow, E. M., and V. K. Jonsson: Absorption and Emission Characteristics of Diffuse Spherical Enclosures, *NASA* TN D-1289, 1962.

12. Nichols, Lester D.: Surface-temperature Distribution on Thin-walled Bodies Subjected to Solar Radiation in Interplanetary Space, *NASA* TN D-584, 1961.

13. Campbell, James P., and Dudley G. Mc Connell: Radiant-Interchange Configuration Factors for Spherical and Conical Surfaces to Spheres, *NASA* TN D-4457, 1968.

14. Grier, Norman T.: Tabulations of Configuration Factors between any Two Spheres and Their Parts, *NASA* SP-3050, 1969.

15. Grier, Norman T., and Ralph D. Sommers: View Factors for Toroids and Their Parts, *NASA* TN D-5006, 1969.

16. Hamilton, D. C., and W. R. Morgan: Radiant-interchange Configuration Factors, *NASA* TN 2836, 1952.

17. Kreith, Frank: "Radiation Heat Transfer for Spacecraft and Solar Power Plant Design," International Textbook Company, Scranton, Pa., 1962.

18. Wiebelt, John A.: "Engineering Radiation Heat Transfer," Holt, Rinehart and Winston, Inc., New York, 1966.

19. Moon, Parry: "The Scientific Basis of Illuminating Engineering," Dover Publications, Inc., New York, 1961.

20. Stevenson, J. A., and Grafton, J. C.: Radiation Heat Transfer Analysis for Space Vehicles, *Rept.* SID-61-91, North American Aviation (AFASD TR 61-119, pt 1), Sept. 9, 1961.

21. Hottel, H. C., and A. F. Sarofim: "Radiation Transfer," McGraw-Hill Book Company, New York, 1967.

22. Hottel, H. C.: Radiant-heat Transmission, in William H. Mc Adams (ed.), "Heat Transmission," 3d ed., pp. 55–125, McGraw-Hill Book Company, New York, 1954.

23. Chung, B. T. F., and P. S. Sumitra: Radiation Shape Factors from Plane Point Sources, *ASME J. Heat Transfer*, vol. 94, no. 3, pp. 328–330, 1972.

24. Wong, H. Y.: "Handbook of Essential Formulae and Data on Heat Transfer for Engineers" Longman Group Ltd., London, 1977.

25. Charles P. Minning: Calculation of Shape Factors between Parallel Ring Sectors Sharing a Common Centerline, *AIAA. J*, vol. 14, no. 6, pp. 813–815, 1976.

26. Leuenberger, H., and R. A. Person: Compilation of Radiation Shape Factors for Cylindrical Assemblies, paper no. 56-A-144, *ASME*. November, 1956.

27. Holchendler, J., and W. F. Laverty: Configuration Factors for Radiant Heat Exchange in Cavities Bounded at the Ends by Parallel Disks and Having Conical Centerbodies, *J. Heat Transfer*, vol. 96, no. 2, pp. 254–257, 1974.

28. Feingold, A, and K. G. Gupta: New Analytical Approach to the Evaluation of Configuration Factors in Radiation from Spheres and Infinitely Long Cylinders, *J. Heat Transfer*, vol. 92, no. 1, pp. 69–76, 1970.

29. Rein. R. G., Jr., C. M. Sliepcevich, and J. R. Welker: Radiation View Factors for Tilted Cylinders, *J. Fire Flammability,* vol. 1, pp. 140–153, 1970.

30. Sparrow, E. M., and E. R. G. Eckert: Radiant Interaction between Fin and Base Surfaces, *J. Heat Transfer,* vol. 84, no. 1, pp. 12–18, 1962.

31. Holcomb, R. S., and F. E. Lynch: Thermal Radiation Performance of a Finned Tube with a Reflector, *Rept.* ORNL-TM-1613, Oak Ridge National Laboratory, April, 1967.

32. Joerg, Pierre, and B. L. Mc Farland: Radiation Effects in Rocket Nozzles, *Rept.* S62-245, Aerojet-General Corporation, 1962.

33. Minning, C. P.: Calculation of Shape Factors between Rings and Inverted Cones Sharing a Common Axis, *ASME J. Heat Transfer,* vol. 99, no. 3, pp. 492–494, 1977. (see also discussion in *ASME J. Heat Transfer,* vol. 101, no. 1, pp. 189–190, 1979).

34. Edwards, D. K.: Comment on "Radiation from Conical Surfaces with Nonuniform Radiosity," *AIAA J.* vol. 7, no. 8, pp. 1656–1659, 1969.

35. Bobco, R. P.: Radiation from Conical Surfaces with Nonuniform Radiosity, *AIAA J.,* vol. 4, no. 3, pp. 544–546, 1966.

36. Cunningham, F. G.: Power Input to a Small Flat Plate from a Diffusely Radiating Sphere, with Application to Earth Satellites, *NASA* TN D-710, 1961.

37. Liebert, Curt H., and Robert R. Hibbard: Theoretical Temperatures of Thin-film Solar Cells in Earth Orbit, *NASA* TN D-4331, 1968.

38. Goetze, Dieter, and Charles B. Grosch: Earth-emitted Infrared Radiation Incident upon a Satellite, *J. Aerospace Sci.,* vol. 29, no. 5, pp. 521–524, 1962.

39. Morizumi, S. J.: Analytical Determination of Shape Factors from a Surface Element to an Axisymmetric Surface, *AIAA J.,* vol. 2, no. 11, pp. 2028–2030, 1964.

40. Robbins, William H., and Carroll A. Todd: Analysis, Feasibility, and Wall-temperature Distribution of a Radiation-cooled Nuclear-Rocket Nozzle, *NASA* TN D-878, 1962.

41. Robbins, William H.: An Analysis of Thermal Radiation Heat Transfer in a Nuclear-rocket Nozzle, *NASA* TN D-586, 1961.

42. Bernard, Jean-Joseph, and Jeanne Genot: "Diagrams for Computing the Radiation of Axisymmetric Surfaces (Propulsive Nozzles)," Office National d'Etudes et de Recherches Aerospatiales, Paris, France, ONERA-NT-185, 1971, (in French).

43. Sommers, Ralph D., and Norman T. Grier: Radiation View Factors for a Toroid: Comparison of Eckert's Technique and Direct Computation, *J. Heat Transfer,* vol. 91, no. 3, pp. 459–461, 1969.

44. Feingold, A.: Radiant-interchange Configuration Factors between Various Selected Plane Surfaces, *Proc. R. Soc. London,* ser. A, vol. 292, no. 1428, pp. 51–60, 1966.

45. Hsu, Chia-Jung: Shape Factor Equations for Radiant Heat Transfer between Two Arbitrary Sizes of Rectangular Planes, *Can. J. Chem. Eng.,* vol. 45, no. 1, pp. 58-60, 1967.

46. Crawford, Martin: Configuration Factor between Two Unequal, Parallel, Coaxial Squares, paper no. 72-WA/HT-16, *ASME,* Nov., 1972.

47. Dummer, R. S., and W. T. Breckenridge, Jr.: Radiation Configuration Factors Program, *Rept.* ERR-AN-224, General Dynamics/Astronautics, February, 1963.

48. Lovin, J. K., and A. W., Lubkowitz: User's Manual for "RAVFAC," a Radiation View Factor Digital Computer Program, *Rept.* HREC-0154-1; Lockheed Missiles and Space Co., Huntsville Research Park, Hunstville, Alabama; LMSC/HREC D148620 (Contract NAS8-30154), November, 1969.

49. Plamondon, Joseph A.: Numerical Determination of Radiation Configuration Factors for Some Common Geometrical Situations, *Tech. Rept.* 32–127, Jet Propulsion Laboratory, California Institute of Technology, July 7, 1961.

50. Tripp, W., C. Hwang, and R. E. Crank: Radiation Shape Factors for Plane Surfaces and Spheres, Circles or Cylinders, *Spec. Rept.* 16, Kansas State University Bulletin, vol. 46, no. 4, 1962.

51. Feingold, A.: A New Look at Radiation Configuration Factors Between Disks, *ASME J. Heat Transfer,* vol. 100, no. 4, pp. 742–744, 1978.

52. Bien, Darl D.: Configuration Factors for Thermal Radiation from Isothermal Inner Walls of Cones and Cylinders, *J. Spacecr. Rockets,* vol. 3, no. 1, pp. 155-156, 1966.

53. Buschman, Albert J., Jr., and Claud M. Pittman: Configuration Factors for Exchange of Radiant Energy between Axisymmetrical Sections of Cylinders, Cones, and Hemispheres and Their Bases, *NASA* TN D-944, 1961.

54. Stephens, Charles W., and Alan M. Haire: Internal Design Considerations for Cavity-type Solar Absorbers, *ARS J.*, vol. 31, no. 7, pp. 896–901, 1961.

55. Cox, Richard L.: "Radiative Heat Transfer in Arrays of Parallel Cylinders," Ph.D. thesis, Tennessee University, Knoxville, 1976.

56. Wiebelt, J. A., and S. Y. Ruo: Radiant-interchange Configuration Factors for Finite Right Circular Cylinder to Rectangular Planes, *Int. J. Heat Mass Transfer*, vol. 6, no. 2, pp. 143–146, 1963.

57. Rea, Samuel N.: Rapid Method for Determining Concentric Cylinder Radiation View Factors, *AIAA J.*, vol. 13, no. 8, pp. 1122–1123, 1975.

58. Masuda, H.: Radiant Heat Transfer on Circular-Finned Cylinders, Report Inst. High Speed Mech., Tohoku University, vol. 27, no. 225, pp. 67–89, 1973.

59. Aleksandrov, V. T.: Determination of the Angular Radiation Coefficients for a System of Two Coaxial Cylindrical Bodies, *Inzh. Fiz. Zh.*, vol. 8, no. 5, pp. 609–612, 1965.

60. Reid, R. L., and J. S. Tennant: Annular Ring View Factors, *AIAA J.*, vol. 11, no. 10, pp. 1446–1448, 1973.

61. Kobyshev, A. A., I. N. Mastiaeva, Iu. A. Surinov, and Iu. P. Iakovlev; Investigation of the Field of Radiation Established by Conical Radiators, *Aviats. Tekh.* vol. 19, no. 3, pp. 43–49, 1976 (in Russian).

62. Watts, R. G.: Radiant Heat Transfer to Earth Satellites, *J. Heat Transfer*, vol. 87, no. 3, pp. 369–373, 1965.

63. Jones, L. R.: Diffuse Radiation View Factors between Two Spheres, *J. Heat Transfer*, vol. 87, no. 3, pp. 421–422, 1965.

64. Juul, N. H.: Diffuse Radiation Configuration View Factors between Two Spheres and Their Limits, *Lett. Heat Mass Transfer*, vol. 3, no. 3, pp. 205–211, 1976.

65. Juul, N. H.: Investigation of Approximate Methods for Calculation of the Diffuse Radiation Configuration View Factor between Two Spheres, *Lett. Heat Mass Transfer*, vol. 3, pp. 513–522, 1976.

66. J. D. Felske: "Approximate Radiation Shape Factors Between Two Spheres", *ASME, J. Heat Transfer*, vol. 100, no. 3, pp. 547–548, 1978.

67. Ballance, J. O., and J. Donovan: Radiation Configuration Factors for Annular Rings and Hemispherical Sectors, *ASME, J. Heat Transfer*, vol. 95, no. 2, pp. 275–276, 1973.

68. Wakao, Noriaki, Koichi Kato, and Nobuo Furuya: View Factor between Two Hemispheres in Contact and Radiation Heat-Transfer Coefficient in Packed Beds, *Int. J. Heat Mass Transfer*, vol. 12, pp. 118–120, 1969.

69. Stasenko, A. L.: Self-irradiation Coefficient of a Moebius Strip of Given Shape, *Akad, Nauk SSSR, Izv. Energetika Transport*, pp. 104–107, July-August, 1967.

70. Toups, K. A.: A General Computer Program for the Determination of Radiant Interchange Configuration and Form Factors: Confac-1, *Rept.* SID-65-1043-1, North American Aviation, Inc. (NASA CR-65256), October, 1965.

71. Chekhovskii, I. R., V. V. Sirotkin, Yu. V. Chu-Dun-Chu, and V. A. Chebanov: Determination of Radiative View Factors for Rectangles of Different Sizes, *High Temperature*, July 1979, Translation of Russian original, vol. 17, no. 1, Jan.-Feb., 1979.

72. Kuroda, Z. and T. Munakata: Mathematical Evaluation of the Configuration Factors Between a Plane and One or Two Rows of Tubes, *Kagaku Sooti* (Chemical Apparatus, Japan) pp. 54-58, Nov., 1979 (in Japanese).

73. Juul, N. H.: Diffuse Radiation View Factors from Differential Plane Sources to Spheres, *ASME, J. Heat Transfer*, vol. 101, no. 3, pp. 558–560, 1979.

CATALOG OF SELECTED CONFIGURATION FACTORS

1	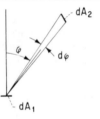	Area dA_1 of differential width and any length, to infinitely long strip dA_2 of differential width and with parallel generating line to dA_1. $$dF_{d1-d2} = \frac{\cos \varphi}{2} d\varphi = \tfrac{1}{2}d(\sin \varphi)$$
2		Area dA_1 of differential width and any length to any cylindrical surface A_2 generated by a line of infinite length moving parallel to itself and parallel to the plane of dA_1. $$F_{d1-2} = \tfrac{1}{2}(\sin \varphi_2 - \sin \varphi_1)$$

3

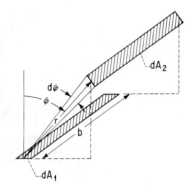

Strip of finite length b and of differential width, to differential strip of same length on parallel generating line.

$$dF_{d1-d2} = \frac{\cos\varphi}{\pi}\, d\varphi \, \tan^{-1}\frac{b}{r}$$

4

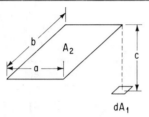

Plane element dA_1 to plane parallel rectangle; normal to element passes through corner of rectangle.

$$X = \frac{a}{c} \qquad Y = \frac{b}{c}$$

$$F_{d1-2} = \frac{1}{2\pi}\left(\frac{X}{\sqrt{1+X^2}}\tan^{-1}\frac{Y}{\sqrt{1+X^2}} + \frac{Y}{\sqrt{1+Y^2}}\tan^{-1}\frac{X}{\sqrt{1+Y^2}}\right)$$

5

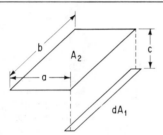

Strip element to rectangle in plane parallel to strip; strip is opposite one edge of rectangle.

$$X = \frac{a}{c} \qquad Y = \frac{b}{c}$$

$$F_{d1-2} = \frac{1}{\pi Y}\left(\sqrt{1+Y^2}\tan^{-1}\frac{X}{\sqrt{1+Y^2}} - \tan^{-1}X + \frac{XY}{\sqrt{1+X^2}}\tan^{-1}\frac{Y}{\sqrt{1+X^2}}\right)$$

6

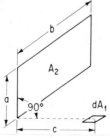

Plane element dA_1 to rectangle in plane $90°$ to plane of element.

$$X = \frac{a}{b} \qquad Y = \frac{c}{b}$$

$$F_{d1-2} = \frac{1}{2\pi}\left(\tan^{-1}\frac{1}{Y} - \frac{Y}{\sqrt{X^2+Y^2}}\tan^{-1}\frac{1}{\sqrt{X^2+Y^2}}\right)$$

7

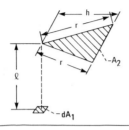

Plane element dA_1 to parallel isosceles triangle of height h and sides r; normal to element passes through vertex of triangle.

$$H = \frac{h}{l} \qquad R = \frac{r}{l}$$

$$F_{d1-2} = \frac{H}{\pi\sqrt{1 + H^2}} \tan^{-1} \sqrt{\frac{R^2 - H^2}{1 + H^2}}$$

8

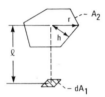

Plane element dA_1 to parallel polygon having n equal sides; normal to element passes through center of polygon.

$$H = \frac{h}{l} \qquad R = \frac{r}{l}$$

$$F_{d1-2} = \frac{nH}{\pi\sqrt{1 + H^2}} \tan^{-1} \sqrt{\frac{R^2 - H^2}{1 + H^2}}$$

9

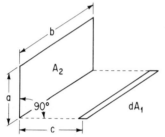

Strip element dA_1 to rectangle in plane 90° to plane of strip.

$$X = \frac{a}{b} \qquad Y = \frac{c}{b}$$

$$F_{d1-2} = \frac{1}{\pi} \left[\tan^{-1} \frac{1}{Y} + \frac{Y}{2} \ln \frac{Y^2(X^2 + Y^2 + 1)}{(Y^2 + 1)(X^2 + Y^2)} - \frac{Y}{\sqrt{X^2 + Y^2}} \tan^{-1} \frac{1}{\sqrt{X^2 + Y^2}} \right]$$

10

Two infinitely long, directly opposed parallel plates of the same finite width.

$$H = \frac{h}{w}$$

$$F_{1-2} = F_{2-1} = \sqrt{1 + H^2} - H$$

11

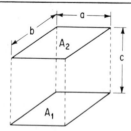

Identical, parallel, directly opposed rectangles.

$$X = \frac{a}{c} \qquad Y = \frac{b}{c}$$

$$F_{1-2} = \frac{2}{\pi XY} \left\{ \ln \left[\frac{(1 + X^2)(1 + Y^2)}{1 + X^2 + Y^2} \right]^{\frac{1}{2}} + X\sqrt{1 + Y^2} \tan^{-1} \frac{X}{\sqrt{1 + Y^2}} \right.$$
$$\left. + Y\sqrt{1 + X^2} \tan^{-1} \frac{Y}{\sqrt{1 + X^2}} - X \tan^{-1} X - Y \tan^{-1} Y \right\}$$

12

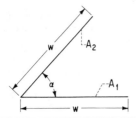

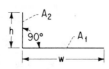

Two infinitely long plates of equal finite width w, having one common edge, and at an included angle α to each other

$$F_{1-2} = F_{2-1} = 1 - \sin\frac{\alpha}{2}$$

13

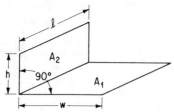

Two infinitely long plates of unequal widths h and w, having one common edge, and at an angle of $90°$ to each other.

$$H = \frac{h}{w}$$

$$F_{1-2} = \tfrac{1}{2}(1 + H - \sqrt{1 + H^2})$$

14

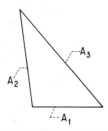

Two finite rectangles of same length, having one common edge, and at an angle of $90°$ to each other.

$$H = \frac{h}{l} \qquad W = \frac{w}{l}$$

$$F_{1-2} = \frac{1}{\pi W}\left(W \tan^{-1}\frac{1}{W} + H \tan^{-1}\frac{1}{H} - \sqrt{H^2 + W^2}\,\tan^{-1}\frac{1}{\sqrt{H^2 + W^2}}\right.$$

$$\left. + \tfrac{1}{4}\ln\left\{\frac{(1 + W^2)(1 + H^2)}{1 + W^2 + H^2}\left[\frac{W^2(1 + W^2 + H^2)}{(1 + W^2)(W^2 + H^2)}\right]^{W^2}\left[\frac{H^2(1 + H^2 + W^2)}{(1 + H^2)(H^2 + W^2)}\right]^{H^2}\right\}\right)$$

15

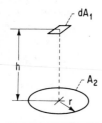

Infinitely long enclosure formed by three plane areas.

$$F_{1-2} = \frac{A_1 + A_2 - A_3}{2A_1}$$

16

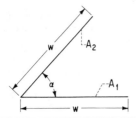

Plane element dA_1 to circular disk in plane parallel to element; normal to element passes through center of disk.

$$F_{d1-2} = \frac{r^2}{h^2 + r^2}$$

17

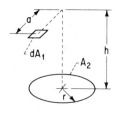

Plane element dA_1 to circular disk in plane parallel to element.

$$H = \frac{h}{a} \qquad R = \frac{r}{a}$$

$$Z = 1 + H^2 + R^2$$

$$F_{d1-2} = \frac{1}{2}\left(1 - \frac{1 + H^2 - R^2}{\sqrt{Z^2 - 4R^2}}\right)$$

18

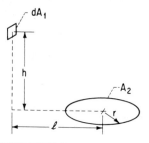

Plane element dA_1 to circular disk; planes containing element and disk intersect at $90°$; $l \geq r$.

$$H = \frac{h}{l} \qquad R = \frac{r}{l}$$

$$Z = 1 + H^2 + R^2$$

$$F_{d1-2} = \frac{H}{2}\left(\frac{Z}{\sqrt{Z^2 - 4R^2}} - 1\right)$$

19

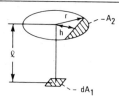

Plane element dA_1 to parallel circular segment; normal to element passes through center of disk containing segment.

$$H = \frac{h}{l} \qquad R = \frac{r}{l}$$

$$F_{d1-2} = \frac{1}{\pi}\left(\frac{R^2}{1 + R^2}\cos^{-1}\frac{H}{R} - \frac{H}{\sqrt{1 + H^2}}\tan^{-1}\sqrt{\frac{R^2 - H^2}{1 + H^2}}\right)$$

20

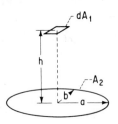

Plane element dA_1 to elliptical plate in plane parallel to element; normal to element passes through center of plate.

$$F_{d1-2} = \frac{ab}{\sqrt{(h^2 + a^2)(h^2 + b^2)}}$$

21

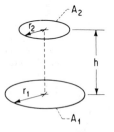

Parallel circular disks with centers along the same normal.

$$R_1 = \frac{r_1}{h} \qquad R_2 = \frac{r_2}{h}$$

$$X = 1 + \frac{1 + R_2^2}{R_1^2}$$

$$F_{1-2} = \frac{1}{2}\left[X - \sqrt{X^2 - 4\left(\frac{R_2}{R_1}\right)^2}\right]$$

22

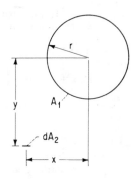

Strip element dA_2 of any length to infinitely long cylinder; $y \geq r$.

$$X = \frac{x}{r} \qquad Y = \frac{y}{r}$$

$$F_{d2-1} = \frac{Y}{X^2 + Y^2} \qquad (Y \geq 1)$$

23

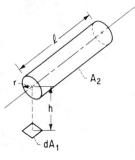

Element of any length on cylinder to plane of infinite length and width.

$$F_{d1-2} = \tfrac{1}{2}(1 + \cos \varphi)$$

24

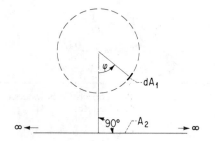

Plane element dA_1 to right circular cylinder of finite length l and radius r; normal to element passes through one end of cylinder and is perpendicular to cylinder axis.

$$L = \frac{l}{r} \qquad H = \frac{h}{r}$$

$$X = (1 + H)^2 + L^2$$

$$Y = (1 - H)^2 + L^2$$

$$F_{d1-2} = \frac{1}{\pi H} \tan^{-1} \frac{L}{\sqrt{H^2 - 1}} + \frac{L}{\pi}\left[\frac{X - 2H}{H\sqrt{XY}} \tan^{-1} \sqrt{\frac{X(H-1)}{Y(H+1)}} - \frac{1}{H} \tan^{-1} \sqrt{\frac{H-1}{H+1}} \right]$$

25

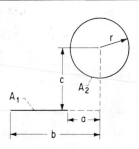

Infinitely long plane of finite width to parallel infinitely long cylinder.

$$F_{1-2} = \frac{r}{b - a}\left(\tan^{-1}\frac{b}{c} - \tan^{-1}\frac{a}{c} \right)$$

26

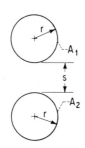

Infinitely long parallel cylinders of the same diameter.

$$X = 1 + \frac{s}{2r}$$

$$F_{1-2} = F_{2-1} = \frac{1}{\pi}\left(\sqrt{X^2 - 1}\right.$$
$$\left. + \sin^{-1}\frac{1}{X} - X\right)$$

27

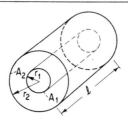

Concentric cylinders of infinite length.

$$F_{1-2} = 1$$

$$F_{2-1} = \frac{r_1}{r_2}$$

$$F_{2-2} = 1 - \frac{r_1}{r_2}$$

28

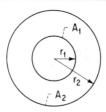

Two concentric cylinders of same finite length.

$$R = \frac{r_2}{r_1} \qquad L = \frac{l}{r_1}$$
$$A = L^2 + R^2 - 1$$
$$B = L^2 - R^2 + 1$$

$$F_{2-1} = \frac{1}{R} - \frac{1}{\pi R}\left\{\cos^{-1}\frac{B}{A} - \frac{1}{2L}\left[\sqrt{(A + 2)^2 - (2R)^2}\cos^{-1}\frac{B}{RA} + B\sin^{-1}\frac{1}{R} - \frac{\pi A}{2}\right]\right\}$$

$$F_{2-2} = 1 - \frac{1}{R} + \frac{2}{\pi R}\tan^{-1}\frac{2\sqrt{R^2 - 1}}{L} - \frac{L}{2\pi R}\left[\frac{\sqrt{4R^2 + L^2}}{L}\sin^{-1}\frac{4(R^2 - 1) + (L^2/R^2)(R^2 - 2)}{L^2 + 4(R^2 - 1)}\right.$$

$$\left. - \sin^{-1}\frac{R^2 - 2}{R^2} + \frac{\pi}{2}\left(\frac{\sqrt{4R^2 + L^2}}{L} - 1\right)\right]$$

where for any argument ξ:

$$-\frac{\pi}{2} \leqslant \sin^{-1}\xi \leqslant \frac{\pi}{2}$$
$$0 \leqslant \cos^{-1}\xi \leqslant \pi$$

29

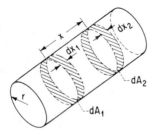

Two ring elements on the interior of a right circular cylinder.

$$X = \frac{x}{2r}$$

$$dF_{d1-d2} = \left[1 - \frac{2X^3 + 3X}{2(X^2 + 1)^{\frac{3}{2}}}\right]dX_2$$

30

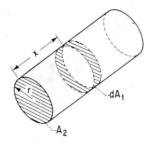

Ring element dA_1 on interior of right circular cylinder to circular disk A_2 at end of cylinder.

$$X = \frac{x}{2r}$$

$$F_{d1-2} = \frac{X^2 + \frac{1}{2}}{\sqrt{X^2 + 1}} - X$$

31

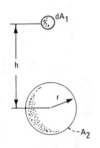

Spherical point source to a sphere of radius r.

$$R = \frac{r}{h}$$

$$F_{d1-2} = \frac{1}{2}(1 - \sqrt{1 - R^2})$$

32

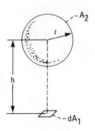

Plane element dA_1 to sphere of radius r; normal to center of element passes through center of sphere.

$$F_{d1-2} = \left(\frac{r}{h}\right)^2$$

33

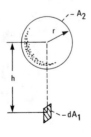

Plane element dA_1 to sphere of radius r; tangent to element passes through center of sphere.

$$H = \frac{h}{r}$$

$$F_{d1-2} = \frac{1}{\pi}\left(\tan^{-1}\frac{1}{\sqrt{H^2 - 1}} - \frac{\sqrt{H^2 - 1}}{H^2}\right)$$

34

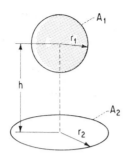

Sphere of radius r_1 to disk of radius r_2; normal to center of disk passes through center of sphere.

$$R_2 = \frac{r_2}{h}$$

$$F_{1-2} = \frac{1}{2}\left(1 - \frac{1}{\sqrt{1 + R_2{}^2}}\right)$$

35

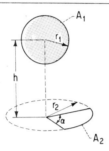

Sphere to sector of disk; normal to center of disk passes through center of sphere.

$$R_2 = \frac{r_2}{h}$$

$$F_{1-2} = \frac{\alpha}{4\pi}\left(1 - \frac{1}{\sqrt{1 + R_2{}^2}}\right)$$

36

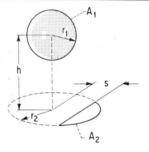

Sphere to segment of disk.

$$R_2 = \frac{r_2}{h} \qquad S = \frac{s}{h}$$

$$F_{1-2} = \frac{1}{8} - \frac{\cos^{-1}(S/R_2)}{2\pi\sqrt{1 + R_2{}^2}} + \frac{1}{4\pi}\sin^{-1}\frac{(1 - S^2)R_2{}^2 - 2S^2}{(1 + S^2)R_2{}^2}$$

37

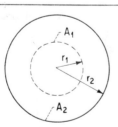

Concentric spheres.

$$F_{1-2} = 1$$

$$F_{2-1} = \left(\frac{r_1}{r_2}\right)^2$$

$$F_{2-2} = 1 - \left(\frac{r_1}{r_2}\right)^2$$

38

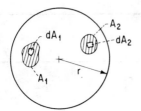

Differential or finite areas on the inside of a spherical cavity.

$$dF_{a1-a2} = dF_{1-a2} = \frac{dA_2}{4\pi r^2}$$

$$F_{d1-2} = F_{1-2} = \frac{A_2}{4\pi r^2}$$

D

RADIATIVE PROPERTIES

Tables of total emissivities, and total absorptivities for incident solar radiation, are provided here for convenience in working problems and to give the reader an indication of the magnitudes to be expected. As discussed in Chap. 5, many factors such as roughness and oxidation can strongly affect the radiative properties. No attempt is made here to describe in detail the condition of the material sample: hence the values given here are only reasonable approximations in some instances. For detailed information on radiative properties including sample descriptions and results from many sources, the reader is referred to the collections in [1-3]. Reference [3] in three volumes is very extensive. Reference [4] provides a limited amount of additional information. As will be seen from these references, for the same material there can sometimes be considerable differences in the property values measured by different investigators.

Normal-total emissivity

Metals	Surface temperature,[a] °F (K)	ϵ_n'
Aluminum:		
highly polished plate	400–1100 (480–870)	0.038–0.06
bright foil	70 (295)	0.04
polished plate	212 (373)	0.095
heavily oxidized	200–1000 (370–810)	0.20–0.33
Antimony, polished	100–500 (310–530)	0.28–0.31
Bismuth, bright	176 (350)	0.34
Brass:		
highly polished	500–700 (530–640)	0.028–0.031
polished	200 (370)	0.09
dull	120–660 (320–620)	0.22
oxidized	400–1000 (480–810)	0.60
Chromium, polished	100–2000 (310–1370)	0.08–0.40
Copper:		
highly polished	100 (310)	0.02
polished	100–500 (310–530)	0.04–0.05
scraped, shiny	100 (310)	0.07
slightly polished	100 (310)	0.15
black oxidized	100 (310)	0.78
Gold:		
highly polished	200–1100 (370–870)	0.018–0.035
polished	266 (400)	0.018
Iron:		
highly polished, electrolytic	100–500 (310–530)	0.05–0.07
polished	800–900 (700–760)	0.14–0.38
freshly rubbed with emery	100 (310)	0.24
wrought iron, polished	100–500 (310–530)	0.28
cast iron, freshly turned	100 (310)	0.44
iron plate, pickled, then rusted red	68 (293)	0.61
cast iron, oxidized at 1100°F	400–1100 (480–870)	0.64–0.78
cast iron, rough, strongly oxidized	100–500 (310–530)	0.95
Lead:		
polished	100–500 (310–530)	0.06–0.08
rough unoxidized	100 (310)	0.43
oxidized at 1100°F	100 (310)	0.63
Magnesium, polished	100–500 (310–530)	0.07–0.13
Mercury, unoxidized	40–200 (280–370)	0.09–0.12

[a] When temperatures and emissivities both have ranges, linear interpolation can be used over these values.

Normal-total emissivity *(Continued)*

Metals	Surface temperature,[a] °F (K)	ϵ'_n
Molybdenum:		
polished	100–500 (310–530)	0.05–0.08
polished	1000–2500 (810–1640)	0.10–0.18
polished	5000 (3030)	0.29
Monel:		
polished	100 (310)	0.17
oxidized at 1100° F	1000 (810)	0.45
Nickel:		
electrolytic	100–500 (310–530)	0.04–0.06
technically pure, polished	440–710 (500–650)	0.07–0.087
electroplated on iron, not polished	68 (293)	0.11
plate oxidized at 1100° F	390–1110 (470–870)	0.37–0.48
nickel oxide	1200–2300 (920–1530)	0.59–0.86
Platinum:		
electrolytic	500–1000 (530–810)	0.06–0.10
polished plate	440–1160 (500–900)	0.054–0.104
Silver, polished	100–1000 (310–810)	0.01–0.03
Stainless Steel:		
Inconel X, polished	−300–900 (90–760)	0.19–0.20
Inconel B, polished	−300–900 (90–760)	0.19–0.22
Type 301, polished	75 (297)	0.16
Type 310, smooth	1500 (1090)	0.39
Type 316, polished	400–1900 (480–1310)	0.24–0.31
Steel:		
polished sheet	−300–0 (90–273)	0.07–0.08
polished sheet	0–300 (273–420)	0.08–0.14
mild steel, polished	500–1200 (530–920)	0.27–0.31
sheet with skin due to rolling	70 (295)	0.66
sheet with rough oxide layer	70 (295)	0.81
Tantalum	2500–5000 (1640–3030)	0.2–0.3
Tin:		
polished sheet	93 (310)	0.05
bright tinned iron	76 (298)	0.043–0.064
Tungsten:		
clean	100–1000 (310–810)	0.03–0.08
filament	80 (300)	0.032
filament	6000 (3590)	0.39
Zinc:		
polished	100–1000 (310–810)	0.02–0.05
galvanized sheet, fairly bright	100 (310)	0.23
gray oxidized	70 (295)	0.23–0.28

[a]When temperatures and emissivities both have ranges, linear interpolation can be used over these values.

Normal-total emissivity *(Continued)*

Dielectrics	Surface temperature,[a] °F (K)	ϵ_n'
Alumina on Inconel	1000–2000 (810–1370)	0.65–0.45
Asbestos:		
paper	100 (310)	0.93
board	100 (310)	0.96
Brick:		
white refractory	2000 (1370)	0.29
fireclay	1800 (1260)	0.75
rough red	100 (310)	0.93
Carbon, lampsoot	100 (310)	0.95
Concrete, rough	100 (310)	0.94
Corundum, emery rough	200 (370)	0.86
Ice:		
smooth	32 (273)	0.966
rough crystals	32 (273)	0.985
Magnesium oxide, refractory	300–900 (420–760)	0.69–0.55
Marble, white	100 (310)	0.95
Mica	100 (310)	0.75
Paint:		
oil, all colors	212 (373)	0.92–0.96
red lead	200 (370)	0.93
lacquer, flat black	100–200 (310–370)	0.96–0.98
Paper:		
roofing	100 (310)	0.91
white	100 (310)	0.95
Plaster	100 (310)	0.91
Porcelain, glazed	70 (295)	0.92
Rokide A on molybdenum	600–1500 (590–1090)	0.79–0.60
Rubber, hard	68 (293)	0.92
Sandstone	100–500 (310–530)	0.83–0.90
Silicon carbide	300–1200 (420–920)	0.83–0.96
Slate	100 (310)	0.67–0.80
Snow	20 (270)	0.82
Soot, candle	200–500 (370–530)	0.95
Water, deep	32–212 (273–373)	0.96
Wood:		
sawdust	100 (310)	0.75
oak, planed	70 (295)	0.90
beech	158 (340)	0.94

[a]When temperatures and emissivities both have ranges, linear interpolation can be used over these values.

Normal-total absorptivity
for incident solar radiation

(Receiving material at 70°F, 295 K)

Metals	α_n'
Aluminum:	
highly polished	0.10
polished	0.20
Chromium, electroplated	0.40
Copper:	
highly polished	0.18
clean	0.25
tarnished	0.64
oxidized	0.70
Galvanized iron	0.38
Gold, bright foil	0.29
Iron:	
ground with fine grit	0.36
blued	0.55
sandblasted	0.75
Magnesium, polished	0.19
Nickel:	
highly polished	0.15
polished	0.36
electrolytic	0.40
Platinum, bright	0.31
Silver:	
highly polished	0.07
polished	0.13
commercial sheet	0.30
Stainless steel #301, polished	0.37
Tungsten, highly polished	0.37

Dielectrics	α_n'
Aluminum Oxide (Al_2O_3)	0.06–0.23
Asphalt pavement, dust-free	0.93
Brick, red	0.75
Clay	0.39
Concrete roofing tile:	
uncolored	0.73
brown	0.91
black	0.91
Earth, plowed field	0.75

Normal-total absorptivity
for incident solar radiation *(Continued)*
(Receiving material at 70° F, 295 K)

Dielectrics	α'_n
Felt, black	0.82
Graphite	0.88
Grass	0.75–0.80
Gravel	0.29
Leaves, green	0.71–0.79
Magnesium oxide (MgO)	0.15
Marble, white	0.46
Paint:	
aluminum	0.55
oil, zinc white	0.30
oil, light green	0.50
oil, light gray	0.75
oil, black on galvanized iron	0.90
Paper, white	0.28
Slate, blue gray	0.88
Snow, clean	0.2–0.35
Soot, coal	0.95
Titanium dioxide (TiO$_2$)	0.12
Zinc oxide	0.15
Zinc sulfide (ZnS)	0.21

REFERENCES

1. Gubareff, G. G., J. E. Janssen, and R. H. Torborg: "Thermal Radiation Properties Survey," 2d ed., Honeywell Research Center, Minneapolis-Honeywell Regulator Co., Minneapolis, 1960.
2. Wood, W. D., H. W. Deem, and C. F. Lucks: "Thermal Radiative Properties," Plenum Press, Plenum Publishing Corporation, New York, 1964.
3. Touloukian, Y. S., et al.: "Thermal Radiative Properties," vol. 7, "Metallic Elements and Alloys," vol. 8, "Nonmetallic Solids," vol. 9, "Coatings," Thermophysical Properties Research Center of Purdue University, Data Series, Plenum Publishing Corp., 1970.
4. Svet, Darii Yakovlevich: "Thermal Radiation, Metals, Semiconductors, Ceramics, Partly Transparent Bodies, and Films," Consultants Bureau, Plenum Publishing Corporation, New York, 1965.

E

ENCLOSURE-ANALYSIS METHOD OF GEBHART

In Chap. 8 the radiative exchange within a diffuse-gray enclosure was analyzed by the method originated by Poljak. A somewhat different viewpoint set forth by Gebhart will be briefly presented here. Additional discussion can be found in [1–3]. The special utility in this formulation is that it yields coefficients that provide the fraction of energy emitted by a surface that is *absorbed* at another surface after reaching the absorbing surface by all possible paths. These coefficients can be of value in formulating some types of problems. After the derivation, a correspondence between the Gebhart and Poljak formulations will be indicated.

As in Chap. 8, an enclosure having N diffuse-gray surfaces is considered, and the same restrictions are imposed here as in Sec. 8-1.2. For a typical surface A_k the net energy loss is the emission from the surface minus the energy that is absorbed by the surface from all incident sources. The emitted energy is $A_k \epsilon_k \sigma T_k^4$. Let G_{jk} be the fraction of the emission from surface A_j that reaches A_k and is *absorbed*. This includes all the paths for reaching A_k; that is, the direct path, paths by means of one reflection, and paths by means of multiple reflections. Thus $A_j \epsilon_j \sigma T_j^4 G_{jk}$ is the amount of energy emitted by A_j that is absorbed by A_k. A heat balance on A_k then gives

$$
\begin{aligned}
Q_k &= A_k \epsilon_k \sigma T_k^4 - (A_1 \epsilon_1 \sigma T_1^4 G_{1k} + A_2 \epsilon_2 \sigma T_2^4 G_{2k} + \cdots + A_j \epsilon_j \sigma T_j^4 G_{jk} \\
&\qquad + \cdots + A_k \epsilon_k \sigma T_k^4 G_{kk} + \cdots + A_N \epsilon_N \sigma T_N^4 G_{Nk}) \\
&= A_k \epsilon_k \sigma T_k^4 - \sum_{j=1}^{N} A_j \epsilon_j \sigma T_j^4 G_{jk}
\end{aligned}
\tag{E-1}
$$

The G_{kk} would generally not be zero since even for a plane or convex surface some of the emission from a surface will be returned to itself by reflection from other surfaces. Equation (E-1) can be written for each surface; this will relate each of the Q's to the surface temperatures in the enclosure. The G factors must now be found.

The quantity G_{jk} is the fraction of energy emitted by A_j that reaches A_k and is absorbed. The total emitted energy from A_j is $A_j\epsilon_j\sigma T_j^4$. The portion traveling by a direct path to A_k and then absorbed is $A_j\epsilon_j\sigma T_j^4 F_{j-k}\epsilon_k$, where for a gray surface ϵ is equal to the absorptivity. All other radiation from A_j arriving at A_k will first undergo one reflection. The emission from A_j that arrives at a typical surface A_n and is then reflected is $A_j\epsilon_j\sigma T_j^4 F_{j-n}\rho_n$. The fraction G_{nk} then reaches A_k and is absorbed. Then all the energy absorbed at A_k originating by emission from A_j is

$$A_j\epsilon_j\sigma T_j^4 F_{j-k}\epsilon_k + (A_j\epsilon_j\sigma T_j^4 F_{j-1}\rho_1 G_{1k} + A_j\epsilon_j\sigma T_j^4 F_{j-2}\rho_2 G_{2k} + \cdots$$
$$+ A_j\epsilon_j\sigma T_j^4 F_{j-k}\rho_k G_{kk} + \cdots + A_j\epsilon_j\sigma T_j^4 F_{j-N}\rho_N G_{Nk})$$

Dividing this energy by the total emission from A_j gives the fraction

$$G_{jk} = F_{j-k}\epsilon_k + F_{j-1}\rho_1 G_{1k} + F_{j-2}\rho_2 G_{2k} + \cdots + F_{j-k}\rho_k G_{kk} + \cdots + F_{j-N}\rho_N G_{Nk}$$

By letting j take on all values from 1 to N, the following set of equations is obtained:

$$G_{1k} = F_{1-k}\epsilon_k + F_{1-1}\rho_1 G_{1k} + F_{1-2}\rho_2 G_{2k} + \cdots$$
$$+ F_{1-k}\rho_k G_{kk} + \cdots + F_{1-N}\rho_N G_{Nk}$$

$$G_{2k} = F_{2-k}\epsilon_k + F_{2-1}\rho_1 G_{1k} + F_{2-2}\rho_2 G_{2k} + \cdots$$
$$+ F_{2-k}\rho_k G_{kk} + \cdots + F_{2-N}\rho_N G_{Nk} \qquad \text{(E-2)}$$

$$\cdots\cdots\cdots\cdots\cdots\cdots\cdots$$

$$G_{Nk} = F_{N-k}\epsilon_k + F_{N-1}\rho_1 G_{1k} + F_{N-2}\rho_2 G_{2k} + \cdots$$
$$+ F_{N-k}\rho_k G_{kk} + \cdots + F_{N-N}\rho_N G_{Nk}$$

Equations (E-2) can be solved simultaneously for $G_{1k}, G_{2k}, \ldots, G_{Nk}$. Equation (E-1) then relates Q_k to the surface temperatures. The k index in Eqs. (E-1) and (E-2) can correspond to any of the surfaces in the enclosure.

At the end of Sec. 8-3.2, it was mentioned that matrix inversion can be applied to (8-19) to yield each Q as a weighted sum of T^4's. The coefficients obtained by the matrix inversion thus correspond to the factors multiplying the T^4's in (E-1). This shows the correspondence between the method described here and that in Chap. 8.

REFERENCES

1. Gebhart, B.: Unified Treatment for Thermal Radiation Transfer Processes—Gray, Diffuse Radiators and Absorbers, paper no. 57-A-34, *ASME,* December, 1957.
2. Gebhart, B.: "Heat Transfer," 2d ed., pp. 150–163, McGraw-Hill Book Company, New York, 1971.
3. Gebhart, B.: Surface Temperature Calculations in Radiant Surroundings of Arbitrary Complexity—for Gray, Diffuse Radiation, *Int. J. Heat Mass Transfer,* vol. 3, no. 4, pp. 341–346, 1961.

EXPONENTIAL INTEGRAL RELATIONS

A summary of some useful exponential integral relations is presented here. Additional relations are given in [1–3].

For positive real arguments, the nth exponential integral is defined as

$$E_n(x) \equiv \int_0^1 \mu^{n-2} \exp\left(\frac{-x}{\mu}\right) d\mu \qquad \text{(F-1)}$$

and only positive integral values of n will be considered here. An alternative form is

$$E_n(x) = \int_1^\infty \frac{1}{t^n} \exp(-xt)\, dt \qquad \text{(F-2)}$$

By differentiating (F-1) under the integral sign, the recurrence relation is obtained:

$$\frac{d}{dx} E_n(x) = -E_{n-1}(x) \qquad n \geq 2$$

$$\frac{d}{dx} E_1(x) = -\frac{1}{x} \exp(-x)$$

(F-3)

Another recurrence relation obtained by integration is

$$n E_{n+1}(x) = \exp(-x) - x E_n(x) = \exp(-x) + x \frac{d}{dx} E_{n+1}(x) \qquad n \geq 1 \quad \text{(F-4)}$$

Also, integration results in

$$\int E_n(x)\, dx = -E_{n+1}(x) \qquad \text{(F-5)}$$

By use of Eq. (F-4), all exponential integrals can be reduced to the first exponential integral given by

$$E_1(x) = \int_0^1 \mu^{-1} \exp\left(\frac{-x}{\mu}\right) d\mu \qquad \text{(F-6)}$$

Alternative forms of $E_1(x)$ are

$$E_1(x) = \int_1^\infty t^{-1} \exp\left(-xt\right) dt = \int_x^\infty t^{-1} \exp\left(-t\right) dt \qquad \text{(F-7)}$$

For $x = 0$ the exponential integrals are equal to

$$E_n(0) = \frac{1}{n-1} \qquad n \geq 2$$
$$\qquad \text{(F-8)}$$
$$E_1(0) = +\infty$$

For large values of x there is the asymptotic expansion

$$E_n(x) = \frac{\exp\left(-x\right)}{x}\left[1 - \frac{n}{x} + \frac{n(n+1)}{x^2} - \frac{n(n+1)(n+2)}{x^3} + \cdots\right] \qquad \text{(F-9)}$$

Therefore, as $x \to \infty$, $E_n(x) \to \exp\left(-x\right)/x \to 0$.

Series expansions are of the form

$$E_1(x) = -\gamma - \ln x + x - \frac{x^2}{2 \times 2!} + \frac{x^3}{3 \times 3!} - \cdots$$

$$= -\gamma - \ln x - \sum_{n=1}^\infty (-1)^n \frac{x^n}{n \times n!}$$
$$\qquad \text{(F-10)}$$

$$E_2(x) = 1 + (\gamma - 1 + \ln x)x - \frac{x^2}{1 \times 2!} + \frac{x^3}{2 \times 3!} - \cdots$$

$$E_3(x) = \tfrac{1}{2} - x + \tfrac{1}{2}(-\gamma + \tfrac{3}{2} - \ln x)x^2 + \frac{x^3}{1 \times 3!} - \cdots$$

where $\gamma = 0.577216$ is Euler's constant. The general series expansion given in [3] is

$$E_n(x) = \frac{(-x)^{n-1}}{(n-1)!}\left[-\ln x + \psi(n)\right] - \sum_{\substack{m=0 \\ (m \neq n-1)}}^\infty \frac{(-x)^m}{(m-n+1)m!} \qquad \text{(F-11)}$$

where $\psi(1) = -\gamma$ and $\psi(n) = -\gamma + \sum_{m=1}^{n-1} \frac{1}{m} \qquad n \geq 2$

Some rough approximations are as follows; from Section 17-5.4,

$$E_3(x) \approx \tfrac{1}{2} \exp(-1.8x)$$

Using (F-3) gives

$$E_2(x) \approx 0.9 \exp(-1.8x)$$

In [4], the approximations are used,

$$E_2(x) \approx \tfrac{3}{4} \exp(-\tfrac{3}{2}x)$$

$$E_3(x) \approx \tfrac{1}{2} \exp(-\tfrac{3}{2}x)$$

Tabulations of $E_n(x)$ are given in [2, 3]. An abridged listing is given in Table F-1 for convenience. Reference [5] presents forms of a generalized exponential integral function that are convenient for numerical computation. The generalized forms include as a special case the $E_n(x)$ discussed here.

Table F-1 Values of exponential integrals [2]

x	$E_1(x)$	$E_2(x)$	$E_3(x)$	$E_4(x)$	$E_5(x)$
0	∞	1.0000	0.5000	0.3333	0.2500
0.01	4.0379	0.9497	0.4903	0.3284	0.2467
0.02	3.3547	0.9131	0.4810	0.3235	0.2434
0.03	2.9591	0.8817	0.4720	0.3188	0.2402
0.04	2.6813	0.8535	0.4633	0.3141	0.2371
0.05	2.4679	0.8278	0.4549	0.3095	0.2339
0.06	2.2953	0.8040	0.4468	0.3050	0.2309
0.07	2.1508	0.7818	0.4388	0.3006	0.2278
0.08	2.0269	0.7610	0.4311	0.2962	0.2249
0.09	1.9187	0.7412	0.4236	0.2919	0.2219
0.10	1.8229	0.7225	0.4163	0.2877	0.2190
0.20	1.2227	0.5742	0.3519	0.2494	0.1922
0.30	0.9057	0.4691	0.3000	0.2169	0.1689
0.40	0.7024	0.3894	0.2573	0.1891	0.1487
0.50	0.5598	0.3266	0.2216	0.1652	0.1310
0.60	0.4544	0.2762	0.1916	0.1446	0.1155
0.70	0.3738	0.2349	0.1661	0.1268	0.1020
0.80	0.3106	0.2009	0.1443	0.1113	0.0901
0.90	0.2602	0.1724	0.1257	0.0978	0.0796
1.00	0.2194	0.1485	0.1097	0.0861	0.0705
1.25	0.1464	0.1035	0.0786	0.0628	0.0520
1.50	0.1000	0.0731	0.0567	0.0460	0.0385
1.75	0.0695	0.0522	0.0412	0.0339	0.0286
2.00	0.0489	0.0375	0.0301	0.0250	0.0213
2.25	0.0348	0.0272	0.0221	0.0185	0.0159
2.50	0.0249	0.0198	0.0163	0.0138	0.0119
2.75	0.0180	0.0145	0.0120	0.0103	0.0089
3.00	0.0130	0.0106	0.0089	0.0077	0.0067
3.25	0.0095	0.0078	0.0066	0.0057	0.0050
3.50	0.0070	0.0058	0.0049	0.0043	0.0038

REFERENCES

1. Chandrasekhar, Subrahmanyan: "Radiative Transfer," Dover Publications, Inc., New York, 1960.
2. Kourganoff, Vladimir: "Basic Methods in Transfer Problems," Dover Publications, Inc., New York, 1963.
3. Abramowitz, Milton, and Irene A. Stegun (eds.): "Handbook of Mathematical Functions with Formulas, Graphs, and Mathematical Tables," *Appl. Math. Ser.* 55, National Bureau of Standards, 1964.
4. Cess, R. D., and S. N., Tiwari: "Infrared Radiative Energy Transfer in Gases," Advances in Heat Transfer, vol. 8, Academic Press, New York, 1972.
5. Breig, W. F., and A. L., Crosbie: Numerical Computation of a Generalized Exponential Integral Function, *Math. Comp.,* vol. 28, no. 126, pp. 575–579, 1974.

INDEX TO INFORMATION IN TABLES AND FIGURES

Subject	Table or figure number
Electromagnetic spectrum	Fig. 1-2
Enclosure-theory equations summary	Table 8-1
Energy and temperature relations for gray medium between gray surfaces:	
differential approximation	Table 15-3
diffusion theory	Table 15-2
exchange-factor relations	Table 17-5
Energy transfer between:	
concentric cylinders	Table 9-1
concentric spheres	Table 9-1
parallel plates	Table 9-1
Energy transfer for gray medium between black parallel plates	Table 14-1, Fig. 14-8
Equation of transfer approximations	Table 15-1
Exponential integral values	Table F-1
Fundamental numerical values	Table A-1
Flame temperatures	Table 17-6
Gas emittance:	
carbon dioxide	Figs. 13-13, 17-11, 17-16
water vapor	Figs. 17-13, 17-17, 17-18
Kirchhoff's-law relations	Table 3-2
Mean-beam-length relations	Table 17-1
Monte Carlo random-number relations:	
surface emission	Table 11-1
radiation in a medium	Table 20-1
Property prediction equations by electromagnetic theory, summary	Table 4-4
Radiation constants	Table A-4
Reflectivity reciprocity relations	Table 3-3
Scattering behavior:	
cross sections	Table 13-1
polarizability	Table 16-5
Soot:	
optical constants	Table 17-7
emittance	Fig. 17-28
Surface properties:	
definitions	Table 3-1
emissivity values	Appendix D
solar-absorptivity values	Appendix D
Water properties:	
absorption coefficient	Table 5-2
solar transmission	Table 5-3

H

GLOSSARY

Absolute temperature—Temperature measured on one of the thermodynamic scales, either Kelvin or Rankine.

Absorbing media—Media that absorb electromagnetic radiation and convert it into internal energy.

Absorptance—The property of a medium that determines the fraction of radiant energy traveling along a path that will be absorbed within a given distance.

Absorption coefficient—The property of a medium that describes the amount of absorption of thermal radiation per unit path length within the medium.

Absorptivity—The property of a material that gives the fraction of energy incident on the material that is absorbed.

Adiabatic—A boundary or material possessing the quality of a perfect insulator, so that no heat is transferred through it.

Albedo for scattering—The ratio of the scattering coefficient to the extinction coefficient.

Angle for total reflection—The angle from the normal of the interface between two materials of differing refractive index at which total reflection of incident intensity occurs.

Band—A spectral interval containing many closely spaced spectral lines, so that radiation within that spectral interval is significantly absorbed.

Bandwidth—The spectral interval containing the major absorption capability of an absorbing band.

Black—Having the property of complete absorption of incident radiation of all wavelengths and from all directions.

Blackbody—Any object or material that absorbs completely all incident radiation. A blackbody emits energy in a manner described by the Stefan-Boltzmann law, Wien's displacement law, and the Planck spectral distribution of energy.

Bouguer's law—The mathematical relation describing the exponential reduction in intensity of radiation as it travels along a path of finite length within a medium.

Brewster's angle—The incidence angle at which there is zero reflection for the parallel polarized component of the radiation incident on an interface.

Broadening—The increase in the spectral width of an absorption line because of any of various mechanisms.

Cavity—Any surface concavity that acts to increase the apparent absorptivity of a surface by increasing the number of reflections (and therefore the absorption) of incident radiation.

Cold-medium approximation—An approximate form of the equation of radiation transfer derived by omitting emission terms in the complete equation. The approximation is justified if the medium is at a low temperature so that its emission can be neglected relative to radiation from boundaries.

Configuration factor—The fraction of uniform diffuse radiant energy leaving one surface that is incident upon a second surface.

Configuration-factor algebra—Mathematical relations between configuration factors.

Cosine law—The mathematical relation describing the variation of emissive power from a diffuse surface as varying with the cosine of the angle measured away from the normal of the surface.

Crossed-string method—A method for easy calculation of the configuration factor between objects having one infinite dimension and a nonvarying cross-sectional geometry.

Cross-section—The apparent projected area of a particle or object relative to its ability to absorb or scatter thermal radiation.

Curtis-Godson approximation—A method for computing radiative transfer through gases that makes use of the limiting absorption characteristics of weakly and strongly absorbing bands to compute absorption in all spectral regions.

Cutoff wavelength—The wavelength at which the absorptive properties of a surface or partially transparent medium undergo a large change, thus imparting spectral selectivity

Dielectric—An electrically insulating material.

Differential approximation—An approximation to the equation of transfer derived by approximating the complete equation with a finite set of moment equations.

Diffuse surface—Surface that emits and/or reflects equal radiation intensity into all directions.

Diffusion approximation—An approximation to the equation of transfer derived by assuming the medium to be optically thick, so that the mean free path for radiation is small.

Draper point—The temperature at which visible radiation emitted by a heated blackbody in darkened surroundings becomes visible to the human eye.

Drude theory—A theoretical approach to predicting the radiative properties of materials, based on electromagnetic theory.

Elsasser model—A model of the absorption properties of an absorption band based on the assumption that the individual lines in the band have a Lorentz shape, and are equally spaced and of the same shape.

Emissive power—The rate of radiative energy emission per unit area from a surface.

Emissivity—The property of a body that describes its ability to emit radiation as compared with the emission from a blackbody at the same temperature.

Emittance—The property of an isothermal material that describes the ability of a given thickness of the material to emit energy as compared to emission by a blackbody at the same temperature.

Enclosure—A concept to account for the incident radiation from all directions of the surrounding space.

Equation of transfer—The mathematical relation describing the variation along a path of the intensity of radiation in an absorbing, emitting and scattering medium.

Exchange areas—Factors accounting for the geometric and absorption effects on radiation between volumes or surfaces separated by attenuating media.

Exchange factor—For radiation exchange between surfaces in a system including specular surfaces, the fraction of diffuse energy leaving one surface that is incident on a second surface by direct paths and by all possible paths of specular reflection.

Exchange-factor approximation—An approximate method for calculating energy transfer in absorbing-emitting media with parallel radiation, convection, and/or conduction. The approximation is based on the assumption that the exchange factors for radiative equilibrium are unchanged by the presence of other heat transfer modes.

Exponential wide-band model—A model of an absorption band constructed of equally spaced Lorentz lines arranged with exponentially decreasing line strengths away from the band center. This model allows accurate description of the band absorption properties at long path lengths because of increasing absorption for that case in the wings of the band.

Extinction coefficient—The property of a medium describing its ability to attentuate intensity per unit of path length. It is composed of the sum of the absorption and scattering coefficients.

Fin efficiency—The ratio of the actual energy lost from a fin to the amount that would be lost if the fin were at a uniform temperature equal to its root temperature.

Fourier's conduction law—The relation between energy flux by conduction heat transfer and the negative of the temperature gradient in the direction of the energy flux.

Fresnel equations—The equations describing the reflection characteristics of electromagnetic radiation at the interface between media with differing refractive indices.

Gray—Having radiative properties that do not vary with wavelength.

Greenhouse effect—The trapping effect for radiation by substances that are transparent to radiation in the visible portion of the spectrum, such as solar energy, but are relatively opaque to radiation in the infrared portion of the spectrum, such as that emitted by low-temperature surfaces.

Hagen-Rubens relations—Relations between the optical, radiative, and electrical properties of materials developed from electromagnetic theory.

Half-width—One-half the spectral width of an absorption line at the point of half the maximum line intensity.

Hemispherical properties—Radiative properties of a surface element that are averaged over all solid angles passing through a hemisphere centered over the surface element.

Hohlraum—A heated cavity constructed in such a way that its opening closely approximates the absorption and emission properties of an ideal blackbody.

Index of refraction—The property of a medium equal to the ratio of the speed of electromagnetic radiation in a vacuum to the speed in the material. In an attenuating material, a complex refractive index is defined.

Induced emission—That portion of thermal radiation emission that is caused by the presence of a radiation field.

Infrared—The part of the electromagnetic spectrum in the wavelength region 0.7–1000 μm.

Intensity—Radiative energy passing through an area per unit solid angle, per unit of the area projected normal to the direction of passage, and per unit time.

Isotropic—Having no dependence on direction or angle.

Jump boundary condition—A boundary condition accounting for the discontinuity between the temperature of a bounding surface and the temperature of an absorbing-emitting medium adjacent to the surface in the absence of heat conduction.

Kernel—That portion of the integrand that contains the dependent variable in an integral equation. If the dependent and independent variables can be interchanged without affecting the value of the kernel, then the kernel is said to be symmetric. Symmetric kernels are necessary for some standard solution methods to be applied to certain integral equations.

Kirchhoff's law—The equality of the directional spectral emissivity and directional spectral absorptivity of a material. Various averaged emissivities and absorptivities are also equal if certain restrictions are met.

Lambert's cosine law—See *Cosine law*.

Line—A very narrow spectral region that absorbs radiation in a medium. The line shape and spectral location are determined by the quantum-mechanical properties of the medium.

Local thermodynamic equilibrium—Having sufficiently approached thermodynamic equilibrium at each location that thermodynamic properties such as temperature can be used to describe the state of each local volume element.

Luminescence—Emission of radiant energy where excitation is by means other than thermal agitation.

Luminous—Emitting radiation in the visible spectrum, usually because of the presence of incandescent particles.

Markov chain—A chain of events, the probability of each event in the chain being independent of all prior events.

Mean absorption coefficient—A spectral absorption coefficient weighted by a spectral energy distribution (for example, the Planck distribution) and averaged over the spectrum.

Mean beam length—An average path length traveled by radiation within a volume of absorbing medium.

Mie scattering—The general scattering theory derived from consideration of the interaction of an electromagnetic wave with a spherical particle, and applying over the entire range of particle diameters.

Milne-Eddington approximation—An approximation to the differential form of the equation of transfer that assumes in a one-dimensional geometry that the radiation traveling with positive direction components is isotropic, while that with negative components is isotropic with a different value.

Monochromatic—Refers to radiation at one wavelength.

Monte Carlo method—A numerical technique based on a probabilistic model of a physical process. For radiation problems, a physical model that simulates the transfer behavior of radiative energy "bundles" is used to calculate energy transmission.

Natural broadening—The increased width of a spectral line due to uncertainty-principle effects on the energy states for the transition process producing the line.

Net-radiation method—A general method for writing a closed set of equations that can be solved for the radiative transfer between surfaces in an enclosure.

Newton-Raphson method—A technique for numerical solution of a set of simultaneous nonlinear algebraic equations.

Opacity—A measure of the ability of a medium to attenuate energy. The opacity is the extinction coefficient integrated over the path length; also called the *optical thickness*.

Optically smooth—See *Optical roughness*.

Optical roughness—The roughness of a surface relative to the wavelength of incident radiation. Surfaces with roughness much smaller than the wavelength of the incident radiation are termed optically smooth; those with roughness greater than the wavelength are optically rough.

Optical thickness—See *Opacity*.

Optically thin—Having an opacity that causes only slight attenuation of radiation along a given path.

Phase function—The mathematical description of a scattering medium that gives the relative amount of intensity that is scattered into various scattering angles from a given angle of incidence.

Planck distribution—The distribution of energy as a function of wavelength that is emitted at a given temperature by a blackbody.

Polarization—The concept of the amplitude of an electromagnetic wave having components that are perpendicular to one another, and thus that can be treated separately and then added vectorially.

Poljak net-radiation method—See *Net-radiation method*.

Prevost's law—The concept that a body emits electromagnetic radiation even when in thermal equilibrium with its surroundings.

Radiative equilibrium—A system in the steady state, with no energy transfer by conduction or convection, is said to be in radiative equilibrium. Every element of such a system must be emitting radiation at the same rate as it absorbs radiation (no internal heat generation).

Radiosity—The rate at which radiant energy leaves a surface by combined emission and reflection of radiation.

Random number—A number chosen at random from a large set of numbers evenly distributed within an interval. The interval is usually 0-1.

Rayleigh-Jeans formula—A distribution of radiant energy with respect to wavelength for a blackbody at a given temperature; derived from the concept of equipartition of energy and valid only at very large values of wavelength times temperature.

Rayleigh scattering—The type of scattering of radiation that occurs when the wavelength of the incident radiation is much larger than the size of the scattering particles. Atmospheric scattering of solar radiation is generally of this type.

Reciprocity—The mathematical relation between the configuration factor for radiation from surface a to b, and that from surface b to a.

Reflectivity—The property of a surface that describes the fraction of incident energy that will be reflected from the surface.

Refraction—The change in direction of radiation upon crossing an interface between materials with differing indexes of refraction.

Refractive index—See *Index of refraction.*

Rosseland diffusion equation—The relation between energy and the gradient in emissive power in an optically thick medium as derived in the diffusion approximation.

Rosseland mean absorption coefficient—The spectral absorption coefficient of a medium averaged over the spectrum as weighted by the derivative of the Planck distribution with respect to total emissive power. Applies to optically thick media.

Scattering coefficient—The property of a medium that describes the amount of scattering of thermal radiation per unit path length for propagation in the medium.

Scattering cross section—The apparent projected area of a particle relative to its ability to scatter radiation.

Schuster-Schwarzchild approximation—An approximation to the differential form of the equation of transfer for one-dimensional planar geometries, based on the assumption that intensities with positive components of direction have one value at all angles, while intensities with negative components have another value. The resulting equations are a simplified form of those derived from moment and discrete-ordinate methods.

Selective surface—A surface with spectral radiation properties that vary widely with wavelength, causing emphasis or suppression of absorption and emission in different spectral regions.

Semigray approximation—A solution technique that assumes that radiant interchange can be treated in two independent spectral regions (usually the solar-dominated and infrared regions) so that the effects of spectrally selective surfaces and transmitting media can be accounted for.

Slip boundary condition—See *Jump boundary condition.*

Snell's law—The mathematical relation between the angles of incidence and refraction at an interface between two media, and the refractive indexes of the media.

Solid angle—Area intercepted on a unit sphere by a conical angle originating at the sphere center.

Source function—The mathematical relation describing the gain in intensity at a location because of both emission and scattering into the direction of the intensity.

Spectral—Having a dependence on wavelength; radiation within a narrow region of wavelength.

Specular—Mirrorlike in reflection behavior.

Spontaneous emission—Emission of radiation because of a spontaneous change in energy level of a substance.

Stefan-Boltzmann constant—The proportionality constant σ between the blackbody hemispherical total emissive power e_b and the fourth power of the absolute temperature; $e_b = \sigma T^4$.

Stefan-Boltzmann law—The relation between blackbody hemispherical total emissive power and the fourth power of the absolute temperature.

Stimulated emission—See *Induced emission*.

Thermal radiation—Radiation detected as heat or light ($\lambda \approx 0.2{-}1000$ μm).

Total—Integrated or summed over all wavelengths.

Transmittance—The property of a material that determines the fraction of energy at the origin of a path that will be transmitted through a given thickness.

Transmittance factor—The fraction of energy leaving one surface that is incident upon a second surface after absorption of some of the energy by an absorbing medium between the surfaces.

Transparent-gas approximation—An approximation to the equation of transfer that contains the assumption that no attenuation of energy occurs between the point of origin of the energy and the point of absorption.

True absorption coefficient—The absorption coefficient of a medium that would be measured in the absence of induced emission.

Unit-sphere method—A graphical method for the determination of configuration factors, based on the fact that if an image of a surface is formed on a unit hemispherical surface and then projected onto the base of the hemisphere, the resulting projected area is simply related to the configuration factor between the original surface and an area element at the center of the hemisphere.

Wien's displacement law—The mathematical relation showing that the product of the wavelength at the peak of a blackbody spectral energy distribution and the blackbody temperature is a constant.

Wien's formula—A spectral distribution of energy emission from an ideal blackbody that is derived from classical thermodynamics. It is quite accurate for most wavelength-temperature ranges of engineering interest.

Zoning method—A method of computing radiant exchange in enclosures containing absorbing-emitting media that is based on dividing the boundary into finite areas, and the medium into finite volumes. The areas and volumes are assumed to be individually uniform, and energy exchange among all elements and volumes is then calculated. The method is suited to computer solution.